RICHARD CALLAS
D0699565

LIGHT'S MANUAL:

Intertidal Invertebrates of the Central California Coast

LIGHT'S MANUAL:
Intertidal Invertebrates of the Central California Coast

THIRD EDITION

edited by

Ralph I. Smith and James T. Carlton

UNIVERSITY OF CALIFORNIA PRESS
BERKELEY · LOS ANGELES · LONDON

University of California Press
Berkeley and Los Angeles, California

University of California Press, Ltd.
London, England

Third Edition
Third printing, with corrections, 1980
ISBN 0-520-02113-4
Library of Congress Catalog Card Number: 70-170726
Printed in the United States of America

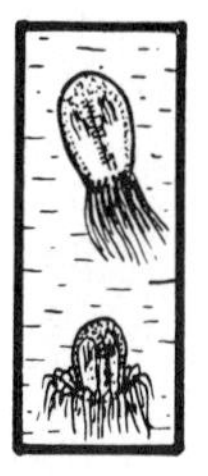

CONTENTS

S. F. Light	ix
Excerpts from Preface to First Edition	xi
Excerpts from Preface to Second Edition	xiii
Preface to Third Edition	xv
Acknowledgments of Illustrations	xviii
Introduction	1
Classification and identification of invertebrates	1
Intertidal collecting	3
Field studies	8
Environmental factors of the intertidal zone	11
A note on the use of this manual	14
General references	14
Introduced Intertidal Invertebrates (*James T. Carlton*)	17
Protozoa (*Zach M. Arnold*)	26
Phylum Porifera (*Willard D. Hartman*)	32
Phylum Cnidaria (Coelenterata) (*Cadet Hand*)	65
Class Hydrozoa (*John T. Rees* and *Cadet Hand*)	65
Class Anthozoa (*Cadet Hand*)	85
Class Scyphozoa	94
Phylum Ctenophora (*Rolf L. Bolin*)	98
Phylum Platyhelminthes (*Eugene C. Haderlie*)	100
Phylum Nemertea (Rhynchocoela) (*Eugene C. Haderlie*)	112
Aschelminthes and Smaller Worm Groups	121
"Mesozoa"; Phyla Gnathostomulida, Rotifera, Gastrotricha, Kinorhyncha, Nematoda, Nematomorpha, Acanthocephala	121
Unsegmented Coelomate Worms (*Mary E. Rice*)	128
Phylum Sipuncula	128
Phylum Echiura	132
Phylum Priapula	133
Phylum Annelida: Introduction and Smaller Groups	135
Class Oligochaeta (*David G. Cook* and *Ralph O. Brinkhurst*)	136

Class Hirudinea 146
Archiannelida (*Colin O. Hermans*) 147
Phylum Annelida: Class Polychaeta (*James A. Blake*) 151
Phylum Arthropoda: Introduction and Lower Crustacea 244
Class Crustacea 245
Subclass Cephalocarida 246
Subclass Branchiopoda 246
Subclass Ostracoda 247
Subclasses Copepoda and Branchiura (*Paul L. Illg*) 250
Phylum Arthropoda: Crustacea, Cirripedia (*William A. Newman*) 259
Phylum Arthropoda: Crustacea, Lower Malacostraca 270
Division Leptostraca 271
Division Peracarida 272
Order Mysidacea 272
Order Cumacea (*William B. Gladfelter*) 273
Phylum Arthropoda: Crustacea, Tanaidacea and Isopoda (*Milton A. Miller*) 277
Order Tanaidacea 277
Order Isopoda 281
Phylum Arthropoda: Crustacea, Amphipoda: Gammaridea (*J. Laurens Barnard*) 313
Phylum Arthropoda: Crustacea, Amphipoda: Caprellidea (*John C. McCain*) 367
Phylum Arthropoda: Crustacea, Eucarida 377
Order Euphausiacea 377
Biology of Decapod Crustacea (*Armand M. Kuris*) 378
Keys to Decapod Crustacea (*James T. Carlton and Armand M. Kuris*) 385
Phylum Arthropoda: Arachnids, Tardigrades, and Insects 413
Class Pycnogonida (*Joel W. Hedgpeth*) 413
Halacaridae (Marine Mites) (*Irwin M. Newell*) 425
Pseudoscorpionida (Chelonethida) 431
Tardigrada 431
Class Insecta: General, Collembola, Hemiptera (*Howell V. Daly*) 432
Class Insecta: Diptera (*Evert I. Schlinger*) 436
Class Insecta: Coleoptera (*John T. Doyen*) 446
Phylum Mollusca: Introduction and Smaller Groups (*Allyn G. Smith*) 453
Class Scaphopoda 455
Class Cephalopoda 455
Class Polyplacophora 457

Phylum Mollusca: Shelled Gastropods (*James T. Carlton* and *Barry Roth*) 467
Phylum Mollusca: Opisthobranchia (*James W. Nybakken*) 515
Key to the Orders of Opisthobranchia (*Terrence M. Gosliner* and *Gary C. Williams*) 517
Order Cephalaspidea 518
Order Anaspidea 521
Order Notaspidea 522
Orders Sacoglossa and Nudibranchia (*Gary R. McDonald*) 522
Phylum Mollusca: Bivalvia (*Eugene V. Coan* and *James T. Carlton*) 543
Phylum Ectoprocta (Bryozoa) (*John D. Soule, Dorothy F. Soule,* and *Penny A. Pinter*) 579
Lesser Lophophorates and Entoprocta 609
Entoprocta (*Richard N. Mariscal*) 609
Phoronida (*Russel L. Zimmer*) 613
Brachiopoda 616
Phylum Hemichordata (*Theodore H. Bullock*) 618
Phylum Echinodermata 620
Class Echinoidea (*John S. Pearse*) 621
Class Asteroidea (*James E. Sutton*) 623
Class Ophiuroidea (*James E. Sutton*) 627
Class Holothuroidea (*James C. Rutherford*) 634
Phylum Chordata: Introduction and Urochordata (*Donald P. Abbott*) 638
Phylum Chordata: Intertidal Fishes (*Rolf L. Bolin,* revised by *Margaret G. Bradbury* and *Lillian J. Dempster*) 656
Intertidal Plants (*Isabella A. Abbott*) 669
The Eelgrasses (Zosteraceae) 669
Conspicuous Intertidal Algae 670
Index 685
Errata and Addenda 717

S. F. LIGHT

The late Professor S. F. Light (1886–1947) was for twenty-two years a member of the Department of Zoology at the University of California, Berkeley. His active interests ranged widely over the field of the invertebrates, and ran the gamut from taxonomy (alcyonarians, scyphozoans, termites, copepods) to the social physiology of termites, their protozoans, symbionts, and caste determination.

Dr. Light gave an extraordinary amount of time and careful thought to his teaching at all levels and exercised a peculiarly pervasive and long-lasting influence on his students' points of view, interests, and habits of thought. His advanced courses were marked by a critical, appreciative, phylogenetic morphology and a critical natural history which insisted on a full realization of the values of sound systematics, keen field observation, and concrete, testable interpretations. The essence of his natural history course was to be found between the lines of its syllabus, which combined a dynamic approach to basic principles (not only of invertebrate zoology but of field biology and scientific methodology) with practical aids to the mastery of a specific fauna. It was the result of more than ten years of active contact, virtually the year around, with that fauna and of continual efforts to perfect a teaching approach which aimed at very high and exacting goals. In the present manual, a conscientious updating of the original published syllabus, the revisers have, I believe, been successful in their aim of retaining the values of Light's approach.

Professor Light played his role in biology and academic life in personal contacts rather than in national or university affairs, and profoundly affected the attitudes of many graduate students and associates. Those who knew him will remember the personal characteristics of modesty—extending to a real underestimation of self—of appreciation of disciplines which lay beyond his own field of study, of exacting criticism in the use of words and ideas—driving him now to caution, now to forward positions—of sincere interest in the human relations of his students and assistants, and of a highly developed aesthetic enjoyment of outdoor beauty. Many have felt that, although they never really knew the inner man, they sensed vividly the goals and standards for which he lived.

THEODORE H. BULLOCK

[An additional, and delightful, reminiscence of S. F. Light by Joel W. Hedgpeth is in the preface of *Between Pacific Tides*, by Ricketts, Calvin, and Hedgpeth (1968). Eds.]

EXCERPTS FROM PREFACE TO FIRST EDITION *

This volume represents accumulations from fifteen years of teaching the natural history of the invertebrates of the central California coast. . . .

In order to study animals in the field it is necessary to be able to identify them, to recognize them, and to know them by name. A considerable part of the time of the course, therefore, is devoted to the study in the laboratory of animals of the various groups with a view to learning the characteristics important in the identification of the species of these groups. . . . No attempt is made to study taxonomy as such . . . Our end is the prosaic one of learning names for the local assemblage as rapidly and simply as possible.

This end, unpretentious as it is, is by no means easily attained. The invertebrate animals of the Pacific Coast are very imperfectly known. For some there is no monographic account. . . .

Finally, modern monographs with keys would not be enough for our purpose. Limited time would still require that these be brought within the range of the study by simplification and limitation of terminology and by limitation of consideration to those species of the various groups significant in our local assemblages. Otherwise it would be impossible for the student to get that familiarity with the fauna as a whole which is one of the greatest values to be obtained in such a study. . . .

The present work, incomplete as it is and continually in process of revision as it is, is the only one known to me which attempts to bring together, in more or less completely illustrated keys and lists, the information necessary for even a tentative identification of the common invertebrates of this area. The work has been enriched by special studies made by students in the class and by graduate students specializing in the invertebrates. . . .

Under the conditions existing with regard to our knowledge of Pa-

* *Laboratory and Field Text in Invertebrate Zoology*, by S. F. Light, published by the Associated Students of the University of California, Berkeley, 1941.

cific Coast invertebrates, changes of name are bound to be the order of the day and a volume such as this is constantly undergoing revision as new works appear or new information is obtained. For its errors, which are numerous, probably beyond even the author's imagination, he accepts full responsibility, consoling himself by the hope that the knowledge of these errors, inevitably forced on the students' attention, may stimulate some of them to undertake corrective investigations. . . .

Berkeley, 1941 S. F. LIGHT

EXCERPTS FROM PREFACE TO SECOND EDITION *

At the time of Professor Light's death in 1947, the first edition of this manual was out of print, and only a small amount of work of revision had been carried out by the author . . . Recognizing that this manual will rarely, if ever, again be used as the text in any course in invertebrate zoology, but rather as an adjunct to standard texts or to an instructor's own syllabus, we have generalized much of the introductory material in the various sections and omitted specific instructions for laboratory procedure in studying particular groups of animals . . . In most instances, keys and discussions were completely rewritten rather than simply emended. In an over-all sense, we edited freely in the effort to achieve a satisfactory and balanced text, reinforced in our efforts by the knowledge of Dr. Light's dissatisfaction with the first edition. Despite these efforts, we may echo Dr. Light's characterization of the work in his preface to the first edition, and state that the second is still "incomplete . . . and continually in process of revision. . . ." If we bring this revision to publication without achieving satisfaction in its completeness, we have, at least, adhered to the tradition of the first.

Especially difficult in the revision of keys has been the problem of where to stop. In attempting to achieve a coverage of intertidal animals adequate for the advanced student of general invertebrate zoology and marine ecology, we have necessarily stopped short of the treatment required for the specialist. At the same time we have gone beyond the needs of the beginning student. Although we have tried not to sacrifice technical exactness, we have frequently resorted to nontechnical descriptions to make the keys usable by persons new to the field. Since these keys cannot include all the animals which may be encountered in the area covered . . . the student may often be un-

* *Intertidal Invertebrates of the Central California Coast: S. F. Light's Laboratory and Field Text in Invertebrate Zoology,* revised by Ralph I. Smith, Frank A. Pitelka, Donald P. Abbott, and Frances M. Weesner, with the assistance of many other contributors, University of California Press, 1954.

able to identify his capture by the aid of this book. While making the keys reasonably inclusive of the common forms, we have tried to make them exclusive of rarer forms, and it is our hope that the student who has in hand a species not included in this manual will recognize that it is not treated here, and betake himself to more specialized works, or send the specimen to a competent authority for identification. . . .

Berkeley, 1954 RALPH I. SMITH

The Light that shines upon our work
Illumines wondrous scenes,
Or would, if we could only tell
Exactly what it means . . .
ANONYMOUS STUDENT (1956)

PREFACE TO THIRD EDITION

We have undertaken this revision of Light's Manual at a critical time. Specialized knowledge of our intertidal fauna has increased greatly in the past two decades, and an expanded manual is necessary to meet the professional need for a means of identifying intertidal invertebrates. But, in the same two decades, threats to our intertidal biota have increased correspondingly, and its very preservation has become a matter of public concern. Professor Light's original manual was directed specifically to the needs and schedule of his own classes; the second edition was revised so as to be useful in various types of courses and in programs of study for preprofessional and professional zoologists.

We are now in a period of intense ecological consciousness, when collecting and identifying no longer claim the share of students' interest and attention they once held, and when our coastal fauna is undergoing serious depredations from commercial and sport collecting, even from student activities, from increased public access, and from all the "improvements" in the form of dredging, diking, draining, filling, and subdividing that we designate as progress. Further, and most significantly, the past decade has witnessed the proliferation of "environmental impact" surveys that depend for their validity on a correct identification of the components of the biota. Without reliable identifications as well as proper estimations of biomass, numbers, and diversity of species, such surveys are useless or, worse, they may lull us into thinking that a scientific evaluation has been made, against which future changes can be measured. There is more than enough evidence that most environmental impact surveys are woefully inadequate in identification and enumeration of intertidal invertebrates.

There is not and perhaps cannot be a single volume that can serve as a comprehensive reference in making identification. Light's Manual has been used and has been useful in such work, despite its shortcomings, and we have no choice in this revision but to attempt to improve its accuracy and coverage. Of necessity it is longer and more detailed, encompassing more species and in general more "professional". While this may make the manual less useful or less convenient in undergraduate classes, its usefulness to graduate students and professional zoologists and ecologists should be correspondingly increased. We cannot go backward, we cannot make it simpler and easier to use, but we hope we have made it better. As in the second edition, we echo Dr. Light in saying that the third is still "incomplete . . . and continually in process of revision . . ." and may it continue so.

As in previous editions, the geographical coverage of this manual is the central California coast approximately from the Monterey Peninsula to Bodega Head and including San Francisco Bay. In functional terms, we are concerned with the area covered by the activities of the marine laboratories at Pacific Grove, Moss Landing, Bolinas, Dillon Beach, and Bodega Bay, and have thus somewhat exceeded the strictly geographical limits. We have not attempted to cover the subtidal fauna of this area; the *Below Pacific Tides* suggested by Hedgpeth in his fourth edition of *Between Pacific Tides* remains to be written. Thus we cannot satisfy the extensive brother- and sisterhood of divers, although much shallow subtidal material is included through the activities of long-armed and water-resistant intertidal collectors. In some instances, as in the sections on barnacles and oligochaetes, it has proved possible to include the intertidal fauna of the entire California coast without using excessive space; unfortunately such coverage has been utterly impractical for the larger groups. Users of this manual in southern California or in Oregon or Washington will find many of their local species represented, but many will be found that will not key out, or will key out wrongly. Unless it is specifically stated, users of this Manual outside the Monterey to Bodega range should exercise even more scepticism than those within the range.

We cannot overemphasize that keys of this sort are only approximating devices for tentative identification, and that identifications made for critical studies should be carefully checked in more complete sources when such exist. The intertidal fauna of the central California coast is still very incompletely known, and this manual should not be used as a primary source of zoogeographic information, since the completeness of coverage and the geographical range covered vary from author to author and from group to group. We have specifically avoided giving geographical ranges of most species because these soon become out of date and may confuse students who find a species outside its reported range. It might as well be stated that some contributors have been more persistent than others in efforts to include species from outside the stated range of the manual! Questions about the actual range of species included in keys and lists should be referred to the authors of the keys rather than to the editors.

In certain parts of the region covered by this manual, the problem of introduced species is becoming increasingly serious as the result of accidental and deliberate introductions by private individuals and state agencies. Especially in San Francisco Bay and other bays that have received overseas shipping and importations of oyster stock, the introduced species constitute a significant or even major part of the fauna. The identification of such new arrivals requires an examination of the problem on a world basis. We have therefore included a new chapter by James Carlton dealing with the introduced fauna. Users of this manual may make valuable contributions to our knowledge of this subject.

As before, we are indebted to many contributors and revisers. Some former contributors have revised their sections, others have passed on and their places have been taken by new contributors. Much of the work of revision has been done at the Bodega Marine Laboratory of the University of California. We are indebted to its director, Dr. Cadet Hand, and to his staff for much help and support.

In a collaborative work of this sort, it is next to impossible to acknowledge every bit of information and the countless small acts of assistance rendered by colleagues and students. The authors of chapters and sections are indicated in the table of contents and in the text; on them has fallen the main burden of the work that has gone into this book. They have worked in a spirit of helpfulness and scientific cooperativeness, and with no material reward; indeed, many have borne considerable personal cost, and we are grateful to them. These authors have acknowledged the assistance of others, as indicated in their respective texts, and to these also the editors are grateful. In the production of this book, which is without royalties, the University of California Press has given support. The encouragement of its staff, especially Mr. Grant Barnes, Miss Susan Peters, Mr. William Snyder, Mr. John Enright, and the careful work of copy-editor Mrs. Diane Hersh, have been invaluable. The actual work of revision has been shared and immeasurably aided by my co-editor and contributor, James T. Carlton, whose knowledge of our coastal fauna has been exceeded only by his boundless enthusiasm and unlimited capacity for work.

Department of Zoology,
University of California,
Berkeley, April, 1974

RALPH I. SMITH

A note on the third printing: The editors have attempted to include all corrections and name changes known to us at this time. When this has resulted in inconsistencies, we accept responsibility. We are especially grateful to University of California Press for accepting these changes, and in particular to Mrs. Barbara Zimmerman for her patience.

Berkeley, May, 1980 R. I. S. and J. T. C.

ACKNOWLEDGMENTS OF ILLUSTRATIONS

Many figures in this manual were executed or redrawn and all figures were numbered and labeled by Mrs. Emily Reid, whose cheerful cooperativeness as well as her care and skill have been indispensable. A number of figures of crustaceans and molluscs taken from J. W. Hedgpeth, *Seashore Life* (1962) are the work of Mrs. Lynn Rudy and are reproduced with the permission of the University of California Press. The figures of algae (Plates 152–156), which were reproduced from the original plates of G. M. Smith's *Marine Algae of the Monterey Peninsula* (1969) for the second edition of this manual, are again used with the permission of Stanford University Press. In addition, we are indebted to Stanford University Press for permission to reproduce figures 229, 230 (Pl. 85), 69 (Pl. 117), and 97 (Pl. 119) from Ricketts and Calvin, *Between Pacific Tides*, 4th edition, revised by Joel W. Hedgpeth, who also drew our title-page figure.

Figure 29 (Pl. 12) is reproduced (modified) from F. M. Bayre and H. B. Owre, *The Free-living Lower Invertebrates* (1968) with permission of Macmillan Publishing Co., Inc. Figures 1 (Pl. 6) and 32 (Pl. 12) are reproduced from A. J. Marshall and W. D. Williams, *Textbook of Zoology: Invertebrates* (1972) with permission of The Macmillan Press Ltd. Figures 11 and 12C (Pl. 9), 28, 30 and 31 (Pl. 12) and 9 (Pl. 16) are reproduced from L. H. Hyman, *The Invertebrates*, vols. I (1940) and III (1951) with permission of McGraw-Hill Book Co., Inc. Figures of molluscs from M. Keen, *Marine Molluscan Genera of Western North America* (1963) and from M. Keen and J. C. Pearson, *Illustrated Key to West North American Gastropod Genera* (1952) are reproduced with permission of Stanford University Press. A figure of *Caecum* from R. T. Abbott, *American Seashells* (1954) is reproduced with permission of D. Van Nostrand Company.

Numerous figures of molluscs, indicated in the corresponding legends, were generously made available to us by Dr. Rudolf Stohler, editor of *The Veliger*, both from his personal collection and from the files of that journal. Figures previously published in scientific journals by authors not named here are acknowledged in the text. Many figures furnished by the authors of sections are original. The decorative figures at the heads of chapters are the work of Donald P. Abbott.

INTRODUCTION

This manual is primarily a guide to the intertidal invertebrate fauna of the central California coast, from the Monterey Peninsula to Bodega Head. It may be useful in other West Coast areas, but its effectiveness will lessen with distance. The bulk of this manual is concerned with identifying and naming animals and only secondarily with taxonomic principles. The following discussion is intended to introduce the student to the problems of animal classification.

CLASSIFICATION AND IDENTIFICATION OF INVERTEBRATES

The purpose of zoological classification is to arrange animals into groups on the basis of fundamental similarities and differences which reflect evolutionary relationships. At least a million species of animals have been described, and 95 percent of these are invertebrates. They include a vast, diverse array of types, including all but one of the classes (Vertebrata) of the animal kingdom. The classification of this assemblage is therefore of cardinal importance in the study of invertebrates. It is also a matter of much controversy.

Zoology students, and biologists from other fields, must be prepared to find that different writers use different systems of classification. These differences are chiefly of two sorts. First, there is the use of different names for the same group; for example, the terms Entoprocta, Kamptozoa, and Calyssozoa refer to the same small group of animals. Second, there is the placing of the same group into different systematic categories, as in the designation of a group as a class by one writer, but as an order by another. Discrepancies of the second sort are so common that it is generally advisable not to have a fixed concept of the taxonomic rank of a particular group, but rather to remember that it is a part of a certain larger group and can be divided into a number of subordinate groups. Thus it is not so important to decide whether the Crustacea represent a subphylum, superclass, or class as it is to know that the group is a major subdivision of the Arthropoda and that it includes the Malacostraca, the Copepoda, and so on.

In contrast to classification, the identification of invertebrates may represent more of empirical techniques than it does of science. However, a necessary preliminary for the study of any animal is the determination of its scientific name. A correct determination is especially important if the animal is to be the subject of a critical investigation.

A scientific name consists of the name of the genus (capitalized), followed by the name of the species (not capitalized), followed by the name of the describer and, if desired, the date of publication of the original description of the species. The name of the describer is placed in parentheses if the generic name now in use differs from that used in the original description, thus *Hemigrapsus nudus* (Dana, 1851) was originally described as *Pseudograpsus nudus* by Dana in 1851. For convenience, the name of the describer may often be omitted in our use of scientific names but should always appear on properly labeled specimens. The generic name may be used alone, the specific name never, unless it has already been used with the generic name on the same page. Both generic and specific names are underlined in manuscript by convention. This aids in picking out scientific names in manuscript and is replaced by italic type in print.

Great pains should be taken to spell scientific names correctly. The correlation between careless, unscientific work and careless use of scientific names is very high.

Common or vernacular names are convenient but have many disadvantages. The rule of priority fixes the scientific name, which is universal. Common names are local. There are no rules to determine which of many such names is the correct name, and the same name may be applied to very different species or types in different regions, or by different persons in the same region. Vernacular names of American birds and certain other groups of vertebrates are relatively uniform owing to the united action of the workers in these fields, but few such standardized common names exist for invertebrates.

It is necessary to make a sharp distinction between these two types of names and to use the vernacular name only after connecting it with the scientific name. Thus, *Hemigrapsus nudus* will be identifiable by all invertebrate zoologists, but "purple shore crab" would have no meaning, or a different meaning, in areas in which *H. nudus* is not found, or even in certain parts of the range of that species.

Few students of zoology realize the difficulties involved in identifying with certainty most species of animals, particularly invertebrates. A relatively few species are readily recognized because of distinctive color, pattern, or structure. Such an animal, for example, is the striped shore crab, *Pachygrapsus crassipes*, which is abundant in rocky crevices above low-tide mark along the Pacific coast. In many other cases, however, identification can only be approximated by the beginner, and in still others even the specialist will find difficulty in making identification. These problems are aggravated on the Pacific coast by the fact that the study of many groups of invertebrates has been greatly neglected. During the period when systematic work was the vogue in zoology, there were few zoologists on this coast. With the change in emphasis in zoology, they have largely abandoned this field, and few others have entered it. For some groups, such as the amphipods and the littoral copepods, the fauna of the Pacific coast is still not well known. In other groups a great amount of work has been accomplished, but much remains to be done.

Some local groups, such as the decapod crustaceans, the marine molluscs, and the echinoderms, are fairly well known to systematists. Even here there is often need for complete and careful systematic revision and monographing. But when all this has been done, still another step will be necessary to make this information available to students of zoology or biology who are not specialists in the particular groups, and also to the intelligent laymen. Manuals of faunas must be produced, containing brief diagnostic descriptions and keys to the species, illustrated, if possible, and accompanied by careful definitions of terms. Such publications that may prove useful in conjunction with the present manual are included in the reference lists of each section.

The keys that form the bulk of this volume represent an attempt to make possible the identification of the common intertidal invertebrates of the central California coast. Many keys are incomplete, and all will need revisions and additions as our knowledge increases. We cannot emphasize too strongly that keys are shortcuts and often very misleading; that their function is merely to clear the way to an approximation; and that identifications made by them, if to be of scientific value, must be reinforced by reference to the original descriptions or by comparisons with descriptions and illustrations in monographs, if such exist, or by comparisons with authentic named specimens, or by submission to a specialist.

INTERTIDAL COLLECTING

With the increasing pressure upon our intertidal fauna resulting from the collecting activities of food-gatherers and large classes of students, not to mention hordes of casual visitors of all ages who collect to no purpose, field studies connected with college classes must have meaningful objectives and rational approaches to the complex and delicate intertidal environment. The days of massive collecting, of competition to secure the greatest list of species, of destructive collecting methods have passed. An intelligently planned field trip aims to show students the invertebrate fauna as an interrelated functional complex, of which a full understanding can be gained only if the components are viewed as parts of a whole.

The approach and spirit of such studies is usually essentially ecological, although limitations in time, training, and equipment may not permit us to utilize the quantitative and experimental methods characteristic of modern ecology. Rather, we make ecological surveys or reconnaissances, on the basis of which we try to set up preliminary hypotheses as to the role and importance of different factors in determining the distribution of animals. These hypotheses are based only on readily observable and relatively obvious facts relating to the structural and behavioral adaptations of organisms, the presence and nature of other biota, and the physical nature of the environment. Such a factor as degree of exposure at a given spot, for example, may be estimated on a relative basis by determining the vertical level in the intertidal zone, the nature and arrangement of the local substrate, and the position with regard to the action of waves, wind, rain, and sun. The importance of the effects of exposure may then, in a preliminary way, be evaluated through a study of the animals present (and of those conspicuously absent), and an appraisal of structural and behavioral adaptations that make life possible under the observed conditions of exposure. Such studies encourage student and teacher alike to think of organisms in ecological terms, and they provide the basis for a degree of insight into the interrelationships between organisms, the variables which limit the distribution of organisms, and the factors which control the formation of characteristic biotic communities. They supplement, but do not replace, the more conventional laboratory studies of invertebrate organisms, and they make these laboratory studies far more meaningful.

No collecting is allowed on the California coast without a fishing license or scientific collecting permit from the California Department of Fish and Game; students and others should obtain a copy of the Fish and Game regulations, available from bait shops and sporting goods stores, for current regulations.

EQUIPMENT RECOMMENDED

Students will need a geologist's pick, or a hammer and cold chisel; putty knife or similar scraper; forceps (most conveniently worn on a string around the neck); plastic bags; jars and vials of various sizes which can be firmly capped; hard-backed notebook and automatic pencil; hand-lens, also on a cord about the neck (6–10 power is ample; higher powers do not permit ready inspection of objects within vials); and a container for equipment, bottles, and specimens (plastic buckets are excellent). Hip-length rubber boots and warm clothing are highly desirable. Shovels and sieves, about one of each per five students, are essential on trips to soft-bottomed areas. Small vials may be more safely and conveniently carried if packed in tins or milk cartons.

SUGGESTIONS FOR COLLECTING

Plan your program beforehand, as every minute of the low tide must be made to count. Go out at once to the water's edge. Follow the tide out and work back as the tide drives you in. On rocky coasts take care not to be trapped on offshore rocks by the incoming tide.

On your first trip to the intertidal, the impact of quantity and variety of animal life will usually lead to more or less excited but aimless collecting of the most obvious and conspicuous species. Get down among the rocks and examine them closely. Extend your search by turning over rocks, lifting algal masses, examining the under surfaces of overhanging ledges, chipping away cracked or soft rock, and so on.

Destroy as little as possible either of animal life or of the natural conditions of the environment. **Replace overturned rocks in their original position!** The fauna living on the lower surface of the rock or beneath the rock cannot survive when exposed to air and wave action; neither can organisms originally on the upper surface survive if turned underneath. Carelessness in this can ruin a good collecting ground within a few years.

On subsequent trips select specific sites that you examine systematically, either to gather data on the aggregation of species occurring together in a given habitat, or to see what representatives of a given systematic group occur there, or both. The most rewarding places for collecting are: (1) deep, shaded tidepools near low-tide level; (2) on and in seaweed, particularly in the holdfasts of larger brown algae, and surf- and eelgrass; (3) in the crevices and cavities of rock and within rock, particularly if the rock is soft, porous, or covered with coralline algae; (4) under rocks; and (5) in sand and mud.

On all field trips, small collections should be made of animals not already known, forms not in laboratory-demonstration sets of specimens (where these are used), and species needed for special studies. It is well to work on common forms first, leaving rarities for later identification and study. Do not collect unless you plan to use the specimens for study or identification or for your class or personal collection. Avoid collecting great numbers of the same species.

Small rocks, especially favorable pieces from larger rocks, and holdfasts of kelp from outer, low-intertidal levels may be brought in and the animals removed and cared for; this requires but one set of notes in the field for many specimens. Care must be taken to segregate certain types of organisms. Most worms, for example, secrete mucus and tend to form a tangled, often inextricable, mass with other organisms. This is particularly true of nemerteans, which should be kept separately. Nudibranchs often liberate quantities of mucus, which entangles small or delicate organisms; and some sponges, when cut or broken, liberate materials deleterious to other animals crowded in the same jar. Certain animals, particularly crabs, sometimes tear others. Voracious animals, such as large isopods, often devour smaller forms, and the spines of echinoderms form a place of escape and entanglement for many smaller, mobile animals. Fragile forms, such as hydroids and arborescent bryozoans, and sensitive forms, such as colonial ascidians, may be rendered unsuitable for study by rough handling, and should be kept only with other inactive forms or by themselves. No jar should ever be crowded or packed with specimens.

Especially in hot weather, problems are raised if animals must be transported any great distance to the laboratory. Packing jars loosely in cold, wet seaweeds in a basket or bucket will maintain the necessary coolness for some time, especially if the container can be kept shaded but exposed to moving air. In a great many cases intertidal animals will survive much longer in jars very loosely packed with damp seaweed than in jars filled with water. Exposure to damp air is not harmful to intertidal animals, and a jar of air contains twenty times as much oxygen as the same jar full of water.

FIELD NOTES

If specimens are taken merely for immediate study or practice in identification, field notes are not necessary. Any specimen to go into a permanent collection, however, *must* be accompanied by certain information to make it of value to any future student. Experience has shown that most collections ultimately are used for study whether that was the original intention of the collector or not, and many troublesome errors and difficulties have arisen owing to lack of information on specimens in such collections. The difficulty of keeping proper field notes and correlating these with a large collection of specimens on a field trip is considerable, especially under the conditions of wet hands, lack of a place to sit, the presence of rain or spray, lack of time, and lack of sufficient containers to segregate specimens. Careful planning is necessary to accomplish the required end.

There are two common methods of keeping these field notes. One, perhaps the safest, is to write the information in pencil on a slip of paper and drop it into the container in the field. When the collection is put into permanent form, the information thus obtained must be copied on the label in (never on) the bottle. The second method is to drop into the container or attach to the specimen a number (written at the time or already prepared), and to record the same number together with the desired information in a field notebook or on cards. This method is easier but involves the risk of loss of the notes (in which case the number means nothing), and the danger of mixing numbers and notes because of haste in the field. Experience will dictate your choice of method.

There is almost no limit to the amount of available information which, if connected with a specimen, might prove valuable to the student of distribution, variation, or ecology. Points commonly noted are: (1) exact **locality,** including county, (2) **date,** (3) **intertidal level,** (4) **environmental situation** or **habitat** (underside of rock, or the like), (5) any **organisms or organic remains apparently associated** with it (e.g., "with polychaete worms, no. 63, in the hole of boring clam"), (6) anything unusual in the **behavior or stage** of the animal, and (7) the **collector's name.** The absolute minimum must include locality, date, habitat, tide level, and collector.

Field notes may have a usefulness far beyond the proper labeling of specimens in collections. Once the investigator has learned the more important species in the fauna, field trips may be devoted not so much to collecting as to making and recording careful observations on specific organisms or habitats. Learning to produce a worthwhile set of field notes in the intertidal takes patience, determination, and practice. The end can be achieved only by limiting the objective, making it sharply definite, and refusing to be deflected from it. Since this in itself requires experience it is generally wise to have earlier field trips devoted to more general objectives, and to confine critical study and recording of ecological features to later trips.

STUDY AND PRESERVATION

On return to the laboratory the first move should be to get material segregated into dishes of fresh, cool sea water, taking care always to maintain the correlation between specimens and field notes. As a routine it is worthwhile to observe under the microscope, while they are still living, any small animals or developmental stages not already studied. Observation of fresh, living material will often bring out much of beauty and interest otherwise lost. If time permits and there is any probability that specimens are to be used for future systematic studies, careful color notes will be of value.

If animals are to be preserved, this should be done while they are still in good condition. Some organisms may be dropped directly into fixative. Others, especially delicate or soft-bodied and contractile organisms, should be first anesthetized. Isotonic magnesium chloride (73 g $MgCl_2 \cdot 6\ H_2O$ per liter of tap water) is the best easily obtainable general anesthetic for marine animals. It may be used straight, or in various proportions with sea water. Other useful anesthetics are: propylene phenoxetol, used in very small quantities; Epsom salts ($MgSO_4$), added directly to dishes of sea water in small amounts; alcohol, arranged to drip slowly into dishes of sea water or fresh water containing animals; menthol, a few crystals floated on the surface of a dish of sea water; or a few shreds of tobacco added to a small amount of the fluid containing animals. Fixation and preservation of museum specimens may be accomplished with 5–10 percent formalin or 70–95 percent alcohol.

FIELD STUDIES

No picture of organisms that ignores their physical and organic environment can be even approximately complete. Studies of dead animals or their parts or even of living animals in the laboratory, indispensable as they are, give but a partial picture. For fuller understanding we must seek firsthand knowledge of living organisms in their natural settings. Field trips are of prime importance in gaining such knowledge and understanding. They make possible a study of the environment itself, of the distribution of organisms within specific habitats, of the behavior and interrelationships of species, and of the influence of physical and biotic factors on the distribution of organisms. Only through information gained in field studies is it possible to establish correlations between the structure and behavior of an organism on the one hand, and its habitat and ecological niche on the other. One purpose of this manual is to make practical the inclusion of extensive field studies of the marine intertidal within a course in the biology of invertebrates for advanced undergraduates or for graduate students.

The practical work in a course in invertebrate zoology normally consists of a number of field trips interspersed with laboratory studies designed to provide knowledge of the animals encountered in the field. Emphasis on field trips will vary with the instructor but, in a summer session at the seashore, field studies can and should constitute the heart of the course. Trips are usually scheduled to coincide with low tides (tide tables for the year can be obtained at most sporting goods or bait stores, or purchased from the Superintendent of Documents, U.S. Government Printing Office, Washington, D.C., 20402) and class programs will therefore vary from year to year.

To be of maximum value the studies should be cumulative and comparative. Each new situation should be compared to others already studied, and similarities and differences should be observed in faunas, environmental conditions, and their interrelationships.

In pursuing field studies it is well to keep in mind certain fundamentals. All animals have similar basic needs or requirements. Ultimately these can be reduced to food, oxygen, protection and the proper conditions for reproduction, the terms food and protection being used in the broadest possible sense. In situations where an animal occurs regularly and in numbers, we may be sure that these needs are met, though the manner in which they are met is not always obvious.

1) **Food,** as used here, includes all substances (except oxygen and water) from the environment necessary to provide animals with energy and body-building materials. The intertidal is a region of abundant light and food. In areas of hard substrate there is often much plant life; this includes not only the larger algae and occasional flowering plants, such as the surfgrass *Phyllospadix,* but also the film of microscopic plant life growing on exposed surfaces of rocks and larger plants. A few organisms (e.g., the kelp crab *Pugettia producta,* the red sea urchin *Strongylocentrotus franciscanus,* the limpets *Notoacmea insessa* and *Collisella instabilis*) graze directly on the attached larger algae. Seaweeds broken loose and washed into crevices and pools or ashore on beaches provide a rich source of food for other forms (e.g., the purple sea urchin *Strongylocentrotus purpuratus,* the beach amphipods *Orchestia* and *Orchestoidea*). Finally, plant material, ground into a fine organic detritus by the action of turbulent waters against the substrate, provides food for a host of forms that feed in many different ways. Detritus suspended in moving waters is taken, together with living plankton, by a variety of particle feeders. Some use mucous nets or webs to trap this food (e.g., the echiuran *Urechis,* the polychaete *Chaetopterus,* the attached gastropod *Petaloconchus*), whereas others use cilia, often in conjunction with sheets or strands of mucus (e.g., most bivalves, brachiopods, bryozoans, tunicates, serpulid and sabellid polychaetes, sand dollars, sponges). Still other feeders on suspended detritus and plankton use combs of fine bristles or setae to catch particles (e.g., the anomurous decapods *Petrolisthes* and *Emerita,* barnacles, and many other lower crustaceans); yet others use tentacles (e.g., the sea cucumber *Cucumaria curata,* terebellid and spionid polychaetes, and ctenophores).

Organic detritus of plant and animal origin accumulates on and in soft substrates, and here it forms food for another complex of organisms. Some of these organisms pick up the surface layer of detritus (e.g., some species of the bivalve genus *Macoma*), others swallow substrate more or less unselectively (e.g., many annelids), and still others burrow and sift the bottom material for edible particles (e.g., the shrimplike anomuran *Callianassa*). Finally, organic detritus tends to cling to exposed surfaces of rocks and plants; here, macroscopic encrusting algae, together with the microscopic plants, bacteria, and protozoa which also occur, serve as food for animals with rasping or scraping organs, particularly the gastropods and chitons.

This great variety of organisms living on plant and animal detritus and plankton provides food in plenty for intertidal predators. Micropredators, like hydroids and the sea anemone *Metridium*, feed upon animal plankton brought in by the tides. Predators on larger organisms (e.g., most starfishes but not *Patiria*, the gastropods *Nucella, Acanthina*, and *Ceratostoma*, crabs, nemerteans) find a rich diet of clams, snails, barnacles, worms, and other forms. Fishes as well as invertebrates are significant predators when the tides are high; shore birds, some small mammals, and man when the waters recede. When studying an organism keep in mind the questions: What does it eat? How does it get its food? What is its role in the food economy of the area?

2) **Oxygen** presents a lesser problem for most intertidal organisms. In most areas continual movement of the shallow waters insures full oxygenation at all times. The oxygen tension in air is the same as in saturated water and, as long as animals remain damp, atmospheric oxygen can readily diffuse across respiratory membranes. The oxygen requirements of sessile and sedentary animals are not large, and some forms at least are capable of withstanding temporary anaerobic conditions. Some intertidal forms, however, do have adaptations to meet specialized needs in respect to oxygen. Species living high in the intertidal zone may show structural modifications enabling them to carry on aerial respiration. Some species of the periwinkle *Littorina* have considerable vascularization of the wall of the mantle cavity, which serves at least to some degree as a lung. Crabs such as *Pachygrapsus* have gills reduced in size and stiff enough for self-support, arrangements for retaining water in the gill cavity, and vascularization of the wall of the cavity itself. Mudflat forms, stranded in the substrate with a limited water and oxygen supply, also have a problem. The echiuran *Urechis* has coelomic corpuscles containing hemoglobin; when tide flats are submerged, the hemoglobin is fully saturated with oxygen and does not function as a respiratory pigment. But at low tide, as the oxygen in the burrow water and in the body tissues becomes depleted, the hemoglobin begins to release its stored oxygen to the tissues. The stored oxygen is sufficient to enable the animal to last until the tide again comes in over the flats, bringing oxygenated water to the burrow.

3) **Reproduction** is somewhat more difficult to study in the field. Many intertidal benthic forms shed eggs and sperm into the sea,

where developing embryos and feeding larvae lead a pelagic existence for a time. Pelagic stages may be taken in plankton tows but otherwise are seldom seen. However, a large number and variety of marine and brackish-water organisms retain, carry, or brood their eggs (e.g., the sea cucumbers *Cucumaria curata* and *Lissothuria nutriens*, the starfish *Leptasterias*, many crustaceans, colonial ascidians, the viviparous polychaete *Neanthes limnicola*). Other invertebrates produce characteristic egg cases (e.g., the snails *Nucella* and *Acanthina*, most nudibranchs and cephalopods, certain polychaete worms). These should be observed and if possible connected with the animals that produced them. Still other animals reproduce asexually and may form extensive colonies or aggregations of individuals (e.g., the aggregating anemone *Anthopleura elegantissima*, colonial ascidians, hydroids, bryozoans, the hydrocoral *Stylantheca*, the alcyonarian *Clavularia*, sponges). Observations of such features in the field or in individuals collected and brought to the laboratory are important in natural-history studies, and may well result in previously unrecorded discoveries.

4) Finally, the questions of **protection** and of the environmental conditions from which organisms require protection are crucial to field studies. Animals persistently present at any particular spot must be adapted to survive the most unfavorable extremes of environmental conditions that occur there. All mechanisms—morphological, physiological, and behavioral—through which an animal preserves internally the conditions for cellular life are here implied in the term protection. In this sense, protection includes the sheltered places sought by motile forms like crabs, the production of anchored fortresses of lime by barnacles, the capacity of brackish-water (and freshwater) organisms to osmoregulate, the ability of the brine shrimp to tolerate a warm and hypersaline medium, and a myriad other adaptations that enable organisms to withstand conditions *externally* which they could not tolerate *internally*.

ENVIRONMENTAL FACTORS OF THE INTERTIDAL ZONE

Owing to the regular rise and fall of the sea, the intertidal zone is a region of great periodic fluctuations in environmental conditions. It is a region of transition between sea and land, but the boundary between aquatic and terrestrial conditions is less sharp and far more complex than that at the shore lines of lakes and ponds. In the intertidal region, animals are subjected to conditions that are alternately marine and semiterrestrial.

Even with their various protective adaptations, animals at the seashore are more or less limited to particular habitats, since each is adapted to a particular way of life under a particular set of conditions. Physical factors of the environment, interacting with specific requirements and adaptations of animals (for food, oxygen, reproduction, and protection), definitely limit the distribution of each species. The most important of the physical factors with which we will be concerned in the field are outlined below.

1) **Nature of the substrate.** The substrate may vary from nearly unbroken cliffs and rocky ledges, through a series of such intergrades as broken rocky reefs, boulders, and pebbles, to the finer substrates of sand, mud, and clay. The nature of the substrate is of cardinal importance in limiting animal and plant distribution, as can be seen at once by comparing the biota of a rocky outcropping with that of an adjacent stretch of sand in a region where both occur. Even the type of rock (e.g., hard granite *vs* relatively soft sandstone or shale) is of great importance, particularly for rock-boring forms.

2) **Degree of wave shock and current action.** The pounding and abrasive action of waves has many effects on both environment and fauna. Where severe, wave action prevents the accumulation of substrates of mud and clay, it restricts the distribution of animals that are not tough, flexible, firmly rooted, or capable of clinging tightly to (or burrowing or boring into) the substrate, and it makes possible the presence of aquatic or semiaquatic forms in a splash zone above the highest level reached by the surface of the sea. Currents may erode soft substrates in one area and redeposit them elsewhere; this shifting of bottom materials and the scouring action of particle-laden water may have important effects on bottom-dwelling organisms present. Ricketts, in *Between Pacific Tides,* clearly recognized the importance of the degree of exposure to wave shock as a factor in limiting animal distribution; his primary ecologic divisions of the Pacific intertidal zone, based on this factor, are "open coast," "protected outer coast," and "bay and estuary," with appropriate subdivisions of each according to the nature of the substrate.

3) **Altered temperatures.** At low tides on sunny summer days and windy winter nights, the temperature extremes for exposed animals markedly exceed those found in the sea. Ability to withstand great changes in temperature, and particularly the higher temperatures, is one important factor in determining the intertidal distribution of organisms.

4) **Desiccation.** Most dwellers in the intertidal zone have come up from the sea; very few are invaders from the land. It follows that withstanding desiccation is a great problem for many intertidal forms, and surviving prolonged submersion is far less commonly a problem. Ex-

posed intertidal organisms tend to dry out, particularly during low tides on hot, sunny days, and the problem is closely related to that of withstanding higher temperatures. Motile forms may show a protective behavior, shifting to positions under rocks or ledges or seeking the slighter protection of depressions and shallow crevices. Sedentary and sessile forms may have protective shells, adherent layers of sand or gravel, tough and often mucus-covered integuments, or large internal stores of water, etc.

5) **Altered salinity.** On hot days at low tide the salinity of high tide pools may rise owing to evaporation. During rains, direct precipitation and runoff from streams and beaches may greatly reduce the salinity in some areas. Organisms on high rocks, in high pools, and near stream mouths must be able to avoid, regulate against, or tolerate at least temporary hypersaline and/or brackish conditions.

6) **Lowered oxygen availability.** Only in certain instances is this an important factor. It becomes important where intertidal forms are desiccated to a degree that exposed surfaces no longer serve as respiratory membranes. It may also be important for dwellers in mud flats exposed by tides.

7) **Predators.** At low tides birds, racoons, and man, at high tides fishes, may invade the intertidal zone, in addition to the large assortment of invertebrate predators.

As a result of all these factors and of many others that usually are less accessible to study, intertidal environments, floras, and faunas may vary greatly, even within relatively restricted regions of the coast. Further, at any one locality one always finds a vertical intertidal zonation of organisms. Plants and animals most resistant to vicissitudes live higher in the intertidal zone; those less tolerant live lower down. Vertical zonation is clearest where the substrate presents uniform, vertical surfaces of rock, as on the walls of deep intertidal channels. It is most complex in areas of irregular and mixed substrates, which provide many different habitats. For a discussion of the general problem of intertidal zonation see Ricketts, Calvin, and Hedgpeth: *Between Pacific Tides*, 4th edition (1968), chapter 14.

The term **ecological niche** is used to refer to the sum total of environmental conditions, physical and biotic, to which a particular species is adapted and which are necessary for its continued existence. **Habitat,** on the other hand, refers to a place or region of a certain kind, characterized by particular major environmental features. A habitat such as a tidepool possesses a characteristic fauna, each species of which has its own niche. Since the distribution of ecological conditions within a habitat may spatially limit an ecological niche we may, for convenience, connect certain niches with places within the habitat, but this is not true of many niches and is not really correct for any niche. To define a niche completely would involve an immense amount of quantitative and experimental investigation of the structure, behavior, and physiology of a species, and a study of the precise physical and biotic conditions of its habitat. However, we are usually able to discern one or more obvious features of each niche which give it objective reality and allow us to give it a preliminary definition and to use it in discussion.

A NOTE ON THE USE OF THIS MANUAL

Subject matter is grouped according to phyla, classes, orders, or other convenient assemblages. For each group there is a general introduction, which may be more or less detailed depending on the needs of the average student and the availability of the information in general texts. Terms necessary for the use of keys are explained and illustrated, and some picture of the group as a whole is attempted.

The keys will be found to vary in their completeness of coverage, geographical range of usefulness, ease of use, and accuracy. Although this is not desirable, it results from differences in numbers of species in various groups, the ease of separating species from each other, the completeness of our knowledge of the group, the professional background of each author, and his success in constructing the key itself. Only use by students and investigators will tell us how suitable a given key may be. The editors and contributors alike welcome criticism and information leading to improvement and revision.

Species lists follow the keys. These lists are of species reliably reported from the central California intertidal zone. Names of species or higher taxa not included in the keys are marked in the lists by asterisks (*). Unless otherwise indicated, the coverage in keys and lists is the intertidal zone from the Monterey Peninsula to Bodega Head, inclusive. Geographical ranges of species are not generally stated because these are often uncertain and constantly subject to revision. Species lists have been annotated with certain general information and references which may be interesting or helpful. Users of this book will doubtless add to these notations, and such marginalia will be of great usefulness in future revisions of this manual.

SOME GENERAL REFERENCES ON INTERTIDAL LIFE

The number of books and papers on the seashore and on intertidal invertebrates and intertidal ecology is so vast that little would be gained here by attempting a complete or even an extensive bibliography.

Rather, we present only a short representative list of significant and pertinent works, in which the reader will find extensive bibliographies as well as general and comprehensive discussions of intertidal life. Several, elementary but of excellent quality, are included for beginning students and newcomers to the seashore; a number of other elementary guides of doubtful quality are on the market. References to particular animal groups will be found at the ends of the respective chapters of this manual. The annotated Systematic Index and General Bibliography in Ricketts and Calvin, *Between Pacific Tides*, is of the utmost value, and no attempt to rival it will be made here.

Gislén, T. 1943–44. Physiographical and ecological investigations concerning the littoral of the northern Pacific. Sections 1–4. Lunds Univ. Årsskr. (N. F.) Avd. 2, 39(5): 64 pp. and 40(8): 92 pp. A perceptive comparison of faunal and environmental aspects of the Pacific Grove intertidal zone with that of Misaki, Japan, at the same latitude.

Glynn, P. W. 1965. Community composition, structure, and interrelationships in the marine intertidal *Endocladia muricata—Balanus glandula* association in Monterey Bay, California. Beaufortia (Zool. Mus., Amsterdam) 12: 1–198. A very detailed account of a common intertidal association.

Hedgpeth, J. W. (ed) 1957. *Treatise on Marine Ecology and Paleoecology*, Vol. 1. *Ecology*. Memoir 67 (1), Geol. Soc. America, 1296 pp. This extensive work, edited by Hedgpeth (who also contributed 7 out of 29 chapters) contains an outstanding and well-edited series of individually documented reviews of topics in marine ecology, including many related to the intertidal zone. Over 200 pages of annotated bibliographies (in the best Hedgpeth tradition) provide a fine selective coverage of all groups of marine organisms. Even if never revised, it will remain the best unified group effort of its kind up to the date of its publication.

Hedgpeth, J. W. 1962. *Introduction to Seashore Life of the San Francisco Bay Region and the Coast of Northern California*. University of California Press, 136 pp. (California Natural History Guides: 9) An interesting, readable, and useful beginners' handbook.

Hinton, S. 1969. *Seashore Life of Southern California*. University of California Press, 181 pp. (California Natural History Guides: 26) Comparable to the small Hedgpeth handbook, but for the region south of Santa Barbara.

Johnson, M. E. and H. J. Snook. 1927. *Seashore Animals of the Pacific Coast*. Macmillan, 659 pp. Reprinted (Dover), 1972. Useful to beginners for identification of common forms, although many parts are out of date and others never were at all complete.

Kozloff, E. N. 1973. *Seashore Life of Puget Sound, the Strait of Georgia, and the San Juan Archipelago*. University of Washington Press, 282 pp. Comparable to the books by Hedgpeth and Hinton on seashore life of the California coast in that it covers both plant and animal life in non-technical language; directed to the beginner and the general public, this well-illustrated book is organized by typical habitats rather than by animal groups. Written by an able biologist, it provides a good general picture of the shallow-water marine life of the Puget Sound area. Detailed keys for the identification of intertidal life of that region have also been published by the author (Univ. Wash. Press, 1974).

MacGinitie, G. E. and N. MacGinitie. 1968. *Natural History of Marine Animals*. Second ed., McGraw-Hill, 523 pp. Contains a great many interesting and useful observations on the biology of West Coast invertebrates, by a husband/wife team of fine observers.

Morris, R. H. and D. P. Abbott. *Marine Invertebrates of California Shores: An Illustrated Guide.* Stanford University Press, about 300 pp. Color photographs of several hundred of the larger invertebrate species by Morris and others, with an extensive text by numerous contributors under the painstaking editorship of Abbott, now in late stages of preparation. There are no keys, but the book contains much more extensive accounts of the biology of many individual species than we could include in an identification manual and, with its illustrations, should be a valuable supplement to the Light Manual.

Ricketts, E. F. and J. Calvin 1968. *Between Pacific Tides.* 4th ed., rev. by J. W. Hedgpeth. Stanford University Press, 614 pp. We urge every user of the present work to buy and read this book, since in many essential respects it supplements the Light Manual. Treatment of intertidal animals is primarily from the standpoint of ecology and natural history; the book is less useful for the identification of any but relatively distinctive forms. An excellent chapter on intertidal zonation was added by Hedgpeth to the third (1952) edition. Although Hedgpeth is still listed as reviser on the title page of the fourth edition, he is in fact a co-author. The fourth edition continues the high standards of, but differs materially from, the third and contains a great deal more that is pure Hedgpeth, especially chapters 13, "Recent Developments," and 15, "Beyond the Tides," but evident also in many other new passages and illustrations. There is an extensively annotated and valuable General Bibliography.

Stephenson, T. A. and A. Stephenson 1972. *Life Between Tidemarks on Rocky Shores.* W. H. Freeman & Co., 425 pp., illus. This represents the study of the biota of rocky shores carried out on a worldwide basis over a period of more than thirty years by T. A. Stephenson, until his death in 1961. This book was completed by his wife and collaborator, Anne Stephenson, and presents in detailed but unified form a comparative picture of plant and animal zonation on rocky shores. Essentially descriptive, it will be of especial value to those who can visit other coasts; to those working in California, the Stephensons' descriptions of the shores of Vancouver Island, Pacific Grove, and La Jolla will be of most immediate interest. It sets a standard of illustration and interpretation that the student will do well to study.

Yonge, C. M. 1949. *The Sea Shore.* London: Collins, 311 pp., illus. Although descriptive of British shores, "This is the finest book of this genre yet published; a *sine qua non* for the bookshelf of all who go to the shore." (Hedgpeth, in Ricketts and Calvin, 1968, p. 565). We agree wholeheartedly; beautifully written and illustrated.

INTRODUCED INTERTIDAL INVERTEBRATES

James T. Carlton
California Academy of Sciences, San Francisco

The identification of invertebrates in estuaries, lagoons, and brackish bays of the central California coast is complicated by the presence of species introduced from other estuarine systems of the world, particularly those of New England and Japan. The large number of introduced species and their impact in our estuaries have long gone unrecognized, partly because no comprehensive review of the known introduced fauna has been previously made, leaving the general impression that only a few large, conspicuous species have been introduced, and partly because many introduced species have been overlooked entirely, redescribed as new species, or misidentified as endemic species.

An **introduced species** is here considered one which was transported to the Pacific coast through man's activities and which has subsequently become established by maintaining a reproducing population. Continued observation of an exotic species may be necessary to determine if it has become established. The result of 130 or more years of introductions has been the establishment on the Pacific coast of perhaps 150 to 200 species of invertebrates, transported from New England, Japan, Australasia, Chile, and other areas, and representing all major invertebrate groups except the Echinodermata. With few exceptions, all of these are restricted to bay, lagoon, and estuarine environments; the open rocky coast or open sandy beach are not recognized as supporting introduced species. This estuarine restriction may be the result, in part, of the lack of mechanisms which might serve to introduce species to non-estuarine areas or, more importantly as discussed below, to the presence here of a mature, endemic, marine fauna and of a relatively immature, endemic, estuarine fauna in our widely separated and geologically rather young bays.

The means by which species have been introduced are summarized below; the first three mechanisms are responsible for the introduction of most of the exotic species. Authoritative identification of an estuarine species established locally as conspecific with an exotic species, its known absence earlier from the fauna, and direct association with one of the mechanisms outlined below are, combined, the primary means to determine the probability of a species being exotic.

1) **Ship Fouling.** The earliest wooden sailing vessels which entered (and occasionally sank in) central California bays may have introduced, as long as two centuries ago, fouling and boring organisms from harbors of the western or southern Pacific and Atlantic. The great influx of ships, many of which were permanently abandoned, into San Francisco Bay in Gold Rush days (1849 and the years following) may have aided in the establishment of numerous species. These include several hydroids (*Obelia* spp., *Syncoryne mirabilis, Tubularia crocea, Garveia franciscana, Clava leptostyla,* and others), the barnacle *Balanus improvisus,* collected as early as 1853 at San Francisco (Carlton and Zullo, 1969), possibly the snail *Ovatella myosotis,* and the entoproct *Barentsia benedeni.* The Australasian tube worm *Mercierella enigmatica,* which appeared in San Francisco Bay at the close of World War I as evidenced by local newspaper accounts (not in the 1930s as has been thought), was also carried to this coast by ships. The New Zealand isopod *Sphaeroma quoyana* and its commensal isopod *Iais californica* (an unfortunate name for an introduced species!) were probably 19th century arrivals in central California, as may have been the Atlantic gribbles *Limnoria tripunctata* and *L. quadripunctata* with their commensal protozoans, flatworms, and copepods. To this list can be added the Atlantic shipworms *Teredo navalis* and *Lyrodus pedicellatus.*

While ship fouling has been reduced in recent years by the development of more efficient anti-fouling paints, this means of introduction may be expected to continue. In the propeller-shaft housing of aircraft carriers returning from Asia, the author has found an abundant fouling fauna of barnacles, tube worms, bryozoans, and other organisms, some of which might become established in warm-water pockets in San Francisco and other bays. Thus, for example, the introduced barnacle *Balanus amphitrite* maintains small populations in San Francisco Bay in areas of shallow warm water.

2) **Ship Ballast.** Ships may also transport organisms in water or shingle ballast. The Oriental shrimp *Palaemon macrodactylus,* which appeared in San Francisco Bay in the early 1950s after the Korean war and is now also known from Los Angeles Harbor, may have been transported in water-ballast tanks by returning ships (Newman, 1963). Infaunal species may also be transported by this method (e.g., the Atlantic polychaete *Neanthes caudata* in southern California harbors), the larval stages being pumped in at a foreign harbor and either the larvae or adults released at the port of arrival. Discharge of shingle ballast (stones, algae, and debris gathered from beaches) by lumber

ships returning from Chile in or before the 1900s was apparently responsible for the introduction of the Chilean beachhopper *Orchestia chiliensis* into San Francisco Bay.

3) **Commercial Oyster Industry.** Extensive importations of adult and seed oysters, *Crassostrea virginica* from the Atlantic coast (1860s to 1910s) and *Crassostrea gigas* from Japan (1920s to the present), into central California bays have been responsible for the introduction of a large number of Atlantic and Japanese estuarine invertebrates. Early shipments of oysters carried abundant epizoics, including sponges, cnidarians, polychaetes, molluscs, crustaceans, bryozoans, and other invertebrates; ironically, many of these species proliferated although the oysters never became established locally. Those species probably introduced with Japanese oysters include the polychaetes *Pseudopolydora kempi, P. paucibranchiata,* and *Spirorbis* sp.; the horn snail *Batillaria attramentaria* and the oyster drill *Ceratostoma inornatum;* the bivalves *Musculus senhousia, Lasaea* sp., and *Tapes japonica;* the nudibranchs *Okenia plana, Eubranchus misakiensis, Trinchesia* sp.; the oyster copepod *Mytilicola orientalis;* the amphipods *Corophium uenoi* and *Grandidierella japonica;* the isopod *Gnorimosphaeroma rayi;* the ectoproct *Victorella pavida;* and other invertebrates. Species introduced with Atlantic oysters include the sponges *Microciona prolifera, Halichondria bowerbanki, Prosuberites* sp. and *Cliona* spp.; cnidarians including the sea anemone *Diadumene leucolena* and apparently several hydroids (such as *Cordylophora lacustris*); the flatworm *Childia groenlandica* and the trematode *Parvatrema borealis;* several polychaetes including *Asychis elongata, Polydora ligni, Streblospio benedicti,* and *Neanthes succinea;* the gastropods *Crepidula convexa, Crepidula fornicata, Crepidula plana, Nassarius obsoletus* (with its schistosome trematode *Austrobilharzia variglandis*), *Busycotypus canaliculatus,* and *Urosalpinx cinerea;* the bivalves *Gemma gemma, Ischadium demissum, Mya arenaria,* and *Petricola pholadiformis;* a number of amphipods but apparently no isopods; the crab *Rhithropanopeus harrisii;* several ectoprocts; and perhaps the tunicate *Molgula manhattensis,* though the latter may have arrived in ship fouling.

Stricter measures to control the accidental introduction of oyster pests are now in effect, but small, more cryptic, encrusting invertebrates are still brought in with shipments of oyster spat, which must continually be imported to maintain the oyster industry on the California coast. In addition, some experimental shipments of other oyster species from Australia, New Zealand, and Europe have been planted in local bays (Hanna, 1966), and small shipments (about 200 bushels per year) of Atlantic oysters from New York State are still imported and planted in California. Some epizoics associated with these oysters may be expected to become established. Barrett (1963) has summarized the history of the California oyster industry, although her introductory sentence, "The California oyster industry is an interesting example of man's adapting an otherwise unproductive part of the landscape to his benefit," remains a classic example of the misunderstanding, not only by geographers, of the importance of mudflats in estuarine ecosystems. Indeed, the aboriginal shell mounds of southern San Francisco Bay alone indicated the great abundance of native oysters (*Ostrea lurida*), and so many bay mussels (*Mytilus edulis*) were gathered by Indians of the bay area that perhaps as much as one-third of the harvest was shipped inland to Indians of the Sacramento-San Joaquin River Delta (Cook, 1946).

4) **Commercial Bait and Fresh Seafood Importation.** Worms (glycerids and nereids) for bait shops and living lobsters (*Homarus americanus*) for restaurants are packed in algae (such as *Ascophyllum* and *Fucus*) and shipped from New England to this coast. The algae, containing numerous invertebrates, may then be discarded in local waters after arrival of the shipment. The presence of small numbers of the Atlantic periwinkle *Littorina littorea* in San Francisco and Newport bays may be attributed to such discarded algae. Other Atlantic species, not yet recognized, are undoubtedly present (see Miller, 1969).

To what extent deliberate, private introductions by individuals of food or bait species have been practiced can never be known, but doubtless such introductions have been and still are going on. The freshwater Asiatic clam *Corbicula manilensis* (the shells of which are common as discarded bait on many ocean and bay beaches) likely owes its initial introduction to attempts by immigrants to establish it. The Atlantic quohog *Mercenaria mercenaria,* often imported by restaurants as well as by individuals, has been found living in Humboldt Bay, San Francisco Bay, and appears to be established in Colorado Lagoon, Long Beach. Certain scientists have been known to culture exotic species in open seawater systems of marine laboratories for experimental purposes, or to place them in convenient, nearby natural habitats with little regard for the possible ecological consequences. Mariculture activities have led to the importation of various species epizoic on the animals being cultured as well as organisms used for feeding the mariculture stocks. Where mariculture operations are maintained in open seawater systems discharging into the sea or into bays, the escape of gametes or larvae is possible. Experimentalists and others need to exercise care and responsibility in such matters.

Misconceptions concerning the distribution of introduced species in California are that oyster epizoics can occur only in areas where oysters have been planted or that cosmopolitan, ship-fouling species can occur only in major ports and harbors. Clearly, once a species has been introduced to this coast, secondary distribution, both natural (as larvae or by rafting) and by man (as by small-craft traffic), can and does occur, and an introduced species may be common in a lagoon or estuary which has had neither oyster culture nor traffic by large ships.

The recognition of a species as exotic, and its correct identification are complicated by several factors. First, the systematics of the endemic species of a group (e.g., protozoans, flatworms, sponges, hydroids, smaller crustaceans, and numerous others) in central California may not be sufficiently advanced to distinguish endemic from exotic species. In these cases, one must rely upon secondary means to establish whether or not a species is exotic. These secondary means include localized occurrence (disjunct distribution) in often widely separated harbors and bays, absence from Indian shellmounds (edible shelled species as well as certain epizoics such as barnacles and bryozoans), and recent appearance in large numbers in an area generally well known faunistically (and from which the species were previously unreported). Disjunct distribution is often obscured in the literature by generalizations of the range of a species; thus we may find a certain species recorded as occurring from "Willapa Bay, Washington to San Francisco Bay" when in fact it may be known *only* from Willapa Bay and San Francisco Bay. Recent arrivals of exotic species must be distinguished from seasonal fluctuations in abundance of endemic animals, such as cephalaspideans, and from temporary extensions of range.

Secondly, the source or "donor" area of an introduced species could conceivably be one of many different estuarine systems of the world, requiring a knowledge, which even most specialists lack, of the world fauna of a given taxon.

Thirdly, the determination of a species as exotic has often rested upon the erroneous criterion that, if it is introduced, the species in question must already be described and be known elsewhere in the world. This assumption leads to the argument that certain taxa must therefore be endemic to this coast if they cannot be referred to any described species elsewhere. In fact, an exotic species could well be undescribed and unknown in the donor area, and this especially holds true for smaller and taxonomically more difficult groups. For example, the tubeworm *Mercierella enigmatica* (abundant in several parts of San Francisco Bay) was first described from France, while undescribed in Australasia where it was endemic; the small, commensal, New Zealand isopod *Iais californica* was first described only after being introduced to California, and the Japanese oyster flatworm *Pseudostylochus ostreophagus* was described after its introduction to Puget Sound (Hyman, 1955).

Examples of species that are probably exotic, but that have not yet been recognized from other areas of the world, are the sea anemone *Diadumene franciscana* and the isopod *Synidotea laticauda*, the former known from Morro Bay and San Francisco Bay, and the latter known only from San Francisco Bay. Both are often associated in San Francisco Bay (their type locality) with introduced estuarine communities. However, Menzies and Miller (1972) have claimed that *S. laticauda* is a relict, warm-water species in San Francisco Bay.

These and other problems have led to the redescription of not a few introduced species as new species, their exotic status being recognized from within a few months to more than fifty years after being described. The Atlantic soft-shelled clam *Mya arenaria* was described in San Francisco Bay as *M. hemphillii;* the Japanese cockle *Tapes japonica* was described as *Paphia bifurcata* in British Columbia; the Japanese oyster copepod *Mytilicola orientalis* as *M. ostrea;* the Atlantic ostracod *Sarsiella zostericola* as *S. tricostata;* the Atlantic polychaete *Streblospio benedicti* as *S. lutinicola;* the cosmopolitan shipworm *Lyrodus pedicellatus* as both *Teredo diegensis* and *T. townsendi*, and *Teredo navalis* as *T. beachi;* the Chilean amphipod *Orchestia chiliensis* as *O. enigmatica;* the New Zealand isopod *Sphaeroma quoyana* as *S. pentodon;* the Atlantic snail *Ovatella myosotis* as *Alexia setifer;* the cosmopolitan tunicate *Styela plicata* as *S. barnharti;* and others. This unnecessary proliferation of new names not only disguises the exotic nature of such species, but should also serve as a warning to zoologists contemplating the description of "new" species or subspecies from any California bay, lagoon, or estuary.

Certain introduced species have been reported from offshore or other non-estuarine habitats. This would seem to contradict the general restriction of introduced species to estuarine areas. The Atlantic gem clam *Gemma gemma* (common in some bays on the Pacific coast) has been reported from 15 and 50 fathoms in Monterey and San Diego bays, respectively. These records may have led Hedgpeth (1968) to conclude that the case for *Gemma* as an introduced species may be moot. The Atlantic slipper shell *Crepidula convexa* was reported from Moss Beach on the open rocky coast of San Mateo County. These records were based on misidentifications of other, endemic, species and should alert workers to pay special attention to identifications from the open coast or offshore waters of introduced species now considered to be restricted to bays and estuaries.

Certain species which are indigenous to this coast, as evidenced by presence in Pleistocene deposits or in Indian shell middens (such as the mussel *Mytilus edulis*), or by occurrence on the open coast or in deep offshore waters where the fauna is generally recognized as indigenous (such as the sea anemone *Metridium senile*, certain caprellids such as *Caprella equilibra* and *C. penantis,* and the sea-squirt *Ciona intestinalis*), are also associated with cosmopolitan ship-fouling faunas. It seems likely that some bay or estuarine populations of these species may represent, at least in part, exotic stocks introduced among endemic populations of the same species, a situation possibly detectable by techniques of population-genetic analysis. Still other species are apparently indigenous to the Pacific coast as far south as Puget Sound, but may have been transported to central California by man; candidates are the bivalve *Macoma balthica* and the isopod *Limnoria lignorum.*

The reasons for the successful introduction of so many exotic species are varied and not fully known. The extensive modification by man of estuarine environments in California and elsewhere on the Pacific coast through filling, dredging, pollution, damming of streams, and construction of wharves, docks, and marinas, has perhaps served to modify certain habitats or to provide new habitats for introduced organisms. Further, the relatively young age of the estuaries of our coast, as compared to the more mature estuarine systems of the Atlantic and Gulf coasts, as well as the fact that large estuarine systems on this coast are few and widely separated, may have resulted in a less diversified endemic, estuarine biota on the Pacific coast, allowing more highly adapted exotic, estuarine species to become established, either occupying "open niches" or successfully competing with what few endemic, estuarine species there are or may have been.

Certain estuarine areas of central California are thus dominated by an introduced fauna which would not be unfamiliar to biologists from Massachusetts or Japan. The Palo Alto Yacht Harbor in south San Francisco Bay furnishes a spectrum of the introduced fauna; here the majority of the macroscopic invertebrates are introduced species. On the pilings are colonies of the hydroids *Obelia* spp., *Syncoryne mirabilis,* and *Tubularia crocea* (the latter with the tiny nudibranch *Tenellia adspersa*); the barnacle *Balanus improvisus;* the ectoproct *Conopeum reticulum,* the entoproct *Barentsia benedeni;* and the tunicate *Molgula manhattensis,* all introduced from the Atlantic; and the little, orange-striped sea anemone *Haliplanella luciae,* possibly from Asia. The isopod *Synidotea laticauda,* its homeland unknown, is abundant, as are several species of the small, cosmopolitan amphipod *Corophium.* Boring into pilings are the gribble *Limnoria tripunctata,* which reached the bay in ships perhaps from the Atlantic, and the New Zealand isopod *Sphaeroma quoyana,* (with its also introduced commensal isopod *Iais californica*) which also forms burrows in mud banks near the docks. Below the floats, on the mud bottom, are great numbers of the mud snail *Nassarius obsoletus* from New England and dense nests of the little mussel *Musculus senhousia* from Japan. Burrowing in the mud are the Japanese cockle *Tapes japonica,* the Atlantic soft-shelled clam *Mya arenaria,* and the Japanese amphipod *Grandidierella japonica.* The Oriental shrimp *Palaemon macrodactylus* swims rapidly about over the bottom and rests in the fouling. In the salt marsh nearby the Atlantic mussel *Ischadium demissum* and snail *Ovatella myosotis* are abundant. The striking faunal changes caused by the introduction of these estuarine wanderers warrants detailed investigations of their distribution and ecological relationships.

I thank Jon Standing, Barry Roth, and Joel Hedgpeth for helpful suggestions.

REFERENCES ON INTRODUCED INTERTIDAL INVERTEBRATES

Abbott, D. P. and J. V. Johnson 1972. The ascidians *Styela barnharti, S. plicata, S. clava,* and *S. montereyensis* in Californian waters. Bull. So. Calif. Acad. Sci. 71: 95–105.

Barnard, J. L. 1950. The occurrence of *Chelura terebrans* Philippi in Los Angeles and San Francisco harbors. Bull. So. Calif. Acad. Sci. 49: 90–97 (mostly taxonomic).

Barnard, J. L. 1961. Relationship of Californian amphipod faunas in Newport Bay and in the open sea. Pac. Nat. 2: 166–186. A brief discussion of introduced faunas.

Barnard, J. L. and D. J. Reish 1959. Ecology of Amphipoda and Polychaeta of Newport Bay, California. Allan Hancock Found. Occ. Pap. 21. Briefly discusses introduced cosmopolitan faunas.

Barrett, E. M. 1963. The California oyster industry. Calif. Dept. of Fish Game, Fish Bulletin 123: 103 pp.

Carlton, J. T. and V. A. Zullo 1969. Early records of the barnacle *Balanus improvisus* Darwin on the Pacific coast of North America. Occ. Pap. Calif. Acad. Sci. no. 75: 6 pp.

Cook, S. F. 1946. A reconsideration of shellmounds with respect to population and nutrition. American Antiquity 11: 50–53.

Elton, C. S. 1958. *The Ecology of Invasions by Animals and Plants.* John Wiley and Sons, 181 pp.

Hanna, G D. 1966. Introduced mollusks of western North America. Occ. Pap. Calif. Acad. Sci. 48: 108 pp.

Hedgpeth, J. W. 1968. Newcomers to the Pacific coast: the estuarine itinerants. *In* Ricketts, E. F. and J. Calvin, *Between Pacific Tides,* 4th ed., rev. by J. W. Hedgpeth, 376–380. Stanford University Press.

Hyman, L. H. 1955. The polyclad flatworms of the Pacific coast of North America: additions and corrections. Amer. Mus. Novitates no. 1704, 11 pp. (*Pseudostylochus ostreophagus,* Puget Sound)

Jones, L. L. 1940. An introduction of an Atlantic crab into San Francisco Bay. Proc. 6th Pac. Sci. Congress 3: 485–486. (*Rhithropanopeus harrisii*).

Lachner, E. A., C. R. Robins, and W. R. Courtenay 1970. Exotic fishes and other aquatic organisms introduced into North America. Smithson. Contrib. Zool. no. 59: 29 pp.

Lindroth, C. H. 1957. *The Faunal Connections between Europe and North America.* John Wiley and Sons, 344 pp.

Menzies, R. J. 1958. The distribution of wood-boring *Limnoria* in California. Proc. Calif. Acad. Sci. (4) 29: 267–272.

Menzies, R. J. and M. A. Miller 1972. Systematics and zoogeography of the genus *Syni-*

dotea (Crustacea: Isopoda) with an account of Californian species. Smithson. Contrib. Zool. no. 102: 33 pp.

Miller, R. L. 1969. *Ascophyllum nodosum:* a source of exotic invertebrates introduced into west coast near-shore marine waters. Veliger 12: 230–231.

Newman, W. A. 1963. On the introduction of an edible oriental shrimp (Caridea: Palaemonidae) to San Francisco Bay. Crustaceana 5: 118–132. (*Palaemon macrodactylus*)

Quayle, D. B. 1964. Distribution of introduced marine Mollusca in British Columbia waters. Jour. Fish. Res. Bd. Canada 21: 1155–1181.

Rotramel, G. 1972. *Iais californica* and *Sphaeroma quoyanum,* two symbiotic isopods introduced to California (Isopoda, Janiridae and Sphaeromatidae). Crustaceana Suppl. III, 193–197.

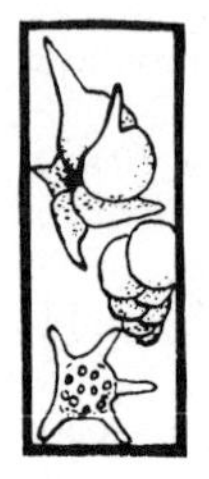

PROTOZOA

Zach M. Arnold
Department of Paleontology,
University of California, Berkeley

(PLATES 1–2)

The zoologist who wishes to enhance his understanding of the marine environment and enrich his general biological experience will find that practically any habitat yielding macroscopic invertebrates also supports a challenging complement of protozoans, once suitable techniques for their recovery and examination are employed. Several types are found attached to metazoans or crawling over them, but others are best procured through plankton tows, epibenthic dredging, centrifuging of sea water, or by sieving washings from algae, rock surfaces, sediments, and invertebrates.

The small size of most flagellates makes prior concentration generally desirable, but such dinoflagellates as the luminescent *Noctiluca scintillans* (fig. 2) may occur in numbers sufficient to discolor tidepools or illuminate footprints along a beach at night. *Gonyaulax polyedra* (fig. 1), whose blooms often produce "red tides," is widely distributed and easily recognized. Though minute, *Oxyrrhis marina* (fig. 3) so often dominates the early phases of successional changes in laboratory dishes as to be commonly encountered.

One of the largest (commonly 2–3 mm in diameter) and most conspicuous intertidal protozoans is *Gromia oviformis* (fig. 4), a brownish spherical or ovoidal sarcodinian that nestles in holdfasts and tufts of algae, on sponges, hydroids, and bryozoans, and in any nook moderately shielded from violent water movement. It is often mistaken for eggs or fecal pellets unless left undisturbed to develop its characteristic web of nongranular pseudopodia.

Foraminifera are among the most easily collected marine protozoa. They may often be seen attached to algal fronds, to the perisarc of hydroids (*Sertularia* and *Obelia*, but not *Aglaophenia*), to tufty bryozoans, to shells of molluscs, to rock surfaces, and even to individual sand grains in relatively quiet water. A useful macroscopic indication of their possible presence is any accumulation of grayish or whitish, silt-sized particles that remains on the surface of attachment following

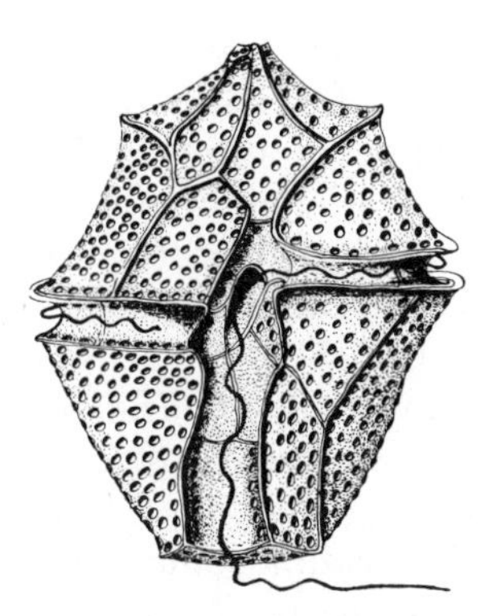

1. *Gonyaulax polyedra*

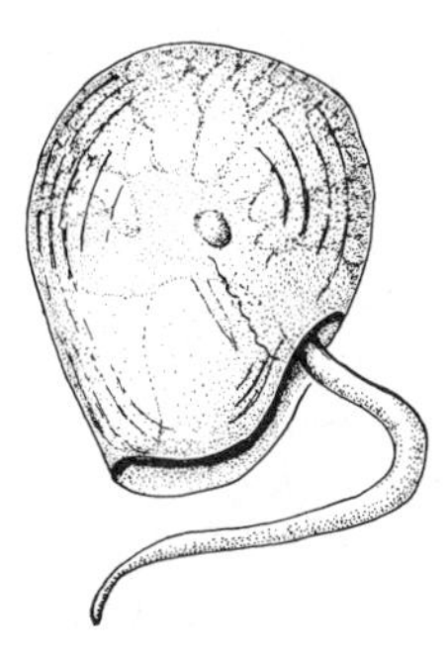

2. *Noctiluca scintillans*

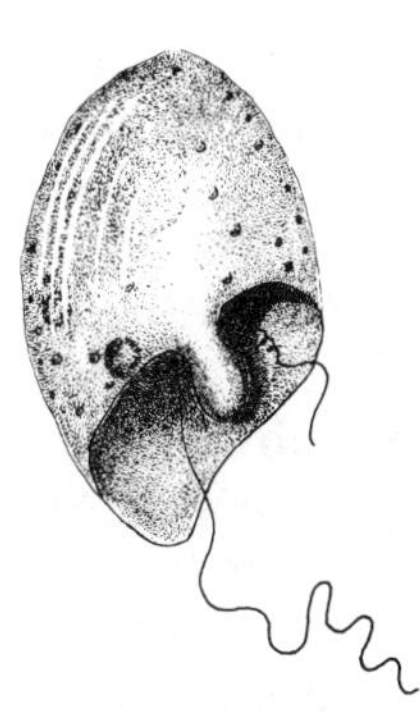

3. *Oxyrrhis marina*

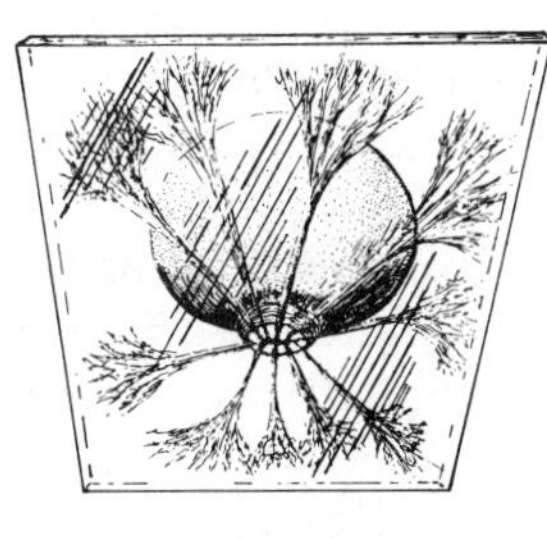

4. *Gromia oviformis*

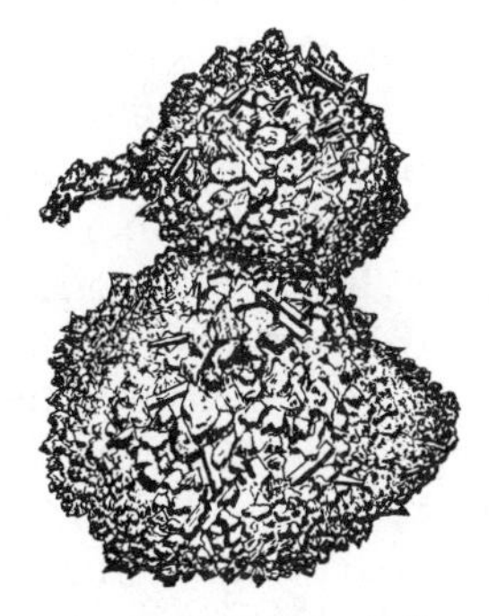

5. *Iridia* cf. *I. serialis*

6. *Haplophragmoides columbiensis* var. *evolutum*

7. *Cornuspira lajollaensis*

8. *Quinqueloculina angulostriata*

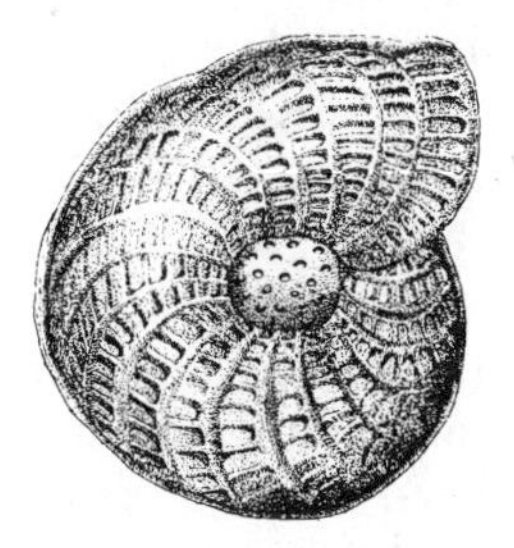

9. *Elphidium crispum*

PLATE 1. **Protozoa** (1). Sizes and useful references for protozoans shown in figures 1–9: 1, 40 µm (Kofoid, 1911); 2, 500 µm to 3 mm (Kofoid and Swezy, 1921); 3, 20–30 µm (Hall, 1925); 4, to 4.5 mm (Jepps, 1926; Hedley, 1962; Arnold, 1972); 5, 500 µm to 5 mm (Le Calvez, 1936); 6, 700 µm to 1 mm (Towe, 1967); 7, 400 µm (Uchio, 1960); 8, to 1.5 mm (Cushman and Valentine, 1930); 9, to 1 mm or more (Jepps, 1942; Myers, 1942). Figures 4–9 are original; the rest in Plates 1 and 2 are modified after various authors.

vigorous swilling under water. Scrubbings from algae, invertebrates, rocks, and debris often yield sediments rich in foraminifera, and strandline deposits occasionally contain good concentrations of their tests (shells). Nondescript accumulations of well-cemented sand grains are often the barely recognizable tests of agglutinated foraminifera. Rocky coasts of granite or sandstone yield good numbers of *Iridia* resembling the Mediterranean *Iridia serialis* (fig. 5), but the multichambered *Haplophragmoides columbiensis* var. *evolutum* (fig. 6) is more easily collected and identified. *Cornuspira lajollaensis* (fig. 7) is a distinctive, planispirally coiled representative of the imperforate calcareous complex. Among the taxonomically troublesome miliolids, *Quinqueloculina angulostriata* (fig. 8) is distinctive. The surface sculpture of its test makes the widely distributed and biologically popular *Elphidium crispum* (fig. 9) simple to recognize, but there is taxonomic disagreement about the complex of variable calcareous-perforate forms to which belongs the common *Rosalina columbiensis* (fig. 11), with its wine red early chambers.

Of the numerous and highly diverse ciliated protozoa, the loricate *Tintinnopsis nucula* (fig. 13) is best taken by plankton tow. Both *Mesodinium pulex* (fig. 12), with its flealike movement, and the graceful "swan animalcule" *Lacrymaria olor* var. *marina* (fig. 16), once seen, are immediately recognizable. Among attached ciliates, the stalked suctorian *Ephelota gemmipara* (fig. 10) has characteristic tentacles of two different types: suctorial and prehensile. Various species of the arborescent colonial genus *Zoothamnium* (fig. 14), their stalks contracting in unison, grow attached to living and nonliving surfaces alike, as do representatives of the solitary sessile *Folliculina* (fig. 15), the "bottle animalcules" with blue or green body and prominent ciliated "wings."

Many protozoans ecto- and endoparasitic on or in local marine invertebrates remain to be extensively studied. Endoparasitic gregarine sporozoans and ecto- and endoparasitic ciliates have been described from a wide variety of intertidal invertebrates from central California; these include species from the shore crab *Pachygrapsus* (Ball, 1938); the burrowing clam *Petricola* (Bush, 1937), the shipworm *Teredo* (Pickard, 1927), the small pulmonate snail *Ovatella* (Kozloff, 1945a), phoronids (Kozloff, 1945b), sea urchins (Lynch, 1929), and *Urechis* (Noble, 1938).

Among the readily available and generally useful reference sources for the identification of protozoans (to genus) are those by Kudo (1966) and Jahn and Jahn (1949).

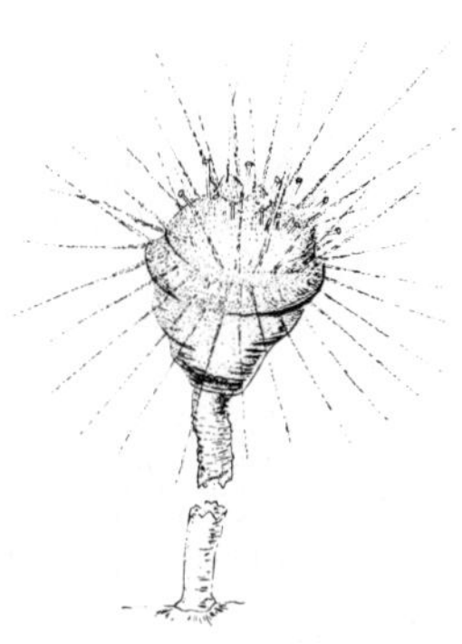

10. *Ephelota gemmipara*

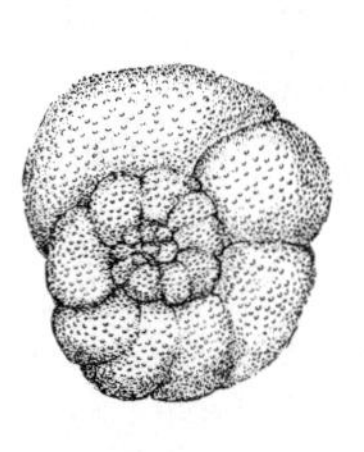

11. *Rosalina columbiensis*

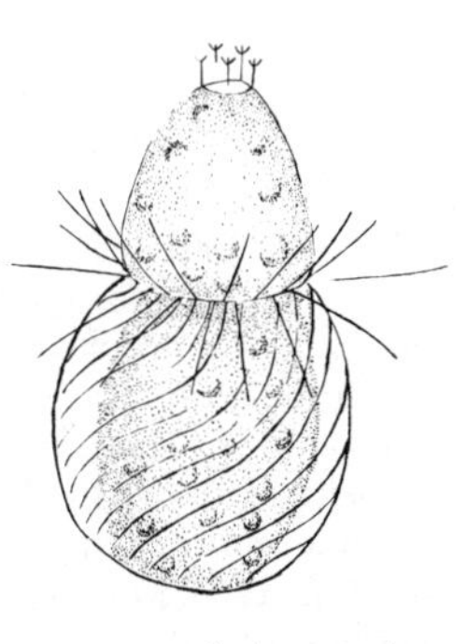

12. *Mesodinium pulex*

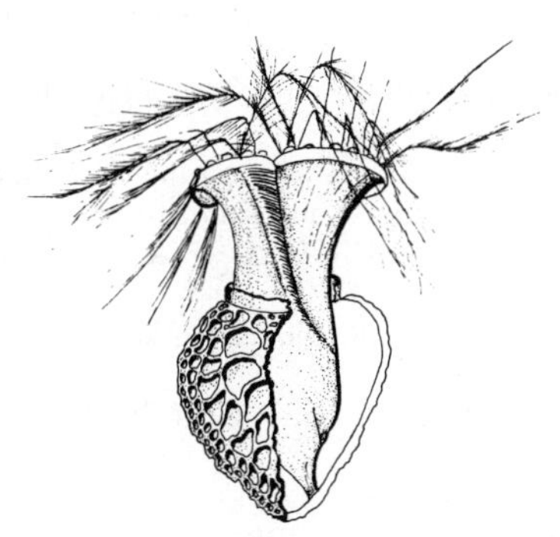

13. *Tintinnopsis nucula*

14. *Zoothamnium* sp.

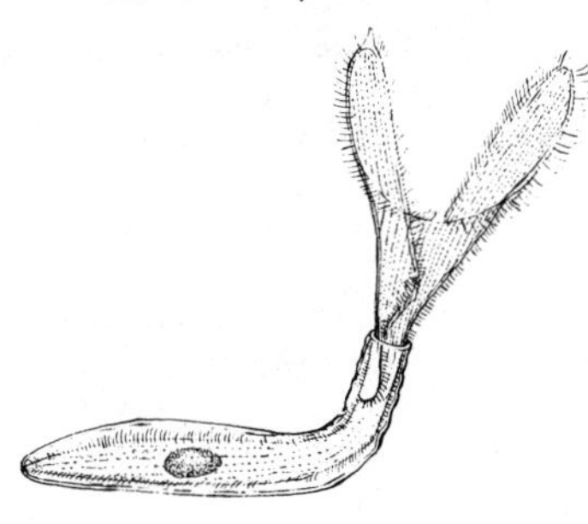

15. *Folliculina* sp.

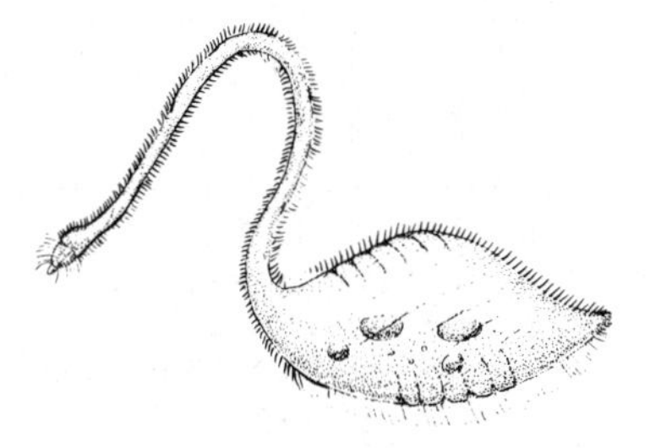

16. *Lacrymaria olor* var. *marina*

PLATE 2. **Protozoa** (2). Sizes and references for protozoans shown in figures 10–16: 10, to 1.5 mm (Grell, 1953); 11, to 500 μm (Angell, 1967); 12, 20–30 μm (Borror, 1963); 13, 100 μm (Campbell, 1926); 14, colony 250 μm or more (Kahl, 1935); 15, 200–500 μm (Andrews, 1921; Hadzi, 1947); 16, 300–500 μm (Kahl, 1935).

REFERENCES ON PROTOZOA

Andrews, E. A. 1921. American folliculinas: taxonomic notes. Amer. Nat. 55: 347–367.

Angell, R. W. 1967. The test structure and composition of the foraminifer *Rosalina floridana*. J. Protozool. 14: 299–307.

Arnold, Z. M. 1972. Observations on the biology of the protozoan *Gromia oviformis* Dujardin. Univ. Calif. Publ. Zool. 100: 1–168.

Ball, G. H. 1938. The life history of *Carcinoecetes hesperus* n. gen., n. sp., gregarine parasite of the striped shore crab, *Pachygrapsus crassipes*, with observations on related forms. Arch. Protistenk. 90: 299–319.

Borror, A. C. 1963. Morphology and ecology of the benthic ciliated Protozoa of Alligator Harbor, Florida. Arch. Protistenk. 106: 465–534.

Bush, M. 1937. *Ancistrina kofoidi* sp. nov., a ciliate in *Petricola pholadiformis* Lamarck, from San Francisco Bay, California. Arch. Protistenk. 89: 100–103.

Campbell, A. S. 1926. The cytology of *Tintinnopsis nucula* (Fol) Laackmann. Univ. Calif. Publ. Zool. 29: 179–236.

Cushman, J. A. 1948. *Foraminifera, their Classification and Economic Use*. 4th ed. Harvard University Press.

Cushman, J. A. and W. W. Valentine 1930. Shallow water Foraminifera from the Channel Islands of Southern California. Stanford Univ. Dept. Geol. Contrib. 1: 1–31.

Grell, K. G. 1953. Die Konjugation von *Ephelota gemmipara* R. Hertwig. Arch. Protistenk. 98: 287–326.

Hadzi, J. 1947. Studien ueber Folliculiniden. Biol. Sbornik Slovens. Akad. vied a umeni. 2: 1–390.

Hall, R. P. 1925. Binary fission in *Oxyrrhis marina* Dujardin. Univ. Calif. Publ. Zool. 26: 281–324.

Hedley, R. H. 1962. *Gromia oviformis* (Rhizopodea) from New Zealand with comments on the fossil Chitinozoa. New Zealand J. Sci. 5: 121–136.

Jahn, T. L. and F. F. Jahn 1949. *How to know the Protozoa*. W. C. Brown Co.

Jepps, M. W. 1926. Contributions to the study of *Gromia oviformis*. Quart. J. Micr. Sci. 70: 701–719.

Jepps, M. W. 1942. Studies on *Polystomella* Lamarck (Foraminifera). J. Mar. Biol. Assoc. U.K. 25: 607–666.

Kahl, A. 1930–1935. Urtiere oder Protozoa I. *In* Dahl, F. *Die Tierwelt Deutschlands*. Teil. 18, 21, 25, 30: 1–886. Jena: G. Fischer.

Kofoid, C. A. 1911. Dinoflagellata of the San Diego region. IV. The genus *Gonyaulax*, with notes on its skeletal morphology and a discussion of its generic and specific characters. Univ. Calif. Publ. Zool. 8: 187–300.

Kofoid, C. A. and O. Swezy 1921. The free-living unarmored Dinoflagellata. Univ. Calif, Memoir 5: 538 pp.

Kozloff, E. N. 1945a. *Cochliophilus depressus* gen. nov., sp. nov., and *Cochliophilus minor* sp. nov., holotrichous ciliates from the mantle cavity of *Phytia setifer* (Cooper). Biol. Bull. 89: 95–102.

Kozloff, E. N. 1945b. *Heterocineta phoronopsidis* sp. nov., a ciliate from the tentacles of *Phoronopsis viridis* Hilton. Biol. Bull. 89: 180–183.

Kudo, R. R. 1966. *Protozoology*. 5th ed. Springfield: C. C. Thomas.

Le Calvez, J. 1936. Observations sur le genre *Iridia*. Arch. Zool. Exp. Gén. 78: 115–131.

Lynch, J. E. 1929. Studies on the ciliates from the intestine of *Strongylocentrotus*, I. Univ. Calif. Publ. Zool. 33: 27–56.

Myers, E. H. 1942. A quantitative study of the productivity of the foraminifera in the sea. Proc. Amer. Phil. Soc. 85: 325–342.

Noble, E. R. 1938. The life cycle of *Zygosoma globosum* sp. nov., a gregarine parasite of *Urechis caupo.* Univ. Calif. Publ. Zool. 43: 41–66.
Pickard, E. A. 1927. The neuromotor apparatus of *Boveria teredinidi* Nelson, a ciliate from the gills of *Teredo navalis.* Univ. Calif. Publ. Zool. 29: 405–428.
Towe, K. M. 1967. Wall structure and cementation in *Haplophragmoides canariensis.* Contr. Cushman Found. Foram. Res. 18: 147–151.
Uchio, T. 1960. Ecology of the living benthonic Foraminifera from the San Diego, California, area. Cushman Found. Foram. Res., Spec. Publ. 5: 72 pp.

PHYLUM PORIFERA

Willard D. Hartman

Department of Biology and Peabody Museum of Natural History, Yale University

(PLATES 3–5)

Sponges are an important component of the intertidal biota of California, but our knowledge of their biology is still rudimentary. Their taxonomy is by no means settled; undescribed species are certain to turn up, and an understanding of their natural relationships is far from being achieved.

The structural plan of sponges is unique among multicellular animals. The body consists of loose aggregations of cells differentiated into ill-defined tissues such as epithelia and mesenchyme, but organs are not formed. A mouth and digestive tract are lacking; instead the surface of sponges is perforated by numerous small incurrent **pores (ostia)** through which water enters. The excurrent apertures (**oscules** or **oscula**) are larger and relatively few in number. Water entering through the ostia passes through a system of canals and chambers lined by flagellated **collar cells (choanocytes)** and leaves the sponge through the oscula. The canal systems are of varying grades of complexity, ranging from those with a single central cavity lined by collar cells (e.g., *Leucosolenia eleanor*) to those with numerous subdivisions of the internal channels and with collar cells restricted to spherical or ellipsoidal chambers along the lengths of the channels (e.g., *Leucandra heathi*, all **Demospongiae**). The latter system is more efficient, causing the water to slow down in its passage through the sponge and thus providing more time for food to be removed. The vast majority of sponges have this complicated plan.

Many sponges have a definite radially symmetrical form. *Leucandra heathi* and *Tethya aurantia* are local examples. Others are irregular in shape and are found encrusting rocks or growing from an encrusting base into irregularly branching forms. Symmetry of form in sponges is correlated with their vertical distribution in the sea. Deep-sea species tend to have regular and definite shapes; intertidal and shallow-water species include a large majority with irregular shapes.

The consistency of sponges varies exceedingly, from hard and stony to friable, rubbery, or gelatinous, depending upon the nature and arrangement of the skeletal elements. All except a few sponges, such as *Halisarca,* possess some type of skeleton. Indeed, the main subdivisions of the phylum are based on skeletal characteristics. The skeleton may consist of calcareous or siliceous spicules alone, of spongin fibers alone, or of a combination of siliceous spicules and spongin fibers.

Some intertidal sponges are easily recognized and have distinctive habitats. *Leucilla nuttingi* is an easily identified, rocky intertidal sponge, occurring in clusters of individuals each with a slender basal stalk and brownish white tube up to an inch tall. *Microciona prolifera,* introduced from the Atlantic coast, forms showy, reddish orange, branched growths in the brackish waters of San Francisco Bay.

Quite resistant to exposure to air, the greenish *Halichondria panicea* and the purple *Haliclona* sp. A are both regularly found in beds of *Mytilus californianus.* The thin, encrusting red *Ophlitaspongia pennata* also occurs commonly in midtidal areas on sides and lower surfaces of rocks. Other species, such as the orange *Tethya aurantia* and the yellow *Polymastia pachymastia,* are common in deeper water and can be collected intertidally on exposed coasts only at the lowest tides.

An interesting family of sponges that has received very little study on the California coast is the Clionidae, which bore into calcareous material such as molluscan shells, corals, and limestone. One local *Cliona* is commonly associated with encrusting calcareous algae, living in a layer under the algal colony and boring up through the alga to the surface in places to allow the exit of contractile tubules on which are borne the ostia and oscula. Other common species (to date no adequate taxonomic study of them has been made) bore into abalone or bivalve shells, and in these the openings, through which the ostial and oscular tubules extend, often form regular circular patterns on the surfaces of the shells. If an abalone shell inhabited by *Cliona* is broken, the yellow sponge can be seen filling extensive galleries which it has bored in the interior of the shell. Shells riddled in this way are greatly weakened. *Cliona* can also be a nuisance on oyster beds.

The yellow nudibranch *Archidoris montereyensis* is commonly seen feeding on *Halichondria panicea* in and around mussel beds, and the red nudibranch *Rostanga pulchra* lives on red sponges which both it and its eggs match closely in color. Some keyhole limpets also feed on sponges, and certain other gastropods, chitons, and fish feed on them.

Many animals find sponges favorable sites in which to live; amphipods, polychaetes, and shrimps are common in certain sponges, and masking crabs, such as *Loxorhynchus crispatus,* often place pieces of sponges on their backs. Other sessile organisms such as hydroids, entoprocts, barnacles, ectoprocts, and ascidians often grow on sponges.

The classification of sponges depends largely upon skeletal characteristics, although these are not always adequate. Embryological and life-history studies, biochemical characteristics, and cytological details have helped in understanding the relationships of some groups. Porifera are divided into four classes on the basis of the chemical composition and geometrical configuration of the skeletal elements, as follows:

Class **Calcarea:** spicules (usually triradiate or monaxonid) of calcium carbonate; spongin absent.

Class **Demospongiae:** skeleton consists of spicules of silicon dioxide (laid down in a hydrated form related to opal) or of the collagenous protein, **spongin,** or of both siliceous spicules and spongin; megascleres are monaxonid or tetraxonid.

Class **Hexactinellida:** skeleton consists of siliceous spicules with three axes (triaxons); not found intertidally, although well represented on the continental shelf and slope of California.

Class **Sclerospongiae:** skeleton consists of a basal mass of crystalline calcium carbonate; siliceous spicules occur in some species; restricted to cryptic tropical reef habitats.

Skeletal characters are the most practical to employ in identification but great care must be exercised in their use. Certain spicules, especially microscleres, may be absent in some members of a population; failure to examine a number of specimens of the same species may lead one astray. For example, toxons are rare or absent in some specimens of the common, red, encrusting sponge *Axocielita originalis;* failure to find them led the describer to place it in a genus and family quite different from that given here. *Penares saccharis* differs from *Penares cortius* in the absence of asters among its microscleres. It is possible that this character is a function of age and that the two are conspecific. Color is useful in separating some species but can be variable (see table, p. 42). Thus the common *Ophlitaspongia pennata* is usually some shade of red, but some specimens are orange, yellow, or buff to light tan. Careful attention to the habitat may also be useful in separating similar species, although relatively little is known about this aspect of sponge biology. For example, *Ophlitaspongia pennata* usually occurs higher in the intertidal than its relative *Axocielita originalis*.

A natural classification demands attention to the biology of sponges in addition to their skeletal morphology. The key presented here gives ample evidence of the need for a taxonomic revision of the sponges of California. Comparative studies of aspects of their biology are likely to provide information useful in carrying out such a revision.

NOTES ON TECHNIQUES

It is important to keep field notes on every specimen collected before fixation. Color in life (best by reference to a color dictionary), oscular sizes in life, notable grouping of pores, surface features, consistency, distinctive odor, tendency to give off mucus, and the presence of buds, gemmules, or larvae are characteristics to look for. Notes on habitat including tidal level are also useful. Some of the substrate should be collected along with encrusting sponges, since certain spicules may be localized on the base, embedded in spongin affixed to the substrate.

Fixation may be done routinely in neutralized 10% formalin (4% formaldehyde) solution in sea water (Baker's calcium formaldehyde is good), followed by transfer to 75% ethyl alcohol within a few days. In certain sponges, such as *Haliclona* sp. B and *Isodictya quatsinoensis*, cells tend to become dissociated from the skeleton in formalin, but most sponges can be fixed successfully in this fluid and are better for histological work than if preserved directly in alcohol. Special fixatives such as Bouin, Zenker, Carnoy, or glutaraldehyde may be necessary for specific histological or ultrastructural studies.

Since skeletal structures are most useful in sponge identification, two types of preparations are needed: (1) sections perpendicular and tangential to the surface and (2) dissociated spicule mounts. Sections of fixed material may be made freehand with a sharp razor blade and stained in 1% basic fuchsin in 95% alcohol, followed by dehydration, clearing, and mounting on a slide in damar or a synthetic resin. Sections may be blotted briefly on a paper towel after each step. It is easier to cut thin, freehand sections of many sponges after embedding a small piece of the sponge in paraffin, but this takes time and is not absolutely necessary for preliminary taxonomic studies. Such stained sections provide information on the arrangement of spicules in the skeletal framework and on the localization of spicule types. Thin microtome sections are necessary for cellular detail. The distribution of spongin is also more clear in thinner sections prepared with a collagen stain.

On the California coast, calcareous sponges may be distinguished from siliceous ones by the presence of triradiate spicules in the former (figs. 15, 39–41). If doubt exists, place a small piece of the sponge on a microscope slide and add a small drop of hydrochloric acid. If the sample fizzes, calcareous matter is present. However, bits of included shell and other calcareous materials may cause fizzing in a siliceous sponge. To make certain of a correct determination, add a cover slip and examine the preparation under the microscope to note whether the spicules are partially dissolved away. Spicules not intrinsic to the sponge in question may adhere to the surface of the sponge or become incorporated into the ectosome or choanosome and hence will turn up on spicule slides. This can cause problems for the beginner as well as the experienced investigator. Examination of a second sample from another part of the sponge or of a second specimen will usually allow one to decide if the spicule type is foreign or proper to the sponge. Sponges with spongin fibers only (order **Dictyoceratida**) often incorporate the spicules of other sponges into their fibers; spicules are usually recognizable as foreign when most are broken or when they represent a random miscellany of spicule types.

Temporary spicule preparations may be made by placing a small piece of the sponge (including both surface and interior tissues) on a slide and adding a few drops of a sodium hypochlorite solution, such as Clorox. After bubbling has subsided, a cover glass should be added and the preparation examined under a compound microscope, with oil immersion objective and ocular micrometer. The use of too much Clorox will lead to diversionary activities of cleaning the microscope. Permanent spicule mounts may be made by boiling a small piece (less than 1 cm^3) of sponge in concentrated nitric acid (or 10% KOH if the sponge has calcareous spicules) under a hood until the organic matter has disappeared. This must be followed by thorough washing away of the acid, best accomplished by transferring the preparation to a centrifuge tube and spinning down the spicules three or four times with a change of water each time. Before the final centrifugation, add 95% alcohol to speed up drying. When decanting the last fluid from the tube, leave a small amount to facilitate removal of the spicules which, after being brought into suspension in the remaining fluid, are poured out on a slide on a warming plate. Check the tube to make certain that all spicules have been transferred to the slide. Allow the slide to dry thoroughly and, before adding mounting medium, add a drop of solvent for that medium to help prevent excessive bubble formation. Now add mounting medium and a long cover slip. Damar, Piccolyte, or other synthetic resins are satisfactory media. As little mounting medium should be added as feasible, since too thick a slide cannot be examined with high power or oil immersion objectives. Such permanent spicule preparations are useful in measuring spicules with an ocular micrometer and in the preparation of drawings. Temporary mounts are sufficient for most routine identification.

In the key, spicule measurements are given as ranges of mean values of several specimens when the overall range of spicule size is relatively low, or as overall ranges of spicule size when the range is great.

DISCUSSION OF SPONGE SPICULES

Some knowledge of the fearsome terminology of sponge spicules is necessary if the taxonomic literature on the group is to be read intelligently. The following discussion of the terms employed is included to aid those who may go further in this field, as well as to clarify certain terms used in the key.

Sponge spicules are formed of calcium carbonate or silicon dioxide, the latter secreted around an axis of organic material. In general we differentiate spicules first on the basis of size: there are large **megascleres** that form the chief supporting framework of the sponge, and smaller **microscleres** scattered throughout the mesenchyme or localized at the surface. Spicules are further subdivided on the basis of the number of axes or rays present. Names for spicules are coined by adding the appropriate numerical prefix to the ending **-axon** (when referring to the number of axes) or **-actine** (when referring to the number of rays or points). For example, an important category of spicules consists of **monaxons,** formed by growth along a single axis. If growth occurs in a single direction, the spicule is a **monactinal monaxon** (figs. 1–4); if in both directions, it is a **diactinal monaxon** (figs. 5–9). Both **monactines** and **diactines** (the latter also called **rhabds**) are usually megascleres, but in some instances the diactines are small and are classed as microscleres. Another important category of spicules includes **tetraxons** (figs. 10–14) with four rays, each pointed in a different direction. The rays may be equal, or one (called the **rhabdome**) may be longer than the others (called **clads**), in which case the spicules are referred to as **triaenes** (figs. 11–14). **Triradiate** (fig. 15) and **quadriradiate** spicules are common in calcareous sponges. **Triaxons** are formed of three axes crossing at right angles, resulting in six rays or points; hence the synonym, **hexactines.** Triaxons occur only in hexactinellid sponges.

Microscleres (figs. 16–25) often have more than four rays; types with numerous equal rays diverging from a central point are known as **asters.** Other types of microscleres are formed by concentric growth around a center and are called **spheres.** Other types of microscleres include C- or S-shaped forms, called **sigmas;** bow-shaped **toxons; chelas,** with recurved hooks, plates, or flukes at each end; and **streptasters,** which are short, spiny rods. There follows a more detailed glossary of spicule types important in identifying sponges. We are indebted to Shirley G. Hartman for drawing figures 26–35 and 38–49.

GLOSSARY OF SPICULE TYPES

(Numerals refer to figures on Plates 3–5; for dimensions of length measurement, see figure 49, page 57)

MEGASCLERES

monactinal monaxons:

styles (fig. 1): rounded at one end, pointed at the other (e.g., *Hymeniacidon ungodon*); the adjectival form is **stylote.**

tylostyles (fig. 2): with a distinct knob at one end; pointed at the other end; resembling a marlinspike (e.g., *Cliona celata*).

subtylostyles (fig. 3): tylostyles with indistinct knob at one end; pointed at the other end (e.g., *Ophlitaspongia pennata*).

acantho- (fig. 4): a prefix denoting that the spicule is covered with thorny processes (e.g., **acanthostyles** in *Microciona microjoanna*); the adjectival form is **acanthose.**

diactinal monaxons:

oxeas (figs. 5, 26–31): pointed at both ends (e.g., *Haliclona* sp. A); the adjectival form is **oxeote.**

strongyles (fig. 6): rounded at both ends (e.g., *Tethya aurantia*).

tornotes (fig. 7): lance-headed at both ends (e.g., *Hymenamphiastra cyanocrypta*).

tylotes (fig. 8): knobbed at both ends (e.g., *Plocamia karykina*).

cladotylotes (fig. 9): tylotes with more or less recurved clads at each end (e.g., *Acarnus erithacus*).

centrotylote (adjective): having the knob near the middle of the shaft of a spicule.

tetraxons: spicules with four rays.

calthrops (fig. 10): having the four rays equal or nearly so (e.g., *Poecillastra tenuilaminaris*, an offshore species).

triaenes: tetraxons having one long ray (the **rhabdome**) and three short rays **(clads).**

orthotriaenes: with clads making an angle of about 90° with the axis of the rhabdome (e.g., *Stelletta clarella*).

plagiotriaenes (fig. 13): with clads directed forward and making an angle of about 45° with the produced axis of the rhabdome (e.g., *Stelletta clarella*).

protriaenes (fig. 12): with three clads directed forward as the tines of a fork. The clads make an angle of less than 45° with the produced axis of the rhabdome (e.g., *Tetilla arb*).

anatriaenes (fig. 11): with the three clads directed backwards (e.g., *Tetilla arb*).

dichotriaenes (fig. 14): with forked clads (e.g., *Stelletta clarella*).

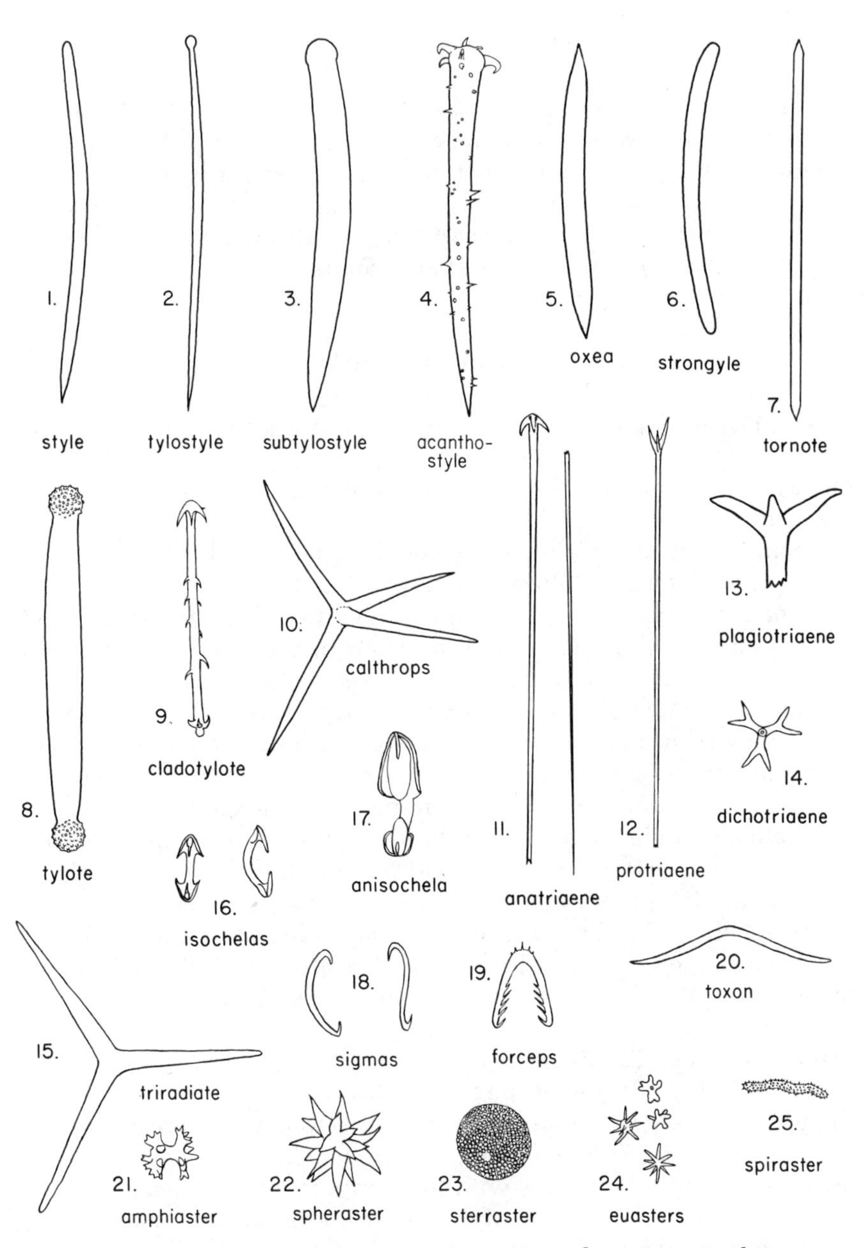

PLATE 3. **Porifera** (1): 1–25, representative types of sponge spicules, not to scale. Hartman, in Light *et al.*, 1954.

diaenes: tetraxons modified through the loss of one clad (e.g., *Tetilla arb*).

monaenes: modified tetraxons with only one clad (e.g., *Tetilla* sp. B). **Anamonaenes** have the clad directed backward like a hook.

triradiates (fig. 15): spicules with three rays more or less in the same plane; common in calcareous sponges such as *Leucosolenia* (figs. 40, 41). They are **sagittal triradiates** when they have two equal angles and one dissimilar angle.

MICROSCLERES

(The following are minute spicules of varied form)

chelas: short, straight or curved rods with ends reflexed and variously elaborated.

isochelas (figs. 16, 35, 38, 46, 48): with equal shovel-shaped ends; of several sorts as follows:

anchorate chelas (fig. 38): shaft straight or slightly curved; each end with thin, winglike extensions of the shaft and a variable number of teeth (e.g., *Plocamissa igzo*).

arcuate chelas (fig. 16): shaft usually strongly curved when viewed from side; the winglike extensions curved forward and indented below to give the appearance of a tooth; a central, narrow tooth present in addition (e.g., *Anaata spongigartina*).

palmate chelas (figs. 17, 48D): shaft straight or slightly curved; winglike extensions large and forming together a triangular or oval plate; tooth broad (e.g., *Microciona prolifera,* fig. 46).

anisochelas (fig. 17): with unequal, shovel-shaped ends; those in local sponges are palmate (e.g., *Mycale macginitiei*).

sigmas (figs. 18, 33, 43, 45): C- or S-shaped (e.g., *Lissodendoryx firma,* fig. 35).

toxons (figs. 20, 42, 46, 48): bow-shaped (e.g., *Ophlitaspongia pennata,* fig. 32); sometimes called **toxas.**

forceps (fig. 19): U-shaped with the arms sometimes crossed; often spined (e.g., *Lissodendoryx firma,* fig. 35).

sigmaspires: spiral spicules with the form of a C or S, usually microspined; characteristic of the order Spirophorida.

asters: microscleres with short, radiating or starlike rays.

euasters (fig. 24): with several equal rays radiating from a central point; e.g., *Stelletta clarella,* in which the ends of the rays vary from pointed (**oxyasters**) to rounded (**strongylasters**).

spherasters (fig. 22): with many rays arising from large central spheres (e.g., *Tethya aurantia*).

sterrasters (fig. 23): spheres covered with many minute rays (e.g., *Geodia mesotriaena*, an offshore species).
amphiasters (fig. 21): short rods with spines at each end (e.g., *Hymenamphiastra cyanocrypta*).
spirasters (fig. 25): spirally twisted, spiny rods (e.g., *Cliona* spp.).
streptasters. Short, spiny rods.

GLOSSARY OF OTHER TERMS USED IN THE KEY

atrium: an exhalant cavity receiving water from one or more exhalant canals or syconoid chambers (in calcareous sponges) and conducting it to one or more usually terminal oscules.
choanosome: interior region of the sponge containing choanocyte chambers.
conulose: surface provided with conical projections, usually caused by the lifting of the pinacoderm by the protruding ends of perpendicular fibers.
cortex: the ectosome, when thick and gelatinous or fibrous (as in *Tethya aurantia*), or packed with spicules of a special type (as in *Geodia*) is called a cortex.
echinate: spicules that project from a spongin fiber either at right or acute angles are said to "echinate" the fiber.
ectosome: peripheral region of sponge, lacking choanocyte chambers.
hastate: in reference to spicules of uniform diameter coming to an abrupt, sharp point.
hispid: surface covered with projecting spicules.
leuconoid: aquiferous system with choanocyte chambers generally small, spherical or ovoid in shape, and usually grouped in the interior of the sponge; most local sponges are of this structure.
mucronate: in reference to spicules, the rounded end produced axially to a sharp point.
pinacoderm: the single layer of flat, contractile cells forming the outer epithelium of a sponge.
syconoid: aquiferous system of closely spaced, elongate choanocyte chambers running radially from the external surface to the central atrial cavity (e.g., *Grantia, Scypha*).
sylleibid: aquiferous system transitional between syconoid and leuconoid conditions, with elongate choanocyte chambers grouped around a common exhalant channel (e.g., *Leucilla nuttingi*).

Editors' note: The sponges of the coast of California present a difficult identification problem, since the fauna is still very imperfectly known, and species are numerous, variable in form and color, in many in-

TABLE I. THE COLORS OF INTERTIDAL SPONGES

	Red	Orange	Yellow	Greenish	Blue	Lavender to Pink	Purple	Gray	Brown to Dark Brown	Buff to Light Brown	Whitish
Acarnus erithacus	X										
Anaata spongigartina	X										
Antho lithophoenix	X	x									
Aplysilla glacialis						X				x	
Aplysilla polyraphis							X				
?*Astylinifer arndti*			x		x	x	x			x	
Axocielita originalis	X	x	x			x			x		
Clathria sp.	X										
?*Clathrina* sp.											X
?*Clathriopsamma pseudonapya*			X								
Cliona ?*celata* var. *californiana*			X						x	x	
Cliona spp.			X							x	
Grantia sp.											x
Halichondria bowerbanki			x							X	
Halichondria panicea		x	x	X						x	
Haliclona sp. A						x	X			x	
Haliclona sp. B		x							x	X	
Halisarca sp.			x							X	
Hymedesanisochela rayae		x							x		
Hymedesmia brepha	X										
Hymedesmia sp. A	X										
Hymedesmia sp. B	X										
Hymenamphiastra cyanocrypta		x			X						
Hymendectyon lyoni		X	x							x	
Hymeniacidon ungodon		x	x				x		x	x	
Hymeniacidon sp.		X								x	
Isodictya quatsinoensis		X									
Leptoclathria asodes			X								
Leucandra heathi											X
Leucilla nuttingi								x		X	x
Leucosolenia eleanor											X

stances separable only by microscopic characteristics. Any simple key, especially one that cannot cover all species, is so likely to lead to erroneous identifications that Dr. Hartman has found it necessary to give much longer descriptions of individual species in the key than is usual in this manual. There has been no comprehensive work on sponges of the central California coast since that of de Laubenfels (1932), and Dr. Hartman's contribution represents a much needed, admittedly preliminary, revision of the sponges of our area. It is our hope that this chapter will provide a basis for badly needed intensive studies of our sponge fauna.

	Red	Orange	Yellow	Greenish	Blue	Lavender to Pink	Purple	Gray	Brown to Dark Brown	Buff to Light Brown	Whitish
Leucosolenia sp.											X
Lissodendoryx firma		x	X							x	
Lissodendoryx topsenti	x	X							x		
Microciona microjoanna	X										
Microciona parthena	X	x							x		
Microciona prolifera	X	x							x		
Mycale macginitiei			x						X	x	
Mycale richardsoni										X	
Ophlitaspongia pennata	X	x	x							x	
?*Pachychalina lunisimilis*						X					
Penares cortius								x	x		
Penares saccharis						x					x
Plocamia karykina	X	x									
Plocamissa igzo	X										
Polymastia pachymastia			X								
Prosuberites sp.										X	
Reniera sp. A						X					
Reniera sp. B					X					x	
Scypha spp.											X
Sigmadocia edaphus										X	x
Sigmadocia sp.						X					
?*Spheciospongia confoederata*						x		x			
Spongia idia								X			
Stelletta clarella						x					X
Suberites sp.		X	x						x	x	
Tedania toxicalis	x	x									
?*Tedanione obscurata*			x							x	
Tethya aurantia var. *californiana*	x	X	x								
Tetilla arb								x		x	
Tetilla sp. B			X							x	
Toxadocia sp.			x						x		
Xestospongia vanilla		x							x	X	
Zygherpe hyaloderma			x							x	

KEY TO INTERTIDAL SPONGES

1. Skeleton present, including calcareous or siliceous spicules or spongin fibers 2
– Skeleton absent (neither spicules nor spongin present); choanocyte chambers large and elongate; consistency soft; color pale tan, yellow-tan or light brown; may be mistaken for a colonial ascidian or the bryozoan *Alcyonidium* .... *Halisarca* sp.

2. Skeleton composed of spongin fibers that may or may not enclose foreign spicules, sand grains, etc. 3
– Skeleton composed of calcareous or siliceous spicules; the latter may be enclosed partly or entirely in spongin fibers .. 5
3. Spongin fibers arise from a basal plate of spongin; fibers may branch but do not anastomose; each fiber has a cortex of concentric layers of spongin surrounding a central pith; choanocyte chambers large (up to 75 μm in greater diameter) 4
– Spongin fibers forming a reticulation; fibers may be cored with foreign spicules or sand grains; choanocyte chambers small, 25 to 30 μm in diameter; surface conulose; massive, slate-gray in color .. *Spongia idia*
4. Thin (1–3 mm), soft, slippery, encrusting; surface conulose; olive-tan, salmon-pink, or coral in color .. *Aplysilla glacialis*
– As above but thicker (up to 5 mm) and of a deep purple color that is given off in large amounts when sponge is collected .... *Aplysilla polyraphis*
5. Spicules siliceous (most local sponges) 6
– Spicules calcareous (few local species; tubular or cuplike; tan or whitish) 61
6. Spicules oxeote megascleres only 7
– Spicules of one or more forms; if one only, not oxeote 14
7. Range of spicule length great in any one specimen, usually greater than 150 μm; spicules of ectosome arranged in tracts that fan out at the surface where they join a more or less reticulate pattern of tracts in the pinacoderm 8
– Range of spicule length small in any one specimen, usually less than 35 μm; spicule arrangement more or less reticulate, or so densely packed as to obscure the basic reticulate pattern 9
8. Surface of sponge (fig. 37) marked by a regular reticulation of multispicular tracts; closely spaced pores occur in spaces between tracts; typically encrusting, sometimes more massive, with rows of oscules borne on ridges; color chrome yellow, yellow-tan, and, especially in well-lighted situations, olive or green at surface; on open coast .... *Halichondria panicea*
– Surface of sponge (fig. 36) marked by loose tracts of spicules or irregularly arranged, isolated spicules; pores tend to be isolated rather than grouped; encrusting or, more often, branching or in flattened lobes; oscules scattered; color yellow-tan, gold, or olive-brown; in bays and harbors .... *Halichondria bowerbanki*
9. Spicules, arranged in a dense reticulate pattern mainly with triangular meshes, form the skeletal matrix; in addition, vertical multispicular tracts or horizontally running spongin fibers, cored with many spicules, may be present 10

– Spicule arrangement reticulate with more or less square meshes; vertical spicular tracts present, joined by horizontally or diagonally placed spicules occurring singly or in groups 11

10. Vertical multispicular tracts, with little spongin, sometimes present, especially in the ectosome; at least some of the oxeas have two bends (fig. 31); oscules small (up to 2 mm), flush with surface or with slightly raised rims; encrusting, usually thin (ca. 5 mm, but up to 15 mm); consistency hard, friable; color buff, tan, or dull orange-brown . . . . *Xestospongia vanilla*

– Anastomosing spongin fibers, up to 300 μm in diameter and filled with oxeas (fig. 30), course horizontally through the skeletal matrix; oscules larger (up to 4 mm), flush with surface or with rims raised up to 2 mm; encrusting, 5–30 mm thick; consistency firm, tough, elastic; color lavender or pale rose-gray . . . . ?*Pachychalina lunisimilis*

11. Vertical tracts composed of a single row of spicules or of a few spicules; cross spicules chiefly horizontal, forming square to polygonal meshes; spicules united at tips with spongin; spicules small (figs. 26, 27), mean length 75–135 μm, mean width 4.5–7.0 μm . . . . *Haliclona* 12

– Vertical tracts multispicular; cross spicules horizontal, diagonal, or irregular in arrangement; spicules larger (figs. 28, 29), mean length 140–165 μm, mean width 6.0–7.5 μm . . . . *Reniera* 13

12. Thinly encrusting (2–3 mm thick); oscules 1–3 mm, closely spaced, almost flush with surface or with raised rims, 1–5 mm high; spicule mean length 75–100 μm, mean width 4.5–6.0 μm; color violet, mauve, or buff; on open coast . . . . *Haliclona* sp. A

– Lobate to tubular, the tubes up to 6.5 cm high with terminal oscules, 3.5–5.0 mm in diameter; spicule mean length 125–135 μm, mean width 7 μm; color light tan to orange-brown or cinnamon; in bays, commonly on floats . . . . *Haliclona* sp. B

13. Encrusting, up to 2 cm thick; surface irregular; oscules less than 2–7 mm in diameter, sometimes arranged in rows on ridges, flush with surface or with raised rims up to 10 mm high; color rose-lavender, lavender, or light gray-brown . . . . *Reniera* sp. A

– Encrusting, up to 5 mm thick; surface smooth; oscules less than 1–2 mm in diameter, barely raised above surface; color gray-blue . . . . *Reniera* sp. B

14. Sponge globular or subglobular in shape, up to 6 cm in diameter, with monaxonid megascleres; surface warty; a thick, fibrous,

contractile cortex present, up to 4 mm in thickness; megascleres (1) fusiform strongyles, 1.25–2.30 mm in length and 20–50 μm in width, arranged in multispicular tracts that radiate from the center of the sponge to the surface where they fan out in the verrucosities, (2) smaller fusiform styles to subtylostyles ranging in length x width 500–1800 μm x 6.0–18.0 μm, occurring chiefly in the cortex; microscleres, asters of several types including (a) spherasters (fig. 22) with rays sometimes forked or spined, less than 30–90 μm in diameter, found chiefly in the cortex, and (b) small asters with pointed, blunt, or subtylote ends, smooth or microspined, 8–25 μm in diameter, chiefly in the cortex; color deep orange, orange-red, or yellow-ochre .... *Tethya aurantia* var. *californiana*

– Sponges encrusting, massive, or boring into calcareous substrate; if globular, with 4-rayed (teraxonid) spicules 15

15. Spicules of one type, either styles or tylostyles 16

– Spicules including two or more categories 23

16. Spicules styles 17

– Spicules tylostyles or subtylostyles 18

17. Encrusting to massive, the surface often provided with low, rounded or irregular processes; styles (fig. 1) arranged in multispicular tracts forming a polygonal network between which are numerous irregularly arranged spicules; subpinacodermal cavities present; spicule mean length 190–200 μm, mean width 6.0–7.0 μm; color bright yellow, yellow-orange, or cinnamon brown; older specimens mahogany-brown or purple with yellow-ochre choanosome; on open coast .... *Hymeniacidon ungodon*

– Pointed processes, up to 2 cm or more high, arise from a thickly encrusting base; spicules arranged in vague plumose tracts with numerous irregularly arranged spicules between; spicule mean length 285 μm, mean width 8.5 μm; color orange-brown or orange-tan; in bays ... . *Hymeniacidon* sp.

18. Encrustation (up to 2 cm or more thick) from which arise numerous rounded to tapering papillae 1–2 cm high; surface smooth to furry; spicules mostly tylostyles of several size categories, the heads small, the shaft narrow at the head end, widest near the middle; long (up to 2.5 mm), thin, subtylostylote to stylote spicules may project from the ectosome, giving rise to the furlike surface; color deep yellow .... *Polymastia pachymastia*

– Prominent papillae absent 19

19. Sponge large, massive, up to 14 cm thick, 70 cm in diameter; spicules tylostyles of quite uniform size, ranging from 200–320 μm (mean 270) x 6–11 μm (mean 9) in length by width; oscules large, up to 1 cm across; pores localized in discrete areas of several cm^2; color lavender-gray, tan choanosome .... ?*Spheciospongia confoederata*

– Sponges encrusting various surfaces or boring into calcareous substrates 20

20. Boring into shells of molluscs, barnacles, or other calcareous substrates; spicules tylostyles (fig. 2), usually with somewhat subterminal heads, mean length by width 230–310 μm x 7–11 μm; color typically bright yellow .... *Cliona ?celata* var. *californiana*

– Free-living, encrusting 21

21. Tylostyles of one size category, mean length by width 230–310 μm x 7–11 μm, usually with somewhat subterminal heads; encrusting the calcareous substrate that it is excavating or encrusting rocks after having completely eroded away its calcareous substrate; encrustations up to 12 mm thick; color light to dark chrome yellow or pale to dark brown .... *Cliona ?celata* var. *californiana*

– Tylostyles of two or more size categories 22

22. Thinly encrusting, less than 1 mm thick; tylostyles of several size categories, varying in length from 175–925 μm, width 7–17 μm; small, upright tylostyles (mean length by width 250 x 10 μm) occur in ectosome; large tylostyles arise from substrate, points up, and form thick tracts running to surface; color gold or hazel; in bays .... *Prosuberites* sp.

– Encrusting to massive, up to 6 cm thick and 10 cm in larger diameter; surface smooth occasionally with low, rounded processes; consistency firm, hard; tylostyles range from 265–460 μm x 8–15 μm in length and width; small tylostyles, arranged vertically with points up, occur in ectosome; larger tylostyles lie in all directions in the choanosome but sometimes form tracts; color light to dark chrome yellow, orange, pinkish-orange, cinnamon-brown, tan, orange-brown; processes, if present, are yellowish .... *Suberites* sp.

23. Two types of megascleres present; microscleres absent 24

– Spicule types otherwise 25

24. Thickly encrusting to massive, up to 7.5 x 5 x 5 cm; megascleres include (1) tylotes (fig. 34A) to subtylotes, sometimes gently sinuous, mean length by width 195–260 μm x 4.5–7 μm, arranged in tracts near and at surface and (2) subtylostyles (fig. 34B) or occasionally styles, mean length by width 235–300 μm x 6.5–9.0 μm, smooth or sometimes sparsely spined most frequently at the head end only, with pointed end usually hastate or mucronate, these spicules arranged in a reticulate pattern in the choanosome; color salmon pink, dull orange-red, dull orange-brown, burnt sienna, or pale terra cotta .... *Lissodendoryx topsenti*

– Thinly encrusting, 1–2 mm thick; megascleres include (1) tylotes to subtylotes, straight or curved, the two ends often subequal, one end being more nearly strongylote, mean length by width 127–162 μm x 2.3–3.4 μm, these spicules arranged in loose tracts running toward the surface and (2) acanthostyles, mean length by width 89–109 μm x 5.3–6.7 μm, oriented vertically with the head end against the substrate; color deep violet-blue, gray-lavender, or yellow-brown .... ?*Astylinifer arndti*

25. Tetraxonid spicules included among the megascleres 56
– Megascleres include one or more monaxonid types; microscleres present 26
26. Megascleres of one type, accompanied by one or more categories of microscleres 27
– Megascleres of two or more types, accompanied by one or more categories of microscleres (rarely, microscleres absent) 35
27. Megascleres diactinal (oxeas or tylotes) 28
– Megascleres monactinal (styles or tylostyles) 31
28. Megascleres tylotes to strongyles with spined ends, mean length by width 220–245 μm x 7–7.5 μm, arranged in vague tracts leading to surface and distributed freely between the tracts; microscleres microspined microxeas, mean length by width 78–88 μm x 1.3–1.7 μm, distributed freely in the flesh; encrusting, up to 12 mm thick; color bright yellow-tan; exudes mucus .... ?*Tedanione obscurata*
– Megascleres oxeas, with one or two categories of microscleres 29
29. Microscleres toxons, mean length 65 μm, scattered in flesh; oxeas, mean length by width 120 x 7 μm, forming a reticulate pattern; encrustation up to 15 mm thick; oscules, 2–3 mm, sometimes on low mounds; color olive-beige, pale gray-brown, or deep chrome yellow .... *Toxadocia* sp.
– Microscleres sigmas .... *Sigmadocia* 30
30. Two categories of sigmas present: (1) C-shaped (fig. 45B), mean length by width 52 x 3.6 μm, (2) strongly curved with unequal arms (fig. 45A), mean length by width of longer arm 87 x 3 μm, oxeas (fig. 45C), mean length by width 275 x 13 μm, arranged in a vague reticulate pattern with some multispicular tracts running to surface; thickly encrusting, up to 2 cm thick; color buff-white .... *Sigmadocia edaphus*
– Single category of sigmas (fig. 43A) present, mean length 33 μm, the shaft often notched outward slightly in the middle; oxeas (fig. 43B), mean length by width 170 to 190 μm x 7 to 9 μm, arranged in a loose reticulate pattern; encrusting, up to 12 mm thick, often overgrowing the sponge *Stelletta clarella;* color rose-lavender .... *Sigmadocia* sp.
31. Megascleres tylostyles 32
– Megascleres subtylostyles or styles 33
32. Microscleres, two categories of streptasters; sponges excavate mollusc shells and other calcareous substrates which they may encrust externally as well; color usually some shade of yellow or tan ... *Cliona* spp.
– Microscleres sigmas (fig. 33B), range of length 15–60 μm, and diancistras (fig. 33A), range of chord lengths 25–40 μm; mean length and width of tylostyles (fig. 33C) 170–185 μm x 5.0–6.0 μm; thinly encrusting, 1–2 mm

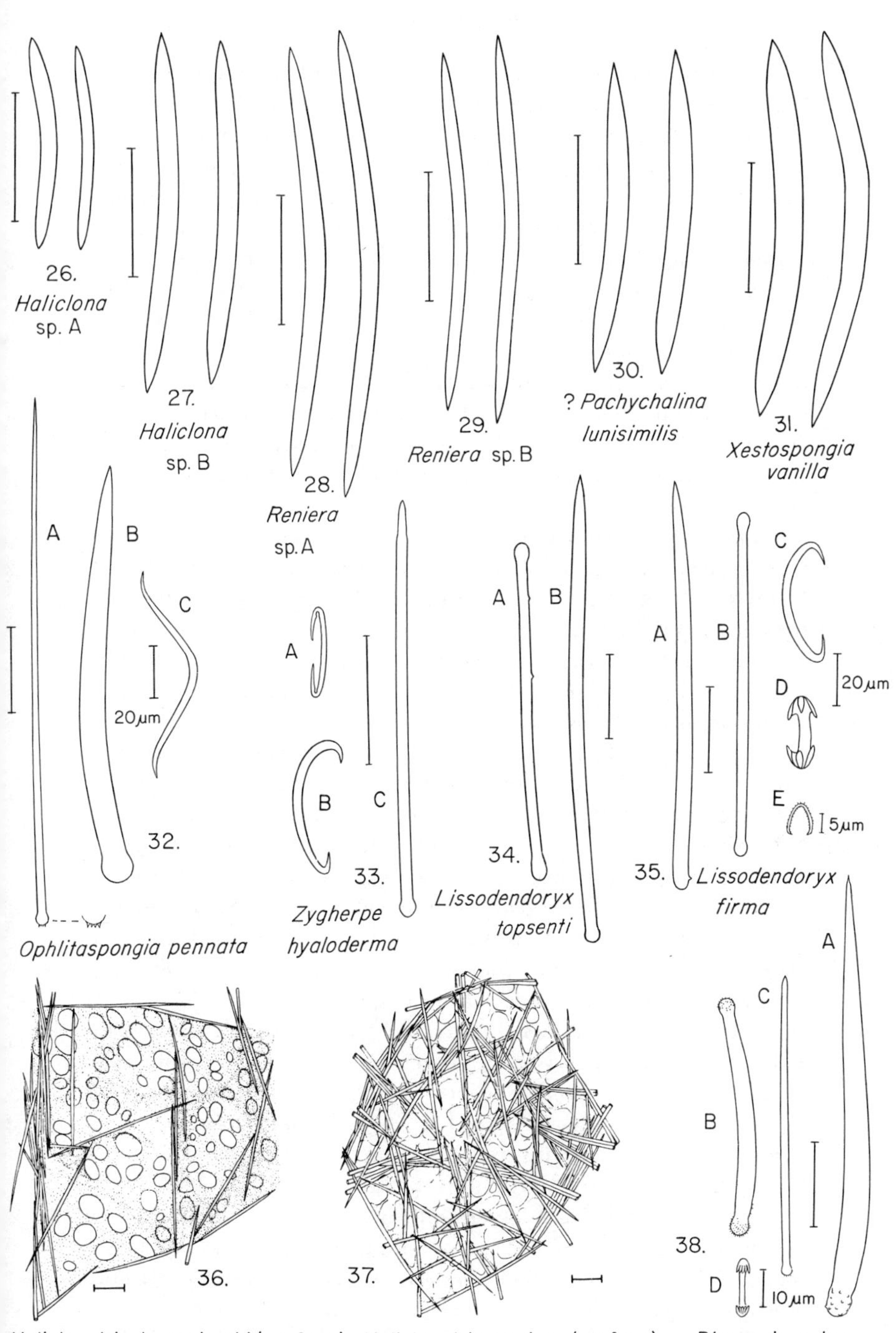

PLATE 4. **Porifera** (2): 36, 37, Hartman, in Smith, 1964; 26–35, 38, original, drawn by Shirley G. Hartman. Scale bars are 50 μm unless otherwise marked.

thick; cinnamon or yellow-ochre . . . . *Zygherpe hyaloderma*

33. Megascleres subtylostyles with proximal 1/3 of spicule (toward head end) thinner than distal 2/3, spicules irregularly sinuous, mean length by width 250–265 μm x 7.5–9.0 μm; microscleres include (1) sigmas, mean length 65–75 μm, (2) toxons, mean length 50–75 μm, (3) larger anisochelas (fig. 17) mean length 30–40 μm, (4) smaller anisochelas, mean length 15–21 μm, encrusting, up to 1 cm thick, often with numerous thin processes, up to 2 cm long, arising from the basal crust; color deep chrome yellow, pale golden brown, light olive-brown, or gray-brown . . . . *Mycale macginitiei*

– Megascleres styles 34

34. Microscleres palmate isochelas, mean length 20–27 μm; styles, mean length by width 135–180 μm x 9.0–11.0 μm; branching subtidal individuals often washed ashore, these up to 17 cm high with large oscules, 7–8 mm across at tips of branches, and smaller oscules (2–3 mm) along sides of branches; occasionally intertidal as short, branching specimens or encrusting to lobate forms with numerous small oscules (1–3 mm) on ridges; color burnt orange . . . . *Isodictya quatsinoensis*

– Microscleres anisochelas, mean length 28–30 μm, and sigmas, mean length 20 μm, sometimes rare or absent; styles arranged in tracts running to surface, tracts interconnected at intervals by a few horizontally placed spicules; styles, mean length by width, 170–175 μm x 7.0–7.5 μm; encrusting, to 4 mm thick; oscules, 1 mm across on low tubules, 1.0–1.5 mm high; color pale gold . . . . *Mycale richardsoni*

35. Megascleres subtylostyles to styles of two size categories . . . 36

– Megascleres otherwise 37

36. Microscleres toxons (fig. 32C), with low arch, range of length by width 26–165 μm x 1.0–6.0 μm; megascleres include (1) thick subtylostyles (fig. 32B), gently curved or bent near middle, rarely with microspined heads, mean length by width 200–240 μm x 14–19 μm, arranged in ascending echinated tracts joined by cross spicules and (2) thin, straight to gently curved subtylostyles with minutely microspined heads (fig. 32A), mean length by width 165–225 μm x 3.0–4.0 μm, localized in groups at surface and scattered within; encrusting, up to 8 mm thick; color usually some shade of scarlet but varying to burnt sienna, deep salmon-red, terra cotta, orange-brown, yellow-tan, or mustard . . . . . *Ophlitaspongia pennata*

– Microscleres of two types, including (1) palmate isochelas (fig. 48D), length 16–19 μm and (2) thin toxons (fig. 48C), with high central arch, range of length by width 46–112 μm x 1.2–1.6 μm, rare or absent in some specimens (both microsclere categories may be absent rarely); megascleres include (1) thick subtylostyles to styles (fig. 48A), usually bent near head end, mean length by width 130–165 μm x 9–12 μm, arranged in ascending echinated tracts joined by cross spicules and (2) thin, straight to gently curved subtylostyles (fig. 48B, E) with smooth or microspined heads, mean length by width 120–155 μm x 3.0–4.0 μm, localized in groups at surface and scattered within; encrusting, usually up to 3 mm, occasionally to 16 mm thick; color scarlet to vermilion, coral, salmon-pink, pale brick red, orange-brown, or pale yellow-ochre .. *Axocielita originalis*

37. Two categories of megascleres present 38
– More than two categories of megascleres present 49
38. One category of megascleres diactinal, the other monactinal (acanthostyles or rarely styles) 39
– Both categories of megascleres monactinal 46
39. Diactinal megascleres tornotes 40
– Diactinal megascleres tylotes 43
40. Microscleres amphiasters (fig. 21), mean length 13–17 μm; megascleres: (1) tornotes (fig. 7), mean length by width 125–160 μm x 3.2 μm and (2) acanthostyles, mean length by width 150–190 μm x 8.0–8.5 μm; thinly encrusting, 4 mm thick; color deep blue or light orange .... *Hymenamphiastra cyanocrypta*
– Microscleres include isochelas 41
41. Microscleres: (1) anchorate isochelas, mean length 21 μm, and (2) sigmas of two size categories, (a) mean chord length 14 μm, (b) mean chord length 28 μm; megascleres: (1) tornotes with microspined ends, mean length by width 150 x 5.0 μm, arranged in vertical tufts at surface, and (2) acanthostyles, sparsely spined, the spines localized at both ends of the spicule and sometimes absent, mean length by width 170 x 9.0 μm, forming a reticulation with some vertical multispicular tracts; thickly encrusting, up to 2.5 cm; color unknown .... *Myxilla agennes*
– Microscleres isochelas only 42
42. Acanthostyles standing erect with heads embedded in basal plate of spongin or forming short tracts arising from substrate; acanthostyles of two size categories: (1) the longer with spines mostly restricted to proximal (toward head) 1/3 or 1/2 of spicule, mean length by width 200–290 μm x 8.0–8.5 μm, (2) the shorter entirely or almost entirely microspined, mean length by width 80–100 μm x 6.5–7.0 μm; tornotes to subtylotes, mucronate, mean length by width 130–160 μm x 3.5 μm, occurring in groups near surface; microscleres arcuate isochelas, mean length 20–30 μm; thinly encrusting, up to 1.5 mm thick; color salmon, salmon-orange, pale terra cotta, brick red or burnt sienna .... *Hymedesmia* sp. A
– Acanthostyles forming a basal reticulation from which arise plumose tracts; acanthostyles of two size categories: (1) the shorter, mean length by width 103–106 μm x 7.0–7.5 μm, forming the basal reticulation, (2) the longer, mean length by width 165–190 μm x 8.5–9.5 μm, occurring in plumose tracts that protrude from the surface; tornotes with mucronate ends, mean length by width 110–130 μm x 3.5–4.5 μm, grouped at

surface; microscleres arcuate isochelas, mean length 22–23 μm; thinly encrusting, up to 2 mm thick; color orange, yellow-ochre, olive-tan, burnt sienna, pale terra cotta .... *Hymendectyon lyoni*

43. Isochelas absent 44
 – Isochelas present 45

44. Microscleres palmate anisochelas with a short "tail" on the smaller end, mean length 18–19 μm; megascleres: (1) acanthostyles with many-spined heads, sparsely spined shafts, and pointed ends mostly free of spines, range of length by width 100–250 μm x 6–13 μm, arranged in vertical echinated tracts and (2) anisotylotes, with one end usually knobbed, the other, thinner, end rounded, mean length by width 157–160 μm x 3.5–4.0 μm, occurring in horizontal clusters at the surface; thinly encrusting, up to 5 mm thick; color light orange to light cinnamon .... *Hymedesanisochela rayae*
 – Microscleres onychaetes (thin, minutely spined rods, tapering to a fine point at one end, thicker at the other end and terminating with a spine) of two size categories, mean length by width, (1) 180 x 1.8 μm, (2) 60 x 2.3 μm; megascleres: (1) styles, straight to curved, mean length by width 300 x 8.5 μm, arranged in vague tracts running to surface, and (2) tylotes to subtylotes with mucronate ends, mean length by width 220 x 4.6 μm, occurring in vertical tufts at surface; encrusting to massive; color brownish orange or brownish red .... *Tedania toxicalis*

45. Thinly encrusting, less than 1 mm thick; microscleres arcuate isochelas only, mean length 35 μm, abundant; megascleres: (1) acanthostyles, mean length by width 160 x 8.0 μm, standing erect with heads embedded in basal plate of spongin and (2) tylotes, mean length by width 200 x 5.5 μm, arranged vertically to form a dense plush at surface and occurring in dense tracts in the choanosome; color salmon .... *Hymedesmia* sp. B
 – Encrusting to massive, up to 4.5 cm thick; surface smooth to rugose; microscleres: (1) arcuate isochelas (fig. 35D), the shaft sometimes spined, mean length 25–35 μm, (2) sigmas (fig. 35C) mean length 32–47 μm, (3) forceps (fig. 35E), mean span of arms 7.0–9.5 μm, frequently absent; megascleres: (1) styles (fig. 35A), straight or slightly curved, smooth or with a few spines usually at the rounded end only, mean length by width 185–250 μm x 7.5–9.5 μm, arranged in vague tracts or sometimes forming an ill-defined reticulation and (2) tylotes (fig. 35B) to subtylotes, mean length by width 170–235 μm x 5.0–5.5 μm, occurring at various angles at the surface and in tracts beneath the surface; color some shade of yellow ....
.... *Lissodendoryx firma*

46. Microscleres toxons, 40 μm in length; megascleres: (1) tylostyles with smooth or microspined heads, range of length by width 230–365 μm x 5.0–13.5 μm, occurring in tracts and scattered without order in the choanosome and scattered tangentially in the ectosome and (2) acanthostyles, of two indistinct size categories, the smaller entirely microspined, with range of length by width 65–85 μm x 5.0–8.0 μm, the larger with spination grading

with increasing size from dense to sparse except on head where spination is generally dense, range of length by width 125–290 μm x 7.0–13.0 μm, echinating the tracts and foreign bodies in the choanosome; encrusting, up to 1 cm thick; color yellow .... *?Clathriopsamma pseudonapya*

– Microscleres isochelas 47

47. Acanthose megascleres echinating vertical, branching spongin fibers; these megascleres acanthostyles, often markedly curved, with spined heads and sparsely spined shafts, the proximal 1/3 (adjacent to the rounded end) of the shaft or the entire shaft of the longest spicules sometimes smooth, range of length by width 80–420 μm x 5.0–25.0 μm; second category of megascleres subtylostyles or occasionally styles or tylostyles, with the rounded ends generally microspined, range of length by width 125–255 μm x 3.5–7.5 μm, arranged vertically or at random at the surface, forming loose tracts near the surface and associated with the main vertical tracts of the skeleton; microscleres, palmate isochelas, range of length 10–17 μm; encrusting, up to 4 mm thick; color bright to deep scarlet .... *Clathria* sp.

– Acanthose megascleres arranged vertically with heads embedded in a basal layer of spongin 48

48. Color yellow; the acanthose megascleres are acanthosubtylostyles without pronounced spination on the heads and of two indistinct size categories: the smaller with the distal (toward pointed end) 1/3 to 1/4 of shaft smooth, range of length by width 92–100 μm x 6.0–11.0 μm, the larger with the distal 2/3 to 1/2 of the shaft smooth, range of length by width 135–470 μm x 6.5–14.5 μm; second category of megascleres straight to gently curved tylostyles to subtylostyles with minutely microspined heads, the pointed ends sometimes truncate and microspined, range of length by width 190–260 μm x 3.5–5.0 μm, occurring in tufts at the surface; microscleres palmate isochelas, range of length 10–14 μm; thinly encrusting, less than 1 mm thick .... *Leptoclathria asodes*

– Color salmon red; acanthose megascleres are acanthosubtylostyles with pronounced spination on the heads and of two size ranges, the smaller with the distal 1/2 to 1/4 of the shaft smooth, range of length by width 105–155 μm x 8.5–12.5 μm, the larger with the distal 3/4 to 1/2 of the shaft smooth, range of length by width 250–470 μm x 11.0–19.5 μm; second category of megascleres mostly straight styles with rounded end smooth or minutely microspined and opposite end hastately pointed, range of length by width 165–230 μm x 2.5–4.0 μm, arranged in choanosomal tracts running to the surface where they end in tufts; microscleres arcuate isochelas, range of length 20–25 μm; thinly encrusting, less than 1 mm thick .... *Hymedesmia brepha*

49. Microscleres arcuate isochelas (fig. 16) of two size categories with mean lengths 20–25 μm and 45–50 μm; megascleres: (1) long acanthostyles, with spines confined to rounded end or, if entirely spined, the spines larger at the rounded end, mean length by width 300–360 μm x 10.5–12.6 μm,

arranged in branching vertical tracts provided with spongin, the pointed ends up and protruding from the tracts at an angle, (2) short acanthostyles, entirely spined, the spines larger at the rounded end, mean length by width 105–117 μm x 7.0–8.5 μm, standing erect in the basal plate of spongin and echinating the vertical tracts, (3) thin tylostyles or subtylostyles with elongate, smooth heads, mean length by width 205–225 μm x 4.0–4.5 μm, occurring in vague tracts near the surface, arranged vertically or at random at the surface, and accompanying the tracts of acanthostyles; encrusting, up to 5 mm thick; color brick red or deep red-brown .... *Anaata spongigartina*

– Microscleres are isochelas and toxons 50

50. Cladotylotes of two sizes included among the megascleres: (1) the larger with smooth shaft, mean length by width 220–230 μm x 10–11 μm, sometimes rare or absent, and (2) the smaller (fig. 9) with spined shaft, mean length by width 95–120 μm x 4–6 μm, always present; chief megascleres smooth, curved styles, mean length by width 290–390 μm x 15–18 μm, arranged in ascending tracts echinated by cladotylotes; ectosomal spicules subtylotes, mean length by width, 190–220 μm x 4.4–4.7 μm; microscleres: (1) palmate isochelas of mean length 16–20 μm, (2) toxons with low, spreading arch and long arms, chord length range 190–440 μm, and (3) toxons with high, rounded arch, length range 15–135 μm; thickly encrusting, up to 5 cm thick, often with oscular chimneys; color bright scarlet to terra cotta .... *Acarnus erithacus*

– Cladotylotes not present 51

51. Thick choanosomal tylotes present 52

– Choanosomal tylotes not present 53

52. Microscleres anchorate isochelas (fig. 38D) with 4 or 5 teeth, mean length 13 μm; main skeleton of ascending, branching, plumose tracts of megascleres including: (1) axial tylotes (fig. 38B) with microspined ends, mean length by width 125 μm x 9.5 μm, and (2) echinating tylostyles or subtylostyles (fig. 38A) with spined heads, range of length by width 130–310 μm x 10.0–17.0 μm; anisotornotes (one end knobbed, the other truncate, pointed, rounded or with smaller knob) with ends minutely microspined (fig. 38C), mean length by width 140 x 4.0μ m, occur at the surface; encrusting, about 1 cm thick; color, carmine .... *Plocamissa igzo*

– Microscleres palmate isochelas, mean length 18–20 μm, and toxons with low, spreading arch, range of length 15–100 μm; chief megascleres straight or usually curved styles with rounded end microspined, heads sometimes smooth, mean length by width 180–215 μm x 15–17 μm, arranged in ascending tracts that penetrate the surface, which is rendered hispid by protruding spicules; tylotes (fig. 8) with microspined heads, mean length by width 210–240 μm x 17–24 μm, interconnecting the vertical tracts; thin tylostyles to subtylostyles with sparsely microspined heads, mean length by width 145–185 μm x 3.3–3.7 μm, occurring mostly near the surface; encrusting, up to at least 2.5 cm thick; color various hues of scarlet to salmon-red or dull salmon-orange; exudes much mucus when injured or collected .... *Plocamia karykina*

53. Chief megascleres acanthostrongyles to acanthostyles arranged in a reticulate pattern, these spicules with mean length by width 130–160 μm x 11.0–12.5 μm; from the nodes of the meshes may arise longer, more sparsely spined acanthostyles, mean length by width 185–210 μm x 11.5–13.0 μm, as well as smooth styles, mean length by width 215–320 μm x 12.5–14.5 μm, these two spicule types, especially the latter, often penetrating the surface to render it hispid; thin styles to subtylostyles, usually with sparsely microspined heads, mean length by width 220–275 μm x 3.5–4.5 μm, stand erect at or near the surface (some of these spicules have microspined truncate ends instead of being pointed); microscleres: palmate isochelas, mean length 22–24 μm, and toxons, range of length 30–260 μm; encrusting, up to 3 cm thick; color bright scarlet, reddish orange or brick red . . . . *Antho lithophoenix*

– Chief megascleres styles or subtylostyles arranged in ascending tracts 54

54. Sponge of anastomosing branches when mature, thinly encrusting when young; megascleres of 3 types: (1) subtylostyles to styles (fig. 46A), with smooth shafts and spined or occasionally smooth heads, arranged in ascending, plumose tracts with terminal tufts that extend beyond the surface and render it hispid, the tracts held together by spongin, mean length by width 180–225 μm x 10.0–12.5 μm, (2) subtylostyles to styles (fig. 46A), mean length by width 155–175 μm x 3.0–3.5 μm, mostly grouped at the surface, and (3) acanthostyles (fig. 46A), mean length by width 80–90 μm x 7.0–8.0 μm, sparsely echinating the tracts; microscleres: (1) palmate isochelas (fig. 46C), mean length 16–20 μm, and (2) toxons (fig. 46B), range of length 15–100 μm; color bright red, varying to shades of orange-brown; introduced into San Francisco Bay . . . . *Microciona prolifera*

– Sponge encrusting; on open coast 55

55. Toxons (fig. 42) with long, often recurved arms, microspined at their tips, range of length 15–325 μm; palmate isochelas, mean length, 13.0–14.5 μm; megascleres of three types: (1) smooth styles, with rounded end often minutely microspined, mean length by width 275–325 μm x 15.0–17.5 μm, arranged in ascending, plumose tracts with terminal tufts that penetrate the surface of the sponge for one spicule length, rendering it hispid, the tracts supported by spongin, (2) subtylostyles with microspined heads, mean length by width 180–220 μm x 3.5 μm, mostly grouped at the surface, and (3) acanthostyles (fig. 4) with heavily spined heads and sparsely spined shafts, mean length by width 105–115 μm x 7.0–8.5 μm, echinating the tracts of styles; sponge encrusting, up to 2 cm thick; color salmon to salmon-red . . . . *Microciona microjoanna*

Toxons (fig. 44) with wide, low arch, the arms not produced and not microspined, range of length 25–100 μm; palmate isochelas, mean length 20–30 μm; megascleres of three types: (1) smooth styles, the rounded end smooth and often having a diameter less than the greatest diameter of shaft, range of length by width 130–450 μm x 10–35 μm, arranged in ascending, plumose tracts provided with spongin, the tracts penetrating the surface for a distance of several spicule lengths, (2) subtylostyles, usually with microspined heads, range of length by width 150–250 μm x 2.5–3.5 μm, mostly grouped at the surface, and (3) acanthostyles, entirely spined, range of length by width 105–170 μm x 8.5–10.5 μm, sparsely echinating the tracts of styles; encrusting, from 3–20 mm thick; color red to orange-brown .... *Microciona parthena*

56\. Microscleres sigmaspires 57
– Microscleres otherwise 59
57\. Sponge club-shaped, up to 22 mm high by 18 mm wide, with conspicuous anchoring root tufts of bundles of thin anatriaenes; on mud flats in San Francisco Bay ... *Tetilla* sp. A
– Sponge spherical or subspherical in shape; on open coast, attached to rocks 58
58\. With a conspicuous crown of spicules up to 5 mm high around the oscules; among the radiating spicule bundles are tracts of anamonaenes; usually provided with a basal mat of long, thin spicules, mostly anamonaenes; color chrome yellow to yellow-tan .... *Tetilla* sp. B
– Without a conspicuous crown of spicules around the oscules; among the radiating spicule bundles are tracts of anatriaenes (fig. 11); young specimens usually without a well-developed basal mat of spicules, although large specimens may have such; color pale tan when young, gray with buff choanosome when large .... *Tetilla arb*
59\. Microscleres are microxeas (fig. 47), often bent once or twice, sometimes indistinctly centrotylote, length by width 68–180 μm x 4.0–10.0 μm, packed densely in the cortex and occurring more sparsely in the choanosome; megascleres: (1) dichotriaenes, the clads spreading out in the cortex and (2) oxeas, length by width 400–890 μm x 10–25 μm, forming vague tracts in the choanosome; encrusting, up to 1.5 cm thick; color dark lavender-brown, or white .... *Penares saccharis*
– Microscleres include asters 60
60\. Microscleres are of 2 types; (1) microstrongyles with 2 bends, sometimes faintly centrotylote, range of length by width 50–160 μm x 3.0–8.0 μm, packed densely in the cortex and occurring more sparsely in the choanosome, and (2) oxyasters, range of diameters 9–25 μm, scattered in the choanosome; megascleres: (1) dichotriaenes with clads spreading out in the cortex and (2) oxeas, mean length by width 400–950 μm x 10–25 μm, forming vague tracts in the choanosome; massive, up to 4 cm thick; color gray to dark brown .... *Penares cortius*
– Microscleres are asters (fig. 24) only, of two sorts; (1) less common, minutely microspined oxyasters, range of diameters 7–15 μm and (2) abundant asters with a tuft of spines at the end of each thick ray, range of diameters 4–5 μm, both types distributed densely at the surface and increasingly sparse in the choanosome; megascleres radially arranged, including

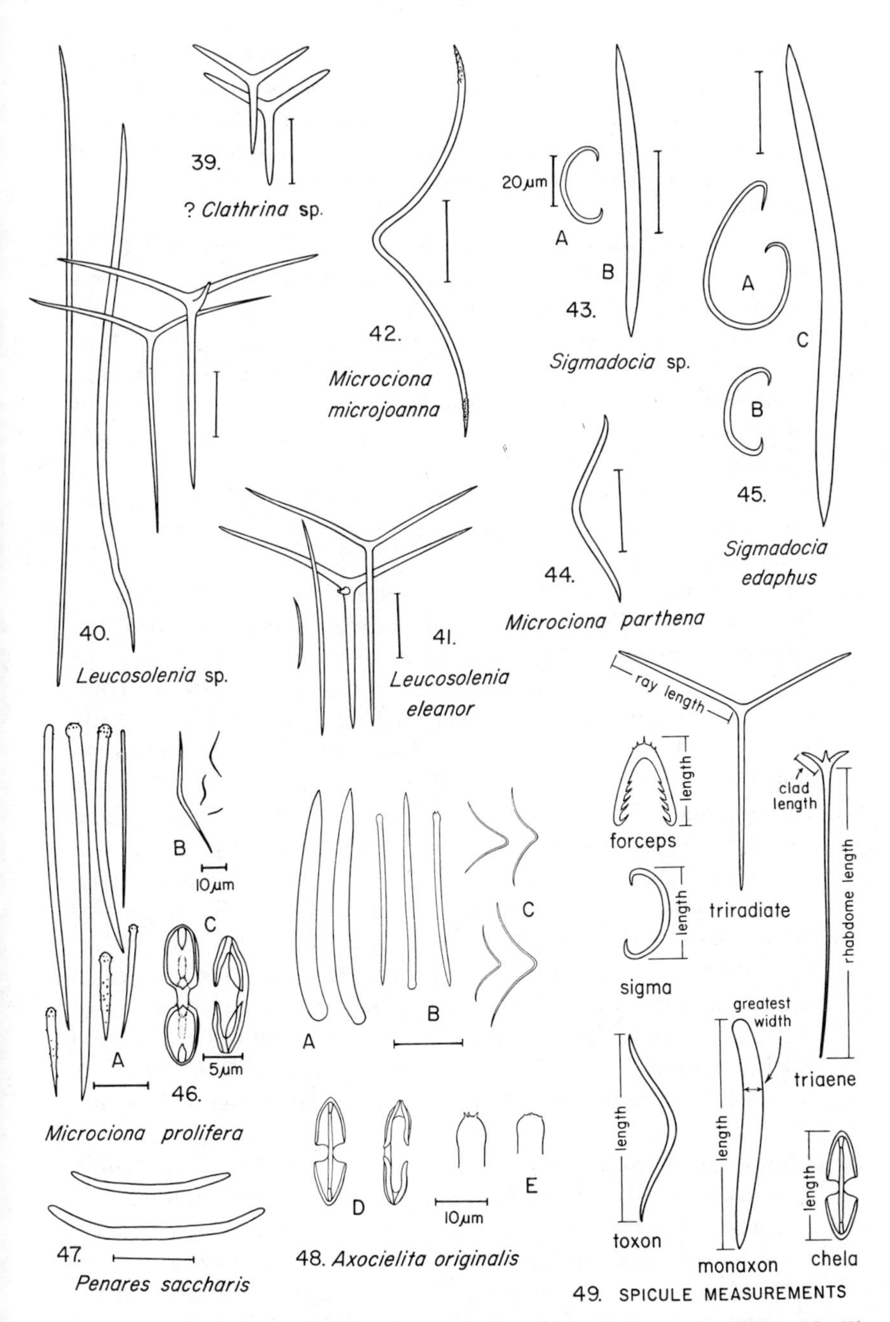

PLATE 5. **Porifera** (3): 46, Hartman, in Smith, 1964; 48, Simpson, 1966; 39–45, 47, 49, original, drawn by Shirley G. Hartman. Scale bars are 50 µm unless otherwise marked.

large oxeas, orthotriaenes, plagiotriaenes (fig. 13), and dichotriaenes (fig. 14), these triaenes with their clads near the surface, and interstitial anatriaenes; encrusting to massive, up to 7 cm thick; color white with buff interior, the surface sometimes tinged with mauve or pink .... *Stelletta clarella*

61. Sponge subspherical to pear-shaped with conspicuous fringe of thin, monaxonid spicules, up to 7 mm high, around the terminal oscule; surface hispid through occurrence of numerous thick, oxeote spicules that lie in the ectosome and project from the surface; a dense layer of vertically oriented microxeas at the surface underlain by a tangential layer of triradiates; choanosome provided with an abundance of triradiates without apparent order; atrium lined by tangentially placed triradiates; choanocyte chamber arrangement leuconoid; up to 9 x 11 cm in height by diameter; whitish .... *Leucandra heathi*
- Sponge shape otherwise 62

62. Sponge tube-shaped, diameter of tube greater than 2 mm when mature 63
- Shape of sponge otherwise 64

63. Sponge vase-shaped or tubular, smooth, up to 1 cm in diameter, narrowing distally, borne on a narrow stalk; commonly occurring in groups; pale tan to grayish white .... *Leucilla nuttingi*
- Sponge tubular, without stalk, with hispid surface; oscule with or without fringe of oxeote spicules; chamber arrangement syconoid; pale tan to whitish .... *Scypha* spp.

64. Sponge sac-shaped, compressed, up to 8 mm wide, without oscular fringe; chamber arrangement syconoid; whitish .... *Grantia* sp.
- Sponge consisting of thin tubes, 4 mm or less in diameter, that branch and may anastomose to form a reticulate mass 65

65. Triradiate spicules (fig. 39) equiangular, with rays equal or subequal, 20–65 μm long (usually 50–60 μm); sponge a tight network of tubes forming masses up to 4 cm across and 1.5 cm high; whitish .... ?*Clathrina* sp.
- Triradiate spicules sagittal, with one ray longer than the other two .... *Leucosolenia* 66

66. Diameter of oscule-bearing tubules usually less than 1 mm, but occasionally up to 1.4 mm; sponge a loose network of anastomosing tubules, may form masses up to at least 7 cm across and 3 cm high; oxeote spicules (fig. 41) from 70–435 μm in length, 3–8 μm in width; buffy or whitish; on open coast .... *Leucosolenia eleanor*
- Diameter of oscule-bearing tubules up to 3.8 mm in greatest width; sponge consists of a basal anastomosis from which arise numerous branching oscule-bearing tubules, varying

greatly in diameter; oxeote spicules (fig. 40) from 175–725 μm in length, 3–9 μm in width; dirty or whitish; in harbors .... *Leucosolenia* sp.

LIST OF INTERTIDAL SPONGES

Class **DEMOSPONGIAE**

Order **Dendroceratida**

HALISARCIDAE

Halisarca sp. Mid to very low intertidal; uncommon. Since spicules are lacking, species identification dependent upon size and shape of choanocyte chambers and details of germ cells and larvae.

APLYSILLIDAE

Aplysilla glacialis (Merejkowsky, 1878). Mid to very low intertidal; moderately common.

Aplysilla polyraphis de Laubenfels, 1930. Intertidal; rare.

Order **Dictyoceratida**

SPONGIIDAE

Spongia idia de Laubenfels, 1932 (=*Leiosella idia*). Very low intertidal; rare.

Order **Haplosclerida**

HALICLONIDAE

Haliclona sp. A [=*H. permollis* (Bowerbank) of second edition]. Mid to low intertidal; common, often in well-lighted situations. The purple, supposedly cosmopolitan, species *H. permollis* needs critical study and the name is abandoned here.

Haliclona sp. B. Occurs on floats in San Francisco Bay and Bodega Harbor. Externally similar to *H. ecbasis* de Laubenfels from southern California but that species has a very different skeletal architecture and probably belongs in the genus *Chalinula*. See Fell, 1970, Pac. Sci. 24: 380–386 (natural history).

RENIERIDAE

?*Pachychalina lunisimilis* (de Laubenfels, 1930) (=*Haliclona lunisimilis*). Low to very low intertidal; rare. Generic placement uncertain.

Reniera sp. A. Low to very low intertidal; common.

Reniera sp. B. Very low intertidal; rare.

Xestospongia vanilla (de Laubenfels, 1930). Low to very low intertidal; common.

ADOCIIDAE

* *Adocia gellindra* (de Laubenfels, 1932). The status of this species is uncertain; it may be identical to *Reniera* sp. A above.

Sigmadocia edaphus de Laubenfels, 1930. Very low intertidal; rare.

Sigmadocia sp. Low to very low intertidal; uncommon.

Toxadocia sp. Low to very low intertidal; uncommon. More than one species may be involved.

Order **Poecilosclerida**

ISODICTYIDAE

Isodictya quatsinoensis (Lambe, 1892). A subtidal species found occasionally in intertidal waters below mean lower low water (tidal zero).

MYCALIDAE

Mycale macginitiei de Laubenfels, 1930 (=*Carmia macginitiei*). Low to very low intertidal; moderately common.

Mycale richardsoni Bakus, 1966. Very low intertidal; rare.

Zygherpe hyaloderma de Laubenfels, 1932. Low to very low intertidal; uncommon.

CLATHRIIDAE

Antho lithophoenix (de Laubenfels, 1927) (=*Isociona lithophoenix*). Mid to very low intertidal; common.

Axocielita originalis (de Laubenfels, 1930) (=*Esperiopsis originalis;* includes *Axocielita hartmani* Simpson, 1966). Very low intertidal; common.

Clathria sp. Very low intertidal; uncommon. May be more mature growth form of *Leptoclathria asodes* (de Laubenfels).

?*Clathriopsamma pseudonapya* de Laubenfels, 1930. Intertidal; rare. Generic assignment uncertain.

Leptoclathria asodes (de Laubenfels, 1930) (=*Eurypon asodes*). Intertidal; rare. Lévi (1960) regards *Leptoclathria* as a synonym of *Microciona*.

* Not in key.

Microciona microjoanna de Laubenfels, 1930. Very low intertidal; uncommon.

Microciona parthena de Laubenfels, 1930. Very low intertidal; rare.

Microciona prolifera (Ellis and Solander, 1786). On pilings and floats in San Francisco Bay, introduced from the Atlantic coast; locally common and conspicuous; the only branching red sponge of the local intertidal fauna.

Ophlitaspongia pennata (Lambe, 1895). Mid to very low intertidal; common. Simpson (1968, Bull. Peabody Mus. Nat. Hist. 25) includes *Ophlitaspongia* in *Microciona* on the basis of cytological characteristics.

MYXILLIDAE

Acarnus erithacus de Laubenfels, 1927. Low to very low intertidal; uncommon.

Anaata spongigartina (de Laubenfels, 1930). Low to very low intertidal; uncommon. Placement in this family uncertain; may be a clathriid but lacks the characteristic palmate isochelas.

?*Astylinifer arndti* de Laubenfels, 1930. Very low intertidal; uncommon. Generic placement uncertain.

Hymedesanisochela rayae Bakus, 1966. Very low intertidal and subtidal; uncommon.

Hymedesmia brepha (de Laubenfels, 1930) (=*Anaata brepha*). Intertidal; rare.

Hymedesmia sp. A. Very low intertidal; moderately common.

Hymedesmia sp. B. Very low intertidal; rare.

Hymenamphiastra cyanocrypta de Laubenfels, 1930. Very low intertidal; uncommon.

Hymendectyon lyoni Bakus, 1966. Mid to very low intertidal; uncommon.

Lissodendoryx firma (Lambe, 1895) (Includes *Lissodendoryx noxiosa* de Laubenfels, 1930). Low to very low intertidal; common.

Lissodendoryx topsenti (de Laubenfels, 1930) (= *Tedania topsenti*). Low to very low intertidal; common. The "microscleres" reported by de Laubenfels (1932) are immature styles.

Myxilla agennes de Laubenfels, 1930. Very low intertidal; rare.

Tedania toxicalis de Laubenfels, 1930. Very low intertidal; rare.

?*Tedanione obscurata* de Laubenfels, 1930. Very low intertidal; rare. Generic placement uncertain.

PLOCAMIIDAE

Plocamia karykina de Laubenfels, 1927. Mid to very low intertidal; common.

Plocamissa igzo (de Laubenfels, 1932). Very low intertidal; uncommon.

Order **Halichondrida**

HALICHONDRIIDAE

Halichondria bowerbanki Burton, 1930. Mid to low intertidal; common on pilings and floats, especially in San Francisco Bay where it may have been introduced with oysters from the Atlantic coast.

Halichondria panicea (Pallas, 1766). Mid to very low intertidal; common on open coast, often in well-lighted situations.

HYMENIACIDONIDAE

Hymeniacidon ungodon de Laubenfels, 1932. Low to very low intertidal; moderately common.

Hymeniacidon sp. Low intertidal; locally common in Tomales Bay.

Order **Hadromerida**

SUBERITIDAE

Prosuberites sp. Low intertidal; locally common in San Francisco Bay, probably introduced from the Atlantic coast with oysters.

Suberites sp. Low to very low intertidal; common; two species may be present.

POLYMASTIIDAE

Polymastia pachymastia de Laubenfels, 1932. Very low intertidal and subtidal.

SPIRASTRELLIDAE

?*Spheciospongia confoederata* de Laubenfels, 1930. Very low intertidal; rare; generic placement uncertain.

CLIONIDAE

Cliona ?*celata* Grant, 1826, var. *californiana* de Laubenfels, 1932. Low to very low intertidal; common; may be a species distinct from *C. celata*.

Cliona spp. Low to very low intertidal; common, boring into molluscan shells. Several microsclere-bearing species occur.

TETHYIDAE

Tethya aurantia (Pallas, 1766) var. *californiana* de Laubenfels, 1932. Low to very low intertidal; moderately common.

Order **Spirophorida**

TETILLIDAE

Tetilla arb de Laubenfels, 1930 (= *Craniella arb*). Very low intertidal; uncommon.

Tetilla sp. A. Formerly found locally on mud flats in San Francisco Bay; status unknown.

Tetilla sp. B. Low to very low intertidal; uncommon.

Order **Choristida**

STELLETTIDAE

Penares cortius de Laubenfels, 1930. Intertidal; rare.

Penares saccharis (de Laubenfels, 1930) (=*Papyrula saccharis*). Very low intertidal; rare. Possibly a juvenile form of the previous species, lacking asters.

Stelletta clarella de Laubenfels, 1930. Very low intertidal; moderately common.

Class **CALCAREA**

Order **Clathrinida**

CLATHRINIDAE

?*Clathrina* sp. Very low intertidal; fairly common.

Order **Leucosoleniida**

LEUCOSOLENIIDAE

Leucosolenia eleanor(?) Urban, 1905. Very low intertidal zone of open coast; moderately common.

Leucosolenia sp. Unidentified species on harbor floats; possibly introduced.

Order **Sycettida**

SYCETTIDAE

Scypha spp. Occurs on floats in harbors as well as on rocks intertidally. Several species are present.

GRANTIIDAE

Grantia sp. Low to very low intertidal zone; uncommon.

Leucandra heathi Urban, 1905 (= *Leuconia heathi*). Very low intertidal zone; locally moderately common. The generic name *Leuconia* is preoccupied by a mollusc.

AMPHORISCIDAE

Leucilla nuttingi (Urban, 1902) (= *Rhabdodermella nuttingi*). Low to very low intertidal zone; common.

REFERENCES ON SPONGES

Bakus, G. J. 1966. Marine poecilosceleridan sponges of the San Juan Archipelago, Washington. J. Zool. Lond. 149: 415–531. Includes descriptions of many species that range southward to California.

Borojevic, R., W. G. Fry, W. C. Jones, C. Lévi, R. Rasmont, M. Sarà, and J. Vacelet 1968. A reassessment of the terminology for sponges. Bull. Mus. National Hist. Nat. (ser. 2) 39: 1224–1235. An attempt at standardization of sponge terminology.

Griessinger, J.-M. 1971. Étude des Réniérides de Méditerranée (Démosponges Haplosclérides). Bull. Mus. National Hist. Nat. (sér. 3) Zool. 3: 97–180. A revision of the difficult families Haliclonidae and Renieridae.

Hyman, L. H. 1940. *The Invertebrates: Protozoa Through Ctenophora*, Vol. I. McGraw-Hill. (Porifera: pp. 284–364.)

Laubenfels, M. W. de 1932. The marine and fresh-water sponges of California. Proc. U.S. Nat. Mus. 81: 1–140. The basic work on California sponges; lists earlier works on their classification.

Lévi, Claude 1960. Les Démosponges des côtes de France. I. Les Clathriidae. Cahiers Biol. Mar. 1: 47–87. A review of an important family of sponges with many representatives in California.

Simpson, T. L. 1966. A new species of clathriid sponge from the San Juan Archipelago. Postilla 103: 1–7. A description of the common clathriid *Axocielita originalis* (de Laubenfels).

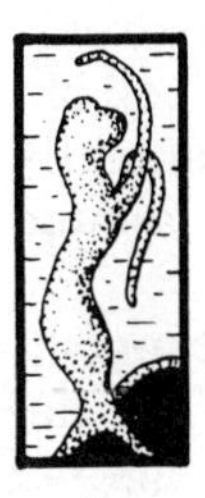

PHYLUM CNIDARIA (COELENTERATA)

Cadet Hand
University of California, Berkeley, and Bodega Marine Laboratory

(PLATES 6–12)

Cnidarians, like sponges, are an ancient group, relatively simple in structural organization, wholly aquatic, and most greatly developed in the sea, where they occur from the shore to abyssal depths. At the seashore, cnidarians are confined with few exceptions to lower tidal levels or below because, like the sponges, ectoprocts, and ascidians, they are not adapted to withstand exposure. But in contrast to these latter groups, which are all filter feeders, cnidarians are primarily predators. Their success seems to be explained by two devices for food-getting and defense, tentacles and nematocysts; by an effective means of distribution, the ciliated planula larva; and in many cases (Scyphozoa, some Hydrozoa) by a free-swimming sexual medusa.

CLASS HYDROZOA

John T. Rees and Cadet Hand
University of California, Berkeley, and Bodega Marine Laboratory

Hydrozoa are abundantly represented in the intertidal zone by the attached **polyp** stages, "hydroids," of the order **Hydroida.** These vary tremendously in form, from tiny individuals to large and showy colonies. The life cycles of hydrozoans (fig. 1) often include a sexual **medusa** stage, which exists free in the plankton. In the intertidal, the medusa stage is more commonly retained upon the polypoid generation as an attached **medusoid** or as an even more reduced "**sporosac.**" The existence of free-living medusa stages has led to difficult problems in taxonomy. In many cases the polyp and the medusa of one species have been described under different generic and specific names, some of which have persisted even after the two forms have been recognized as stages in the life cycle of one species. It is necessary to have a separate key for the attached polypoid or "hydroid" forms (page 72), which are chiefly encountered in intertidal collecting, and one for the Hydromedusae (page 77), which are generally taken in plankton tows, by dip-netting in pools, in harbors around floats, or among *Zostera*.

In addition to the **Hydroida,** which includes almost all local species of hydrozoans, members of other orders may be encountered. The calcareous "hydrocorals" are represented intertidally by one member of the order **Stylasterina,** the lavender, encrusting *Stylantheca porphyra;* species of *Allopora* occur subtidally as pink, encrusting or branching growths. The order **Trachylina** consists of typical hydromedusans with the polyp stage lacking in the life cycle; trachyline medusae are occasionally taken in our plankton, but they are essentially oceanic forms. Another order, **Chondrophora,** is often abundantly represented on our beaches by the blue "by-the-wind sailor" *Velella* (fig. 30), which may be blown ashore in vast numbers. Chondrophorans consist of a float beneath which is a single complex polyp comparable to that of *Tubularia.* The polyp possesses a central mouth surrounded by cycles of blastostyles bearing medusa buds. The blastostyles in turn are surrounded by a marginal ring of tentacles. Chondrophorans are recognized as distinct from members of the order **Siphonophora,** which have a much more complex organization (see Totton, 1965). Each siphonophore colony has a distinct form, size, and arrangement of its members. The colony forms a complex array of polyps and medusoids specialized for feeding, swimming, reproduction, or other functions and may be supported by a common float. Onshore currents and winds carry both chondrophorans and siphonophores to our coast but both are more characteristic of oceanic waters.

The order **Hydroida** includes many local intertidal species of such varied form and structure that a detailed account is desirable. Certain terms are widely used for the parts of the hydroid colony. Unfortunately, they by no means always have the same meaning nor are they always consistently applied. The only monographic treatment for the West Coast is that of Fraser (1937), but the distinctions he makes are often obscure, his terminology is complex, and in his work little or no attention is paid to the medusoid stages.

The hydroid colony (fig. 1) is a continuous, often branching, cellular tube, the **coenosarc.** The coenosarc consists of a layer of ectoderm separated by a thin layer of noncellular **mesoglea** from an inner layer of endoderm which surrounds a continuous central cavity, the **coelenteron** or digestive cavity. Partly or completely surrounding the coenosarc is a thin, chitinous, noncellular, nonliving layer, the **perisarc.**

Hydroid polyps show considerable polymorphism, and the zooids

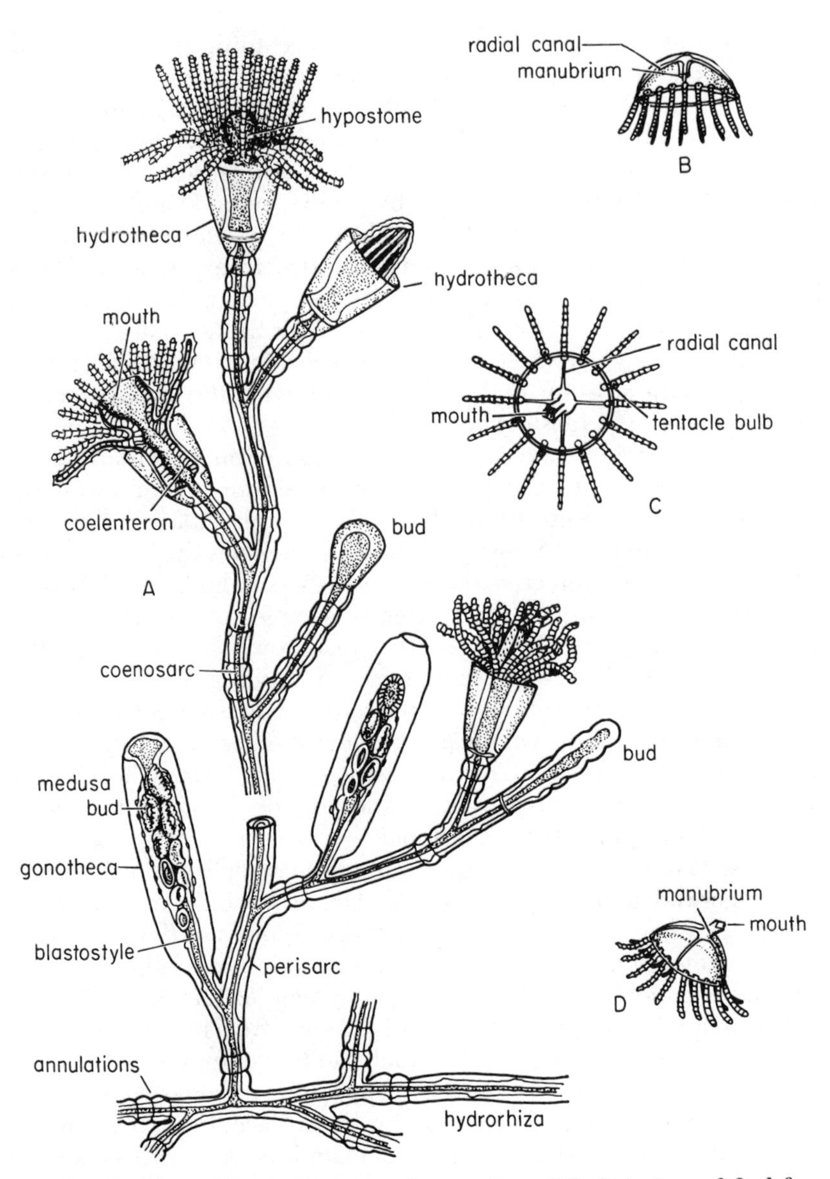

PLATE 6. **Cnidaria** (1). Hydrozoan polymorphism (*Obelia*): 1, modified from Hand in Marshall and Williams, 1972; used with permission of The Macmillan Company. A, the polypoid colony; B, C, immature medusae; D, medusa swimming inside out. (Note that *Obelia* lacks the typical hydromedusan velum.)

may be of several different types, named according to their specialized function: nutritive **gastrozooids,** generative **gonozooids,** or defensive **dactylozooids.** The term **hydranth** is used to designate the terminal part of a nutritive zooid but does not include associated perisarcal structures such as the **hydrotheca.** The hydranth is therefore entirely coenosarcal, consisting of the body, hypostome, mouth, and tentacles.

The term **gonosome** is used to include all the specialized generative zooids of the colony and the perisarcal structures associated with them; the term **trophosome** refers to the rest of the colony. In thecate hydroids (fig. 1) the gonosome includes the asexual generative zooids or **blastostyles,** which produce sexual zooids—**medusae, medusoids,** or **sporosacs**—by budding, together with the **gonothecae** or cases enclosing the blastostyles.

The sexual zooids are termed **gonophores** by some, or are referred to in general as the "medusoid" stage or generation, in contrast to the "hydroid" or polyp(oid) stage. However, Fraser (1937) uses gonophore as a synonym of blastostyle, but often includes the budding sexual zooids and the protective theca as well. We and others call this assemblage a **gonangium,** a term Fraser uses to mean gonotheca. Such a reproductive element of a colony (blastostyle with buds and protective covering, if any) is often spoken of as a "fruiting body." The term gonophore is used, therefore, with the most diverse meanings, and we need to know with which of these meanings it is used in each case. Dr. Light has remarked that, "This necessity of using terms whose meanings differ with the author, while annoying for the moment, affords very excellent intellectual experience."

The sexual zooids produce gametes from which, by fertilization, arise zygotes that develop into **planula larvae.** Each larva can give rise to a new colony. All the zooids of a given colony are derived from a single zygote; hence the sexual zooids of a colony are all of the same sex, and we speak of the colony as being male or female.

The generalized hydromedusan (fig. 2) is a free-swimming animal consisting of a gelatinous "bell" which can range from bell- to saucer-shaped with all gradations between. The outer surface is known as the exumbrellar, the inner as the subumbrellar surface. From the center of the subumbrellar surface hangs the **manubrium,** which can be of various lengths and, in some species, is mounted upon a gelatinous **peduncle.** The oral opening is terminal on the manubrium. It frequently carries lobes (often spoken of as "lips"), frills, or tentacles, which are liberally provided with nematocysts. Where the manubrium joins the bell or peduncle there is usually a gastric cavity. Radial canals arise from the gastric cavity and course through the bell to the margin, where they join the ring canal. There are usually four radial canals, but other numbers occur (six, eight, numerous). In a few forms the radial canals are branched (figs. 14, 28), while in others centripetal canals arise from the ring canal but may not reach the stomach.

Hydromedusae are typically "craspedote," that is, they possess a **velum** or membrane that partly closes off the subumbrellar space at the level of the bell margin. The velum is occasionally lacking, as in *Obelia* (fig. 1).

The bell margin is usually simple and unscalloped. Tentacles usually arise from the bell margin and may be simple, few or many in number, occurring singly or in groups, or they may be branched (e.g., *Cladonema*, fig. 4B) or rudimentary. The margin may also be provided with specialized sense organs. Chief among these are **ocelli** and **statocysts.** Ocelli occur as dark pigmented spots, usually one on each tentacle bulb. Statocysts are vesicles or open pits containing one or more concretions known as **statoliths.**

Medusae are almost always of separate sexes. The gonads are epidermal structures on the radial canals, peduncle, or manubrium.

Many of the characteristics customarily used in the classification of hydrozoans are now recognized as varying markedly with environmental conditions and developmental stage, with the result that many species are now being or should be re-examined and their validity established (e.g., West and Renshaw, 1970). In addition, for many species the complete life cycle is not known, resulting in a curious double taxonomy, in which polyp and medusa of the same animal have been described under separate names. Thus, the polyp originally named *Lar* (fig. 15) is now known to give rise to the hydromedusan *Proboscidactyla* (fig. 14), and when such life cycles are established, the older name takes precedence. In many species only the polyp is known, but not the medusa, or vice versa.

Positive identification of many hydrozoan polyps cannot be made unless the sexually mature medusae or the fixed gonophores associated with them are known. Forms such as these are at present best keyed out only to genus. Fraser (1937), Russell (1953), Naumov (1960), and Kramp (1961) (particularly the latter three) can be of help for those who wish to pursue the taxonomy of the group.

Medusae are best examined alive, but it is frequently necessary to anesthetize them. For this, a solution of magnesium chloride (73.2 g of $M_gCl_2 \cdot 6\ H_2O$ per liter of fresh water) is recommended in the proportion of 30 to 40 percent added to the water containing the animals. Preservation should be in 10 percent formalin, which tends to dissolve the statoliths but leaves the structure of the statocyst intact. For histological work, Bouin's fixative is recommended.

For hydromedusae that cannot be identified by the following key, the most useful reference is Russell (1953). The serious student will also find Kramp (1961) of great assistance, since that monograph de-

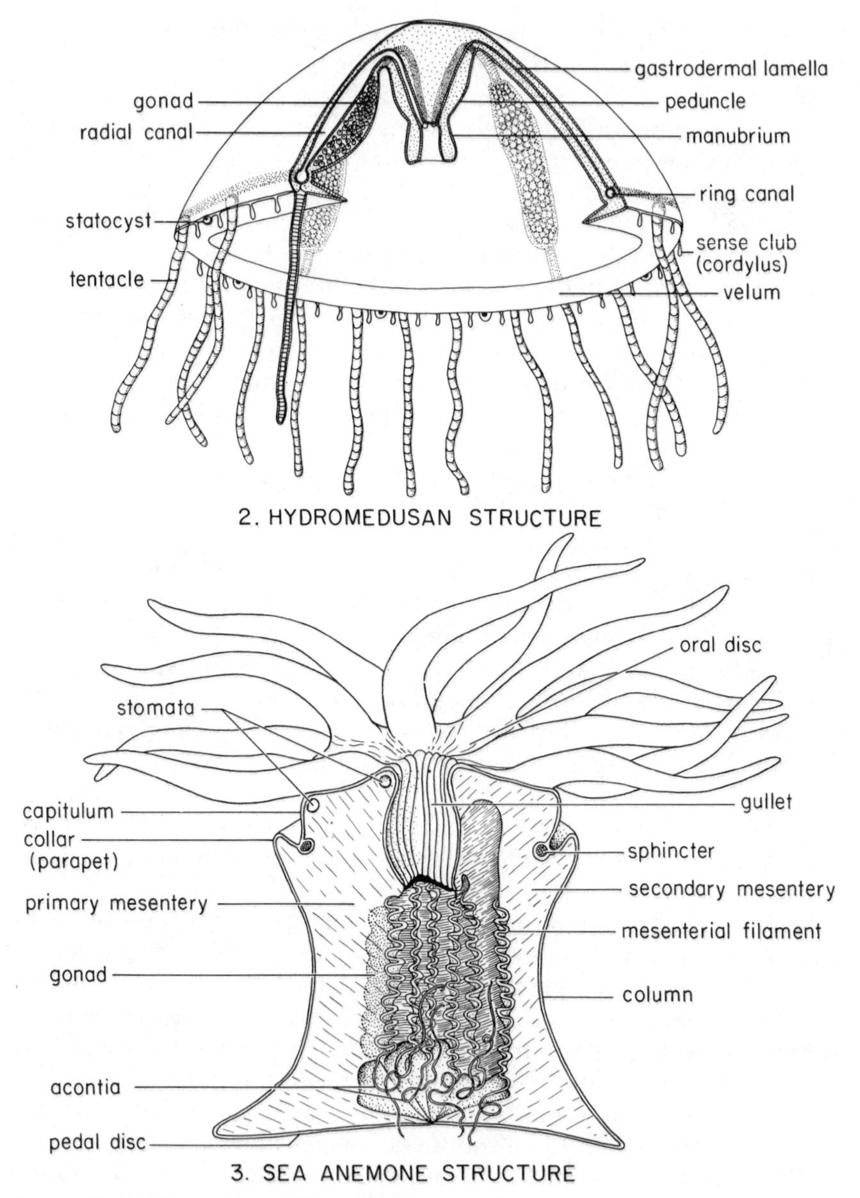

PLATE 7. **Cnidaria** (2). Hydromedusan and sea anemone structure: 2, hydromedusan structure, diagrammatic, a section of bell removed; 3, sea anemone structure, animal shown as if cut in half.

fines all known genera of medusae and gives a brief diagnosis of species.

The help of Dennis Lees in the preparation of the key to polyp stages of Hydrozoa is gratefully acknowledged, as is that of Vicki B. Pearse in respect to the Anthozoa. In the unified species list which follows the key to hydromedusae, we have attempted to integrate the confused, separate taxonomies of hydroid polyps and their medusae.

GLOSSARY OF HYDROZOA

aboral: pertaining to the region away from or most distant from the mouth.
annulated: having a serially ringed appearance (fig. 1).
capitate: knobbed at the tip (figs. 4, 7A).
cirri: small, tentaclelike organs on the bell margin of medusae between the true marginal tentacles; characteristic of Leptomedusae.
exumbrella: in the medusa, the outer surface of the bell.
gonophore: that structure on the hydrozoan polyp which (1) gives rise directly to the sexual products, or (2) gives rise to an attached or free medusa which in turn releases gametes, or (3) the sexual medusa itself.
filiform: threadlike.
hydranth: that portion of the polyp possessing tentacles and mouth, not including the stem or stalk.
hydrotheca or **theca:** the cuplike structure surrounding the polyp of most Leptomedusae.
internode: that portion of the stem or branch between noticeable joints or nodes; characteristic of the Plumulariidae.
manubrium: a tubular structure bearing the mouth at its free end.
nematophore: a polyp modified as a tiny dactylozooid, encased in a protective **nematotheca** (figs. 16B, 18D); characteristic of the Plumulariidae.
operculum: flap or lid covering the hydrotheca.
peduncle: in the hydromedusae, cone of jelly on which the manubrium and stomach are situated.
perisarc: chitinous covering surrounding the stolons and polyps.
polymorphism: polyps becoming modified to perform specific functions; in *Hydractinia* (fig. 9), for example, reproductive gonozooids, feeding gastrozooids, and defensive dactylozooids.
radial canals: in medusae, the canals extending from the stomach to the ring canal.
ring canal: in medusae, the canal around the margin of the bell.
statocyst: small vesicles found on the ring canals of Leptomedusae; possibly organs of orientation.

stolon: tubelike structure connecting the polyps of a hydrozoan colony.
subalternate: not quite opposite, yet not regularly alternate.
theca: see hydrotheca.
velum: flap of tissue within the rim of the bell of hydromedusae.
zooid: the individual member of a colonial organism.

For the student who needs to go beyond the level of this manual or to use Fraser (1937), the use of a good standard or zoological dictionary is advised.

KEY TO ATTACHED (POLYPOID) STAGES OF HYDROIDA

1. Hydranths when retracted encased partially or wholly within a theca 18
– Hydranths when retracted not encased within a theca 2
2. Hydranth with only filiform tentacles (figs. 5A, 8, 9, 10A, 13A) 3
– Hydranth with at least some capitate tentacles (figs. 4A, 6A, 7A) 11
3. Hydranth with 2 whorls of filiform tentacles; hydranth large, pink and showy, with stalk 2 cm or more in height; gonophores borne among lower whorl of tentacles .... *Tubularia* 4
– Filiform tentacles otherwise arranged 5
4. Colony unbranched, up to 3 cm in length; distal tentacles less numerous than proximal; in rocky intertidal of open coast .... *Tubularia marina*
– Colony branched, growing in large, bushy clusters, up to 15 cm in length; proximal and distal tentacles nearly equal in number (20–24); on pilings and floats in bays and estuaries .... *Tubularia crocea*
5. Hydranth with 2 tentacles (figs. 15A, B, C); on rims of tubes of sabellid worms .... *Proboscidactyla*
– Hydranth with more than 2 tentacles; on rocks, algae, etc. . 6
6. Hydranth with tentacles in a single whorl 7
– Hydranth with scattered tentacles 17
7. Stalks of hydranths without perisarc; colony a crowded mass of pink, unbranched gastrozooids, gonozooids, and sometimes

PLATE 8. **Cnidaria** (3). **Anthomedusae (Gymnoblastea, Athecata):** 4A, *Cladonema* polyp with medusa bud; 4B, medusa; 5A, *Leuckartiara* polyps and medusa bud; 5B, medusa, showing only 4 of 16 tentacles; 6A, *Hydrocoryne* polyp; 6B, medusa, optical section showing only 2 of 4 tentacles; 7A, *Stauridiosarsia japonica* polyp with medusa buds; 7B, "*Sarsia*," the medusa of 7A; 8, *Cordylophora* with gonangium; 9, *Hydractinia*. 4–7, Rees, original; 8–9 after Smith, 1964.

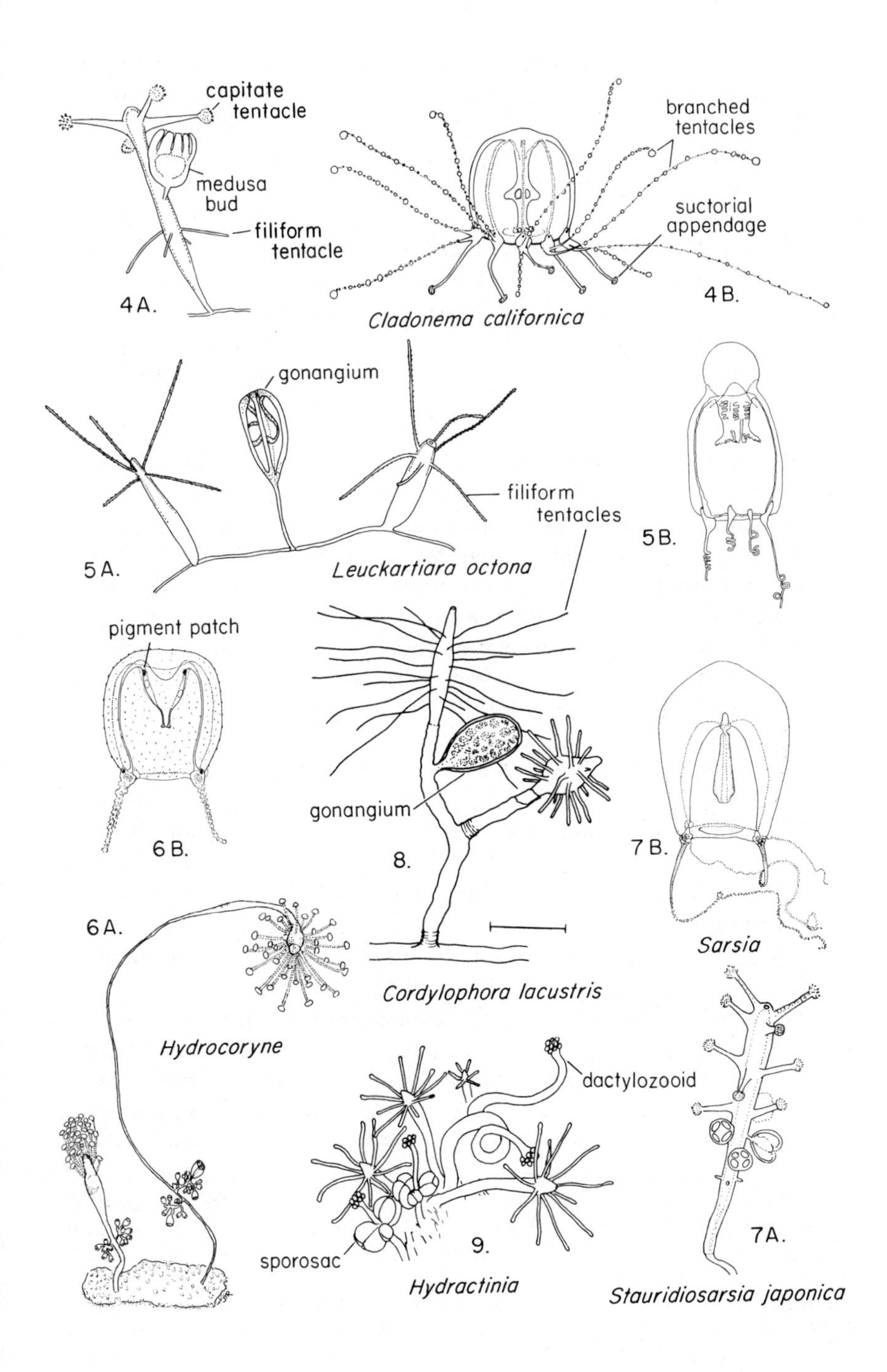
capitate tentacle
medusa bud
filiform tentacle
4A.
branched tentacles
suctorial appendage
4B.
Cladonema californica
gonangium
filiform tentacles
5A.
Leuckartiara octona
5B.
pigment patch
6B.
gonangium
8.
7B.
6A.
Sarsia
Cordylophora lacustris
Hydrocoryne
dactylozooid
sporosac
9.
Hydractinia
7A.
Stauridiosarsia japonica

dactylozooids arising from a basal network of stolons (fig. 9) .... *Hydractinia*
– Stalks of hydranth with a perisarc; colony usually branched 8
8. Hydranth with trumpet-shaped manubrium; perisarc stiff, brown, closely and sharply annulated .... *Eudendrium californicum*
– Hydranth with a conical manubrium; perisarc usually not sharply annulated, or if so only on the smaller branches ... 9
9. Colony small, less than 1 cm high; stems not at all or very slightly branched (fig. 5A) .... *Leuckartiara octona*
– Colony larger, up to several cms in length; stems branched 10
10. Hydranths bright orange; possesses fixed gonophores .... *Garveia annulata*
– Hydranths colorless; releases medusae ...... *Bougainvillia*
11. Hydranths with filiform and capitate tentacles 12
– Hydranth with capitate tentacles only 13
12. Hydranth with circlet of 4 capitate tentacles and a basal whorl of 4 filiform tentacles; in bays (fig. 4A) *Cladonema californica*
– Hydranth with scattered capitate tentacles and a basal whorl of 3–5 small, filiform tentacles (fig. 7A) .... *Stauridiosarsia*
13. Hydranths 1 cm or more long; perisarc reduced or absent .. 14
– Hydranths never more than 3 mm long; perisarc well developed 15
14. Tentacles 60–70, arranged in 5–6 circlets; releases free medusae; (fig. 6A) .... *Hydrocoryne*
– Tentacles up to 500, irregularly arranged, fixed gonophores .... *Candelabrum*
15. Hydranth does not produce free medusae; gonangia with net-like pattern .... *Coryne*
– Hydranth releases free medusae; medusa buds with tentacle bulbs and ocelli 16
16. Colony much branched; perisarc irregularly annulated .... *Syncoryne eximia*
– Colony unbranched or slightly branched; perisarc not annulated .... *Syncoryne mirabilis*

PLATE 9. **Cnidaria** (4). **Leptomedusae (Calyptoblastea, Thecata) and Limnomedusae:** 10A, *Eutonina* polyp with gonangium; 10B, medusa; 11, *Aequorea;* 12A, *Phialidium* colony; 12B, hydrotheca; 12C, medusa; 13A, *Phialella* polyp; 13B, medusa; 14, *Proboscidactyla* medusa; 15A, *Proboscidactyla* colony on rim of worm tube; 15B, young polyp; 15C, mature polyps with medusa-buds. 10, 13, Rees, original; 12A after West and Renshaw, 1970 (as *Clytia attenuata*); 14, 15 after Hand, 1954; 11, 12C from L. H. Hyman, 1940, used with permission of McGraw-Hill Book Company.

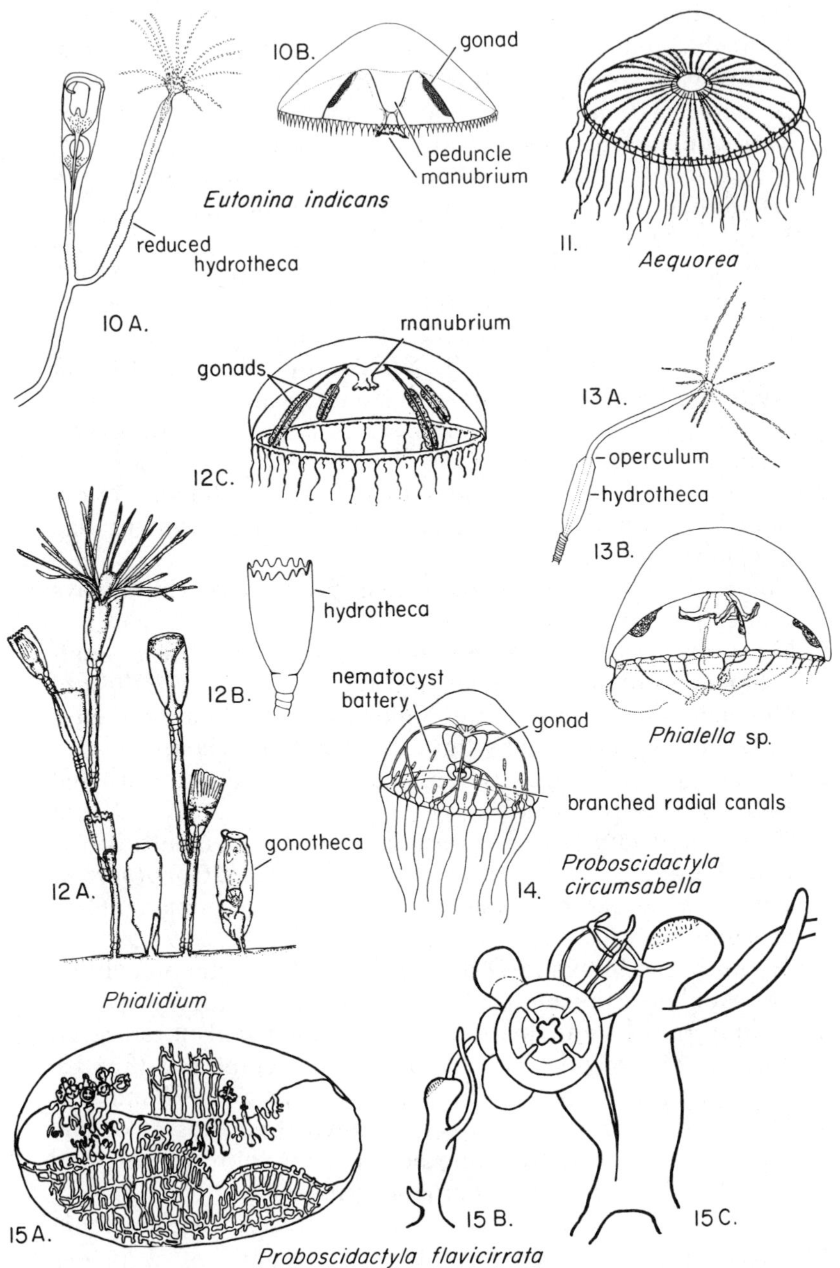
10B.
gonad
peduncle
manubrium
Eutonina indicans
reduced
hydrotheca
10 A.
11.
Aequorea
manubrium
gonads
13 A.
operculum
hydrotheca
12C.
13B.
hydrotheca
12B.
nematocyst
battery
gonad
Phialella sp.
branched radial canals
gonotheca
Proboscidactyla
circumsabella
12 A.
14.
Phialidium
15A.
15 B.
15 C.
Proboscidactyla flavicirrata

17. (*Note 3 choices*) Marine; colony regularly branched; produces free medusae .... *Turritopsis nutricula*
 – In fresh or brackish water; colony irregularly branched; possesses fixed gonophores (fig. 8) .... *Cordylophora lacustris*
 – In bays; hydranths pink, often in masses; perisarc absent; fixed gonophores pink in male, purple in female .... *Clava leptostyla*
18. Hydrotheca with a more or less distinct stalk 19
 – Hydrotheca sessile, without a stalk 23
19. (*Note 3 choices*) Hydrothecae tubular with pointed operculum of converging segments or folds; colony simple or branched, with small hydranths 1–2 mm in height (fig. 13A) ..... families CAMPANULINIDAE and PHIALELLIDAE
 – Hydrothecae reduced, saucer-shaped, too small to contain retracted hydranths (fig. 19) .... *Halecium*
 – Hydrothecae bell- or wineglass-shaped 20
20. Colony small, generally less than 2 cm in height, rarely branching; when branching, each stem bearing not more than 3–4 polyps 21
 – Colony larger, generally more than 2 cm in height, always branching, sometimes profusely 22
21. Gonangia release free medusae; young medusae hemispherical, with 4–8 tentacles (fig. 12) .... *Phialidium*
 – Gonangia do not release free medusae, but retain the medusoids on the blastostyle within the gonotheca *Campanularia*
22. Gonangia release free medusae; young medusae with flat bell, with 12–16 tentacles (fig. 1) .... *Obelia*
 – Gonangia retain sexual zooids in a cluster at mouth of the gonotheca .... *Gonothyraea*
23. Hydrothecae on one side of branch only; nematophores present 24
 – Hydrothecae in two rows on opposite sides of the branch; no nematophores present 25
24. Colony translucent; hydrothecae alternate with internodes; margins of hydrothecae not toothed or sculptured (fig. 18) .... *Plumularia*
 – Colony stiff, brown or greenish brown; branches of colony resemble feathers; hydrothecae close-set with toothed and sculptured margins; no internodes present (fig. 16) .... *Aglaophenia*
25. Hydrothecae opposite each other; operculum of 2 pieces (fig. 20) .... *Sertularia*
 – Hydrothecae alternate or subalternate 26

26. Hydrotheca with smooth margin; operculum of one piece; colony plumelike, with numerous side branches in one plane .... *Abietinaria*
- Hydrotheca with toothed margin; operculum of 3 or 4 pieces; colony without side branches (fig. 17) .... *Sertularella*

KEY TO HYDROMEDUSAE

1. Gonads borne on the radial canals, or on the radial canals associated with the peduncle (figs. 2, 10B, 12C, 28) 2
- Gonads borne on the manubrium only, not associated with radial canals (fig. 14) 12
2. Tentacles arising in a single whorl around bell margin 6
- At least some tentacles arising above the bell margin; tentacles above bell margin with distinct endodermal core running to the ring canal 3
3. Tentacles without suckerlike adhesive disc 4
- Tentacles with suckerlike adhesive disc 5
4. (*Note 3 choices*) In fresh water; pendant, pouch-like gonads .... *Craspedacusta sowerbii*
- Marine; gonads hang like a folded curtain .... *Aglauropsis*
- Marine; gonads numerous, fingerlike, pendant; peduncle well developed, broadly conical, and extending to the velum; large, to 85 mm high .... *Scrippsia pacifica*
5. Four of tentacles *terminating* in suckerlike adhesive disc; other tentacles with adhesive discs located other than terminally .... *Vallentinia adherens*
- Adhesive discs located on aboral surface of all tentacles but none terminal .... *Gonionemus vertens*
6. Gonads fingerlike, pendant .... *Polyorchis*
- Gonads not fingerlike; extending along radial canals 7
7. Radial canals numerous, more than 24, simple; mouth large, surrounded by numerous frilled lips (fig. 11) .... *Aequorea*
- Four radial canals 8
8. Bell flat; medusae with numerous tentacles, lacking a velum; often swim "inside out;" statocysts 8 (2 per quadrant), borne on basal bulbs of tentacles (fig. 1) .... *Obelia*
- Bell hemispherical; not normally seen swimming inside out 9
9. Peduncle well developed, reaching the level of the ring canal, numerous small tentacles; with 8 statocysts (2 per quadrant) (fig. 10B) .... *Eutonina indicans*
- Peduncle poorly developed or absent 10
10. Statocysts lacking; with cirri and sensory clubs (cordyli) between tentacles .... *Laodicea*

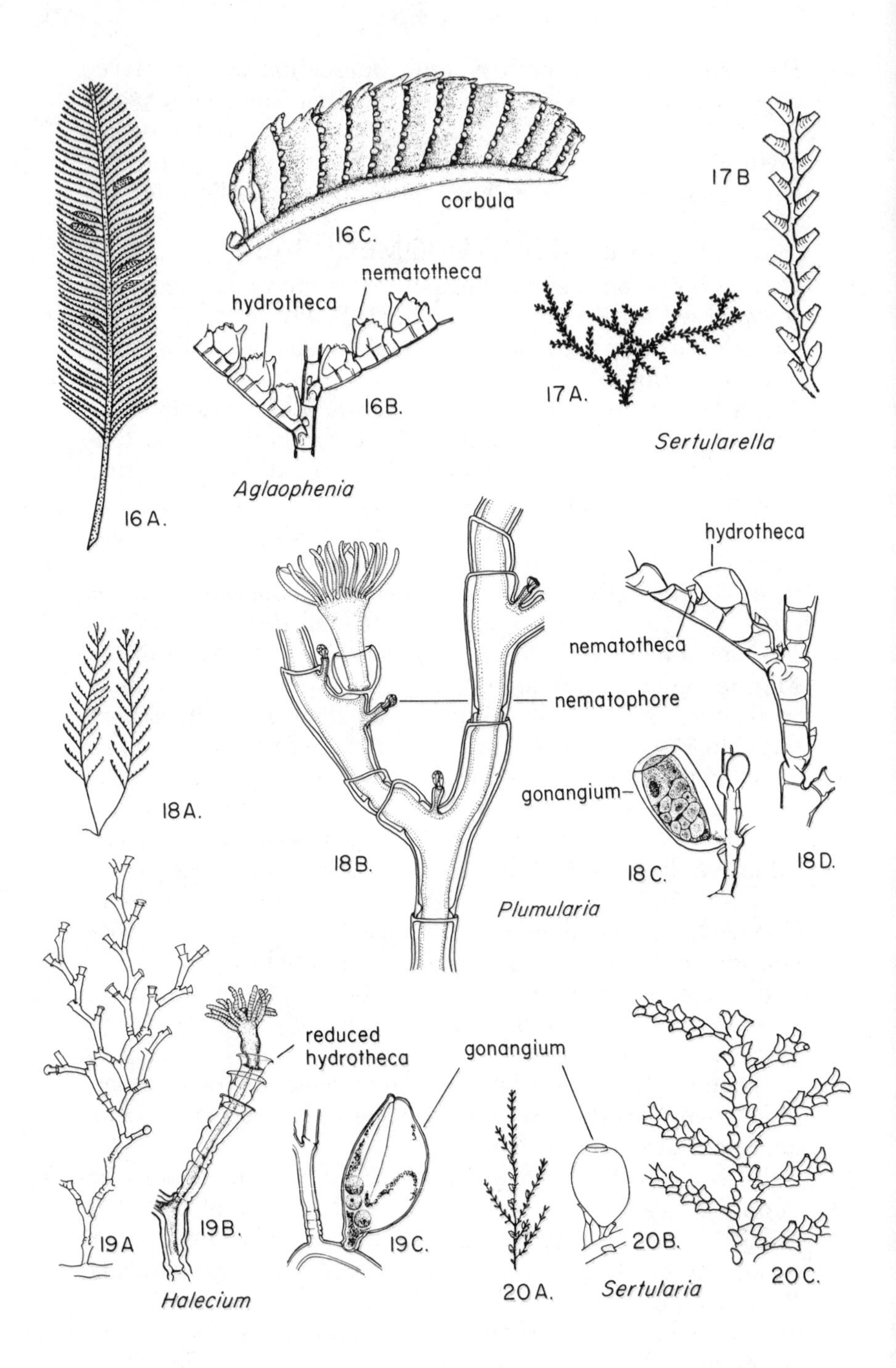
corbula
16 C.
nematotheca
hydrotheca
16 B.
Aglaophenia
16 A.
17 B
17 A.
Sertularella
hydrotheca
nematotheca
nematophore
gonangium
18 A.
18 B.
18 C.
18 D.
Plumularia
reduced
hydrotheca
gonangium
19 A
19 B.
19 C.
Halecium
20 A.
20 B.
20 C.
Sertularia

– Statocysts present 11
11. Two statocysts per quadrant; 40–50 tentacles (fig. 13B) .... *Phialella*
– Four or more statocysts per quadrant (fig. 12C) *Phialidium*
12. Tentacles arise in 4 clusters at distal ends of the 4 radial canals .... *Bougainvillia*
– Tentacles arise singly; may be simple or branched 13
13. Tentacles branched in adult; radial canals and tentacles usually 9 in number, with a single, well-developed suctorial appendage borne orally on each tentacle (fig. 4B) .... *Cladonema californica*
– Tentacles not branched 14
14. Manubrium long, extending 2–3 times the height of the bell beyond velum; gonads in 2–3 distinct rings ....*Dipurena*
– Manubrium short, not extending beyond velum, or only slightly so 15
15. Radial canals branched in older specimens; nematocyst batteries on exumbrella (fig. 14) ....*Proboscidactyla*
– Radial canals simple even in sexually mature specimens ... 16
16. Gonads in a ring around manubrium; 4 equally developed tentacles 17
– Four gonads, complexly folded; medusa "double-decked" in appearance, with a more or less conspicuous apical projection 18
17. Manubrium tubelike; no pigment at the proximal portion of the manubrium (fig. 7B) .... *Sarsia*
– Manubrium conical; ochre pigment patches more or less pronounced at the proximal portion of the manubrium (fig. 6B) .... *Hydrocoryne*
18. Two opposed, well-developed tentacles, others rudimentary .... *Stomotoca*
– More than 2 well-developed tentacles 19
19. Tentacles 12–24, and in addition about 16 rudimentary tentacles (fig. 5B) .... *Leuckartiara octona*
– Tentacles 30 or more; no rudimentary tentacles ..*Neoturris*

PLATE 10. **Cnidaria** (5). **Leptomedusae (Calyptoblastea, Thecata):** 16A, *Aglaophenia* colony; 16B, hydrothecae; 16C, corbula; 17A, *Sertularella* colony; 17B, hydrothecae; 18A, *Plumularia* colony; 18B, D, hydranths and nematophores; 18C, gonangium; 19A, *Halecium* colony; 19B, hydrotheca; 19C, gonangium; 20A, *Sertularia* colony; 20B, gonangium; 20C, hydrothecae. 16A,C, 17A, 18A, 20A after Fraser, 1937; 16B, 17B, 18C,D, 19A,B,C, 20B,C are New Zealand species, after Ralph, 1957–61.

UNIFIED LIST OF HYDROZOA

Since this list includes both polyps and medusae, the following abbreviations are used:

P, M—Polyp and/or Medusa reported from central California.

(P), (M)—Polyp and/or Medusa present in life cycle but not reported from central California intertidal zone.

P′—only the Polyp stage present in life cycle.

Order **HYDROIDA**

Suborder **Anthomedusae** (= **Athecata, Gymnoblastea**)

BOUGAINVILLIIDAE

Bimeria, see *Garveia*.

Bougainvillia spp. P, M Local species probably include *B. ramosa* (Van Beneden, 1844), a possible introduction in bays and harbors, and *B. mertensi* Agassiz, 1862.

Garveia annulata Nutting, 1901. P′ Rocky intertidal zone of open coast; a conspicuous, bright orange hydroid.

* *Garveia franciscana* (Torrey, 1902) (Formerly *Bimeria franciscana*). P′ In bays and estuaries, apparently introduced and cosmopolitan; other species may be present.

HYDRACTINIIDAE

Hydractinia spp. P′ Commonly found as pink patches on undersides of rocks and ledges.

CANDELABRIDAE

Candelabrum sp. P′ See Hand and Gwilliam, 1951 J. Wash. Acad. Sci. 41: 206–209.

CLADONEMIDAE

Cladonema californica Hyman, 1947. P, M Both polyp and medusa occur on *Zostera* and algae in bays and harbors.

CLAVIDAE

Clava leptostyla Agassiz, 1862. P′ Large pink patches on pilings in estuaries; probably introduced from Atlantic.

* *Cordylophora lacustris* Allman, 1844. P′ On pilings in waters of low salinity, or in fresh water, as in Lake Merced; introduced.

Turritopsis nutricula McCrady, 1857. P (M) In estuaries and bays; possibly introduced.

* Not in key.

CORYNIDAE

Coryne spp. P′ A *Coryne* found intertidally on algae may be *C. brachiata* Torrey, 1901.

Dipurena sp. (P) M Uncommon in inshore waters; polyp subtidal, generally associated with sponges.

Sarsia spp. P, M Several species of *Sarsia* medusae occur in plankton; most are small and difficult to identify; the name "*Sarsia*" refers to the medusae produced by corynid polyps, including *Syncoryne* and *Stauridiosarsia.*

Stauridiosarsia spp. P, M *S. japonica* Nagao, 1962 occurs intertidally on rocks and shells; other species occur in Bodega Harbor.

Syncoryne eximia (Allman, 1864). P, M In rocky intertidal zone; releases medusae with tentacles held in a characteristic hooked stance.

Syncoryne mirabilis (Agassiz, 1862). P, M In bays and harbors; introduced from Atlantic.

EUDENDRIIDAE

Eudendrium spp. P′ *E. californicum* Torrey, 1902 appears to be the common *Eudendrium* of the open rocky coast; other species are present.

HYDROCORYNIDAE

Hydrocoryne sp. P, M The medusa occurs in plankton in summer and fall; polyp chiefly subtidal, found occasionally on intertidal algae.

PANDEIDAE

Leuckartiara octona (Fleming, 1823) [=*Perigonimus repens* (Wright, 1858)]. P, M The polyp occasionally occurs on the shells of living gastropods, including *Olivella biplicata* and *Nassarius mendicus.*

Neoturris spp. (P) M Oceanic; uncommon inshore.

Perigonimus, see *Leuckartiara.*

Stomotoca spp. (P) M Oceanic; uncommon inshore.

POLYORCHIDAE

Polyorchis spp. (P?) M There is considerable morphological variation among the species of *Polyorchis* (see Skogsberg, 1948). The common *Polyorchis* with diverticula on the radial canals is *P. penicillatus* (Eschscholtz, 1829). A variety or possibly distinct species, *P. montereyensis* Skogsberg, 1948, also occurs. Those without diverticula and with a well-developed peduncle are *P. haplus* Skogsberg, 1948, but this species is uncommon. The polyp stage is probably unknown, despite the unconfirmed claim of Brinckmann-Voss, 1977, Can. J. Zool., 55:93–96.

Scrippsia pacifica (Torrey, 1909) (Formerly *Polyorchis*). (P) M Largest hydromedusan in central California; often washed up on beaches in summer.

TUBULARIIDAE

Tubularia crocea (Agassiz, 1862). P′ Forms large, conspicuous clusters on pilings in estuaries; introduced from Atlantic.

Tubularia marina (Torrey, 1902). P′ More delicate and sparse, on and under rocks in the low rocky intertidal zone.

ZANCLEIDAE

* *Zanclea costata* Gegenbauer, 1856. P, M A very small polyp with capitate tentacles; associated with cheilostome bryozoans on pilings.

Suborder **Leptomedusae (= Thecata, Calyptoblastea)**

AEQUOREIDAE

Aequorea spp. (P), M In open water plankton; very luminescent.

CAMPANULARIIDAE

Campanularia spp. P′ Although Naumov (1960) has synonymized *Eucopella* and *Clytia* in this genus thus combining species with and without free medusae, we retain the usage that *Campanularia* does not release medusae, but retains medusoids within the gonotheca. *Campanularia (=Eucopella) compressa* (Clark, 1876) is common on larger hydroids; other species are present.

Gonothyraea spp. P′ The sexual zooids or sporosacs are extruded as a mass held at mouth of the gonotheca. *Gonothyraea clarki* (Marktanner-Turneretscher, 1895) is often very common on harbor floats.

Obelia spp. P, M Although Naumov (1960) synonymizes *Gonothyraea* with *Obelia*, we retain the usage that *Obelia* releases small, flat medusae with 12–16 tentacles. Local species include *O. geniculata* (Linnaeus, 1767) with a characteristic zigzag, central stem, and *O. longissima* (Pallas, 1766), a large (up to 60 cm) fouling species common in harbors; other species are present.

Phialidium spp. P, M Traditionally, *Phialidium* has been the medusa produced by the hydroid *Clytia*, and opinions differ on which name should prevail. We follow Russell in using *Phialidium;* the medusa of *P. gregarium* (Agassiz, 1862) occurs in local plankton and the small polyps occur subtidally and on floats. However, West and Renshaw (1970) use *Clytia* for a southern California species that is clearly of the same genus.

CAMPANULINIDAE

* *Calycella syringa* (Linnaeus, 1767). P′ Very tiny, inconspicuous polyps, with thecae closed by an operculum of pointed flaps.

EUTIMIDAE

Eutonina indicans (Romanes, 1876). P, M Medusae abundant in summer plankton; the polyp is subtidal.

HALECIIDAE

Halecium spp. P′ In exposed rocky intertidal.

LAODICEIDAE

Laodicea spp. (P) M Medusae in open-water plankton.

PHIALELLIDAE

Phialella sp. P, M Polyp occurs on pilings and mussels in harbors.

PLUMULARIIDAE

Aglaophenia spp. P′ "Ostrich-plume hydroids," conspicuous in the rocky intertidal; species include *A. struthionides* (Murray, 1860), probably the largest local *Aglaophenia*, and *A. inconspicua* Torrey, 1904, a small, delicate species found on algae; other species are present.

Plumularia spp. P′ *Plumularia plumularoides* (Clark, 1876) and *P. setacea* (Ellis, 1755) appear to be common intertidal *Plumularia;* other species probably present.

SERTULARIIDAE

Abietinaria spp. P′ Common in rocky intertidal.

Sertularella spp. P′ *Sertularella turgida* (Trask, 1857) is the most common; other species probably present.

Sertularia spp. P′ *Sertularia furcata* Trask, 1857 is very common intertidally on algae, in large patches; other species probably present.

Suborder **Limnomedusae**

OLINDIASIDAE

Aglauropsis sp. P, M Medusa an open-water form.

Craspedacusta sowerbii Lankester, 1880. P, M This introduced *fresh-water* medusa occurs sporadically in late fall, often in artificial impoundments; the polyp ("*Microhydra*") has been reported only once from California.

Gonionemus vertens A. Agassiz, 1862. (P), M Medusa common among eelgrass (*Zostera*) in Puget Sound region; rare locally.

Vallentinia adherens Hyman, 1947. (P), M Medusa taken among floating algae, such as *Macrocystis*.

PROBOSCIDACTYLIDAE

Proboscidactyla spp. P, M *Proboscidactyla circumsabella* Hand, 1954 occurs in central California; the polyp stage is found on rims of tubes of the sabellid polychaete *Pseudopotamilla occelata;* other species (such as *Proboscidactyla flavicirrata* Brandt, 1835, fig. 15A–C) are common in the Puget Sound region.

Order **CHONDROPHORA**

* *Velella velella* (Linnaeus, 1758) (=*V. lata* Chamisso and Eysenhardt, 1821). P, M Often washed ashore in large numbers in spring and summer. Fields and Mackie (1971) consider chondrophores members of the superfamily Tubularioidea.

Order **STYLASTERINA**

* *Stylantheca porphyra* Fisher, 1931 (=*Allopora porphyra*). Vivid purple, low intertidal and subtidal encrustations, sometimes bearing small, paired holes of a burrowing spionid polychaete. Although described as *Stylantheca* by Fisher and transferred to *Allopora* by him in 1938, most subsequent workers have retained the original generic assignment.

REFERENCES ON HYDROZOA

CNIDARIA (general)

Hyman, L. H. 1940. *The Invertebrates: Protozoa through Ctenophora,* Vol. I. McGraw-Hill, pp. 365–661.

HYDROZOA

Fields, W. G. and G. O. Mackie 1971. Evolution of the Chondrophora: evidence from behavioral studies on *Velella.* J. Fish. Res. Bd. Canada 28: 1595–1602.

Fisher, W. K. 1938. Hydrocorals of the north Pacific Ocean. Proc. U.S. Nat. Mus. 84: 493–554.

Fraser, C. McL. 1937. *Hydroids of the Pacific Coast of Canada and the United States.* University of Toronto Press, 207 pp.

Hand, C. 1954. Three Pacific species of "*Lar*" (including a new species), their hosts, medusae, and relationships (Coelenterata, Hydrozoa). Pacific Sci. 8: 51–67.

Hyman, L. H. 1947. Two new Hydromedusae from the California coast. Trans. Amer. Micr. Soc. 66: 262–268.

Kramp, P. L. 1961. Synopsis of the medusae of the world. J. Mar. Biol. Assoc. U.K. 40: 1–469.

Kramp, P. L. 1968. *The Hydromedusae of the Pacific and Indian Oceans.* Dana-Report no. 72. Copenhagen: Carlsberg Foundation, 200 pp.

Mayer, A. G. 1910. *Medusae of the World,* Vols. I and II. *The Hydromedusae.* Carnegie Inst. Wash. Publ. no. 109: 1–498, i–xv.

Naumov, D. V. 1960. *Hydroids and Hydromedusae of the USSR* (in Russian). Zool. Inst. Akad. Nauk, SSSR, no. 70, 585 pp., pl. I–XXX. English translation, 1969, by Israel Program for Scientific Translation, available from U.S. Dept. of Commerce, Clearinghouse for Sci. and Tech. Information, Springfield, Va. 22151.

Nutting, C. C. American hydroids. Spec. Bull. U.S. Nat. Mus. No. 4 (Parts 1–3), 736 pp., 102 plates.

Part 1: 1900. The Plumulariidae. 285 pp., 35 pls.

Part 2: 1904. The Sertulariidae. 325 pp., 41 pls.
Part 3: 1905. The Campanulariidae and Bonneviellidae. 126 pp., 27 pls.
Russell, F. S. 1953. *The Medusae of the British Isles.* Vol. I: *Anthomedusae, Leptomedusae, Limnomedusae . . .* xiii + 530 pp. Vol. II. 1970. *Pelagic Scyphozoa with a Supplement to the First Volume on Hydromedusae.* xii + 284 pp. Cambridge University Press. These are excellent accounts of "jellyfish," both Hydromedusae and Scyphomedusae, useful even on this coast.
Skogsberg, T. 1948. A systematic study of the family Polyorchidae (Hydromedusae). Proc. Calif. Acad. Sci. (4) 26: 101–124.
Totton, A. K. 1965. A synopsis of the Siphonophora. London: Publ. British Mus. Nat. Hist., 230 pp.
West, D. L., and R. W. Renshaw. 1970. The life cycle of *Clytia attenuata.* Mar. Biol. 7: 332–339.

CLASS ANTHOZOA

Cadet Hand
University of California, Berkeley, and Bodega Marine Laboratory

Anthozoa, the largest class of cnidarians, number more that 6,000 species, many of which are massive. The group reaches its greatest and most diversified development in the coral-reef areas of tropical and subtropical seas, where several orders not found in our region occur. Locally, the class is chiefly represented by a large number of sea anemones of the subclass **Zoantharia** (or **Hexacorallia**), which (at least in this area) has numerous simple, unbranched, hollow tentacles.

1) Order **Actiniaria:** Sea anemones exemplify the major distinctive features of the Zoantharia, except that they are always solitary rather than colonial, and lack a skeleton. Figure 3 illustrates the anatomy of a typical sea anemone. Most are attached by a broad **pedal disc,** except for some burrowing types in which the base is replaced by a bulbous, burrowing structure, the **physa** (fig. 25). The broad oral disc is surrounded by circlets of hollow tentacles used in food-getting, and bears the slitlike mouth, opening into a deep **gullet (actinopharynx).** The digestive cavity is divided by numerous radially placed, vertical partitions known as **mesenteries** (sometimes called **septa,** although this term may also be used for the calcareous sclerosepta of corals). The edges of the mesenteries are thickened into glandular and digestive regions known as **mesenterial filaments,** and may at their lower ends possess threadlike **acontia** (of taxonomic importance but present only in certain sea anemones).
2) Order **Corallimorpharia:** Although commonly regarded as a sea

anemone, the well-known *Corynactis* has certain distinctive features, including its very large nematocysts, and is placed by systematists in a separate order; it is, as it were, "a coral without a skeleton."

3) Order **Madreporaria:** The stony corals, represented in the local intertidal only by the solitary, orange cup-coral *Balanophyllia elegans*.
4) Order **Ceriantharia:** Includes a few very elongate, sea-anemone-like animals with two rings of tentacles, only rarely intertidal in this area.

In contrast to the Zoantharia, the subclass **Alcyonaria** or **Octocorallia** are distinguished by having eight pinnately branched tentacles, and are always colonial:

1) Order **Pennatulacea:** The sea pens are practically always subtidal in this area.
2) Order **Stolonifera:** Small, sessile polyps connected by a network of stolons. We have one or two low-intertidal representatives (fig. 31).

No trace of the medusoid stage remains in the life cycle of any anthozoan. Gametes are produced by endodermal gonads on the mesenteries, and a planula larva is developed which, after wandering, settles down to grow into a new polyp. In the sea anemone *Epiactis*, tiny young anemones are protected in a groove around the outside of the column of the parent, where they look as if they were being budded off asexually. Although this is not true in *Epiactis*, a number of sea anemones do reproduce asexually. *Metridium senile*, as it moves about, often leaves behind small mounds of tissue from the edges of the pedal disc, a process called "pedal laceration." From these bits of tissue, small anemones develop. The dense, clonal aggregations of *Anthopleura elegantissima* also arise asexually, but in this case by longitudinal fission of the entire animal.

KEY TO ANTHOZOA

1. Tentacles simple, neither branched nor pinnate, usually numerous .... Subclass **Zoantharia (Hexacorallia)** 2
– Tentacles pinnately branched and 8 in number .... Subclass **Alcyonaria (Octocorallia)** 25
2. With septate calcareous skeleton .... stony corals, Order **Madreporaria**
The only local intertidal representative is the bright, transparent, orange solitary cup-coral *Balanophyllia elegans*.
– Without skeleton 3

3. With few to numerous marginal tentacles; no oral inner ring of tentacles; abundant and diverse intertidally .... sea anemones, Orders **Corallimorpharia** and **Actiniaria** 4

– With many very long tentacles, and with an inner oral ring in addition to the marginal ring of tentacles; inhabit slimy tubes; few species and rarely intertidal ....Order **Ceriantharia**

Represented locally by *Pachycerianthus fimbriatus.*

4. Tentacles capitate (knobbed at tip); body color variable (white, pink, red, brown, lavender) ....

– .... *Corynactis californica*

Tentacles not capitate 5

5. Base attached to substrate or solid object; 24 or more tentacles (except in very juvenile specimens) 8

– Base not attached; usually more or less buried in soft substrate, with oral disc and tentacles protruding; small, slender, and elongated 6

6. With 10 tentacles; up to 6 cm long by 6 mm in column diameter ....*Halcampa decemtentaculata*

– With 12–18 tentacles 7

7. (*Note 3 choices*) With 12–18 tentacles, usually 16; nematosomes (spherical, ciliated bodies) circulating in coelenteron; column 10–15 mm long; generally transparent when expanded (fig. 25) ....*Nematostella vectensis*

– Usually with 12 tentacles; no nematosomes; column divisible into physa, scapus, scapulus, and capitulum; scapus with brownish, well-developed cuticle; column up to 5 cm long, but only 1–3 mm in diameter ....*Edwardsia* sp.

– With 12 tentacles; opaque, creamy white column with crinkled, cuticular appearance and attached sand grains; column to 5 cm long and 5–6 mm diameter ......*Halcampa crypta*

8. With 24 tentacles; column with cuticular sheath; tentacles orange, sometimes with brown markings *Cactosoma arenaria*

– With more than 24 tentacles 9

9. With acontia (white, threadlike filaments extruded through cinclides (pores) or breaks in the body wall when animal is detached or handled roughly) 10

– Without acontia 16

10. Column elongate, vermiform; to 46 cm long by 1.5 cm in diameter; burrowing in sandy mud but with base attached to shells or pebbles; column translucent or pale white (fig. 24) ....*Flosmaris grandis*

– Neither burrowing nor vermiform 11

11. Margin of tentacle-bearing disc usually deeply frilled or lobed when extended; tentacles short and very numerous;

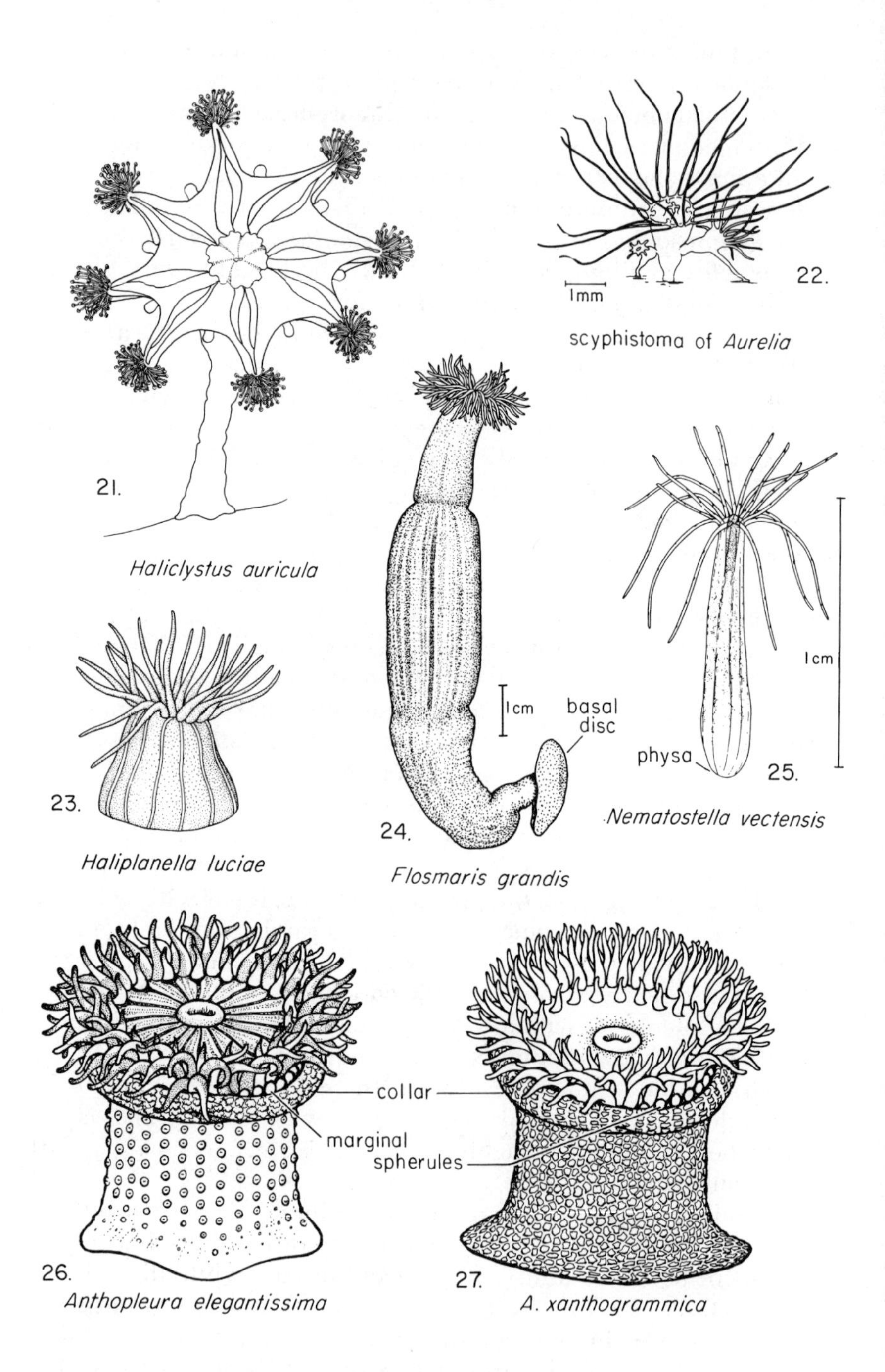
21.
Haliclystus auricula
1mm
22.
scyphistoma of Aurelia
1cm
basal
disc
physa
1cm
25.
Nematostella vectensis
23.
Haliplanella luciae
24.
Flosmaris grandis
collar
marginal
spherules
26.
Anthopleura elegantissima
27.
A. xanthogrammica

very little tentacle-free area around the mouth .... *Metridium senile*

– Margin of disc circular and not frilled or lobed when extended; rarely more than 100 tentacles present; a fairly large area of disc tentacle-free 12

12. The tentacles (directives) closest to each end of mouth-slit with yellow bases; color of the column variable (cream, gray, light green); column transparent when extended and frequently with vertical white stripes *Diadumene franciscana*

– Directive tentacles not marked differently than others 13

13. Column transparent olive green, brownish green, or other shade of green when extended; sometimes orange gonads are visible in the lower parts of the column; vertical stripes of orange, yellow, or white frequently present on column (fig. 23) .... *Haliplanella luciae*

– Column flesh-colored, yellow, orange, or reddish; not as above 14

14. Column and tentacles flesh to pale salmon-colored; upper parts of column may show tints of green; long and slender when extended .... *Diadumene leucolena*

– Column yellow, orange, or reddish 15

15. In extension, column about as tall as wide or slightly taller than wide; up to 1 cm in diameter; orange or yellow, sometimes reddish .... *Metridium exilis*

– In extension, column at least twice as tall as wide; up to 5 mm in diameter; transparent light orange or yellow .... *Diadumene lighti*

16. Column with tubercles or diffuse adhesive properties, capable of holding bits of sand, shells, etc. 17

– Column smooth, unable to hold debris 22

17. With white to yellow, inconspicuous marginal spherules (figs.

PLATE 11. **Cnidaria** (6). Various polypoid cnidarians: 21, *Haliclystus,* from False Bay near Friday Harbor, Wash.; 22, scyphistoma of *Aurelia;* 23, *Haliplanella;* 24, *Flosmaris;* 25, *Nematostella;* 26, *Anthopleura elegantissima;* 27, *A. xanthogrammica.* The latter two are drawn semidiagrammatically to emphasize differences. In *A. elegantissima,* tentacles generally multicolored and variously spotted, often pink-tipped; disc usually with radial lines; tubercles on column round (except on collar), in longitudinal rows, sparse near base; in *A. xanthogrammica,* tentacles and disc generally uniformly colored, disc seldom with conspicuous radial lines; tubercles on column small, irregular, closely and uniformly distributed (not in rows except on collar). 21, Gwilliam, 1954; 22, Bush, in Smith, 1964; 23, after Hargitt, 1914, redrawn by Emily Reid; 24, Hand and Bushnell, 1967; 25, Crowell, 1946, redrawn by Emily Reid; 26, 27, Vicki B. Pearse, original, redrawn by Mildred Waltrip.

26, 27) at top of column just outside tentacles; column green, gray, yellow, or white 18

– No marginal spherules; column some shade of red, sometimes interrupted by green or brownish green patches 20

18. Tubercles on upper $^2/_3$ of column only; column white or pink on lower $^2/_3$, usually black or gray on upper $^1/_3$; tentacles frequently brightly colored; body capable of great elongation; lives buried in sand or gravel and attached below the surface to rocks .... *Anthopleura artemisia*

– Whole column covered with tubercles; column usually green 19

19. Tubercles irregular, compound, not conspicuously in longitudinal rows; disc and tentacles uniform in color, not pink-tipped; disc usually without radial lines; large, solitary anemones (fig. 27) .... *Anthopleura xanthogrammica*

– Tubercles round, arranged in longitudinal rows; tentacles often tipped with pink, purple, or variously marked; disc usually with radial lines; column green to yellow or white; small to medium-sized anemones, commonly densely massed on rocks in sand; solitary individuals may be larger (fig. 26) .... *Anthopleura elegantissima*

20. Column densely covered with tubercles which are strongly adherent; usually buried in gravel or shell debris; column covered with adhering gravel, shells, etc. .. *Tealia coriacea*

– Column not densely covered with tubercles and with or without adhering foreign material 21

21. (*Note 3 choices*) Column bright scarlet; tubercles white, few, in regular longitudinal rows .... *Tealia lofotensis*

– Column red, frequently with irregular patches of green or brownish green, sometimes to point of appearing mostly green; tubercles absent or very few, weak, and scattered .... *Tealia crassicornis*

– Column dull red to brown; edge of base may be marked with light striations; very flat when contracted; commonly with scattered adherent sand grains .... *Cnidopus ritteri*

22. Diameter of column greater than height; edge of base may have white lines; oral disc with white lines or dots at bases of tentacles 23

– Column as long as broad or longer; edge of base without white lines 24

23. Young anemones usually attached on column of the adult; conspicuous, radiating white lines at edge of base and on oral disc .... *Epiactis prolifera*

– Never with attached young; basal lines present or absent, oral disc without lines; basal area usually a darker, different or more intense color than column .... *Cnidopus ritteri*

24. Column several times longer than broad (up to 18 cm long by 1.5 cm in diameter); burrowing in mud or sand, attached to stones, worm tubes, or shells .... *Zaolutus actius*
 – Column as tall as broad; color red or red varied with green or greenish brown patches; large (up to about 9 cm diameter) .... *Tealia crassicornis*

25. Low, encrusting, ramifying growths on rocks, shells, etc.; individual polyps usually contracted to small, pinkish mounds, tentacles rarely visible in the field; sometimes mistaken for small ascidians .... Order **Stolonifera**
 There are two local intertidal forms, of which the more common appears to be a species of *Clavularia* (fig. 31).
 – Erect, featherlike; rooted in soft substrates by a fleshy base; usually subtidal but occasionally washed up or seen in protected, shallow-water areas; the sea pens .... Order **Pennatulacea** 26

26. Distinctly thick and fleshy .... *Ptilosarcus gurneyi*
 – Elongate and slender 27

27. Rough to the touch; axis strong and brittle .... *Stylatula elongata*
 – Smooth to the touch; axis thin and flexible .... *Acanthoptilum gracile*

LIST OF ANTHOZOA

Subclass **ZOANTHARIA (=HEXACORALLIA)**

Order **Actiniaria**

ACTINIIDAE (see Hand, 1955a)

Anthopleura artemisia (Pickering in Dana, 1848) (formerly *Evactis*). Open coast and in estuaries; generally attached to stones below surface of muddy sand, giving the appearance of being a burrower.

Anthopleura elegantissima (Brandt, 1835) (formerly *Bunodactis* or *Cribrina*). Produces clonal aggregations by division; bay-dwelling individuals may be largely buried in substrate. See Ford, 1964; Francis, 1937a,b.

Anthopleura xanthogrammica (Brandt, 1835). The "great green anemone" of tide pools.

Cnidopus ritteri (Torrey, 1902). Under and on rocks in protected situations on outer coast; viviparous, embryos and small anemones brooded in coelenteron; see Hand and Dunn, 1974.

Epiactis prolifera Verrill, 1869. Ubiquitous; in bays and on outer coast; young brooded on outside of column of parent.

Tealia coriacea (Cuvier, 1798). Mid to low intertidal of outer coast and penetrating into bays; buried in sand or gravel.

Tealia crassicornis (Müller, 1776). Mid to low intertidal of outer coast; on undersides of large rocks and in protected pools.

Tealia lofotensis (Danielssen, 1890). Low intertidal on vertical rock faces and subtidal.

DIADUMENIDAE (see Hand, 1955b)

Diadumene franciscana Hand, 1955. San Francisco Bay; sporadically abundant in Aquatic Park, Berkeley; on pilings in Lake Merritt in summer, etc.

Diadumene leucolena (Verrill, 1866) (=*Sagartia leucolena*). Probably introduced; on floats, pilings, stones, and oyster shells, in bays.

Diadumene lighti Hand, 1955. In sand among algal holdfasts along edges of rocky-shore tidal channels.

EDWARDSIIDAE

Edwardsia sp. A small, undescribed or unidentified species occurs in Bodega Harbor, among *Phoronopsis* on muddy sand flats.

Nematostella vectensis Stephenson, 1935. In *Salicornia* marshes in San Francisco and Tomales Bays; see Hand, 1957.

HALCAMPIDAE (see Hand, 1954)

Cactosoma arenaria Carlgren, 1931. On open coasts; frequently on kelp holdfasts.

Halcampa crypta Siebert and Hand, 1974. A twelve-tentacled species, found in muddy shale gravel of inshore pools of Duxbury Reef, Bolinas.

Halcampa decemtentaculata Hand, 1954. This ten-tentacled species is found among root masses of *Phyllospadix*, holdfasts of laminarians, and gravelly pools of low rocky intertidal. Both *Halcampas* occur also in the Friday Harbor area.

HALIPLANELLIDAE (see Hand, 1955b)

Haliplanella luciae (Verrill, 1898) (= *Sagartia luciae*). In bays and estuaries, under rocks and on pilings; may be introduced.

ISANTHIDAE (see Hand, 1955a)

Zaolutus actius Hand, 1955. Referred to from Elkhorn Slough as *Harenactis attenuata* Torrey by MacGinitie, 1935, Amer. Midl. Nat. 16: 629–765; apparently rare in this area, although common subtidally in the south; burrowing in muddy sand.

ISOPHELLIIDAE

Flosmaris grandis Hand and Bushnell, 1967. Burrowing in sand or sandy mud, San Francisco Bay.

METRIDIIDAE (see Hand, 1955b)

Metridium exilis Hand, 1955. Under rocks or ledges on open coast.

Metridium senile (Linnaeus, 1767). The American Pacific coast population is considered a subspecies, *M. senile fimbriatum* (Verrill, 1865) (see Hand, 1955b); common on pilings, floats, and jetties of bays and harbors, as well as subtidally.

Order **Madreporaria** (= **Scleractinia**)

Balanophyllia elegans Verrill, 1864. More common subtidally.

Order **Corallimorpharia**

Corynactis californica Carlgren, 1936. Possesses very large nematocysts, excellent for class study; low rocky intertidal and subtidal; in clonal groups; see Hand, 1954.

Order **Ceriantharia**

Pachycerianthus fimbriatus McMurrich, 1910 (= *P. torreyi* Arai, 1965 and *P. plicatulus* Carlgren, 1924). Generally subtidal, occasionally in low intertidal in very soft mud; the large, golden brown tentacular crown is very conspicuous when expanded; the thick, tough, soft, black, slimy tubes go down to a depth of 3 feet or more; see Arai, 1971, J. Fish. Res. Bd. Canada 28: 1677–1680.

Subclass **ALCYONARIA** (= **OCTOCORALLIA**)

Order **Stolonifera**

Clavularia sp. Under rocks and ledges of low intertidal zone on open coast.

Order **Pennatulacea**

Acanthoptilum gracile (Gabb, 1863). Shallow subtidal of bays to 100 m or more.

Ptilosarcus gurneyi (Gray, 1860) (formerly *Leioptilus* spp.). Occasionally washed ashore; see Batie, 1972, Northwest Sci. 46: 290–300 (taxonomy).

Stylatula elongata (Gabb, 1863). Low intertidal to 60 or more meters; commonly taken by commercial fishermen.

REFERENCES ON ANTHOZOA

Carlgren, O. 1949. A survey of the Ptychodactaria, Corallimorpharia and Actiniaria, with a Preface by T. A. Stephenson. K. Svenska Vetenskapsakad. Handl. (ser. 4) 1: 1–121.

Carlgren, O. 1952. Actiniaria from North America. Arkiv för Zool. (ser. 2) 3: 373–390.
Ford, C. E. 1964. Reproduction in the aggregating sea anemone, *Anthopleura elegantissima*. Pacific Sci. 18: 138–145.
Francis, L. 1973a. Clone specific segregation in the sea anemone, *Anthopleura elegantissima*. Biol. Bull. 144: 64–72.
Francis, L. 1973b. Intraspecific aggression and its effect on the distribution of *Anthopleura elegantissima* and some related sea anemones. Biol. Bull. 144: 73–92.
Hand, C. 1954. The sea anemones of central California. Part I. The corallimorpharian and athenarian anemones. Wasmann J. Biol. 12: 345–375.
Hand, C. 1955a. The sea anemones of central California. Part II. The endomyarian and mesomyarian anemones. Wasmann J. Biol. 13: 37–99.
Hand, C. 1955b. The sea anemones of central California. Part III. The acontiarian anemones. Wasmann J. Biol. 13: 189–251.
Hand, C. 1957. Another sea anemone from California and the types of certain Californian anemones. J. Wash. Acad. Sci. 47: 411–414.
Hand, C. and R. Bushnell 1967. A new species of burrowing acontiate anemone from California (Isophelliidae: *Flosmaris*). Proc. U.S. Nat. Mus. 120: 1–8.
Hand, C. and D. F. Dunn 1974. Redescription and range extension of the sea anemone *Cnidopus ritteri* (Torrey)(Coelenterata: Actiniaria). Wasmann J. Biol. 32: 187–194.
Kükenthal, W. 1913. Über die Alcyonarienfauna Californiens und ihre tiergeographischen Beziehungen. Zool. Jahrb. Abt. f. Syst. 35: 219–270.
Nutting, C. C. 1909. Alcyonaria of the California Coast. Proc. U.S. Nat. Mus., 35: 681–727. (Kükenthal is considered more reliable.)
Siebert, A. E., Jr. and C. Hand 1974. A description of the sea anemone *Halcampa crypta* new species. Wasmann J. Biol. 32: 327–336.
Stephenson, T. A. 1928, 1935. *The British Sea Anemones*. London: Ray Society. Vols. I, 148 pp., and II, 426 pp.

CLASS SCYPHOZOA

Scyphozoa chiefly comprise the so-called "true" jellyfish, which may generally be distinguished from hydromedusae by their larger size, fringed mouth lobes, scalloped margins, absence of a velum, and complex pattern of radial canals (fig. 29). Hydromedusae usually are small, glassy clear, and possess a velum; most have four simple radial canals. For an account of more fundamental morphological differences, the student is referred to Hyman, 1940.

Scyphomedusae, being pelagic, are not encountered in the intertidal zone as adults except when cast ashore, but may be often seen in harbors. Exceptions are the curious stauromedusans (fig. 21), which possess the basic structure of scyphomedusans, but are attached by an aboral stalk to eelgrass, rocks, or algae. In addition, the white, soft, attached "scyphistoma" stages of certain scyphomedusae may be encountered under floats in harbors or marinas (fig. 22). For extensive reviews of the literature, see Kramp (1961) and Russell (1970); the latter also covers several local species in detail.

KEY TO THE MORE COMMON SCYPHOMEDUSAE

1. Free-swimming medusae, often large (fig. 29) .... Order **Semaeostomeae** 2
– Small, inconspicuous, attached forms with 8 marginal clusters of knobs or short, knobbed tentacles (fig. 21) .... Order **Stauromedusae** (see note in species list)
2. Bell dish-shaped; marginal tentacles very small and numerous; 4 horseshoe-shaped gonads; 4 moderately fringed mouth arms .... *Aurelia* 3
– Bell flat or domed; marginal tentacles large and well developed; mouth arms extremely fringed and prominent 4
3. Bell margin scalloped into 8 lobes; up to 40 cm diameter .... *Aurelia aurita*
– The 8 lobes of bell margin secondarily notched so as to appear as 16; up to 30 cm diameter .... *Aurelia labiata*
4. Marginal tentacles arranged singly; bell domed; mouth lobes prominent and long 5
– Marginal tentacles arranged in clusters; bell flatter; mouth lobes extensively fringed 6
5. Eight long marginal tentacles; bell radially patterned in deep purplish brown; up to 70 cm diameter ...... *Pelagia colorata*
– Twenty-four marginal tentacles; bell with radial, yellow brown lines; up to 30 cm diameter ... *Chrysaora melanaster*
6. Marginal tentacles in 16 linear groups; 16 marginal sense organs; up to 50–60 cm diameter *Phacellophora camtschatica*
– Marginal tentacles in 8 crescentic groups; 8 marginal sense organs; up to 1 m diameter .... *Cyanea capillata*

LIST OF SCYPHOZOA

Order **Semaeostomeae**

Aurelia aurita (Linnaeus, 1758) (occasionally as *Aurellia*). See Russell, 1970.

Aurelia labiata Chamisso and Eysenhardt, 1821.

Chrysaora melanaster Brandt, 1838.

Cyanea capillata (Linnaeus, 1758). See Russell, 1970.

Pelagia colorata Russell, 1964. Previously called *P. noctiluca* (Forskål) and *P. panopyra*, an Atlantic species. See Russell 1964, 1970.

Phacellophora camtschatica Brandt, 1838.

Order **Stauromedusae:** This group is poorly known on the West Coast. G. F. Gwilliam (1956) in an unpublished thesis (University of California, Berkeley) lists seven species for western North America. One of these, a new species of the genus *Manania,* is found intertidally and subtidally in central California. This is probably the form listed in Ricketts and Calvin (1968) as *Haliclystus stejnegeri.* The genus *Haliclystus* is represented in northern California by *H. auricula* (Rathke, 1806) (fig. 21), found rarely at low tides on the exposed rocky coast. *H. auricula* (= *H. sanjuanensis* of several authors) is common on the eelgrass *Zostera* in the Friday Harbor area, Washington, and *H. salpinx* Clark, 1863 occurs locally in that area on laminarians. *H. stejnegeri* Kishinouye, 1899 is known only from Alaska and Japan. *Manania* sp. is to be looked for on coralline algae and algal holdfasts in low intertidal pools. Still another species, representing an undescribed genus, has been taken in beds of floating *Macrocystis* near Monterey.

REFERENCES ON SCYPHOZOA

Gwilliam, G. F. 1956. Studies on West Coast Stauromedusae. Doctoral thesis, University of California, Berkeley (unpublished).

Hyman, L. H. 1940. *The Invertebrates: Protozoa through Ctenophora,* Vol. I. McGraw-Hill (Scyphozoa, pp. 497–538; see also Vol. V, 1959, Retrospect, pp. 718–729).

Kramp, P. L. 1961. Synopsis of the Medusae of the World. J. Mar. Biol. Assoc. U.K. 40: 1–469.

Mayer, A. G. 1910. *Medusae of the World.* Vol. III. The Scyphomedusae. Carnegie Inst. Wash. Publ. 109: 499–735.

Russell, F. S. 1970. *The Medusae of the British Isles.* Vol. II. *Pelagic Scyphozoa . . .* Cambridge University Press, xii + 284 pp. Includes a detailed index to literature on Scyphomedusae.

Russell, F. S. 1964. On Scyphomedusae of the genus *Pelagia.* J. Mar. Biol. Assoc. U.K. 44: 133–136.

PLATE 12. **Cnidaria** (7) **and Ctenophora.** 28, *Polyorchis penicillatus* (Hydrozoa); 29, Scyphomedusan structure, diagrammatic; 30, *Velella* (Hydrozoa, Chondrophora); 31, *Clavularia* (Anthozoa, Stolonifera), colony; 32, 33, **Ctenophora:** 32, *Beroe;* 33, *Pleurobrachia.* 28, 30, 31, from L. H. Hyman, 1940, used with permission of McGraw-Hill Book Company; 29, modified after Naumov from Bayer and Owre: *The Free-living Lower Invertebrates,* 1968; used with permission of The Macmillan Company; 32, after Chun from Hand in Marshall and Williams: *Textbook of Zoology: Invertebrates,* 1972; used with permission of The Macmillan Company.

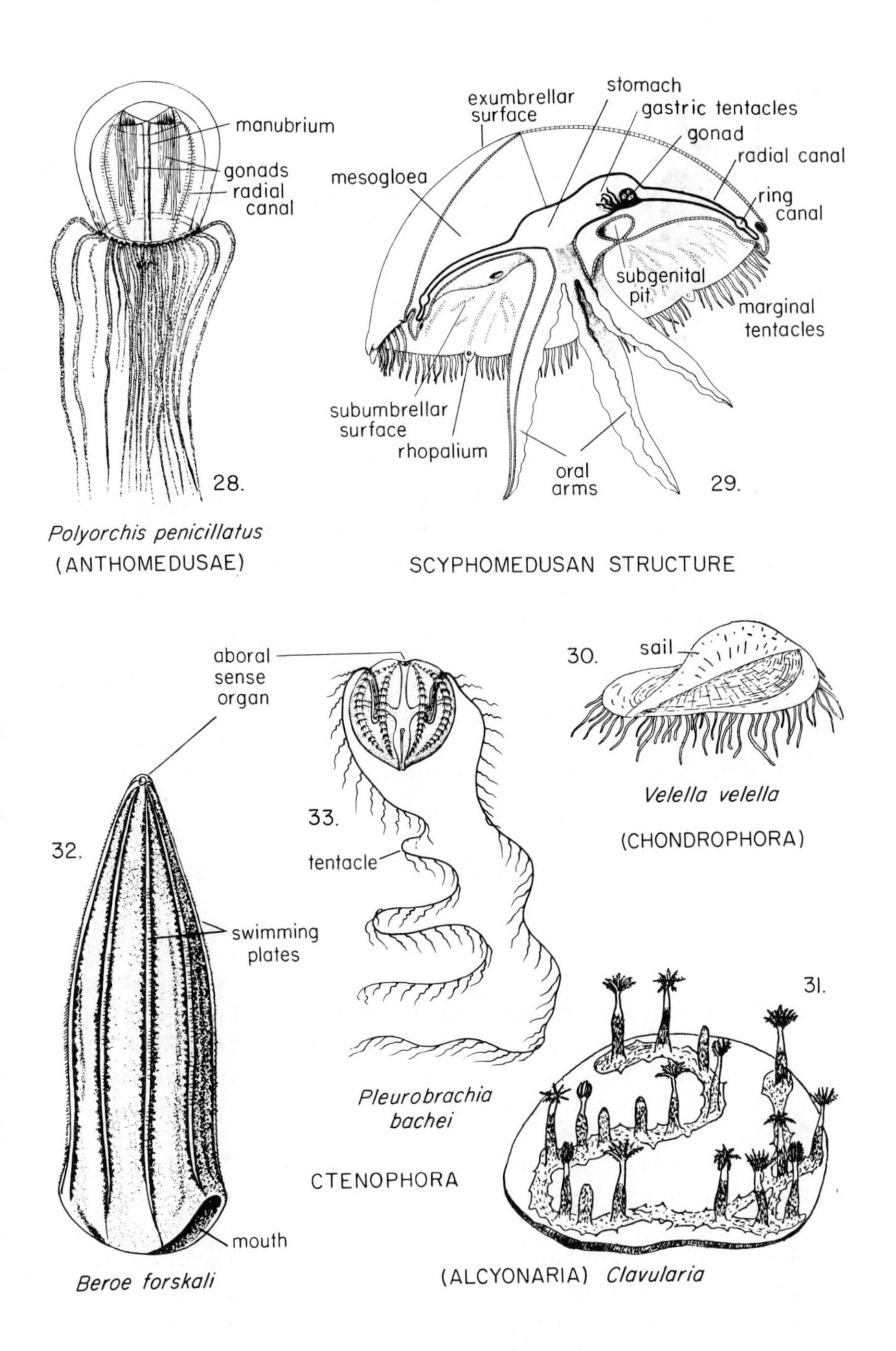
manubrium
gonads
radial canal
28.
Polyorchis penicillatus
(ANTHOMEDUSAE)
stomach
exumbrellar surface
gastric tentacles
gonad
radial canal
mesogloea
ring canal
subgenital pit
marginal tentacles
subumbrellar surface
rhopalium
oral arms
29.
SCYPHOMEDUSAN STRUCTURE
aboral sense organ
30.
sail
Velella velella
(CHONDROPHORA)
33.
32.
tentacle
swimming plates
31.
Pleurobrachia bachei
CTENOPHORA
mouth
Beroe forskali
(ALCYONARIA) Clavularia

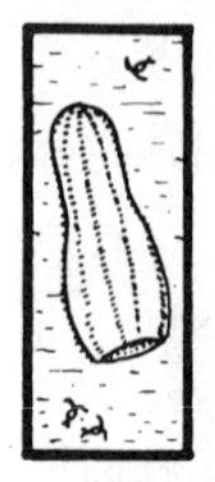

PHYLUM CTENOPHORA

The late Rolf L. Bolin
Hopkins Marine Station of Stanford University, Pacific Grove

(PLATE 12)

The ctenophores or comb jellies (figs. 32, 33) are transparent animals belonging to a small and entirely marine phylum which resemble medusae in appearance, but which are set apart by a number of important differences. These include the absence of nematocysts and of tentacles about the mouth, the possession of a unique aboral sense organ, a marked biradial symmetry, and locomotion by means of eight meridional rows of ciliary platelets (the comb rows) which give the phylum its name. All ctenophores are predators, often capturing their prey by **colloblasts** (glue cells) on long, paired tentacles. Some members of this group glow brilliantly in darkness. The commonest local species is *Pleurobrachia bachei,* often cast up on beaches and variously called sea gooseberries, sea walnuts, or cats' eyes.

KEY TO CTENOPHORA

1. Body resembling empty sack; neither tentacles nor oral lobes present .... Class **NUDA** Order **Beroida** 2
– Body not sacklike; tentacles (in some species minute and almost impossible to detect) and sometimes oral lobes present .... Class **TENTACULATA** 3
2. Side branches of meridional canals forming anastomosing network; body basically triangular in shape, widest at mouth (fig. 32) .... *Beroe forskali*
– Side branches of meridional canals ending in blind twigs; body basically ovoid in shape, mouth somewhat constricted .... *Beroe cucumis*
3. Oral lobes (liplike flaps on either side of mouth) well developed and conspicuous .... Order **Lobata** 4
– Oral lobes not developed .... Order **Cydippida** 5
4. General body surface smooth and colorless .... *Bolinopsis microptera*

- – Surface of body bearing large, conspicuous papillae tipped with orange or brown; may be 15–30 cm long *Leucothea* sp.
- 5. Body much flattened laterally .... *Mertensia ovum*
- – Body subcircular in cross-section 6
- 6. Body almost spherical (fig. 33) ....*Pleurobrachia bachei*
- – Body oval, markedly longer than wide 7
- 7. Ciliary plates ending about midway between oral pole and equator; tentacular sheaths in contact with stomach throughout most of their length ....*Hormiphora* sp.
- – Ciliary plates ending much nearer oral pole than equator; tentacular sheaths not in contact with stomach ....*Euplokamis californiensis*

LIST OF CTENOPHORA

Class **NUDA**

Beroe cucumis Fabricius, 1780. Pronounced *Beroë*.

Beroe forskali Milne-Edwards, 1841. Luminesces beautifully when disturbed after a period in darkness.

Class **TENTACULATA**

Bolinopsis microptera (A. Agassiz, 1865).

Euplokamis californiensis Torrey, 1904.

Hormiphora sp.

Leucothea sp.

Mertensia ovum (Fabricius, 1780).

Pleurobrachia bachei A. Agassiz, 1860. By far the commonest species. See Hirota, 1974, Fishery Bull. 72: 295–335 (natural history)

REFERENCES

Easterly, C. O. 1914. A study of the occurrence and manner of distribution of the Ctenophora of the San Diego region. Univ. Calif. Publ. Zool. 13: 21–38.

Hyman, L. H. 1940. *The Invertebrates: Protozoa through Ctenophora,* Vol. I. McGraw-Hill (Ctenophora: pp. 662–696; see also Vol. V, 1959, Retrospect, pp. 730–731).

Mayer, A. G. 1911. Ctenophores of the Atlantic coast of North America. Carnegie Inst. Wash. Publ. 162: 1–58.

Torrey, H. B. 1904. The ctenophores of the San Diego Region. Univ. Calif. Publ. Zool. 2: 45–51.

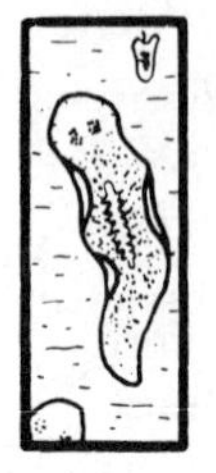

PHYLUM PLATYHELMINTHES

Eugene C. Haderlie
Naval Postgraduate School, Monterey

(PLATES 13–15)

The Platyhelminthes is the lowest of the truly bilateral phyla. Two of the three classes, the **Trematoda** or flukes and the **Cestoda** or tapeworms, are parasitic, particularly in vertebrates, and will not be discussed here. The third class, **Turbellaria,** is composed of herbivorous or predaceous free-living forms with a few commensal or parasitic members, in five groups or orders:

(1) **Acoela** are small marine forms without an intestine; the most conspicuous local intertidal acoel is the orange *Polychoerus carmelensis* (fig. 3) of the *Ulva*-filled, high tidepools of the Monterey region; the smaller *Childia groenlandica* has been found in San Francisco Bay.

(2) **Rhabdocoela** are small, typically narrow forms in which the intestine is a single, straight tube; common in fresh water, and represented in the local intertidal by at least three symbiotic species: *Syndisyrinx franciscanus* (fig. 1) in the intestine of sea urchins; *Syndesmis dendrastrorum* in sand dollars; and *Collastoma pacifica* in the sipunculan *Themiste pyroides.* Minute, free-living species will also be encountered.

(3) **Alloeocoela** are mostly marine forms, sometimes lumped with rhabdocoels, from which they differ in having a more complex intestine and other features; they are small and commonly overlooked (fig. 2). Alloeocoels are common interstitially in mud, sand, and gravel but can also be found on seaweed and in tidepools. Two dozen species from the central California coast have been described by Karling (1962a, 1962b, 1964, 1966). Because of their minute size and the difficulty of identification, no key is attempted here; students should consult Karling's papers for any serious study.

(4) **Tricladida,** which include the familiar freshwater and terrestrial planarians, have a three-branched gut and a protrusible pharynx. Although common in fresh water and in certain damp situations on land, triclads are uncommon in the sea. One local intertidal triclad, *Nexilis*

epichitonius (fig. 4), is commensal in the mantle cavity of the chiton *Mopalia hindsii*, and Holmquist and Karling (1972) have described two interstitial species, *Pacificides psammophilus* (fig. 5) and *Oregoniplana opisthopora.*

(5) **Polycladida** are a large marine group, characterized by an intestine with many branches (figs. 6, 7). Abundant locally, some polyclads attain large size, and many are active and strikingly colored. There are many species of polyclads on the Pacific coast, often abundant under boulders and on pilings. But, except for highly colored or distinctively marked species, polyclads are difficult to identify. Exact determination is often impossible without studying sagittal serial sections of the copulatory apparatus. Eye arrangement is sometimes clear in living worms, but in darkly pigmented species eyes cannot be seen except in fixed, dehydrated, and cleared specimens. Characteristic and often diagnostic features, such as the nature of the penis stylet, Lang's vesicle (seminal bursa), and the spermiducal bulbs, may sometimes be seen by carefully compressing living animals between slides and examining by transmitted light. Color and color patterns are often distinctive, but are often variable and may depend upon the food. The sizes given in the key are averages; many specimens collected will be larger or smaller.

For more details of Pacific coast polyclads and for species not treated in the following key, the student should consult the valuable papers of Hyman (1953, 1955, 1959). The following key is a revised version of that prepared by the late Dr. Libbie Hyman for the second edition of this manual. John J. Holleman of Merritt College, Oakland, and Eugene N. Kozloff of the Friday Harbor Laboratories of the University of Washington have given valuable advice in the preparation of this chapter.

GLOSSARY

cerebral eyes: eyespots, usually of dark color and in paired clusters over the brain area.

common genital pore: single pore to outside from both male and female reproductive systems.

marginal eyes: eyespots in band along whole or anterior part of body margin.

marginal tentacles: tentacles on anterior margin often formed from folds in margin.

nuchal tentacles: tentacles located over the lateral brain area, well back from anterior margin.

penis: muscular projection that terminates the male copulatory system and is employed by simple protrusion to the outside.

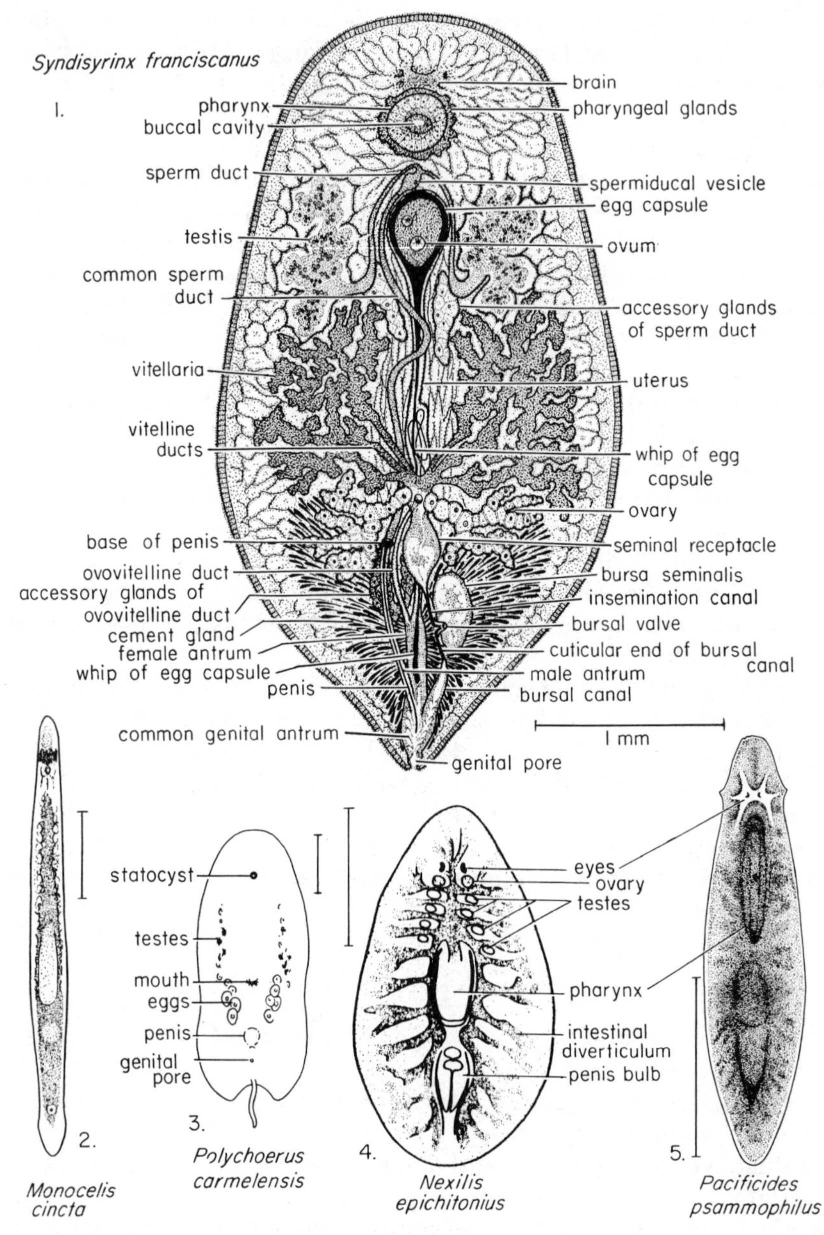

PLATE 13. **Flatworms** (1). **Smaller Turbellaria:** Scale marks are 1 mm. 1, after Lehman, 1946; 2, Karling, 1966; 3, Bush, in Smith, 1964; 4, Holleman and Hand, 1962; 5, Holmquist and Karling, 1972.

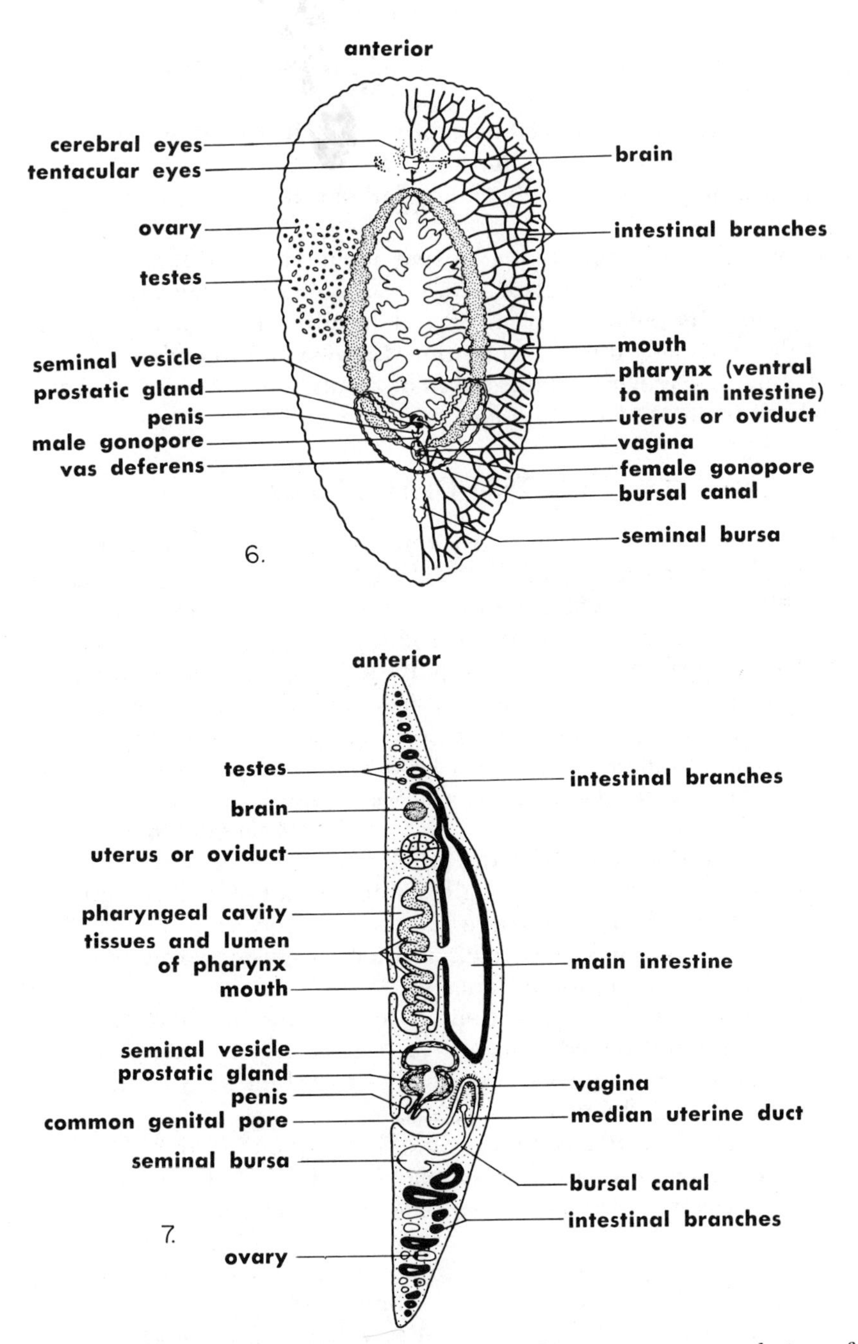

PLATE 14. **Flatworms** (2). **Polyclad anatomy:** 6, diagrammatic ventral view of *Leptoplana timida* (redrawn from Bresslau, after Heath and McGregor); 7, diagrammatic median longitudinal section of *Stylochoplana agilis* (redrawn from Bresslau, after Lang).

penis stylet: thorn or stylet on penis; may be short or long, straight or curved.

Lang's vesicle or seminal bursa: a receptacle for sperm storage at upper end of vagina and bursal canal.

spermiducal bulbs: heavily muscularized terminations of sperm ducts.

sucker: a glandulo-muscular adhesive organ on the midventral line at various levels in polyclads, forming a noticeable elevation or depression. Its presence or absence determines the primary division of the polyclads into the suborders **Cotylea** and **Acotylea.**

tentacular eyes: eyespots located in or around the tentacles or at sites where tentacles would be located if present.

KEY TO POLYCLADIDA

1. Ventral surface without a sucker; tentacles when present of nuchal type (figs. 8, 12) 2
– With sucker on ventral surface (figs. 9, 10); tentacles when present situated at anterior margin 17
2. Dorsal surface covered with pointed papillae; pair of conspicuous, pointed, conical nuchal tentacles; tentacular eyes in rings at base of tentacles, cerebral eyes in 2 clusters medial and anterior to tentacles; shape oval to round, fairly thick; color orange-red to tan with dark spots on dorsum; living on encrusting ectoprocts; to 10 mm long (fig. 12) .... *Hoploplana californica*
– Dorsal surface not papillate 3
3. Band of eyes present along all or anterior part of body margin; eyes also present elsewhere 4
– Without marginal eyes; eyes limited to clusters over brain region or in and around tentacles 9
4. Marginal band of eyes limited to anterior 1/4–1/2 of body margin; with nuchal tentacles 5
– Marginal band of eyes completely encircling body; with or without nuchal tentacles 6
5. Small, oval, to 10 mm long; pair of inconspicuous nuchal tentacles; marginal eyes limited to anterior 1/4 of body margin; tentacular eyes inside tentacles; few cerebral eyes; color creamy with minute tan specks .... *Stylochus franciscanus*
– Large, oval, to 50 mm long, 25 mm wide; conspicuous, conical nuchal tentacles well back from anterior margin; marginal eyes extending back from anterior end 1/4–1/2 body length; tentacles with eyes; cerebral eyes numerous in 2 elongated clusters between tentacles; color buff, with short brown streaks (fig. 8) .... *Stylochus tripartitus*
6. With prominent nuchal tentacles 7

– Nuchal tentacles wanting in adult, present in young; adult very large, to 60–100 mm long, 25 mm wide; thick; elongate oval in shape with somewhat pointed anterior end; marginal eyes numerous forward, thinning posteriorly; color dark brown with closely set dark spots .... *Stylochus atentaculatus*

7. Very large, 50–150 mm long; oval; no posterior notch 8

– Small, 7 mm long; oval; distinct posterior median notch; cerebral eyes few, in a line on either side of midline forward of tentacles; color buff to brown .... *Stylochus exiguus*

8. Elongate oval in shape, narrowing at ends; rather thin but very large, commonly 50–100 mm or more; color gray with inconspicuous darker spots .... *Stylochus californicus*

– Broadly oval in shape with conspicuous contractile tentacles; thick and tough; very large, to 100 mm long, 70 mm wide; color tan or brown heavily marked with dark brown dashes or spots .. *Kaburakia excelsa*

9. With nuchal tentacles 10

– Without nuchal tentacles 12

10. Small, up to 12 mm long; cuneate (wedge-shaped), expanded anteriorly, tapering to bluntly pointed posterior; pointed nuchal tentacles; 4–6 eyes in tentacular clusters in and around tentacles; 10–15 cerebral eyes scattered in two longitudinal tracts; color variable, pale brownish yellow or buff, translucent yellow or golden tan; ventral surface white (fig. 11) .... *Stylochoplana gracilis*

– Larger, oval in shape 11

11. To 40 mm long, thick and firm; nuchal tentacles nipplelike and contractile; tentacular eyes fill the tentacles and occur diagonally before and behind tentacle bases; cerebral eyes in elongated areas broadening anteriorly; color transparent bluish green or light olive with branches of digestive tract showing as zigzag chocolate lines radiating from the central main intestine to the periphery .... *Alloioplana californica*

– Similar to *A. californica* in size and shape, but color light brown, speckled or reticulated, and digestive branches not obvious .... *Pseudostylochus burchami*

12. Penis armed with straight, relatively short stylet; body elongate with expanded anterior end tapering to bluntly pointed posterior end; to 30 mm long; tentacular eye clusters of up to 13 large eyes, cerebral clusters elongate, each with about 50 small eyes; color light brown or pale tan, dotted with reddish brown spots .... *Notoplana inquieta*

– Without penis stylet 13

13. With highly developed spermiducal bulbs 14

– Without spermiducal bulbs 15

14. With large common genital pore and large, conical, unarmed penis; body shape varies from elongate oval to wedge-shaped; to 40 mm long; color highly variable, may be uniform reddish brown with darker

streaks outlining pharynx and copulatory apparatus, or mottled dark brown with darker stripe middorsally, or with dark brown to nearly black blotches on nearly white background ....*Freemania litoricola*

– With separate, inconspicuous genital pores; penis small; body elongate oval, rounded anteriorly, tapering posteriorly, thick and firm; to 40 mm long, 6 mm wide; eyes in distinct cerebral and tentacular groups of up to 50 eyes each; color speckled gray or light tan mottled with green and brown ....*Leptoplana chloranota*

15. Lang's vesicle short or reduced 16

– Lang's vesicle very long; body expanded anteriorly, tapering posteriorly; to 35 mm long, 15 mm wide; up to 30 eyes in each tentacular and cerebral cluster; color pale with pink or red tinge, brown along middorsal region, color faint at margins ....*Notoplana rupicola*

16. Lang's vesicle short and plump; body widest anteriorly, tapering to pointed tail; to 60 mm long when extended and crawling; tan or pale gray shading to patchy brown middorsally (fig. 13) ....*Notoplana acticola*

– Lang's vesicle small or much reduced; penis exceptionally long and slender; body elongate, not expanded anteriorly, sides nearly parallel along anterior 1/3 then tapering gradually to blunt posterior end; to 14 mm long, 4 mm wide; yellowish gray in color ....*Notoplana saxicola*

17. Marginal tentacles formed from upturned folds of anterior margin; pharynx ruffled 18

– Marginal tentacles not obvious folds; pharynx tubular 20

18. Color basically black and white, sometimes also some red . 19

– Color not black and white but pale tan or brown with darker brown flecks; oval or rounded, of firm consistency; to 28 mm long; sucker central in position, very large, oval, and much folded (fig. 10) ....
.................................*Pseudoceros canadensis*

19. Elongate oval with rounded ends and ruffled margin; to 50 mm long; thin, delicate; pronounced middorsal ridge extending most of body length; tentacles broad, flaplike; cerebral eyes in single cluster like inverted V; eyes on tentacle flaps; sucker well anterior to middle; white with middorsal black stripe along center of ridge, stripe forks anteriorly between tentacular flaps and forks continue to anterior margin ....
....................................*Pseudoceros luteus*

– Oval, with gracefully ruffled margin; 40–90 mm long; coloration striking, basically white with darker markings, black stripe with tinges of red on middorsal ridge from just behind tentacles to posterior 1/3 of body, similar stripe encircles entire margin of animal except anterior edge between tentacles, this marginal band is just inward of edge, leaving outer margin white; elongate spots scattered over entire dorsal surface, mostly black but with some small, wine red spots and a few indistinct, white spots; tentacles with dark band and dark spot between them ....
.............................*Pseudoceros montereyensis*

20. Marginal tentacles well developed, long and pointed 21
– Marginal tentacles rudimentary, short and blunt; body elliptical, to 9 mm long; about 80 small eyes at base of each tentacle; ground color of dorsum orange with darker shade along middorsal line from eyes to posterior end of midgut, laterally this color lighter and, near the margin, alternates with bright yellow in raylike expansions; minute white specks scattered over dorsum .... *Stylostomum lentum*
21. (*Note 3 choices*) Orange-red, yellowish pink, or salmon with minute white specks and pink streak along middorsal line; another color variation is yellowish white with very yellow margin, orange anterior end and numerous white stripes running longitudinally; body oval, thick; sucker small, rounded, somewhat posterior to middle; to 30 mm long .... *Eurylepta aurantiaca*
– Color grayish white with pure white border, white middorsal ridge and irregularly distributed small white spots; this pattern is crisscrossed with black lines, usually terminating in red tips that invade the white border; a few short, scattered, disconnected red streaks; a dark mark at base of tentacles, distally red, proximally black and in between is a small black dash; to 20 mm long (fig. 9) .... *Eurylepta californica*
– Striking coloration of alternating black and white longitudinal stripes of varying width with middorsal orange stripe and marginal orange band; black stripes often show orange spots, white stripes sometimes tinged with red; ventral surface white with orange margin; tentacles black and flaring; to 40 mm long .... *Prostheceraeus bellostriatus*

LIST OF TURBELLARIA

(only Polycladida are keyed out)

Order **ACOELA**

Polychoerus carmelensis Costello and Costello, 1938. Often abundant on Monterey Peninsula on *Ulva* and stones in high tidepools; excellent for class study; see Armitage, 1961, Pac. Sci. 15: 203–210 (biology).

Childia groenlandica (Levinsen, 1879). From bottom mud in San Francisco Bay; see Hyman, 1959.

Order **RHABDOCOELA**

Collastoma pacifica Kozloff, 1953. In gut of the sipunculan, *Themiste pyroides;* see Kozloff, 1953, J. Parasitol. 39: 336–340.

Syndesmis dendrastrorum Stunkard and Corliss, 1951. In intestine of sand dollar *Dendraster.*

Syndisyrinx franciscanus Lehman, 1946 (=*Syndesmis francis-*

cana). Common in intestine of sea urchin *Strongylocentrotus;* good class material (see Lehman, 1946).

Order **ALLOEOCOELA**

About 25 species in 13 genera are known from this area; all but one of these species were described by Karling (1962a, 1962b, 1964, 1966). Dr. Karling says *Monocelis cincta* is the luminous *Monocelis* mentioned in Ricketts and Calvin (4th ed., p. 73). Most alloeocoels occur on algae or interstitially in coarse sand.

Order **TRICLADIDA**

Nexilis epichitonius Holleman and Hand, 1962. Commensal in mantle cavity of the chiton *Mopalia hindsii;* also associated with other invertebrates; see Holleman, 1972.

Oregoniplana opisthopora Holmquist and Karling, 1972. In sandy substrate of *Zostera* meadows and on rocks in surf zone at low-tide level.

Pacificides psammophilus Holmquist and Karling, 1972. Interstitial on moderately exposed, sandy beaches.

Procerodes pacifica Hyman, 1954. Described from stranded *Macrocystis* near San Diego, and reported from Friday Harbor (Holleman, 1972); probably occurs in our area.

Order **POLYCLADIDA**

Suborder **Acotylea**

Alloioplana californica (Heath and McGregor, 1912). Abundant under boulders in gravel and in crevices at mean tide and above.

Freemania litoricola (Heath and McGregor, 1912). Beneath stones below midtide level.

Hoploplana californica Hyman, 1953. Common on encrusting ectoprocts, especially *Celleporaria brunnea,* on pilings.

Kaburakia excelsa Bock, 1925. Under rocks, on boat bottoms, and on pilings around mussels.

Leptoplana chloranota (Boone, 1929). Under stones in low intertidal.

Notoplana acticola (Boone, 1929). Extremely common under rocks from high- to low-tide zone.

Notoplana inquieta (Heath and McGregor, 1912). Found under stones at low water; also associated with *Macrocystis.*

Notoplana rupicola (Heath and McGregor, 1912). Under rocks, low intertidal.

Notoplana saxicola (Heath and McGregor, 1912). On masses of algae in tidepools and on algal holdfasts.

Pseudostylochus burchami (Heath and McGregor, 1912). Mainly subtidal but may be found in low intertidal pools.

Stylochoplana gracilis Heath and McGregor, 1912. On blades and stipes of *Macrocystis*, occasionally on pilings.

Stylochus atentaculatus Hyman, 1953. In rocky crevices and under stones.

Stylochus californicus Hyman, 1953. Occurs in association with bivalve molluscs, on which it feeds.

Stylochus exiguus Hyman, 1953. In burrows of mudflat dwellers.

Stylochus franciscanus Hyman, 1953. In very low intertidal and subtidally in San Francisco Bay.

Stylochus tripartitus Hyman, 1953. On kelp stipes and holdfasts.

Suborder **Cotylea**

Eurylepta aurantiaca Heath and McGregor, 1912. Under stones and crawling about on bottom of tidepools; sluggish, clings tenaciously to rocks.

Eurylepta californica Hyman, 1959. Among coralline algae on rocky shores.

Prostheceraeus bellostriatus Hyman, 1953. On wharf pilings.

Pseudoceros canadensis Hyman, 1953. In semiprotected rocky areas or jetties at low-tide level; generally northern.

Pseudoceros luteus (Plehn, 1898). In low tidepools.

Pseudoceros montereyensis Hyman, 1953. Underside of rocks in midintertidal.

Stylostomum lentum Heath and McGregor, 1912. Among rocks, low intertidal.

REFERENCES ON FLATWORMS

Boone, E. S. 1929. Five new polyclads from the California coast. Ann. Mag. Nat. Hist. (ser. 10) 3: 33–46.

Costello, H. M. and D. P. Costello 1938. A new species of *Polychoerus* from the Pacific coast. Ann. Mag. Nat. Hist. (ser. 11) 1: 148–155.

Heath, H. and E. A. McGregor 1912. New polyclads from Monterey Bay, California. Proc. Acad. Nat. Sci. Philadelphia 64: 455–488, Pl. xii–xviii.

Holleman, J. J. 1972. Marine turbellarians of the Pacific coast. I. Proc. Biol. Soc. Wash. 85: 405–411.

Holleman, J. J. and C. Hand 1962. A new species, genus, and family of marine flatworms (Turbellaria: Tricladida, Maricola) commensal with mollusks. Veliger 5: 20–22.

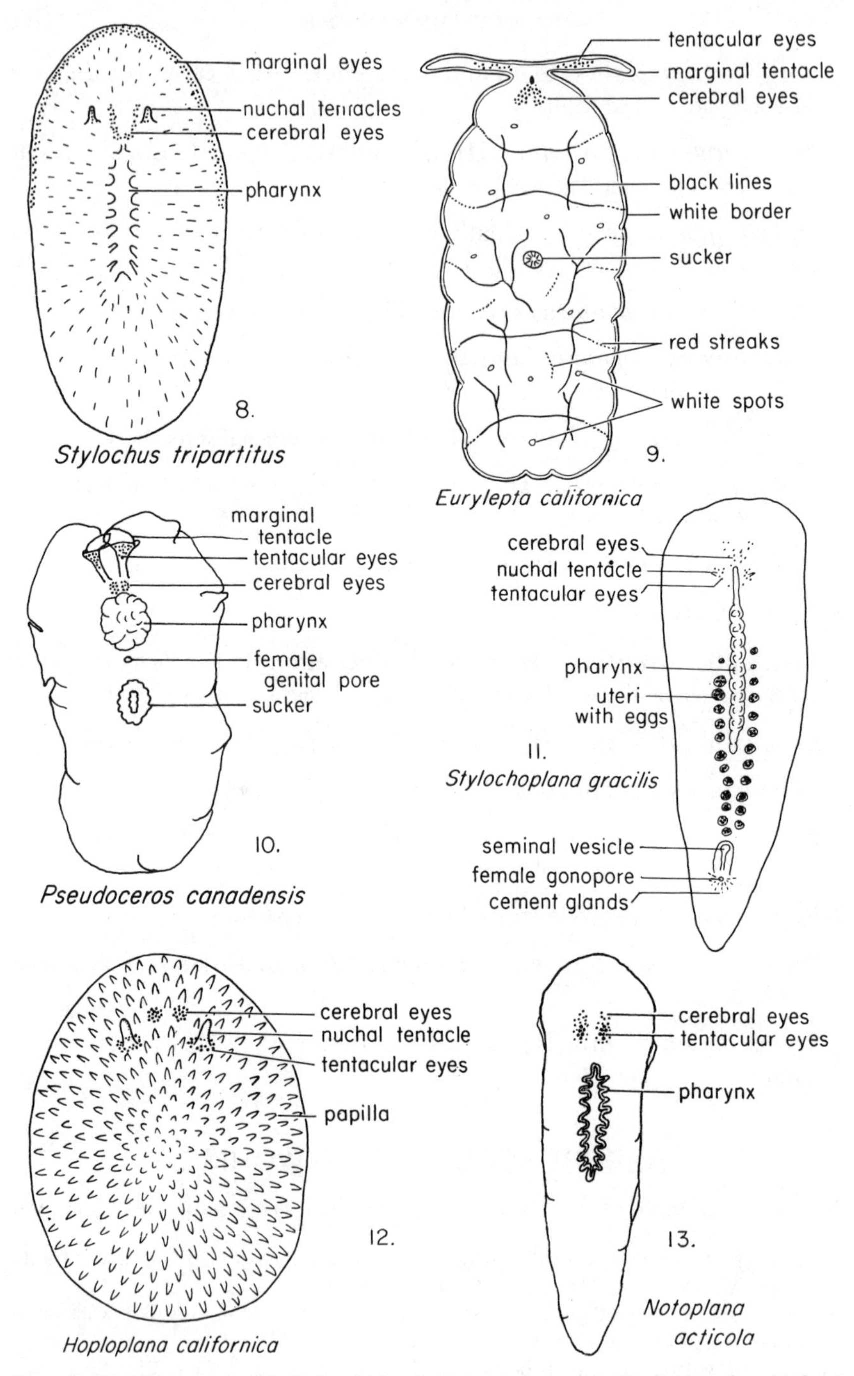

PLATE 15. **Flatworms** (3). **Polyclads:** Scales various. 12, Haderlie, from life; 8, 10, 11, 13 redrawn from Hyman; 9, redrawn from a sketch by D. H. Montgomery.

Holmquist, C. and T. G. Karling 1972. Two new species of interstitial marine triclads from the North American Pacific coast, with comments on evolutionary trends and systematics in Tricladida (Turbellaria). Zool. Scripta 1: 175–184.

Hyman, L. H. 1951. *The Invertebrates: Platyhelminthes and Rhynchocoela,* Vol. II. McGraw-Hill.

Hyman, L. H. 1953. The polyclad flatworms of the Pacific coast of North America. Bull. Amer. Mus. Nat. Hist. 100: 267–392.

Hyman, L. H. 1954. A new marine triclad from the coast of California. Amer. Mus. Novit. no. 1679: 5 pp.

Hyman, L. H. 1955. The polyclad flatworms of the Pacific coast of North America: Additions and corrections. Amer. Mus. Novit. no. 1704: 11 pp.

Hyman, L. H. 1959. Some Turbellaria from the coast of California. Amer. Mus. Novit. no. 1943: 17 pp.

Karling, T. G. 1962a. Marine Turbellaria from the Pacific Coast of North America. I. Plagiostomidae. Arkiv f. Zool. 15: 113–141.

Karling, T. G. 1962b. II. Pseudostomidae and Cylindrostomidae. Arkiv f. Zool. 15: 181–209.

Karling, T. G. 1964. III. Otoplanidae. Arkiv f. Zool. 16: 527–541.

Karling, T. G. 1966. IV. Coelogynoporidae and Monocelididae. Arkiv f. Zool. 18: 493–528.

Kozloff, E. N. 1965. New species of acoel turbellarians from the Pacific Coast. Biol. Bull. 129: 151–166.

Lehman, H. E. 1946. A histological study of *Syndisyrinx franciscanus,* gen. et sp. nov., an endoparasitic rhabdocoel of the sea urchin, *Strongylocentrotus franciscanus.* Biol. Bull. 91: 295–311.

PHYLUM NEMERTEA (RHYNCHOCOELA)

Eugene C. Haderlie
Naval Postgraduate School, Monterey

(PLATE 16)

Nemerteans or ribbon worms are readily recognized by their soft, elongated, contractile, ciliated, and unsegmented bodies. The intestine is usually straight, with a posterior anus. A distinctive feature is the long, eversible proboscis, which is usually not connected with the digestive tract, but is an infolding of the anterior body wall. This proboscis may be caused to evert by adding alcohol dropwise to a small amout of sea water containing the worms or by putting them into fresh water, and is often everted when the worm is disturbed or injured. Like polyclads, nemerteans are predators with a wide range of prey. For this purpose the proboscis with its sticky secretions or needlelike stylets is an efficient weapon.

Although most nemerteans are free-living, bottom-dwelling, and marine, a few species are found in other situations. *Prostoma rubrum* inhabits fresh water such as Lake Temescal in Oakland; *Geonemertes* is found in damp places on land; and a few, such as *Nectonemertes*, swim or float in mid- or deep water. *Malacobdella grossa* lives commensally in the mantle cavity of bivalve molluscs, and *Carcinonemertes epialti* lives among the gills and egg masses of *Hemigrapsus oregonensis* and other crabs.

Nemerteans are handled with difficulty because of their extreme contractility, their secretion of large quantities of mucus, and the tendency of many to break into fragments. They should be kept separated from other animals and must be anesthetized carefully before preservation. For most purposes preservation in 10% formalin is satisfactory but, since acids or formalin destroy the stylets of hoplonemerteans, members of this group should be preserved in 80% alcohol.

Identification of the orders of nemerteans by the use of even such helpful references as those of Coe is difficult because the division into orders is based in large part upon the arrangement of muscle layers in the body wall. The general descriptive features given below may help in deciding the probable order of an unknown specimen.

GLOSSARY

ocelli: eye spots, usually of dark color, particularly on lateral margins of head.
stylet: tiny, needlelike harpoon on end of proboscis (fig. 3).
accessory stylets: spare stylets, usually stored in pouches (fig. 3).
cephalic grooves: furrows or slits running longitudinally along the lateral sides of the head (fig. 5).
caudal cirrus: thin, tapering, tail-like appendage at posterior end of worm (fig. 5); often lost in rough handling.

CLASSIFICATION

Class **ANOPLA.** Mouth posterior to brain; proboscis not armed with stylets.

Order **Paleonemertea.** In general, paleonemerteans are slender, soft, and extensile, heads somewhat blunt, bodies not much flattened, ocelli and longitudinal cephalic grooves lacking.

Order **Heteronemertea.** Heads characteristically rather snakelike, with marked lateral cephalic grooves (*Zygeupolia* is an exception); a small caudal cirrus is found in *Cerebratulus* (fig. 5), *Micrura,* and *Zygeupolia.*

Class **ENOPLA.** Mouth anterior to brain; proboscis armed with one or more stylets except in Bdellonemerteans.

Order **Hoplonemertea.** The stylets are diagnostic and can be seen by flattening the anterior end of an animal carefully between microscope slides and examining by transmitted light (fig. 3).

Order **Bdellonemertea.** Commensal in mantle cavities of marine bivalves; the leechlike form and posterior sucking disc are distinctive (fig. 4).

KEY TO NEMERTEA

1. (*Note 3 choices*) Commensal in mantle cavity of marine bivalve molluscs; body short, flattened, leechlike, with large, rounded, muscular, posterior sucker and anterior notch; white, occasionally orange; 3–4 cm long (fig. 4) .... *Malacobdella grossa*
– Commensal among egg masses of various crabs; body minute, slender; lateral grooves lacking; bright orange, reddish, or yellow; head paler; 4–6 mm long .... *Carcinonemertes epialti*
– Free-living; body usually elongate and never with sucker . 2

2. Entire body or dorsal surface solid color; no conspicuous lines or rings of contrasting color 3
– With longitudinal lines, transverse rings, or spots of color sharply contrasting with ground color 16
3. Approximately same color or shade dorsally and ventrally . 4
– Darker dorsally, distinctly lighter ventrally 11
4. Small and/or of slender build; whitish, yellowish, pink, or pale flesh colored 5
– Usually larger, distinctive coloration (red, dark brown, brown, orange) 7
5. (*Note 3 choices*) Very slender, opaque white, sometimes with pink or yellowish tinge; brain pink, intestinal canal often brown; ocelli numerous, in two groups on each side of flattened head; with 2 or 3 pouches of accessory stylets; 2.5–5 cm .... *Amphiporus imparispinosus*
– Similar to *A. imparispinosus* but with 6–12 pouches of accessory stylets (fig. 3); 10–30 cm .... *Amphiporus formidabilis*
– White or whitish in color; neither ocelli nor stylets 6
6. Minute, slender; somewhat translucent; may have indistinct dorsal and ventral bands of pale yellow or orange; in delicate, parchment-like tubes; to 2.5 cm .... *Tubulanus pellucidus*
Elongate, body rounded anteriorly; long, very flattened posterior region in which internal organs are visible as transverse dark yellow or reddish lines in lateral areas; shape of head constantly changes; worm tends to coil posterior region of body into spiral; to 20 cm .... *Carinoma mutabilis*
7. (*Note 3 choices*) Without cephalic grooves or slits on side of head 8
– With cephalic grooves and a small caudal cirrus (note: this may be lost by rough handling) 9
– With cephalic grooves but lacking a caudal cirrus; row of 4–8 small ocelli on each side of head; color variable, usually brownish green, dark brown, or reddish brown, commonly paler ventrally; body contracts by shortening and thickening, not by coiling; brackish water; to 20 cm .... *Lineus ruber*
8. Lacking caudal cirrus 10
– With conspicuous caudal cirrus; body slender, head very long and acutely pointed; ocelli wanting; body cylindrical, flattened posteriorly; head marked off from body by shallow annular constriction; color pinkish anteriorly, deep red tinged with yellow posteriorly; to 8 cm .... *Zygeupolia rubens*
9. Deep cephalic grooves; anterior body firm and stout; posterior flattened and easily fragmented; color exceedingly variable, pale yellow, buff, light brown or chocolate brown; lateral margins thin and often pale; lat-

eral nerve cords reddish; to 1 m or more (fig. 5) .... *Cerebratulus californiensis*

– Shallow cephalic grooves; head tending to be pointed; body flattened posteriorly; anterior body flesh colored, rosy, or gray shading into pink, with yellow or white on head; sometimes a cream-colored median ventral stripe; to 60 cm .... *Micrura alaskensis*

10. Body very slender, long, extensible (1–3 meters); color red, orange, or bright vermillion ...... *Tubulanus polymorphus*

– Body thicker and less extensible (to 60 cm); color deep red, orange-red, mahogany or red-brown, sometimes with whitish bloom; anterior of head with broad spot of slightly darker color bordered by white, terminally and laterally; head broad, rounded, withdrawn into body when worm is contracted .... *Baseodiscus punnetti*

11. Head distinctly darker or lighter than rest of body 12

– Shade of head not distinctly different from rest of body 15

12. Head darker than rest of body, or with prominent dark patch 13

– Head lighter, but with dark markings 14

13. Head with 2–8 small ocelli on each anterolateral margin; body very slender and elongate; contracts by coiling or spiraling; head with pale gray frontal or lateral margin; body dusky or brownish green, dark or reddish brown, paler ventrally; to 15 cm .... *Lineus vegetus*

– Head with 4 distinct ocelli in square about dark square, triangular or shield-shaped patch; body small, slender; dorsally purple or brown with white longitudinal line, or reddish with brown flecks; sometimes with narrow, transverse markings; to 7 cm .. *Tetrastemma nigrifrons*

14. Head with 4 ocelli around a dark patch; small, slender; to 7 cm .... (dark variety of) *Tetrastemma nigrifrons*

– Head light, with 2 conspicuous, dark brown or black oval or triangular spots; 12–20 ocelli in irregular clusters on each side; dorsal surface of body brown or reddish, ventrally pale reddish, orange, or flesh colored; to 15 cm .... *Amphiporus bimaculatus*

15. Body very long and slender; worms often in tangled groups; dorsal surface greenish, ventral pale green to white, tip of head paler; 2 clusters of 10–15 ocelli on each side of head; very long, curved stylets; to 50 cm or more .... *Emplectonema gracile*

– Body stouter; purplish to dark brown or orange brown dorsally to ventrolaterally; yellowish to white ventally; narrow, conspicuous, V-shaped, white marking back of head; pair of yellowish spots on each lateral margin in front of brain; to 25 cm *Paranemertes peregrina*

16. Small (to 15 cm) or very slender 17

– Larger (20 cm or more) and stouter 20

17. Head with long cephalic grooves 18

– Without cephalic grooves 19

18. Head with 6–15 ocelli on each side; body small (to 3 cm), flattened, with caudal cirrus; dorsal surface pale brown or yellow, nearly covered with small black or brown spots often elongated and in irregular lines or grouped into rectangular areas separated by narrow, transverse lines ....*Micrura pardalis*
 – Head narrower than body, tip with narrow white border enclosing oval orange spots in area of lemon yellow; body larger (to 15 cm), somewhat flat, soft, often twisted or knotted, fluted longitudinally and constricted transversely when contracted; body deep brown, chestnut, or slaty with tinge of green or red, and velvety sheen, posterior very pale; usually transverse and longitudinal lemon-yellow markings dorsally with diamond-shaped thickenings in median line; 7–15 fine, longitudinal lines ... . *Lineus pictifrons*
19. Row of 2–8 ocelli on each side of head; body very slender; dusky or brownish green, dark or reddish brown, paler ventrally; frontal and lateral margins of head white or pale gray, brain region rosy; body with up to 20 or more very narrow, inconspicuous, light rings; fine, longitudinal, light lines often present on sides of body (see 13 for those without lines or rings); to 15 cm ....*Lineus vegetus*
 – Two large ocelli on each side of head; dorsal surface dull white with tinge of brown and two narrow, deep brown lines the length of body; ventral white, gray, or pale flesh color; to 15 cm ..... *Nemertopsis gracilis*
20. Body deep brown with numerous narrow white rings and 5 or 6 longitudinal, white lines; lives in thin, tough, transparent tubes open at each end; average length 20 cm, but can extend to 1 m .... *Tubulanus sexlineatus*
 – General body color ivory white with wide, deep purple or wine-colored, rectangular dorsal areas separated by narrow, white bands; tip of head red or flesh colored, dorsal side of head deep orange or vermillion; ventral surface white; small caudal cirrus (often lost); to 20 cm or more ....*Micrura verrilli*

LIST OF NEMERTEA

Class **ANOPLA**

Order **Paleonemertea**

Carinoma mutabilis Griffin, 1898. In sand, sandy mud, mud, and clay.

Tubulanus pellucidus (Coe, 1895). In delicate tubes under stones and among algae, in sand and mud.

Tubulanus polymorphus Renier, 1804. Under heavy boulders bedded in gravel, among mussels, in muddy situations.

Tubulanus sexlineatus (Griffin, 1898). In thin, transparent tubes among algae, mussels, etc. on rocks and pilings.

Order **Heteronemertea**

Baseodiscus punnetti (Coe, 1904). Among red algae, corallines, and other growths near low-water mark.

Cerebratulus californiensis Coe, 1905. In sand and mud flats of bays and harbors.

Lineus pictifrons Coe, 1904. In crevices of rocks and among algae exposed to surf; also in mud of harbors and bays.

Lineus ruber (O. F. Müller, 1771). Beneath stones and among algae, in sandy and muddy situations; brackish water, see Jennings and Gibson, 1969 (nutrition).

Lineus vegetus Coe, 1931. Beneath stones and among corallines exposed to full surf from midtide to low-tide levels; has remarkable ability to regenerate from small fragments.

Micrura alaskensis Coe, 1901. Beneath stones and in crevices, in sand and mud.

Micrura pardalis Coe, 1905. Among corallines and other growths on rocks and pilings; in tidepools exposed to full surf.

Micrura verrilli Coe, 1901. Beneath stones in sandy mud and among algae and roots of surf grass *Phyllospadix* at low-tide level.

Zygeupolia rubens (Coe, 1895). In sand of harbors and bays.

Class **ENOPLA**

Order **Hoplonemertea**

Amphiporus bimaculatus Coe, 1901. In crevices of rocks and pilings, or beneath stones near low-water line; secretes vast quantities of mucus.

Amphiporus formidabilis Griffin, 1898. In mussel-bed areas, often crawling over wet, exposed rock or mussels.

Amphiporus imparispinosus Griffin, 1898. Among corallines or beneath stones on exposed rocky shores, among growths on pilings, etc.

Carcinonemertes epialti Coe, 1902. In egg masses of *Hemigrapsus oregonensis, Cancer magister, Pugettia producta* (rarely), and other crabs; see Humes, 1942.

Emplectonema gracile (Johnston, 1837). In *Endocladia* zone in holdfasts and under barnacles, and in masses among mussels and barnacles on rocks and pilings.

* *Geonemertes* sp. We are informed by Dr. J. W. Hedgpeth that a species of this terrestrial genus occurs under boards and stones in salt marshes, as in Tomales Bay and Elkhorn Slough.

Nemertopsis gracilis Coe, 1904. In *Endocladia* holdfasts and empty barnacle shells, and under large boulders bedded in coarse sand.

Paranemertes peregrina Coe, 1901. Common and widely distributed; among mussels and corallines on exposed rock, also crawling on surface of mudflats; see Roe, 1970, Biol. Bull. 139: 80–91 and Gibson, 1970, Biol. Bull. 139: 92–106 (feeding and nutrition).

* *Prostoma rubrum* (Leidy, 1850). In fresh water; introduced; a small (to 2 cm), pinkish worm, with 2 to 4, usually 3, pairs of ocelli.

Tetrastemma nigrifrons Coe, 1904. Among algae, bryozoans, and other growths on rocks and pilings.

Order **Bdellonemertea**

Malacobdella grossa (O. F. Müller, 1776). Uncommon; in mantle cavity of many species of bivalves, including *Macoma secta* and *Siliqua patula.* See Gibson and Jennings, 1969, J. Mar. Biol. Assoc. U.K. 49: 17–32 (nutrition); Guberlet, 1925, Publ. Puget Sound Biol. Sta. 5: 1–14.

* Not in key.

REFERENCES ON NEMERTEA

Coe, W. R. 1901. Papers from the Harriman Alaska Expedition, 20. The Nemerteans. Proc. Wash. Acad. Sci. 3: 1–111.

Coe, W. R. 1904. The Nemerteans. Harriman Alaska Exped. 11: 1–220.

PLATE 16. **Nemertea, "Mesozoa", Aschelminthes.** 1–5, **Nemertea:** 1, *Micrura verrilli,* its white caudal cirrus has been lost, after Coe; 2, *Amphiporus bimaculatus,* after Coe; 3, *A. formidabilis,* stylets in a flattened preparation, drawn by Emily Reid from a photograph by John Wourms; 4, *Malacobdella grossa,* after Verrill; 5, *Cerebratulus californiensis,* after Coe; 6, **Rotifera,** *Philodina;* 7, *Seison,* both after Thane-Fenchel, 1969; 8, **Gastrotricha,** *Turbanella mustela,* combined dorsal and ventral view, after Wieser, 1957, Puget Sound intertidal; 9, **Kinorhyncha,** unidentified, from life, Puget Sound, after Hyman, 1951, used with permission of McGraw-Hill Book Company; 10, **Gnathostomulida,** *Gnathostomula,* after Riedl, 1971; 11, **Nematoda,** *Pseudocella triaulolaimus,* female, Dillon Beach, in *Egregia* holdfast, after Hope, 1967; 12, **Orthonectida,** *Ciliocincta sabellariae,* after Kozloff, 1965; 13, **Dicyemida,** *Dicyema apollyoni,* Moss Beach, after McConnaughey, 1949; 4, 5, 10, 11 redrawn by Emily Reid.

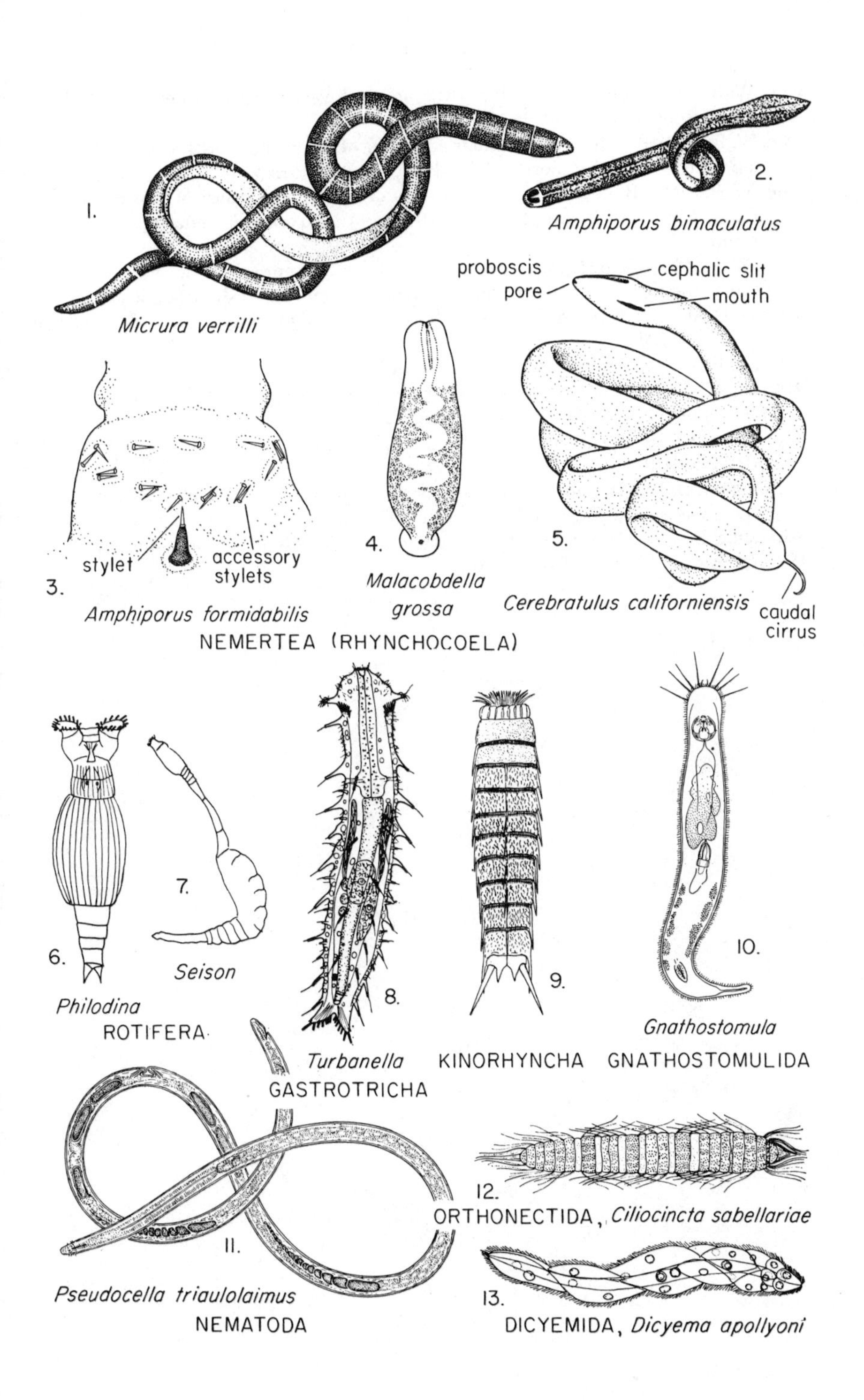
1.
Micrura verrilli
2.
Amphiporus bimaculatus
proboscis
pore
cephalic slit
mouth
3.
stylet
accessory
stylets
Amphiporus formidabilis
4.
Malacobdella
grossa
5.
Cerebratulus californiensis
caudal
cirrus
NEMERTEA (RHYNCHOCOELA)
6.
Philodina
7.
Seison
ROTIFERA
8.
Turbanella
GASTROTRICHA
9.
KINORHYNCHA
10.
Gnathostomula
GNATHOSTOMULIDA
11.
Pseudocella triaulolaimus
NEMATODA
12.
ORTHONECTIDA, Ciliocincta sabellariae
13.
DICYEMIDA, Dicyema apollyoni

Coe, W. R. 1905. Nemerteans of the west and northwest coasts of America. Bull. Mus. Comp. Zool. Harvard 47: 1–319.

Coe, W. R. 1940. Revision of the nemertean fauna of the Pacific coasts of North, Central, and northern South America. Allan Hancock Pacific Exped. 2: 247–323.

Coe, W. R. 1943. Biology of the nemerteans of the Atlantic coast of North America. Trans. Conn. Acad. Arts Sci. 35: 129–328. An excellent general account of the group.

Coe, W. R. 1944. Geographical distribution of the nemerteans of the Pacific coast of North America, with descriptions of two new species. J. Wash. Acad. Sci. 34: 27–32.

Corrêa, D. D. 1964. Nemerteans from California and Oregon. Proc. Calif. Acad. Sci. (ser. 4) 31: 515–558.

Gibson, R. 1973. Nemerteans. Hutchinson University Library, 224 pp.

Humes, A. G. 1942. The morphology, taxonomy, and bionomics of the nemertean genus *Carcinonemertes*. Illinois Biol. Monogr. 18: 105 pp.

Hyman, L. H. 1951. *The Invertebrates: Platyhelminthes* and *Rhynchocoela,* Vol. II. McGraw-Hill, 550 pp. (Rhynchocoela: pp. 459–531).

Jennings, J. B. and R. Gibson 1969. Observations on the nutrition of seven species of rhynchocoelan worms. Biol. Bull. 136: 405–443.

Kirsteuer, E. 1967. Marine, benthonic nemerteans: how to collect and preserve them. Amer. Mus. Nov. no. 2290: 1–10.

ASCHELMINTHES AND SMALLER WORM GROUPS

Anyone who examines freshly collected material microscopically will find hosts of tiny animals, adult as well as larval, attached to or creeping on and among other animals and the substrate. Identification of such animalculae, even to major group, is often difficult, and many interesting types may be passed over. In this chapter are gathered several groups, which in common are individually generally small, often very numerous as individuals, usually members of quite large groups and, on the California coast, little studied. Of these, many are members of the meiofauna, interstitial in sand or mud habitats (see Hulings, 1971, and Hulings and Gray, 1971); mesozoans and acanthocephalans are parasitic. The groups treated here are not closely related; Mesozoa are of uncertain position, Gnathostomulida are acoelomates, and the aschelminth complex and Acanthocephala are pseudocoelomates.

"MESOZOA"

(PLATE 16)

Mesozoans are small, wormlike parasites of marine invertebrates, consisting of an outer layer of ciliated somatic cells enclosing internal reproductive cells; their uncertain phylogenetic position has evoked much discussion. The **Dicyemida** (fig. 13) are common parasites in the kidneys of cephalopods (see McConnaughey, 1949, 1951, for central California species from the squid *Loligo* and various intertidal octopuses); recent work indicates that an intermediate invertebrate host may not be required. The **Orthonectida** are not closely related to dicyemids and, according to Kozloff (1969), placing them in the Mesozoa is inappropriate. Orthonectids are internal parasites of a number of intertidal invertebrates, including flatworms, nemerteans, polychaetes, bivalves, and brittle stars. *Ciliocincta* (fig. 12), described from *Sabellaria cementarium* in the San Juan Archipelago, Washington (Kozloff, 1965, 1971), may be sought for in that polychaete locally, while *Rhopalura* (Kozloff, 1969) occurs locally in the brittle star *Amphipholis* and in other invertebrates. The biology, ecology, and distribution of mesozoans on our coast are still poorly known, and alert students can make important contributions to knowledge of the group.

PHYLUM GNATHOSTOMULIDA

(PLATE 16)

Gnathostomulida are small (several mm long), interstitial marine worms, characterized by a ciliated epidermis and a complex muscular pharynx usually provided with paired jaws and an unpaired cuticular basal plate. The phylum is of uncertain position, but has affinities with Platyhelminthes and Aschelminthes. Although known for several decades, the first species were not described until 1956; Riedl (1969) first recorded the group from North America, and species of *Gnathostomula* (fig. 10) are known from the Pacific coast (Riedl, 1971). Gnathostomulids may be abundant (and in some samples even outnumber nematodes) in deeper, black, fine-sediment (sand) layers with a high organic content (often hydrogen-sulfide rich); intertidally, they may be sought for in sheltered situations, such as harbor sand flats; other species, however, occur in clean sand, among surfgrass roots, or in mud. Hulings and Gray (1971) and Sterrer (1971) give techniques for extraction and preparation, and Sterrer (1972) has summarized the literature.

ASCHELMINTH COMPLEX

In this complex of groups (united by Hyman, 1951, in the Phylum Aschelminthes) are small, mostly vermiform, pseudocoelomate animals with a cuticle and a posterior anus. Some move by cilia, others lack cilia altogether. They include the familiar rotifers, the tiny gastrotrichs and kinorhynchs (echinoderids), the huge group of nematodes, and the gordiaceans or horsehair worms (with a single marine genus, *Nectonema*).

PHYLUM ROTIFERA

(PLATE 16)

Rotifers, the "wheel animalculae" of marine, freshwater, and semiterrestrial habitats, are minute (generally under 0.5 mm), usually free-swimming or creeping animals; sessile, tubiculous, and colonial forms are known, as well as ecto- and endoparasitic species. Rotifers are characterized by anterior ciliated lobes (the corona), a pharynx with complex jaws (mastax), flame-cell nephridia, and an annulated cuticle that may give them the appearance of being segmented. The slender, posterior part of the body, called a "foot," characteristically bears two "toes." *Rotatoria* and *Philodina* (fig. 6), common in fresh water, have marine representatives and are typical of the group. Free-swimming rotifers are readily recognized by the corona, by which they swim, and which may resemble rotating wheels; occasionally abundant in the plankton, they might be confused with veliger larvae of molluscs. *Proales* occurs in or on other organisms (such as oligochaetes, hydroids, hermit crabs, or the yellow-green alga *Vaucheria,* in which gall-like swellings are produced); the wormlike *Seison* (fig. 7) occurs on the primitive malacostracan crustacean *Nebalia*. Other species have been reported on sea cucumbers and preying upon the sessile ciliate *Folliculina*. Living specimens are best for identification, and permanent slides of the mastax may be necessary for identification. Hyman (1951) gives a good general account of rotifers, and Thane-Fenchel (1968) provides a key to marine and brackish-water genera.

PHYLUM GASTROTRICHA

(PLATE 16)

Gastrotrichs are a small phylum of minute, marine and freshwater animals occurring interstitially, often in coarse sand or in mud. They have a scaly or spiny cuticle, lack the corona of rotifers, but possess tufts or bands of ventral cilia and frequently a forked posterior end (caudal furca) (fig. 8). They are easily distinguished at higher magnifications from turbellarians and gnathostomulids by possession of a terminal mouth, an anterior pharynx, and an anus, and by lack of lateral and dorsal locomotory cilia and statocyst. They commonly glide by means of ventral cilia, but some have a leechlike motion. Little is known of the phylum on the California coast. For collecting and mounting techniques, see Hulings and Gray (1971, pp. 39–40). Hyman (1951), Hummon (1971), and D'Hondt (1971) give good general accounts and summarize the literature; see Hummon (1966, 1969, 1972) and Wieser (1957) for Pacific coast species.

PHYLUM KINORHYNCHA (ECHINODERA)

(PLATE 16)

The exclusively marine kinorhynchs are seldom sought for; like rotifers and most gastrotrichs they are less than 1 mm in length. The yellow brown cuticle of the adult is divided into joints or annulations resembling segments, and is characteristically spiny (fig. 9); there are no external cilia (as there are in rotifers and gastrotrichs), and the retractible head is armed with a circlet of spines. Kinorhynchs may be abundant in surface mud layers in shallow water, but also occur less abundantly on muddy sand or algal substrates. Higgins (1971) and Hyman (1951) provide reviews and literature; Zelinka (1928) is still the most important reference. Higgins (1960, 1961) has described several species from the San Juan Archipelago, and Kozloff (1972) has studied the development of one species in that area. See Hulings and Gray (1971, pp. 40–41) and Higgins (1971) for collecting, extracting (an aerating technique brings the animals to the surface trapped on minute air bubbles), and procedures for study.

PHYLUM NEMATODA

(PLATE 16)

The nematodes or roundworms are an enormous group, found in practically all habitats, and may be the most abundant metazoans in marine sediments. Although the best known are parasitic in animals or in crops, the majority of species are free-living in soil and in marine and fresh water. Nematodes are easily recognized by the elongated, cylindrical or fusiform body (fig. 11), usually small and pointed at the ends; by the unsegmented smooth cuticle; and by the characteristic, stiff, whipping motion. Large nematodes (to 1 or 2 cm or more) occur commonly in intertidal sponges, among bryozoans, in mussel beds and kelp holdfasts, and in detritus. Identification is a task for the specialist; the works of Allgén, Chitwood, Hope, Murphy, and others (some of which are cited below) should be consulted by any student seriously interested in the local intertidal nematode fauna. Hyman

(1951) provides a good general account; Hulings and Gray (1971, pp. 41–42) outline extraction and preparation techniques.

PHYLUM NEMATOMORPHA

Nematomorphans include the well-known "horsehair snakes" or gordiacean worms, the larvae of which are parasitic in insects and crustaceans. Adults have a unique, nonfunctional alimentary tract and are commonly seen as very long, cylindrical, dark brown worms in freshwater ponds, streams, or ditches; they may reach lengths of 1 m or more. The one marine genus, *Nectonema,* is pelagic as an adult, and passes its juvenile life as an internal parasite of shrimps, hermit crabs, and brachyurans (see Born, 1967 for references, and general account by Hyman, 1951). *Nectonema* is unreported on this coast, but should it turn up it will likely be first seen as a striking, white horsehair worm swimming at a harbor night light.

PHYLUM ACANTHOCEPHALA

Acanthocephalans or spiny-headed worms (so-called from the recurved spines on the retractible proboscis) are elongate, cylindrical endoparasites, lacking digestive, circulatory, and respiratory systems, requiring invertebrate intermediate hosts as larvae, and occurring as adults in the digestive tract of vertebrates. Local acanthocephalans may be sought in the intestines of marine mammals (harbor seals or sea lions washed ashore may provide abundant and fresh material), aquatic birds, and fishes. Intermediate hosts may include amphipods, hermit crabs, or other crustaceans. See Hyman (1951) for a general account, and recent parasitology texts for later literature. The group has received little attention in our area.

REFERENCES ON ASCHELMINTHES AND SMALLER WORM GROUPS

Allgén, C. A. 1947. West American marine nematodes. Vid. Medd. Dansk. naturhist. Foren. i København 110: 65–219 (see also 113: 366–411).

Born, J. W. 1967. *Palaemonetes vulgaris* (Crustacea, Decapoda) as host for the juvenile stage of *Nectonema agile* (Nematomorpha). J. Parasitol. 53: 793–794. (Woods Hole)

Chitwood, B. G. 1960. A preliminary contribution to the marine nemas (Adenophorea) of northern California. Trans. Amer. Micr. Soc. 79: 347–384. Includes species from intertidal substrates in San Francisco Bay and Tomales Bay, *Egregia* holdfasts at Dillon Beach, and the Moss Landing sand dunes.

D'Hondt, J.-L. 1971. Gastrotricha. *In* H. Barnes, ed., Ocean. Mar. Biol. Ann. Rev., London 9: 141–191.

Higgins, R. P. 1960. A new species of *Echinoderes* (Kinorhyncha) from Puget Sound. Trans. Amer. Micr. Soc. 79: 85–91.

Higgins, R. P. 1961. Three new homalorhage kinorhynchs from the San Juan Archipelago, Washington. J. Elisha Mitchell. Sci. Soc. 77: 81–88.

Higgins, R. P. 1971. A historical overview of kinorhynch research. Smithson. Contr. Zool. 76: 25–31.

Hope, W. D. 1967. Free-living marine nematodes of the genera *Pseudocella* . . . *Thoracostoma* . . . and *Deontostoma* . . . (Nematoda: Leptosomatidae) from the west coast of North America. Trans. Amer. Micr. Soc. 86: 307–334. Includes species from Dillon Beach, Bolinas, Pacific Grove, and Point Pinos.

Hope, W. D. and D. G. Murphy 1972. A taxonomic hierarchy and checklist of the genera and higher taxa of marine nematodes. Smithson. Contrib. Zool. 137: 101 pp.

Hulings, N. C., ed. 1971. Proceedings of the First International Conference on Meiofauna. Smithson. Contrib. Zool. 76: 205 pp.

Hulings, N. C. and J. S. Gray 1971. A manual for the study of meiofauna. Smithson. Contrib. Zool. 78: 83 pp.

Hummon, W. D. 1966. Morphology, life history, and significance of the marine gastrotrich, *Chaetonotus testiculophorus* n. sp. Trans. Amer. Micr. Soc. 85: 450–457 (Puget Sound).

Hummon, W. D. 1969. *Musellifer sublittoralis*, a new genus and species of Gastrotricha from the San Juan Archipelago, Washington. Trans. Amer. Micr. Soc. 88: 282–286.

Hummon, W. D. 1971. The marine and brackish-water Gastrotricha in perspective. Smithson. Contr. Zool. 76: 21–23.

Hummon, W. D. 1972. Dispersion of Gastrotricha in marine beaches of the San Juan Archipelago, Washingon. Mar. Biol. 16: 349–355.

Hyman, L. H. 1951. *The Invertebrates: Acanthocephala, Aschelminthes, and Entoprocta. The Pseudocoelomate Bilateria,* Vol. III. McGraw-Hill, 572 pp. (Acanthocephala, pp. 1–52; Rotifera, pp. 59–151; Gastrotricha, pp. 151–170; Kinorhyncha, pp. 170–183; Nematoda, pp. 197–455; Nematomorpha, pp. 455–472; see also Vol. V, 1959, Retrospect).

Kozloff, E. N. 1965. *Ciliocincta sabellariae* gen. and sp. n., an orthonectid mesozoan from the polychaete *Sabellaria cementarium* Moore. J. Parasit. 51: 37–44.

Kozloff, E. N. 1969. Morphology of the orthonectid *Rhopalura ophiocomae*. J. Parasit. 55: 171–195.

Kozloff, E. N. 1971. Morphology of the orthonectid *Ciliocincta sabellariae*. J. Parasit. 57: 585–597.

Kozloff, E. N. 1972. Some aspects of development in *Echinoderes* (Kinorhyncha). Trans. Amer. Micr. Soc. 91: 119–130.

McConnaughey, B. H. 1949. Mesozoa of the family Dicyemidae from California. Univ. Calif. Publ. Zool. 55: 1–34.

McConnaughey, B. H. 1951. The life cycle of the dicyemid Mesozoa. Univ. Calif. Publ. Zool. 55: 295–336.

McConnaughey, B. H. 1963. The Mesozoa. *In* E. C. Dougherty, ed., *The Lower Metazoa,* University of California Press, pp. 151–168.

McConnaughey, B. H. 1968. The Mesozoa. *In* M. Florkin and B. T. Scheer, eds., *Chemical Zoology,* vol. II, Academic Press, pp. 557–570.

Riedl, R. J. 1969. Gnathostomulida from America. Science 163: 445–452.

Riedl, R. J. 1971. On the genus *Gnathostomula* (Gnathostomulida). Intern. Revue Ges. Hydrobiol. 56: 385–496.

Sterrer, W. 1971. Gnathostomulida: problems and procedures. *In* Hulings, 1971, pp. 9–15.

Sterrer, W. 1972. Systematics and evolution within the Gnathostomulida. Syst. Zool. 21: 151–173.

Stunkard, H. W. 1954. The life history and systematic relations of the Mesozoa. Quart. Rev. Biol. 29: 230–244.

Stunkard, H. W. 1972. Clarification of taxonomy in the Mesozoa. Syst. Zool. 21: 210–214.

Thane-Fenchel, A. 1968. A simple key to the genera of marine and brackish-water rotifers. Ophelia 5: 299–311.

Wieser, W. 1957. Gastrotricha Macrodasyoidea from the intertidal of Puget Sound. Trans. Amer. Micr. Soc. 76: 372–381.

Wieser, W. 1959. *Free-living nematodes and other small invertebrates of Puget Sound beaches.* University of Washington Press, 179 pp.

Zelinka, C. 1928. *Monographie der Echinodera.* Leipzig: Wilhelm Engelmann, 396 pp.

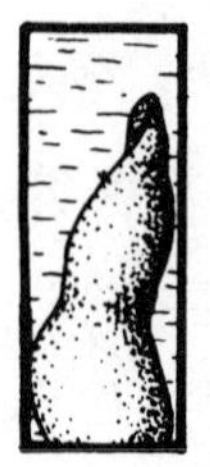

UNSEGMENTED COELOMATE WORMS

Mary E. Rice
National Museum of Natural History, Smithsonian Institution, Washington, D.C.

In this chapter are grouped three types of marine worms bearing some superficial resemblance in having an apparently unsegmented, stout, cylindrical body and a spacious body cavity or coelom. They were for a century (up to about the 1940s) called a class, "Gephyrea," of the Annelida, but this concept is now discarded (Hyman, 1959, p. 611). These groups show such fundamental differences that they are now generally placed in the separate phyla, Sipuncula, Echiura, and Priapula, although Hyman, with some reservations, places the Priapula in the Aschelminthes and the Echiura in or close to the Annelida.

PHYLUM SIPUNCULA

(PLATE 17)

The commonest local sipunculans are known as "peanut worms" because of their appearance when contracted, although the term would not be descriptive for many other species. The body is usually cylindrical and elongate when relaxed, with a slender anterior part. Upon contraction this anterior introvert is pulled back into itself, and the worm becomes more ovoid, with the body very firm and turgid. The muscular body wall encases a spacious coelom and a long, coiled intestine. The anus opens dorsally at the base of the introvert (in contrast to its posterior position in the echiurans). Sipunculans represent a distinct phylum, differing from annelids in the lack of segmentation and setae, and from echiurans in the position of the anus and in other characteristics. The introverted anterior body wall, which can be extended for feeding (fig. 3–5), has tentacles about the mouth at its tip and is quite unlike the proboscides of either annelids or echiurans.

Sipunculans are marine or estuarine and commonly, in this area, inhabit rock crevices, empty shells, algal holdfasts, or other protected situations along the open coast. Others occur in firm sand or mud or among eelgrass roots, although they are absent from the shifting sands of exposed beaches. The rocky intertidal forms apparently collect detritus with their tentacles; burrowing forms may swallow large quantities of mud and sand.

In order to identify sipunculans it is best to let them relax naturally until the tentacles are exposed and, if possible, to anesthetize them in an extended condition. A good relaxant for many species is 10% ethanol in sea water. If during anesthetization the anterior end is retracted, then a slight pressure on the body wall will often cause extension of the introvert. Because of their tendency to pull in the introvert and tentacles when placed in the preservative, it is well to grasp the animals with forceps just behind the tentacles, and to hold them until killing is completed. Fisher's monograph (1952) covers local sipunculans, although some of his names have been altered by subsequent workers.

KEY TO SIPUNCULA

1. Tentacles conspicuous when extended, branching (fig. 3) ... 2
– Tentacles inconspicuous, fingerlike (figs. 4B, 5) 5
2. Introvert armed with small, black or brown spines; tentacles arising on 4 stems (fig. 3) .... *Themiste pyroides*
– Introvert devoid of spines; tentacles arising on 6 stems 3
3. Body fusiform (spindle-shaped) to pyriform (pear-shaped); resembling *T. pyroides;* collar (at base of tentacles) reddish purple .... *Themiste dyscrita*
– Body cylindrical; collar not obviously reddish or purplish .. 4
4. Intestinal coils numbering up to 30 .....*Themiste zostericola*
– Intestinal coils numbering up to 100 ...*Themiste perimeces*
5. Muscles of body wall divided into separate longitudinal bands 6
– Muscles of body wall without trace of bands 8
6. Adults medium-sized (commonly 5–7 cm or up to 12 cm fully extended); skin thick, rough, with prominent papillae largest in anal region and posterior extremity; introvert with anterior rows of small hooks and transverse, brownish bands; body often spotted with black, brown, or purple (figs. 4A, B) .... *Phascolosoma agassizii*
– Adults large (12–50 cm extended); skin smooth without prominent papillae on trunk, but may be grooved with longitudinal and circular furrows; no hooks on introvert 7
7. Introvert short, with scalelike papillae and sharply marked off from body; tentacular fold surrounding mouth with lobulate tentacles .... *Sipunculus nudus*
– Introvert without scalelike papillae, not sharply marked off from body; tentacles arranged in numerous longitudinal series,

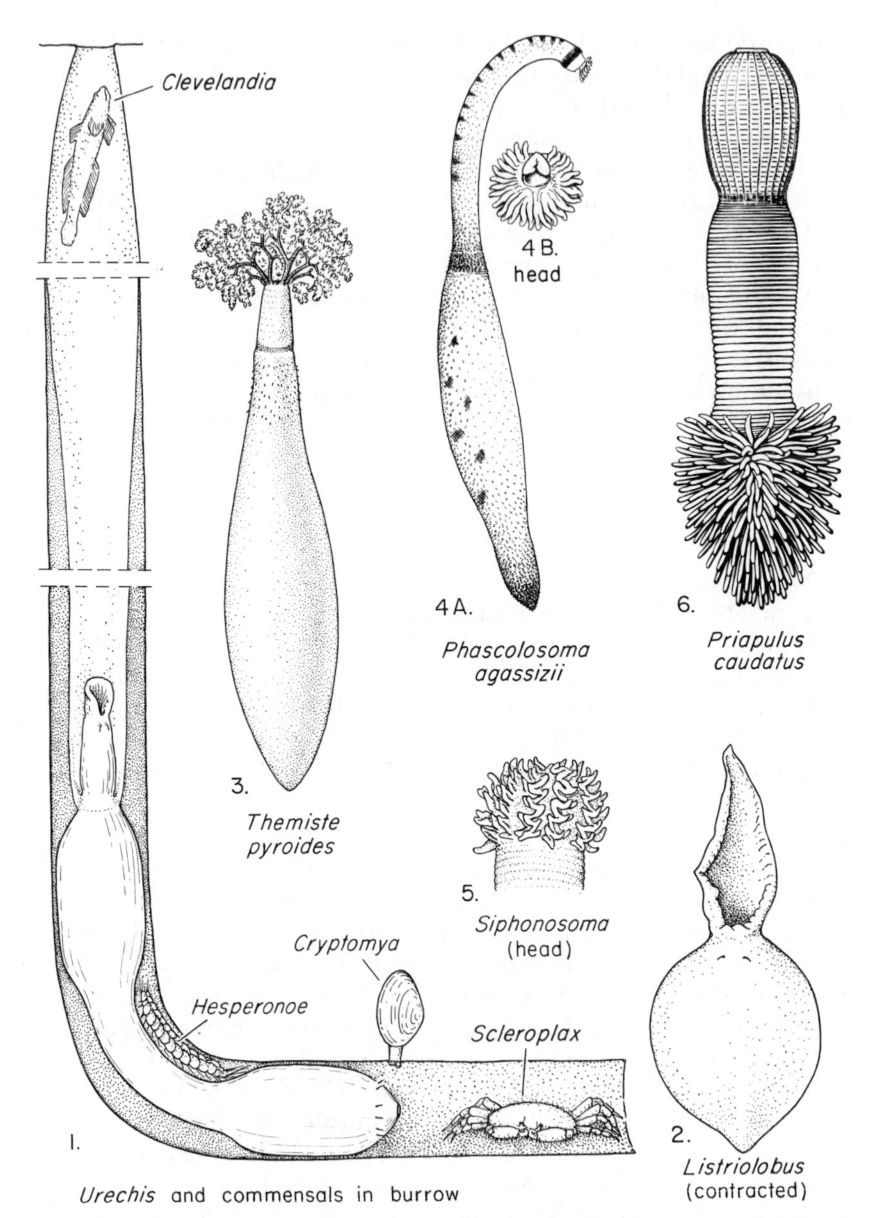

PLATE 17. Unsegmented coelomate worms: **Sipuncula, Echiura, and Priapula** (not to scale). 1, *Urechis* with mucous feeding net, in burrow with commensals; 2, *Listriolobus*, contracted; 3, *Themiste;* 4A, *Phascolosoma* with introvert extended; 4B, head with tentacles, oral view; 5, *Siphonosoma*, head; 6, *Priapulus*. 1, 2, 3, 5 after Fisher, 1946, 1952, redrawn by Emily Reid; 4, 6, original, drawn by Carolyn B. Gast.

forming a sort of head (fig. 5) .... *Siphonosoma ingens*

8. Body small, slender, threadlike; introvert 6–8 times length of trunk with rows of small hooks anteriorly; intestine attached posteriorly by spindle muscle ....*Golfingia hespera*

– Body small, cylindrical; introvert short, length usually less than trunk, no hooks; intestine not attached posteriorly .... *Golfingia margaritacea californiensis*

LIST OF SIPUNCULA

GOLFINGIIDAE

Golfingia hespera (Chamberlain, 1919). The curious generic name was coined by E. Ray Lankester to commemorate a holiday spent golfing at St. Andrews in 1885. In tubes of ceriantharians and among tubes of the polychaete *Mesochaetopterus;* also among *Zostera* on mudflats.

Golfingia margaritacea californiensis Fisher, 1952. In rock crevices.

Themiste dyscrita (Fisher, 1952). In sand among rocks and in pholad burrows. This and following species of *Themiste* were formerly *Dendrostomum.*

Themiste perimeces (Fisher, 1928). In mudflats.

Themiste pyroides (Chamberlain, 1919) (=*Dendrostomum petraeum* Fisher, 1928). Generally in rock crevices; look for the small rhabdocoel *Collastoma pacifica* in gut (see p. 107).

Themiste zostericola (Chamberlain, 1919). Among *Zostera,* in sand and mud under boulders, low intertidal. A southern species; see Peebles and Fox, 1933.

PHASCOLOSOMATIDAE

Phascolosoma agassizii Keferstein, 1867 (formerly *Physcosoma*). The commonest sipunculan of our intertidal; in mud, rock crevices, shells, holdfasts, and *Mytilus* beds. See Towle and Giese, 1967; Rice, 1973.

SIPUNCULIDAE

Siphonosoma ingens (Fisher, 1947). In muddy sand and among *Zostera* roots.

Sipunculus nudus Linnaeus, 1766. Subtidal; occasionally washed in.

REFERENCES ON SIPUNCULA

Fisher, W. K. 1952. The sipunculid worms of California and Baja California. Proc. U.S. Nat. Mus. 102: 371–450.

Hyman, L. H. 1959. *The Invertebrates: Smaller Coelomate Groups,* Vol. V. McGraw-Hill (Sipunculida, pp. 610–696).

Peebles, F. and D. L. Fox 1933. The structure, functions, and general reactions of the marine sipunculid worm *Dendrostoma zostericola*. Bull. Scripps Inst. Oceanog., Tech. Ser., 3: 201–224.
Rice, M. E. 1967. A comparative study of the development of *Phascolosoma agassizii*, *Golfingia pugettensis*, and *Themiste pyroides* with a discussion of developmental patterns in the Sipuncula. Ophelia 4: 143–171.
Rice, M. E. 1973. Morphology, behavior, and histogenesis of the pelagosphera larva of *Phascolosoma agassizii* (Sipuncula). Smithson. Contrib. Zool. no. 132, 51 pp.
Rice, M. E. and M. Todorović, editors. 1975. *Proc. Internat. Symposium on the Biology of the Sipuncula and Echiura.* Naučno Delo, Belgrade.
Stephen, A. C. 1964. A revision of the classification of the phylum Sipuncula. Ann. Mag. Nat. Hist. Ser. 13, 7: 457–462.
Stephen, A. C. and S. J. Edmonds 1972. *The Phyla Sipuncula and Echiura.* London, British Museum (Natural History), viii + 528 pp.
Towle, A. and A. C. Giese 1967. The annual reproductive cycle of the sipunculid *Phascolosoma agassizii.* Physiol. Zool. 40: 229–237.

PHYLUM ECHIURA

(PLATE 17)

Echiurans resemble annelids in a general way, but are unsegmented as adults. The body is usually sausage-shaped, consisting of a muscular wall about a spacious coelom, and contains a very long, looped intestine, opening by a posterior anus. A characteristic solid, extensible proboscis just anterior to the mouth has given the group the name of "spoon worms," from its shape when contracted. There are usually one or more pairs of setae placed ventrally behind the mouth, and one or two posterior rings of setae.

Echiurans are entirely marine. Most burrow in sand or mud, or inhabit rock crevices, empty shells, sand-dollar tests, pholad holes, and so on, and swallow large quantities of bottom material or lighter detritus gathered by the long proboscis. However, the most common echiuran of this area, *Urechis caupo* (fig. 1), has the specialized habit of collecting fine-particulate material, including bacteria, in a net of mucus through which it pumps a flow of water, periodically consuming both net and collected food. The natural history of the "innkeeper" *Urechis*, so called for the various commensals sharing its burrow, is interestingly described by Fisher and MacGinitie (1928b). Fisher's monograph (1946) has a good general account of the group.

KEY TO ECHIURA

1. Size large; a ring of conspicuous setae at posterior end of flesh colored body; preoral proboscis very short; forms perma-

nent burrows in muddy sand; (fig. 1) ....
.... *Urechis caupo* Fisher and MacGinitie, 1928

– Size small to medium; body globose, green to gray violet, no posterior setae (fig. 2); proboscis yellow, elongate, soft, easily lost; rare, in mud among *Zostera* and in sandy mud, to off-shore depths .... *Listriolobus pelodes* Fisher, 1946.

REFERENCES ON ECHIURA

Fisher, W. K. 1946. Echiuroid worms of the North Pacific Ocean. Proc. U.S. Nat. Mus. 96: 215–292.

Fisher, W. K., and G. E. MacGinitie 1928a. A new echiuroid worm from California. Ann. Mag. Nat. Hist. (10) 1: 199–204, Pl. IX.

Fisher, W. K., and G. E. MacGinitie 1928b. The natural history of an echiuroid worm (*Urechis*). Ann. Mag. Nat. Hist. (10) 1: 204–213, Pl. X.

Newby, W. W. 1940. The embryology of the echiuroid worm *Urechis caupo.* Mem. Amer. Phil. Soc. 16: 1–213.

Stephen, A. C. and S. J. Edmonds 1972. *The Phyla Sipuncula and Echiura.* London: British Museum (Natural History), viii + 528 pp.

PHYLUM PRIAPULA

(Plate 17)

Members of this phylum bear a superficial resemblance to sipunculans or echiurans, but the current tendency is to place them among or near the Aschelminthes (Hyman, 1951) together with nematodes, rotifers, and other groups, or as a separate phylum (Shapeero, 1961). The priapulan body is muscular and cylindrical, up to several centimeters long, with an anterior, bulbous, spiny proboscis and one or two posterior, hollow appendages, the cavity of which is continuous with the perivisceral space of the body. There are only eight species in the phylum, all in marine or brackish water. *Priapulus caudatus* Lamarck, 1816 (fig. 6) occurs in muddy or sandy bottom in Tomales Bay. It apparently does not construct a permanent burrow, but moves about through the substrate. According to Lang (1948a), *Priapulus* is an active predator on polychaetes and even other priapulans, which it swallows whole. The species is known also from Puget Sound, northern Japan, and the North Atlantic. Its early embryology, which might shed light on its affinities, is not known, but its rotiferlike larval form has been described (Lang, 1948b).

REFERENCES ON PRIAPULA

Hyman, L. H. 1951. *The Invertebrates: Acanthocephala, Aschelminthes, and Entoprocta,* Vol. III. McGraw-Hill (Priapulida, pp. 183–197).

Lang, K. 1948a. Contribution to the ecology of *Priapulus caudatus* Lam. Arkiv f. Zool. 41A (5): 1–12.

Lang, K. 1948b. On the morphology of the larva of *Priapulus caudatus* Lam. Arkiv f. Zool. 41A (9): 1–8.

Shapeero, W. L. 1961. Phylogeny of Priapulida. Science 133: 879–880.

Shapeero, W. L. 1962a. The distribution of *Priapulus caudatus* Lam. on the Pacific coast of North America. Amer. Midl. Nat. 68: 237–241.

Shapeero, W. L. 1962b. The epidermis and cuticle of *Priapulus caudatus* Lamarck. Trans. Amer. Micr. Soc. 81: 352–355.

Van der Land, J. 1970. Systematics, zoogeography, and ecology of the Priapulida. Zool. Verhandl. no. 112: 118 pp.

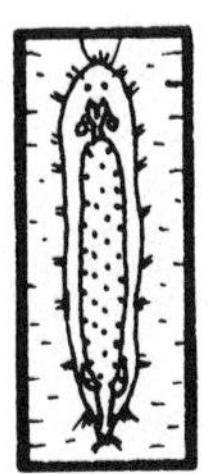

PHYLUM ANNELIDA: INTRODUCTION AND SMALLER GROUPS

The annelids or segmented worms are characterized by an elongated body, divided into segments and formed on the plan of a tubular jacket of muscle surrounding a fluid-filled coelom. Although lacking a rigid internal skeleton, annelids can use the hydrostatic pressure of coelomic fluid, acted upon by the muscular body wall, as a "fluid skeleton," aiding in extension of the body and in burrowing. Locomotion in polychaetes is also aided by numerous setae (or chaetae) which project from the sides of the body. The annelid body plan has proved to be extremely plastic and adaptable in an evolutionary sense, so that we find a great diversity of form, habitat, and mode of life within the phylum.

Three classes are generally recognized: (1) **Polychaeta** have many setae (hence the name) and parapodia, as well as a diversity of tentacles and gill-like devices. Polychaetes, to be discussed in the next chapter (p. 151), are almost all marine, exceedingly diverse in form, and very numerous, comprising about 60 families and some 1600 genera. **Archiannelida** (p. 147), a small and poorly known marine group, may be regarded as an appendage to the Polychaeta rather than as a distinct class. Archiannelids are small forms, showing a curious assortment of features, including weak segmentation, frequent absence of parapodia and setae; some are hermaphroditic. Some of the foregoing features may be primitive, others reduced. The group is probably an assortment of degenerate and rather specialized animals related to widely diverse, better-known families. (2) **Oligochaeta** (p. 136) have no parapodia and few setae (as the name implies). They include the familiar earthworms, as well as marine and freshwater representatives. (3) **Hirudinea** or leeches (p. 146) have no setae (except in one order) or parapodia, and are found mainly in fresh water.

CLASS OLIGOCHAETA

David G. Cook
Great Lakes Biolimnology Laboratory,
Canada Centre for Inland Waters, Burlington, Ontario, Canada

Ralph O. Brinkhurst
Fisheries Research Board of Canada, Biological Station,
St. Andrews, New Brunswick, Canada

(PLATE 18)

The class **Oligochaeta** contains the familiar earthworms and about ten families of small to minute, often obscure, predominantly aquatic worms; the best known representatives of the latter group (known collectively as microdriles) are the tubificids. These red, threadlike worms, often sold as fish food, may occur in great abundance in organically polluted habitats. On the Pacific coast of North America, representatives of three families occur in salt or brackish water, namely the **Tubificidae, Naididae** and **Enchytraeidae.** Unlike all other microdrile families, the enchytraeids are predominantly terrestrial or semi-aquatic, and include the familiar white worms associated with compost heaps and other decaying vegetable matter. Naididae are small, easily overlooked worms that reproduce regularly by simple fission and lead a predatory or browsing existence on plant or sediment surfaces; this mode of life, rather than the burrowing or interstitial habit of Tubificidae and Enchytraeidae probably accounts for their relatively high degree of cephalisation (see below).

Oligochaetes are hermaphroditic annelids, lacking parapodia, palps, tentacles, and jaws, and possessing comparatively few setae; the reproductive function is confined to a few anterior segments where, at sexual maturity, a glandular **clitellum** is formed that secretes the cocoon within which one or a few eggs develop directly into juveniles resembling small adults. A microdrile oligochaete consists of a long, thin, segmented, cylindrical body with a prostomium, an anterior ventral mouth, and a terminal anus (fig. 1). The gut, ventral nerve cord, and the main dorsal and ventral blood vessels (except in Enchytraeidae) extend the length of the body, but all are modified to some degree in anterior segments. The taxonomically significant alimentary canal begins at the mouth as a narrow, thin-walled esophagus. In the region of the second segment, the esophagus opens into the pharynx, consisting of a tube with thickened dorsal and dorsolateral gland cells; associated with the pharynx are the pharyngeal or septal glands, which are more or less discrete masses of glandular cells whose long, secretory processes penetrate the pharyngeal wall. In some Enchytraeidae, a pair of tubular peptonephridia (fig. 2) arise at the esophageal-pharyngeal junction; these and the pharyngeal glands probably both serve a combination of digestive, lubricative, and food-collecting functions.

Other structures are segmentally arranged and include the setae, which usually occur in two dorsolateral and two ventrolateral bundles on all segments except the **peristomium** (segment I), and a pair of excretory organs, the nephridia, which are located ventrolaterally in most postclitellar segments. The reproductive organs in marine microdriles consist of one pair of testes associated with paired male genitalia, one pair of ovaries associated with a pair of small, ventrolateral, female funnels that conduct eggs to the exterior, and a pair of ectodermal spermathecae that receive and store sperm after copulation. Ovaries and female funnels occur in the segment immediately posterior to the testes segment; that is, in microdriles they are located in the atrial or male genital segment (fig. 11).

TAXONOMIC CRITERIA

In the microdrile families under consideration, the most important taxonomic characters are: the morphology and arrangement of the male genitalia and spermathecae; the number, form, and detailed structure of the setae; and, in Enchytraeidae, the anatomy of the anterior digestive system.

Genitalia (figs. 11, 12): In Tubificidae the male genitalia consist of a pair of funnels, on the anterior face of septum X/XI, that collect sperm and conduct them into a pair of narrow, tubular vasa deferentia, which open into the storage and copulatory apparatus; the latter consist typically of a pair of muscular **atria** that each open to the exterior on segment XI either as a simple pore in the body wall, or, more usually, as some form of modified penis or pseudopenis; true penes often possess a sheath of thickened cuticle around them. Usually the atrium bears a stalked, glandular organ, the prostate gland, which probably provides nutrients and lubrication for the sperm prior to copulation; in some tubificid genera the prostate glands are diffuse masses of cells more or less covering the atria. After copulation, sperm from the sexual partner are stored in spermathecae which, in tubificids, are located in segment X; that is, in the segment immediately anterior to the atria. Spermathecae are simple invaginations of the body wall, each consisting of a narrow tubular duct and a voluminous storage ampulla (fig. 11).

The naidid genital system is basically the same as in tubificids, but the various elements are located in segments V and VI and the prostate glands are always diffuse cell masses. The comparative rarity of naidids in a fully sexual condition is explained by the high incidence of asexual reproduction by simple fission in this family.

Enchytraeidae differ considerably from other microdrile families in many respects, including their genital systems; spermathecae, whose ducts often bear gland cells and whose ampullae are often in open communication with the gut, are located in segment V, and the male genitalia are situated in segments XI and XII. Large, more or less tubular male funnels, lying freely in the body cavity of segment XI, connect with the very narrow, often tortuously coiled vasa deferentia which lie in segment XII; the latter open to the exterior simply on penial papillae or in association with glandular penial bulbs; atria, in the naidid and tubificid sense, are lacking in this family.

In some Tubificidae and Naididae, the ventral setae of the atrial or spermathecal segments, or both, may be modified at sexual maturity as genital (spermathecal and penial) setae.

Setae (figs. 3–8): With a few comparatively rare exceptions, all microdriles possess four bundles of chitinous setae implanted in the body wall of all segments except the peristomium (segment I). Two basic types occur in the families considered here, namely **crochets** and **hair setae.** In the Naididae and Tubificidae, crochets are usually slender, sigmoid structures (fig. 4) possessing a more or less median thickening (the **node** or **nodulus**); the distal ends may be simple-pointed or bifid (forked), and the latter are sometimes further modified by one or a series of fine intermediate teeth between the outer major teeth (**pectinate** setae) (fig. 7). Enchytraeid crochets are simple-pointed or rounded (with one rare exception) and range in form from slender, sigmoid setae with a nodulus to the more usual condition of stout, straight setae without nodal thickenings (fig. 5).

Hair setae occur only in the dorsal bundles of some naidids and

---

PLATE 18. **Oligochaeta.** 1, tubificid oligochaete (from life); 2, diagrammatic optical section of anterior end of *Enchytraeus;* 3, diagrammatic transverse section through midsection of generalized microdrile (left side) and *Peloscolex* (right side) showing papillate body wall; 4, sigmoid seta, with nodulus; 5, proximally curved (enchytraeid) seta, without nodulus; 6, distal end of bifid seta; 7, distal end of pectinate seta; 8, distal end of seta of *Chaetogaster limnaei;* 9, spermathecal duct of *Lumbricillus vancouverensis;* 10, sperm bundle of *Lumbricillus georgiensis;* 11, diagrammatic longitudinal section through genitalia of *Peloscolex* spp.; 12, diagrammatic longitudinal section through male genitalia of *Monopylephorus irroratus;* 13, penis sheath of *Limnodrilus hoffmeisteri;* 14, penis sheath of *Peloscolex* spp. (generalized, diagrammatic). All by Cook, original. a, Prostomium; b, Clitellum; c, Peptonephridium; d, Pharyngeal glands; e, Pharynx; f, Dorsal blood vessel; g, Ventral nerve cord; h, Hair seta; i, Dorsal crochets; j, Ventral setae (crochets); k, Spermatheca; l, Testis; m, Ovary; n, Vas deferens; o, Atrium; p, Prostate gland; q, Penis, r, Nodulus; s, Intermediate teeth.

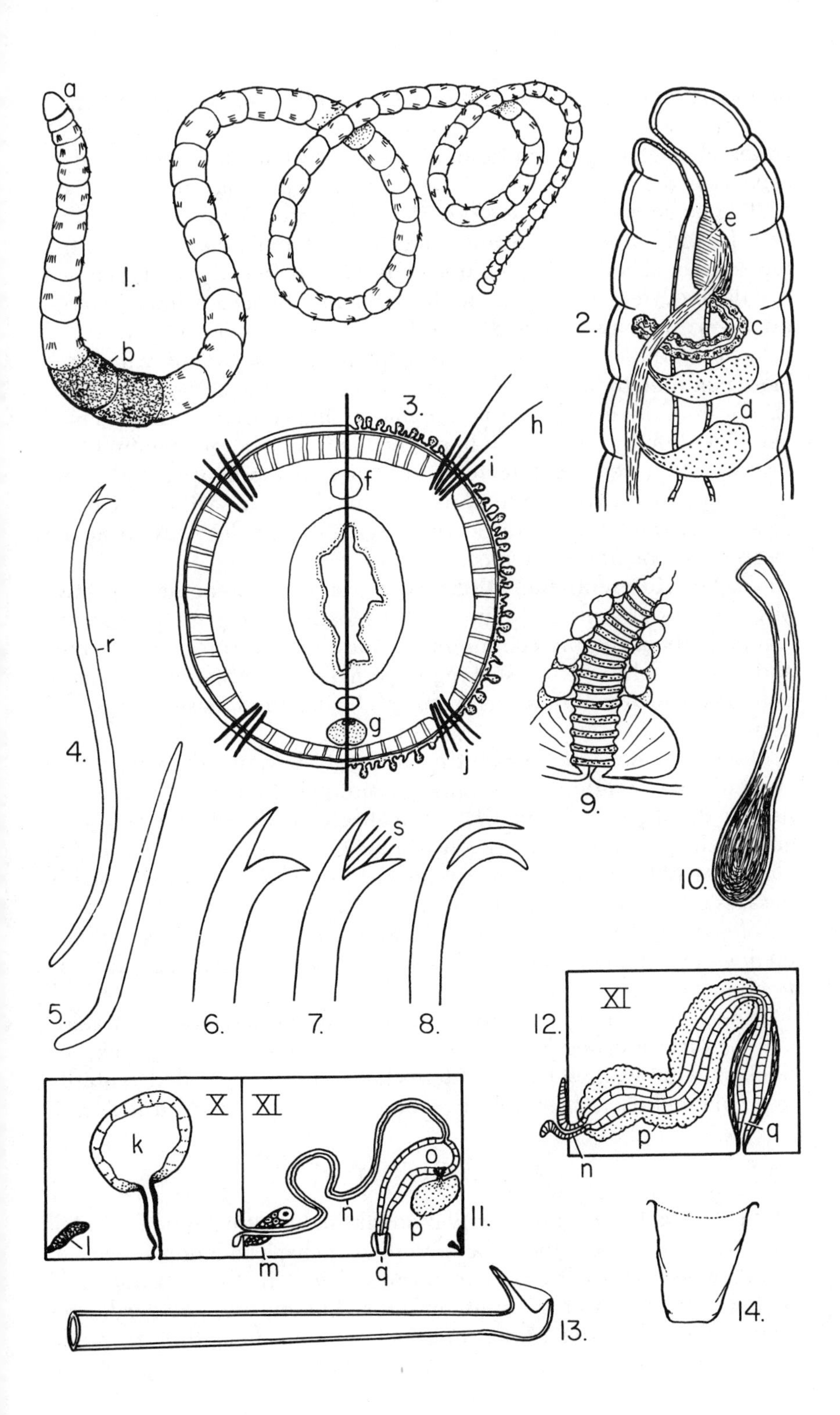
a
b
1.
e
c
2.
d
3.
h
i
f
g
j
r
4.
5.
9.
10.
s
6.
7.
8.
XI
12.
p
q
n
X
XI
k
o
l
m
n
p
q
11.
13.
14.

tubificids, and consist of an elongate, slender, simple-pointed shaft without a nodulus.

Other characters: The anterior digestive system in Enchytraeidae, especially the presence or absence of **peptonephridia,** are major diagnostic characters and have been covered above. For purposes of this key the only other significant character in Tubificidae concerns the presence or absence of large, free cells within the body cavity, known as **coleomocytes;** these occur in the subfamily Rhyacodrilinae in large numbers and may completely fill the coelom of some segments.

Most Naididae are characterized by anterior segments which are more modified than in other microdrile families; dorsal setae are often absent from the first four or five segments and sensory structures may be present on the peristomium or prostomium. The prostomium may bear an elongated tactile extension known as a **proboscis** and many naidids possess eyes on the peristomium which are visible as paired, crescent-shaped masses of opaque pigment granules located at the bases of the photosensitive cells.

Characters differentiating Oligochaeta from other annelids: Because oligochaetes are often obscure and, to many biologists, unexpected components of marine communities, and because some polychaetes and archiannelids bear a striking superficial resemblence to them, a note on characters differentiating oligochaetes from these groups will be useful.

Mature oligochaetes present no difficulty because their elaborate genitalia and glandular clitellum are unique. Immature worms can best be distinguished from those polychaetes with reduced parapodia and small numbers of setae, such as some Capitellidae, by detailed examination of the setae; in these polychaetes, the bifid crochets usually possess a thin, membranelike hood around their distal ends. In some species the hood is reduced to a small, membranous keel connecting the lower tooth to the setal shaft; such ornamentation is totally lacking in oligochaetes.

The few Archiannelida which could be confused with enchytraeids may be distinguished by examining the structure of the pharynx; the thickened gut wall which constitutes the pharynx is situated ventrally in archiannelids and capitellids, whereas in oligochaetes it is dorsal.

ECOLOGY

Marine oligochaetes are intertidal or subtidal benthic animals which live within, and feed upon, the bottom deposits. The larger species of Tubificidae and Enchytraeidae burrow freely in the substrate and probably feed indiscriminately on the sediment; the very small species of those families (down to about 0.1 mm diameter) are meiobenthic (interstitial) worms which inhabit the interstices between substrate particles and feed on fine, organic debris: free burrowing species tend to live in silts and poorly sorted, fine sands; meiobenthic worms are more restricted to coarser sands. In general oligochaetes are particularly abundant in areas of organic enrichment, often to the exclusion of other groups because they are able to withstand the low oxygen tensions usually associated with this condition. In the intertidal zone, for example, many Enchytraeidae are found in or beneath masses of decaying seaweed, and Tubificidae often occur in large numbers under stones which are partly buried in the sediment. Subtidally, in areas of organic pollution, very large populations of a few species of tubificids may occur; for example, in the San Francisco Bay system Brinkhurst and Simmons (1968) found that three species of *Peloscolex* formed up to 97.8 percent of the total bottom fauna at some stations.

Naididae generally are confined to sediment or plant surfaces, browsing off algae and organic debris, but one species (*Chaetogaster limnaei*) appears to be an obligate commensal in the mantle cavity and kidney of some pulmonate gastropod molluscs.

Very little data is available on the distribution of marine oligochaetes, especially on the Pacific coast, but it is probable that some are cosmopolitan (*Enchytraeus albidus, Chaetogaster limnaei, C. diaphanus, Paranais litoralis, Limnodrilus hoffmeisteri*) and others at least Amphiamerican (*Peloscolex apectinatus, P. gabriellae, P. nerthoides*). From our experience with the Atlantic coast fauna, it seems likely that many species remain to be described from the Pacific coast.

COLLECTION, PRESERVATION AND EXAMINATION PROCEDURES

Oligochaetes may be extracted from sediments in which they live by any combination of seiving, flotation, or manual sorting techniques. Intertidal sediments may be collected by digging or scooping material into containers, and any coring device, grab, dredge, or fine-mesh hand net is suitable for obtaining subtidal samples.

Material to be killed, either in the sediment or after extraction, is best narcotized in 0.015% propylene phenoxetol, killed and fixed in 10% formaldehyde solution for 48 hours, and stored in 70 to 90% ethyl alcohol.

In order to identify microdriles, it is necessary to make temporary or permanent microscopical preparations of whole or dissected animals. The following key uses readily seen setal characters wherever possible; these and cuticular penis sheaths can be observed by mounting worms temporarily in glycerol or Ammans lactophenol (40 g carbolic acid, 400 ml lactic acid, 800 ml glycerol, 400 ml water). Both mountants, expecially the latter, are good for handling large, routine collections, but both will eventually destroy soft tissues and neither will accept nuclear-stained animals. Therefore, for more critical work, it is recommended that worms should be dehydrated in absolute ethyl alcohol, cleared in xylol, and mounted in Canada Balsam. Using this method, the genitalia and pharyngeal structures of animals very lightly stained in acid hematoxylin may be observed.

Enchytraeidae can be more readily identified from living material; worms should be mounted in a small volume of sea water on a microscope slide and immobilized by gentle pressure of a cover slip. Larger, more vigorous individuals can be stopped by mounting them in a drop of gelatine solution which is just ready to set.

KEY TO MARINE AND BRACKISH-WATER OLIGOCHAETA

The marine oligochaetes of the Pacific coast are virtually unknown. It is probable that many species are distributed more widely than is now suspected, and that many species remain to be discovered. Therefore, the geographic coverage of this manual has been expanded for the oligochaetes to include all salt-water species recorded from Vancouver Island to the southern California border. The key is designed so that most worms can be keyed to family with a reasonable degree of accuracy, but it must be stressed that the remainder of the key is artificial and applies only to species recorded in the literature. New or previously unrecorded oligochaetes will not necessarily key out near their closest relatives, and it is strongly suggested that species keyed out be checked against complete descriptions (see references).

1. Some or all crochets bifid, sigmoid; hair setae present or absent; male genitalia situated in segment VI or XI 2
– All crochets single-pointed or rounded, straight, or curved proximally, or sigmoid; hair setae always absent; male genitalia situated in segment XII .... ENCHYTRAEIDAE 12
2. Dorsal setae absent from segments II to IV or V, or totally absent; eyes present or absent; male genitalia in segment VI .... NAIDIDAE 3
– Dorsal setae present from segment II rearwards; eyes always absent; male genitalia in segment XI .... TUBIFICIDAE 7
3. Dorsal setae totally absent .... *Chaetogaster* 4
– Dorsal setae present from segment V or VI 5
4. Commensal on pulmonate gastropods; setal teeth strongly curved to about 90° to setal shaft (fig. 8) .... *Chaetogaster limnaei*

– Free-living; setal teeth only moderately curved .... *Chaetogaster diaphanus*
5. Hair setae present; dorsal setae begin on segment VI .... *Nais communis*
– Hair setae absent; dorsal setae begin on segment V .... *Paranais* 6
6. All setae with upper tooth longer than lower .... *Paranais frici*
– Posterior setae with equal teeth .... *Paranais litoralis*
7. Hair setae present in some or all dorsal bundles 8
– Hair setae absent 11
8. Hair setae twisted distally; coelomocytes present, very numerous; prostate glands diffuse (fig. 12) .... *Monopylephorus irroratus*
– Hair setae straight or bent, never twisted; coelomocytes absent; prostate glands discrete, stalked (fig. 11) 9
9. Hair setae short, bent, present behind segment V; body wall smooth .... *Illyodrilus frantzi capillatus*
– Hair setae long, straight, present in anterior bundles; body wall usually with raised papillae, or with accumulations of foreign particles in body-wall ridges, especially in posterior segments (fig. 3, right) 10
10. Anterior dorsal crochets with 1 or 2 intermediate teeth; hair setae smooth .... *Peloscolex nerthoides*
– Anterior dorsal crochets simple bifids; hair setae serrate ... . *Peloscolex apectinatus*
11. Long, thick, tubular penis sheaths with reflexed hood present in segment XI (fig. 13); body wall smooth .... *Limnodrilus hoffmeisteri*
– Penis sheaths short, thin, truncated cone-shaped (fig. 14); body wall papillate or with accumulations of foreign particles .... *Peloscolex gabriellae*
12. Very small worms (up to 6 mm long, 0.3 mm diameter); dorsal setae absent; ventral setae 2 per bundle on all segments ... . *Marionina subterranea*
– Larger worms; dorsal setae present; ventral setae more than 2 per bundle 13
13. Setae straight or bent proximally (fig. 5); paired peptonephridia present .... *Enchytraeus* 14
– Setae sigmoid or, if bent proximally, living worms are green; peptonephridia absent .... *Lumbricillus* 17
14. One or 2 diverticula present at junction of spermathecal duct and ampulla 15
– Spermathecal diverticula absent 16
15. Each setal bundle associated with a bundle of nonprotuber-

ant, supplementary setae whose tips converge; 2 or 3 ventral setae per regular (protuberant) bundle; segments without secondary annuli .... *Enchytraeus cryptosetosus*

– Without supplementary setae; 5 ventral setae per bundle; each segment with a number of secondary annuli .... *Enchytraeus multiannulatus*

16. Penial bulb with 15–20 accessory glands .... *Enchytraeus albidus*

– Penial bulb simple, without accessory glands .... *Enchytraeus pugetensis*

17. Setae bent proximally; living worms green .... *Lumbricillus mirabilis*

– Setae sigmoid; living worms white, pink, or red 18

18. Proximal half of spermathecal duct with distinct, spiral ridges (fig. 9) .... *Lumbricillus vancouverensis*

– Spermathecal duct simple 19

19. Sperm in spermathecae in discrete, spoon-shaped bundles (fig. 10) ....*Lumbricillus georgiensis*

– Sperm in spermathecae in random masses or spirals, never in bundles 20

20. Sperm funnels about 1.5 times as long as wide .... *Lumbricillus qualicumensis*

– Sperm funnels 3–4 times as long as wide *Lumbricillus belli*

LIST OF OLIGOCHAETA

ENCHYTRAEIDAE

Enchytraeus albidus Henle, 1837. A cosmopolitan species with wide ecological tolerances, usually found under decaying seaweed or stones near high-tide line; also inhabits salt marshes, compost heaps, sewage beds, and effluents.

Enchytraeus cryptosetosus Tynen, 1969. Under decaying seaweed near high-tide line.

Enchytraeus multiannulatus Altman, 1936. Under decaying plant material at edge of salt marshes; sexually mature in spring.

Enchytraeus pugetensis Altman, 1931. In decaying seaweed on gravel and shell-fragment substrates; intertidal.

Lumbricillus belli Tynen, 1969. Intertidal in coarse sand under decaying seaweed.

Lumbricillus georgiensis Tynen, 1969. Under decaying seaweed in sand and small shingle near high-tide line.

Lumbricillus mirabilis Tynen, 1969. Under decaying seaweed in mixture of sand and pebbles near high-tide line.

Lumbricillus qualicumensis Tynen, 1969. Under decaying seaweed in coarse sand near high-tide line.

* *Lumbricillus santaeclarae* Eisen, 1904. Estuarine, under debris and decaying driftwood along the shore; Lake Merritt, Oakland. (J. T. Carlton).

Lumbricillus vancouverensis Tynen, 1969. Under decaying seaweed in sand and small shingle near high-tide line.

Marionina subterranea Knöllner, 1935. A very small worm, reported interstitially in sand near high-tide level, Half Moon Bay (R. E. Mesick), but probably widely distributed.

NAIDIDAE

Chaetogaster diaphanus (Gruithuisen, 1828). Sometimes in brackish water; probably predatory on smaller worms.

Chaetogaster limnaei von Baer, 1827. A cosmopolitan species found in fresh and brackish water; commensal in pulmonate gastropods, sometimes in the kidney, but more usually in the mantle cavity; probably feed on cercariae.

Nais communis Piguet, 1906. Sometimes in brackish water.

Paranais frici Hrabe, 1941. In brackish water.

Paranais litoralis (Müller, 1784). In brackish to fully marine habitats, intertidal and subtidal.

TUBIFICIDAE

Illyodrilus frantzi capillatus Brinkhurst and Cook, 1966. Sometimes subtidal in brackish water.

Limnodrilus hoffmeisteri Claparède, 1862. A cosmopolitan and very abundant tubificid, predominantly in fresh water but often in brackish habitats; subtidal.

Monopylephorus irroratus (Verrill, 1873). Brackish water and marine, intertidal and subtidal; often associated with decaying plant material and freshwater seepages.

Peloscolex apectinatus Brinkhurst, 1965. A marine subtidal species, apparently fairly intolerant of brackish water, but tolerant of organic pollution.

Peloscolex gabriellae Marcus, 1950. A broadly defined and problematical species of uncertain specific limits; probably amphiamerican, brackish water and marine, intertidal, subtidal and abyssal; tolerant of organically polluted water.

Peloscolex nerthoides Brinkhurst, 1965. Subtidal, brackish water and marine; tolerant of organic pollution.

* Not in key.

REFERENCES ON OLIGOCHAETA

Altman, L. C. 1936. Oligochaeta of Washington. Univ. Wash. Publ. Biol. 4: 1–137.

Brinkhurst, R. O. 1964. Studies on the North American aquatic Oligochaeta I: Naididae and Opistocystidae. Proc. Acad. Nat. Sci. Philadelphia 116: 195–230.

Brinkhurst, R. O. 1965. Studies on the North American aquatic Oligochaeta II. Tubificidae. Proc. Acad. Nat. Sci. Philadelphia 117: 117–172.

Brinkhurst, R. O. and D. G. Cook 1966. Studies on the North American aquatic Oligochaeta III: Lumbriculidae and additonal notes and records of other families. Proc. Acad. Nat. Sci. Philadelphia 118: 1–33.

Brinkhurst, R. O. and B. G. M. Jamieson 1971. *Aquatic Oligochaeta of the World.* Edinburgh: Oliver & Boyd, 860 pp.

Brinkhurst, R. O. and M. L. Simmons 1968. The aquatic Oligochaeta of the San Francisco Bay system. Calif. Fish Game 54: 180–194.

Cook, D. G. 1971. The Tubificidae (Annelida, Oligochaeta) of Cape Cod Bay, II: Ecology and systematics, with the description of *Phallodrilus parviatriatus* nov. sp. Biol. Bull. 141: 203–221.

Cook, D. G. 1974. The systematics and distribution of marine Tubificidae (Annelida, Oligochaeta) in the Bahia de San Quintin, Baja California, with descriptions of five new species. Bull. So. Calif. Acad. Sci. 73.

Nielsen, C. O. and B. Christensen 1959. The Enchytraeidae; critical revision and taxonomy of European species. Natura Jutl. 8–9: 7–160.

Sperber, C. 1948. A taxonomical study of the Naididae. Zool. Bid. Uppsala 28: 1–296.

Stephenson, J. 1930. *The Oligochaeta.* Oxford, 978 pp.

Tynen, M. J. 1969. New Enchytraeidae (Oligochaeta) from the east coast of Vancouver Island. Canad. J. Zool. 47: 387–393.

CLASS HIRUDINEA

Hirudinea or leeches are highly specialized annelids which have neither setae nor parapodia, but are equipped with an anterior and a larger posterior sucker for attachment. The external body divisions (annuli) do not reflect the true number of internal segments. The body is usually elongated, highly muscular, and rather solid, and the coelom is reduced to a network of sinuses.

Most leeches occur in fresh water in ponds, lakes, or streams, under vegetation and rocks, although there are amphibious, terrestrial (largely in the tropics), and brackish water and marine species. Locally, leeches may be common on marine fishes and elasmobranchs (the MacGinities have reported *Branchellion* and *Pontobdella* on sting rays and skates in Elkhorn Slough), although crustaceans (such as shrimp) and other invertebrates may also serve as hosts; leeches may occasionally also be encountered in the intertidal zone. Nearly all are predators, specialized for piercing their prey and feeding upon blood or soft tissues.

REFERENCES ON HIRUDINEA

Knight-Jones, E. W. 1962. The systematics of marine leeches. *In* K. H. Mann, *Leeches (Hirudinea) Their Structure, Physiology, Ecology and Embryology*, Pergamon, Appendix B, pp. 169–186.

Meyer, M. C. and A. A. Barden, Jr. 1955. Leeches symbiotic on Arthropoda, especially decapod Crustacea. Wasmann J. Biol. 13: 297–311.

ARCHIANNELIDA

Colin O. Hermans
California State College, Sonoma

(PLATE 19)

Archiannelids are small polychaetes (less than 1 mm to over 20 mm), with parapodia either reduced or lacking, with a ventral band of cilia for locomotion (except in Polygordiidae), and often lacking setae. They are characteristic of the interstitial habitat, which comprises the spaces between sand grains, algal debris, etc., and hence are part of the diverse assemblage of small forms making up the "meiobenthos." Such animals, because of their small size and obscure habitat, are generally overlooked in ordinary collecting, but with proper methods are found in large numbers.

The five families here considered as archiannelids appear to be more closely related to each other than to other polychaete groups, and show a combination of characteristics which represent adaptations to small size and the interstitial habitat. Taxonomically they can be regarded either as a group equivalent in rank to the Polychaeta Errantia and Sedentaria, or to the orders recognized by Dales (1967).

The archiannelid fauna of the central California intertidal zone is very poorly known, but representatives of all five families can be expected to be found if a proper search is made. The basic problem in collecting is to separate the tiny meiobenthic animals from their substratum; methods for handling meiobenthos are discussed by Huling and Gray (1971). Representative genera are shown in figures 1–13.

KEY TO FAMILIES OF ARCHIANNELIDA

1. Without parapodia 2
– With parapodia 4
2. Without setae or ventral ciliation; 2 stiff tentacles lacking ampullary apparatus at base; cuticle thick and iridescent; body turgid and nematode–like .... **Polygordiidae**
(example: *Polygordius*, fig. 1)

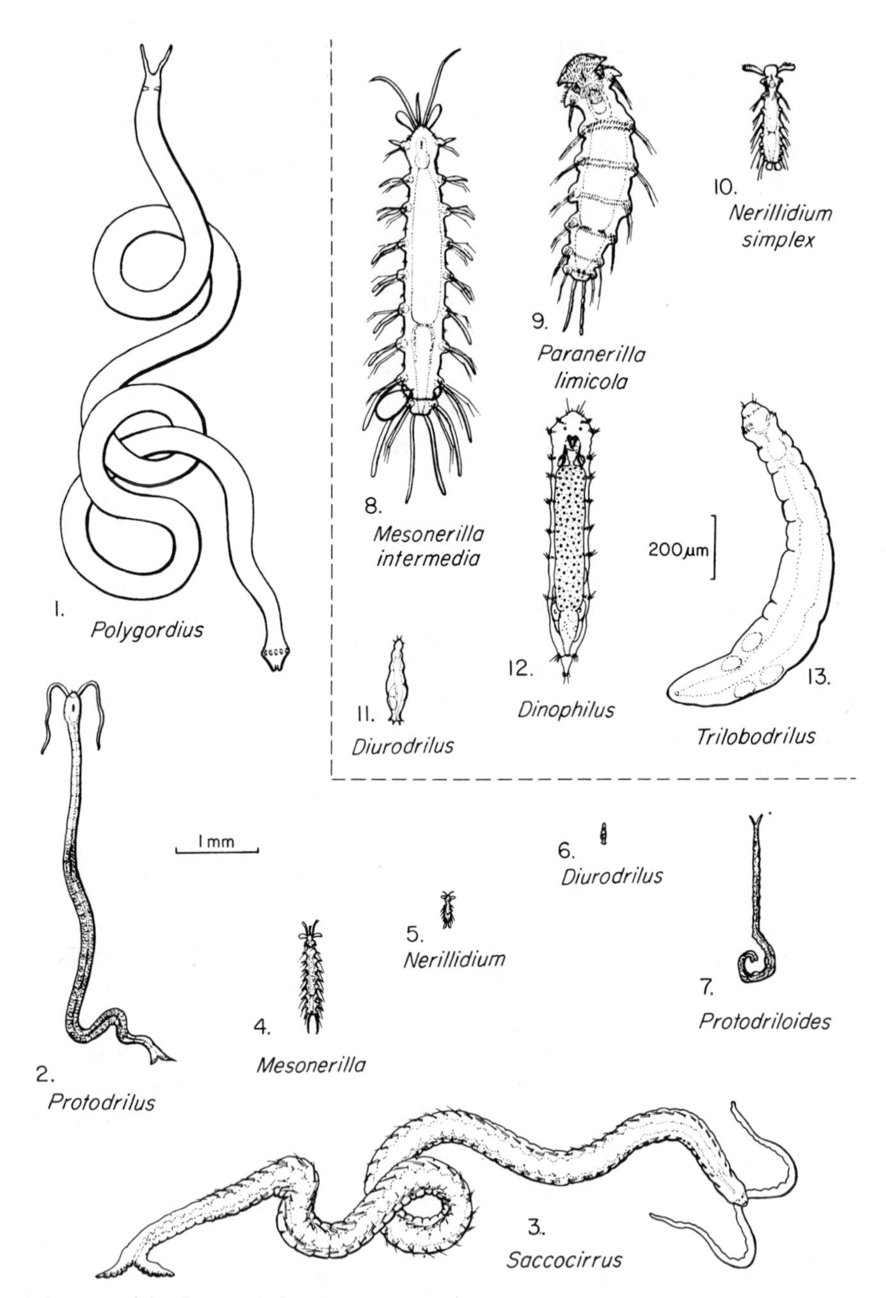

PLATE 19. **Archiannelida.** Representative genera, to scales indicated: 1–11, 13 after Jouin, 1971; 12 after Bush in Smith, 1964.

– With ventral ciliation; buccal diverticulum muscular; ventral nerve bands (VNB) separate 3
3. Most without setae; 2 tentacles, most with ampullae; VNB unsegmented; segmental gonoducts only in males, these without penes .... **Protodrilidae**
(examples: *Protodrilus*, fig. 2; *Protodriloides*, fig. 7)
– Lacking setae or tentacles; VNB segmented, gonoducts in 1 or 2 pairs or unpaired; genital pores of male median ventral unpaired, with penis ....**Dinophilidae**
(examples: *Dinophilus*, fig. 12; *Diurodrilus*, figs. 6, 11; *Trilobodrilus*, fig. 13)
4. With more than 20 segments; with retractile, stubby parapodia containing a single bundle of setae; 2 tentacles with ampullae; with or without muscular buccal diverticulum (BD); ventral nerve bundles separate, segmented; segmental gonoducts present, in males with seminal vesicles and eversible penes, in females with seminal receptacles ....**Saccocirridae**
(example: *Saccocirrus*, fig. 3)
– With less than 10 segments; with short parapodia bearing 2 setal bundles and usually 1 cirrus; 2–5 tentacles, 2 of these spoon-shaped, without ampullae; BD muscular, sometimes with jaws; VNB separate, unsegmented; gonoducts of females opening in a pair (and males with up to 3 pairs) of ventral genital pores .... **Nerillidae**
(examples: *Nerilla; Mesonerilla* (figs. 4, 8); *Paranerilla*, fig. 9; *Nerillidium* figs. 5, 10)

LIST OF ARCHIANNELIDA

Dinophilus gyrociliatus Schmidt, 1857. In coarse sediment and algal films along Sonoma County coast and in marine aquaria.

Nerilla antennata Schmidt, 1848. Occasional in coarse sediments, but mainly known from marine aquaria.

Polygordius sp. Larvae occur in local plankton.

Protodrilus sp. Often found in coarse sandy beaches with *Saccocirrus sonomacus*.

Saccocirrus sonomacus Martin, 1977. Common in coarse sandy beaches; known at present only from north of San Francisco.

Several other genera and species are known from the Puget Sound region, and may well occur on the California coast; see Wieser, 1957.

REFERENCES ON ARCHIANNELIDA

Dales, R. P. 1967. *Annelids.* 2nd ed. London: Hutchinson University Press.

Hermans, C. O. 1969. The systematic position of the Archiannelida. Systematic Zool. 18: 85–102.

Huling, N. C. and J. S. Gray 1971. A manual for the study of meiofauna. Smithson. Contrib. Zool. 78: 1–84. Numerous references and methods.

Jouin, C. 1971. Status of the knowledge of the systematics and ecology of Archiannelida. *In* Proceedings of the First Internat. Conf. on Meiofauna, Smithson. Contr. Zool. 76: 47–56.

Martin, G. G. 1977. *Saccocirrus sonomacus,* N. Sp., a new archiannelid from California. Trans. Amer. Micr. Soc. 96: 97–103.

Remane, A. 1932. Archiannelida. In *Tierwelt der Nord – und Ostsee,* 6(a): 1–36.

Wieser, W. 1957. Archiannelids from the intertidal of Puget Sound. Trans. Amer. Micr. Soc. 76: 275–285.

PHYLUM ANNELIDA: CLASS POLYCHAETA

James A. Blake
Pacific Marine Station, University of the Pacific, Dillon Beach

(PLATES 20–53)

Polychaeta ("many setae") are numerous, diverse, almost entirely marine, and often constitute a major component of benthic communities.

There is no generally accepted scheme for dividing the class into orders. Some recognize an order of more or less sedentary forms (**Polychaeta Sedentaria**) and another of mostly free-living types (**Polychaeta Errantia**). Such a division is, however, highly artificial since polychaete families represent a wide spectrum of adaptations, from entirely free-swimming, through crawlers, burrowers, temporary tube builders, to those which construct permanent tubes. Dales (1967) has developed a scheme of orders (based largely on the pharyngeal apparatus) which, with some exceptions, is gaining acceptance. In practice, however, it is more useful to recognize the principal family types such as nereid, polynoid, spionid, syllid, sabellid, serpulid, and so on.

GENERAL MORPHOLOGY

In order to make clear the basic structure of polychaetes, the morphology of free-living forms is treated first. The structure of free-living polychaetes is perhaps best understood by taking a typical example, describing its structure, and pointing out how other families deviate from this plan.

The body is generally elongated with numerous segments, and consists of a **prostomium** (anterior cephalic lobe), a **metastomium** (the following body segments) and a **pygidium** (the last segment).

The heads of polychaetes are exceedingly diverse. As an example of the head of an errant polychaete, see *Nereis* (figs. 1, 2). The head, in this genus, consists of a preoral prostomium, provided at its anterior margin with a pair of small **antennae,** and at its sides with paired, fleshy, biarticulated **palps.** The prostomium of *Nereis* bears two pairs of eyes; other polychaetes may have one pair or none; in some, numerous eyespots are scattered on the peristomium or even on the tentacles or sides of the body.

The segment just behind the prostomium is the **peristomium** (in *Nereis*, a fusion of two segments). It bears, in *Nereis*, four pairs of **tentacular cirri** on short stalks at its anterior margin. The usage of the terms palp, antenna, tentacle, and cirrus varies greatly. **Antennae,** unless otherwise specified, are usually dorsal or marginal on the prostomium. **Palps** are usually associated with the mouth, and tend to be lateral or ventral to the prostomium and border the anterior margin of the mouth. However, certain dorsolateral structures, especially if these are large, elongated, grooved, or prehensile as in the spionids, are frequently called palps. The term **cirrus** is usually applied to structures arising dorsally or ventrally on the parapodia, whereas comparable structures on the anterior part of the body, if elongated, may be designated tentacular cirri (or peristomial tentacles). **Tentacle** is a very general term used to signify any of a variety of elongated sensory or feeding structures, usually on the head.

In most free-living polychaetes the pharyngeal region may be everted to form a **proboscis,** which in *Nereis* bears stout **jaws** and small, horny teeth **(paragnaths).** The proboscis of *Nereis* is divisible into two external regions (figs. 1, 2): (1) an **oral ring** external to the mouth (internal when proboscis is retracted) and divisible into six areas, numbered V to VIII; and (2) a second ring, the **maxillary ring** with stout, horny jaws at its outer end, and divisible into six areas, numbered I to IV. Even-numbered areas are paired; odd-numbered areas lie in the dorsal and ventral midlines. A study of the arrangement of paragnaths is necessary in taxonomic work on nereids; in other families, the proboscis may be smooth or covered with soft papillae.

The jaw pieces of some polychaetes are complex (e.g., Eunicidae, Arabellidae, Lumbrineridae, Onuphidae, Dorvilleidae) and composed of several parts, each of which may have numerous small teeth (figs. 5, 6). These are important characteristics in taxonomic work.

The body segments bear conspicuous **parapodia.** The parapodia of *Nereis* are paired locomoter appendages, each of which is composed of an upper lobe, the **notopodium,** and a lower **neuropodium** (fig. 3). Each lobe typically contains a bundle of slender, projecting, chitinous **setae** and a larger, dark, internal spine or **aciculum** (fig. 7). The shape, size, number, and position of setae are important in classification. Arising from the bases of the notopodium above and the neuropodium below, are often slender, flexible outgrowths, the **dorsal cirrus** (pl., cirri) and **ventral cirrus,** respectively. The notopodium consists of a dorsal and a middle lobe, between which are an aciculum and projecting setae; the neuropodium consists of a neuroacicular lobe, provided

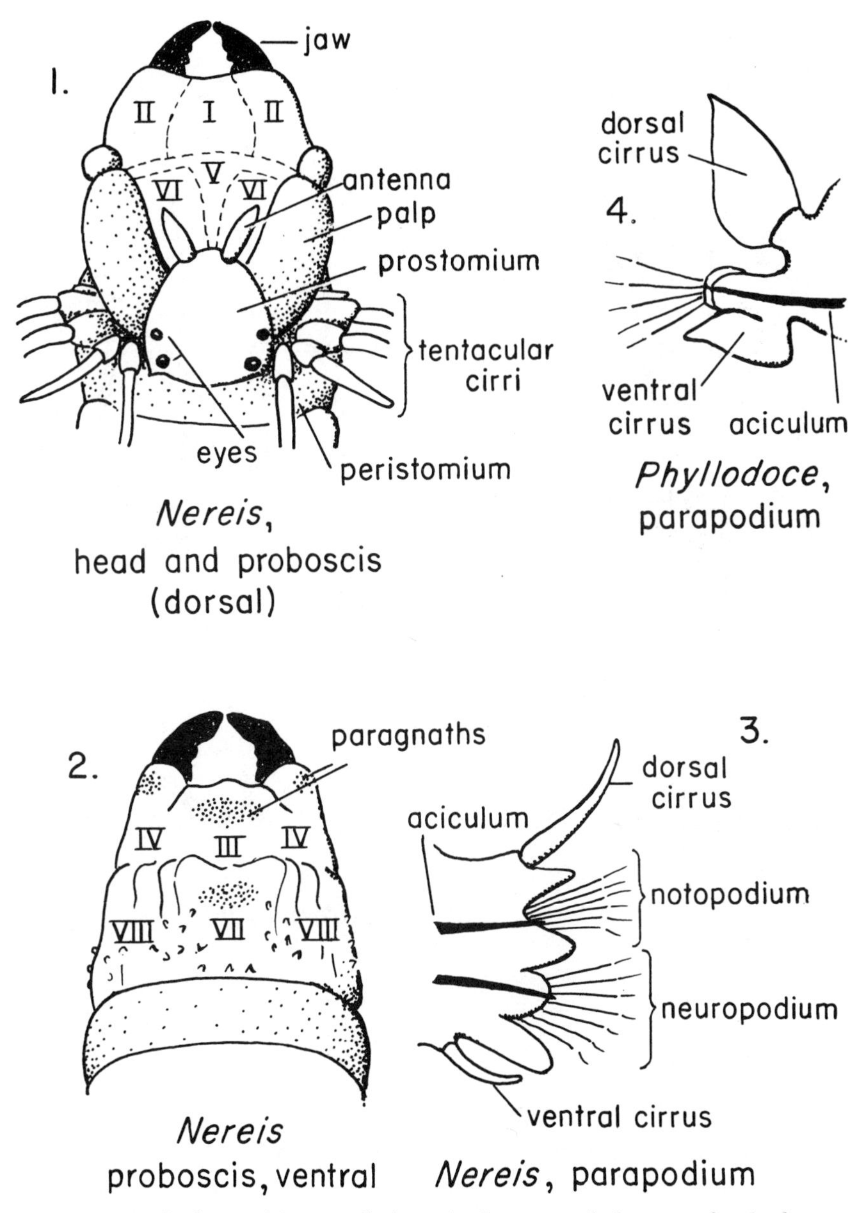

PLATE 20. **Polychaeta** (1). Morphology (unless stated, figures of polychaetes are by Blake, original): 1, head of *Nereis* with everted proboscis, dorsal; 2, same, ventral; 3, parapodium, *Nereis;* 4, parapodium, *Phyllodoce.*

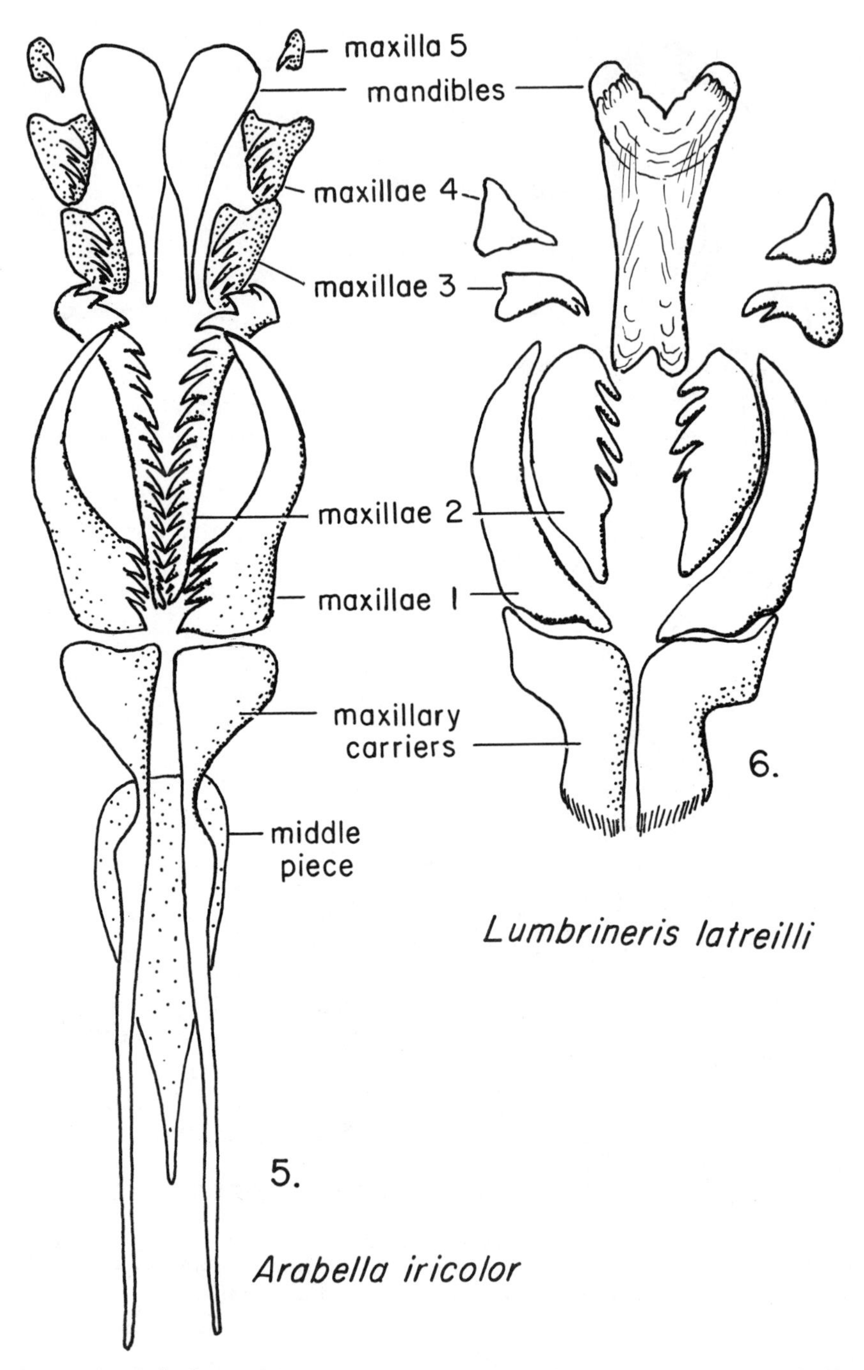

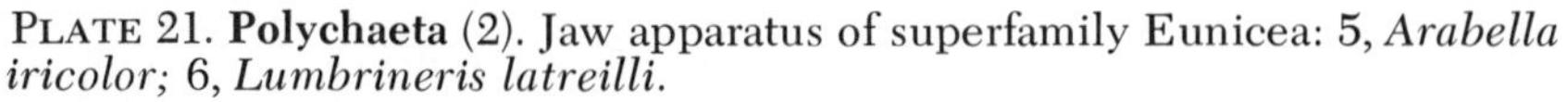
PLATE 21. **Polychaeta** (2). Jaw apparatus of superfamily Eunicea: 5, *Arabella iricolor;* 6, *Lumbrineris latreilli.*

with setae and an aciculum, and a ventral lobe. In many polychaetes, dorsal gills or **branchiae,** made conspicuous by red or green blood within, arise from the bases of the parapodia in certain parts of the body.

The first parapodia of *Nereis* are borne on the segment behind the peristomium. The first two pairs of parapodia are **uniramous,** that is, each contains a single aciculum and **setal fascicle** (bundle); all succeeding parapodia are **biramous.** In other polychaetes the parapodia vary: some are only uniramous (notopodium lacking) (fig. 4); others only biramous; some with both types; some with parapodia greatly reduced or absent **(apodous).**

Setae vary widely in form and furnish precise characteristics for determination of species. Because microscopic examination is necessary, the use of setal characteristics has been avoided where possible in the keys, but is necessary in some cases. Common setal types and their names should be recognized. Figures 7–48 give some idea of the diversity of setal form. We may distinguish **simple setae** (figs. 8–26, 35–48) from **composite** or **jointed setae** (figs. 27–34). Long, slender, simple setae are called **capillary setae.** The tips of setae, whether simple or composite, may be **entire, bifid** (fig. 26), or **trifid.** If bent like a sickle they are **falcate;** if flattened like an oar blade, **limbate** (fig. 15) or **bilimbate** (fig. 16). Some simple setae have stubby, bent, usually bifid ends called **hooks** or **crotchets** (figs. 35–38). These are usually relatively stout and grade into short, broadened types known as **uncini** (figs. 39–48), usually set in close rows, which are especially characteristic of Sedentaria.

Composite setae may be multiarticulate, as in the long bristles of some flabelligerids (fig. 27), but are characteristically two-jointed, composed of a shaft and a blade (fig. 28–34). This blade in turn may have various shapes and may itself be a hook (fig. 34). The blade rests in a notch in the end of the shaft. If the two sides of the notch are equal, it is **homogomph** (fig. 33); if unequal, **heterogomph** (fig. 32). Finally setae, either simple or compound, may be embedded at their tips in a clear matrix and are spoken of as hooded. Thus, we may have **simple hooded hooks** (figs. 26, 35) or **composite hooded hooks** (fig. 34).

Sedentary polychaetes depart widely from the body form of errant types. Prostomium and eyes are often reduced, proboscis and jaws usually absent, and the anterior end, especially in types dwelling in fixed tubes, greatly elaborated for feeding and respiration. In sabellids and serpulids the peristomial tentacles form a branchial crown of featherlike "gills," which serves both for feeding and respiration. Cilia pass water between branches of the plumes and transport food, entangled in mucus, down to the mouth. In other forms such as terebellids, the peristomial tentacles are long, filamentous, and prehen-

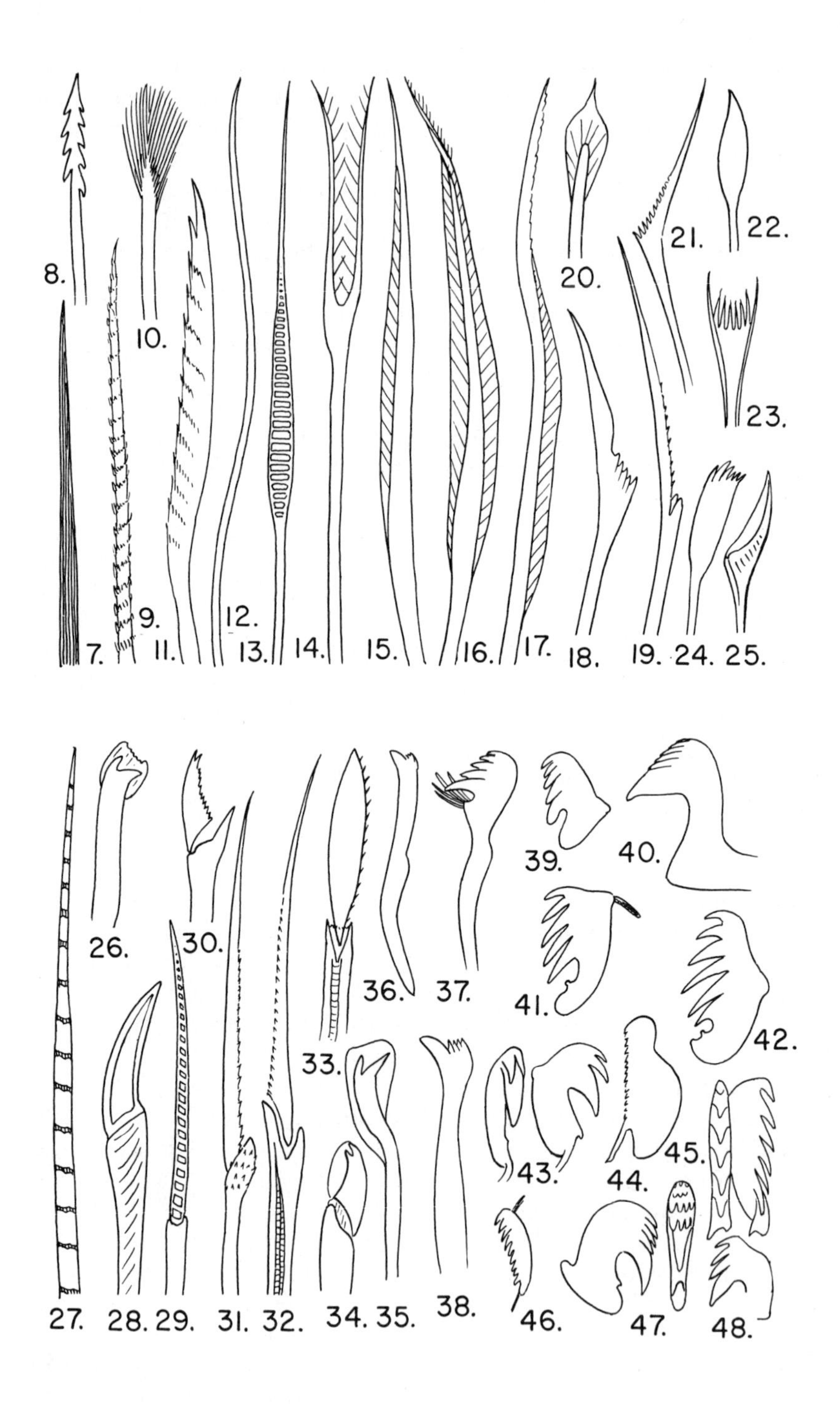
8.
10.
9.
7.
11.
12.
13.
14.
15.
16.
17.
18.
19.
20.
21.
22.
23.
24.
25.
26.
27.
28.
29.
30.
31.
32.
33.
34.
35.
36.
37.
38.
39.
40.
41.
42.
43.
44.
45.
46.
47.
48.

sile, serving to carry food by ciliary action in a groove running along each filament. Just behind the head, tufted, blood-filled branchiae may be present. Parapodia in tubicolous polychaetes tend to be small and are provided with rows of uncini for gripping the sides of the tube. Special glands may secrete tube-forming material. In serpulids, which form a rigid calcareous tube, one or more peristomial cirri may be modified to a plug-like operculum that can block the tube entrance. The body in tube-dwelling forms is often divisible into a more anterior and specialized **thorax** and a less specialized, posterior **abdomen.** The thorax may bear anteriorly a **collar,** which may be extended back to the posterior end of the thorax as a pair of folds, the **thoracic membranes.** The abdomen may be followed by a caudal region. The preceding accounts are of selected, free-living and tubicolous types, and can convey but a poor idea of the actual diversity of pattern of the numerous annelid families.

COLLECTION AND PRESERVATION

Polychaetes occur in a variety of habitats including sand and mud, under rocks, in algal holdfasts, amongst barnacles and mussels, on floats and wharves, and may even be found as borers in rocks and shells or as commensals of other invertebrates. When sorting polychaetes from sediments, samples should be washed through a 0.5–1.0 mm mesh screen to retain very small species.

Polychaetes should be relaxed or anesthesized prior to fixation. Magnesium chloride (7.5%) has commonly been used. However, propylene phenoxetol (Gallard-Schlesinger, New York) has been used with success in benthic studies (McKay and Hartzband, 1970, Trans. Amer. Micr. Soc. 89: 53–4). A stock solution of 1.5% propylene phenoxetol is made up in warm tap water. A working solution of 0.15% is prepared by mixing one part stock solution with nine parts sea water.

PLATE 22. **Polychaeta** (3). Types of setae (after Fauvel, 1923). Fig. 7, aciculum; Figs. 8–27, simple setae: 8, barbed; 9, spinous capillary; 10 brushlike; 11, serrated; 12, simple capillary; 13, camerated capillary; 14, furcate or forked; 15, limbate or winged; 16, bilimbate; 17, seta of *Apomatus;* 18–19, bayonet setae, Serpulidae; 20, spatulate (palea), Sabellidae; 21, geniculate or bent, Serpulidae; 22, lanceolate or styliform, Chaetopteridae; 23, pectinate, Eunicidae; 24, 25, paleae, Sabellariidae; 26, bifid acicular seta; 27, multiarticulate, Flabelligeridae; 28–34, compound setae: 28, Flabelligeridae; 29, Sigalionidae; 30, Syllidae; 31, Phyllodocidae; 32, heterogomph spiniger, Nereidae; 33, homogomph falciger, Nereidae; 34, Eunicidae; 35–48, simple hooks and uncini: 35, *Polydora;* 36, *Arenicola;* 37, Maldanidae; 38, *Trichobranchus;* 39–48, uncini: 39, *Serpula;* 40, avicular uncinus of Sabellidae; 41–45, Serpulidae; 46, *Chaetopterus;* 47, *Amphitrite;* 48, Sabellidae.

Worms relax immediately in this solution. They may be revived simply by returning them to sea water.

Fixation of free-living polychaetes is best in formalin (5–20% depending on the use of the material). Tubicolous worms generally have a softer cuticle and should be fixed in Bouin's fluid, Kahle's fixative, or a potassium dichromate fixative. These fixatives harden the cuticle and impart an excellent appearance to the body, making study easy.

During fixation, care should be taken to ensure a straight specimen. Coiled worms are difficult to study and break easily when manipulated. Straightening may be achieved by placing the living worm on a piece of paper towel or filter paper and adding the fixative with a pipette. The same results are achieved by using a plastic petri dish, adding the fixative a drop at a time and manipulating the specimen with forceps. If the worm everts its proboscis, this should be held with forceps to prevent inversion until the worm is fixed. After a suitable time in the fixative, worms are preserved in 70% ethyl or isopropyl alcohol.

DISSECTION OF JAWS AND MOUNTING OF PARAPODIA

For taxonomic purposes it is often necessary to dissect out the proboscis to examine jaws or other structures. Parapodia must always be examined under the microscope. The proboscis of a large worm may be dissected out with a pair of fine scissors; smaller worms require a pair of iris scissors, and minute specimens require the use of forceps and a microscalpel. The specimen is held with forceps while the razor-sharp scalpel cuts out the proboscis. The jaws of eunicid-like polychaetes may be removed along with the entire pharyngeal apparatus. A medial incision is made dorsally from about setigers 4–8, or wherever the jaws can be seen through the cuticle. The pharynx is then removed with forceps. Musculature is removed or trimmed under a dissecting microscope.

Jaws, pharyngeal structures, and parapodia must be mounted on slides for proper study. A simple technique is to place the structure to be mounted in 100% alcohol until dehydrated (5–10 minutes) and to mount it in a drop of Diaphane or Permount. This will clear the specimen and make a permanent slide. Diaphane can also be used with 95% alcohol. Small worms may be mounted whole by this method.

GENERAL COMMENTS ABOUT THE FAUNA AND THE KEYS

Approximately 275 species in 44 families occur in intertidal and shallow subtidal areas of northern and central California. The Spionidae is the largest family, with 35 species. Other large families and approximate numbers of species are: Syllidae, 24; Phyllodocidae, 19; Sabellidae, 18; Nereidae, 16; Polynoidae, 13; Terebellidae, 13; Serpulidae, 11; and Opheliidae, 10.

Although the polychaete fauna of central and northern California appears to be large, the northern sector of California is not as well known as comparable areas in the southern part of the state (Hartman, 1968). Microhabitats of the rocky intertidal zones have only recently been explored; bays and estuaries of the central and northern coasts are poorly known; and the interstitial fauna of sand and mud has been essentially ignored. It is probable that numerous additional species will be found with increased study. Some unidentified species have been included as "sp." with comments on their occurrence.

The keys to Polychaeta begin with a key to families. Since the ability to recognize families is of great practical value to the zoologist, the student should early learn the common family types. Each family is then considered separately. The families are arranged in the so-called natural order, beginning with errant families (scale-bearing families first) and concluding with the sedentary families (serpulids last). This arrangement is considered more desirable than an alphabetical sequence because it places related families together and makes comparisons of figures easier. Families may be located by referring to the page number following the family name in the Family Key, or by referring to the alphabetical Index to Polychaete Families on p. 160. Under each family are listed the species, and in most cases there are keys to genera and species. For more complete accounts of California polychaetes, the student is referred to Hartman (1968, 1969). Additional important general references include Berkeley and Berkeley (1948, 1952), Day (1967), Fauvel (1923, 1927), and Hartman (1959, 1965). For planktonic polychaetes, see Dales (1957).

GLOSSARY OF TERMS NOT DEFINED IN TEXT

aristate seta: a stout seta with smooth shaft and terminal tuft of fine hairs.

buccal: (adjective) of or pertaining to the mouth.

caruncle: a sensory lobe which is a posterior projection of the prostomium.

cephalic cage: long, forwardly directed setae which enclose and protect the head.

ceratophore: the basal joint of an antenna.

cirrophore: a basal projection on which a cirrus is mounted.

elytron: a dorsal, scalelike structure; plural is **elytra.**

epitoke: swimming sexual stage of a polychaete; usually with highly modified body structures.

esophageal caeca: lateral diverticula of esophagus as in arenicolids.
falciger: compound seta with a stout, curved end.
felt: matted hairs or fine setae, produced by notopodia in Aphroditidae; usually with entangled detritus.
foliaceous: leaflike.
nuchal epaulette: a raised and elongated sensory organ projecting posterolateral to the prostomium.
nuchal organ: a sensory organ on the prostomium or extending back from it, usually in form of a groove or ciliated ridge.
occipital: pertaining to the posterior part of the prostomium.
palea (-ae): a broad, flattened type of seta.
radiole: one of the main radii or tentacles of the branchial crown of a sabellid or serpulid; each radiole normally bears two rows of side branches or pinnules.
setiger: a seta-bearing segment of the body.
spatulate: flattened, bladelike seta with blunt end; may bear a terminal mucron (point) in some species.
spiniger: a compound seta whose blade tapers to a fine tip.
tubercle: an elongated papilla present in some scale worms.

INDEX TO POLYCHAETE FAMILIES

AMPHARETIDAE, 231
AMPHINOMIDAE, 177
APHRODITIDAE, 174
ARABELLIDAE, 203
ARENICOLIDAE, 226
CAPITELLIDAE, 225
CHAETOPTERIDAE, 216
CHRYSOPETALIDAE, 177
CIRRATULIDAE, 218
COSSURIDAE, 221
CTENODRILIDAE, 221
DORVILLEIDAE, 202
EUNICIDAE, 201
EUPHROSINIDAE, 178
FLABELLIGERIDAE, 221
GLYCERIDAE, 194
GONIADIDAE, 194
HESIONIDAE, 183
LUMBRINERIDAE, 202
MAGELONIDAE, 216
MALDANIDAE, 228
NEPHTYIDAE, 197
NEREIDAE, 190
NOTOPHYCIDAE, 178
ONUPHIDAE, 199
OPHELIIDAE, 222
ORBINIIDAE, 205
OWENIIDAE, 228
PARAONIDAE, 207
PECTINARIIDAE, 230
PEISIDICIDAE, 176
PHYLLODOCIDAE, 178
PILARGIIDAE, 184
POLYNOIDAE, 174
SABELLARIIDAE, 228
SABELLIDAE, 235
SCALIBREGMIDAE, 222
SERPULIDAE, 239
SIGALIONIDAE, 176
SPIONIDAE, 208
STERNASPIDAE, 224
SYLLIDAE, 184
TEREBELLIDAE, 232
TROCHOCHAETIDAE, 216

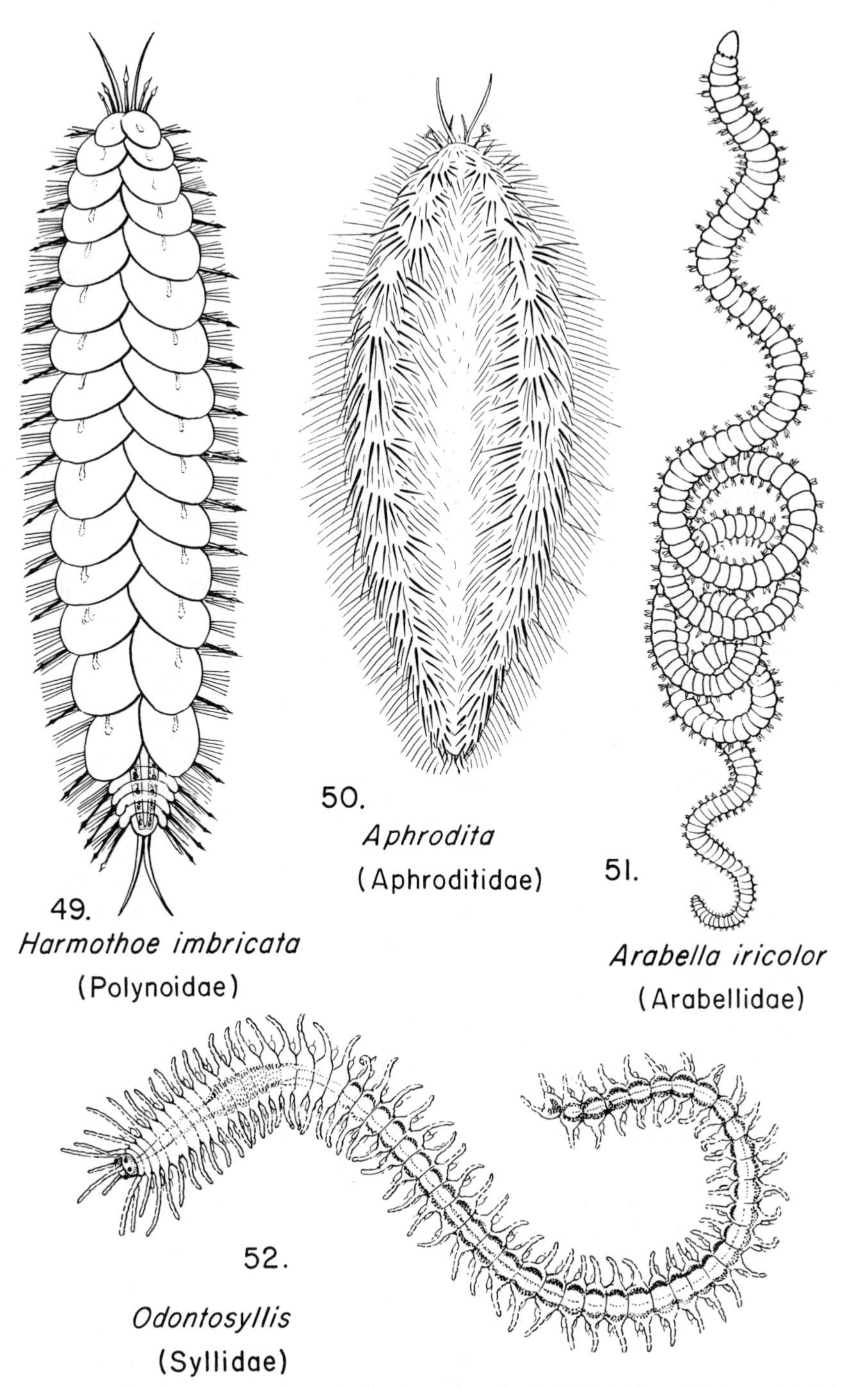

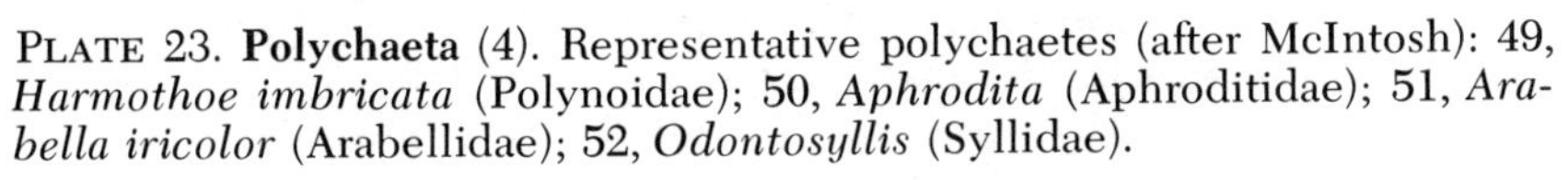

PLATE 23. **Polychaeta** (4). Representative polychaetes (after McIntosh): 49, *Harmothoe imbricata* (Polynoidae); 50, *Aphrodita* (Aphroditidae); 51, *Arabella iricolor* (Arabellidae); 52, *Odontosyllis* (Syllidae).

KEY TO FAMILIES OF POLYCHAETA

1. Dorsal surface more or less covered with overlapping elytra (scales), paleae, or felt 2
– Dorsal surface not covered with elytra, paleae, or felt 5
2. Dorsal surface more or less concealed by felt; body short, oval; sea mice (fig. 50) .... APHRODITIDAE (p. 174)
– Dorsal surface more or less concealed by paleae (fig. 86); animal usually minute .........CHRYSOPETALIDAE (p. 177)
– Dorsal surface more or less concealed by elytra 3
3. Compound neurosetae present 4
– Setae all simple, elytra and dorsal cirri alternate regularly from setiger 4 to about 23; thereafter every 2 elytra are followed by a dorsal cirrus (fig. 49) ... POLYNOIDAE (p. 174)
4. Elytra and dorsal cirri alternate regularly on all segments from setiger 5 to posterior end .... PEISIDICIDAE (p. 176)
– Elytra and dorsal cirri alternate in anterior segments, but in posterior segments elytra occur on all segments and completely replace cirri .... SIGALIONIDAE (p. 176)
5. Pharynx well developed, muscular, often armed with jaws or teeth; parapodia well developed, commonly bear compound setae, setal lobes supported by internal acicula; prostomium usually with sensory appendages (except in Lumbrineridae, Arabellidae, and *Exogone* of the Syllidae) 6
– Pharynx without jaws or teeth, usually saclike, not muscular; parapodia reduced, simple setae predominate, no acicula; prostomium seldom with sensory appendages and often fused with the peristomium, which may bear grooved palps, buccal cirri, or a branchial crown 20
6. Notosetae in transverse rows across dorsum (fig. 88) ....EUPHROSINIDAE (p. 178)
– Notosetae not in rows across dorsum 7
7. Prostomium extended posteriorly as a prominent caruncle (fig. 87) ........................AMPHINOMIDAE (p. 177)
– Caruncle not prominent 8
8. (*Note 3 choices*) Dorsal and ventral cirri flattened, leaflike, paddlelike, or globular; prostomium with 4 frontal antennae (fig. 53) and sometimes a medial one as well; tentacular cirri 2–4 pairs; parapodia uniramous (figs. 96, 100); setae compound .... PHYLLODOCIDAE (p. 178)
– Dorsal and ventral cirri short, attached laterally to large, bulbous pads; prostomium with 2 pairs of long, lateral antennae (fig. 89); parapodia biramous (fig. 90); setae compound ....NOTOPHYCIDAE (p. 178)

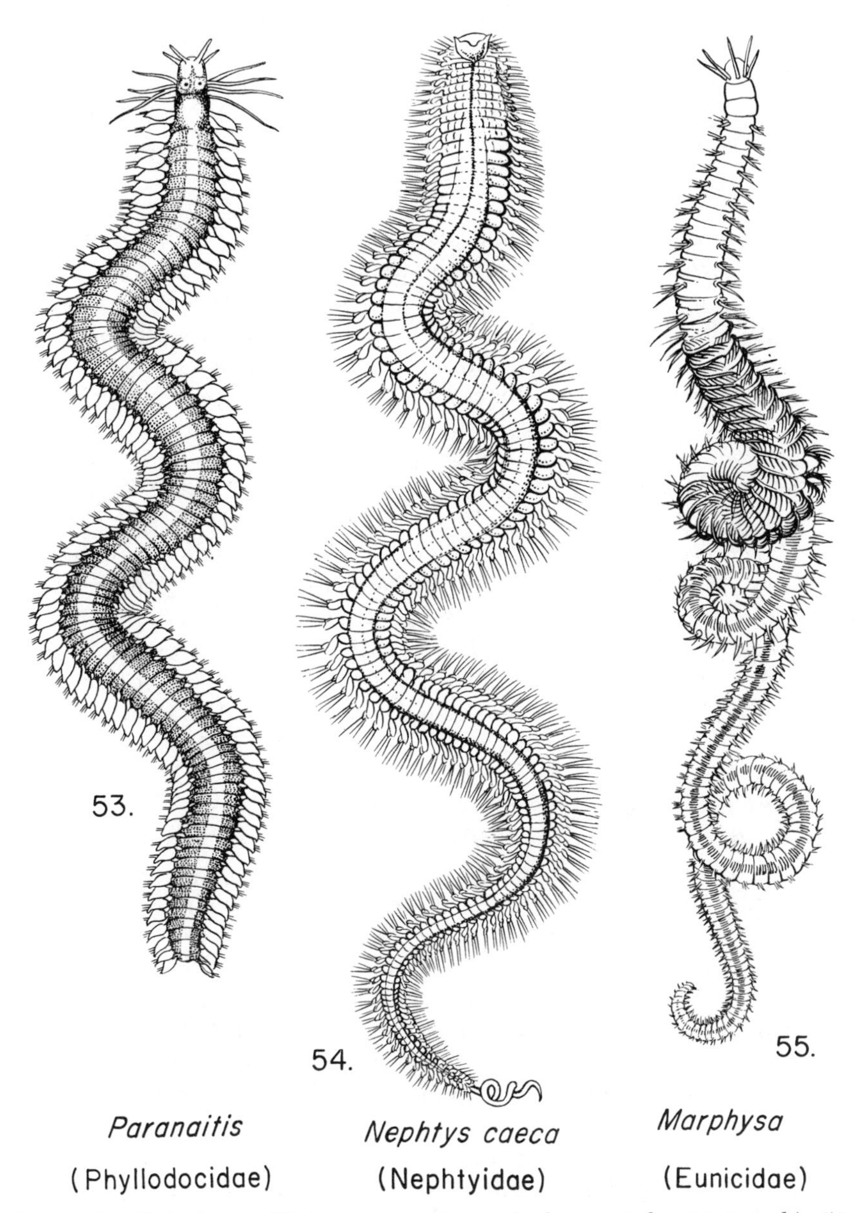

PLATE 24. **Polychaeta** (5). Representative polychaetes (after McIntosh): 53, *Paranaitis* (Phyllodocidae); 54, *Nephtys caeca* (Nephtyidae); 55, *Marphysa* (Eunicidae).

- Dorsal and ventral cirri if present, not leaflike or globular 9

9. Prostomium a pointed, annulated cone terminating distally in 4 minute antennae (figs. 151–153); peristomium fused with prostomium, without tentacular cirri; a large, powerful proboscis; burrows in sand or mud 10

- Prostomium otherwise 11

10. Body with parapodia similar throughout, either uniramous or biramous; dorsal cirri small, ventral cirri larger, conical; proboscis with 4 horny jaws with supports (fig. 151) .... GLYCERIDAE (p. 194)

- Body divided into 2–3 regions: 1) anterior region with uniramous parapodia; 2) transitional region in which notopodia gradually develop (or this region may be lacking); 3) posterior region with well-developed biramous parapodia; both dorsal and ventral cirri conical to fingerlike; proboscis with 2 large, toothed jaws and a circle of denticles (figs. 152, 153) .... GONIADIDAE (p. 194)

11. Prostomium flattened, pentagonal, with 4 small antennae (figs. 54, 165–169); body subrectangular in cross-section; biramous parapodia with rami well separated and with long cilia along the interramal border; notosetae and neurosetae arranged in fan-shaped fascicles, with more or less developed presetal and postsetal lamellae; burrow in sand or mud .... NEPHTYIDAE (p. 197)

- Prostomium, body, and parapodia otherwise 12

12. With an elaborate jaw apparatus consisting of a pair of ventral mandibles and dorsal maxillae consisting of few to numerous paired pieces (figs. 5, 6); with 1–2 achaetous and apodous tentacular or buccal segments, without tentacular cirri, or with only a single short, laterodorsal pair 13

- Jaws absent or otherwise; with 1–8 pairs of tentacular cirri (except in *Exogonella*) 17

13. Prostomium simple, conical or suboval, without antennae or distinct palps; parapodia without dorsal or ventral cirri; first 2 segments achaetous and apodous, without tentacular cirri; body smooth, elongate, cylindrical, resembling an earthworm; burrowing, carnivorous 14

- Prostomium suboval, with 1–7 antennae, 2 palps; parapodia with dorsal and ventral cirri; body otherwise 15

14. Neurosetae consisting of : 1) limbate setae with fine tips and 2) hooded hooks; jaw apparatus with 2 short, broad maxillary carriers (fig. 6), no median piece; eyes absent .... LUMBRINERIDAE (p. 202)

- Neurosetae consisting of limbate setae and with or without

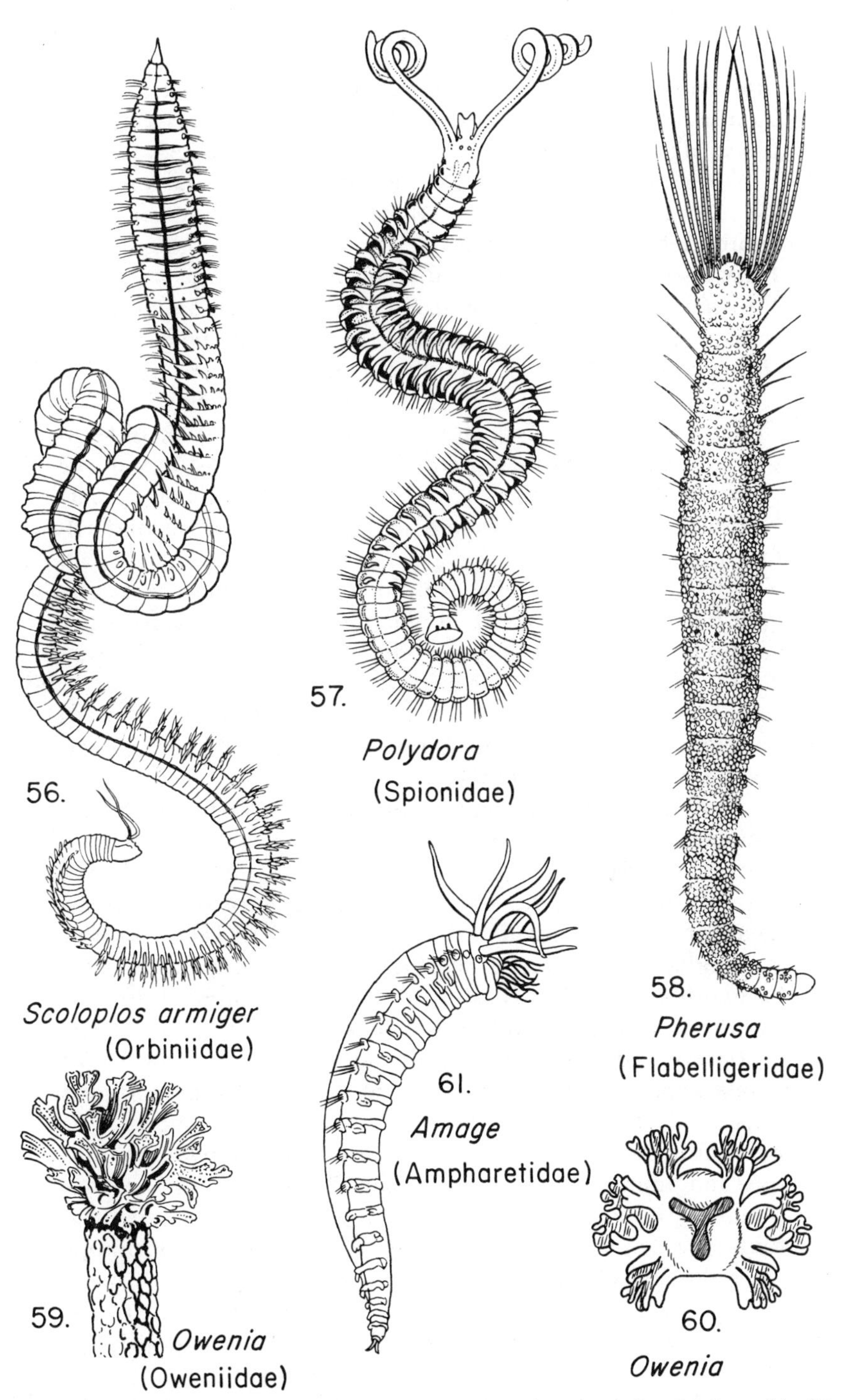

PLATE 25. **Polychaeta** (6). Representative polychaetes (after McIntosh): 56, *Scoloplos armiger* (Orbiniidae); 57, *Polydora* (Spionidae); 58, *Pherusa* (Flabelligeridae); 59, 60, *Owenia* (Oweniidae); 61, *Amage* (Ampharetidae).

projecting acicular setae; without hooks or crotchets; jaw apparatus with 2 long, slender maxillary carriers plus a median piece (fig. 5); eyes present (fig. 51) ARABELLIDAE (p. 203)

15. First segment apodous and achaetous; 7 prostomial antennae (5 long occipital, 2 short frontal, fig. 171); paired ventral palps short, globular; tube dwelling .... ONUPHIDAE (p. 199)
- First 2 segments apodous and achaetous 16

16. Prostomium with a pair of articulated antennae and a pair of long, curved ventral palps (figs. 183, 185) or both short in *Ophryotrocha* (fig. 184); crawlers and burrowers .... DORVILLEIDAE (p. 202)
- Prostomium with 1–5 occipital antennae and a pair of short, globular ventral palps more or less fused to prostomium (fig. 55); tube dwelling .... EUNICIDAE (p. 201)

17. Neurosetae compound (some may have blades secondarily fused to shaft) 18
- Neurosetae and notosetae simple, not compound (notosetae may be stout or hooked); tentacular segment apodous and achaetous, more or less fused with prostomium, usually with 2 pairs of small tentacular cirri (figs. 111, 112) .... PILARGIIDAE (p. 184)

18. Parapodia biramous or sub-biramous; notopodia at least represented by internal acicula 19
- Parapodia uniramous (may be biramous in sexual epitokes); tentacular segment apodous and achaetous, with 1–2 pairs of tentacular cirri; prostomium suboval with 3 antennae, 2 palps (fig. 52) (palps may be reduced or fused; prostomial appendages absent in *Exogonella*) .... SYLLIDAE (p. 184)

19. Parapodia with varying degrees of development of extra tonguelike lobes or ligules; prostomium suboval to subpyriform, with 2 frontal antennae and 2 biarticulate palps; proboscis with a pair of distal, dentate, hooked jaws; with single apparent tentacular segment bearing 3–4 pairs of cirri (figs. 1, 2); notosetae compound ...... NEREIDAE (p. 190)
- Parapodia without ligules; prostomium suboval to subquadrangular, with 2–3 antennae, 2 palps (may be biarticulate) (fig. 91); proboscis without jaws; with 1–4 achaetous tentacular segments and 2–8 pairs of tentacular cirri; notopodia reduced, with notosetae simple or lacking .... HESIONIDAE (p. 183)

20. Body short and stout; with a tuft of filamentous anal gills (fig. 65) .... STERNASPIDAE (p. 224)
- Body elongate; no anal gills 21

21. Head modified by development of frilly membranes, buccal tentacles, or a branchial crown of feathery tentacles around

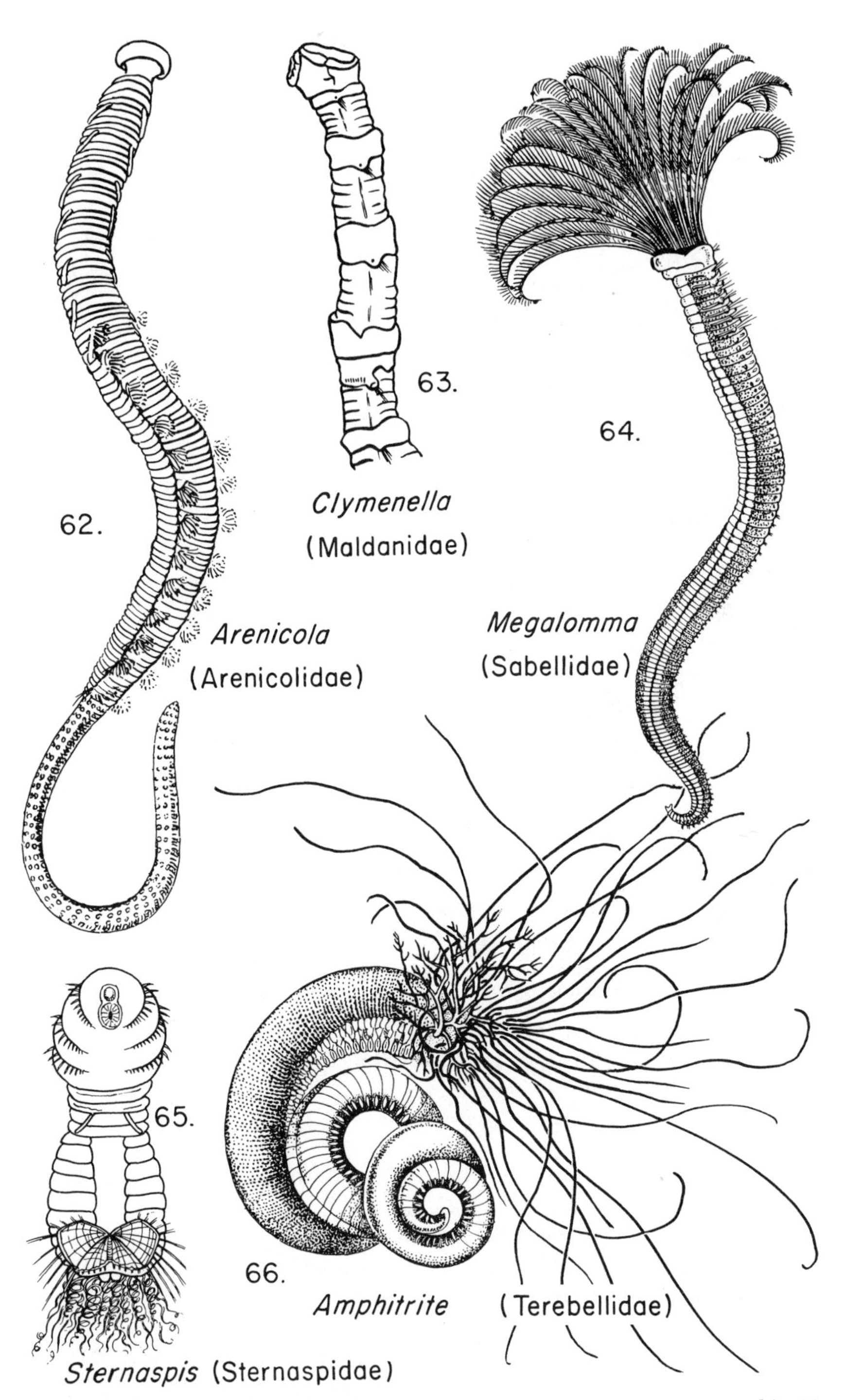

PLATE 26. **Polychaeta** (7). Representative polychaetes (after McIntosh): 62, *Arenicola* (Arenicolidae); 63, *Clymenella* (Maldanidae); 64, *Megalomma* (Sabellidae); 65, *Sternaspis* (Sternaspidae); 66, *Amphitrite* (Terebellidae).

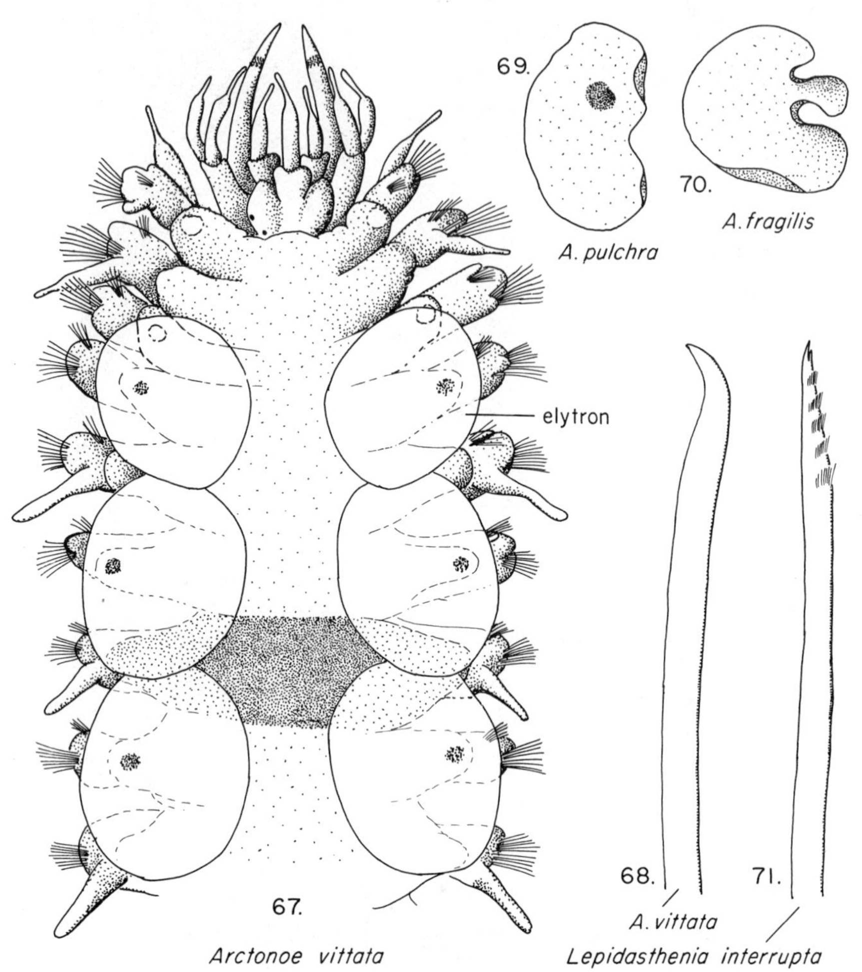

PLATE 27. **Polychaeta** (8). Polynoidae: 67, *Arctonoe vittata;* 68, neuroseta; 69, *A. pulchra,* elytron; 70, *A. fragilis,* elytron; 71, *Lepidasthenia interrupta,* neuroseta; 69 after Johnson.

the mouth; prostomium often reduced and indistinguishable from buccal segment 35

– Head not greatly modified; prostomium usually well developed and obvious; buccal segment sometimes with parapodia and may bear a pair of adhesive palps or a few grooved tentacles 22

22. Buccal segment with tentacles retractile into mouth (fig. 61) .... AMPHARETIDAE (p. 231)

– Buccal segment otherwise 23

23. Buccal segment with a pair of adhesive palps or several grooved tentacles located on anterior setigers 24
– Buccal segment without appendages 28
24. Anterior end with a pair of papillose, adhesive palps (fig. 254); head flattened and spadelike; gills absent ....MAGELONIDAE (p. 216)
– Palps, if present, not papillose or adhesive; head not flattened; gills often present 25
25. Body divided into 2 or 3 distinct regions 26
– Body not divided into distinct regions 27
26. Body divided into 3 distinct regions; prostomium reduced; peristomium with a large lip; setiger 4 bears large, modified setae; inhabit distinctive tubes CHAETOPTERIDAE (p. 216)
– Body divided into 2 regions; prostomium large, with a single occipital tentacle; 2 anterior pairs of parapodia directed forward, covering sides of prostomium (fig. 259); neuropodia of setigers 2 and 3 bear stout acicular setae; not in distinctive tubes ......................TROCHOCHAETIDAE (p. 216)
27. Prostomium with 2 dorsolaterally grooved palps, often long and coiling (fig. 57); neuropodia and/or notopodia of posterior setigers bear hooded hooks; some, none, or many segments with paired branchiae ....SPIONIDAE (p. 208)
– Prostomium usually lacking appendages; first setigerous segment often bearing 1 pair of large palpi or numerous tentacular filaments; numerous long, filamentous gills present on body setigers .....................CIRRATULIDAE (p. 218)
28. Multidentate hooks (sometimes with hoods) present at least in posterior setigers (fig. 37) 29
– Unique serrated setae present (fig. 275); body minute; usually only 9–11 segments (fig. 274) . . CTENODRILIDAE (p. 221)
– Setae not of this type 31
29. Multidentate hooks with hoods; body resembles an earthworm .... CAPITELLIDAE (p. 225)
– Multidentate hooks without hoods; body not resembling an earthworm 30
30. Body segments elongated, with body appearing jointed, but never annulate (fig. 63); gills rare; construct sand- or mud-covered tubes .... MALDANIDAE (p. 228)
– Body segments not elongated, always annulated; gills present (fig. 62); form L- or U-shaped burrows in mud .... ARENICOLIDAE (p. 226)
31. A single long, filiform gill arising from dorsum of setiger 2 or 3 (fig. 273) .... COSSURIDAE (p. 221)
– Gills, if present, in pairs along the body segments 32

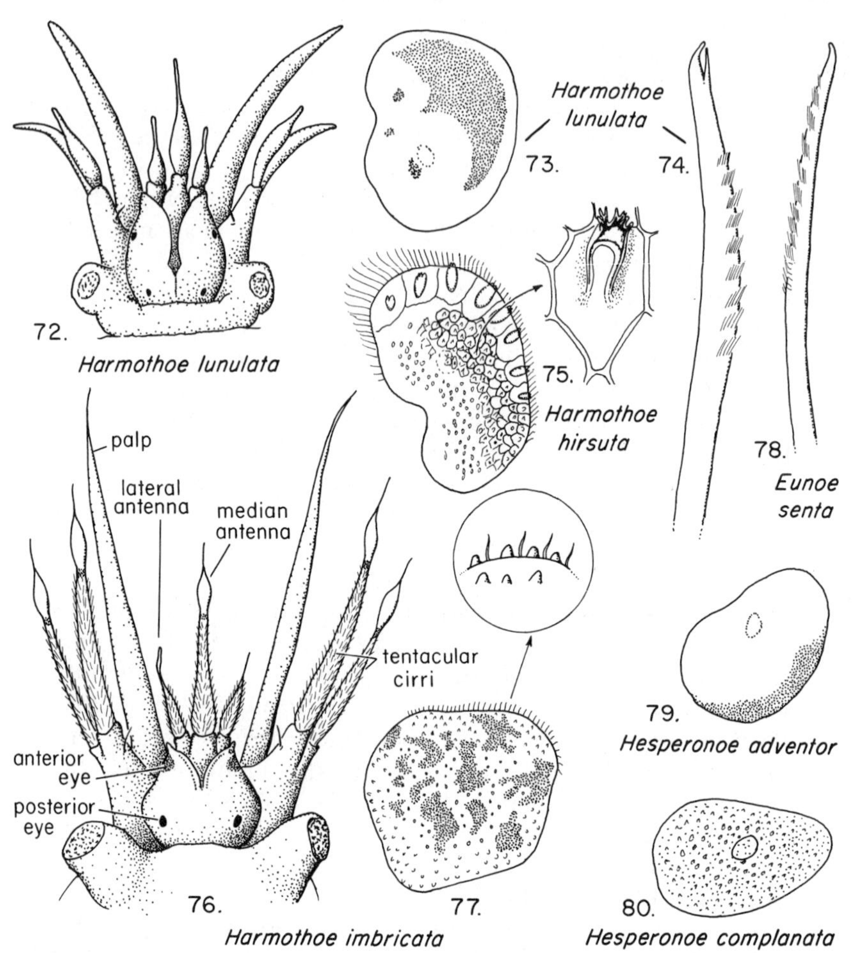

PLATE 28. **Polychaeta** (9). Polynoidae: 72–74, *Harmothoe lunulata:* 72, anterior end in dorsal view, elytra omitted; 73, elytron; 74, neuroseta; 75, *H. hirsuta* elytron, inset shows detail of tubercle; 76–77, *H. imbricata:* 76, anterior end in dorsal view, elytra omitted; 77, elytron, inset shows detail of margin; 78, *Eunoe senta,* neuroseta; 79, *Hesperonoe adventor,* elytron; 80, *H. complanata,* elytron; 75, after Johnson; 78, after Hartman; 79 after Skogsberg.

32. Body sleek, ventral groove often present (fig. 279); segmental eyes sometimes present; prostomium a sharply tapered cone; a pair of evaginable nuchal organs present ....OPHELIIDAE (p. 222)
– Body otherwise; segmental eyes absent; prostomium conical, blunt, or lobed, not sharply tapered; nuchal organs not evaginable 33
33. Prostomium notched or lobed on frontal margin; body often with a rough appearance, swollen anteriorly, branchiae if

present branched and restricted to anterior segments (fig. 278) ....................SCALIBREGMIDAE (p. 222)

– Prostomium conical or rounded; body not rough in appearance or swollen anteriorly; branchiae dorsally directed and distributed over a long body region 34

34. Parapodia with internal acicula, lobes well developed; often elongate; branchiae continue to posterior end of body; capillary setae crenulated, often arranged in palisades; prostomium without medial antenna (fig. 56) .... ORBINIIDAE (p. 205)

– Parapodia without internal acicula, lobes reduced; branchiae absent from posterior end of body; setae all smooth or faintly striated, not crenulated; prostomium with medial antenna in some genera; small, threadlike worms (fig. 253) .... PARAONIDAE (p. 207)

35. Head terminates in a frilled membrane (figs. 59, 60); body enclosed in tube of closely fitting sand grains .... OWENIIDAE (p. 228)

– Head has tentacles, palps, or branchial crown of feathery tentacles around the mouth 36

36. Head bearing heavy setae or paleae 37

– Head without setae or paleae 39

37. Body regions indistinct; multiarticulate composite setae present; capillary setae appearing annulated; cephalic cage often present (fig. 58) .... FLABELLIGERIDAE (p. 221)

– Body regions well defined; composite setae absent 38

38. Paleae in a single row; caudal region short and flattened; tube free, conical, formed of close fitting sand grains (figs. 303–306) ..................PECTINARIIDAE (p. 230)

– Paleae form 2–3 rows; caudal region long and cylindrical; rigid sand tubes attached to rocks or shells, often in dense colonies (figs. 293–302) .... SABELLARIIDAE (p. 228)

39. Head with soft tentacles for deposit feeding; gills often present on anterior segments; setal types not inverted in posterior region 40

– Head with a crown of bipinnate radioles; gills absent in body segments; setal types inverted in posterior region 41

40. Tentacles retractile into mouth, either grooved or papillose (fig. 61) .... AMPHARETIDAE (p. 231)

– Tentacles not retractile into mouth, grooved but never papillose (fig. 66) .... TEREBELLIDAE (p. 232)

41. Tubes leathery or of mucus, sand or mud, operculum absent (fig. 64) .... SABELLIDAE (p. 235)

– Tubes calcareous; stalked operculum often present among radioles (figs. 344–354) .... SERPULIDAE (p. 239)

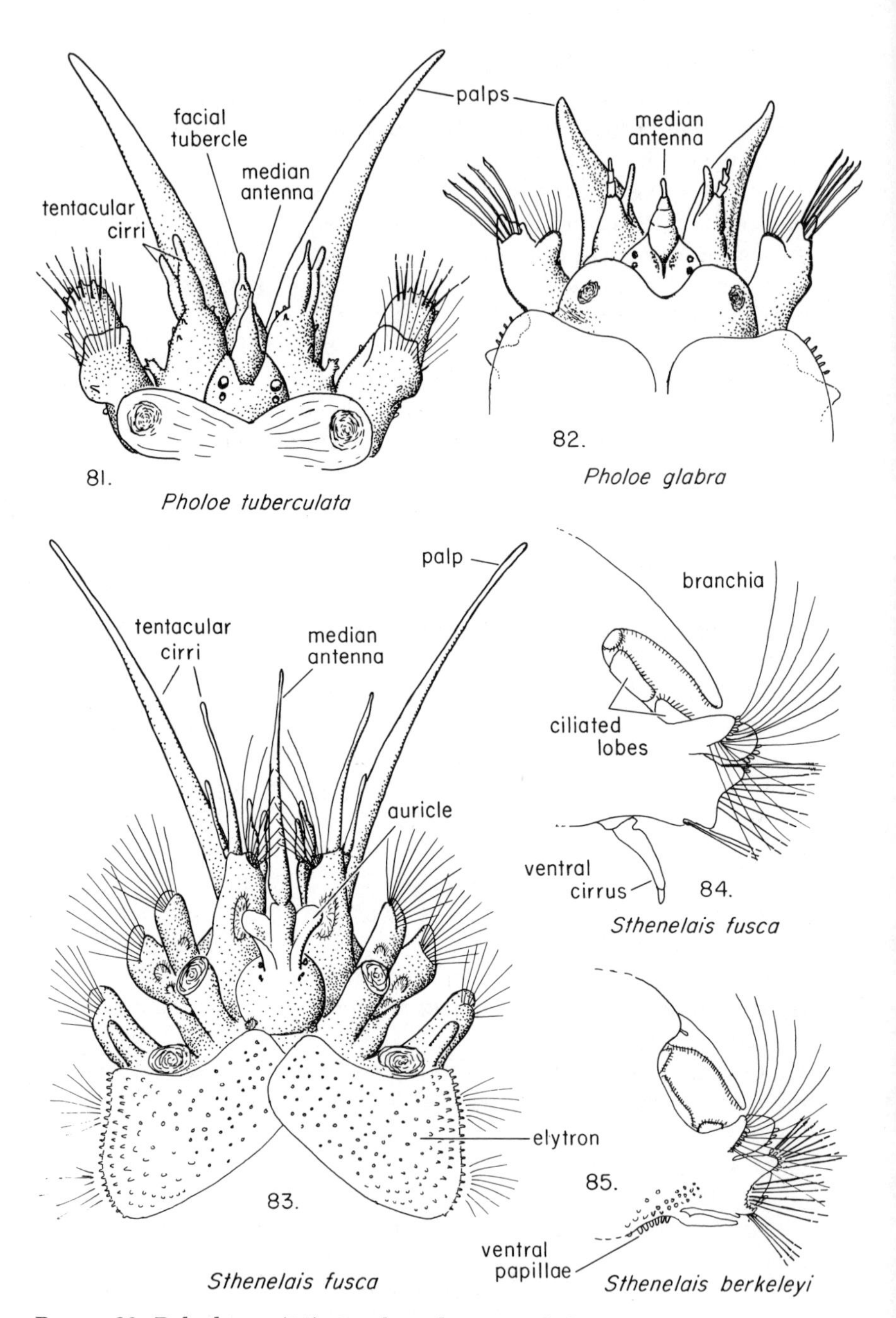

PLATE 29. **Polychaeta** (10). Sigalionidae: 81, *Pholoe tuberculata;* 82, *P. glabra;* 83, *Sthenelais fusca;* 84, *S. fusca,* parapodium; 85, *S. berkeleyi,* anterior parapodium; 82, after Hartman; 84–85 after Pettibone.

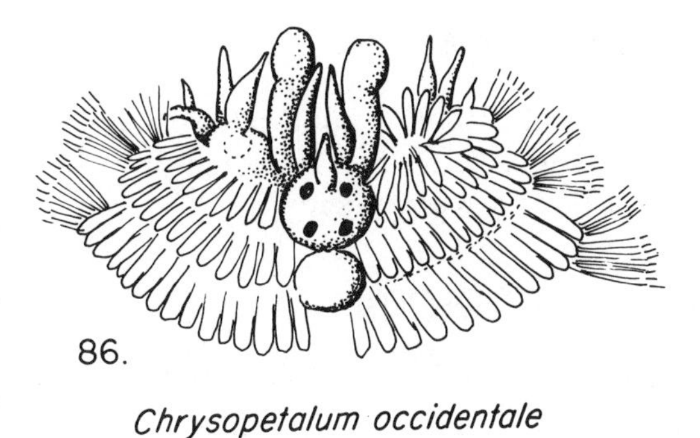

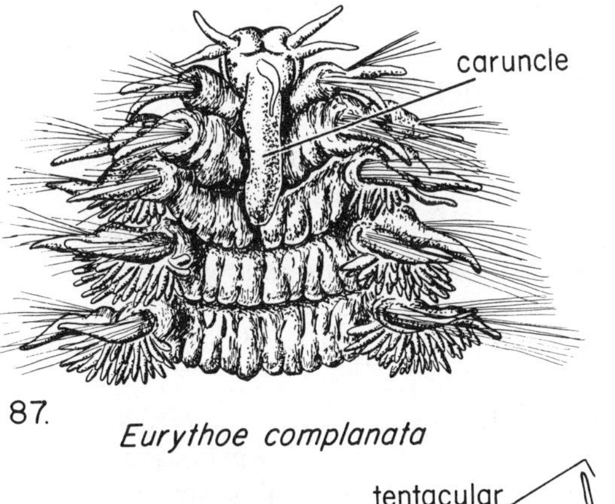

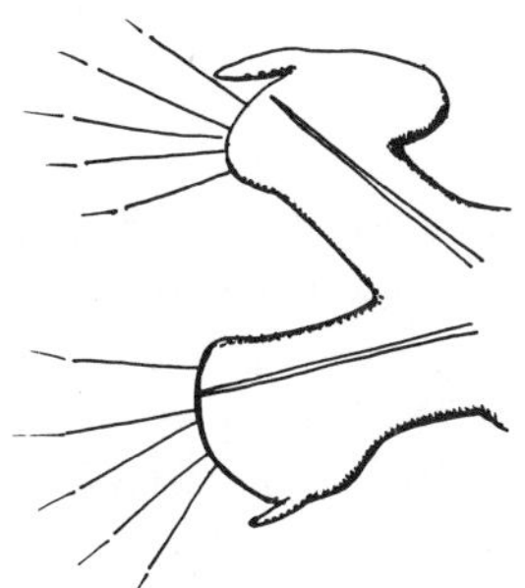

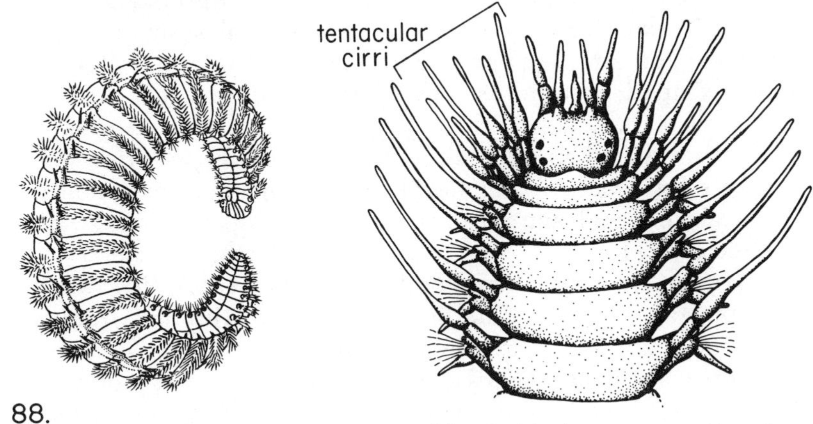

PLATE 30. **Polychaeta** (11). 86, *Chrysopetalum occidentale* (Chrysopetalidae); 87, *Eurythoe complanata*, (Amphinomidae); 88, *Euphrosine* (Euphrosinidae); 89, *Phyllodocella bodegae* (Notophycidae); 90, *P. bodegae*, 7th parapodium in anterior view; 91, *Ophiodromus pugettensis* (Hesionidae); 86 after Imajima and Hartman; 87 after Hartman; 88 after McIntosh; 89–90 after Fauchald and Belman.

APHRODITIDAE

Several species of *Aphrodita* occur in shallow subtidal areas off the California coast; they may be dredged and are occasionally washed ashore. See Hartman (1968, p. 14) for key to species.

LIST OF APHRODITIDAE

* *Aphrodita armifera* Moore, 1910. Dredged on silty or rocky bottoms.

* *Aphrodita castanea* Moore, 1910. Dredged on shallow bottoms.

* *Aphrodita refulgida* Moore, 1910. Dredged on shallow bottoms.

POLYNOIDAE

1. With 12 pairs of elytra; lateral antennae inserted terminally on anterior prolongations of the prostomium .... *Lepidonotus squamatus*

–. With 15 or more pairs of elytra; lateral antennae inserted ventral to the median antenna 2

2. With 18 pairs of elytra .... *Halosydna brevisetosa*

– With 15 pairs of elytra 5

– With 20 or more pairs of elytra 3

3. Elytra cover at least 2/3 of the dorsum on anterior segments (fig. 67), those on posterior segments noticeably smaller than those on anterior segments; dorsal cirri and antennae club- or pear-shaped; neuropodial setae falcate, sides smooth (fig. 68) .... *Arctonoe* 4

– Elytra reduced in size throughout, those in posterior segments not noticeably smaller than those in anterior segments; dorsal cirri and antennae cirriform; neuropodial setae serrated (fig. 71) .... *Lepidasthenia interrupta*

4. Elytra strongly frilled or folded at external margins (fig. 70); in ambulacral grooves of starfish .... *Arctonoe fragilis*

– Elytral margin smooth; broad, transverse, dark band across anterior segments (fig. 67); in branchial grooves of molluscs, with other invertebrates or free-living ..... *Arctonoe vittata*

– Elytral margin slightly undulate (fig. 69); commensal with holothurians .... *Arctonoe pulchra*

5. Notopodial and neuropodial setae each of one kind 7

– Notopodial and neuropodial setae each of two kinds, one stout and the other slender .... *Hesperonoe* 6

* Not in key.

6. Elytral surface smooth or with very few minute, low papillae (fig. 79); commensal with *Urechis* ... *Hesperonoe adventor*
– Elytra with numerous scattered, low papillae (fig. 80); commensal with *Callianassa* .... *Hesperonoe complanata*
7. Neuropodial setae entire at the distal end (fig. 78) 10
– Some neuropodial setae with bifid tips (fig. 74) *Harmothoe* 8
8. Anterior eyes far forward, under the prostomial peaks (fig. 76); elytral margin lightly fringed (fig. 77) *Harmothoe imbricata*
– Anterior eyes near the middle of the prostomium, clearly visible dorsally (fig. 72); elytra smooth or strongly fringed 9
9. Elytra strongly fringed along outer margins, and with large, hexagonal areas adorned with characteristic spines (fig. 75) .... *Harmothoe hirsuta*
– Elytra smooth, with a characteristic dark patch along the inner border (fig. 73) .... *Harmothoe lunulata*
10. Neuropodial setae long and slender, distal ends hairlike ... . *Antinoella* sp.
– Neuropodial setae with distal ends thicker, not hairlike (fig. 78); elytra conspicuously covered with furcated spines .... *Eunoe senta*

LIST OF POLYNOIDAE

See Hartman, 1939, 1968.

Antinoella sp. Two deep-water species are reported in Hartman (1968); an intertidal species has occasionally been referred to as *Antione* sp.

† *Arctonoe fragilis* (Baird, 1863). Commensal in ambulacral grooves of asteroids.

Arctonoe pulchra (Johnson, 1897). Commensal with asteroids, holothurians, molluscs, and terebellid polychaetes. See Dimock and Davenport 1971, Biol. Bull. 141: 472–484.

Arctonoe vittata (Grube, 1855). Commensal with asteroids, molluscs, or free-living. See Gerber and Stout 1968, Physiol. Zool. 41: 169–179.

† *Eunoe senta* (Moore, 1902). Rare.

Halosydna brevisetosa Kinberg, 1855 [= *H. johnsoni* (Darboux, 1899)]. The most common scaleworm in central and northern California; in rocky habitats; commensal with terebellids or molluscs as well as free-living. See Gaffney, 1973, Syst. Zool. 22: 171–175.

† In the characteristic "-oe" ending of the names of many polynoids and other worms, the "e" is pronounced, as indicated by the diacritic marks (-oë) formerly used but unfortunately eliminated in current nomenclatorial practice; thus, Hăr mo thó ē.

† *Harmothoe hirsuta* Johnson, 1897. More southern in distribution; in rocky habitats.

Harmothoe imbricata (Linnaeus, 1767). A cosmopolitan species usually free-living in rocky habitats. See Daly, 1972, Mar. Biol. 12: 53–56 (maturation and breeding biology in England); Cazaux 1968, Arch. Zool. Expér. Gen. 109: 477–543 (larval development).

Harmothoe lunulata (delle Chiaje, 1841). Free-living or commensal with holothurians or polychaetes; see Cazaux, 1968, above.

† *Hesperonoe adventor* (Skogsberg, 1928). With *Urechis* and *Upogebia*.

Hesperonoe complanata (Johnson, 1901). With *Callianassa* and *Urechis*.

Lepidasthenia interrupta (Marenzeller, 1902). Rare, free-living or commensal in maldanid tubes; see Cazaux, 1968, above.

Lepidonotus squamatus (Linnaeus, 1767). A cosmopolitan species free-living in rocky habitats.

PEISIDICIDAE

Peisidice aspera Johnson, 1897 is the only local member. This rare species is best known from Monterey, where it is found at low-tide mark crawling over stones or hiding in crevices. The species has been classified both as a sigalionid and polyodontid; the present tendency is to regard the genus as comprising a distinct family. The body is short (to 12 mm); prostomium lacks median antenna; neuropodia have composite setae; and the elytra have a heavily fimbriated margin. See Johnson, 1897, Proc. Calif. Acad. Sci. (3) 1: 153–198.

SIGALIONIDAE

1. Body small to minute; prostomium with one antenna (fig. 81); branchiae absent 2
– Body long; prostomium with one large median antenna and 2 lateral auricles (fig. 83); a cirriform branchia on most parapodia 3
2. Elytra not completely covering the dorsum, leaving a medial gap; a prominent "facial tubercle" present (fig. 81); ventrum papillated .... *Pholoe tuberculata*
– Elytra completely cover the dorsum; facial tubercle absent; ventral papillae limited to mouth area (fig. 82) *Pholoe glabra*
3. Ventral surface thickly papillated (fig. 85) .... *Sthenelais berkeleyi*

– Ventral surface smooth; first 2–3 pairs of elytra light to transparent, the rest a mottled green; head red (figs. 83–84) .... *Sthenelais fusca*

LIST OF SIGALIONIDAE

Pholoe tuberculata Southern, 1914. Probably = *P. minuta* (Fabricius, 1780). Occurs on sand-mud substrates in Tomales Bay; low intertidal to subtidal.

Pholoe glabra Hartman, 1961. Silt and mud bottoms.

Sthenelais berkeleyi Pettibone, 1971. Intertidal to 110 meters; recorded from Puget Sound and southern California. See Pettibone, 1971, J. Fish. Res. Bd. Canada 28: 1393–1401.

Sthenelais fusca Johnson, 1897. Common under rocks, among algal holdfasts, and rhizomes of *Phyllospadix*. See Pettibone, 1971, above; Hartman, 1939.

CHRYSOPETALIDAE

1. Paleae of 1 kind, narrow, not longitudinally serrated; color in life pale rust to yellowish (fig. 86) .... *Chrysopetalum occidentale*

– Paleae of 2 kinds; dorsal broad ones, longitudinally serrated, and laterally directed narrower ones, not serrated; color in life glistening white or greenish .... *Paleanotus bellis*

LIST OF CHRYSOPETALIDAE

See Hartman, 1940.

Chrysopetalum occidentale Johnson, 1897. Rocky habitats; more southern in distribution.

Paleanotus bellis (Johnson, 1897). A small species (to about 3 mm), common among barnacles, bryozoans, and sponges on pilings and elsewhere; see Banse and Hobson, 1968.

AMPHINOMIDAE

1. Caruncle reduced, extends posteriorly to middle of second setigerous segment; smaller, to 50 mm long .... *Pareurythoe californica*

– Caruncle larger, extends posteriorly behind the third setigerous segment and conceals much of the prostomium (fig. 87); larger, to about 35 cm long ...... *Eurythoe complanata*

LIST OF AMPHINOMIDAE

See Hartman, 1940.

Eurythoe complanata (Pallas, 1766). Under rocks; more common in southern California.

Pareurythoe californica (Johnson, 1897). Under rocks.

EUPHROSINIDAE

Euphrosine is represented by several species, such as *Euphrosine aurantiaca* Johnson, 1897. In life they are bright red to orange and may occur in algal holdfasts or crevices of rocks. Typically subtidal, they are occasionally washed ashore. See Hartman (1940, 1968) for key to species.

NOTOPHYCIDAE

A single species, *Phyllodocella bodegae* Fauchald and Belman, 1972, has been found in a thick, gelatinous sac attached to the hydroid *Obelia* in Bodega Harbor. Figures 89 and 90 are from Fauchald and Belman, 1972, Bull. So. Calif. Acad. Sci. 71: 107–108.

(Note: This has since been determined to be a species of *Micronereis* (Nereidae) [K. Fauchald, personal communication]

PHYLLODOCIDAE

1. With 2 pairs of tentacular cirri 2
– With 4 pairs of tentacular cirri 6
2. Tentacular cirri and prostomial antennae short ..... *Eteone* 3
– Tentacular cirri and prostomial antennae long .... *Hesionura* sp.
3. Dorsal cirri broadly rounded (fig. 94); body with irregularly spaced black spots; more than 50 mm long *Eteone pacifica*
– Dorsal cirri otherwise; without black spots; smaller, less than 50 mm long 4
4. Prostomium broader than long (fig. 92); first segment dorsally reduced; dorsal cirri distally pointed, approximately triangular in shape (fig. 93) .... *Eteone lighti*
– Prostomium as long as broad or longer; first segment fully developed dorsally 5
5. Prostomium trapezoidal, longer than wide (fig. 95); dorsal cirri subrectangular (fig. 96); body pale green .... *Eteone dilatae*
– Prostomium semicircular, as long as wide (fig. 97); dorsal cirri

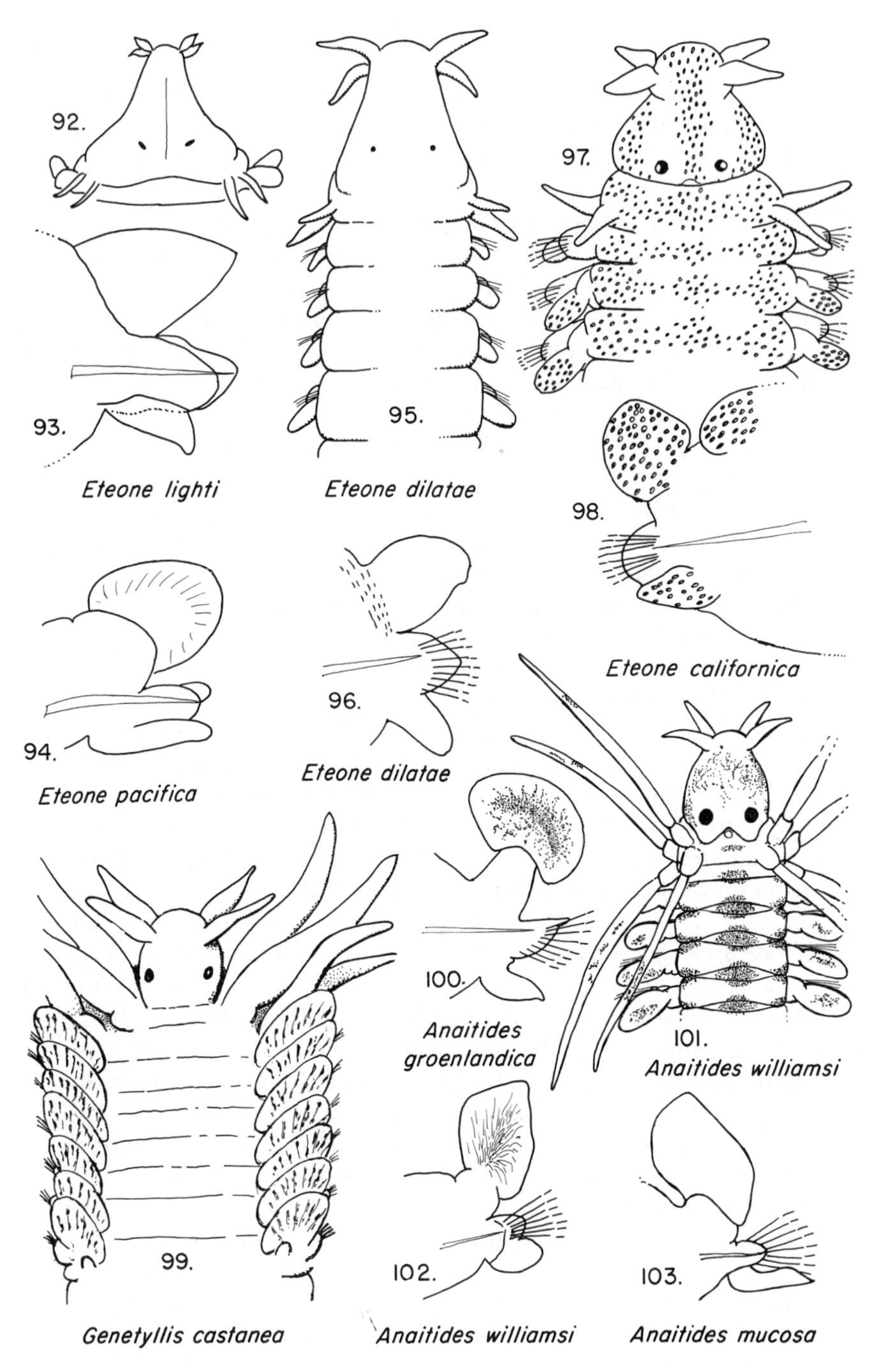

PLATE 31. **Polychaeta** (12). Phyllodocidae: 92, *Eteone lighti;* 93, parapodium; 94, *E. pacifica,* parapodium; 95, *E. dilatae;* 96, parapodium; 97, *E. californica;* 98, parapodium; 99, *Genetyllis castanea;* 100, *Anaitides groenlandica,* parapodium; 101, *A. williamsi;* 102, parapodium; 103, *A. mucosa,* parapodium; 91–93 after Hartman.

as broad as long, inflated (fig. 98); body brownish due to small, granular, pigment spots .... *Eteone californica*

6. Median antenna absent; nuchal papilla may be present 7
– Median antenna present 12
7. Ventral tentacular cirrus of segment 2 foliaceous, others cirriform .... *Sige montereyensis*
– All tentacular cirri cirriform 8
8. Proboscis with papillae arranged in longitudinal rows near the base; nuchal papilla present .... *Anaitides* 10
– Proboscis with papillae in dispersed arrangement; nuchal papilla present or absent 9
9. Nuchal papilla absent; body greenish with reddish-brown on the dorsal cirri and dorsum (fig. 99) .....*Genetyllis castanea*
– Nuchal papilla present ....*Phyllodoce* sp.
10. Dorsal cirri wider than long, with darkly pigmented center and clear around the margin (fig. 100); pharyngeal papillae number 12–15 in a row .... *Anaitides groenlandica*
– Dorsal cirri longer than wide 11
11. (*Note 3 choices*) Each body segment with 3 pigmented areas across the dorsum, forming 3 longitudinal lines down the body (figs. 101–102); pharyngeal papillae number up to 9 in a row ....*Anaitides williamsi*
– Body uniformly iridescent purplish brown; pharyngeal papillae number 10–12 in a row ....*Anaitides medipapillata*
– Body pale, dorsum with dark punctate spots; pharyngeal papillae number up to 12 in a row (fig. 103) ....*Anaitides mucosa*
12. (*Note 3 choices*) First segment dorsally reduced (fig. 106) ... .*Eumida* 13
– Tentacular segments form complete rings (figs. 105, 108, 109) 14
– First segment dorsally and ventrally reduced; second segment dorsally reduced; tentacular cirri clavate; body with granular red pigment on tentacular cirri, dorsal cirri, and segments (fig. 104) ....*Clavodoce splendida*
13. Dorsum crossed by dark green transverse bands; dorsal cirri ovoid (figs. 106, 107) .... *Eumida bifoliata*
– Dorsum without bands; dorsal cirri pointed at tips ....*Eumida sanguinea*
14. Ventral tentacular cirrus of segment 2 foliaceous and asymmetrical, other tentacular cirri cirriform (fig. 105) ....*Steggoa californiensis*
– Tentacular cirri all cirriform (fig. 109) ....*Eulalia* 15
15. First segment somewhat fused to the prostomium; color pale yellow to greenish with a pair of lateral, longitudinal brown

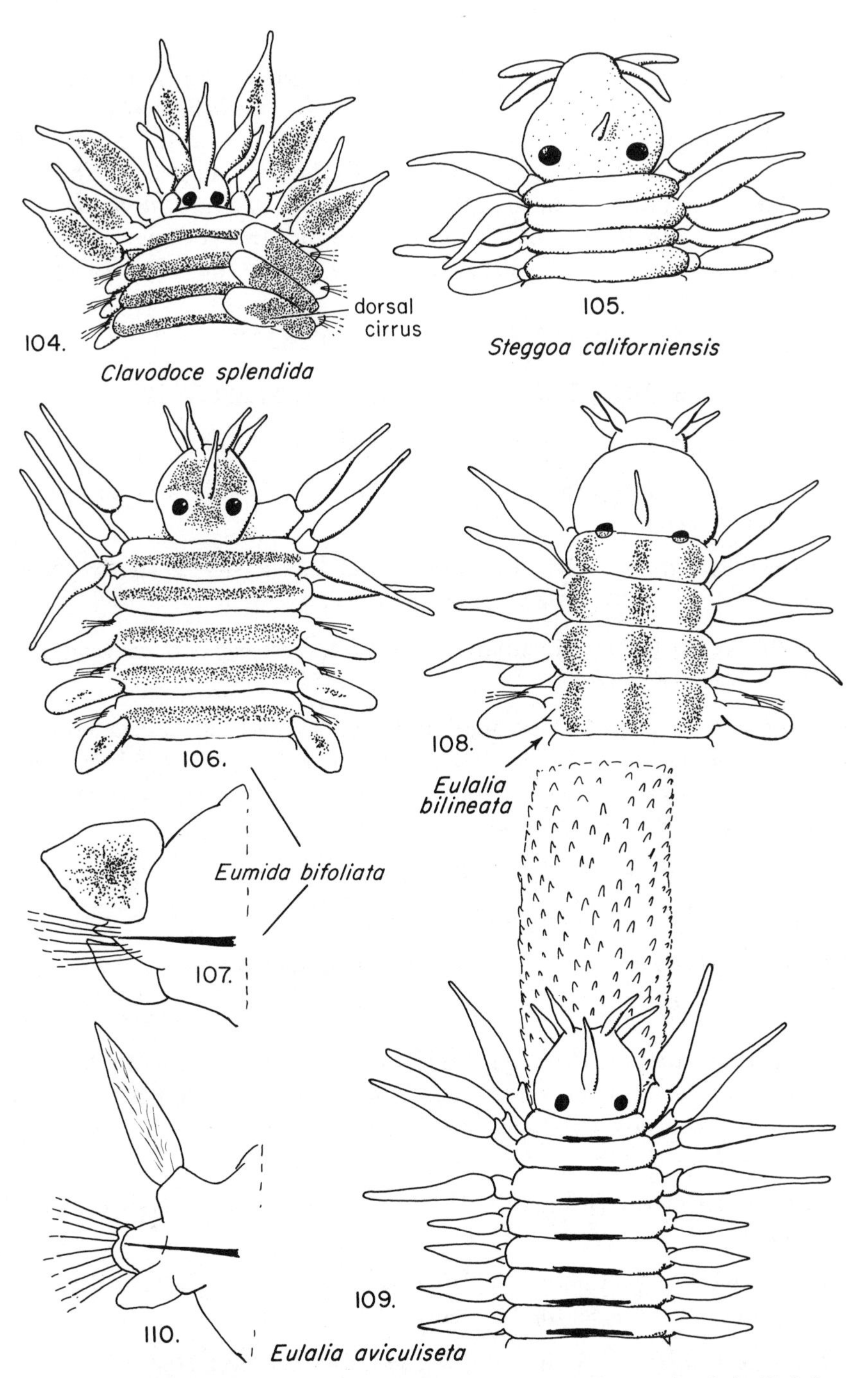

PLATE 32. **Polychaeta** (13). Phyllodocidae: 104, *Clavodoce splendida* (left dorsal cirri missing); 105, *Steggoa californiensis;* 106, *Eumida bifoliata;* 107, parapodium; 108, *Eulalia bilineata;* 109, *E. aviculiseta;* 110, parapodium.

stripes or bands running down the length of the body (fig. 108) .... *Eulalia bilineata*
– First segment clearly set off from the prostomium 16
16. Body pale to light green with or without brown spots on the dorsal cirri and on bases of parapodia ...... *Eulalia viridis*
– Body greenish with black intersegmental grooves (figs. 109, 110) .... *Eulalia aviculiseta*

LIST OF PHYLLODOCIDAE

See Hartman 1936, Univ. Calif. Publ. Zool. 41: 117–132.

Anaitides groenlandica (Oersted, 1843). Silt and mud.

Anaitides medipapillata Moore, 1909. Rocky habitats; more common in southern California.

Anaitides mucosa (Oersted, 1843). Sand and silt. See Cazaux 1969, Arch. Zool. Expér. Gen. 110: 145–202 (larval development).

Anaitides williamsi Hartman, 1936. Mixed sand and mud sediments; lays eggs on blades of *Zostera*.

† *Clavodoce splendida* Hartman, 1936. In sand flats with *Zostera* and *Ulva* (Hartman, 1936); in algal holdfast, Doran Beach Jetty, Bodega Harbor; known only from Bodega Bay region, rare occurrence probably due to its being most common at the lowest tide level and subtidally.

† *Eteone californica* Hartman, 1936. Sandy mud.

Eteone dilatae Hartman, 1936. Clean sand beaches.

Eteone lighti Hartman, 1936. Muddy sediments of estuaries.

Eteone pacifica Hartman, 1936. Silt and mud; see Banse, 1972.

Eulalia aviculiseta Hartman, 1936. One of the most common polychaetes of northern California; occurs among mussels and barnacles, in algal holdfasts, and under rocks in debris. This species may be a synonym of *E. quadrioculata* Moore, 1906; see Banse, 1972.

Eulalia bilineata (Johnston, 1840). Common in algal holdfasts and under rocks.

Eulalia viridis (Linnaeus, 1767). A cosmopolitan species, appears to be rare locally; under rocks and in algal holdfasts.

Eumida bifoliata (Moore, 1909). Sometimes mistaken for *Eulalia aviculiseta*, this species is common among algal holdfasts and debris in rocky habitats of the Bodega Bay region; also known from sandy-silt sediments and shell bottoms.

† Terminal "e" in names of this sort is pronounced.

Eumida sanguinea (Oersted, 1843). A cosmopolitan species; in northern California it occurs among algae and bryozoans on rocks; see Cazaux, 1969, above.

Genetyllis castanea (Marenzeller, 1879). Common in algal holdfasts and other microhabitats in rocky areas.

Hesionura sp. Known from Moss Landing; in coarse sand; may be related to *H. coineaui difficilis* (Banse, 1963) from the state of Washington.

† *Phyllodoce* sp. Among rocks, Bodega Bay region.

† *Sige montereyensis* Hartman, 1936. Known only as a single specimen from Monterey; in shallow water.

Steggoa californiensis Hartman, 1936. Sandy mud.

HESIONIDAE

1. Minute (not over 4–5 mm long), interstitial; 6 pairs of tentacular cirri; body pale with brown marks along the sides of segments and paired spots on the posterior margins of each segment .... *Microphthalmus sczelkowii*
– Larger and robust (more than 5 mm long); 6–8 pairs of tentacular cirri; color dark or pale, but never spotted 2
2. With 6 pairs of tentacular (peristomial) cirri (forward of cirri on setigerous segments, but not including prostomial cirri, fig. 91); parapodia biramous but with notopodia reduced almost to a cirrus with inconspicuous setae; body reddish brown to purple; free-living on muddy bottoms or commensal in ambulacral grooves of starfishes ..... *Ophiodromus pugettensis*
– With 8 pairs of tentacular cirri; parapodia uniramous; color pale; free-living in shallow muddy bottoms .... *Gyptis brevipalpa*

LIST OF HESIONIDAE

Gyptis brevipalpa (Hartmann-Schroeder, 1959) [= *G. arenicola glabra* (Hartman, 1961)]. Silt and mud; principally subtidal; see Banse and Hobson, 1968.

* *Gyptis brunnea* Hartman, 1961. Larval stages have been taken from plankton in Tomales Bay and reared in the laboratory; juveniles grow well in a sand substrate; adults have not been found outside southern California.

Microphthalmus sczelkowii Mecznikow, 1865. Clean sand beaches; Dillon Beach (previously unrecorded from Pacific coast; see Pettibone, 1963).

Ophiodromus pugettensis (Johnson, 1901) (= *Podarke pugettensis*). The most common intertidal hesionid of California; on silty bottoms and in ambulacral grooves of asteroids, especially *Patiria miniata*. See Lande and Reish, 1968, Bull. So. Calif. Acad. Sci. 67: 104–111.

PILARGIIDAE

1. Prostomial antennae and peristomial cirri short (fig. 112); setae short, inconspicuous .... *Pilargis* 2
– Prostomial antennae and peristomial cirri long (fig. 111); setae conspicuous .... *Sigambra bassi*
2. Dorsum heavily papillated; notopodia with subglobular base and terminal clavate process (figs. 112, 113) .... *Pilargis berkeleyi*
– Dorsum lightly papillated; notopodia with a broad, quadrate base and tapering short cirri (fig. 114) .... *Pilargis maculata*

LIST OF PILARGIIDAE

See Hartman, 1947b; all occur in sandy sediments.

Pilargis berkeleyae Monro, 1933.

Pilargis maculata Hartman, 1947. Pettibone, 1966, Proc. U.S. Natl. Mus. 118: 155–208, regards *P. berkeleyi* and *P. maculata* as synonyms.

Sigambra bassi (Hartman, 1945).

SYLLIDAE

The syllids are poorly known on the California coast; there may be fifty or more undescribed intertidal species. Several species not in Hartman (1968) have been assigned to appropriate genera in the following key. Some well-known names may be simply "catch-all" terms for endemic species as yet undescribed. Because of the great need for taxonomic study of the Syllidae, the following keys are obviously incomplete and the student should be prepared to find new species and range extensions.

1. Ventral cirri absent (fig. 116); dorsal cirri smooth; two nuchal epaulettes extend from the hind margin of the prostomium through the first few segments (fig. 115) ..... *Autolytus* spp.
– Ventral cirri present; dorsal cirri smooth or articulated 2
2. Dorsal cirri articulated 3
– Dorsal cirri smooth or wrinkled 11
3. Pharynx with a single medial tooth on the distal end; body not flattened 5

- – Pharynx denticulate on the distal end; body flattened .... *Trypanosyllis* 4
- 4. Color of dorsum pale, crossed with dark transverse lines; composite falcigers with subdistal tooth (fig. 117) .... *Trypanosyllis gemmipara*
- – Color ivory to tawny; cirri purplish brown; composite falcigers with smooth cutting edge (fig. 118) .... *Trypanosyllis ingens*
- 5. (*Note 3 choices*) Setae entirely simple .... *Haplosyllis spongicola*
- – Setae partly simple, partly composite .... *Syllis* 6
- – Setae all composite, except in posteriormost segments where single simple setae may occur in superior and inferior positions in setal fascicles .... *Typosyllis* 7
- 6. Median parapodia with thick, Y-shaped seta (fig. 119) in addition to composite setae .... *Syllis gracilis*
- – Median parapodia with 2–3 projecting acicular setae (fig. 120) in addition to composite setae .... *Syllis elongata*
- 7. Composite falcigers distally entire; paired antennae inserted ventrally .... *Typosyllis adamanteus*
- – Some or all composite falcigers distally bifid (fig. 122) 8
- 8. Median antenna inserted at or near the middle of prostomium (fig. 121) 9
- – Median antenna inserted at frontal margin of prostomium .. 10
- 9. Paired antennae inserted at frontal margin of prostomium .. .. *Typosyllis pulchra*
- – Paired antennae inserted in front of anterior eyes (figs. 121, 122) .... *Typosyllis aciculata*
- 10. Dorsal cirri thick (fig. 123) .... *Typosyllis armillaris*
- – Dorsal cirri slender (fig. 124) .... *Typosyllis hyalina*
- 11. Palps fused along medial line; typically small, 4 mm or less .. 12
- – Palps free or only partly fused; animal usually larger than 4 mm 17
- 12. (*Note 3 choices*) Two pairs of peristomial cirri (fig. 125); dorsal cirri long and filiform .... *Brania* sp.
- – Peristomial cirri and prostomial antennae absent; dorsal cirri short .... *Exogonella* sp.
- – One pair of peristomial cirri; dorsal cirri short and ovoid .. 13
- 13. Body papillate; peristomial cirri prominent .. *Sphaerosyllis* 14
- – Body smooth; peristomial cirri minute .... *Exogone* 15
- 14. Palps elongate; a special capsule of rods (rhabdites) in most parapodia (fig. 126) .... *Sphaerosyllis hystrix*
- – Palps broad; without special capsule in parapodia (fig. 127). .... *Sphaerosyllis californiensis*

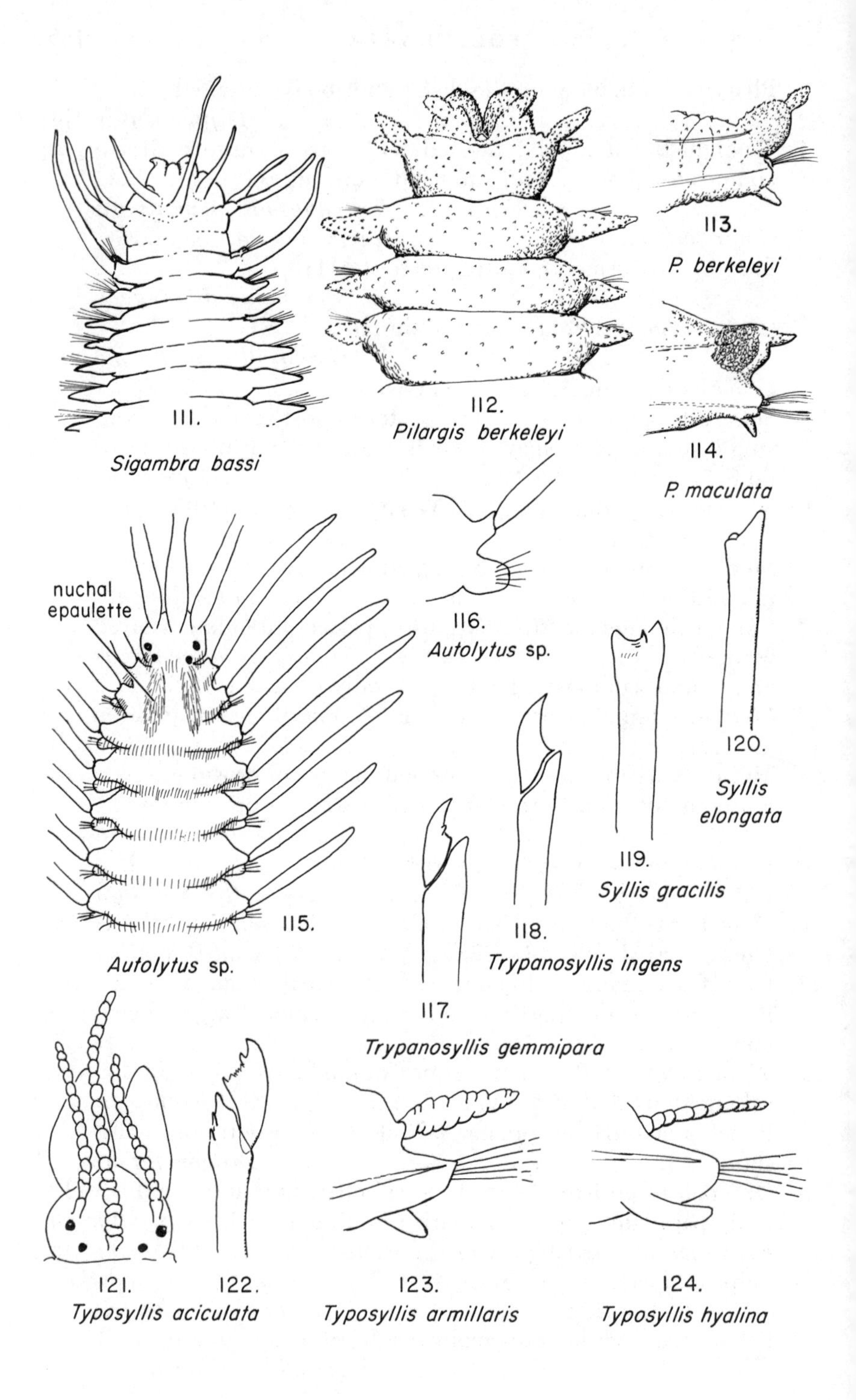
111.
Sigambra bassi
112.
Pilargis berkeleyi
113.
P. berkeleyi
114.
P. maculata
nuchal
epaulette
115.
Autolytus sp.
116.
Autolytus sp.
117.
Trypanosyllis gemmipara
118.
Trypanosyllis ingens
119.
Syllis gracilis
120.
Syllis
elongata
121.
122.
Typosyllis aciculata
123.
Typosyllis armillaris
124.
Typosyllis hyalina

15. (*Note 3 choices*) Composite setae include both spinigers and falcigers, with falcigers all distally entire (figs. 131, 132) 16
– Composite setae all bifid falcigers (fig. 128) .... *Exogone gemmifera*
– Composite setae all entire falcigers (fig. 129) .... *Exogone verugera*
16. Median antenna shorter than prostomium, and hardly larger than lateral antennae .... *Exogone uniformis*
– Median antenna longer than prostomium, and much longer than lateral antennae (figs. 130–132) .... *Exogone lourei*
17. Prostomium covered posteriorly by a nuchal hood (fig. 133). .... *Odontosyllis* 18
– Prostomium without nuchal hood 19
18. Dorsal cirri in median region short, thick, no longer than half the width of the body; dorsum relatively unpigmented .... *Odontosyllis parva*
– Dorsal cirri in median region long, slender and tapering to a fine tip; dorsum dark with pale transverse bands (fig. 133) .... *Odontosyllis phosphorea*
19. Proboscis armed with bi- or tri-cusped teeth; peristomium with a pair of long, linear nuchal epaulettes which extend back over the first setiger; body short with few segments .. .. *Amblyosyllis* sp.
– Proboscis terminates in 9–10 marginal papillae; nuchal processes absent; body large .... *Pionosyllis gigantea*

LIST OF SYLLIDAE

Amblyosyllis sp. Rocky habitat, open coast, associated with siliceous sponges.

Autolytus spp. The species of *Autolytus* remain to be worked out in California. At least four species have been taken in plankton where the sexual stages are found. See Gidholm, 1965, Zool. Bidr. Uppsala 37: 1–44 (reproduction); 1966, Ark. Zool. 19: 157–213 (systematics) of the sub-family Autolytinae.

Brania sp. In sandy mud.

Exogone gemmifera Pagenstecher, 1862. Sandy muds; with kelp.

PLATE 33. **Polychaeta** (14). Pilargiidae, Syllidae: 111, *Sigambra bassi;* 112, *Pilargis berkeleyae;* 113, parapodium; 114, *P. maculata,* parapodium; 115, *Autolytus* sp.; 116, parapodium; 117, *Trypanosyllis gemmipara;* 118, *T. ingens,* composite falciger; 119, *Syllis gracilis,* Y-shaped seta; 120, *Syllis elongata,* acicular seta; 121, *Typosyllis aciculata;* 122, composite falciger; 123, *T. armillaris,* parapodium; 124, *T. hyalina,* parapodium; 111–114 after Hartman; 115–116 after Gidholm; 117, 119 after Imajima; 118 after Johnson; 121–122 after Reish; 123–124 after Fauvel.

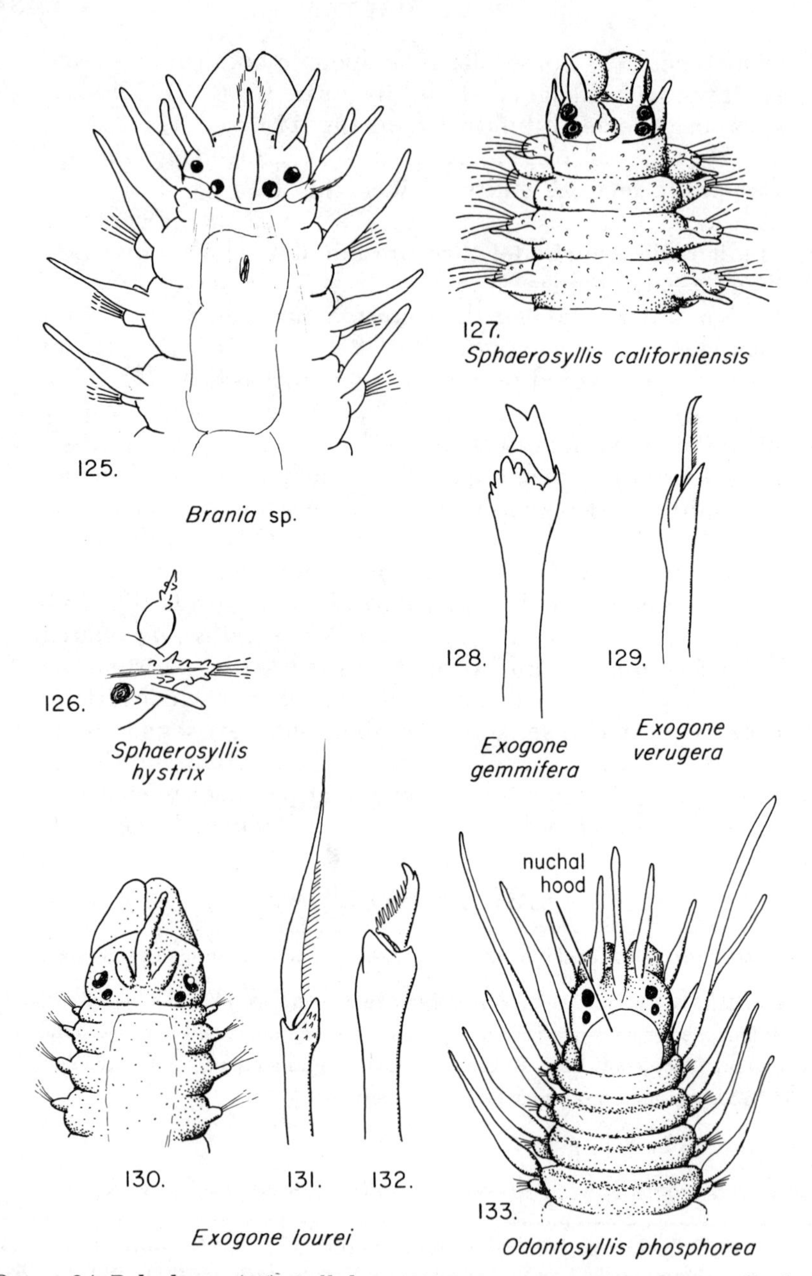

PLATE 34. **Polychaeta** (15). Syllidae: 125, *Brania* sp.; 126, *Sphaerosyllis hystrix,* parapodium; 127, *S. californiensis;* 128, *Exogone gemmifera,* composite falciger; 129, *E. verugera,* composite falciger; 130, *E. lourei;* 131, composite falciger; 132, composite spiniger; 133, *Odontosyllis phosphorea;* 126 after Berkeley and Berkeley; 127 after Hartman; 128–129 after Imajima.

Exogone lourei Berkeley and Berkeley, 1938. Sandy mud; algal holdfasts; see Banse and Hobson, 1968; Banse, 1972.

Exogone uniformis Hartman, 1961. Southern California; sand and mixed sediments; see Banse, 1972.

Exogone verugera (Claparède, 1868). Sandy sediments.

Exogonella sp. Sandy mud of Tomales Bay.

Haplosyllis spongicola (Grube, 1855). Among sponges and bryozoans in rocky habitats.

Odontosyllis parva Berkeley, 1923. Among algae, mussels, bryozoans, and sponges; see Banse, 1972.

Odontosyllis phosphorea Moore, 1909. Among algal holdfasts and *Zostera;* see Banse, 1972.

Pionosyllis gigantea Moore, 1908. Subtidal in rocky and muddy sediments.

Sphaerosyllis californiensis Hartman, 1966. Silt and mud sediments.

Sphaerosyllis hystrix Claparède, 1863. Among algal holdfasts; muddy bottoms.

* *Sphaerosyllis pirifera* Claparède, 1868. See Hartman's Atlas.

Syllis elongata (Johnson, 1901). Common among algal holdfasts on rocky shores; see Banse, 1972.

Syllis gracilis Grube, 1840. With algal holdfasts and in crevices in rocky habitats.

* *Syllis* spp.

Trypanosyllis gemmipara Johnson, 1901. Among bryozoans, sponges, and tunicates in exposed rocky habitats.

Trypanosyllis ingens Johnson, 1902. Among rocks and algal holdfasts; see Berkeley and Berkeley, 1952, J. Fish. Res. Bd. Canada 8: 488–496.

Typosyllis aciculata Treadwell, 1945. Rocky habitats; see Reish, 1950, Amer. Mus. Nov. no. 1466 (redescription).

Typosyllis adamanteus (Treadwell, 1914) (= *Syllis spenceri* Berkeley and Berkeley, 1938). Among algae and barnacle clumps; see Banse, 1972.

Typosyllis armillaris (Müller, 1771).

Typosyllis hyalina (Grube, 1863). Rocky habitats.

Typosyllis pulchra (Berkeley and Berkeley, 1938). Among mussel and algal holdfasts; see Banse, 1972.

NEREIDAE

1. Peristomium enlarged, forming a collar around and under the prostomium (fig. 134); commensal with hermit crabs ....*Cheilonereis cyclurus*
– Peristomium not so enlarged 2
2. Posterior notopodia with simple, dark brown, hooked setae (fig. 135); peristomial cirri greatly elongated ....*Platynereis bicanaliculata*
– Posterior notopodia without such setae 3
3. Paragnaths absent from either the maxillary or oral ring 4
– Paragnaths present on both rings 5
4. Paragnaths present only on the oral ring (areas VII and VIII) as 6 brown cones set in a transverse row, 2 cones to each area; neuropodial lobes of posterior setigers as a large, saclike lobe (fig. 136) ....*Eunereis longipes*
– Paragnaths present only on the maxillary ring (area I with none, II with 3 minute cones, III with 1 minute cone, IV with 3 small cones in a transverse row); anterior parapodia with large dorsal and ventral lobes (fig. 137) ....*Ceratonereis tunicatae*
5. Posterior notopodia with some homogomph falcigers in addition to spinigers (fig. 147) 6
– Posterior notopodia without homogomph falcigers 10
6. Posterior notopodial lobes greatly elongate, straplike (fig. 138) ....*Nereis vexillosa*
– Posterior notopodial lobes otherwise 7
7. Proboscis with many tiny paragnaths over both oral and maxillary rings 8
– Proboscis with paragnaths larger and restricted to certain areas 9
8. Jaws with 8–9 teeth (fig. 141); eyes small; inhabits sediments (figs. 139, 140) ....*Nereis procera*
– Jaws with 3–5 teeth (fig. 142); eyes large; rocky intertidal .. ..*Nereis eakini*
9. (*Note 3 choices*) Parapodial lobes typically dark (fig. 143); no difference in size of median and posterior notopodial lobes; proboscis lacks teeth in area V *Nereis pelagica neonigripes*
– Parapodial lobes not dark (fig. 144); notopodial lobe in posterior region increases in size to subrectangular lobe about twice as long as broad (fig. 145); body bright green to brown; proboscis lacks teeth in area V ....*Nereis grubei*
– Parapodial lobes not dark; posterior notopodial lobes not larger (figs. 146, 147), body pale, dorsum with bars of brown

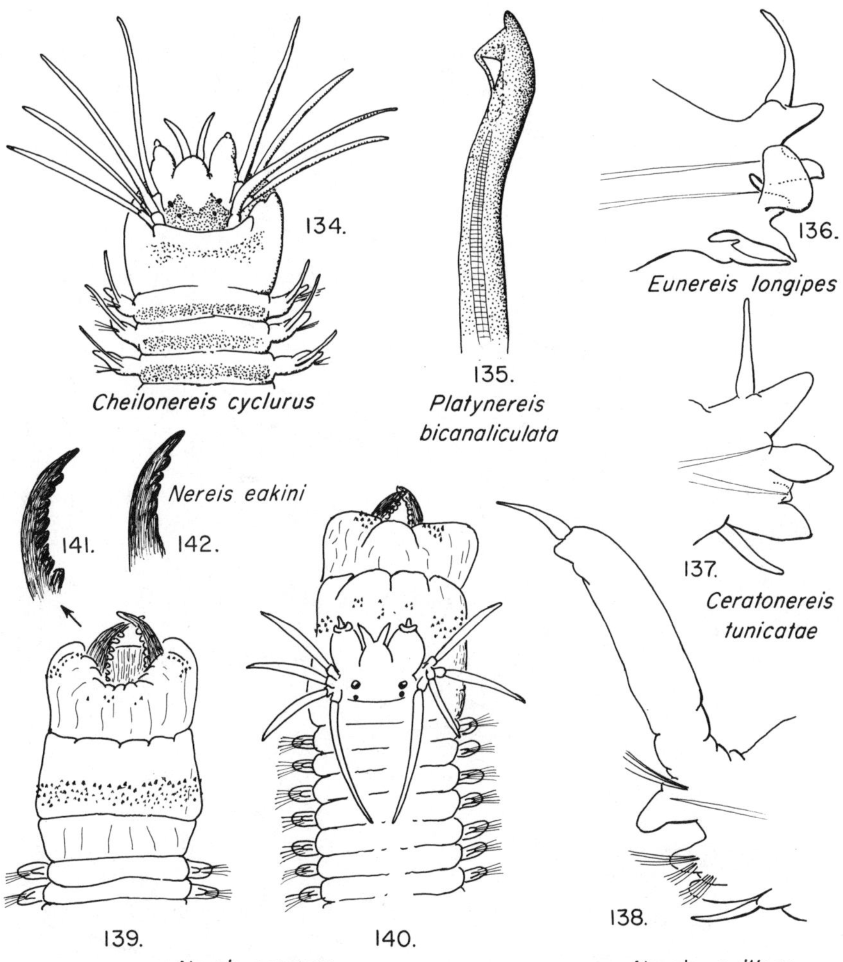

PLATE 35. **Polychaeta** (16). Nereidae: 134, *Cheilonereis cyclurus;* 135, *Platynereis bicanaliculata,* posterior notopodial hook; 136, *Eunereis longipes,* posterior parapodium; 137, *Ceratonereis tunicatae,* anterior parapodium; 138, *Nereis vexillosa,* posterior parapodium; 139, *N. procera,* anterior end, ventral view; 140, anterior end, dorsal view; 141, *N. procera,* jaw, showing number of teeth; 142, *Nereis eakini,* jaw.

or rust-colored pigment; proboscis with a single tooth in area V .... *Nereis latescens*

10. Area VI of proboscis with transverse paragnaths, body pale with dark, quadrate bars .... *Perinereis monterea*

– Area VI of proboscis with conical paragnaths; pigment bars absent 11

11. Posterior notopodial lobe becomes broadly oval, foliose, with

minute dorsal cirrus (fig. 148); marine; see note in species list .... *Neanthes brandti* and *Neanthes virens*

– Posterior parapodial lobes very elongate (fig. 149); estuarine .... *Neanthes succinea*

– Posterior parapodial lobes short (fig. 150); in coastal estuaries,

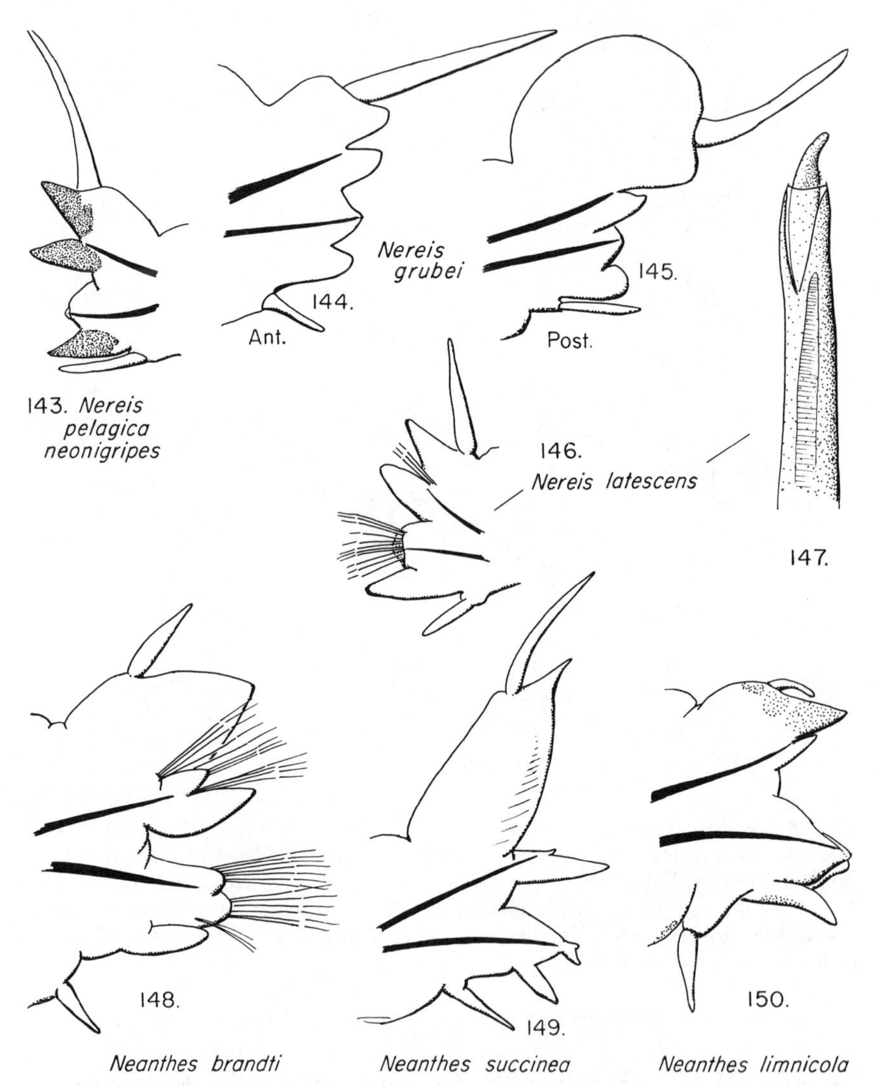

PLATE 36. **Polychaeta** (17). Nereidae: 143, *Nereis pelagica neonigripes*, posterior parapodium; 144, *N. grubei*, anterior parapodium; 145, posterior parapodium; 146, *N. latescens*, posterior parapodium; 147, homogomph falciger; 148, *Neanthes brandti*, posterior parapodium; 149, *N. succinea*, posterior parapodium; 150, *N. limnicola*, posterior parapodium.

lagoons, and lakes; see note in species list .... *Neanthes limnicola*

LIST OF NEREIDAE

See Hartman, 1936, Proc. U.S. Nat. Mus. 83: 467–480; Hartman, 1940.

Ceratonereis tunicatae Hartman, 1936. In ascidians.

Cheilonereis cyclurus (Harrington, 1897). Commensal with hermit crabs; dredged.

Eunereis longipes Hartman, 1936. Rock crevices; rare.

* *Lycastopsis pontica neapolitana* La Greca, 1950. From peat-clay banks in Tomales Bay; see Hartman, 1959, Bull. Mar. Sci. 9: p. 163. This species is a cosmopolitan representative of a subfamily (Namanereinae) of very small worms with reduced nereid features, inhabiting fresh or brackish water at high intertidal or spray zone levels, and usually overlooked.

Neanthes brandti (Malmgren, 1866) and *Neanthes virens* (Sars, 1835). These species are separated with difficulty. The former has many paragnaths on both rings of the proboscis while the second has few. Because of the overlap in geographic range and great similarities in morphology, especially of the parapodia, there is a possibility that these species may not be genetically isolated. Both are known from intertidal areas, including sand and rock.

Neanthes limnicola (Johnson, 1901) (=*N. lighti* Harman, 1938). This species is sometimes referred to as *N. diversicolor* (Müller, 1776) to which it is closely related. The two species, however, have greatly different reproductive patterns and need not be confused. *Neanthes limnicola* is viviparous, while *N. diversicolor* of Europe and eastern North America has a free-living, nonplanktonic larva; see Smith, 1950, J. Morph. 87: 417–465 (embryology); 1958, Syst. Zool. 7: 60–73 (taxonomy); Pettibone, 1963. Inhabits brackish or even fresh water of estuarine streams, coastal lagoons, and Lake Merced (San Francisco).

Neanthes succinea (Frey and Leuckart, 1847). Sandy mud to muddy sediments of bays.

Nereis eakini Hartman, 1936. Rocky habitats.

Nereis grubei (Kinberg, 1866) (=*N. mediator* Chamberlin, 1918). Among algae on rocky shores. See Reish, 1954, Occ. Pap. no. 14, Allan Hancock Found. (life history, ecology); Schroeder, 1968, Pac. Sci. 22: 476–481 (life history).

Nereis latescens Chamberlin, 1919. Common in algal holdfasts and among rocks and debris.

* *Nereis natans* Hartman, 1936. Perhaps a sexual stage of another central California species in this list.

Nereis pelagica neonigripes Hartman, 1936. Rocky habitats.

Nereis procera Ehlers, 1868. Sandy mud to muddy sediments; low intertidal to subtidal.

Nereis vexillosa Grube, 1851. Common on pilings with barnacles and mussels; see Johnson, 1943, Biol. Bull. 84: 106–114 (life history).

Perinereis monterea (Chamberlin, 1918). Rocky habitats; rare.

Platynereis bicanaliculata (Baird, 1863). Abundant in protected rocky habitats among algal holdfasts; builds tubes and aggregates.

GONIADIDAE

1. Proboscis with soft papillae over the surface and with a set of chevrons on each side at the base (fig. 152) *Goniada brunnea*
– Proboscis with numerous hard, yellow, chitinized spines (figs. 153, 154) .... *Glycinde* 2
2. Dorsal cirrus incised near the tip (fig. 156); proboscidial organs of area V rudimentary or lacking .... *Glycinde polygnatha*
– Dorsal cirrus entire (fig. 155); proboscidial organs of area V conspicuous, resembling those of area II (fig. 154) .... *Glycinde armigera*

LIST OF GONIADIDAE

See Hartman, 1940, 1950.

Glycinde armigera Moore, 1911. In sandy mud sediments.

Glycinde polygnatha Hartman, 1950. In sandy mud sediments.

Goniada brunnea Treadwell, 1906. In sandy mud; low intertidal to deep water.

GLYCERIDAE

1. Parapodia uniramous; setae all composite 7
– Parapodia biramous; notosetae simple, neurosetae composite 2
2. Parapodia with 1 postsetal lobe; branchiae absent 3
– Parapodia with 2 postsetal lobes; branchiae present 4
3. Parapodia with 2 presetal lobes throughout (fig. 157); proboscidial organs without ridges .... *Glycera capitata*
– Parapodia with 1 presetal lobe in posterior setigers (fig. 158);

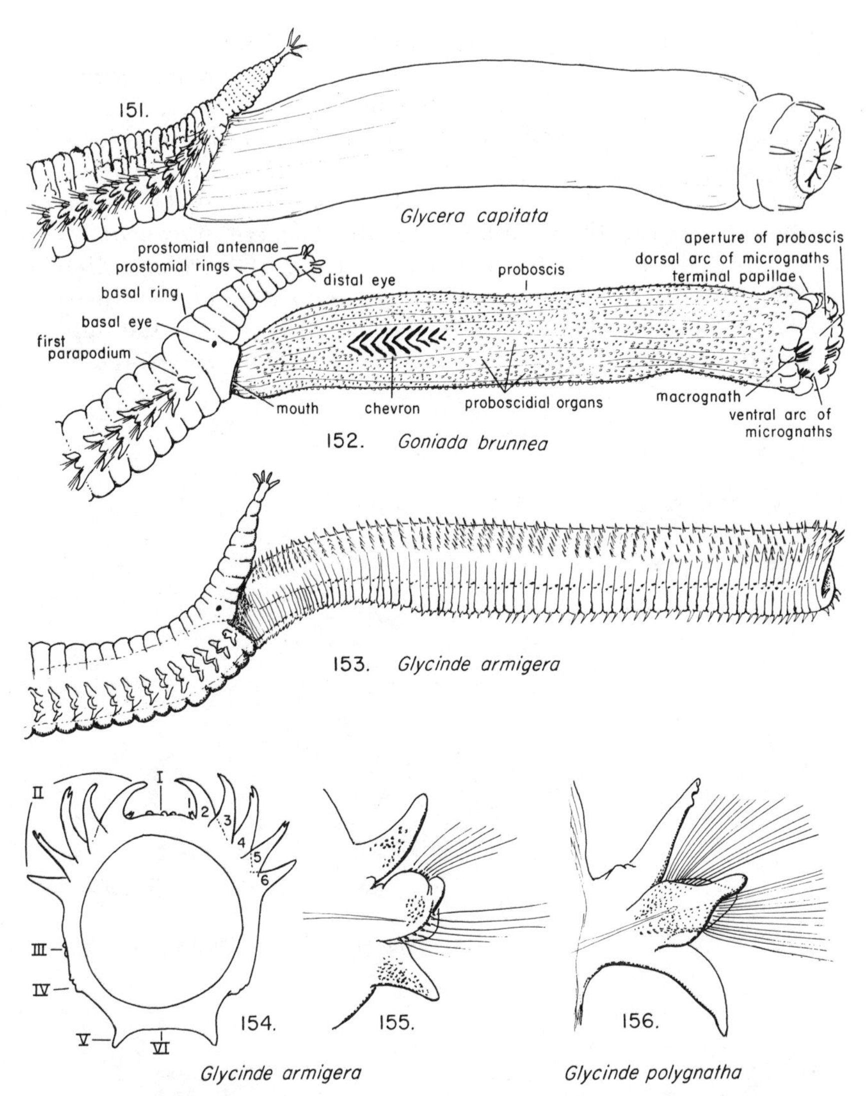

PLATE 37. **Polychaeta** (18). Glyceridae and Goniadidae: 151, *Glycera capitata*, proboscis everted; 152, *Goniada brunnea*, proboscis everted; 153, *Glycinde armigera*, proboscis everted; 154, cross-section through middle of everted proboscis showing distribution of proboscidial armature; 155, anterior parapodium; 156, *Glycinde polygnatha*, anterior parapodium. 152–154, 156 after Hartman.

proboscidial organs with transverse ridges ... *Glycera tenuis*

4. Parapodia with blisterlike branchiae on dorsoposterior sides (fig. 159) .... *Glycera robusta*
– Parapodia with long, fingerlike branchiae 5
5. Branchiae retractile, emerging from the posterodorsal side of

the parapodia, forming branched lobes (fig. 160) .... *Glycera americana*
– Branchiae not retractile, not branched 6
6. Parapodia with a single branchia located above the setal lobe (fig. 161) .... *Glycera convoluta*
– Parapodia with 2 fingerlike branchiae, one located above and the other below the setal lobe (fig. 162) *Glycera dibranchiata*
7. Parapodial lobes short, wider than long (fig. 163); color in life light green .... *Hemipodus californiensis*
– Parapodial lobes longer than wide (fig. 164); color in life dull or bright red .... *Hemipodus borealis*

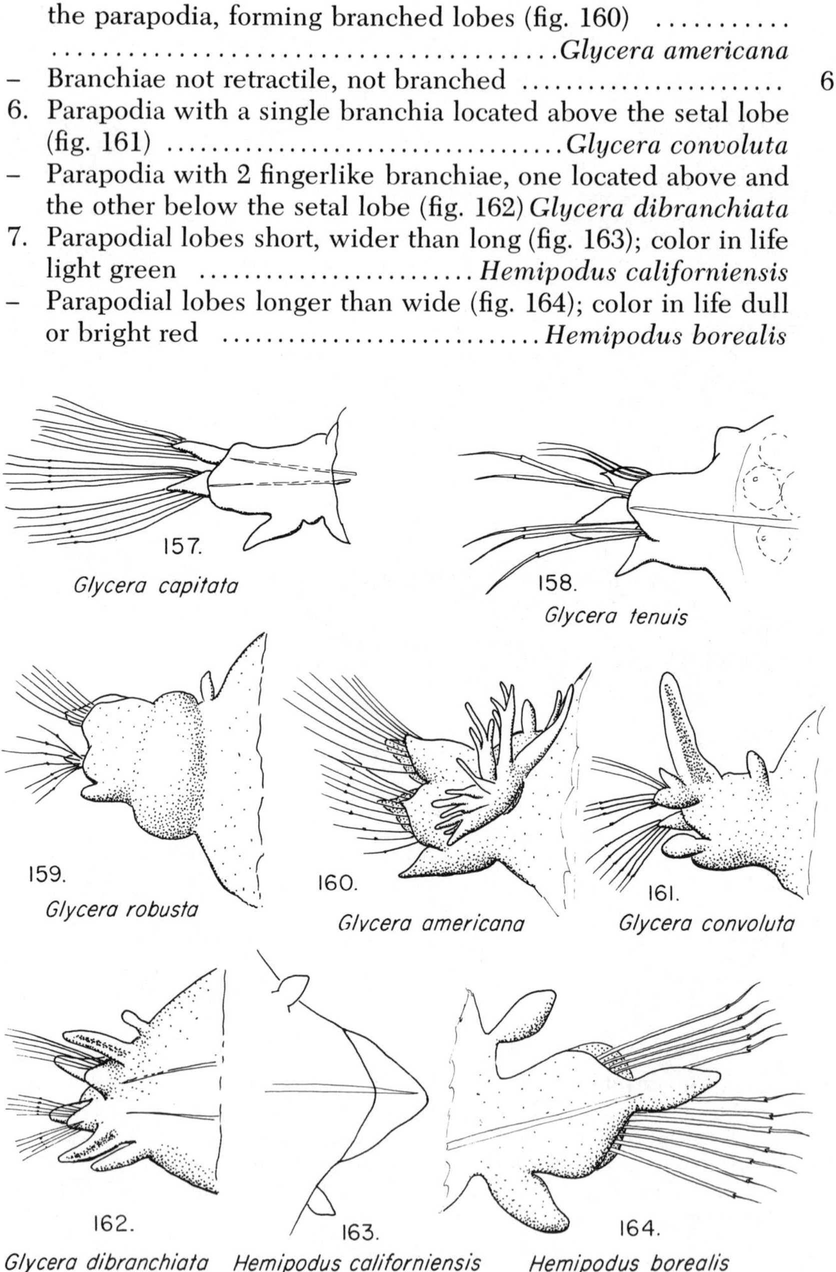

PLATE 38. **Polychaeta** (19). Glycerid parapodia: 157, *Glycera capitata;* 158, *G. tenuis;* 159, *G. robusta;* 160, *G. americana;* 161, *G. convoluta;* 162, *G. dibranchiata;* 163, *Hemipodus californiensis;* 164, *H. borealis;* 157, 158, 163 after Hartman.

LIST OF GLYCERIDAE

See Hartman, 1940, 1950

Glycera americana Leidy, 1855. Sandy mud sediments.

Glycera capitata Oersted, 1843. Sandy mud to mud.

Glycera convoluta Keferstein, 1862. Sandy sediments.

Glycera dibranchiata Ehlers, 1868. Sandy mud to mud.

Glycera robusta Ehlers, 1868. Sand and cobble sediments.

Glycera tenuis Hartman, 1944. Clean sands.

Hemipodus borealis Johnson, 1901. Sands to sandy muds.

Hemipodus californiensis Hartman, 1938. Muddy sediments of estuaries.

NEPHTYIDAE

1. Interramal cirri begin on setiger 3; prostomium with spread-eagle pigment pattern (fig. 167); first few segments pale dorsally; inhabits clean, sandy beaches . . *Nephtys californiensis*
– Interramal cirri begin on setiger 4 2
– Interramal cirri begin on setigers 5–6 3
2. Prostomium and first few segments with characteristic dark pigment pattern on the dorsal side (fig. 165); inhabits muddy sand flats . . . . *Nephtys caecoides*
– Prostomium without pigment pattern . . . . *Nephtys parva*
3. Posterior prostomial antennae bifurcate (fig. 168); interramal cirri begin on setiger 5 . . . . *Nephtys cornuta franciscana*
– Posterior prostomial antennae not bifurcate; interramal cirri begin on setigers 5–6 . . . . *Nephtys caeca*

LIST OF NEPHTYIDAE

See Hartman, 1938, Proc. U.S. Nat. Mus. 85: 143–158; Hartman, 1940, 1950.

Nephtys caeca (Fabricius, 1780). Sandy sediments; rare.

Nephtys caecoides Hartman, 1938. Sandy muds of bays and lagoons; one of the most common nephtyids in California. See Clark and Haderlie, 1962, J. Anim. Ecol. 31: 339–357.

Nephtys californiensis Hartman, 1938. Clean sandy beaches; very common. See Clark and Haderlie, above.

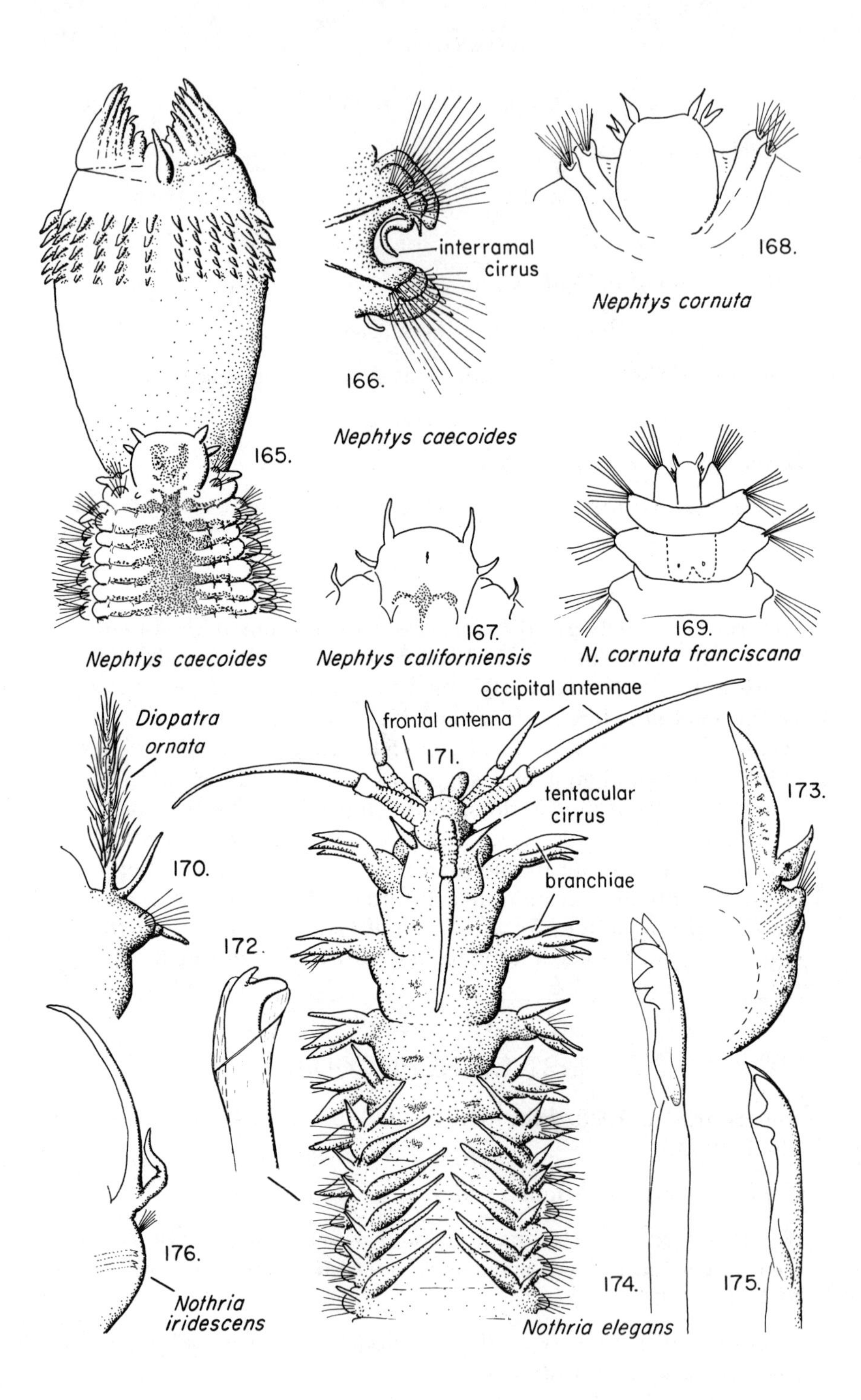
165.
Nephtys caecoides
interramal
cirrus
166.
Nephtys caecoides
168.
Nephtys cornuta
167.
Nephtys californiensis
169.
N. cornuta franciscana
Diopatra
ornata
170.
occipital antennae
frontal antenna
171.
tentacular
cirrus
branchiae
172.
173.
176.
Nothria
iridescens
174.
175.
Nothria elegans

Nephtys cornuta franciscana Clark and Jones, 1955. In muddy sediments of San Francisco Bay; the subspecies *franciscana* may only represent the juvenile stage of the stem species; figure 168 shows the stem species from British Columbia and figure 169 the subspecies from San Francisco.

Nephtys parva Clark and Jones, 1955. Muddy sediments of San Francisco Bay. See Clark and Jones, 1955, J. Wash. Acad. Sci. 45: 143–146.

ONUPHIDAE

1. Peristomium without tentacular cirri .... *Hyalinoecia* sp.
– Peristomium with tentacular cirri (fig. 171) 2
2. (*Note 3 choices*) Branchial filaments spiralled on the main stalk (fig. 170); pectinate setae with numerous teeth; ventral cirri padlike from the 6th segment .... *Diopatra ornata*
– Branchial filaments simple, cirriform (fig. 173) ..... *Nothria* 3
– Branchial filaments pinnately divided ....*Onuphis* spp.
3. Branchiae present from the first parapodia 4
– Branchiae lacking from the first 19 or more segments .... *Nothria stigmatis*
4. Branchiae on setigers 10–80 are thick (fig. 173); subacicular hooks usually begin on setiger 9 (fig. 172); pseudocompound hooks usually tridentate but 3rd tooth may be very small, usually on setigers 1–4 (figs. 174, 175) ....*Nothria elegans*
– All branchiae slender (fig. 176); subacicular hooks usually begin on setigers 11–14; pseudocompound hooks tridentate .... *Nothria iridescens*

LIST OF ONUPHIDAE

Onuphids occur in sandy sediments, low intertidal to subtidal. See Hartman, 1944; Hobson, 1971; Fauchald, 1968, Allan Hancock Monogr. Mar. Biol., no. 3.

Diopatra ornata Moore, 1911.

Hyalinoecia sp. Generally offshore.

PLATE 39. **Polychaeta** (20). Nephtyidae and Onuphidae: 165, *Nephtys caecoides*, proboscis everted; 166, 30th parapodium, in anterior view; 167, *N. californiensis;* 168, *N. cornuta;* 169, *N. cornuta franciscana;* 170, *Diopatra ornata,* anterior parapodium; 171, *Nothria elegans;* 172, subacicular seta from setiger 60; 173, 6th parapodium; 174, tridentate pseudocompound hook from setiger 2; 175, tridentate pseudocompound hook from setiger 2 with small 3rd tooth; 176, *N. iridescens,* 25th parapodium; 167, after Hartman; 168, after Berkeley; 169, after Clark and Jones; 176 after Hobson.

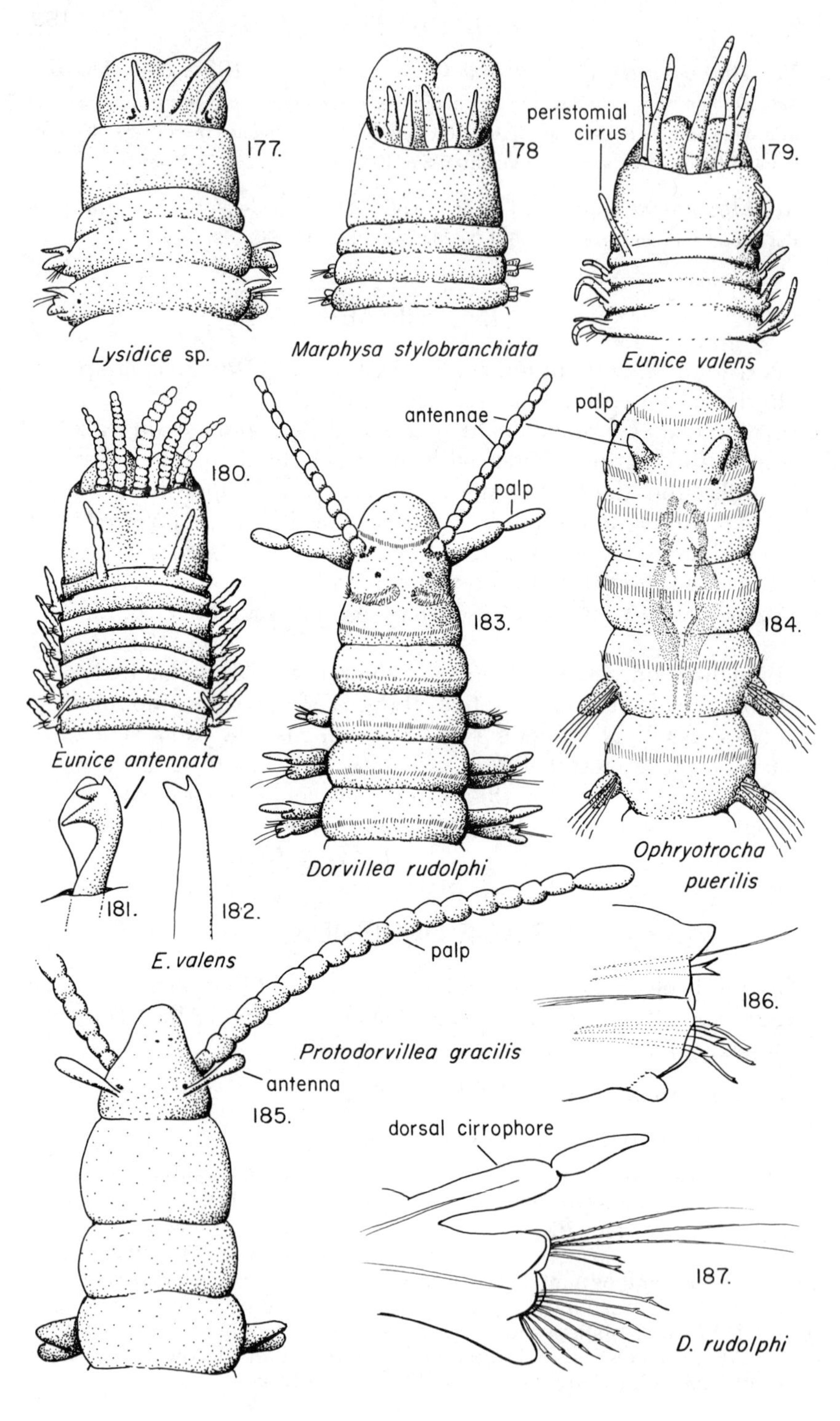

177.
Lysidice sp.
178
Marphysa stylobranchiata
peristomial cirrus
179.
Eunice valens
180.
Eunice antennata
181.
182.
E. valens
antennae
palp
183.
Dorvillea rudolphi
palp
184.
Ophryotrocha puerilis
palp
antenna
185.
Protodorvillea gracilis
186.
dorsal cirrophore
187.
D. rudolphi

Nothria elegans (Johnson, 1901).

Nothria iridescens (Johnson, 1901).

Nothria stigmatis (Treadwell, 1922).

Onuphis spp. See Hartman's Atlas.

EUNICIDAE

1. With 3 prostomial antennae (fig. 177) .... *Lysidice*
– With 5 prostomial antennae 2
2. Peristomial cirri present 3
– Peristomial cirri absent (fig. 178) *Marphysa stylobranchiata*
3. Subacicular hooks (figs. 181, 182) and pectinate setae present .... *Eunice* 4
– Subacicular hooks and pectinate setae absent .... *Palola*
4. Prostomial antennae moniliform (fig. 180); branchiae begin on setigers 5–6 .... *Eunice antennata*
– Prostomial antennae with faint articulations (fig. 179), not moniliform; branchiae begin on setiger 3 .... *Eunice valens*

LIST OF EUNICIDAE

See Fauchald, 1970; Hartman, 1944.

Eunice antennata (Savigny, 1820). Rocky intertidal and shallow subtidal; rare.

Eunice valens (Chamberlin, 1919). See Fauchald, 1969, Smithson. Contr. Zool. 6: 10–12, fig. 5; Akesson, 1967, Acta Zool. 48: 141–192 (embryology, as *E. kobiensis*).

Lysidice sp. Rocky intertidal, Tomales Point; rare, this may prove to be a juvenile *Marphysa*.

Marphysa stylobranchiata Moore, 1909. Rocky habitats; intertidal to the continental slope.

* *Palola paloloides* (Moore, 1909).

PLATE 40. **Polychaeta** (21). Eunicidae and Dorvilleidae: 177, *Lysidice* sp.; 178, *Marphysa stylobranchiata;* 179, *Eunice valens;* 180, *E. antennata;* 181, subacicular hook; 182, *E. valens,* subacicular hook; 183, *Dorvillea rudolphi;* 184, *Ophryotrocha puerilis;* 185, *Protodorvillea gracilis;* 186, parapodium; 187, *D. rudolphi,* parapodium; 179, 182 after Fauchald; 186, after Hartman.

DORVILLEIDAE

1. Animal minute (2–8 mm); prostomium with minute, inconspicuous antennae (fig. 184) .... *Ophryotrocha puerilis*
– Animal larger (greater than 8 mm); prostomium with well-developed antennae 2
2. Parapodia with dorsal acicula and dorsal cirrophores (fig. 187) 3
– Parapodia without dorsal acicula or dorsal cirrophores (fig. 186); antennae short, palps long with many articulations (fig. 185) .... *Protodorvillea gracilis*
3. Prostomium with a prominent nuchal papilla posteriorly; palps longer than antennae .... *Dorvillea moniloceras*
– Prostomium without nuchal papilla; antennae longer than palps (figs. 183, 187) .... *Dorvillea rudolphi*

LIST OF DORVILLEIDAE

See Hartman, 1944; Fauchald, 1970; Jumars, 1974, Zool. J., Linnean Soc. 54: 101–135, has recently revised the dorvilleids.

* *Dorvillea annulata* (Moore, 1906). Fauchald (1970) has shown that *D. annulata* is separable from *D. rudolphi* in the structure of the jaw apparatus. Since *D. annulata* is known from Washington (type locality) and western Mexico, it probably will turn up in our fauna.

Dorvillea moniloceras (Moore, 1909). Rocky intertidal; rare.

Dorvillea rudolphi (delle Chiaje, 1828) (=*D. articulata* of Hartman's Atlas, 1968, p. 817). Common in harbors, in mud on floats and pilings. See Richards, 1967, Mar. Biol. 1: 124–133 (reproduction and development).

Ophryotrocha puerilis Claparède and Mecznikow, 1869. In mud and detrital masses in harbors; a contaminant in aquaria; cosmopolitan, principally in southern California.

Protodorvillea gracilis (Hartman, 1938). Sand or muddy sand sediments; intertidal and shelf depths; see Hobson, 1971.

LUMBRINERIDAE

1. Some anterior parapodia with composite hooded hooks (fig. 193) and simple capillary setae 2
– Composite setae absent 3
2. Acicula yellow .... *Lumbrineris latreilli*
– Acicula black .... *Lumbrineris japonica*
3. Posterior parapodia with prolonged postsetal lobes (figs. 190, 191) 4

– Posterior parapodia with postsetal lobes only slightly longer than presetal (fig. 188, 189); color in life orange red on body and iridescent pigment on surface ...... *Lumbrineris zonata*
4. Simple hooded hooks (falcigers) begin setiger 30–40; postsetal lobes stand erect on posterior parapodia (fig. 190) .... *Lumbrineris erecta*
– Simple hooded hooks (falcigers) begin on setiger 1; posterior postsetal lobes do not stand erect (figs. 191, 192) .... *Lumbrineris tetraura*

LIST OF LUMBRINERIDAE

See Hartman, 1944; Fauchald, 1970.

Lumbrineris erecta (Moore, 1904). Sandy mud sediments.

Lumbrineris japonica (Marenzeller, 1879). Sandy mud sediments; not common.

Lumbrineris latreilli Audouin and Milne Edwards, 1834. Sandy mud sediments; rock crevices.

Lumbrineris tetraura (Schmarda, 1861). Sand and mixed sediments; Bodega Bay.

Lumbrineris zonata (Johnson, 1901). The most common lumbrinerid in northern California; common in flats of Tomales Bay and Bodega Harbor; sand or mixed sand-mud sediments.

ARABELLIDAE

1. Parapodia with conspicuous projecting acicular setae .... *Drilonereis* 2
– Acicular setae not heavy nor projecting conspicuously from parapodia (figs. 195–197) .... *Arabella* 3
2. Mandibles (ventral jaw pieces) absent; maxillae prominent (fig. 198) .... *Drilonereis nuda*
– Mandibles present (fig. 199); maxillae prominent .... *Drilonereis falcata*
3. Parapodial lobes short throughout (fig. 195); prostomium usually with 4 eyespots in a transverse row (fig. 194); color in life dark red to green and highly iridescent *Arabella iricolor*
– Postsetal lobes of posterior parapodia elongate and directed upward (fig. 197); eyespots absent or obscure in adult; color in life pale gray or with 3 dorsal rows of dark bluish spots, and with an iridescent epithelium ....*Arabella semimaculata*

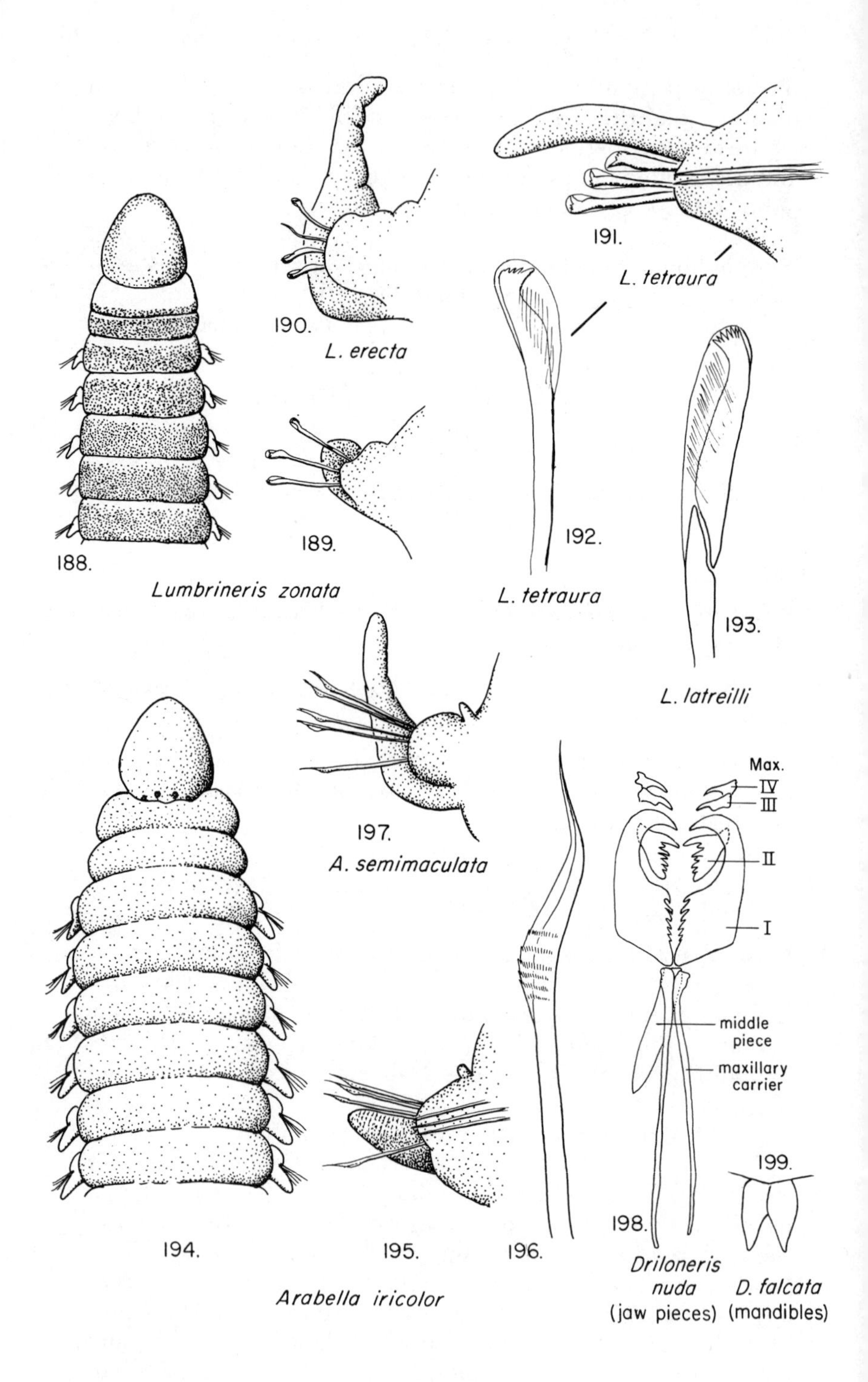

191.
L. tetraura
190.
L. erecta
192.
L. tetraura
193.
L. latreilli
188.
189.
Lumbrineris zonata
197.
A. semimaculata
Max.
IV
III
II
I
middle piece
maxillary carrier
199.
194.
195.
196.
198.
Driloneris nuda (jaw pieces)
D. falcata (mandibles)
Arabella iricolor

LIST OF ARABELLIDAE

See Hartman, 1944; Fauchald, 1970.

Arabella iricolor (Montagu, 1804). Common in rocky habitats, in crevices, with algae and debris; also in mixed sediments.

Arabella semimaculata (Moore, 1911). In rocks and mixed sediments.

Drilonereis falcata Moore, 1911. Mud and sand.

Drilonereis nuda Moore, 1909. Rocky bottoms with kelp.

ORBINIIDAE

1. Anterior margin of the prostomium truncate or broadly rounded (fig. 201) 2
– Anterior margin of the prostomium pointed (fig. 200) 3
2. Two achaetous segments anterior to the first setigerous segment; small, less than 6 mm .... *Protoaricia* sp.
– One achaetous segment anterior to the first setigerous segment (fig. 201); animal larger .... *Naineris dendritica*
3. Thoracic region with ventral papillae (fig. 200) .... *Orbinia johnsoni*
– Thoracic region without ventral papillae 4
4. Thoracic neuropodia with slender pointed setae only (fig. 204) .... *Haploscoloplos elongatus*
– Thoracic neuropodia with blunt spines (fig. 203) and pointed setae 5
5. Neuropodia of some thoracic setigers with 2 postsetal lobes (fig. 205) .... *Scoloplos armiger*
– Thoracic neuropodia with a single postsetal lobe (fig. 202) .... *Scoloplos acmeceps*

LIST OF ORBINIIDAE

See Hartman, 1957; Pettibone, 1957. J. Wash. Acad. Sci. 47: 159–167.

PLATE 41. **Polychaeta** (22). Lumbrineridae, Arabellidae: 188, *Lumbrineris zonata;* 189, posterior parapodium; 190, *L. erecta,* posterior parapodium; 191, *L. tetraura,* posterior parapodium; 192, compound seta; 193, *L. latreilli,* compound seta; 194, *Arabella iricolor;* 195, posterior parapodium; 196, simple seta; 197, *A. semimaculata,* posterior parapodium; 198, *Driloneris nuda,* maxillary apparatus; 199; *D. falcata,* mandibles; 198 after Hartman; 199 after Moore.

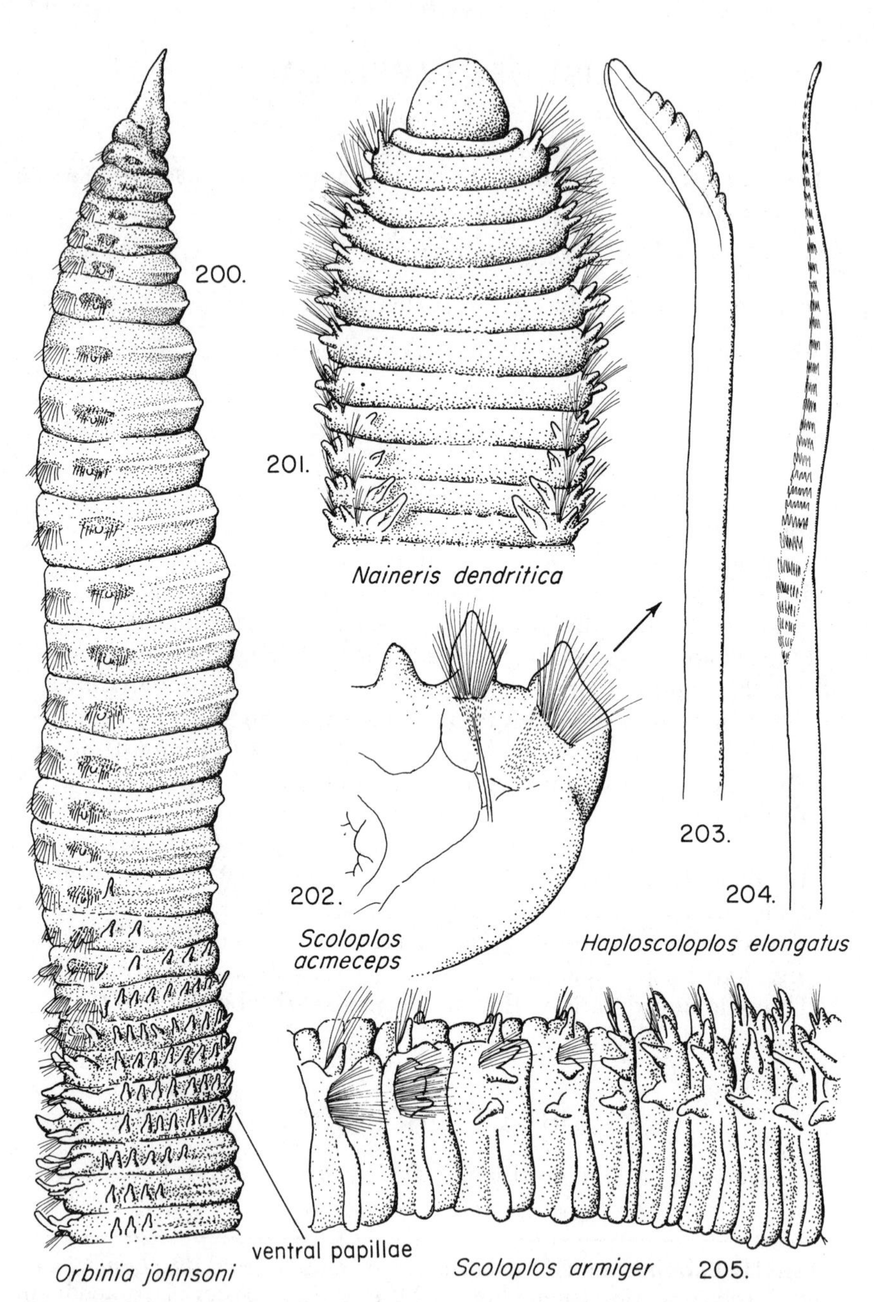

PLATE 42. **Polychaeta** (23). Orbiniidae: 200, *Orbinia johnsoni;* 201, *Naineris dendritica;* 202, *Scoloplos acmeceps,* 18th parapodium; 203, thoracic neuropodial uncinus; 204, *Haploscoloplos elongatus,* thoracic neuropodial seta; 205, *S. armiger,* lateral view of transitional region, segments 16–24; 200, 202–205 after Hartman.

Haploscoloplos elongatus (Johnson, 1901). The most common orbiniid in bays and estuaries, also in subtidal sediments in offshore areas; inhabits sands and muds.

Naineris dendritica (Kinberg, 1867). The most common orbiniid in rocky intertidal areas; found in algal holdfasts, bryozoan masses, and debris under and between rocks.

Orbinia johnsoni (Moore, 1909). Sand sediments of protected beaches.

Protoaricia sp. In rocky intertidal; may be confused with *Naineris dendritica.*

Scoloplos acmeceps Chamberlin, 1919. Mud, algal holdfasts, roots of *Zostera.*

Scoloplos armiger (Müller, 1776). A cosmopolitan species in mixed sediment types; rare in northern California; largely subtidal.

PARAONIDAE

Paraonids are small, threadlike worms often overlooked in routine collecting (fig. 253). Numerous species occur in California, principally in subtidal zones. Because there have been few reports of paraonids in northern and central California, the following key to genera reported from California is provided to stimulate interest should these animals be encountered. See Hartman (1969) for keys to species.

1. Prostomium with median antenna 2
– Prostomium without median antenna (fig. 253) 3
2. Posterior neuropodia with modified setae ....*Aricidea*
– Posterior notopodia with modified setae ....*Cirrophorus*
– Modified setae absent ....*Aedicira*
3. Posterior neuropodia with modified setae .... *Paraonis*
– Posterior notopodia with modified setae ....*Paraonides*

LIST OF PARAONIDAE

See Hartman, 1944b, 1957; Hobson, 1972, Proc. Biol. Soc. Wash. 85: 549–556

* *Aricidea suecica* (Eliason, 1920). Sandy mud; shallow subtidal to deep water.

* *Paraonis gracilis* (Tauber, 1879). Silt and mud; shallow subtidal to deep water.

SPIONIDAE

1. Setiger 5 modified and provided with specialized setae 2
– Setiger 5 not modified 21
2. Branchiae first present anterior to setiger 5 ...... *Boccardia* 3
– Branchiae first present after setiger 5 8
3. Major spines of setiger 5 of one type 4
– Major spines of setiger 5 of two types 5
4. Recurved posterior spines present (fig. 213); pygidium with 2 broad, ventral lappets, each with a short process .... *Boccardia hamata*
– Posterior spines absent; pygidium saucerlike .... *Boccardia truncata*
5. Heavy spines of setiger 5 are (1) falcate and (2) tridentate (figs. 207, 208); branchiae anterior to setiger 5 small and inconspicuous (fig. 206) .... *Boccardia tricuspa*
– Heavy spines of setiger 5 are (1) falcate and (2) bristle-topped (figs. 209, 212); branchiae anterior to setiger 5 larger, conspicuous 6
6. Notosetae absent on setiger 1, acicular setae present on posterior notopodia (fig. 210); bristle-topped spines of setiger 5 with a small accessory tooth (fig. 209) .... *Boccardia berkeleyorum*
– Notosetae present on setiger 1; no acicular setae in posterior notopodia; bristle-topped spines of setiger 5 lack an accessory tooth 7
7. Notosetae of setiger 1 long and directed forward in a fanshaped fascicle (fig. 211; body, fig. 212) .... *Boccardia columbiana*
– Notosetae of setiger 1 short ....*Boccardia proboscidea*
8. Setiger 5 only slightly modified, with a prominent postsetal lobe (figs. 214, 216); heavy spines of setiger 5 inconspicuous, arranged in a J-shaped, double row (fig. 218); hooded hooks

---

PLATE 43. **Polychaeta** (24). Spionidae: 206, *Boccardia tricuspa;* 207–208, modified spines of setiger 5; 209, *B. berkeleyorum,* modified spine of setiger 5; 210, posterior acicular seta; 211, *B. columbiana;* 212, modified spine of setiger 5; 213, *B. hamata,* posterior spine; 214, *Pseudopolydora paucibranchiata;* 215, posterior end; 216, *P. kempi;* 217, posterior end; 218, lateral view of setigers 3–6; 219, *Polydora commensalis;* 220, heavy spine from setiger 5; 221, *P. brachycephala,* modified spine of setiger 5; 222, *P. pygidialis,* modified spine from setiger 5; 223, *P. pygidialis,* posterior end in lateral view; 224, *P. nuchalis;* 225, modified spine of setiger 5; 226, companion seta from setiger 5; 227, *P. ligni,* companion seta from setiger 5; 228, modified spine from setiger 5; 229, *P. elegantissima;* 230, modified spine from setiger 5; 206, 224–226 after Woodwick.

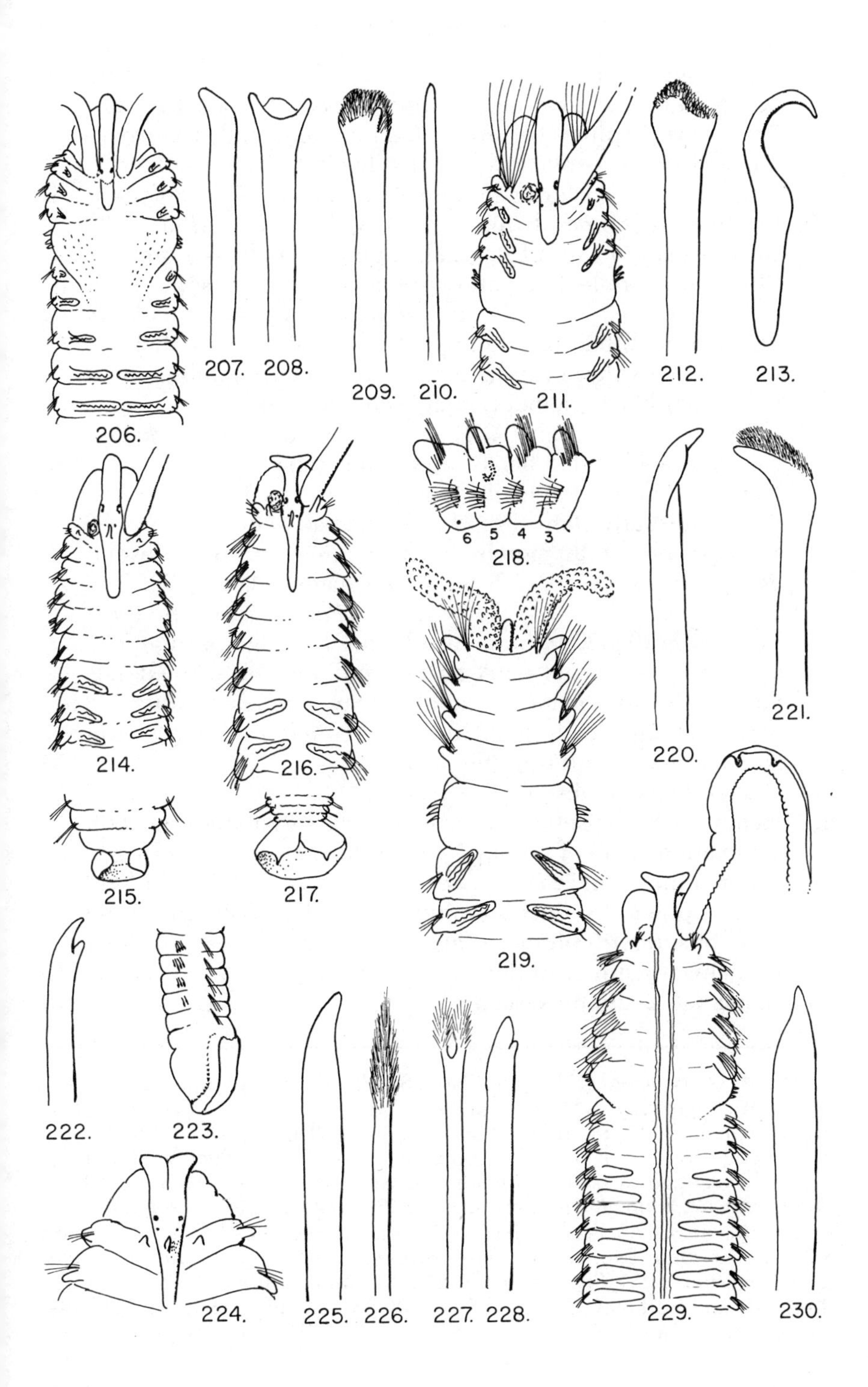

206.
207.
208.
209.
210.
211.
212.
213.
214.
215.
216.
217.
218.
6
5
4
3
219.
220.
221.
222.
223.
224.
225.
226.
227.
228.
229.
230.

begin on setiger 8 .... *Pseudopolydora* 9

– Setiger 5 greatly modified, without postsetal lobes (figs. 229, 231); heavy spines of setiger 5 prominent and usually arranged in a crescentic row; hooded hooks begin on setigers 7, 9, 10, or more posterior setigers .... *Polydora* 10

9. Prostomium rounded (fig. 214); pygidium saucerlike, with smooth margin (fig. 215); small (approximately 4–6 mm); in life, exhibits reflective yellow pigment spots on palps .... *Pseudopolydora paucibranchiata*

– Prostomium bifid (fig. 216); pygidium cup-shaped with 2 dorsal processes (fig. 217); larger (up to 28 mm); exhibits intersegmental black pigment on anterior segments .... *Pseudopolydora kempi*

10. Neuropodial hooded hooks begin on setiger 7 12

– Neuropodial hooded hooks begin on setigers 10–17 11

11. Palps unusually short (fig. 219); branchiae begin on setiger 6 and are long with membranous margins; caruncle absent; heavy spines of setiger 5 with large lateral flange (fig. 220) .... *Polydora commensalis*

– Palps long (fig. 229); branchiae begin on setiger 8 (rarely 7), are short and do not meet at the midline; caruncle extends posteriorly over many segments; heavy spines of setiger 5 with a lateral sheath (fig. 230) ....*Polydora elegantissima*

12. Nuchal tentacle present (fig. 224) 13

– Nuchal tentacle absent 14

13. Heavy spines of setiger 5 have an accessory tooth (fig. 228); companion setae of setiger 5 featherlike, closely adhering to major spines (fig. 227) ....*Polydora ligni*

– Heavy spines of setiger 5 simple, falcate (fig. 225); companion setae of setiger 5 plumose (fig. 226) .... *Polydora nuchalis*

14. Branchiae begin on setiger 7 15

– Branchiae begin on setigers 8–10 19

PLATE 44. **Polychaeta** (25). Spionidae: 231, *Polydora convexa;* 232, modified spine of setiger 5; 233, *P. spongicola,* modified spine of setiger 5; 234, *P. limicola,* modified spine from setiger 5; 235–236, *P. websteri,* modified spines from setiger 5; 237, *P. socialis,* companion seta from setiger 5; 238, heavy spine from setiger 5; 239, *Spiophanes missionensis;* 240, *S. bombyx;* 241, *S. fimbriata;* 242, *Rhynchospio arenincola;* 243, *Streblospio benedicti;* 244, *Pygospio californica;* 245, *P. elegans;* 246, *Laonice cirrata,* head; 247, lateral view of genital region; 248, *Paraprionospio pinnata;* 249, *Prionospio pygmaeus;* 250, *P. cirrifera;* 251, *Spio filicornis;* 252, *Nerinides acuta;* 233, 241, 246 after Berkeley and Berkeley; 239, 242, 244, 248, 251, 252 after Hartman; 247 after Wesenberg-Lund.

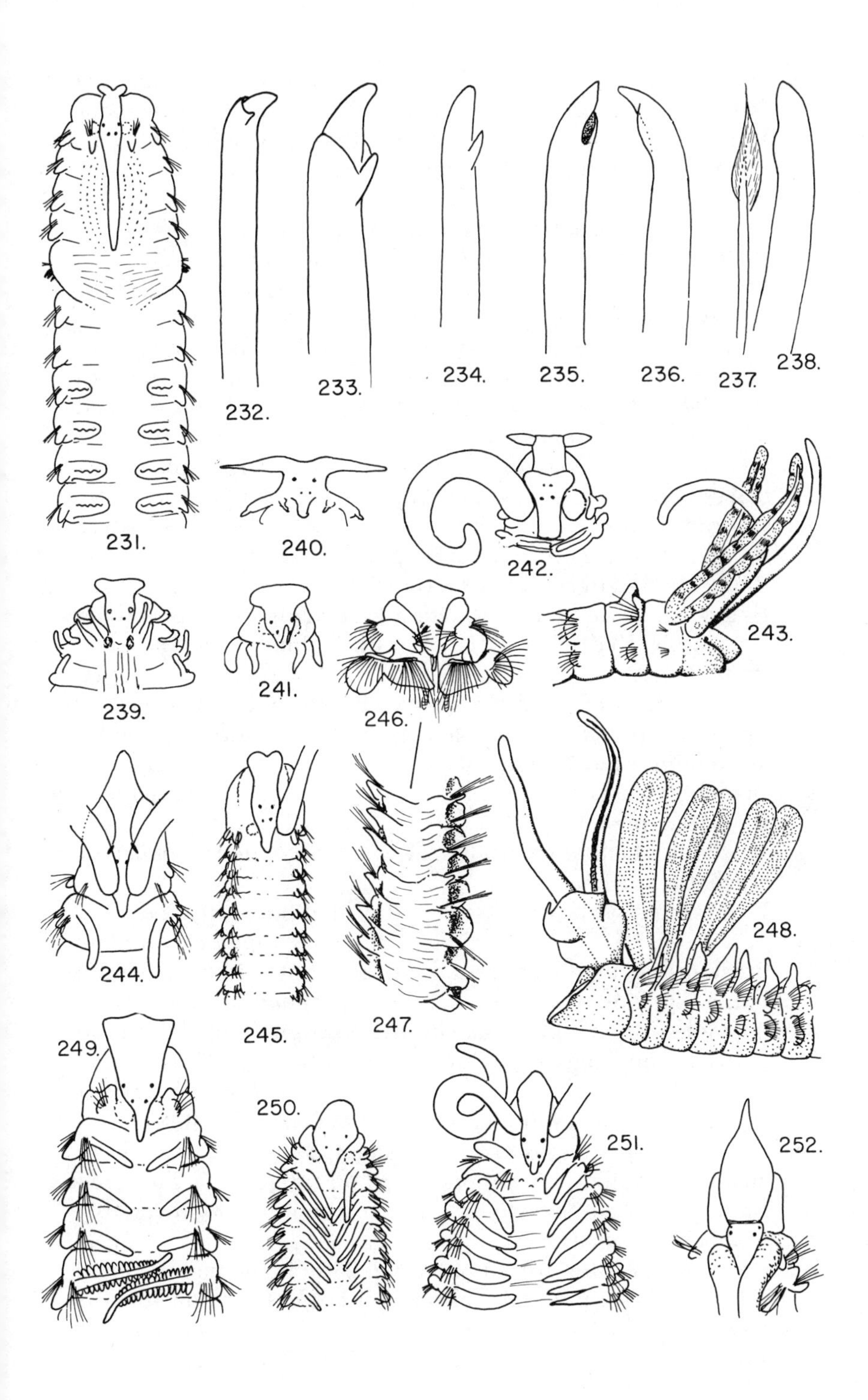
231.
232.
233.
234.
235.
236.
237.
238.
240.
242.
243.
239.
241.
246.
244.
245.
247.
248.
249.
250.
251.
252.

15. Heavy spines of setiger 5 falcate with bristled tops (fig. 221) .... *Polydora brachycephala*
- Heavy spines of setiger 5 falcate without bristled tops 16
16. Prostomium entire on anterior margin 17
- Prostomium incised 18
17. Heavy spines of setiger 5 falcate with a large, lateral accessory tooth (fig. 222); pygidium large, scoop-shaped (fig. 223) .... *Polydora pygidialis*
- Heavy spines of setiger 5 falcate with the terminal end surrounded by a collar (fig. 233) .... *Polydora spongicola*
18. Heavy spines of setiger 5 falcate with a small accessory tooth (fig. 234); often forms massive colonies on rocks, wharves, and other exposed surfaces .... *Polydora limicola*
- Heavy spines of setiger 5 falcate with an accessory flange attached along one side (figs. 235, 236); bores into calcareous substrates .... *Polydora websteri*
19. Branchiae begin on setigers 9–11 .... *Polydora giardi*
- Branchiae begin on setiger 8 (fig. 231) 20
20. Heavy spines of setiger 5 with a broad collar on the convex side of the shaft (figs. 231, 232); bores into calcareous substrates .... *Polydora convexa*
- Heavy spines of setiger 5 simple, falcate with no accessory structures (fig. 238; also fig. 237); dorsal surface usually flecked with black pigment .... *Polydora socialis*
21. Branchiae absent .... *Spiophanes* 22
- Branchiae present 24
22. Nuchal tentacle present (fig. 241) ... *Spiophanes fimbriata*
- Nuchal tentacle absent 23
23. Prostomium with well-developed frontal horns (fig. 240) ... . *Spiophanes bombyx*
- Prostomium broad at frontal margin, but without definite horns (fig. 239) .... *Spiophanes missionensis*
24. A single pair of branchiae inserted posterior to the paired palps; second segment with a dorsal collar (fig. 243) .... *Streblospio benedicti*
- Branchiae occur in numerous pairs; dorsal collar absent ... 25
25. Branchiae begin posterior to setiger 10, males may have small, elongated pair of dorsal cirri on setiger 2 .. *Pygospio* 26
- Branchiae begin on setiger 1 or 2 and continue for a variable number of segments 27
26. Prostomium anteriorly bilobed (fig. 245); branchiae begin on setigers 11–13; neuropodial hooded hooks begin on setigers 8–9 .... *Pygospio elegans*
- Prostomium anteriorly conical (fig. 244); branchiae begin on

setiger 19; neuropodial hooks begin on setiger 23 .... *Pygospio californica*

27. Prostomium with articulated lateral horns (fig. 242); branchiae begin on setiger 2 .... *Rhynchospio arenincola*
- Prostomium otherwise 28

28. Branchiae present only on anterior setigers; interramal (genital) pouches may be present in middle and posterior segments 29
- Branchiae present throughout length of the body 32

29. Hooded hooks in neuropodia only, bidentate; interramal pouches present in middle region (figs. 246, 247) .... *Laonice cirrata*
- Hooded hooks in both neuropodia and notopodia, multidentate; interramal pouches present or absent 30

30. Branchiae all pinnate, begin on setiger 1; peristomium fused dorsally and laterally with the first achaetous segment, forming prominent peristomial wings; palps long, having a broad membrane on the outer margin .... *Paraprionospio pinnata*
- Branchiae begin on setiger 2 and may be cirriform, pinnate, or both 31

31. Branchiae number 4 pairs, with 1–3 cirriform and 4 pinnate (fig. 249); .... *Prionospio pygmaeus*
- Branchiae number 6–12 pairs (fig. 250), all cirriform; interramal pouches sometimes present, beginning on setigers 5–7 and extending for 20 setigers. .... *Prionospio cirrifera*

32. Branchiae begin on setiger 2; prostomium acutely pointed (fig. 252) 33
- Branchiae begin on setiger 1; prostomium rounded (fig. 251) .... *Spio filicornis*

33. Hooded hooks present in both neuropodia and notopodia of posterior setigers; neuropodial lamellae bilobed in posterior setigers .... *Scolelepis squamatus*
- Hooded hooks present only in neuropodia; neuropodial lamellae entire .... *Nerinides* 34

34. Nuchal tentacle present .... *Nerinides tridentata*
- Nuchal tentacle absent (fig. 252) .... *Nerinides acuta*

LIST OF SPIONIDAE

See Hartman, 1936, Univ. Calif. Publ. Zool. 41: 45–52; Hartman, 1941a; Blake, 1969, 1971; Blake and Evans, 1973; Blake and Woodwick, 1971, Bull. So. Calif. Acad. Sci. 70: 31–42 (review of *Boccarida*) and 1972, *Ibid.* 70: 72–79 (new species of *Polydora*).

Boccardia berkeleyorum Blake and Woodwick, 1971. Bores in coralline algae, hermit-crab shells, and the jingle shell *Pododesmus*.

Boccardia columbiana Berkeley, 1927. Bores in mollusc shells and coralline algae. See Woodwick, 1963, Bull. So. Calif. Acad. Sci., 62: 132–139.

Boccardia hamata (Webster, 1879) (= *B. uncata* Berkeley, 1927). Mudflats with oysters; algal holdfasts. See Blake, 1966, Bull. So. Calif. Acad. Sci. 65: 176–184 (taxonomy); Dean and Blake, 1966, Biol. Bull. 130: 316–330 (life history).

Boccardia proboscidea Hartman, 1940. High intertidal rock pools, in crevices; sandy mudflats; occasionally in debris in rocky intertidal; see Woodwick, 1963 (above).

Boccardia tricuspa (Hartman, 1939). Bores in mollusc shells and coralline algae. See Woodwick, 1963, Proc. Biol. Soc. Wash. 76: 209–216.

Boccardia truncata Hartman, 1936.

Laonice cirrata (Sars, 1851). Silty mud.

Nerinides acuta (Treadwell, 1914). Clean sand beaches.

Nerinides tridentata Southern, 1914. Specimens near this species have been taken from mud in Tomales Bay.

Paraprionospio pinnata (Ehlers, 1901) (= *Prionospio pinnata*). Silty mud; principally subtidal.

* *Polydora alloporis* Light, 1970. Subtidal, in *Allopora californica* and *A. venusta*. See Light, 1970, Proc. Calif. Acad. Sci. (4) 37: 459–471.

Polydora brachycephala Hartman, 1936. Sandy mud sediments, low intertidal.

Polydora commensalis Andrews, 1891. Commensal in hermit-crab shells; see Blake, 1969, 1971.

Polydora convexa Blake and Woodwick, 1972. Bores in hermit-crab shells, and the jingle shell *Pododesmus;* also with algal holdfasts, *Balanus*, bryozoans, and sponges.

Polydora elegantissima Blake and Woodwick, 1972. Bores in hermit-crab and Pismo-clam (*Tivela*) shells.

Polydora giardi Mesnil, 1896. Bores in coralline algae and mollusc shells. This species in California is nearly identical to *P. tridenticulata* Woodwick from the Marshall Islands; it is not clear which name should be applied. See Woodwick, 1964, Pac. Sci. 18: 146–159.

Polydora ligni Webster, 1879. Mud flats of bays and estuaries; possibly introduced; see Blake, 1969, 1971.

Polydora limicola Annenkova, 1934. Forms large aggregations on rocks, wharves, and ship bottoms.

Polydora nuchalis Woodwick, 1953. Mudflats of estuaries and bays. See Woodwick, 1953, J. Wash. Acad. Sci. 43: 381–383 and 1960, Pac. Sci. 14: 122–128 (larval development).

Polydora pygidialis Blake and Woodwick, 1972. Bores in mollusc shells and ectoprocts.

Polydora socialis (Schmarda, 1861). Sandy mud to mud sediments; also algal holdfasts, occasionally boring in hermit-crab shells; see Blake, 1969, 1971.

Polydora spongicola Berkeley and Berkeley, 1950. Commensal with sponges. See Woodwick, 1963, Proc. Biol. Soc. Wash. 76: 209–216.

Polydora websteri Hartman, 1943. Bores in mollusc shells and other calcareous materials; see Blake, 1969, 1971.

Pseudopolydora kempi (Southern, 1921). Sandy mud sediments of bays. See Okuda, 1937, J. Fac. Sci. Hokkaido Imp. Univ. 5: 217–254; Imajima and Hartman, 1964.

Pseudopolydora paucibranchiata (Okuda, 1937). A recent introduction with *P. kempi* in California bays and estuaries. In Tomales Bay *P. paucibranchiata* may be the dominant spionid polychaete on many sand flats; prefers finer sediments than *P. kempi.* Both species have probably been introduced with Japanese oyster (*Crassostrea gigas*) imports; see Okuda, 1937 (above); Imajima and Hartman, 1964.

Prionospio cirrifera Wirén, 1883. Silty mud.

Prionospio pygmaeus Hartman, 1961. Silty mud.

Pygospio californica Hartman, 1936. Intertidal sand flats.

Pygospio elegans Claparède, 1863. With the sabellid *Chone ecaudata* in sand and tube aggregations in rocky intertidal areas; sand flats, high intertidal.

Rhynchospio arenincola Hartman, 1936. Sand and mud sediments. See Banse, 1964, Proc. Biol. Soc. Wash. 76: 203–204.

Scolelepis squamatus (Müller, 1806). Clean sand sediments.

Spio filicornis (Müller, 1766). Sand flats.

Spiophanes bombyx (Claparède, 1870). Sandy mud.

Spiophanes fimbriata Moore, 1923. Mud; rare.

Spiophanes missionensis Hartman, 1941. Sandy mud sediments.

Streblospio benedicti Webster, 1879. Mud flats of estuaries and tributaries;

see Dean, 1965, Biol. Bull. 128: 67–76 (reproduction and larval development; based on Atlantic material).

MAGELONIDAE

1. Lateral pouches present between successive parapodia of abdominal segments (fig. 254); setae of modified 9th setiger of two types, one limbate (fig. 256) other mucronate (fig. 255) .... *Magelona sacculata*
– Lateral pouches absent; setae of modified 9th setiger limbate (fig. 258) .... *Magelona pitelkai*

LIST OF MAGELONIDAE

See Hartman, 1944b; Jones, 1968, Biol. Bull. 134: 272–297.

Magelona pitelkai Hartman, 1944. The most common magelonid in central and northern California; clean sand to sandy mud.

Magelona sacculata Hartman, 1961. Fine sands; rare

TROCHOCHAETIDAE (= DISOMIDAE)

Trochochaeta (=*Disoma*) *franciscanum* (Hartman, 1947) (fig. 259) is known from shallow mud in San Francisco Bay, and is probably present in other northern California localities; larvae have been taken in plankton from Tomales Bay. See Hartman, 1947, J. Wash. Acad. Sci. 37: 160–169; Banse and Hobson, 1968.

CHAETOPTERIDAE

1. With one pair of head appendages (palps) 2
– With two pairs of head appendages, a pair of small cirri located behind the longer pair of palps (fig. 261); tubes very slender (up to 1 mm in diameter), often in clusters, with faint or obscure annulations .... *Phyllochaetopterus prolifica*
2. Palps shorter than the anterior region (fig. 263); middle region highly modified, with 12th pair of notopodia long and winglike,

PLATE 45. **Polychaeta** (26). Paraonidae, Magelonidae, Trochochaetidae, Chaetopteridae: 253, *Paraonis gracilis;* 254, *Magelona sacculata;* 255, mucronate seta from setiger 9; 256, limbate seta from setiger 9; 257, *Magelona pitelkai;* 258, winged seta from setiger 9; 259, *Trochochaeta franciscanum;* 260, *Mesochaetopterus taylori;* 261, *Phyllochaetopterus prolifica;* 262, *Chaetopterus variopedatus* in tube; 263, *C. variopedatus;* 253–259 after Hartman; 260 after Potts; 263 after MacGinitie.

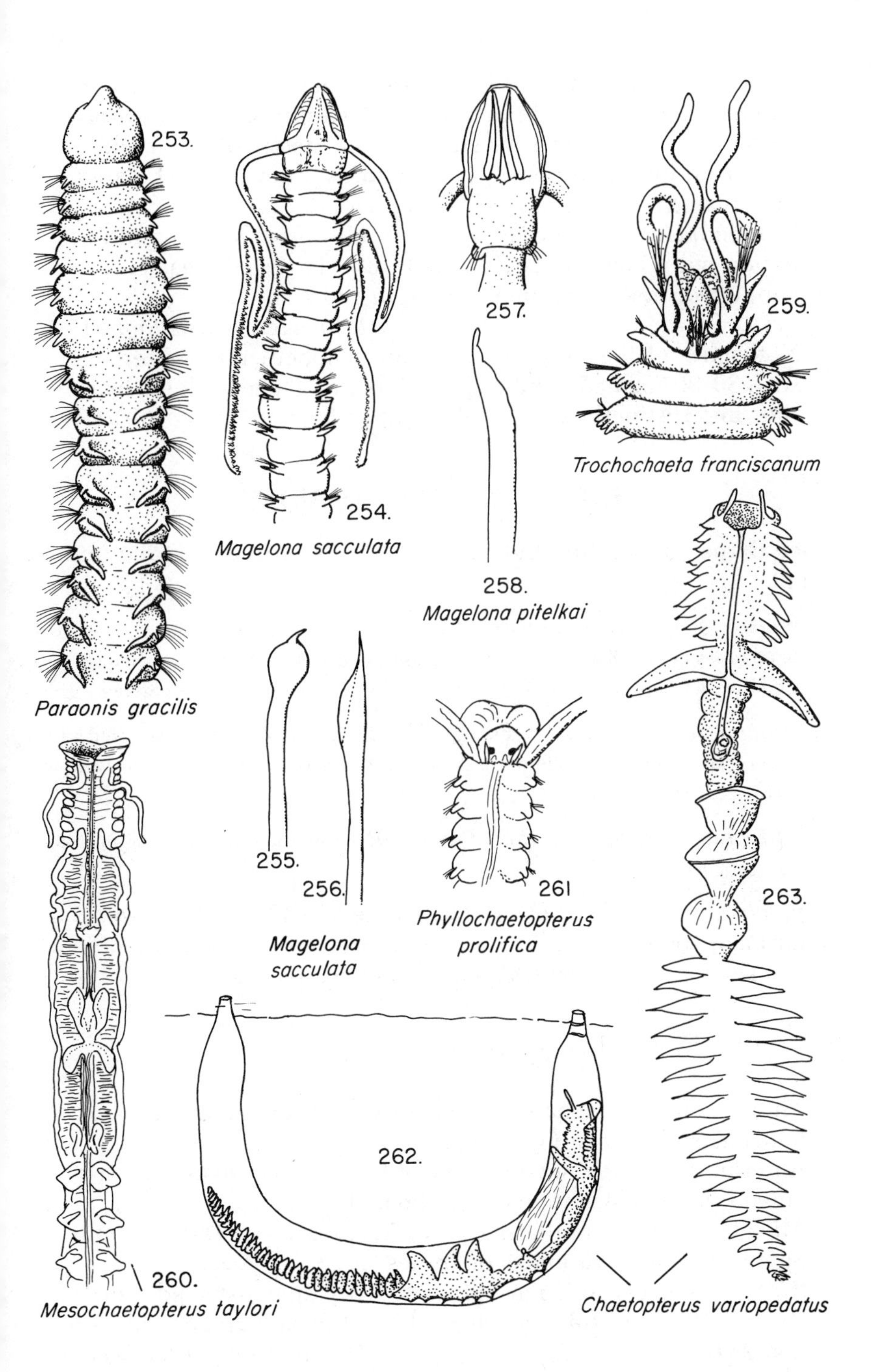
253.
Paraonis gracilis
254.
Magelona sacculata
257.
258.
Magelona pitelkai
259.
Trochochaeta franciscanum
255.
256.
Magelona
sacculata
261
Phyllochaetopterus
prolifica
263.
260.
Mesochaetopterus taylori
262.
Chaetopterus variopedatus

and others on somites 14–16 forming fans; tubes U-shaped parchmentlike (fig. 262); in sand .. *Chaetopterus variopedatus*
– Palps longer than the anterior region; notopodia of middle region unilobed, bilobed, or trilobed, but never fused across the dorsum 3
3. Notopodia of the middle region are simple lobes; middle region with 3 segments (fig. 260); tube large and coarse, not annulated; in sand or among *Zostera* roots .... *Mesochaetopterus taylori*
– Notopodia of the middle region are trilobed; middle region with 30–90 segments; tube about 2 mm in diameter, long, hyaline, and distinctly annulated ... *Spiochaetopterus costarum*

LIST OF CHAETOPTERIDAE

See Barnes, 1965, Biol. Bull. 129: 217–233 (feeding and tube-building).

Chaetopterus variopedatus (Renier, 1804). Cosmopolitan; forms U-shaped tubes in sand flats of bays and estuaries; mucous-bag feeding mechanism described by MacGinitie, 1939, Biol. Bull. 77: 115–118; rare north of Monterey.

Mesochaetopterus taylori Potts, 1914. In sand flats, usually associated with *Zostera;* rare; northern California may be the southern limit; known from Dillon Beach and Humboldt Bay.

Phyllochaetopterus prolifica Potts, 1914. Intertidal to about 100 m; common; rocky habitats.

Spiochaetopterus costarum (Claparède, 1870) (as *Telepsavus* in Hartman, 1969). Fine sands; cosmopolitan; our species has been named *S. costarum pottsi* (E. Berkeley, 1927). See Gitay, 1969, Sarsia 37: 9–20.

CIRRATULIDAE

1. Anterior end with a pair of thick, grooved palps in addition to tentacular filaments (figs. 264, 271) 2
– Anterior end with several grooved tentacular filaments arranged in a transverse row (figs. 266, 268) 4
2. All setae slender capillaries .... *Tharyx* spp.
– Some setae acicular hooks or spines 3
3. (*Note 3 choices*) Acicular spines distally spoon-shaped; constructs calcareous tube masses; color in life dark green to black (fig. 264) .... *Dodecaceria fewkesi*

- Acicular spines in posterior segments distally entire, forming nearly complete rings about the body (fig. 267) .... *Chaetozone setosa*
- Acicular spines in posterior segments distally bifid (fig. 272) .... *Caulleriella* spp.

4. Transverse row of tentacular filaments arise on setiger 1; first branchiae appear on the same segment as the tentacular filaments .... *Cirratulus cirratus*
- Transverse row of tentacular filaments arise on setiger 5 (7th segment); first branchiae appear anterior to the tentacular filaments (fig. 268) .... *Cirriformia* 5

5. Posterior parapodia with one or few black spines (fig. 270) .. .. *Cirriformia luxuriosa*
- Posterior parapodia with numerous slender, yellow spines (fig. 269) .... *Cirriformia spirabrancha*

LIST OF CIRRATULIDAE

Caulleriella spp. Several species of this genus are known from California; see Hartman (1969, p. 226) for key to species.

Chaetozone setosa Malmgren, 1867. Cosmopolitan; silty sediments.

Cirratulus cirratus (Müller, 1776). Cosmopolitan, rocky habitats; three subspecies have been reported from the area: *C. cirratus cingulatus* Johnson, 1901; *C. cirratus spectabilis* (Kinberg, 1866); and *C. cirratus cirratus;* see Hartman (1969) for more information.

Cirriformia luxuriosa (Moore, 1904). Rocky habitats.

Cirriformia spirabrancha (Moore, 1904). Forms extensive beds in *Zostera* communities; sandy mud sediments; estuaries and bays.

Dodecaceria fewkesi Berkeley and Berkeley, 1954. Common in rocky habitats, forming calcareous masses; the more cosmopolitan *D. concharum* Oersted has been reported in our fauna, but despite examination of hundreds of specimens of *Dodecaceria,* I have seen none which can be referred to *D. concharum*. See Berkeley and Berkeley 1954. J. Fish. Res. Bd. Canada 11: 326–334 (life history).

* *Tharyx multifilis* Moore, 1909. *T. multifilis* and *T. parvus* are so similar that no attempt has been made to separate them in the key. *T. multifilis* occurs in marine environments; mud; also in debris in rocky habitats. See Hartman, 1969, pp. 263–266; Banse and Hobson, 1968.

* *Tharyx parvus* Berkeley, 1929. Estuarine habitats; fine, silty muds.

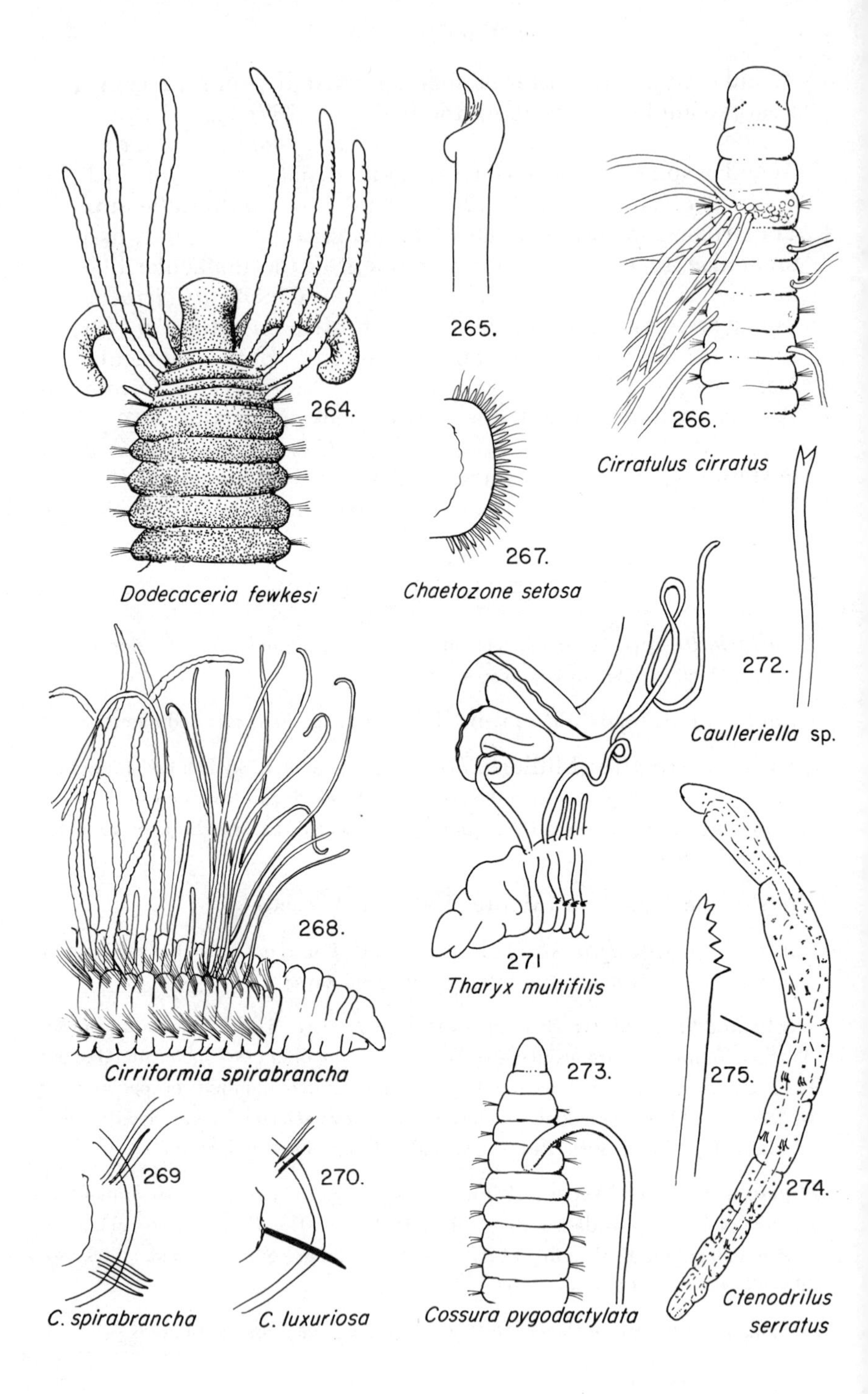
264.
Dodecaceria fewkesi
265.
267.
Chaetozone setosa
266.
Cirratulus cirratus
272.
Caulleriella sp.
268.
Cirriformia spirabrancha
271
Tharyx multifilis
269
C. spirabrancha
270.
C. luxuriosa
273.
Cossura pygodactylata
275.
274.
Ctenodrilus serratus

COSSURIDAE

Cossura is rare, small, and threadlike, inhabiting sandy subtidal mud. *Cossura pygodactylata* Jones, 1956 (fig. 273) was described from San Francisco Bay at 1–10 m, and has been found in Tomales Bay. See Jones, 1956, J. Wash. Acad. Sci. 46: 127–130.

CTENODRILIDAE

Ctenodrilus serratus (Schmidt, 1857) (fig. 274) is cosmopolitan in distribution. It is small, inconspicuous, and occurs intertidally in clumps of debris; also commonly found in seawater systems of marine laboratories. The species is small (2.5–6.0 mm in length); purplish with a speckled surface (fig. 274); serrated setae are characteristic (fig. 275); rapid asexual reproduction is prevalent. See Hartman, 1944b.

FLABELLIGERIDAE

1. Body smooth, encased in a transparent, gelatinous tube; free-living, color in life dark green . . *Flabelligera infundibularis*
– Body papillated, without gelatinous sheath 2
2. Neurosetae both simple and composite; body densely papillated; may be commensal with *Strongylocentrotus purpuratus;* color in life, dorsum reddish purple and ventrum orange yellow with greenish cast . . . . *Flabelliderma commensalis*
– Neurosetae all simple; papillae dense or thinly dispersed . . . 3
3. Cephalic cage absent; body short; branchiae numerous . . . . *Brada* sp.
– Cephalic cage present; body long; 4 or many pairs of retractile branchiae 4
4. Surface papillae form a fringe at the anterior end of each segment (fig. 276); branchiae numerous . . . . *Pherusa inflata*
– Surface papillae uniformly distributed (fig. 277); 4 pairs of branchiae . . . . *Pherusa papillata*

PLATE 46. **Polychaeta** (27). Cirratulidae, Cossuridae, and Ctenodrilidae: 264, *Dodecaceria fewkesi;* 265, spoon-shaped acicular seta; 266, *Cirratulus cirratus;* 267, *Chaetozone setosa,* posterior segment; 268, *Cirriformia spirabrancha,* head; 269, posterior parapodium; 270, *C. luxuriosa,* posterior parapodium; 271, *Tharyx multifilis* (after Moore); 272, *Caulleriella* sp., bifid acicular spine; 273, *Cossura pygodactylata;* 274, *Ctenodrilus serratus;* 275, serrated seta.

LIST OF FLABELLIGERIDAE

* *Brada villosa* (Rathke, 1843). Shallow, dredged.

Flabelliderma commensalis (Moore, 1909). Rocky intertidal; free-living or commensal on the surface of the purple sea urchin *Strongylocentrotus purpuratus;* not known north of Monterey Bay. See Spies, 1973, J. Morph. 139: 465–490 (circulatory system).

Flabelligera infundibularis Johnson, 1901. Muddy bottoms; mainly subtidal.

Pherusa inflata (Treadwell, 1914). Rocky habitats. See Hartman, 1952, Pac. Sci. 6: 71–74.

Pherusa papillata (Johnson, 1901). Rocky sediments.

SCALIBREGMIDAE

Representatives of this family are characteristic of subtidal muddy sediments and are rarely encountered in the intertidal. An exception in California is *Oncoscolex pacificus* (Moore, 1909) which may occasionally be taken in the low rocky intertidal (fig. 278).

OPHELIIDAE

1. Ventral groove present and well defined (fig. 279) 2
– Ventral groove absent; a pair of setal fascicles occur anterior to the mouth; posterior parapodia with large lobes or lappets; body stout; characteristic foul smell .... *Travisia gigas*
2. Ventral groove extends entire length of the body 3
– Ventral groove limited to posterior region 5
3. Branchiae absent .... *Polyophthalmus pictus*
– Cirriform branchiae present on some parapodia 4
4. Some parapodia with lateral eyespots ...... *Armandia brevis*
– Lateral eyespots absent .... *Ammotrypane aulogaster*
5. Anterior region, including setigers 1 and 2, set off from thorax by a constriction (fig. 280) 7
– Anterior region not set off by a constriction 6
6. (*Note 3 choices*) Body with 33 setigers .... *Ophelia assimilis*

PLATE 47. **Polychaeta** (28). Flabelligeridae, Scalibregmidae, Opheliidae, and Capitellidae: 276, *Pherusa inflata;* 277, *P. papillata;* 278, *Oncoscolex pacificus;* 279, *Ophelia limacina;* 280, *Euzonus mucronata;* 281, branchia; 282, *E. dillonensis*, branchia; 283, *E. williamsi*, branchia; 284, *Mediomastus californiensis;* 285, *Capitella capitata;* 276–277, 281–284 after Hartman; 278 after Imajima; 280, after Treadwell; 285 after Berkeley and Berkeley.

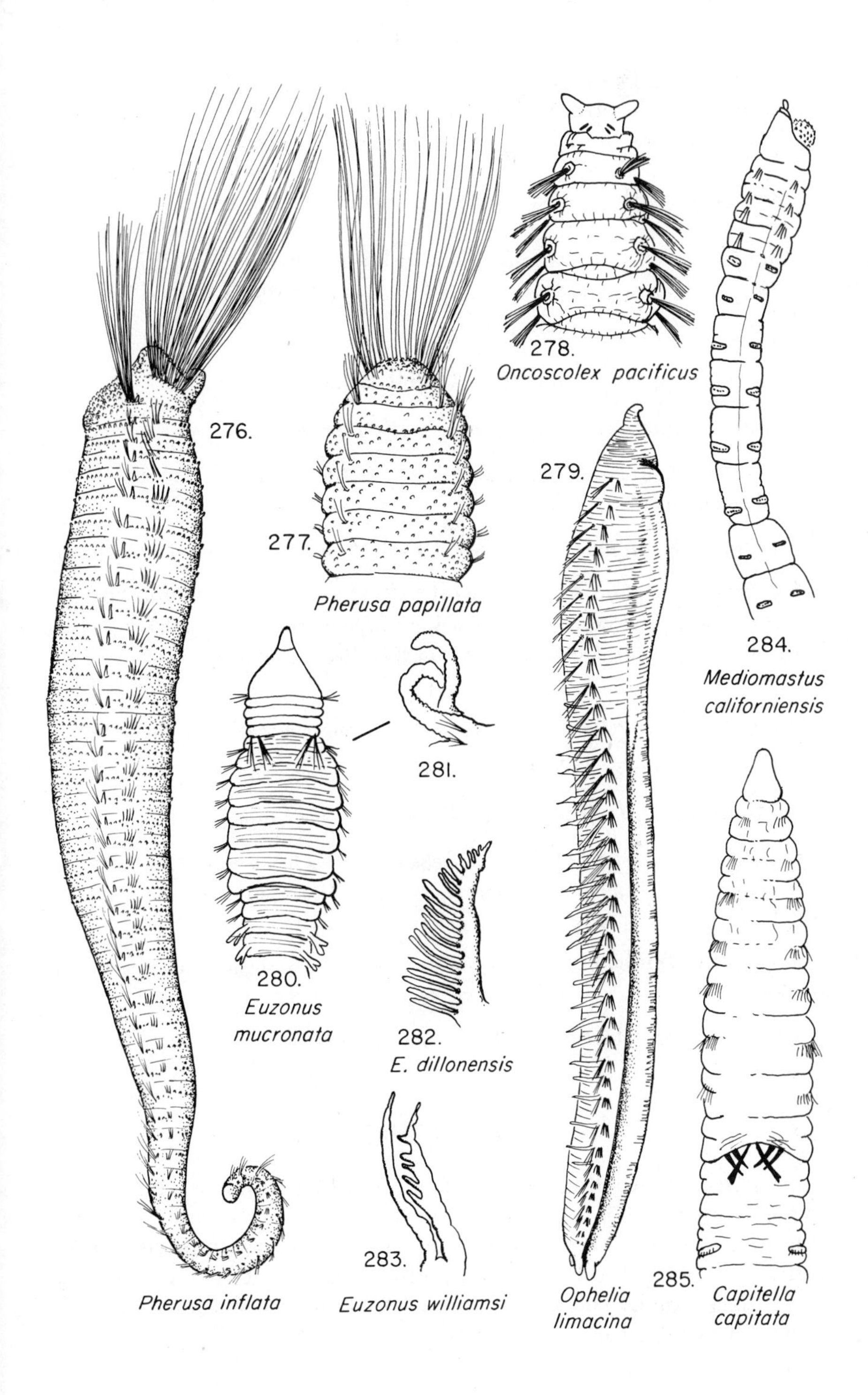
276.
277.
Pherusa papillata
278.
Oncoscolex pacificus
279.
284.
Mediomastus californiensis
280.
Euzonus mucronata
281.
282.
E. dillonensis
283.
Euzonus williamsi
285.
Pherusa inflata
Ophelia limacina
Capitella capitata

- Body with 38 setigers .... *Ophelia pulchella*
- Body with 39 setigers (fig. 279) .... *Ophelia limacina*

7. Branchiae not branched, with pectinate divisions (fig. 282) .. .. *Euzonus dillonensis*

- Branchiae with 2 or 3 branches and a few lateral pinnules (fig. 283) .... *Euzonus williamsi*
- Branchiae with 2 branches and without pinnules (fig. 281) .. .. *Euzonus mucronata*

LIST OF OPHELIIDAE

Ammotrypane aulogaster Rathke, 1843. Muddy sediments.

Armandia brevis (Moore, 1906) (= *A. bioculata* Hartman, 1938). Sandy mud and silt.

Euzonus dillonensis (Hartman, 1938). Clean sand beaches; Dillon Beach.

Euzonus mucronata (Treadwell, 1914) (=*Thoracophelia mucronata*). Clean sand beaches. See Dales, 1952, Biol. Bull. 102: 232–252 (ecology, larval development); McConnaughey and Fox, 1949, Univ. Calif. Publ. Zool. 47: 319–340.

Euzonus williamsi (Hartman, 1938). Clean sand beaches; Dillon Beach.

Ophelia assimilis Tebble, 1953. Sand beaches.

Ophelia limacina (Rathke, 1843). Sand beaches; cosmopolitan.

Ophelia pulchella Tebble, 1953. Sandy mud sediments; shallow-dredged in Bodega Bay. See Tebble, 1953, Ann. Mag. Nat. Hist. (12) 6: 361–368 (review of *Ophelia*).

Polyophthalmus pictus (Dujardin, 1839). Rocky habitats associated with algae.

Travisia gigas Hartman, 1938. Sandy mudflats.

STERNASPIDAE

Sternaspis fossor Stimpson, 1854 is a small, swollen worm with an ovoid shape (fig. 65). It has very few segments and burrows head first into fine, silty sediments. Hard anal plates cover the entrance to the burrow allowing the terminal branchial filaments to extend into the water. It is usually subtidal and may be taken with shallow grabs; occasionally washed in.

CAPITELLIDAE

1. Thorax with 9 setigers; notopodia of setigers 8–9 of male with heavy genital spines (fig. 285); females with large genital pore .... *Capitella capitata*
– Thorax with 10 or more setigers; genital setae absent 2
2. Thorax with 10 setigers; setigers 1–4 with capillary setae, setigers 5–10 with hooks (fig. 284) ... *Mediomastus californiensis*
– Thorax with 11 or more setigers 3
3. Thorax with 11 setigers 5
– Thorax with 13 or 16 or more setigers 4
4. Thorax with 13 setigers, all with capillary setae *Dasybranchus* 8
– Thorax with more than 16 setigers, all with capillary setae .. .. *Anotomastus gordioides*
5. Thoracic setigers 1–5 with capillary setae, 6–11 with hooks .... *Heteromastus* 6
– All thoracic setigers with capillary setae .... *Notomastus* 7
6. Branched branchiae in posterior abdominal notopodia .... *Heteromastus filobranchus*
– Distinct branchiae absent .... *Heteromastus filiformis*
7. Branchiae absent; nephridial pores limited to thorax .... *Notomastus tenuis*
– Branchiae present in median and posterior segments as tufts of small retractile filaments; nephridial pores conspicuous on abdomen .... *Notomastus magnus*
8. Thoracic segments relatively smooth; branchiae with 2–3 lobes .... *Dasybranchus glabrus*
– Thoracic segments with deep lines; branchiae large, branched .... *Dasybranchus lumbricoides*

LIST OF CAPITELLIDAE

See Hartman, 1947a.

Anotomastus gordioides (Moore, 1909). Mud; rare.

Capitella capitata (Fabricius, 1780). Cosmopolitan in mud flats; common.

Dasybranchus glabrus Moore, 1909. Fine sands.

Dasybranchus lumbricoides Grube, 1878. Mud.

Heteromastus filiformis (Claparède, 1864). Cosmopolitan in fine silty sediments; common.

Heteromastus filobranchus Berkeley and Berkeley, 1932. Common, in bays.

Mediomastus californiensis Hartman, 1944. Mud; common.

Notomastus magnus Hartman, 1947. Sandy muds.

Notomastus tenuis Moore, 1909. Sandy muds; common.

ARENICOLIDAE

1. Neuropodia of branchial segments long, extending ventrally to near the midline (fig. 288); with a single pair of esophageal caeca .... *Arenicola* 2
– Neuropodia of branchial segments short throughout and widely separated ventrally (fig. 287); with more than one pair of esophageal caeca .... *Abarenicola* 3
2. Branchiae, 11 pairs; color in life, light yellow with a purplish cast .... *Arenicola brasiliensis*
– Branchiae, 16–18 pairs; color in life rich dark green anteriorly and brown posteriorly .... *Arenicola cristata*
3. Nephridial pores present setigers 5–9, each uncovered; esophageal caeca include one large anterior pair and 3–6 smaller pairs (fig. 289); see also fig. 286 .... *Abarenicola pacifica*
– Nephridial pores present setigers 5–9, each covered with a hood; esophageal caeca include one large anterior pair and 7–9 smaller pairs .... *Abarenicola claparedii oceanica*

LIST OF ARENICOLIDAE

See Wells, 1962, Proc. Zool. Soc. London 138: 331–353.

Abarenicola claparedii oceanica Healy and Wells, 1959 (= *A. vagabunda oceanica*). Sand flats; Humboldt Bay and north. See Wells, 1963, *Speciation in the Sea,* Syst. Assoc. Publ. 5, 79–98.

Abarenicola pacifica Healy and Wells, 1959 Sand flats; Humboldt Bay and north. See Healy and Wells, 1959, Proc. Zool. Soc. London 133: 315–335.

Arenicola brasiliensis Nonato, 1958. Reported locally from San Francisco Bay.

Arenicola cristata Stimpson, 1856. In sand flats.

* *Branchiomaldane vincentii* Langerhans, 1881. Rare; may occur among rhizomes of *Phyllospadix.*

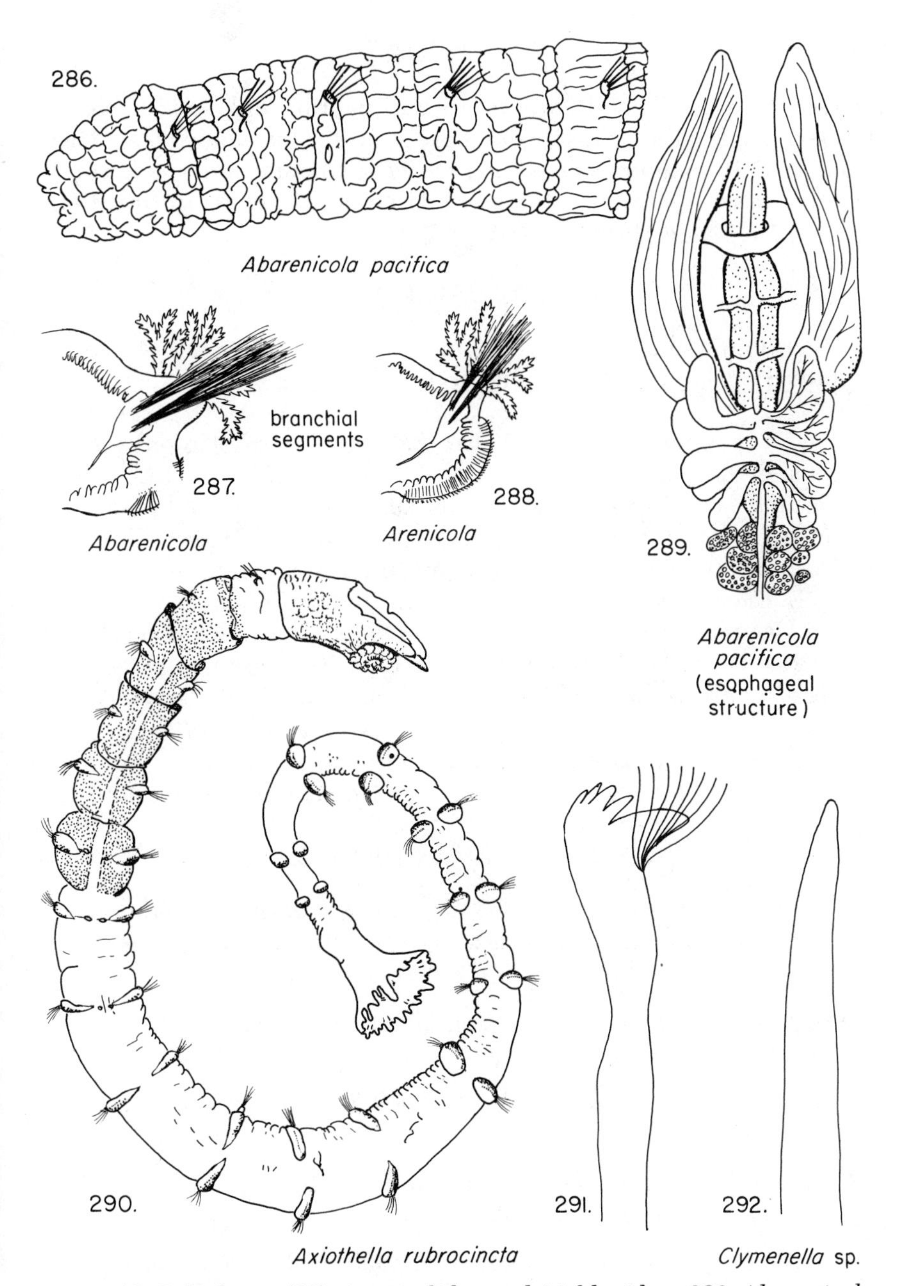

PLATE 48. **Polychaeta** (29). Arenicolidae and Maldanidae: 286, *Abarenicola pacifica;* 287, *Abarenicola*, branchial segment; 288, *Arenicola*, branchial segment; 289, *Abarenicola pacifica*, esophageal structure; 290, *Axiothella rubrocincta;* 291, rostrate uncinus; 292, *Clymenella* sp., acicular neuropodial spine of setiger 1; 289, after Okuda; 290, after Spies.

OWENIIDAE

Oweniids (figs. 59, 60) are tubicolous worms encased in tightly formed, sandy tubes. Most species are subtidal. *Owenia collaris* Hartman, 1955 is common in shallow, sandy sediments and was formerly referred to in our fauna as the more cosmopolitan *O. fusiformis* delle Chiaje, 1841. See Hartman, 1955, 1969.

MALDANIDAE

1. Pygidium without anal cirri, but formed as slanting plate with a dorsal anal opening; no neurosetae on setiger 1 .... *Asychis elongata*
– Pygidium encircled by anal cirri (fig. 290); neurosetae present on setiger 1 2
2. Neurosetae all rostrate hooks (fig. 291); green with red bands on segments 4–8 (fig. 290) .... *Axiothella rubrocincta*
– Neurosetae of first 3–4 setigers are thick acicular spines (fig. 292), those of succeeding setigers are rostrate hooks 3
3. Collar present on anterior margin of setiger 4 ... *Clymenella* 4
– Segments without collars .... *Praxillella affinis pacifica*
4. Head with well-developed, raised margin; 22–27 setigers; mudflats .... *Clymenella californica*
– Head without a raised margin; 21 setigers; shaly rocks .... *Clymenella complanata*

LIST OF MALDANIDAE

Asychis elongata (Verrill, 1873) (= *A. amphiglypta* of authors). San Francisco Bay, introduced from Atlantic. See Light 1974, Proc. Biol. Soc. Wash. 87: 175–184.

Axiothella rubrocincta (Johnson, 1901). Fine sand sediments; very common in bays and estuaries.

Clymenella complanata Hartman, 1969. In crevices of shale rocks in San Mateo County.

Clymenella californica Blake and Kudenov, 1974. Tomales Bay in muds at extreme low intertidal.

Praxillella affinis pacifica Berkeley, 1929. Mud of harbors and bays; see Banse and Hobson, 1968.

SABELLARIIDAE

1. Opercular paleae form a black cone (fig. 297); outer opercular paleae with a distal plume (fig. 296); colonies often form mas-

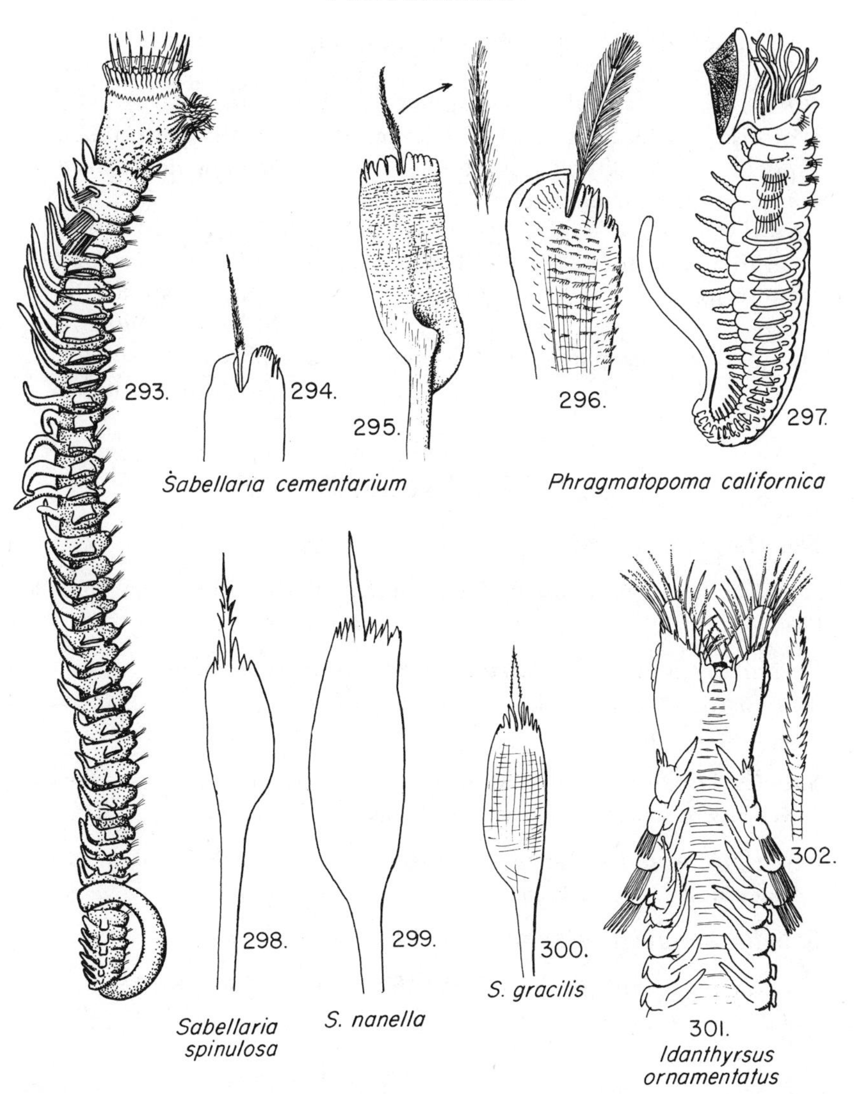

PLATE 49. **Polychaeta** (30). Sabellariidae: 293, *Sabellaria cementarium;* 294–295, outer opercular paleae; 296, *Phragmatopoma californica,* outer opercular palea; 297, *P. californica;* 298, *Sabellaria spinulosa;* 299, *S. nanella,* outer opercular palea; 300, *S. gracilis,* outer opercular palea; 301, *Idanthyrsus ornamentatus;* 302, *I. ornamentatus,* opercular palea; 293, 295 after Okuda; 296, 299, 300 after Hartman.

sive reefs .... *Phragmatopoma californica*

– Operculum formed by diverging yellow spines (fig. 293) 2

2\. Opercular paleae form 2 visible rows (fig. 301); outer opercular paleae pinnate (fig. 302) .... *Idanthyrsus ornamentatus*

– Opercular paleae form 3 visible rows; outer opercular paleae are broad plates with some type of terminal spine or arista .. .. *Sabellaria* 3

3. Outer opercular paleae terminate in a flat plate, with a distal spinose arista (figs. 294, 295); opercular stalk with black speckles (fig. 293) .... *Sabellaria cementarium*

– Outer opercular paleae and stalk otherwise 4

4. (*Note 3 choices*) Outer opercular paleae with longest spine marginally serrated (fig. 298) .... *Sabellaria spinulosa*

– Outer opercular paleae with longest spine a smooth spike (fig. 299) .... *Sabellaria nanella*

– Outer opercular paleae with longest spine a plumed arista (fig. 300) .... *Sabellaria gracilis*

LIST OF SABELLARIIDAE

See Hartman, 1944b.

Idanthyrsus ornamentatus Chamberlin, 1919. Rocky habitats, low intertidal.

Phragmatopoma californica (Fewkes, 1889). Constructs large colonies of honeycombs of tubes on rocks; may form massive reefs. See Dales, 1952, Quart. J. Micr. Sci. 93: 435–452.

Sabellaria cementarium Moore, 1906. Rocky habitats; tubes of sand and gravel, may be solitary or colonial, often forms reefs; common.

Sabellaria gracilis Hartman, 1944. Rocky habitats; tubes made of fine sand, colonies smaller, more delicate, seldom conspicuous; common.

Sabellaria nanella Chamberlin, 1919. San Francisco; rare.

Sabellaria spinulosa Leuckart, 1849. San Francisco Bay; probably introduced as a fouling organism.

PECTINARIIDAE

1. Cephalic spines long and tapered (figs. 303, 304); tubes straight, formed of fine, reddish brown sand (fig. 305) .... *Pectinaria californiensis*

– Cephalic spines short and blunt; tubes curved and formed of coarse, black and white sand grains (fig. 306) .... *Cistenides brevicoma*

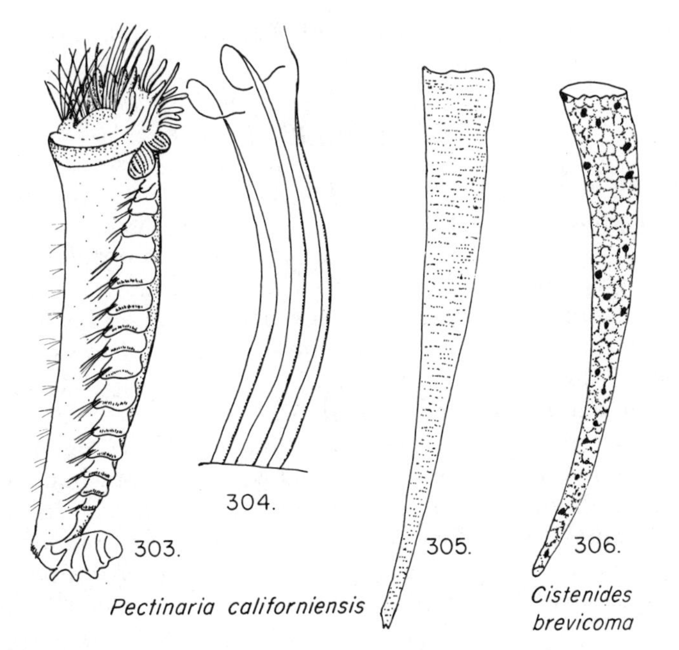

PLATE 50. **Polychaeta** (31). Pectinariidae: 303, *Pectinaria californiensis;* 304, 3 cephalic spines; 305, tube of *P. californiensis;* 306, tube of *Cistenides brevicoma.*

LIST OF PECTINARIIDAE

See Hartman, 1941b.

Cistenides brevicoma (Johnson, 1901). Gravel and coarse sand.

Pectinaria californiensis Hartman, 1941. Sand.

AMPHARETIDAE

1. Branchiae smooth (fig. 307); thorax with 14 setigers .... *Ampharete labrops*
– Branchiae of 2 kinds, pinnate and smooth (fig. 308); thorax with 15 setigers .... *Schistocomus hiltoni*

LIST OF AMPHARETIDAE

Ampharete labrops Hartman, 1961. Fine sands; Tomales Bay and Morro Bay.

* *Amphicteis* spp. See Hartman's Atlas.

Schistocomus hiltoni Chamberlin, 1919. In sand- and debris-covered tubes under rocks and in crevices on open coast.

TEREBELLIDAE

1. Thoracic neuropodia and uncini absent; prostomial lobe greatly prolonged as a long flap with 3 lobes (fig. 309) .... *Amaeana occidentalis*
– Thoracic uncini present 2
2. (*Note 3 choices*) Thoracic uncini arranged in single rows ... 3
– Thoracic uncini arranged in double or alternating rows from setigers 7–11 4
– Thoracic uncini arranged in single rows on setigers 2–5, then in a depressed ring on succeeding thoracic setigers (fig. 310); notosetae begin on second branchial segment; branchiae are tufts of unbranched coiling filaments ...... *Thelepus crispus*
3. With 3 pairs of branchiae; notosetae begin on first branchial segment .... *Streblosoma crassibranchia*
– Branchiae absent; notosetae begin on second segment .... *Polycirrus* spp.
4. Branchiae absent; peristomium with transverse rows of minute eyespots; thoracic notosetae include long and short, curved spines with flaring denticulate tips and thickened shafts, and bear numerous spinelets (figs. 311, 312); with over 30 thoracic setigers .... *Spinosphaera oculata*
– Branchiae present 5
5. With 1 pair of slightly branched branchiae; peristomium with many small eyespots behind tentacular bases; notosetae limbate, without denticles (fig. 313); with 13 thoracic setigers; color in life light red .... *Ramex californiensis*
– With 2 or 3 pairs of branchiae 6
6. With 2 pairs of branchiae 7
– With 3 pairs of branchiae; thoracic notosetae limbate; with 17 thoracic setigers 8
7. Anterior branchiae larger; ventral scutes (scalelike structures) prominent, continuous through 16 setigers, thereafter reduced; thoracic notosetae with smooth shafts and denticulate

---

PLATE 51. **Polychaeta** (32). Ampharetidae and Terebellidae: 307, *Ampharete labrops;* 308, *Schistocomus* sp.; 309, *Amaeana occidentalis;* 310, *Thelepus crispus,* arrangement of thoracic uncini; 311–312, *Spinosphaera oculata,* thoracic notosetae; 313, *Ramex californiensis,* thoracic notoseta; 314, *Terebella californica,* thoracic notoseta; 315, *Pista elongata,* long-shafted thoracic uncinus; 316, *P. elongata,* short uncinus; 317, *P. elongata;* 318, *P. pacifica;* 319, tube; 320, *Loimia medusa,* thoracic uncinus; all after Hartman.

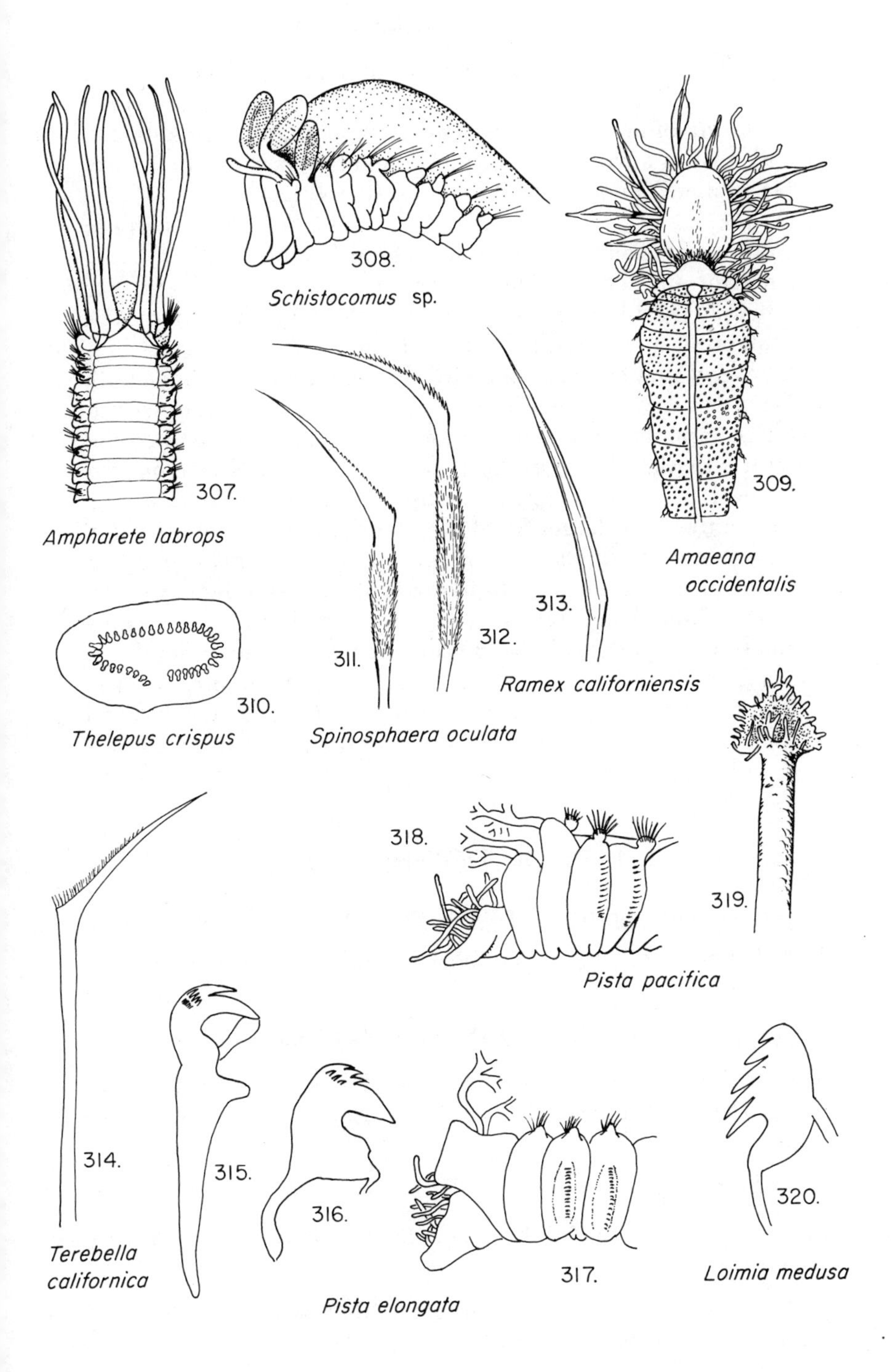
307.
Ampharete labrops
308.
Schistocomus sp.
309.
Amaeana occidentalis
310.
Thelepus crispus
311.
312.
313.
Spinosphaera oculata
Ramex californiensis
314.
315.
Terebella californica
316.
317.
Pista elongata
318.
319.
Pista pacifica
320.
Loimia medusa

tips (fig. 314); with 23–28 or more thoracic setigers .... *Terebella californica*

– Posterior branchiae larger; lateral lappets on segments 1–3 and smaller auricular lobes on segments 4–6; thoracic notosetae limbate, smooth; 17 thoracic setigers .... *Pista brevibranchiata*

8. Anterior end with conspicuous lateral lappets 9

– Anterior end without lateral lappets 11

9. Uncini of first few setigers avicular with long shafts (fig. 315), remaining uncini with short shafts (fig. 316) 10

– Uncini of first few setigers with short base and pectinate with 5–6 teeth in a row (fig. 320) .... *Loimia medusa*

10. Large lappets on second branchial segment (fig. 317); tube with reticulated top; under rocks .... *Pista elongata*

– Large lappets on second and third branchial segments (fig. 318); tube with large, hoodlike, overlapping membrane (fig. 319); sandy mudflats .... *Pista pacifica*

11. Color in life green with light tan tentacles and deep reddish brown branchiae; notosetae with smooth tips; numerous small dark eyespots behind tentacular bases; tubes sandy .. .. *Eupolymnia crescentis*

– Color in life brown with numerous white tentacles and red branchiae; notosetae with denticulate tips; eyes absent; tubes muddy .... *Neoamphitrite robusta*

LIST OF TEREBELLIDAE

Amaeana occidentalis (Hartman, 1944). Mudflats of bays and estuaries.

Eupolymnia crescentis Chamberlin, 1919. Sandy mud sediments of bays and estuaries.

Loimia medusa (Savigny, 1818). Sandy mud sediments.

Neoamphitrite robusta (Johnson, 1901). In mud and under rocks. See Brown and Ellis, 1971, J. Fish. Res. Bd. Canada 28: 1433–1435 (tube-building, feeding).

Pista brevibranchiata Moore, 1923. In mud.

Pista elongata Moore, 1909. In rocky habitats, forms muddy tubes in crevices and under rocks.

Pista pacifica Berkeley and Berkeley, 1942. Sandy mud sediments; forms characteristic tubes which extend deep into the substratum.

Polycirrus spp. The systematics of *Polycirrus* are confused and no names have been assigned to our local species. *Polycirrus* is common in rocky habitats and also in mud in tires and other habitats on floats and docks. Several species may be present.

Ramex californiensis Hartman, 1944. Rocky habitats on exposed outer coast.

Spinosphaera oculata Hartman, 1944. Rocky habitats on exposed outer coast.

Streblosoma crassibranchia Treadwell, 1914. Sandy mud.

Terebella californica Moore, 1904. Rocky habitats; algal holdfasts.

Thelepus crispus Johnson, 1901. Tube attached to undersides of rocks in exposed rocky habitats; one of the most common intertidal terebellids.

SABELLIDAE

1. Body covered with a thick sheath of slimy mucus; radioles united with a membrane for most of their length .... Subfamily **Myxicolinae:** *Myxicola infundibulum*
– Body not covered with mucoid sheath; may be enclosed in tube of mucus or of cemented particles; radioles partly united by a membrane or radioles free 2
2. Thoracic neurosetae include a row of avicular uncini (fig. 327) and sometimes an additional row of pennoned setae (fig. 328); tubes permanent, with most species incapable of building another; tube firm, tough, or leathery, formed of mucus and cemented sediment particles .... Subfamily **Sabellinae** 3
– Thoracic neurosetae include a single row of long-shafted hooks (fig. 341); tubes temporary, animals capable of leaving tube and forming another; tubes generally simple, fragile, formed of mucus with adhering sediment particles .... Subfamily **Fabricinae** 11
3. Radioles of branchial crown divided several times (fig. 322); color in life variable, but often light tan with deep red crown .... *Schizobranchia insignis*
– Radioles not divided 4
4. Composite eyes spiraled around distal end of some radioles (fig. 323), especially the dorsal pairs .... *Megalomma splendida*
– Composite eyes if present, not in such a position 5
5. Thoracic collar with 4 lobes (fig. 324); thoracic notosetae of 2 different types: limbate (fig. 329) and spatulate (figs. 330) .. 6
– Thoracic collar with 2 lobes (fig. 331); thoracic notosetae differing gradually among themselves, long and shortened bilimbate (figs. 333, 334) 10

6. Each base of branchial crown conspicuously spiraled .... *Eudistylia* 7
– Each base of branchial crown forms a semicircle .... *Pseudopotamilla* 8
7. Dorsal edge of branchial crown deeply cleft (fig. 325), color variable, radioles often deep maroon with orange tips .... *Eudistylia polymorpha*
– Dorsal edge of branchial crown not cleft (fig. 324); color variable, radioles often crossed by alternating bars of red and white .... *Eudistylia vancouveri*
8. Dorsal edges of branchial crown deeply cleft (fig. 326); dorsal collar lobes short or obscure; eyespots conspicuous with several on a radiole; tentacular crown brilliantly colored, transversely banded with deep red ... *Pseudopotamilla occelata*
– Dorsal edges of branchial crown not cleft; dorsal collar lobes prominent; eyespots present or absent 9
9. Large, to 58 mm; collar membrane with high dorsal lobes; eyespots absent .... *Pseudopotamilla intermedia*
– Small, to 25 mm; collar membrane with low dorsal lobes, widely separated from lateral lobes by a deep cleft; eyespots not conspicuous, usually one or few on a radiole; pale colored in life .... *Pseudopotamilla socialis*
10. Tentacular radioles with paired eyespots; radioles with deep red bands in region of paired eyes ..... *Sabella crassicornis*
– Radioles without eyespots; color pale, with radioles reddish and mottled with white .... *Sabella media*
11. Radioles partly united by a membrane .... *Chone* 15
– Radioles free or united only at the base 12
12. Radioles without pinnules, undivided (fig. 335); inhabits fresh water .... *Manayunkia speciosa*
– Radioles pinnately divided; inhabits marine or brackish water 13
13. With 3 abdominal setigers; abdominal uncini short *Fabricia* 14
– With 7–8 abdominal setigers; abdominal uncini short .... *Oriopsis gracilis*
14. Thoracic collar short, divided into 2 short dorsal lobes and a ventral plaque (figs. 336, 337) .... *Fabricia brunnea*
– Thoracic collar prominent, encircles peristomium (fig. 338). .... *Fabricia berkeleyi*
– Thoracic collar absent; with pair of short palps (fig. 339)... .... *Fabricia sabella*
15. In rocky intertidal; inhabits shell and coarse sand covered tubes; thoracic spatulate setae with distal mucron (fig. 340); minute, to 10 mm; with 7–8 pairs of radioles . .*Chone ecaudata*

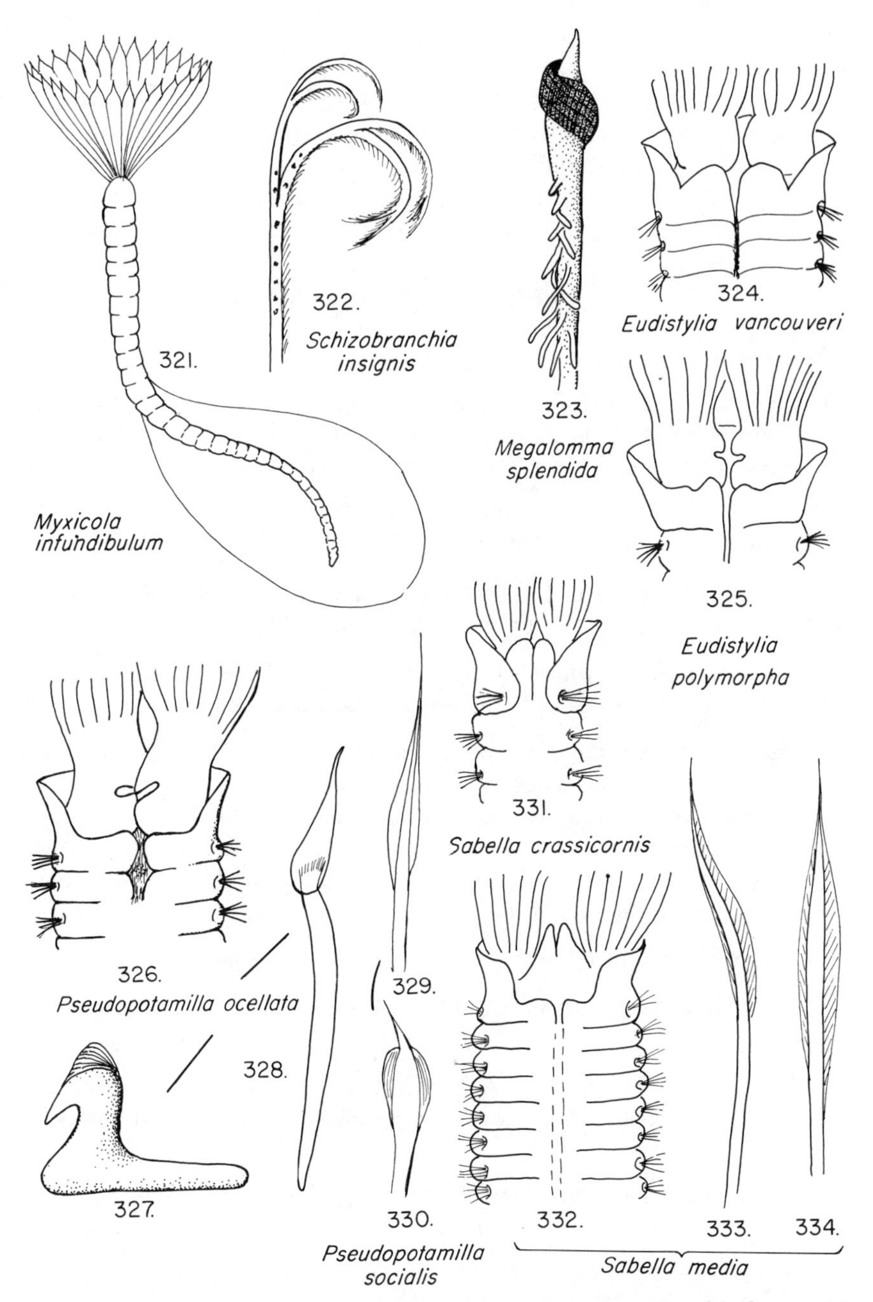

PLATE 52. **Polychaeta** (33). Sabellidae: 321, *Myxicola infundibulum;* 322, *Schizobranchia insignis,* radiole; 323, *Megalomma splendida,* radiole with eye; 324, *Eudistylia vancouveri;* 325, *E. polymorpha;* 326, *Pseudopotamilla occelata;* 327, thoracic avicular uncinus; 328, thoracic pennoned seta; 329, *P. socialis,* thoracic bilimbate notoseta; 330, thoracic spatulate notoseta; 331, *Sabella crassicornis;* 332, *S. media;* 333–334, thoracic notosetae; 322, 324 after Berkeley and Berkeley; 329–330, 332–334 after Hartman.

- In sandy flats; inhabits thin, silt-covered tubes; thoracic spatulate setae without distal mucron (fig. 342); larger, to 63 mm; with up to 15 pairs of radioles .... *Chone mollis*
- Generally subtidal in sandy mud; resembling *C. mollis* but with distal mucron on thoracic spatulate setae *Chone gracilis*

LIST OF SABELLIDAE

See Hartman, 1951.

Chone ecaudata (Moore, 1923) (=*C. minuta* Hartman, 1944; see Banse, 1972). Rocky intertidal, amongst algae, in crevices; may accumulate sand by forming tube masses which may be reef-like, associated with *Pygospio elegans* in such situations.

Chone gracilis Moore, 1906. Sandy mud sediments; subtidal.

Chone mollis (Bush, 1904). Common in sand-mud sediments. See Bonar, 1972, J. Exp. Mar. Biol. Ecol. 9: 1–18 (tube-building, feeding).

Eudistylia polymorpha (Johnson, 1901). Rocky habitats, also in harbors on floats and wharves; may be confused with *E. vancouveri* (see below).

Eudistylia vancouveri (Kinberg, 1867). Common in harbors on floats and wharves, also in sandy mudflats; specimens exhibiting characteristics of both *E. polymorpha* and *E. vancouveri* have been found, suggesting possibility of hybridization. The copepod *Gastrodelphys dalesi* is reported on *E. vancouveri* at Tomales Point; see Dudley, 1964, Amer. Mus. Nov. no. 2194 on symbiotic gastrodelphyid copepods, largely from Washington.

Fabricia berkeleyi (Banse, 1956). Mudflats of estuaries.

Fabricia brunnea Hartman, 1969. Rocky intertidal, in sandy bottoms of tidepools.

Fabricia sabella (Ehrenberg, 1837). In mud in estuaries; often among algae and barnacles on pilings. See Lewis, 1968, J. Linn. Soc. London 47: 515–526 (ecology).

Manayunkia speciosa Leidy, 1858. In freshwater streams, canals, and lakes along Pacific coast. See Pettibone, 1953, Biol. Bull. 105: 149–153 (morphology).

Megalomma splendida (Moore, 1905). Rocky bottoms, rare intertidal.

Myxicola infundibulum (Renier, 1804). Slime masses of these worms are often found in harbors, attached to ropes or other sunken objects; cosmopolitan; well-known experimental animal.

Oriopsis gracilis Hartman, 1969. Rocky intertidal; in tide pools with sand; form tube clusters with sponges and tunicates.

Pseudopotamilla intermedia Moore, 1905. Rocky or mixed bottoms; rare.

Pseudopotamilla occelata Moore, 1905. Rocky habitats, open coast; common. Look for small, commensal, two-tentacled hydroid *Proboscidactyla* on rims of tubes.

Pseudopotamilla socialis Hartman, 1944. Rocky habitats, open coast; associated with masses of sponge and tunicates.

Sabella crassicornis Sars, 1851. Rocky and mixed bottoms.

Sabella media (Bush, 1904). Rocky habitats, in crevices; Hartman (1969) places this species in *Demonax*.

Schizobranchia insignis Bush, 1904. Rocky habitats; in harbors on floats and wharves; common.

SERPULIDAE

1. Tube minute, neatly coiled; body asymmetrical 7
– Tube larger, straight (fig. 345), twisted but not coiled; body symmetrical 2
2. Operculum absent; 3 pairs of branchiae; forms small, fragile tube masses; color in life red ...... *Salmacina tribranchiata*
– Operculum present 3
3. Operculum bladderlike with smooth stalk and terminating in a long, curved spine (fig. 347) ..... *Vermiliopsis multiannulata*
– Operculum funnel-shaped 4
4. (*Note 3 choices*) Operculum with up to 160 crenulations, stalk smooth (fig. 348); tubes hard, white, cylindrical with a mid-dorsal ridge .... *Serpula vermicularis*
– Operculum bordered with 26–30 crenulations, stalk terminates with 3 knobbed processes below the funnel (fig. 349); tubes thick, white .... *Crucigera zygophora*
– Operculum with spines 5
5. Spines of operculum form a single expanded disc (fig. 346); tubes white, in aggregations, each with a series of annulations (fig. 345; body, fig. 344) .... *Mercierella enigmatica*
– Spines of operculum form a two-storied disc 6
6. Distal series of spines with lateral projections (fig. 350); tube white, slender, erect, crossed by delicate growth lines .... *Hydroides pacificus*
– Distal series of spines with smooth sides (fig. 351); tube solitary with long ridges and cross lines .... *Eupomatus gracilis*
7. Tube dextral (figs. 352, 353); aperture facing to right when tube is placed with opening below .... *Spirorbis spirillum*

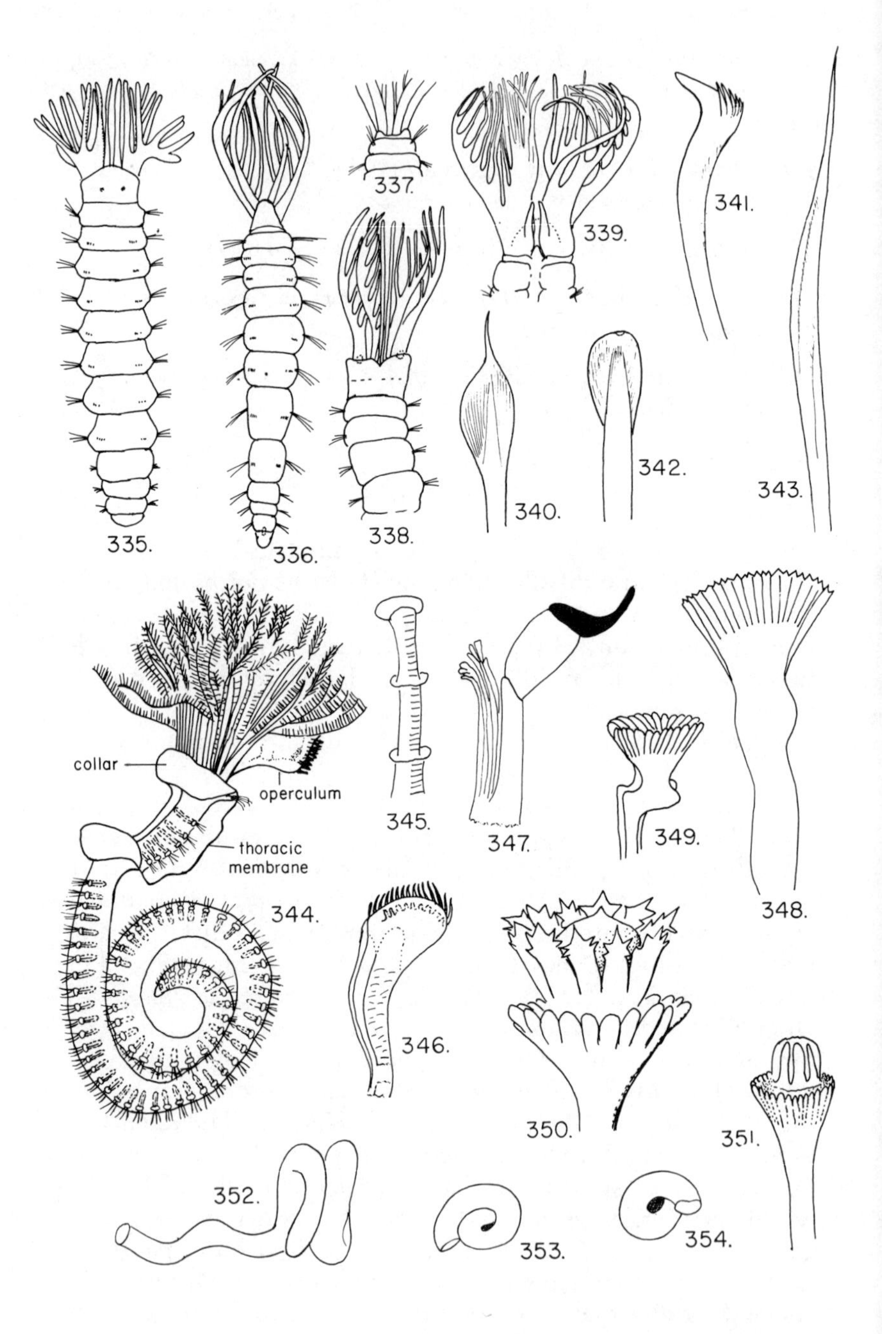
335.
336.
337.
338.
339.
340.
341.
342.
343.
collar
operculum
thoracic membrane
344.
345.
346.
347.
348.
349.
350.
351.
352.
353.
354.

- Tube sinistral (fig. 354); aperture facing to left when tube so placed 8
8. Tube conspicuously corrugated on surface .. *Spirorbis eximius*
- Tube smooth on the surface 9
9. Larvae incubated in operculum .... *Spirorbis moerchi*
- Larvae incubated in tube (fig. 354) .... *Spirorbis borealis*

LIST OF SERPULIDAE

Crucigera zygophora (Johnson, 1901). Rocky habitats.

Eupomatus gracilis Bush, 1904. Rare, more southern in distribution; rocky surfaces.

Hydroides pacificus Hartman, 1969. This is probably *H. elegans* Haswell, 1883 (=*H. norvegicus* of authors). A fouling species in harbors and marinas of southern California. See Zibrowius, 1971, Tethys 2: 691–746.

Mercierella enigmatica Fauvel, 1923. Abundant in San Francisco Bay in areas (such as Aquatic Park, Berkeley and Lake Merritt, Oakland) of reduced salinity and little wave action; may form massive aggregations which block water courses and channels; an introduced cosmopolitan fouling species.

Salmacina tribranchiata (Moore, 1923). Sheltered rocky habitats.

Serpula vermicularis Linnaeus, 1767. Attaches to most types of hard surfaces on exposed shores; also in harbors on floats and wharves; cosmopolitan.

Spirorbis borealis Daudin, 1800. Attaches to hard surfaces or algae; cosmopolitan.

Spirorbis eximius Bush, 1904. Attaches to hard surfaces.

Spirorbis moerchi Levinsen, 1883. Attaches to shells.

Spirorbis spirillum (Linnaeus, 1758). Attaches to rocks and algae; cosmopolitan.

PLATE 53. **Polychaeta** (34). Sabellidae and Serpulidae: 335, *Manayunkia speciosa;* 336, *Fabricia brunnea*, ventral view; 337, *F. brunnea*, dorsal view; 338, *F. berkeleyi;* 339, *F. sabella;* 340, *Chone ecaudata*, thoracic spatulate notoseta; 341, thoracic long-shafted neuropodial hook; 342, *C. mollis*, thoracic spatulate notoseta; 343, superior thoracic notoseta; 344, *Mercierella enigmatica;* 345, tube; 346, operculum; 347, *Vermiliopsis multiannulata*, operculum; 348, *Serpula vermicularis*, operculum; 349, *Crucigera zygophora*, operculum; 350, *Hydroides pacificus*, operculum; 351, *Eupomatus gracilis*, operculum; 352–353, *Spirorbis spirillum*, dextral tubes; 354, *S. borealis*, sinistral tube; most after Hartman; 352–354 after Smith, 1964.

"Spirorbis" spp. *Note added in 3rd printing:* The taxonomy of *"Spirorbis"* on our coast is not worked out; there are numerous species not in this key, and it is best to apply no names to local species until revisionary work has been done. See Bailey, 1969. J. Linnean Soc. (Zool.) 48: 387–407 (brood protection, taxonomy).

Vermiliopsis multiannulata (Moore, 1923). Intertidal to subtidal rocky habitats.

REFERENCES ON POLYCHAETA

Note: For literature before 1950 not cited below, see Hartman's bibliography of polychaetes (1951a).

Banse, K. 1972. Redescription of some species of *Chone* Dröher (Sabellidae, Polychaeta). Fishery Bull. 70: 459–495.

Banse, K. and K. D. Hobson 1968. Benthic polychaetes from Puget Sound, Washington, with remarks on four other species. Proc. U. S. Nat. Mus. 125: 1–53.

Berkeley, E. and C. Berkeley 1935. Some notes on the polychaetous annelids of Elkhorn Slough, Monterey Bay, California. Amer. Mid. Nat. 16: 766–775.

Berkeley, E. and C. Berkeley 1941. On a collection of Polychaeta from southern California. Bull. So. Calif. Acad. Sci. 40: 16–60.

Berkeley, E. and C. Berkeley 1948. Canadian Pacific fauna 9. Annelida 9b(1). Polychaeta Errantia. Fish. Res. Bd. Canada, pp. 1–100.

Berkeley, E. and C. Berkeley 1952. Canadian Pacific fauna 9. Annelida 9b(2). Polychaeta Sedentaria. Fish. Res. Bd. of Canada, pp. 1–139. (See Pettibone, 1967, Proc. U. S. Natl. Mus. 119: 1–23, for review of the Berkeley type material.)

Blake, J. A. 1969. Reproduction and larval development of *Polydora* from northern New England (Polychaeta: Spionidae). Ophelia 7: 1–63.

Blake, J. A. 1971. Revision of the genus *Polydora* from the east coast of North America (Polychaeta: Spionidae). Smithson. Contrib. Zool. 75: 1–32.

Blake, J. A. and J. W. Evans 1973. *Polydora* and related genera as borers in mollusk shells and other calcareous substrates. Veliger 15: 235–249.

Dales, R. P. and G. Peter, 1972. A synopsis of the pelagic Polychaeta. J. Nat. Hist. 6: 55–92.

Dales, R. P. 1967. *Annelids*. 2nd ed. London: Hutchinson, 200 pp. A useful introduction, devoted mainly to polychaetes.

Day, J. H. 1967. A monograph on the Polychaeta of southern Africa. Brit. Mus. Nat. Hist. London Publ. no. 656: 1–878, 2 volumes.

Fauchald, K. 1970. Polychaetous annelids of the families Eunicidae, Lumbrineridae, Iphitimidae, Arabellidae, Lysaretidae and Dorvilleidae from western Mexico. Allan Hancock Monogr. Mar. Biol. 5: 1–335.

Fauvel, P. 1923. Polychètes errantes. Faune de France 5: 1–488.

Fauvel, P. 1927. Polychètes sédentaires. Addenda aux Errantes, Archiannélides, Myzostomaires. Faune de France 16: 1–494.

Hartman, Olga 1939. Polychaetous annelids. Part I. Aphroditidae to Pisionidae. Allan Hancock Pac. Exped. 7: 1–156.

Hartman, O. 1940. Polychaetous annelids. Part II. Chrysopetalidae to Goniadidae. Allan Hancock Pac. Exped. 7: 173–287.

Hartman, O. 1941a. Polychaetous annelids. Part III. Spionidae. Some contributions to the biology and life history of Spionidae from California. Allan Hancock Pac. Exped. 7: 289–323.

Hartman, O. 1941b. Polychaetous annelids. Part IV. Pectinariidae. Allan Hancock Pac. Exped. 7: 325–345.
Hartman, O. 1944. Polychaetous annelids. Part V. Eunicea. Allan Hancock Pac. Exped. 10: 1–238.
Hartman, O. 1944a. Polychaetous annelids from California including the descriptions of two new genera and nine new species. Allan Hancock Pac. Exped. 10: 239–310.
Hartman, O. 1944b. Polychaetous annelids. Part VI. Paraonidae, Magelonidae, Longosomidae, Ctenodrilidae, and Sabellariidae. Allan Hancock Pac. Exped. 10: 311–390.
Hartman, O. 1947a. Polychaetous annelids. Part VII. Capitellidae. Allan Hancock Pac. Exped. 10: 391–482.
Hartman, O. 1947b. Polychaetous annelids. Part VIII. Pilargidae. Allan Hancock Pac. Exped. 10: 483–524.
Hartman, O. 1950. Goniadidae, Glyceridae and Nephtyidae. Allan Hancock Pac. Exped. 15: 1–182.
Hartman, O. 1951a. *Literature of the polychaetous annelids.* Los Angeles: privately published, 290 pp.
Hartman, O. 1951b. Fabricinae (Feather-duster polychaetous annelids) in the Pacific. Pac. Sci. 5: 379–391.
Hartman, O. 1954. The marine annelids of the San Francisco Bay and its environs, California. Allan Hancock Found., Pub. Occ. Pap. 15: 1–20. Somewhat out of date.
Hartman, O. 1955. Endemism in the north Pacific ocean with emphasis on the distribution of marine annelids, and descriptions of new or little known species. Essays in the natural sciences in honor of Captain Allan Hancock on the occasion of his birthday July 26, 1955. University of Southern California Press, pp. 39–60.
Hartman, O. 1956. Polychaetous annelids erected by Treadwell, 1891 to 1948, together with a brief chronology. Bull. Amer. Mus. Nat. Hist. 109: 245–310.
Hartman, O. 1957. Orbiniidae, Apistobranchidae, Paraonidae, and Longosomidae. Allan Hancock Pac. Exped. 15: 211–394.
Hartman, O. 1959, 1965. Catalogue of the polychaetous annelids of the world. Parts I and II. Allan Hancock Found. Pub. Occ. Pap. 23: 1–628. See also Hartman, 1965, Supplement 1960–1965 and Index, Allan Hancock Found. Publ., Occ. Pap. 23: 197 pp.
Hartman, O. 1961. Polychaetous annelids from California. Allan Hancock Pac. Exped. 25: 226 pp.
Hartman, O. 1968. *Atlas of the errantiate polychaetous annelids from California.* Los Angeles: Allan Hancock Found. Univ. South. Calif., 828 pp.
Hartman, O. 1969. *Atlas of the sedentariate polychaetous annelids of California.* Los Angeles: Allan Hancock Found. Univ. South. Calif., 812 pp.
Hartman, O. and D. J. Reish 1950. The marine annelids of Oregon. Ore. State College Monogr. Zool. 6: 1–64.
Hobson, K. D. 1971. Some polychaetes of the superfamily Eunicea from the North Pacific and North Atlantic oceans. Proc. Biol. Soc. Wash. 83: 527–544.
Imajima, M. and O. Hartman 1964. Polychaetous annelids of Japan. Allan Hancock Found., Pub. Occ. Pap. 26: 1–462 (2 vols.).
Pettibone, M. H. 1953. *Some scale-bearing polychaetes of Puget Sound and adjacent waters.* University of Washington Press, 89 pp.
Pettibone, M. H. 1963. Marine polychaete worms of the New England region. I. Aphroditidae through Trochochaetidae. Bull. U. S. Nat. Mus. 227: 1–356.
Reish, D. J. and J. L. Barnard 1967. The benthic Polychaeta and Amphipoda of Morro Bay, California. Proc. U. S. Nat. Mus. 120 (3565): 1–25.

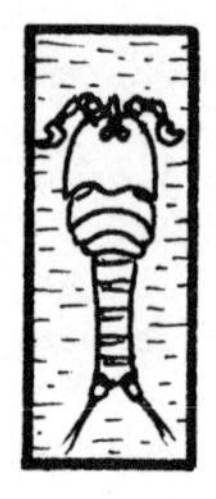

PHYLUM ARTHROPODA: INTRODUCTION AND LOWER CRUSTACEA

The arthropods are a diverse and successful group of segmented animals which have developed a firm, jointed exoskeleton and jointed appendages. The parts of their body armor are movable upon each other by discrete, strandlike, striated muscles. The arthropod body plan is radically different from that of the segmented annelids, in which the body wall is a soft, "dermo-muscular tunic" of smooth muscle surrounding a "fluid skeleton," which contributes rigidity by being under pressure from the muscular body wall. Arthropods, unlike most of the annelids, have a greatly reduced coelom, and have opened out the vascular system into a system of blood-filled spaces, the **haemocoel.** The hearts of arthropods (and of the arthropod-like Onychophora) possess inlet valves (**ostia**) by which the blood enters. The eggs are yolky, and cleavage and embryology are highly modified. Only in the case of the crustacean nauplius do we find anything comparable to a simple larval form, and even this larva shows marked structural complication. Probably as a consequence of the relatively impermeable exoskeleton, the arthropods have been the most successful phylum in adapting to a terrestrial existence. Several large arthropod groups occur only on land or reenter the water secondarily.

Since the arthropods constitute by far the largest animal phylum, it is necessary that the student have a clear, over-all grasp of the major groups, not only those which are well represented in the intertidal zone, but others which will not be encountered there. The phylum may be divided, for our purposes, into two main lines or subphyla (excluding Tardigrada, the extinct Trilobita, and certain other groups of doubtful affinities).

1. Subphylum **MANDIBULATA:** The first pair of appendages are antennae (**antennules**), the second pair are also antennae or are missing, the third pair form jaws (**mandibles**) which give the group its name.
   a) Class **Crustacea.** The first *two* pairs of appendages are antennae. Crustaceans are predominately aquatic and, although extremely

diverse, they preserve to a large extent the more primitive features of arthropods in general.

b) Class **Labiata.** Only the first pair of antennae are present. The bases of the second maxillae are fused to form a lower lip (**labium**) which gives the group its name. Included are the great subclass **Insecta** and several smaller subclasses, including centipedes and millipedes, collectively referred to as **Myriapoda.** Intertidal insects are treated on pages 432–452.

2. Subphylum **CHELICERATA:** The first pair of appendages are usually pincerlike (**chelicerate**); this subphylum includes the arachnids and their allies. Although predominately a terrestrial group, the Chelicerata include the ancient marine "horseshoe crabs," *Limulus* (order **Xiphosura**), not found on this coast, the "sea spiders" (class or subphylum **Pycnogonida,** page 413), the marine mites (Halacaridae, page 425), and the littoral pseudoscorpions (page 431).

CLASS CRUSTACEA

A general classification of the Crustacea demonstrates the diverse nature of the group. The student of invertebrate zoology should be able to place almost all crustaceans "on sight" into the classes and orders summarized below.

Subclass **Cephalocarida** (page 246): primitive, minute, interstitial forms.

Subclass **Branchiopoda** (page 246): includes *Artemia* the brine shrimp and the freshwater fairy shrimps (Order **Anostraca**), tadpole shrimps (**Notostraca**), clam shrimps (**Conchostraca**), and water fleas (**Cladocera**); all but *Artemia* and a few cladocerans are freshwater animals.

Subclass **Ostracoda** (page 247): small, short-bodied, bivalved forms common in marine and fresh waters.

Subclass **Copepoda** (page 250): numerous and important in all waters; mostly free-living, but with many symbiotic species.

Subclass **Branchiura** (page 253): a small group of flattened ectoparasites of fish; formerly classified with the copepods.

Subclass **Cirripedia** (page 259): sessile and stalked barnacles, the parasitic rhizocephalans, and acrothoracicans.

Subclass **Malacostraca** (page 270): diverse forms, including:

- Division **Leptostraca** (page 271): a very small, primitive, shrimplike marine group.
- Division **Peracarida** (page 272): mysids, cumaceans, tanaidaceans (cheliferans), isopods, and amphipods.
- Division **Eucarida** (page 377): euphausiids and decapods, includ-

ing the familiar shrimps, true crabs, hermit crabs, various lobsters, crayfishes, and others.

Division **Hoplocarida (Stomatopoda):** the mantis shrimps; marine forms with massive raptorial claws, not found in the local intertidal zone.

GENERAL REFERENCES ON CRUSTACEA

Green, J. 1963. *A biology of Crustacea.* Chicago: Quadrangle Books, 180 pp.

Lockwood, A. M. 1967. *Aspects of the physiology of Crustacea.* W. H. Freeman & Co., 328 pp.

Schmitt, W. L. 1965. *Crustaceans.* University of Michigan Press, 204 pp.

Snodgrass, R. E. 1965. Crustacean metamorphoses. Smith. Miscell. Colls. 131: 78 pp.

Waterman, T. H., ed. 1960. *The physiology of Crustacea.* Vol. I, *Metabolism and growth,* 670 pp.; 1961. Vol. II, *Sense organs, integration and behavior.* 681 pp. Academic Press.

Whittington, H. B. and W. D. I. Rolfe, eds. 1963. *Phylogeny and evolution of Crustacea.* Mus. Comp. Zool., Harvard Univ., Spec. Publ. 192 pp.

SUBCLASS CEPHALOCARIDA

Cephalocarids, the most primitive living crustaceans, are tiny (about 3 mm long), subtidal, interstitial crustaceans with a horseshoe-shaped head, a thorax of nine segments, and an abdomen of ten; only the thorax bears appendages, which are triramous (rather than biramous as in other crustaceans). The telson bears a pair of large processes forming a caudal fork **(furca).** *Lightiella serendipita* Jones, 1961 has been found in shallow water on a muddy sand bottom in San Francisco Bay, but is probably more widely distributed.

REFERENCES ON CEPHALOCARIDA

Jones, M. L. 1961. *Lightiella serendipita* gen. nov., sp. nov., a cephalocarid from San Francisco Bay, California. Crustaceana 3: 31–46.

Sanders, H. L. 1955. The Cephalocarida, a new subclass of Crustacea from Long Island Sound. Proc. Natl. Acad. Sci. 41: 61–66.

Sanders, H. L. 1957. The Cephalocarida and crustacean phylogeny. Syst. Zool. 6: 112–129.

Sanders, H. L. 1963. Cephalocarida. Functional morphology, larval development, comparative external anatomy. Mem. Connecticut Acad. Arts Sci. 15: 1–80.

SUBCLASS BRANCHIOPODA

(PLATE 54)

Branchiopods, with few exceptions, are found in fresh waters. The order **Anostraca** (which have stalked eyes and lack the carapace characteristic of other branchiopods) are represented by the reddish brine shrimp *Artemia salina* (Linnaeus, 1758) (fig. 1), locally abundant in the concentrated (hypersaline) sea water of evaporation pools for commercial salt production around San Francisco and other bays, as well as in inland saline lakes. Brine shrimp are commonly used as food for aquarium fish, and can be obtained cheaply from aquarium stores, or raised from dried eggs. Other anostracans, commonly known as fairy shrimps, occur in seasonal freshwater ponds. The **Notostraca,** represented by the tadpole shrimps, have a broad, shieldlike carapace notched posteriorly, and sessile eyes, and occur in seasonal freshwater pools or in rice fields (Linder, 1952; Longhurst, 1955). The entirely freshwater **Conchostraca** (with thin, hinged, bivalved carapace covering the entire body and head, numerous appendages, and sessile eyes) are known as clam or claw shrimps, are widely found in vernal pools, and are often mistaken for small, bivalved molluscs. The **Cladocera** or water fleas have an apparently bivalved, but not hinged, carapace which does not cover the head, a shortened body with relatively few appendages, and a single prominent compound eye. Most cladocerans, such as the familiar *Daphnia*, are abundant in fresh water, but two genera (*Evadne* and *Podon*) are common in marine plankton (see Baker, 1938). The Conchostraca and Cladocera are sometimes placed in a single order, **Diplostraca.** Pennak (1953) and Ward and Whipple (Edmondson, 1959) provide good, if somewhat dated, keys and detailed discussions of the freshwater branchiopods.

REFERENCES ON BRANCHIOPODA

Baker, H. M. 1938. Studies on the Cladocera of Monterey Bay. Proc. Calif. Acad. Sci. (4) 23: 311–365.

Edmondson, W. T., ed. 1959. Ward and Whipple's *Fresh-Water Biology*. 2nd ed., John Wiley & Sons, 1248 pp.

Linder, F. 1952. Contributions to the morphology and taxonomy of the Branchiopoda Notostraca, with special reference to the North American species. Proc. U.S. Nat. Mus. 102: 1–69.

Littlepage, J. L. and M. N. McGinley 1965. A bibliography of the genus *Artemia* (*Artemia salina*) 1812–1962. San Francisco Aquarium Society, Spec. Publ. 1: 73 pp.

Longhurst, A. R. 1955. A review of the Notostraca. Bull. Brit. Mus. (Nat. Hist.) Zool. 3: 1–57.

Pennak, R. W. 1953. *Fresh-water Invertebrates of the United States*. Ronald Press, 769 pp.

SUBCLASS OSTRACODA

(PLATE 54)

The minute ostracods occur in nearly all collections of freshwater, brackish, and marine plants, bottom material, or clumps of hydroids and bryozoans. They present a characteristic appearance, with a bivalved, often bean-shaped shell, from which project several appendages (figs. 2A, B). The bivalved carapace (closed by an adductor muscle) entirely encloses both head and body; there is no visible body segmentation and the body is extremely shortened, having at most seven pairs of appendages. Local intertidal species have not been studied in detail, and only scattered work has been done. Watling (1970, 1972) has described several species from shallow water and the intertidal of the Tomales region; *Sarsiella zostericola* Cushman, a possible introduction from the Atlantic, has been found in shallow mud in San Francisco Bay (Jones, 1958, J. Wash. Acad. Sci. 48: 48–52, 238–239; Kornicker, 1967, Proc. U.S. Nat. Mus. 122: 1–46), while species of *Cyprideis* may be found in lagoons and estuaries. Skogsberg (1928, 1950) has described several intertidal species from Monterey. Those interested in the taxonomy and biology of ostracods are referred to the *Treatise on Invertebrate Paleontology*, which deals extensively with Recent forms, to the symposium volume by Neale, and to Sars' classic monograph. Hulings (1971) has given a review of the status of the taxonomy and ecology of benthic marine ostracods.

REFERENCES ON OSTRACODA

Benson, R. H., *et al.* 1961. Ostracoda. In *Treatise on Invertebrate Paleontology*, R. C. Moore, ed., Part A, Arthropoda 3, Crustacea, Ostracoda, xxiv + 442 pp. Univ. Kansas Press and Geol. Soc. Amer.

Hulings, N. C. 1971. Summary and current status of the taxonomy and ecology of benthic Ostracoda including interstitial forms. *In* N. C. Hulings, ed., Proc. First Internat. Conf. Meiofauna, Smithson. Contrib. Zool. 76: 91–96.

Neale, J. W., ed. 1969. *The Taxonomy, Morphology, and Ecology of Recent Ostracoda.* Edinburgh: Oliver & Boyd.

Sars, G. O. 1922–1928. *Ostracoda. An Account of the Crustacea of Norway.* Vol. 9. Bergen Mus., 277 pp.

Skogsberg, T. 1928. Studies on marine ostracods. Part II, External morphology of the genus *Cythereis* with descriptions of twenty-one new species. Occ. Pap. Calif. Acad. Sci. 15: 155 pp. (Includes species from Pacific Grove.)

Skogsberg, T. 1950. Two new species of marine Ostracoda (Podocopa) from California. Proc. Calif. Acad. Sci. (4) 26: 483–505. (Pacific Grove.)

Watling, L. 1970. Two new species of Cytherinae (Ostracoda) from central California. Crustaceana 19: 251–263. (Tomales Bay.)

Watling, L. 1972. A new species of *Acetabulastoma* Shornikov from central California with a review of the genus. Proc. Biol. Soc. Wash. 85: 481–488. (Low intertidal coralline zone, Tomales Point.)

PLATE 54. **Various Crustacea.** 1, **Anostraca:** *Artemia salina*, male, Hedgpeth, 1962; 2A,B, **Ostracoda:** A, *Cythereis aurita*, from Pacific Grove, male right valve, Skogsberg, 1928; B, *Cylindroleberis* sp., after Sars; 3A,B, **Leptostraca:** A, *Nebalia pugettensis*, head of male, after Clark, 1932; B, *Nebalia bipes*, female, after Claus, from Calman; 4, **Mysidacea:** *Mysis* sp. from Calman; 5A,B,C, **Cumacea:** *Cumella vulgaris;* A, male lateral; B, dorsal; C, female lateral, by W. B. Gladfelter; 6, **Euphausiacea:** *Euphausia pacifica* Hansen, 1911, after Boden, Johnson and Brinton, 1955. 2A,B, 3A,B, 4 and 6 redrawn by Emily Reid.

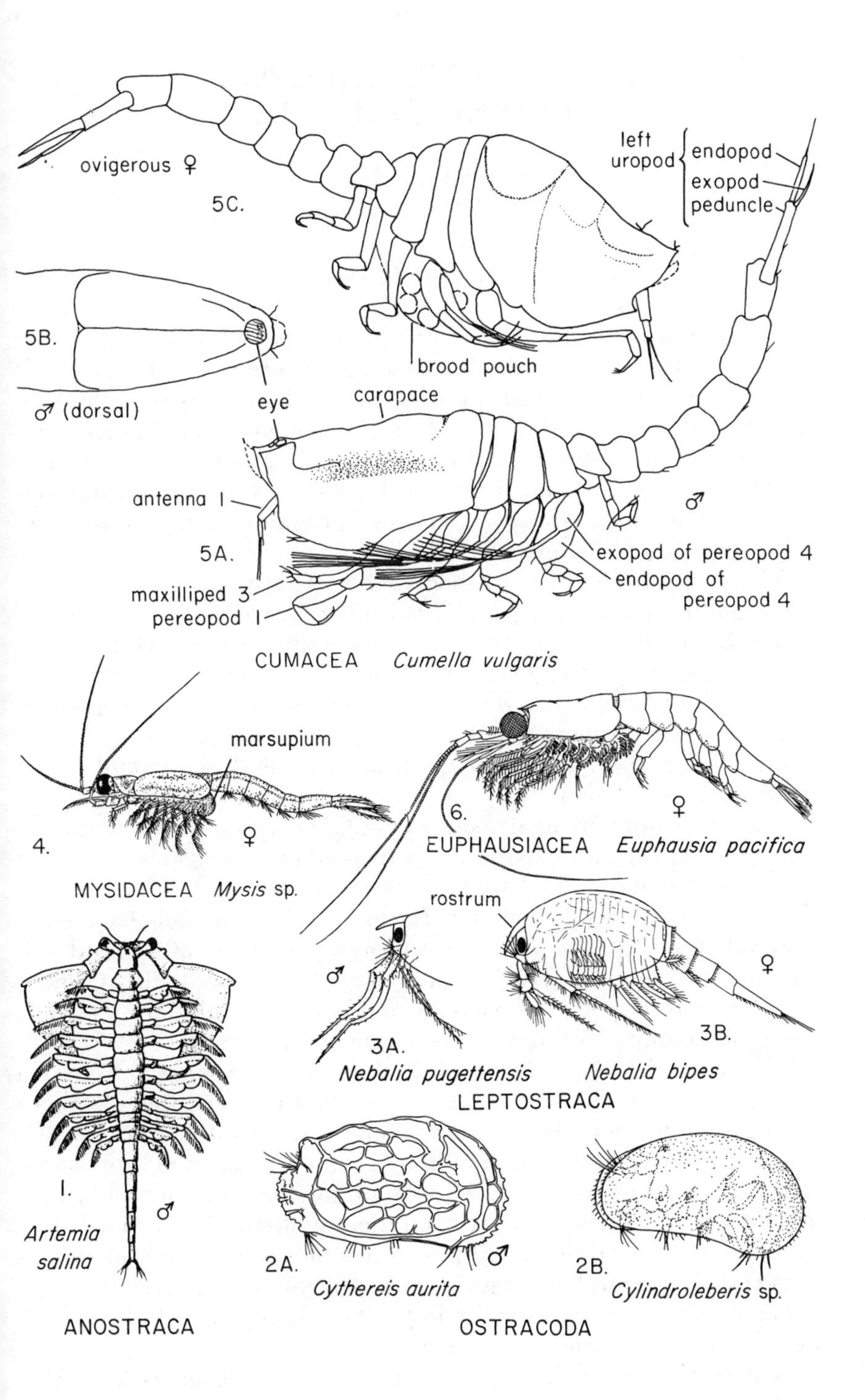
ovigerous ♀
5C.
left uropod
endopod
exopod
peduncle
5B.
♂ (dorsal)
brood pouch
eye
carapace
antenna I
♂
5A.
exopod of pereopod 4
endopod of pereopod 4
maxilliped 3
pereopod I
CUMACEA Cumella vulgaris
marsupium
6.
♀
EUPHAUSIACEA Euphausia pacifica
4.
♀
MYSIDACEA Mysis sp.
rostrum
♂
♀
3A.
3B.
Nebalia pugettensis
Nebalia bipes
LEPTOSTRACA
1.
♂
Artemia salina
2A.
♂
Cythereis aurita
2B.
Cylindroleberis sp.
ANOSTRACA
OSTRACODA

SUBCLASSES COPEPODA AND BRANCHIURA

Paul L. Illg
University of Washington, Seattle

(PLATES 55–57)

There are great difficulties in dealing with copepods of the California coast, since there are hundreds of species. There exist a fair roster of marine planktonic species; a report on San Francisco Bay plankton (Esterley, 1924); a masterly sampling of just a few of the harpacticoid copepods of tidepools, interstitial waters of various shore deposits, and the algal zone (Lang, 1965); and a few papers on parasites and other symbionts of a variety of marine vertebrates and invertebrates and even of a few marine plants.

The diversity of habitat and mode of life of copepods is expressed in exuberant morphological variety, but it is characteristic of this group to express adaptations to specific living conditions even while retaining deeply conservative features. Copepods are therefore notable for convergent body forms in diverse lineages. These facts are reflected in the complex taxonomy.

In terms of populations, copepods are extremely important in the marine plankton, where they occur in astronomical numbers and occupy a key position in many food chains. Copepods are also significant both quantitatively and qualitatively in another setting: the interstitial fauna (meiobenthos), in which copepods are exceeded in numbers and biomass only by nematodes. In this habitat copepods have been reported at over 60 species per association and over 150,000 individuals per square meter (Hulings and Gray, 1971; Noodt, 1971). There are thousands of species of parasitic or otherwise symbiotic copepods, over 2,500 being listed from fishes (Yamaguti, 1963).

The copepod body is divided into an anterior **prosome** in front of a movable articulation, and a smaller, posterior **urosome** (fig. 1). These terms are better than thorax and abdomen, about the limits of which opinions differ. The last or anal body segment bears a pair of **caudal rami.** The uniramous first antennae (antennules) are large and conspicuous, particularly in plankters, reflecting their use in swimming; there are smaller, often biramous, second antennae. A series of mouthparts variously expresses adaptive specializations, and there are up to five pairs of thoracic (swimming) legs. These legs, particularly the last, are often of systematic importance.

About 90 percent of all copepods fall in the orders **Calanoida, Cyclo-**

poida, **Harpacticoida,** and **Caligoida.** The first three include the common free-living copepods of fresh water and many marine forms, both free-living and symbiotic.

Calanoida (figs. 1–6) usually have obvious planktonic adaptations, the antennules very long with up to twenty-five articles, and the urosome sharply distinct from the much longer and broader prosome. Among the common calanoids of San Francisco Bay is *Acartia tonsa* (fig. 5), a very small plankter with almost worldwide distribution in inshore waters. A widespread oceanic genus sometimes found in the bay close to the Golden Gate is *Calanus* (fig. 2) about which there is a tremendous literature, most of it under *Calanus finmarchicus* (cf. Marshall and Orr, 1955). Sometimes found in the bay and common offshore is *Epilabidocera longipedata* (=*E. amphitrites* of most previous accounts) (fig. 4), which is large and anatomically complex (Park, 1966). In the inner reaches of the bay are two interesting calanoids: in brackish water, *Eurytemora hirundoides* (fig. 3) can form large populations; in hypersaline salt pans, *Pseudodiaptomus euryhalinus* (fig. 6) is often abundant. In almost any freshwater body, particularly in temporary ponds, occur several species of *Diaptomus*, some of which reach large size, often conspicuous by bright red coloration.

Cyclopoida are common in all marine and fresh waters. Planktonic cyclopoids are small, the antennules short, usually with less than seventeen articles, and the urosome and prosome distinctly articulated and of about the same length. However, cyclopoid parasites and commensals, which are often encountered in intertidal collecting, may reach large size and show bizarre modifications. A series of "semiparasitic" symbionts of various invertebrates, from cnidarians and flatworms to tunicates, retain the cyclopoid body form. A common form is *Clausidium vancouverense* (fig. 7), found as obvious red specks on the surfaces of *Callianassa* and *Upogebia* (Light and Hartman, 1937). Several species originally described in a now-fractionated genus, *Paranthessius* (fig. 13), occur in *Tresus, Panopea, Tivela, Saxidomus,* and other clams, lying very inconspicuously on the gills. *Mytilicola orientalis* (fig. 17) is a reddish cyclopoid of bizarrely modified, worm-like form found in the rectum of *Mytilus edulis* and other bivalves. It is a pest of mussels and commercial oysters, and there is much current literature on the group. An interesting series of cyclopoid symbionts (figs. 14–16) occur in the branchial cavities of simple ascidians and in the matrix or zooids of compound ascidians (Illg, 1958; Dudley, 1966).

Harpacticoida includes thousands of littoral species. They are usually very small, the antennules are short with few segments, and the urosome is not conspicuously set off from the prosome. The most minute copepods occur among the large assemblage of harpacticoid species living in the meiofauna. Certain to be encountered is the "red

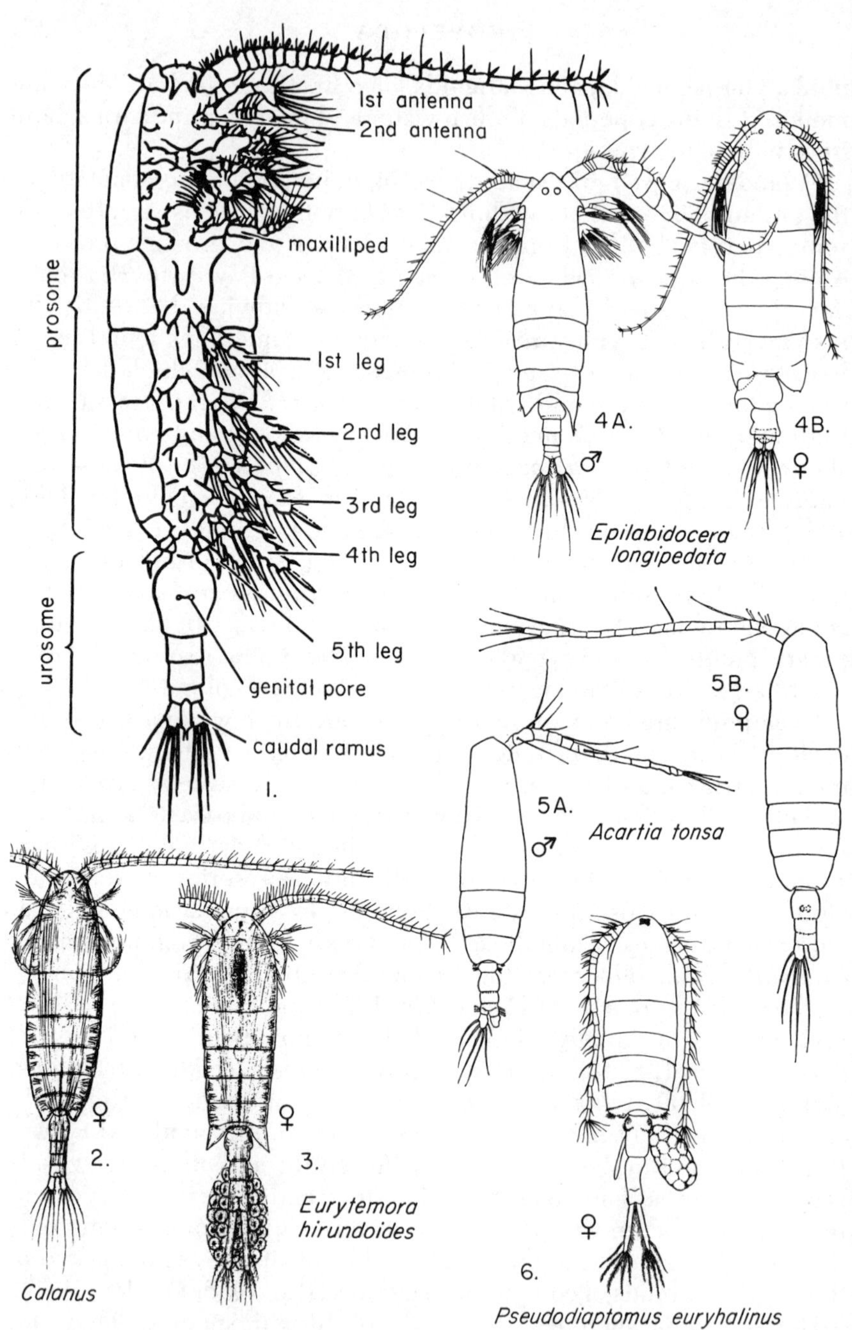

PLATE 55. **Copepoda** (1). **Calanoida:** 1, Diagrammatic figure of a female calanoid copepod, after Giesbrecht and Schmeil; 2, 3, after Sars, 1901; 4, after Park, 1966; 5, after Gonzáles and Bowman, 1965; 6, after Johnson, 1939.

high-tidepool copepod" *Tigriopus californicus* (fig. 8). In equally distinctive habitats occur *Diarthrodes cystoecus* (fig. 11), which burrows in the familiar red alga *Halosaccion*, and *Porcellidium* spp. (fig. 10), minute sowbug-like animals obvious as mobile, pepperlike specks between the blades of eelgrass or surfgrass. On all sorts of marine surfaces and particularly obvious as inhabitants of temporary or permanent seawater systems, are *Tisbe furcata* (fig. 9) and other species (cf. Johnson and Olson, 1948).

Caligoida include rather large copepods, predominately parasites of marine fishes. The basic body form is obviously copepodan (fig. 18), but very modified forms make up some of the important subdivisions of this order.

Because of the scattered records and the enormous number of local species, it is impractical to provide a key to intertidal copepods of this region. The illustrations are provided to show representative copepod types, and must not be relied upon for identification. For freshwater forms, Ward and Whipple (Edmondson, 1959) present excellent keys and brief diagnoses. Any body of fresh water, and particularly temporary ponds, yields typical calanoids, cyclopoids, and harpacticoids. The marine plankton, for which there is no local key, offers predominately calanoids, but a few small cyclopoids also usually occur. Most copepod parasites of fishes can be keyed out in the volume by Yamaguti (1963). Lang (1965) presents a digest of the few species dealt with in all preceding papers on Pacific-coast harpacticoids and adds information on 98 California species, 81 of these new. He includes pertinent keys, but for more complete coverage of the harpacticoids it is necessary to refer to the comprehensive keys of his *Monographie der Harpacticiden* (1948).

SUBCLASS BRANCHIURA

Branchiura or **Arguloidea** are sometimes treated as an order of copepods, but form a separate subclass, since they differ from copepods proper in the possession of compound eyes and in other characteristics. They are predators ("ectoparasites") of both freshwater and marine fish. The two species most commonly encountered in this region are *Argulus japonicus* Thiele, 1900 on freshwater goldfish and *Argulus pugettensis* Dana, 1852 (fig. 19) on various marine perch and sea-run salmonid fish. Wilson (1944) and Yamaguti (1963) treat branchiurans along with parasitic copepods of fishes.

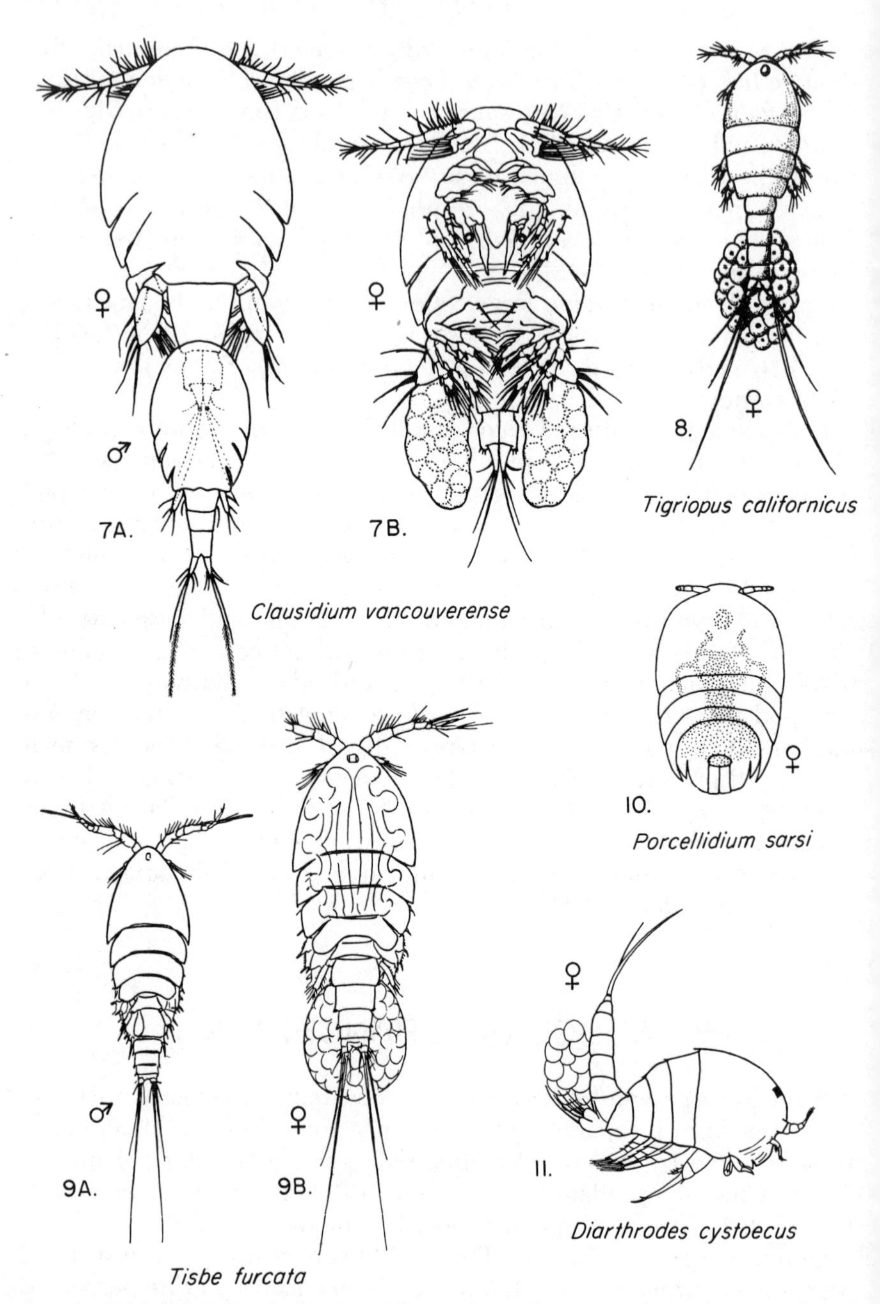

PLATE 56. **Copepoda** (2). **Cyclopoida:** 7, *Clausidium vancouverense:* A, dorsal view of clasping pair; B, female with egg masses, in ventral view; 8–11, **Harpacticoida:** 8, after Hedgpeth, 1962; 9, after Lang, 1948; 10, after Battaglia, 1954; 11, after Fahrenbach, 1962.

LIST OF SOME LOCAL MARINE COPEPODA

M: Marine
B: Brackish water
H: Hypersaline
C: Commensal
P: Parasitic

Order **Calanoida**

Acartia tonsa Dana, 1849. MB.

Calanus spp. M.

Epilabidocera longipedata (Sato, 1913) (=*E. amphitrites;* see Park, 1966). M.

Eurytemora hirundoides (Nordquist, 1888). B. Common in brackish waters of San Francisco Bay area.

Pseudodiaptomus euryhalinus Johnson, 1939. BH. In San Francisco Bay area is especially common in hypersaline waters and salt pans.

Order **Cyclopoida**

Clausidium vancouverense (Haddon, 1912). MC. On *Callianassa* and *Upogebia;* see Light and Hartman, 1937.

Mytilicola orientalis Mori, 1935. MP. In the intestine of *Mytilus edulis* and other bivalves. See Humes, 1954, J. Parasitol. 40: 186–194.

Hemicyclops thysanotus Wilson, 1935. MC. On *Upogebia, Callianassa,* and *Hermissenda.* See Gooding, 1960, Proc. U.S. Nat. Mus. 112: 159–195.

Paranthessius spp. MC. On gills of *Tresus, Protothaca, Saxidomus,* and other bivalves; see Illg, 1949.

Pholeterides furtiva Illg, 1958. MC. In *Aplidium* and other ascidians.

Pygodelphys aquilonaris Illg, 1958. MC. In *Pyura, Cnemidocarpa, Molgula,* and other simple ascidians.

Pythodelphys acruris Dudley and Solomon, 1966. MC. From the matrix of the compound ascidian, *Archidistoma.*

Order **Harpacticoida**

Diarthrodes cystoecus Fahrenbach, 1954. MC. In the intertidal red alga *Halosaccion.* See Fahrenbach, 1954, J. Wash. Acad. Sci. 44: 326–329.

Porcellidium spp. M. Between the blades of eelgrass (*Zostera*) and surfgrass (*Phyllospadix*).

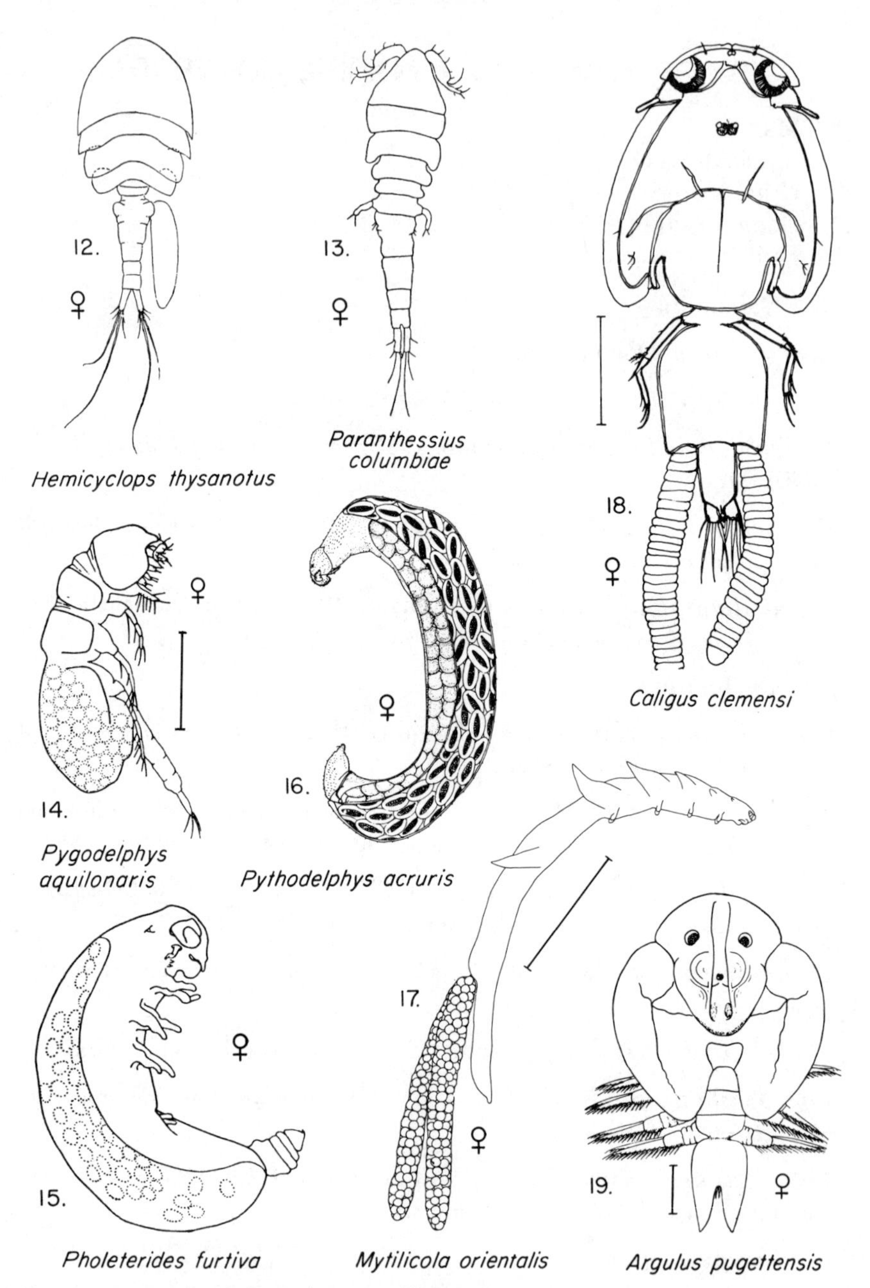

PLATE 57. **Copepoda** (3); **Branchiura:** 12–17, **Cyclopoida:** 12, after Gooding, 1960; 13, after Illg, 1949; 14, 15, after Illg, 1958; 16, after Dudley and Solomon, 1966; 17, after Humes, 1954; 18, **Caligoida:** after Parker and Margolis, 1964; 19, **Branchiura:** after Wilson, 1912. Scale bars = 1 mm; rest not to scale.

Tigriopus californicus (Baker, 1912) (=*T. triangulus* Campbell). M. Characteristic of higher tide pools, often in very high salinity.

Tisbe furcata (Baird, 1837). MB. See Johnson and Olson, 1948.

Order **Caligoida** (Ectoparasites of marine fishes)

Caligus clemensi Parker and Margolis, 1964. MP. Known from many host species and also taken free in plankton. Other caligoids are present. See Margolis *et al.* 1975.

REFERENCES ON COPEPODA

Davis, C. C. 1949. The pelagic Copepoda of the northeastern Pacific Ocean. Univ. Wash. Publ. Biol. 14: 1–117. (Key.)

Dudley, P. L. 1966. Development and systematics of some Pacific marine symbiotic copepods. The biology of the Notodelphyidae. Univ. Wash. Publ. Biol. 21: v + 282 pp.

Esterly, C. O. 1924. The free-swimming copepoda of San Francisco Bay. Univ. Calif. Publ. Zool. 26: 81–129.

Fahrenbach, W. H. 1962. The biology of a harpacticoid copepod. La Cellule 62: 301–376. (On *Diarthrodes cystoecus.*)

Fraser, J. H. 1936a. The occurrence, ecology, and life history of *Tigriopus fulvus* (Fischer). J. Mar. Biol. Assoc. U.K. 20: 523–536.

Fraser, J. H. 1936b. The distribution of rock pool copepods according to tidal level. Animal Ecol. 5: 23–38.

Hulings, N. C. and J. S. Gray, eds. 1971. A Manual for the Study of Meiofauna. Smithson. Contrib. Zool. 78: 83 pp.

Illg, P. L. 1949. A review of the copepod genus *Paranthessius* Claus. Proc. U.S. Nat. Mus. 99: 391–428.

Illg, P. L. 1958. North American copepods of the family Notodelphyidae. Proc. U.S. Nat. Mus. 107: 463–649.

Johnson, M. W. 1939. *Pseudodiaptomus* (*Pseudodiaptallous*) *euryhalinus*, a new subgenus and species of Copepoda, with preliminary notes on its ecology. Trans. Amer. Micr. Soc. 58: 349–355.

Johnson, M. W. 1948. The postembryonic development of the copepod, *Pseudodiaptomus euryhalinus* Johnson, and its phylogenetic significance. Trans. Amer. Micr. Soc. 67: 319–330.

Johnson, M. W. and J. B. Olson. 1948. The life history and biology of a marine harpacticoid copepod, *Tisbe furcata* (Baird). Biol. Bull. 95: 320–332.

Kaestner, A. 1970. *Invertebrate Zoology. Crustacea.* Vol. III. Translated by H. W. and L. R. Levi. Interscience, John Wiley & Sons.

Lang, K. 1948. *Monographie der Harpacticiden.* Stockholm: H. Ohlsson. 2 vols., 1682 pp.

Lang, K. 1965. Copepoda Harpacticoidea from the California Pacific Coast. K. Svenska Vetenskapsakad. Handlingar, 4th Ser., Vol. 10, no. 2.

Light, S. F., and O. Hartman 1937. A review of the genera *Clausidium* Kossmann and *Hemicyclops* Boeck (Copepoda, Cyclopoida). Univ. Calif. Publ. Zool. 41: 173–188.

Margolis, L., Z. Kabata, and R. R. Parker, 1975. Catalogue and synopsis of *Caligus*, a genus of Copepoda (Crustacea) parasitic on fishes. Bull. Fish. Res. Bd. Canada 192: 117 pp.

Marshall, S. M. and A. P. Orr. 1955. *The Biology of a Marine Copepod.* Edinburgh: Oliver & Boyd. (Reprint, 1972. Berlin: Springer.)

Monk, C. R. 1941. Marine harpacticoid copepods from California. Trans. Amer. Micr. Soc. 60: 75–99.

Noodt, W. 1971. Ecology of the Copepoda. In Proceedings of the First International Conference on Meiofauna. Smithson. Contrib. Zool. 76: 97–102.

Park, T. S. 1966. The biology of a calanoid copepod *Epilabidocera amphitrites* McMurrich. La Cellule 66: 131–251.

Ward, H. B. and G. C. Whipple 1959. *Fresh-water Biology.* Edited by W. T. Edmondson. 2nd ed. John Wiley & Sons.

Wilson, C. B. 1932. The copepods of the Woods Hole region, Massachusetts. Bull. U.S. Nat. Mus. 158: 1–635 (Includes both marine and freshwater forms.)

Wilson, C. B. 1935. Parasitic Copepoda from the Pacific Coast. Amer. Midl. Nat. 16: 776–797.

Wilson, C. B. 1944. Parasitic copepods in the United States National Museum. Proc. U.S. Nat. Mus. 94: 529–582. (Contains bibliography of earlier work.)

Yamaguti, S. 1963. *Parasitic Copepoda and Branchiura of Fishes.* Interscience, John Wiley & Sons.

PHYLUM ARTHROPODA: CRUSTACEA, CIRRIPEDIA

William A. Newman
Scripps Institution of Oceanography, La Jolla

(PLATES 58–61)

Cirripeds are specialized crustaceans in that they undergo two metamorphic changes during development and all adults are attached. Most are planktotrophic setose feeders. However, a number are parasitic and so highly modified that they are recognizable as crustaceans only by their larval stages. The subclass includes five orders, three of which occur locally.

The principal order, **Thoracica,** encompasses the stalked and sessile barnacles, and representatives are common along the California coast. Their shell remains are so numerous in some situations that Charles Darwin suggested that the present geological period might be known in the fossil record as the "Age of Barnacles." The order to which these forms belong will be taken up in more detail below.

The second order, **Acrothoracica,** contains relatively small, shell-less forms found burrowing in calcareous substrates such as molluscan shells, coral, and limestone. The order reaches its greatest diversity in coral seas. Locally the minute *Trypetesa lateralis* makes its burrows in the shells of *Tegula* occupied by hermit crabs. Occasionally other snail shells are similarly occupied. The female excavates a small burrow with a slitlike opening on the inside of the body whorl of the shell. Tiny dwarf males, as cyprid larvae, locate the burrows and reside with the female.

The third order, **Rhizocephala,** includes the most specialized parasites among the crustaceans. Many species start life as typical **nauplii** which after four molts metamorphose into **cyprid larvae.** The cyprid searches out a host, usually a recently molted anomuran or brachyuran decapod, and attaches by a first antenna to the base of a seta. It then undergoes a metamorphosis in which the legs and eyes are cast and the remaining tissue is extruded through the antenna into the host. This tissue migrates around the digestive tract and sends out nutritive rootlets that pervade the host's body. Eventually a globular reproductive body erupts beneath the abdomen of the host. The solitary, globose *Heterosaccus californicus* is fairly common on older *Pugettia producta*, the kelp crab, and a small percentage of hermit crabs bear the gregarious, fingerlike *Peltogasterella gracilis* (see Reinhard, 1944).

In a typical stalked barnacle, such as the goose barnacle *Lepas* (fig. 1), the capitulum is armored with five calcareous plates: a pair of large **scuta** (sing., **scutum**), a pair of **terga** (sing., **tergum**) and a narrow median **carina** along the dorsal margin. Together the terga and scuta close the aperture through which the cirri are extended during feeding. In the stalked barnacle *Pollicipes* (fig. 2) a sixth plate, the **rostrum,** has been added, and numerous accessory platelets are found around the basal margin of the capitulum. The rostrum and carina, and probably some of the plates of the accessory whorl, are homologous with the wall plates of sessile barnacles. The terga and scuta of sessile barnacles have become separated from the wall and specialized to form the **operculum.**

In sessile or acorn barnacles, such as most species of *Balanus* (fig. 3), there are three calcareous regions: (1) the **basis,** attached to the substratum; (2) the **wall,** formed of rigidly articulated plates; and (3) a movable portion, the **opercular valves,** closing the aperture. The wall of most species of *Balanus* consists of six plates, each known as a **compartment** (fig. 5). The carina, furthest from the head, is dorsal or posterodorsal; the rostrum, with **rostrolateral plates** fused to it, forms the opposite end of the wall. The **carinolaterals** lie on either side and overlap the carina. Between them and the rostrum lie the **laterals,** overlapping the carinolaterals and overlapped by the rostrum. In *Tetraclita* there are only four plates, the carinolaterals having been eliminated. In most species of *Tetraclita* and *Chthamalus* and in some species of *Balanus*, the basis is membranous. There are also six plates in *Chthamalus* (fig. 4), but the course of evolution has been different and they are not all the same as in *Balanus*. The most obvious difference is that the carinolaterals have dropped out and the rostrolaterals have been retained as free plates. Thus the rostrum is overlapped by adjacent plates rather than overlapping them.

The exposed median triangular portion of each compartment is the **paries** (pl., **parietes**). The edge of a compartment that is overlapped by an adjacent compartment is an **ala** (pl., **alae**). If the overlapping edge is marked off from the paries by a definite change in the direction of growth lines, it is known as a **radius.** Radii are present on all but the most primitive balanids. The carina, rostrum, and rostrolaterals of *Chthamalus* are symmetrical; the first two always having alae, the last radii. On the other hand, the rostrum of *Balanus* is distinguished in having radii rather than alae. The carinolaterals and laterals in *Balanus*, and the laterals in *Chthamalus*, are asymmetrical,

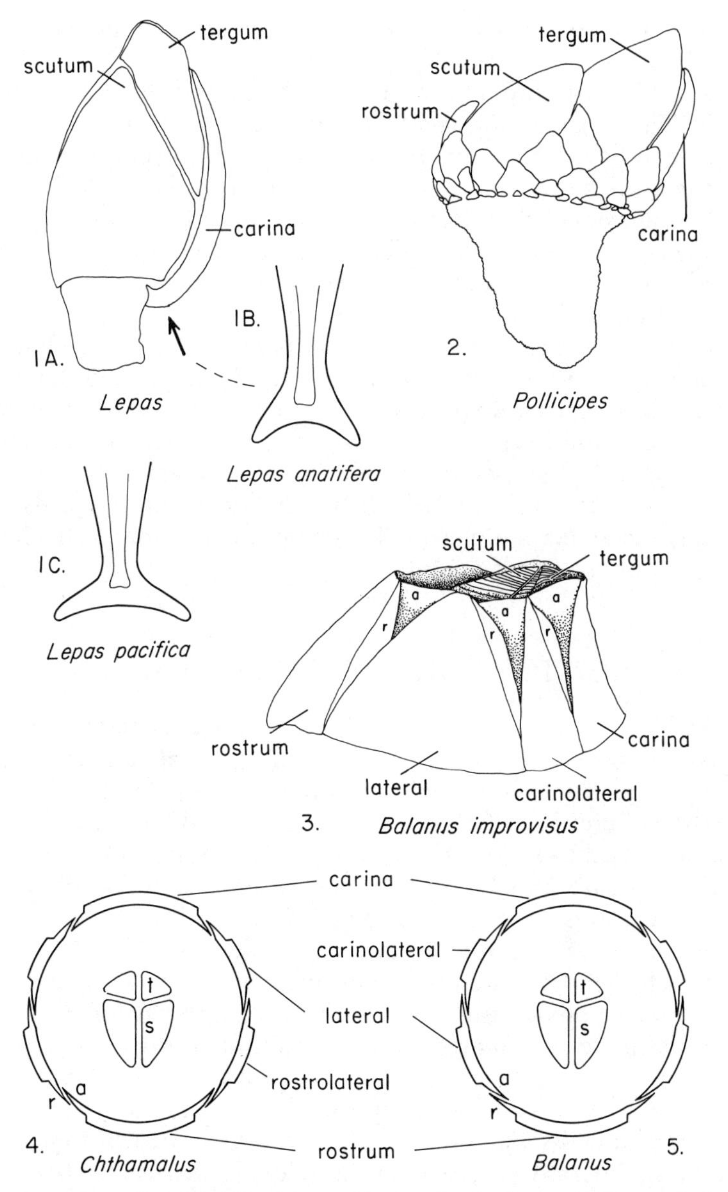

PLATE 58. **Cirripedia** (1). Types of barnacles and terminology of plates: 1A, goose barnacle (*Lepas*) with 5 plates, from right; 1B, base of carina of *L. anatifera;* 1C, of *L. pacifica;* 2, stalked barnacle (*Pollicipes polymerus*) with numerous plates from right; 3, sessile barnacle (*Balanus improvisus*), from right, showing how radii overlap alae; 4–5, schematic cross-sections to show arrangement of plates of *Chthamalus* and *Balanus* (r = radius, a = ala, t = tergum, s = scutum). Newman, original.

having alae on the anterior (rostral) margins and radii on the posterior (carinal) margins.

The movable operculum is formed by the paired terga and scuta (figs. 3–7). The terga are intricately articulated with the scuta of each side; each pair is provided with muscles that open and close the aperture and control the position of the operculum in the orifice of the shell. The interiors of the valves are marked by ridges, grooves, and pits. Likewise, the exteriors of the valves in some species are ornamented in various distinctive ways. These characteristics are useful in identification (figs. 10–25).

Most barnacles are cross-fertilized hemaphrodites. Eggs are laid in the mantle cavity of the adult where they are fertilized and develop into larvae. Although a few species produce and release cyprid larvae, most produce nauplii (fig. 8), which are released into the plankton where they develop into cyprid larvae (fig. 9). Usually there are six naupliar stages, the last five of which are planktotrophic. Cyprids are nonfeeding and select the place of attachment where, once attached, they metamorphose into the adult form.

KEY TO THORACIC CIRRIPEDIA

1. Stalked forms .... **Lepadomorpha** 2
– Sessile forms .... **Balanomorpha** 7
2. Capitular plates more than 5 in number, surrounded basally by whorl of imbricated plates; exposed intertidal on rocks and artificial structures (fig. 2) .... *Pollicipes polymerus*
– Capitular plates 5 or less in number, without basal whorl of imbricate plates 3
3. Capitular plates 5 in number 4
– Capitular plates 2 in number, reduced to a pair of Y-shaped scuta, on scyphomedusae .... *Alepas pacifica*
4. Capitular plates completely covering capitulum (fig. 1A) .... *Lepas* 5
– Capitular plates reduced to narrow slips; occurring in gill chambers of crabs and lobsters .....*Octolasmis californiana*
5. Capitulum laterally compressed; individuals always attached to floating objects, not forming float of their own 6
– Capitulum more or less globular; plates thin, papery; individuals forming floats of their own, initial attachment to feathers, *Velella*, or floats of adults .... *Lepas fascicularis*
6. Plates thin but not papery, pigment of hypodermis showing through scuta; base of carina laterally expanded like whale flukes (fig. 1C) .... *Lepas pacifica*

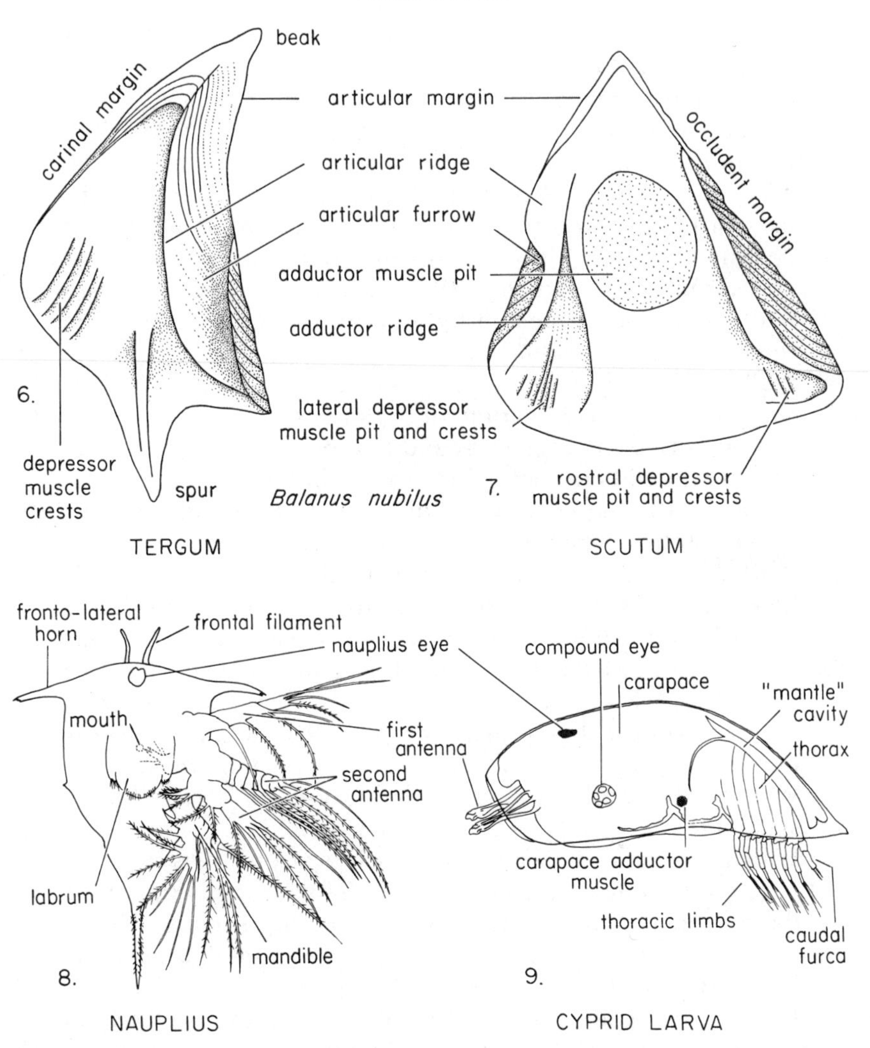

PLATE 59. **Cirripedia** (2). Terminology of interior of opercular valves of *Balanus nubilus:* 6, right tergum; 7, right scutum. Larval forms: 8, second stage nauplius of *B. amphitrite* seen from below, right appendages deleted; 9, cyprid of *Trypetesa* seen from left, modified after Kuhnert (1934). All but 9, Newman, original.

– Plates thick, pigment of hypodermis not showing through scuta except in very small specimens; base of carina forming a blunt fork (figs. 1A,B) .... *Lepas anatifera*

7. Wall composed of 4 plates, permeated by numerous longitudinal tubes usually filled with calcareous reddish material exposed externally by erosion; interior of opercular valves

heavily sculptured (fig. 25) *Tetraclita squamosa rubescens*
– Wall composed of 6 plates 8
8. Rostrum overlapped by adjacent plates (fig. 4); small, brown, or gray green .... CHTHAMALIDAE 9
– Rostrum overlapping adjacent plates (fig. 5)..BALANIDAE 10
9. Scutum with long, strong, adductor ridge, with lateral depressor muscle crests (fig. 11) .... *Chthamalus dalli*
– Scutum with short, strong, adductor ridge, without lateral depressor muscle crests (fig. 10) .... *Chthamalus fissus*
10. Wall solid, not permeated by longitudinal tubes (except for some young *B. glandula*) 11
– Wall permeated by longitudinal tubes 14
11. Free-living, on wide variety of substrates 12
– Living buried in gorgonians and hydrocorals 13
12. Wall white, ribbed; scutum with pit on either side of adductor ridge (fig. 19); immature individuals may have small, irregular unfilled tubes in the wall ....*Balanus glandula*
– Wall white, smooth or weakly ribbed; scutum with callus forming several ridges extending from the articular ridge to the adductor ridge (fig. 24) ..*Balanus hesperius laevidomus*
13. In gorgonians; basis boat-shaped, tergum apically truncated (fig. 12) .... *Balanus galeatus*
– In hydrocorals; basis more or less flat, tergum apically pointed (fig. 13) .... *Balanus nefrens*
14. Wall permeated by a single row of more or less uniformly spaced tubes 15
– Wall permeated by many irregularly spaced tubes not in a single row; outer surface white, irregularly ribbed, having thatched appearance; tergum with beak (fig. 14) ....*Balanus cariosus*
15. Exterior of wall white, without colored markings 18
– Exterior of wall ornamented by conspicuous reddish or maroon, more or less longitudinal markings 16
16. Radii solid, not permeated by transverse tubes 17
– Radii permeated by transverse tubes; exterior of scutum without longitudinal striations; tergum with faint depressor muscle crests (fig. 22) ......*Balanus tintinnabulum californicus*
17. (*Note 3 choices*) Exterior of scutum marked by one or more longitudinal rows of pits; tergal spur nearly 1/2 width of basal margin (fig. 23) ....*Balanus trigonus*
– Exterior of scutum with numerous longitudinal striations; tergal spur about 1/3 width of basal margin (fig. 17) ....*Balanus pacificus*
– Exterior of scutum marked by simple transverse growth lines,

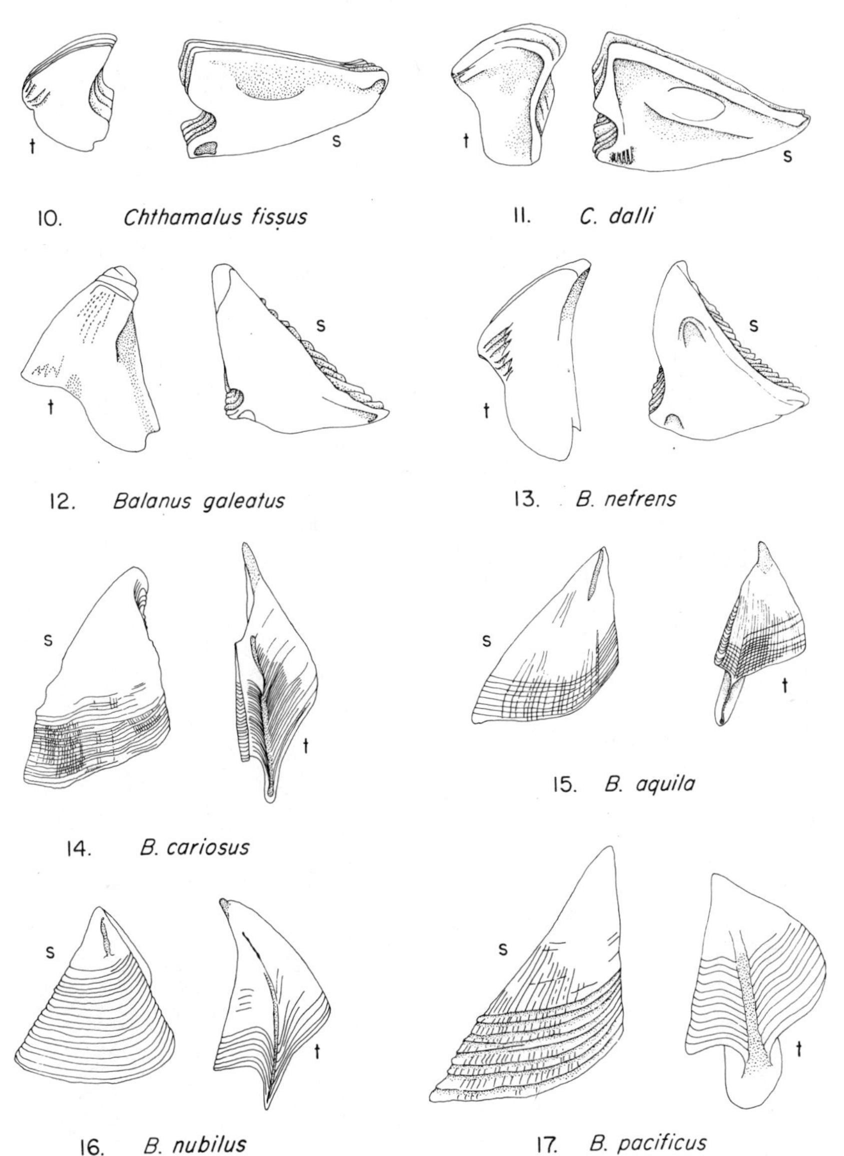

PLATE 60. **Cirripedia** (3). Right-hand terga and scuta of sessile barnacles: 10–13, interior; 14–17, exterior. Newman, original.

without pits or longitudinal striations; tergal spur about 1/3 width of basal margin (fig. 21); in bays and harbors .... *Balanus amphitrite amphitrite*

18. Tergum beaked 19
– Tergum not beaked 20

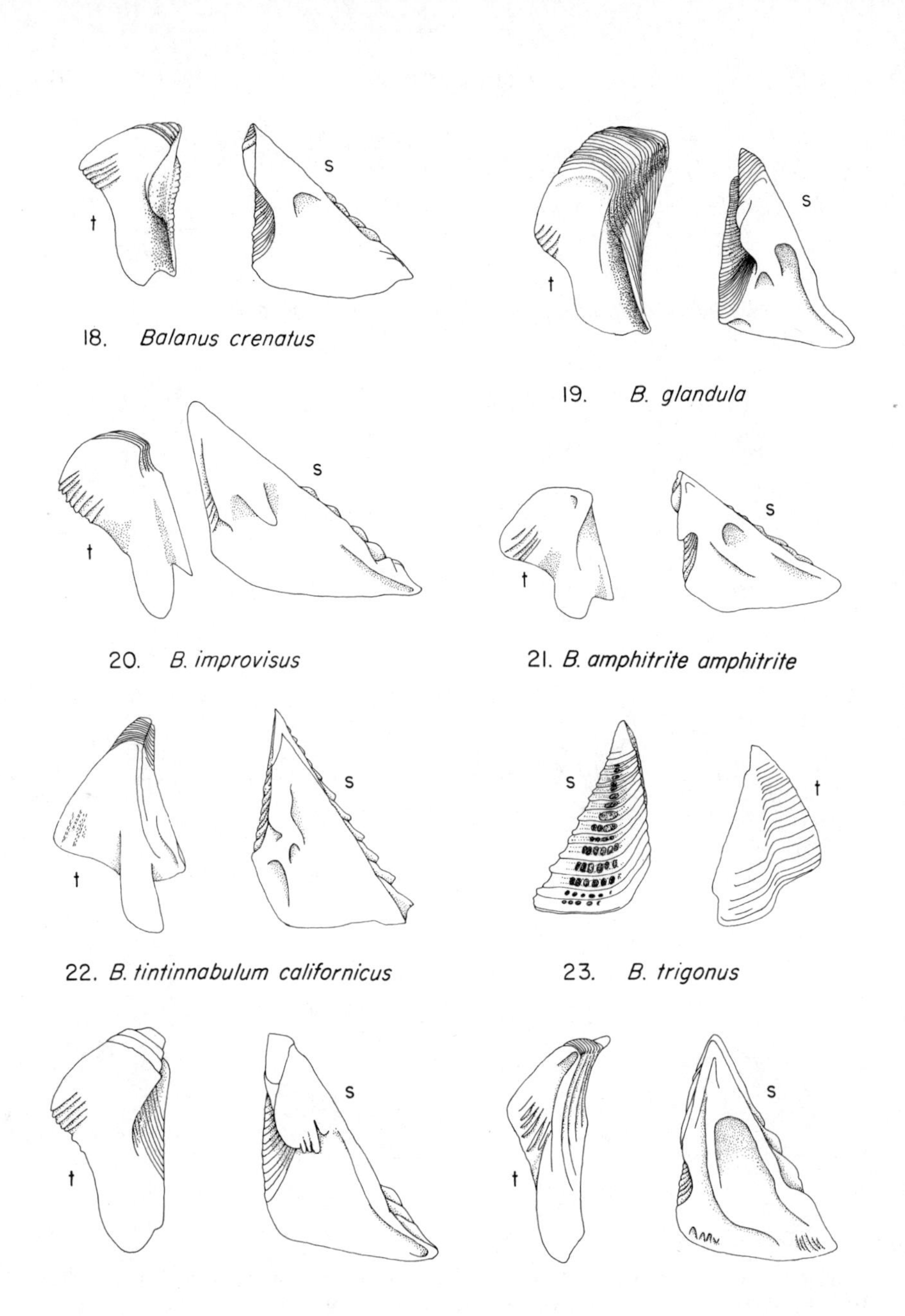

PLATE 61. **Cirripedia** (4). Right-hand terga and scuta of sessile barnacles: 23, exterior; rest, interior. Newman, original.

19. Aperture of shell relatively small, not flaring; exterior of scutum and tergum marked by longitudinal striations (fig. 15) .... *Balanus aquila*
 – Aperture of shell relatively large, flaring; exterior of scutum and tergum without longitudinal striations (figs. 6, 7, 16) ... . *Balanus nubilus*
20. Scutum without adductor ridge; tergal spur wider than long, occupying at least 1/2 of basal margin (fig. 18) .... *Balanus crenatus*
 – Scutum with strong adductor ridge; tergal spur longer than wide, occupying less than 1/2 of basal margin; only in brackish water (figs. 3, 20) .... *Balanus improvisus*

LIST OF CIRRIPEDIA

Order **Acrothoracica**

* *Trypetesa lateralis* Tomlinson, 1953. Burrowing inside gastropod shells inhabited by hermit crabs. See Tomlinson, 1953, J. Wash. Acad. Sci. 43: 373–381; 1955, J. Morph. 96: 97–114 (morphology).

Order **Rhizocephala**

* *Heterosaccus californicus* Boschma, 1933. Occurs on spider crabs; solitary.

* *Peltogasterella gracilis* (Boschma, 1927). Occurs gregariously on hermit crabs. See Yanagimachi, 1961, Crustaceana 2: 183–186 (*Peltogasterella* life cycle).

Order **Thoracica**

Suborder **Lepadomorpha**

Alepas pacifica Pilsbry, 1907. Attached to scyphomedusae, usually *Phacellophora.*

Lepas (Dosima) fascicularis Ellis and Solander, 1786. Like other goose barnacles, but forming gas-filled float.

Lepas (Lepas) anatifera Linnaeus, 1758. Goose barnacle, pelagic, washed ashore on floating timber and other objects.

Lepas (L.) pacifica Henry, 1940. On floating timber and other objects.

Octolasmis californiana Newman, 1960. A southern species, ranging north to Monterey Bay; inhabiting gill chambers of crabs and spiny lobsters.

* Not in key.

Pollicipes polymerus Sowerby, 1833 (formerly *Mitella*). Frequently forming dense stands with *Mytilus californianus* on mid-tidal rocks exposed to heavy wave action. See Barnes and Reese, 1959, Proc. Zool. Soc. Lond. 132: 569–585 (feeding); 1960, J. Anim. Ecol. 29: 169–185 (behavior, ecology); Hilgard, 1960, Biol. Bull. 119: 169–188 (reproduction).

Suborder **Balanomorpha**

Balanus (Armatobalanus) nefrens Zullo, 1963. Subtidal southern species ranging north to Monterey Bay, embedded in the hydrocorals *Allopora californica* and *Errinopora pourtalesia*.

Balanus (Balanus) amphitrite amphitrite Darwin, 1854. Introduced; generally restricted to warmer portions of bays and harbors; low intertidal and subtidal.

Balanus (Balanus) aquila Pilsbry, 1907. Large, common in low intertidal and subtidal on pilings and rocks in bays; rare north of San Francisco.

Balanus (Balanus) crenatus Bruguière, 1789. Low intertidal and subtidal in protected situations.

Balanus (Balanus) glandula Darwin, 1854. Common; high intertidal, outer coast and in bays; forms dense stands with *Chthamalus*. See Barnes and Barnes, 1956, Pac. Sci. 10: 415–422 (biology).

Balanus (Balanus) improvisus Darwin, 1854. Introduced; in brackish water of bays on shells, pilings, etc.; low intertidal and subtidal.

Balanus (Balanus) nubilus Darwin, 1854 (=*B. flos* Pilsbry, 1907). Often mis-spelled *nubilis*, following an error by Pilsbry (1916); Darwin's original spelling was *nubilus*. Very large; low intertidal on pilings and rocks.

Balanus (Balanus) pacificus Pilsbry, 1916. Subtidal southern form extending north to Monterey Bay; usually on other organisms or man-made structures.

Balanus (Balanus) trigonus Darwin, 1854. Low intertidal and subtidal southern form extending north to Monterey Bay.

Balanus (Conopea) galeatus (Linnaeus, 1771). Southern form extending as far north as Monterey Bay, occurring on gorgonians; only balanid in this region with complemental males.

Balanus (Megabalanus) tintinnabulum californicus Pilsbry, 1916. Low intertidal and subtidal southern form, scarce north of Monterey Bay; often found in large clusters on buoys and pilings.

Balanus (Semibalanus) cariosus (Pallas, 1788). Intertidal northern species ranging to somewhat south of San Francisco Bay.

Balanus (Solidobalanus) hesperius laevidomus Pilsbry, 1916. Subtidal northern form ranging south to San Francisco Bay, frequently on crabs and mollusc shells; low intertidal in laminarian holdfasts, etc.

Chthamalus dalli Pilsbry, 1916. High-intertidal northern species, generally with and above *Balanus glandula.*

Chthamalus fissus Darwin, 1854 (=*C. microtretus* Cornwall, 1937). Southern species occurring with *C. dalli* in central California.

Tetraclita (Tetraclita) squamosa rubescens Darwin, 1854. Commonly intertidal, singly or in clusters; peltate white variety *elegans* on crabs and molluscan shells; rare north of San Francisco.

REFERENCES ON CIRRIPEDIA

Cornwall, I. E. 1951. The barnacles of California (Cirripedia). Wasmann J. Biol. 9: 311–346.

Crisp, D. J. and P. S. Meadows 1962. The chemical basis of gregariousness in cirripedes. Proc. Roy. Soc. Lond. B. 156: 500–520.

Darwin, C. *A Monograph of the sub-class Cirripedia.* Part I, *Lepadidae,* 1851; Part II, *Balanidae,* 1854. London: Ray Society (reprinted, Cramer, 1964).

Gruvel, A. 1905. *Monographie des Cirrhipèdes.* Paris: Masson, 472 pp. (reprinted, A. Asher, 1965).

Henry, D. P. 1940a. Notes on some pedunculate barnacles from the North Pacific. Proc. U.S. Nat. Mus. 88: 225–236.

Henry, D. P. 1940b. The Cirripedia of Puget Sound with a key to the species. Univ. Wash. Publ. Oceanogr. 4: 1–48.

Henry, D. P. 1942. Studies on the sessile Cirripedia of the Pacific coast of North America. Univ. Wash. Publ. Oceanogr. 4: 95–134.

Newman, W. A. 1960. *Octolasmis californiana,* a new pedunculate barnacle from the gills of the California Spiny Lobster. Veliger 3: 9–11.

Newman, W. A. 1967. On physiology and behavior of estuarine barnacles Proc. Symposium on Crustacea, Part III, pp. 1038–1066, Mar. Biol. Soc. India.

Newman, W. A. and A. Ross 1976. Revision of the balanomorph barnacles, including a catalog of the species. San Diego Soc. Nat. Hist. Memoir 9: 108 pp. (This alters several names in the present key.)

Newman, W. A., V. Zullo, and T. H. Withers 1969. Cirripedia. In *Treatise on Invertebrate Paleontology,* R. C. Moore, ed., Part R. *Arthropoda.* 4 (1): R206–295. Geol. Soc. Amer. and Univ. Kansas Press.

Pilsbry, H. A. 1916. The sessile barnacles (Cirripedia) contained in the collections of the U.S. National Museum; including a monograph of the American species. Bull. U.S. Nat. Mus. 93: xi + 366 pp.

Reinhard, E. G. 1944. Rhizocephalan parasites of hermit crabs from the Northwest Pacific. Jour. Wash. Acad. Sci. 34: 49–58.

Reischman, P. O. 1959. Rhizocephala of the genus *Peltogaster* from the coast of the State of Washington and the Bering Sea. Nederl. Akad. Wetensch. Proc. C. 62: 409–435.

Tomlinson, J. T. 1969. The burrowing barnacles (Cirripedia: order Acrothoracica). Bull. U.S. Nat. Mus. 296: 162 pp.

Zullo, V. 1963. A review of the subgenus *Armatobalanus* Hoek (Cirripedia, Thoracica), with the description of a new species from the California coast. Ann. Mag. Nat. Hist. (ser. 13) 6: 587–594.

PHYLUM ARTHROPODA: CRUSTACEA, LOWER MALACOSTRACA

The great subclass **Malacostraca** includes most of the large, common, and conspicuous crustaceans. Although many of its members are well known, the group as a whole shows a diversity of features that makes it rather hard to define, and that makes it necessary to recognize a number of orders. Among the characteristics that unite the various orders of malacostracans, perhaps the most striking is the division of the body into well-marked thoracic and abdominal regions: the thorax (often to some extent fused with the head) of eight somites and bearing appendages variously adapted for swimming, feeding, or walking; and an abdomen of six somites, all of which usually bear appendages except when the abdomen is reduced, as in crabs; the last pair of these appendages forms, with the telson, a caudal fan. Other features include a carapace covering part or all of the thorax (but lacking in amphipods and isopods) and biramous first antennae. The *more primitive* members of the various orders show a basic unity in an approach to a common, shrimplike form said to exhibit the "caridoid facies" (Pl. 91, fig. 1). The mysids or the carideans perhaps most closely approach this generalized concept of a malacostracan. Such types as amphipods and isopods diverge widely in appearance, although study reveals the malacostracan ground plan in them also.

KEY TO DIVISIONS OF LOCAL MALACOSTRACA

1. Seven abdominal segments; prominent posterior caudal furca; hinged carapace with adductor muscle and movable, hinged rostrum .... **LEPTOSTRACA** (page 271)
– Six abdominal segments; no caudal furca; carapace, if present, not hinged and without adductor muscle 2
2. Carapace, if present, not fusing with more than 4 thoracic segments; oöstegites present (plates on ventral side of female

which form a brood pouch for hatching of eggs and carrying of young) .... **PERACARIDA** (page 272)
– Carapace fusing with all thoracic segments; no oöstegites ... . **EUCARIDA** (page 377)

The primitive Leptostraca include only one reported local intertidal species. The Peracarida include the orders Mysidacea, Cumacea, Tanaidacea, Isopoda, and Amphipoda; the Eucarida include the Euphausiacea and Decapoda, the latter the familar shrimps, true crabs, hermit crabs, lobsters and crayfish, porcellain crabs, and thalassinids. One more division, **HOPLOCARIDA,** the mantis shrimps (not keyed out), may be mentioned but is not represented in the local intertidal.

DIVISION LEPTOSTRACA, ORDER NEBALIACEA

(PLATE 54)

Nebaliaceans are small, primitive, filter-feeding malacostracans, occasionally abundant in pools and in or under *Ulva, Enteromorpha,* and the eelgrass *Zostera* on mudflats. They differ from other malacostracans in the unique possession of seven, rather than six, abdominal somites, the last lacking appendages; in having the abdomen end in a caudal furca; and in having a hinged carapace closed by an adductor muscle. The head is covered by a conspicuous, movable, hinged rostrum. Eggs are brooded on the thoracic legs of the female and juveniles emerge as postlarvae, the nauplius and metanauplius stages having been completed before hatching.

Our local species is apparently *Nebalia pugettensis* (Clark, 1932), the female of which is said to be nearly identical to the Atlantic *Nebalia bipes* (fig. 3B); the male, however, possesses prominent, sickle-shaped antennae (fig. 3A). The wormlike rotifer *Seison* (page 123, Pl. 16, fig. 7) has been found on *Nebalia.*

REFERENCES ON LEPTOSTRACA

Cannon, H. G. 1927. On the feeding mechanism of *Nebalia bipes.* Trans. Roy. Soc. Edinburgh 55: 355–369.

Cannon, H. G. 1960. Leptostraca. *In* H. G. Bronn, *Klassen und Ordnungen das Tierreichs,* 5, Abt. 1, Buch 4, 81 pp.

Clark, A. E. 1932. *Nebaliella caboti* n. sp. with observations on other Nebaliacea. Trans. Roy. Soc. Canada, (3) 26: 217–235.

Menzies, R. J. and J. L. Mohr 1952. The occurrence of the wood-boring crustacean *Limnoria* and of Nebaliacea in Morro Bay, California. Wasmann J. Biol. 10: 81–86.

DIVISION PERACARIDA

The abundant and successful peracaridan crustaceans comprise most of the so-called lower malacostracans. Members of the orders **Mysidacea** and **Cumacea** (Pl. 54) have a shrimplike or "caridoid" body form with a distinct carapace covering (although not attached to) the thorax. Members of the orders **Tanaidacea (Chelifera)** (Pl. 62), **Isopoda** (Pls. 63–69) and **Amphipoda** (Pls. 70–90) either lack or have a very small carapace, have more compact bodies, and are less shrimplike. Although the body form of peracaridans differs greatly, all adult females possess a brood pouch, in which the eggs are carried and the young developed, formed of plates (**oöstegites**) on the thoracic limbs.

KEY TO ORDERS OF PERACARIDA

1. Body shrimplike, with elongate abdomen and distinct carapace over thorax 2
– Body with thorax and abdomen not sharply distinct; carapace very small or absent 3
2. Eyes stalked; carapace covering all or most of thorax (Pl. 54, fig. 4) .... **Mysidacea** (page 272)
– Eyes sessile; carapace covering only 3 or 4 thoracic somites and inflated into a branchial chamber on each side (Pl. 54, figs. 5A, B, C) .... **Cumacea** (page 273)
3. Carapace very small, covering first 2 thoracic somites; resemble small isopods, but have 1st pair of legs (gnathopods) chelate, and eyes borne on anterolateral stalks (Pl. 62) .... **Tanaidacea** (page 277)
– Carapace absent; eyes sessile 4
4. Body usually dorsoventrally flattened; thoracic legs (except for maxillipeds) essentially alike (Pl. 63) .... **Isopoda** (page 281)
– Body usually laterally compressed; thoracic appendages of more than one form, with gnathopods (pereopods 1 and 2) prehensile (adapted for grasping) (Pls. 70, 86) .... **Amphipoda** (pages 313, 367)

ORDER MYSIDACEA

(PLATE 54)

Mysids or opossum shrimps, so-called from the brood pouch of the females (fig. 4), are small, translucent, shrimplike animals commonly encountered in marine, brackish, and fresh waters. They are filter feeders and scavengers and, although most mysids are small, some deep-sea species may reach a length of 35 cm. On exposed sandy beaches mysids are best found by digging shallow pools in the low beach and puddling the sand; sometimes only their shadows reveal the transparent, swimming shrimps.

A number of local species may occasionally be abundant: *Archaeomysis grebnitzkii* Czerniavsky, 1882 (see Holmquist, 1975), and various species of *Acanthomysis*, including A. *sculpta* (Tattersall, 1933), are common in the surf zone on open sandy beaches. Other species of *Archaeomysis* and *Acanthomysis* occur in sand pockets, tide pools, and kelp beds. *Neomysis mercedis* (Holmes, 1897) (= *N. awatschensis* of authors; see Holmquist, 1973) occurs in Lake Merced, San Francisco, is often abundant in swarms in brackish water of bays, estuaries, and sloughs, and is important in the diet of striped bass and other fishes. Other species of *Neomysis* may occur. The works of Tattersall and Banner should be consulted for detailed descriptions, keys, and figures.

REFERENCES ON MYSIDS

Banner, A. H. A taxonomic study of the Mysidacea and Euphausiacea (Crustacea) of the northeastern Pacific. Part 1, 1948, Trans. Roy. Canad. Inst. 26: 345–399 (includes *Archaeomysis*); Part 2, 1948, *Ibid.* 27: 65–125 (includes *Neomysis* and *Acanthomysis*); Part 3, 1950, *Ibid.* 28: 63 pp. (includes key).

Banner, A. H. 1954. New records of Mysidacea and Euphausiacea from the northeastern Pacific and adjacent areas. Pac. Sci. 8: 125–139 (includes *Archaeomysis, Neomysis,* and *Acanthomysis*).

Holmquist, C. 1973. Taxonomy, distribution, and ecology of the three species *Neomysis intermedia* (Czerniavsky), *N. awatschensis* (Brandt) and *N. mercedis* (Holmes) (Crustacea, Mysidacea). Zool. Jahrb. (Abt. syst. Ökol. Geogr. Tiere) 100: 197–222.

Holmquist, C. 1975. A revision of the species *Archeomysis grebnitzkii* Czerniavsky and *A. maculata* (Holmes) (Crustacea, Mysidacea) Zool. Jahrb. Syst. 101: 51–71.

Tattersall, W. M. 1932. Contributions to a knowledge of the Mysidacea of California. Univ. Calif. Publ. Zool. 37: 301–347 (La Jolla and San Francisco Bay).

Tattersall, W. M. 1951. A review of the Mysidacea of the United States National Museum. U.S. Nat. Mus. Bull. 201, 292 pp. (See also Supplement, by Banner, 1954, Proc. U.S. Nat. Mus. 103: 575–583.)

Tattersall, W. M. and O. S. Tattersall 1951. *The British Mysidacea.* Ray Society, London, 460 pp.

ORDER CUMACEA

William B. Gladfelter

Pacific Marine Station, University of the Pacific, Dillon Beach

(PLATE 54)

Cumaceans are small peracaridans, usually one to several mm long. The head is fused with the first three or four thoracic somites to form a cephalothorax, which is enveloped by an inflated carapace, forming a spacious branchial chamber on each side. The tubular, flexible abdomen is distinctly set off from the thoracic somites. The telson is free or fused to the last segment. Behind the two pairs of antennae and two pairs of maxillae are three pairs of maxillipeds, the third pair serving as a floor to the branchial chambers. The first two pairs of pereopods, and sometimes the third and fourth (but never the last pair), have well-developed exopods. Biramous pleopods are present only in the males, in which different species have from none to five pairs. The uropods are slender, cylindrical, and biramous. The exoskeleton is weakly to strongly mineralized; pigment, either diffuse or in chromatophores, is often present in the integument. A single, sessile, median compound eye is present, with only three to eleven ommatidia (fig. 5B).

Sexual dimorphism is pronounced in adults: the males (fig. 5A) are often markedly more slender (including the carapace) and, in species in which the female lacks the compound eye, it may be present in the male. The second antennae of the male are long and well developed in contrast to the abbreviated ones of the female; males often have pleopods, whereas females never do; mature females have a brood pouch (fig. 5C). Adult male characters develop gradually during several molts. Eggs hatch in the brood pouch of the female and the young leave after three molts, resembling the adult female, but lacking the last pair of pereopods.

Cumaceans inhabit the surface layer of sediments from mid-intertidal to great depths, in densities as high as 15,000 per m^2 (Lie, 1969). Intertidal species can be found in the surface layer of sand flats or protected beaches at low tide. Many species, including intertidal forms, undergo diurnal vertical migrations, rising to the water surface at night; they can be attracted and captured by night-lighting from a dock. Burrowing into the substrate is rapid, and is effected by the last three or four pairs of pereopods (Dixon, 1944). Swimming is effected by the thoracic exopods, abdominal flexures, or by pleopods when present (Foxon, 1936). Feeding is by manipulation and cleaning of individual sediment grains (when the animals are in coarse sediment), by predation (for example, on Foraminifera), or by surface deposit feeding in those species living in fine mud. A respiratory current is maintained by pumping water into the branchial chamber ventrally and out anteriorly. A number of species have structural adaptations to particular ranges of sediment size and amount of organic deposits in the sediment (see Dixon, 1944; Foxon, 1938; Gladfelter, 1975), such as the mesh size of the setal filter at the entrance of the branchial chamber.

The cumacean fauna of central California is virtually unknown;

much of what is known is based on studies by Gladfelter (1975). Almost all records are subtidal, and although only *Cumella vulgaris* is common intertidally, it is possible that some species which occur in the shallow sublittoral zone (less than 6 m) may also occasionally be intertidal.

KEY TO SHALLOW-WATER CUMACEA

1. Telson free and distinct 2
– Telson fused to last segment 5
2. Adult male with 2 pairs of pleopods; telson much shorter than peduncle of uropod 3
– Adult male with no pleopods; telson about as long or longer than peduncle of uropod 4
3. Carapace long and cylindrical, smooth dorsally as seen in profile; telson about half as long as peduncle of uropod .... *Diastylopsis dawsoni*
– Carapace arched, not smooth dorsally as seen in profile; telson more than 1/2 as long as peduncle of uropod .... *Anchicolurus occidentalis*
4. Lateral spines present on telson ... *Lamprops quadriplicata*
– Lateral spines absent on telson ....*Lamprops* sp.
5. Adult male without pleopods; inner margin of peduncle of uropod in female with only 1 spine (figs. 5A, B, C) .... *Cumella vulgaris*
– Adult male with pleopods; inner margin of peduncle of uropod in female with several spines 6
6. Adult male with 2 pairs of pleopods; female with well-developed exopod on 3rd pereopod ......*Eudorella pacifica*
– Adult male with 5 pairs of pleopods; female without well-developed exopod on 3rd pereopod .... *Cyclaspis* sp.

LIST OF SHALLOW-WATER CUMACEA

Anchicolurus occidentalis (Calman, 1912) (= *Colurostylis occidentalis*).

Cumella vulgaris Hart, 1930. The common intertidal species in this area; on harbor flats.

Cyclaspis sp.

Diastylopsis dawsoni Smith, 1880.

Eudorella pacifica Hart, 1930. See Barnard and Given, 1961, Pac. Nat. 2: 153–165.

Lamprops quadriplicata Smith, 1879.

Lamprops sp.

REFERENCES ON CUMACEA

Calman, W. T. 1912. The Crustacea of the order Cumacea in the collection of the United States National Museum. Proc. U.S. Nat. Mus. 41: 603–676.

Dixon, A. Y. 1944. Notes on certain aspects of the biology of *Cumopsis goodsiri* (Van Beneden) and some other cumaceans in relation to their environment. J. Mar. Biol. Assoc. U.K. 26: 61–71.

Foxon, G. E. H. 1936. Notes on the natural history of certain sand-dwelling Cumacea. Ann. Mag. Nat. Hist. (10) 17: 377–393.

Gladfelter, W. B. 1975. Quantitative distribution of shallow-water Cumacea from the vicinity of Dillon Beach, California, with the description of five new species. Crustaceana 29: 241–251.

Hart, J. F. L. 1930. Some Cumacea of the Vancouver Island region. Contr. Canad. Biol. Fish. 6: 1–18.

Lie, U. 1969. Cumacea from Puget Sound and off the northwestern coast of Washington, with descriptions of two new species. Crustaceana 17: 19–30 (see also 21: 33–36).

Pike, R. B. and R. F. Le Sueur 1958. The shore zonation of some Jersey Cumacea. Ann. Mag. Nat. Hist. (13) 1: 515–523.

Wieser, W. 1956. Factors influencing the choice of substratum in *Cumella vulgaris* Hart (Crustacea, Cumacea). Limnol. Oceanogr. 1: 274–285.

Zimmer, C. 1936. California Crustacea of the order Cumacea. Proc. U.S. Nat. Mus 83: 423–439.

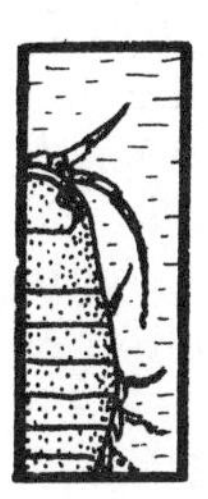

PHYLUM ARTHROPODA: CRUSTACEA, TANAIDACEA AND ISOPODA

Milton A. Miller
University of California, Davis

Tanaidacea (Chelifera) and **Isopoda** are closely related orders, similarly adapted for bottom dwelling, and commonly taken together in benthic samples. Their general similarity led earlier authors, notably Richardson (1905), to classify Chelifera as a superfamily or suborder of isopods. Attaching greater significance to certain basic differences (detailed below), however, modern carcinologists regard isopods and tanaidaceans as separate orders.

Anyone working on Pacific coast isopods and tanaidaceans owes a debt to Robert J. Menzies, co-author of the section on these groups in the second edition of this manual. Thanks are also due Ernest W. Iverson for many suggestions, to Emily Reid for help with illustrations, and to others who have helped in various ways, notably John E. Bodle, Jerry Brill, Nicholas K. Temnikow, and Robert Tufft.

ORDER TANAIDACEA (CHELIFERA)

(PLATE 62)

Tanaidaceans are a small order compared to Isopoda and Amphipoda. They are tiny, cryptic creatures, a few mm long, commonly found with isopods and other small, benthic crustaceans crawling awkwardly on hydroids, bryozoans, or algae, or hiding in crevices, empty worm tubes, or shells. Species of the aptly named genus *Pagurapseudes* resemble hermit crabs, living with the abdomen coiled inside tiny snail shells and the chelae protruding—a remarkable example of convergent evolution.

Tanaidaceans resemble isopods in having: (1) a dorsoventrally flattened (depressed) body; (2) a long, distinctly segmented thorax bearing seven pairs of legs; (3) a single pair of **maxillipeds;** and (4) a relatively short, usually segmented, abdomen **(pleon)** with a series of biramous lamellar pleopods and a single pair of **uropods.**

They differ from isopods in the following important features: (1) The first pair of legs (**gnathopods**) bears pincers or true chelae (figs. 3, 5, 6); those of isopods are simple or, at most, subchelate. (2) The first two thoracic somites are fused to the head and covered with a small carapace, leaving six free thoracic somites (**pereonites**) forming the **pereon** (fig. 4); in isopods only the *first* thoracic segment is fused to the head, a carapace is lacking, and the pereon is thus composed of seven free somites. (3) The eyes of tanaidaceans are borne on short stalks, usually separated from the head by a suture; the eyes of isopods are sessile. (4) In tanaidaceans, at least the inner branch (**endopod**) of the uropods is jointed; in isopods, the uropodal rami are uniarticulate.

Sexual characteristics are of major importance in distinguishing higher tanaidacean taxa and, at the species level, marked sexual dimorphism may cause taxonomic difficulty. Hence, when making determinations, it is advisable to have mature adults of both sexes, preferably from the same population. Males of dimorphic species can readily be distinguished from females by their more strongly developed gnathopods, which sometimes bear grotesque chelae; by having longer first antennae, often with many more joints in the flagellum and heavier setation; by having larger eyes; and, in some genera, by having metamorphosed (fused) mouthparts.

Lang (1956) divides the Tanaidacea into two suborders, primarily on the basis of whether the sperm ducts terminate separately in a pair of conical penial processes (suborder **Dikonophora**) or in a single genital cone (suborder **Monokonophora**). The penes in both suborders are located between the last pair of legs and are often difficult to see. Tanaidacean families are distinguished, in part, by the number of pairs of oöstegites that form the brood pouch (**marsupium**) in ovigerous females. Oöstegites are thin plates that project medially from the basal joint of one or more pairs of legs. Only six tanaidacean genera, each represented by one species, are known from central California.

KEY TO TANAIDACEA (CHELIFERA)

1. First antenna with biramous flagellum (fig. 2); mandibles with palp; males with single penial process; marsupium formed by 4 pairs of oöstegites in female Suborder MONOKONOPHORA 2
– First antenna with uniramous flagellum; mandibles without palp; males with double penial process; marsupium formed by 1 or 4 pairs of oöstegites .... Suborder DIKONOPHORA 3
2. Pleon with 6 somites including pleotelson; pleopods present; living in tiny snail shells .... *Pagurapseudes*
– Pleon with 3 somites including pleotelson; pleopods absent (figs. 1–3) .... *Synapseudes intumescens*

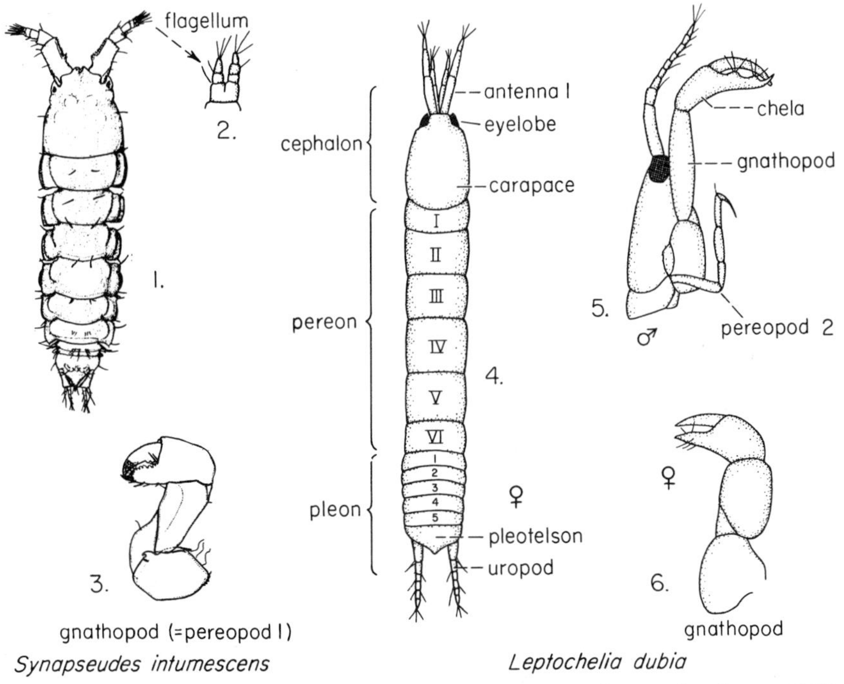

PLATE 62. **Tanaidacea.** 1–3 after Menzies, 1949; 4–6 after Richardson, 1905.

3. Pleon with 6 somites and 5 pairs of pleopods; uropods biramous (outer branch inconspicuous); marsupium formed by 4 pairs of oöstegites issuing from bases of 2nd–5th legs; mouthparts of male metamorphosed, either totally fused or only maxillipeds distinguishable (figs. 4–6) .... *Leptochelia dubia*
– Pleon with 3–6 somites and 3 pairs of pleopods; uropods uniramous; marsupium formed by 1 pair of oöstegites from 5th legs; no sexual dimorphism in mouthparts 4
4. Pleon with 3 somites plus telson, bearing 3 pairs of pleopods, the third rudimentary .... *Pancolus californiensis*
– Pleon with 5–6 somites and 3 normal pairs of pleopods 5
5. Pleon with 6 somites .... *Anatanais normani*
– Pleon with 5 somites .... *Tanais*

LIST OF TANAIDACEA

Suborder **Monokonophora**

METAPSEUDIDAE

Synapseudes intumescens Menzies, 1949. Exposed rocky intertidal, on holdfasts of brown algae, bryozoans, *Mytilus* beds (Menzies, 1949).

PAGURAPSEUDIDAE

Pagurapseudes sp. In snail shells, like tiny, slow-moving hermit crabs; rocky intertidal. See Howard, 1952.

Suborder **Dikonophora**

PARATANAIDAE

Leptochelia dubia (Krøyer, 1842). Cosmopolitan; abundant among algae in pools and on mudflats.

TANAIDAE

Anatanais normani (Richardson, 1905).

Pancolus californiensis Richardson, 1905. In high-intertidal *Cladophora;* see Lang, 1961.

Tanais sp. In bays and estuaries; a tanaid common in San Francisco Bay is tentatively identified as *Tanais vanis* Miller, 1940, described from Hawaii.

REFERENCES ON TANAIDACEA

Hansen, H. J. 1913. Crustacea Malacostraca. II. IV. The Order Tanaidacea. Danish Ingolf Exped. 3: 1–145.

Hatch, M. H. 1947. The Chelifera and Isopoda of Washington and adjacent regions. Univ. Wash. Publ. Biol. 10: 155–274.

Howard, A. D. 1952. Molluscan shells occupied by tanaids. Nautilus 65: 75–76.

Lang, K. F. 1949. Contributions to the systematics and synonymics of the Tanaidacea. Ark. Zool. 42A: 1–14.

Lang, K. F. 1950. The genus *Pancolus* Richardson and some remarks on *Paratanais euelpis* Barnard (Tanaidacea). Ark. Zool. (ser. 2) 1: 357–360.

Lang, K. F. 1956. Neotanaidae nov. fam., with some remarks on the phylogeny of the Tanaidacea. Ark. Zool. (ser. 2) 9: 469–475.

Lang, K. F. 1957. Tanaidacea from Canada and Alaska. Contr. Depart. Pech. Quebec 52: 1–54.

Lang, K. F. 1961. Further notes on *Pancolus californiensis* Richardson. Ark. Zool. (ser. 2) 13: 573–577.

Lang, K. F. 1970. Taxonomische und phylogenetische Untersuchungen über die Tanaidaceen. Ark. Zool. (ser. 2) 22: 595–626.

Menzies, R. J. 1949. A new species of apseudid crustacean of the genus *Synapseudes* from northern California (Tanaidacea). Proc. U.S. Nat. Mus. 99: 509–515.

Menzies, R. J. 1953. The apseudid Chelifera of the Eastern Tropical and North Temperate Pacific Ocean. Bull. Mus. Comp. Zool. Harvard 107: 443–496.

Nierstrasz, H. F. 1913. Die Isopoden der Siboga-Expedition. I. Isopoda and Chelifera. Siboga Expedition Monograph 32a: 1–56.

Richardson, H. 1905. Monograph on the isopods of North America. U.S. Nat. Mus. Bull. 54: 727 pp. Includes the Tanaioidea or Chelifera as an isopodan superfamily.

ORDER ISOPODA

(PLATES 63–69)

Isopods are a large, diverse order with eight suborders, all but one of which (**Phreatoicidea**) occur in California. Although best represented in the sea, they are also found in fresh and brackish water and in moist habitats on land (suborder **Oniscoidea**). Isopoda compare in ecological importance with the related Amphipoda, notably as intermediate links in food chains. Isopods are predominantly benthic and cryptic, living under rocks, in crevices, empty shells and worm tubes, and among sessile and sedentary organisms, such as algae, sponges, hydroids, ectoprocts, mussels, barnacles, and ascidians. Some burrow in natural substrates including mud, sand, sandstone, and driftwood, while a few burrowers, such as the "gribble" (*Limnoria*) and *Sphaeroma*, do extensive damage to pilings and wooden boats. Several isopod groups are facultative or obligate parasites on other crustaceans or fishes, and some live commensally with various invertebrates. Several species are important scavengers on wrack or dead animals. Certain worm parasites, notably acanthocephalans, use isopods as intermediate hosts.

Isopods can be distinguished from other crustaceans by a combination of the following characteristics: (1) flattened (depressed) body shape; (2) compact head (**cephalon**) with sessile compound eyes, two pairs of antennae, and mouthparts comprising a pair of **mandibles,** two pairs of **maxillae,** and a pair of **maxillipeds;** (3) a long thorax of eight **thoracomeres,** the first fused with the head and bearing the maxillipeds, the remaining seven (called **pereonites**) being free and collectively comprising a body division called the **pereon**; (4) seven (rarely 5) pairs of similar walking legs or **pereopods** (hence "isopod"), never chelate, and (5) a relatively short abdomen (**pleon**) composed of six somites (**pleonites**), at least one of which is fused to the terminal anal plate (**telson**) to form a **pleotelson**; (6) six pairs of biramous pleonal appendages, including five pairs of respiratory (sometimes also natatory) pleopods with platelike branches and a single pair of fanlike or sticklike **uropods** with uniarticulate (unjointed) branches (for details and variations, see figs. 7–14).

Identification of isopods often requires dissection and microscopic examination of appendages and other structures by use of needle knives and fine-pointed jewelers forceps under a dissecting binocular. Dissected parts may be mounted on microscope slides in glycerin or a more permanent medium for observation under a compound microscope.

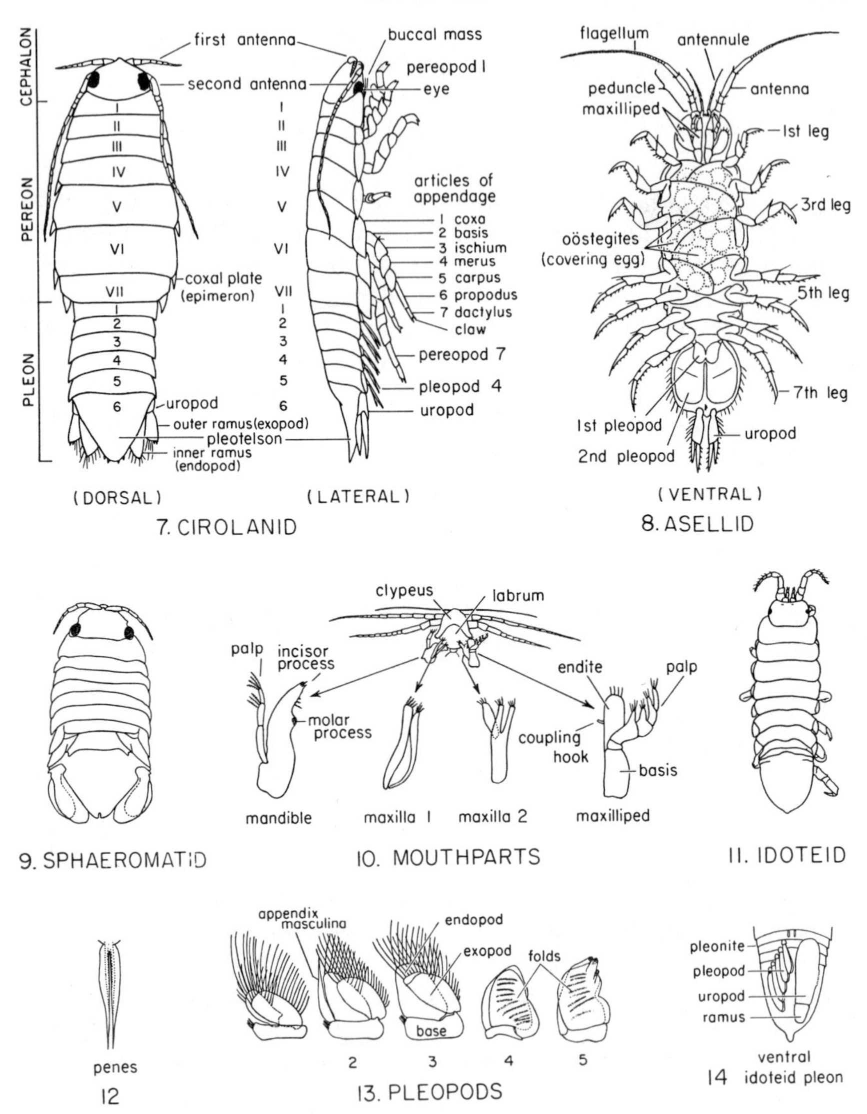

PLATE 63. **Isopoda** (1). Isopod anatomy in representative families: 7 after Menzies and Frankenberg, 1966; 8 after Van Name, 1936; 9, 10, 12, 13 after Menzies and Glynn, 1968; 14 after Eales.

KEY TO SUBORDERS OF ISOPODA

1. Only 5 pairs of walking legs; head and first 2 thoracomeres fused, leaving 6 free thoracomeres (pereonites), the last reduced and legless; adult males with protruding, sickle-shaped

mandibles; ovigerous females with pereonites 3–5 swollen with developing eggs .... **Gnathiidea** (p. 284)

– Normally 7 pairs of pereopods, usually ambulatory; only 1st (maxillipedal) thoracomere fused with head, leaving 7 free pereonites; ovigerous females with subthoracic brood pouch formed by overlapping plates (oöstegites) 2

2. Parasitic only on Crustacea; females often asymmetrical, often with segmentation and appendages reduced or absent; legs when present prehensile; uropods when present simple, small, terminal; males diminutive, segmented, symmetrical, often on body of female .... **Epicaridea** (p. 284)

– Free-living or parasitic on fish, but not on Crustacea (caution: some non-epicarids, e.g., asellotes, may be ectocommensal on crustaceans) 3

3. Uropods hinged laterally or ventrally (figs. 7, 9, 11, 14) 4

– Uropods terminal, styliform (figs. 8, 54–61, 64–67) 6

4. Uropods ventral, not visible dorsally, forming opercular plates or valves that fold together, like cabinet doors, beneath pleopods (fig. 14); [caution: **Tylidae** (not covered here) of the terrestrial suborder Oniscoidea have somewhat similar opercular uropods] ....**Valvifera** (p. 287)

– Uropods lateral, visible dorsally, fanlike (sometimes lacking) 5

5. Body at least 7 times as long as wide, subcylindrical; uropodal exopod arching over spatulate telson; pereopod 1 larger than others, subchelate (2nd and 3rd may also be subchelate) .... **Anthuridea** (p. 291)

– Body usually less than 5 times as long as wide, depressed; uropodal exopod not arching over telson; pereopod 1 not enlarged or subchelate ....**Flabellifera** (p. 291)

6. Terrestrial; first antennae vestigial; pleon consists of 5 free pleonites plus short pleotelson (figs. 58–67) **Oniscoidea** (p. 302)

– Aquatic 7

7. Body narrow, elongate (about 7 times longer than wide), tiny (less than 2 mm), eyeless, albino; pereopod 1 subchelate (as in Anthuridea, above); pleon composed of 3 subequal divisions (fig. 41); interstitial in sand .... **Microcerberidea** (p. 297)

– Body generally narrow, but much less elongate than in Microcerberidea (above) at least in shallow-water forms, and generally much larger; pereopod 1 subchelate or simple; first or first 2 pleonites short and free, the rest fused with telson to form a large, shieldlike pleotelson (figs. 42, 45, 48, 51, 54–57) .... **Asellota** (p. 298)

SUBORDER GNATHIIDEA

Seven species of this aberrant suborder have been found in California, all subtidal and all south of Point Conception, except *Gnathia crenulatifrons* which has been taken in benthic samples at Monterey. Only one family (Gnathiidae) and one genus (*Gnathia*) are represented. See Schultz (1969) for a key to North American species.

SUBORDER EPICARIDEA

Three of the four families of this parasitic suborder (Bopyridae, Cryptoniscidae and Entoniscidae) are represented in central California; the Dajidae have not been reported north of Point Conception. The epicaridean fauna of this area is imperfectly known; a systematic search in local crustaceans may reveal many new or unreported species. Moreover, no life cycle of any California species has been completely worked out. The generalized life history (based on a few known cases) comprises a series of molts producing a sequence of three larval stages and highly modified adults. The eggs develop in the brood pouch (marsupium) of the female from which the young emerge as tiny (2–3.5 mm), pelagic larvae. This first larval stage, the "epicaridium," has six free pereonites, each bearing a pair of prehensile subchelate pereopods with well-developed claws with which it attaches to some plankter, usually a copepod. On the intermediate host it molts to the "microniscium" larval stage ("microniscus" of some authors). A subsequent molt produces the third larval stage, the "cryptoniscium" (or "cryptoniscus"), which is longer (to 7 mm) and adds a seventh pereonite with another pair of legs. The cryptoniscium detaches from the planktonic host, becomes benthic, finds its definitive crustacean host, and metamorphoses into a more or less highly modified female or transforms with slight alteration into a diminutive male. The adult bopyrid is called a "bopyridium;" the early adult entoniscid is known as an "asticot." The larval stages in the four epicaridean families are superficially indistinguishable, whereas the adults, especially the females, are quite distinctive. The following key is based mainly on the adult female.

KEY TO EPICARIDEA

1. Body of female segmented (sometimes indistinctly), not simply an egg sac; antennae and mouthparts present, but rudimentary or modified for sucking; pereonal legs developed, rudimentary, or absent 2

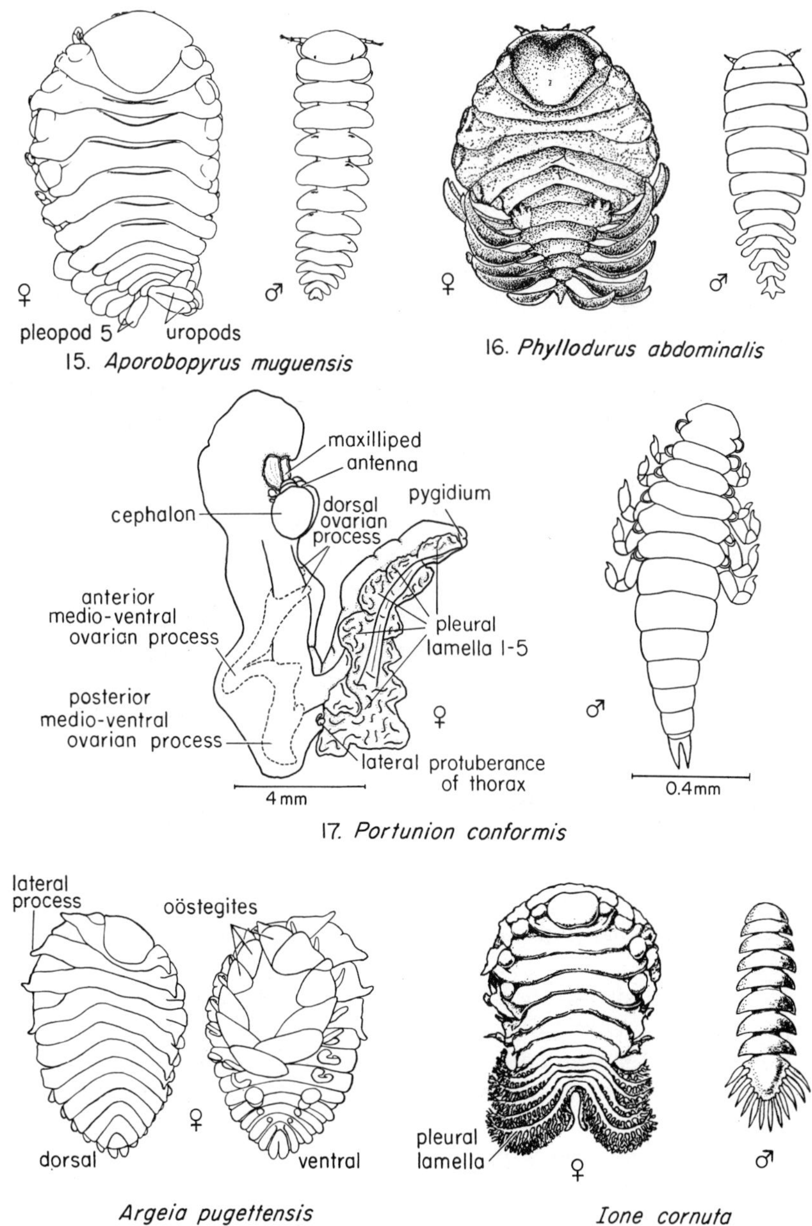

PLATE 64. **Isopoda** (2)1 **Epicaridea:** 15 after Shiino, 1964; 17 after Muscatine, 1956; 16, 18, 19 after Richardson, 1905. Males, diminutive compared to females, are here drawn to a much larger scale.

- Female unsegmented or with segmentation feebly indicated, transformed into a rotund or lobular egg sac; antennae and mouthparts absent; legs rudimentary or absent .... CRYPTONISCIDAE

(One intertidal species is known in California, *Cryptothir balani* (= *Hemioniscus balani*), parasitic in barnacles, *Balanus* and *Chthamalus*. *Cryptothir* can be distinguished from other cryptoniscid genera by the lobulated body of the female.)

2. Body of female without indication of rigid exoskeleton, seemingly undifferentiated, but body divisions and segmentation present; pereonites expanded laterally into thin plates; maxillipeds are only recognizable mouthparts; legs stubby or absent; internal parasites of other crustaceans ENTONISCIDAE

(One Californian species, *Portunion conformis* (fig. 17), in body cavity of *Hemigrapsus* spp.)

- Female distinctly segmented but more or less asymmetrical; antennae and mouthparts rudimentary; pereopods prehensile, 7 present on one side, but all except 1st may be absent on the other side; parasitic on decapod crustaceans .. BOPYRIDAE 3

3. Pleon with lateral plates (pleural lamellae) elongate, those of female fringed on one side with long, branched processes (fig. 19, female), those of male without digitations (fig. 19, male); in branchial cavity of *Callianassa* .... *Ione cornuta*

- Pleon in both sexes with pleural lamellae rudimentary or absent; (caution: lateral pleopods may be mistaken for pleural processes, even though at least the 1st 3 are biramous) 4

4. Both branches of pleopods in adult female similar in size and shape; posterolateral margins of pereon not produced in processes; pleonites in male distinct 5

- Branches of pleopods in adult female dissimilar in size and shape 6

5. Dorsolateral surface of pleonite 1 on each side bears a large, papillose process; pleopods biramous with long, narrow branches extending laterally from narrow pleotelson (fig. 16); eyes absent; among pleopods of *Upogebia pugettensis* .... *Phyllodurus abdominalis*

- Dorsolateral surface of pleonite 1 smooth; pleopods biramous with short branches, broad anteriorly but narrowing posteriorly; uniramous, lobose uropods extend posteriorly from terminus of small pleotelson; eyes present but minute; in branchial chamber of *Pachycheles rudis* (fig. 15) .... *Aporobopyrus muguensis*

6. Exopod of pleopods oval and small; endopod narrow and elongate; posterolateral margins of pereon not produced into

narrow processes; male with pleonites distinct *Pseudione* sp.
- Exopod of pleopods narrow and elongate; endopod oval and small; pereon with posterolateral margins produced into narrow processes; male with pleonites fused ....*Argeia* 7

7. Lateral processes on all pereonites; head not lobate; inner branches of all pleopods present (fig. 18) *Argeia pugettensis*
- Lateral processes absent on some anterior pereonites; head bilobate; inner branches present only on first 3 pairs of pleopods ....*Argeia pauperata*

SUBORDER VALVIFERA

Two families of this suborder, Arcturidae and Idoteidae, are represented in California. Valviferans are larger and more diverse in colder waters.

KEY TO VALVIFERA

1. Body narrow, elongate, subcylindrical; anterior 4 pereopods unlike posterior 3, being smaller, setose, and nonambulatory; head fused with 1st pereonite, leaving 6 free thoracic somites .... *Idarcturus hedgpethi*
- Body elongate and depressed; legs nearly alike and ambulatory; 7 free thoracic somites (fig. 11) ....IDOTEIDAE 2

2. Pereonites lack dorsally visible epimeral sutures; all pleonites completely coalesced except for a single, incomplete suture or incision on each side near base of consolidated pleon; maxillipedal palp with 3 articles 3
- All pereonites except 1st with distinct, dorsally visible epimeral sutures; pleon with 2 complete and 1 partial intersegmental suture dividing it into 3 divisions—2 small anterior pleonites and a large, shieldlike pleotelson with an incompletely fused pleonite near its base; maxillipedal palp with 4 or 5 articles .... *Idotea* 9

3. Flagellum of antenna 2 rudimentary ... *Edotea sublittoralis*
- Flagellum of antenna 2 with many joints ....*Synidotea* 4

4. Body smooth; head without preocular horns or other projections 5
- Body with tubercles or ridges; head with preocular horns and other sculpture 6

5. Pleon less than 1/3 longer (in midline) than greatest width; frontal margin of head transverse or slightly concave with slight median excavation (fig. 20) ...... *Synidotea laticauda*

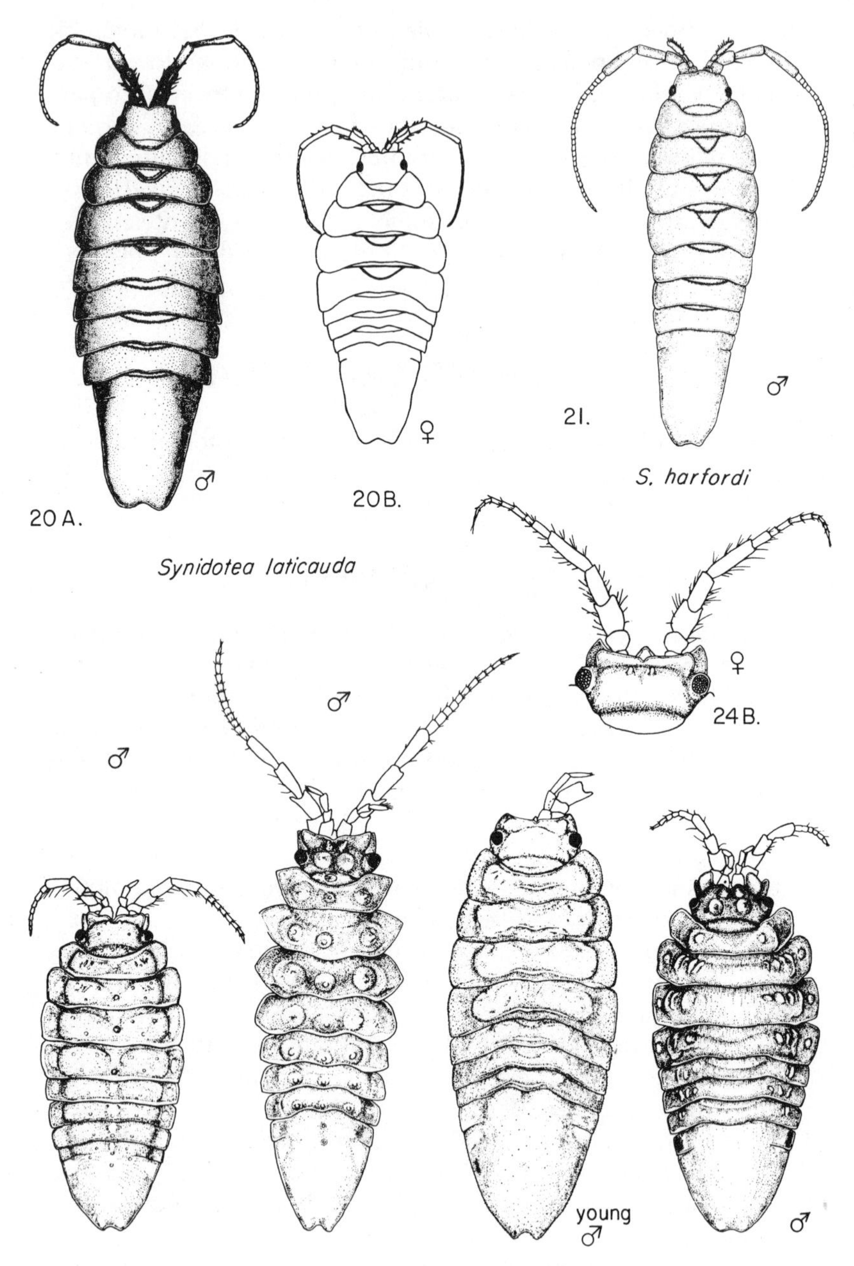

PLATE 65. **Isopoda** (3). **Valvifera:** 20–25, *Synidotea*, after Menzies and Miller, 1972.

– Pleon at least 1/3 longer than broad; frontal margin of head transverse or slightly convex with no median emargination (fig. 21) .... *Synidotea harfordi*
6. Dorsal surface of each pereonite with a transverse carina, but no tubercles (fig. 24) .... *Synidotea bicuspida*
– Dorsal surface of body with tubercles; preocular horns large, extending to or beyond frontal border of head 7
7. Mid-dorsal line of pereon generally lacking tubercles; preocular horns project forward (fig. 25) ..... *Synidotea ritteri*
– Mid-dorsal line of pereon with at least one tubercle per somite; preocular horns project laterally 8
8. Lateral borders of first 4 pereonites generally acute; each pereonite with a transverse row of 3 prominent tubercles (fig. 23) .... *Synidotea pettiboneae*
– Lateral borders of first 4 pereonites blunt; pereonites beset with small tubercles (fig. 22) ...... *Synidotea berolzheimeri*
9. Maxillipedal palp with 4 articles .... *Idotea* (*Idotea*) 10
– Maxillipedal palp with 5 articles .... *Idotea* (*Pentidotea*) 12
10. Superior margin of clypeus (1st frontal lamina) with pronounced median concavity (fig. 26) *Idotea* (*Idotea*) *urotoma*
– Superior margin of clypeus evenly rounded 11
11. Frontal process of head apically blunt or notched; posterior margin of telson somewhat concave *Idotea* (*Idotea*) *rufescens*
– Frontal process apically pointed; posterior margin of telson with a pronounced median tooth or projection (fig. 27) .... *Idotea* (*Idotea*) *fewkesi*
12. Apex of frontal process of head entire; maxilliped with 1 coupling hook; eyes not markedly transversely elongate 13
– Apex of frontal process with a median notch; maxilliped with 1 or 2 coupling hooks; eyes oval or transversely elongate .. 17
13. Frontal process blunt or widely angulate, not extending beyond frontal lamina 1; frontal lamina 1 triangulate in dorsal view 14
– Frontal process narrow, pointed, exceeding considerably the forward extent of semicircular frontal lamina 1 16
14. Posterolateral margin of epimeral plate of 7th (last) pereonite evenly convex, not acute; eyes somewhat pear-shaped .... *Idotea* (*Pentidotea*) *schmitti*
– Posterolateral margin of last pereonite acute; eyes kidney-shaped (reniform) or oval 15
15. First pleonite with acute lateral borders; eyes reniform (fig. 29) .... *Idotea* (*Pentidotea*) *wosnesenskii*
– First pleonite with blunt lateral borders; eyes oval 18
16. Posterior margin of telson deeply concave with acute pos-

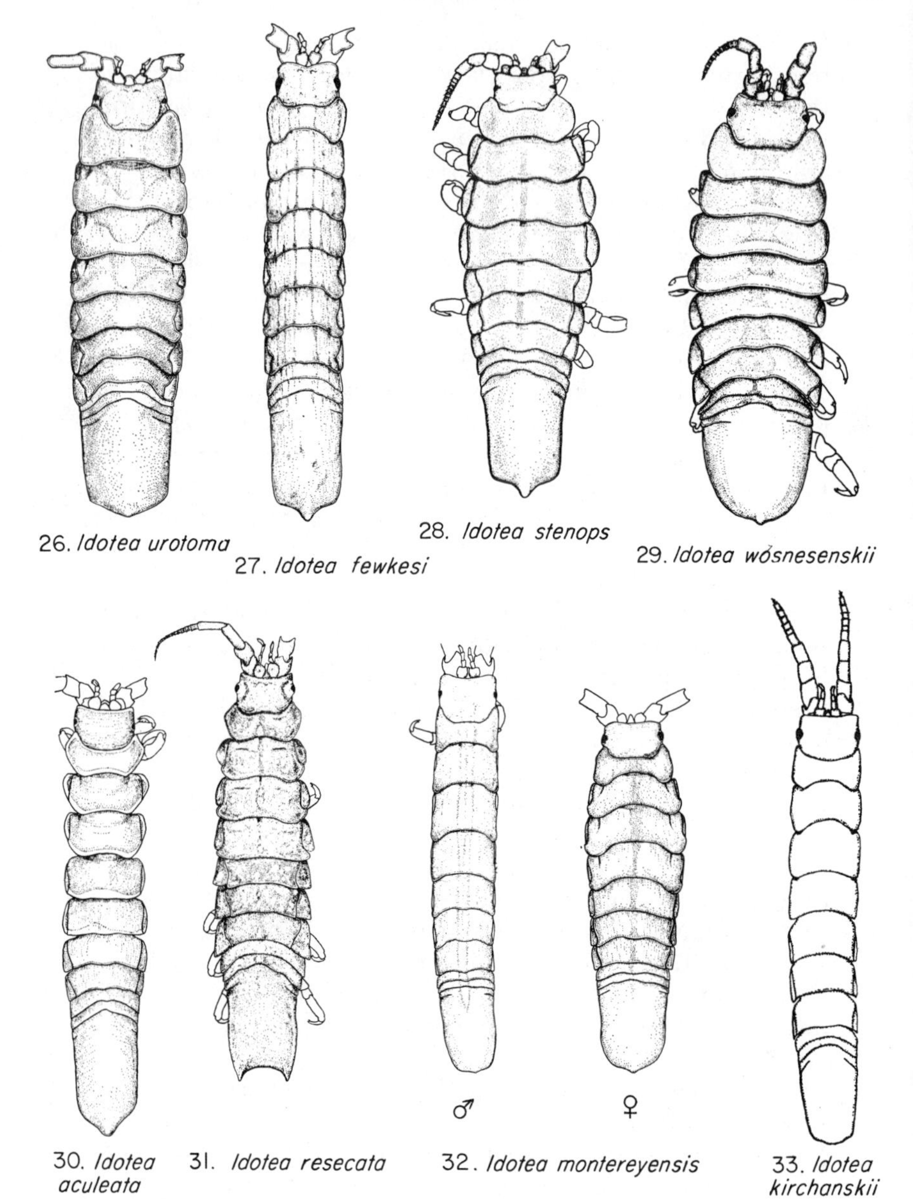

PLATE 66. **Isopoda** (4). **Valvifera:** 26–33; *Idotea:* 26–32 after Menzies, 1950; 33 after Miller and Lee, 1970.

terolateral angles; each angle with a small keel (fig. 31); often on eelgrass *Zostera;* also common on *Macrocystis* .... *Idotea (Pentidotea) resecata*

– Posterior margin of telson usually convex, with small but distinct median tooth; if concave, only slightly so and lacking acute posterolateral angles and any dorsal keel above each angle (fig. 32); often on surf grass *Phyllospadix* .... *Idotea (Pentidotea) montereyensis*

17. Maxilliped with 2–3 coupling hooks; eyes narrow, elongate (fig. 28) .... *Idotea (Pentidotea) stenops*
– Maxilliped with 1 coupling hook; eyes oval (fig. 30) .... *Idotea (Pentidotea) aculeata* (in part)

18. Pereonites distinctly separated laterally; epimera visible dorsally on pereonites 2–7; sides of head straight; antenna 2 reaching almost to posterior margin of 4th pereonite (fig. 30) .... *Idotea (Pentidotea) aculeata* (in part)
– Pereonites not sharply separated laterally; epimera visible dorsally only on pereonites 5–7 (fig. 33) .... *Idotea (Pentidotea) kirchanskii*

SUBORDER ANTHURIDEA

This suborder is represented by one family, Anthuridae, formerly placed in the **Flabellifera.**

KEY TO ANTHURIDEA

1. Free part of maxilliped consists of 1 article; mandibular palp lacking; only 6 pereonites readily visible in dorsal view, the 7th being greatly reduced (fig. 34) *Colanthura squamosissima*
– Free part of maxilliped consists of 2 or 3 articles; mandibular palp present; all 7 pereonites visible in dorsal view 2

2. Maxilliped with 2 free articles; mouthparts adapted for piercing and sucking; pleonites not fused, pleon distinctly segmented (fig. 36) .... *Paranthura elegans*
– Maxilliped with 3 free articles; mouthparts adapted for biting and chewing; anterior 5 pleonites coalesced (fig. 35) .... *Cyathura munda*

SUBORDER FLABELLIFERA

The Flabellifera is the largest and most diverse isopodan suborder, represented in California by eight families, three of which (Anuropidae, Excorallanidae, and Serolidae) have not been reported north of Point Conception. Because of the great diversity of this suborder, it is more convenient to key the families first and then the species in each family.

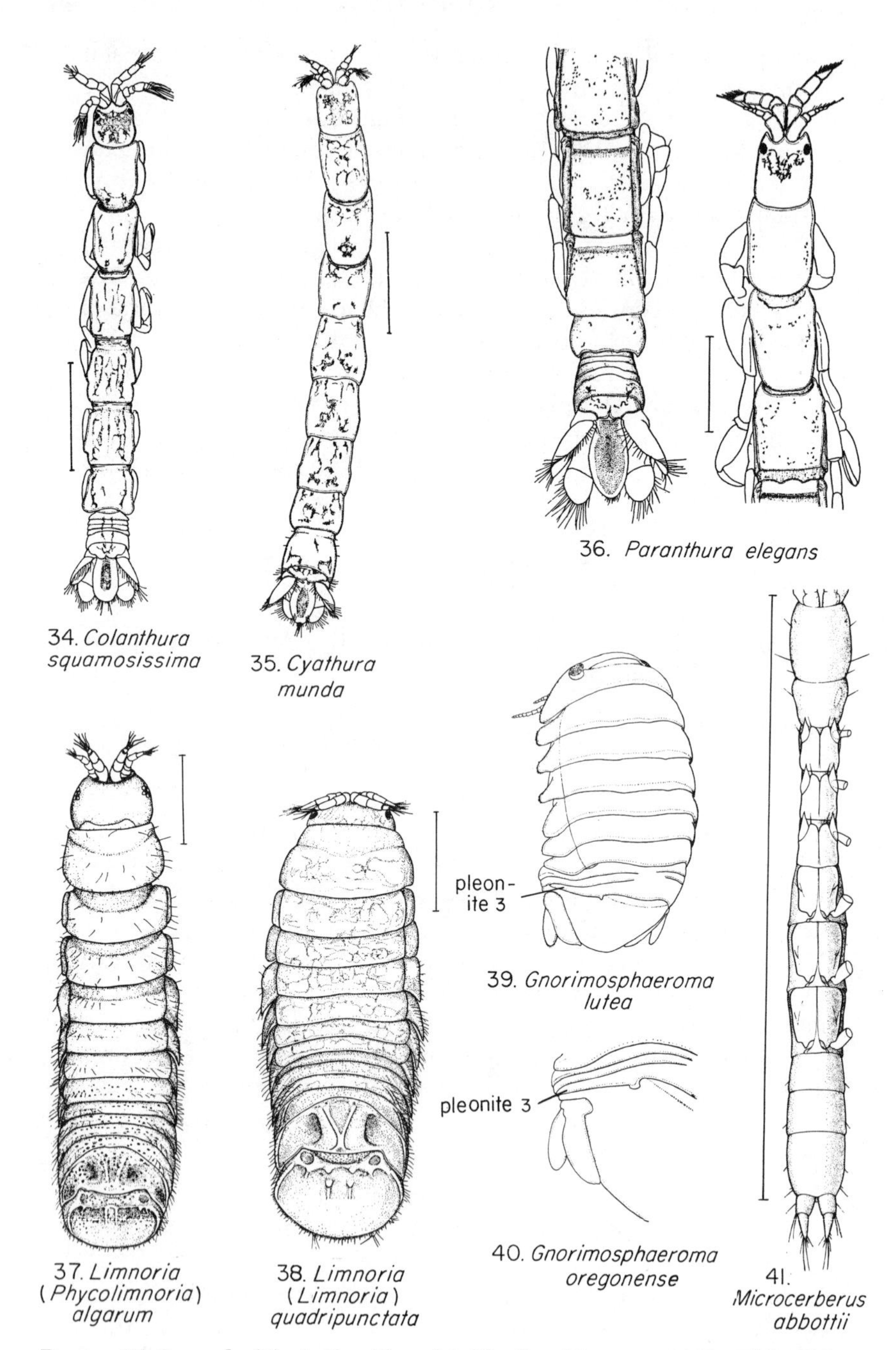

PLATE 67. **Isopoda** (5). **Anthuridea:** 34–36 after Menzies, 1951; **Flabellifera:** Limnoriidae: 37, 38 after Menzies, 1957; **Flabellifera:** Sphaeromatidae: 39, 40 after Menzies, 1954; **Microcerberidea:** 41 after Lang, 1961. Scale bars = 1 mm.

KEY TO FAMILIES OF FLABELLIFERA

1. Pleon comprises 3 divisions: the 1st concealed under the last pereonite; the 2nd composed of several coalesced pleonites, often indicated by 3 partial lateral sutures; the 3rd forming a large pleotelson (fig. 9); most are able to roll into a ball .... SPHAEROMATIDAE (p. 293)
– Pleon comprises 5 free pleonites plus a pleotelson; generally incapable of rolling into a ball 2
2. Uropods greatly reduced, with small, clawlike exopod and slightly larger styliform endopod with blunt apex; borers in wood or kelp holdfasts (figs. 37, 38) LIMNORIIDAE (p. 295)
– Uropods well developed, with similar platelike branches ... 3
3. Maxillipeds with broad, free, setose palp, never furnished with hooks (fig. 7) .... CIROLANIDAE (p. 296)
– Maxillipedal palp embracing other mouthparts distally, not setose apically but furnished with outwardly curved hooks, at least in males and in females without eggs 4
4. Antennae without clear distinction between peduncle (base) and flagellum; maxillipedal palp composed of 2 articles; plumose setae absent on pleopodal margins .... CYMOTHOIDAE (p. 296)
– Antennae with well-defined peduncle and flagellum; maxillipedal palp with 2 or 5 articles; plumose setae on margins of first 2 pairs of pleopods .... AEGIDAE (p. 297)

KEY TO SPHAEROMATIDAE

Sphaeromatid isopods can easily be recognized by their compact, convex bodies, usually capable of rolling into a ball (conglobation); by their pleon which is consolidated into two–three divisions; and by their lateral uropods in which the endopod is rigidly fused to the basal article and the exopod (if present) is movable. In their ability to roll up, sphaeromatids resemble certain terrestrial isopods, called pill-bugs—a striking example of parallel evolution. Identification of genera and species is often difficult because of marked sexual dimorphism and parallel adaptations. Hence it is advisable, when making determinations, to have a representative sample including adults of both sexes.

In the following key, the Sphaeromatidae are partitioned among five major groups: **Platybranchiata, Hemibranchiata,** and **Eubranchiata** (Hansen, 1905); **Colobranchiata** (Richardson, 1909a) and **Pentadibranchiata** (Bodle, 1969). This scheme is based primarily on pleopod structure, especially the presence or absence of transverse, accordionlike folds or pleats on one or both branches of the last two pairs (pleopods 4 and 5).

1. Both branches of pleopods 4 and 5 thin or somewhat fleshy, but lacking transverse folds 2
– Transverse folds (fig. 13) present on 1 or both rami of pleopods 4 and/or 5 6
2. First pair of pleopods small, widely separated at base; 2nd pair much larger, operculate; plumose setae on margin of 1st pleopod only; uropod with a single slender branch (Group **Colobranchiata**) .... *Bathycopea daltonae*
– Pleopods 1 and 2 similar in size, 1st pair not widely separated, 2nd pair not operculate; marginal plumose setae on rami of 1st 3 pairs of pleopods; uropods biramous (Group **Platybranchiata**) .... *Gnorimosphaeroma* 3
3. Basal articles of antennae 1 contiguous in midline ....
.... *Gnorimosphaeroma noblei*
– Basal articles of antennae 1 separated from each other by rostrum 4
4. Only 1st 2 pleonites of 2nd (1st visible) pleonal division form its lateral margin; pointed lateral edge of 3rd pleonite usually not reaching lateral pleonal border; body 1.7–2 times longer than wide (fig. 39) .... *Gnorimosphaeroma lutea*
– All 3 pleonites comprising 2nd (1st visible) pleonal division reach and form its lateral margin; body broader, 1.5–1.75 times longer than wide (fig. 40) 5
5. Basis (1st free joint) of pereopod 1 with only 1 seta; ischium (2nd free joint) with rows of long setae ....
.... *Gnorimosphaeroma oregonense*
– Basis of pereopod 1 with tuft of 7–9 setae; sternal crest of ischium with 2–3 setae .... *Gnorimosphaeroma rayi*
6. Transverse folds present only on endopods of pleopods 4 and 5; exopods more or less membranous (Group **Hemibranchiata**) 7
– Transverse folds present on both branches of pleopod 5, but either present or absent on both rami of pleopod 4 14
7. Telson apically notched (fig. 9) or with a ventral, subterminal groove .... *"Dynamenella"* 8
– Telson entire 10
8. Pleotelson smooth; exopod of 3rd pleopod biarticulate .....
.... *"Dynamenella" glabra*
– Pleotelson with 4 longitudinal ridges or confluent rows of tubercles; exopod of 3rd pleopod uniarticulate 9
9. Outer or both pairs of longitudinal ridges on pleotelson composed of fused tubercles; a row of 2 tubercles lateral to these

near anterior margin of pleotelson *"Dynamenella" sheareri*

– Both pairs of longitudinal ridges on pleotelson solid; no additional sculpturing on pleotelson *"Dynamenella" benedicti*

10. Maxillipedal palp with articles 2–4 not produced into fingerlike lobes; uropodal exopod normally with 5 prominent teeth on lateral margin, but number of teeth variable .... *Sphaeroma pentodon*

– Maxillipedal palp with articles 2–4 produced into fingerlike lobes (fig. 10); uropodal exopod not toothed *Exosphaeroma* 11

11. Telson apically pointed; uropodal rami in males extraordinarily large and broad; pereonites with transverse rows of tubercles ....*Exosphaeroma amplicauda*

– Telson not apically pointed; uropodal rami in both sexes of moderate size; pereon smooth 12

12. Telson apically produced into a rhomboid process; lateral margins of pleotelson deeply indented for uropods ....*Exosphaeroma rhomburum*

– Pleotelson apically rounded, laterally convex 13

13. Pleon with 8 tubercles, 2 anterior to and 6 on pleotelson .. ..*Exosphaeroma octoncum*

– Pleon without tubercles ....*Exosphaeroma inornata*

14. Transverse folds on endopod and exopod of pleopod 5 only; first 2 pairs of legs subchelate in male; only 1st pair subchelate in female (Group **Pentadibranchiata**) *Tecticeps convexus*

– Transverse folds on both branches of pleopods 4 and 5; all legs ambulatory (Group **Eubranchiata**) 15

15. Basal article of antenna 1 dilated; dorsal surface of pleotelson with 3 longitudinal ridges ....*Dynamenella dilatata*

– Basal article of antenna 1 not dilated; dorsal surface of pleotelson with a transverse row of 3 large tubercles (bifid in male) plus a median tubercle anterior to terminal notch; males differ markedly from females in having (1) pleotelson produced posteriorly with a much larger terminal excavation that is partially subdivided by 3 pairs of teeth into a tandem of 3 connected sinuses, and (2) uropodal exopod elongated, incurved, and armed with 3–4 spines ....*Paracerceis cordata*

KEY TO LIMNORIIDAE

1. Incisor process of mandibles simple, lacking rasp or file; algal holdfast borers (fig. 37) ..*Limnoria (Phycolimnoria) algarum*

– Incisor of right mandible with filelike ridges, that of left with rasplike sclerotized plates; wood borers .... Subgenus *Limnoria* 2

2. Dorsal surface of pleotelson with a median Y-shaped keel at base; lateral and posterior borders smooth ....*Limnoria (Limnoria) lignorum*
– Dorsal surface of pleotelson with symmetrically arranged tubercles at anterior end; lateral and posterior borders smooth or tuberculate 3
3. Four anterior tubercles on telson, thus (::) (fig. 38); posterior and lateral margins not tuberculate ....*Limnoria (Limnoria) quadripunctata*
– Three anterior tubercles on telson, thus (.˙.); posterior and lateral borders tuberculate .. *Limnoria (Limnoria) tripunctata*

KEY TO CIROLANIDAE

1. Posterior border of telson armed with about 26 stout, non-plumose setae; frontal margin of head with inconspicuous median rostrum ....*Cirolana harfordi*
– Posterior margin of telson fringed with long, narrow, plumose setae; head with long, narrow, spatulate rostrum 2
2. Telson broadly rounded and crenulate posteriorly; peduncle (base) of antenna 1 with 2nd and 3rd articles subequal in length .... *Excirolana linguifrons*
– Telson angulate or obtusely rounded posteriorly; peduncle of antenna 1 with 3rd article longer than 2nd 3
3. Telson obtusely rounded; spatulate frontal process almost as long as broad; distance between eyes about diameter of eye .... *Excirolana chiltoni*
– Telson obtusely angulate; spatulate frontal process longer than broad; distance between eyes greater than diameter of eye .... *Excirolana kincaidi*

KEY TO CYMOTHOIDAE

(based mainly on adult females)

1. Head posteriorly produced in 3 lobes, not at all immersed in 1st pereonite; last (7th) pereopod with a few setae (spines) on inferior margin of carpus and propodus *Nerocila californica*
– Head not posteriorly trilobate, more or less immersed in 1st pereonite; last leg lacking setae on inferior margin of carpus and propodus 2
2. Telson in adult female nearly twice as broad as long; juveniles with diffuse dark pigmentation on uropodal exopod and anterolateral areas of pleotelson; eyes medium-sized and widely separated; anterior border of head broadly rounded or trun-

cate; antenna 2 with 10–11 articles .... *Lironeca vulgaris*
- Telson in adult female about as broad as long; juveniles with pigment granules concentrated in melanophores, lacking distinct color pattern; eyes large, close-set medially; anterior border of head strongly produced, apically blunt; antenna 2 with 8–9 articles .... *Lironeca californica*

KEY TO AEGIDAE

1. Eyes large (except in *Aega microphthalma*), occupying 1/2 or more of dorsal area of head; maxillipedal palp composed of 5 articles; propodus of pereopods 1–3 simple, cylindrical, not expanded .... *Aega* 2
- Eyes of moderate size; maxillipedal palp composed of 2 articles; propodus of pereopods 1–3 expanded and armed with spines .... *Rocinela* 5
2. Eyes small, round .... *Aega microphthalma*
- Eyes large 3
3. Peduncle of antenna 1 with first 2 articles enlarged, dilated, and flattened .... *Aega lecontii*
- Peduncle of antenna 1 with first 2 joints not enlarged, but compressed and rounded 4
4. Eyes contiguous medially; posterior margin of telson smooth .... *Aega tenuipes*
- Eyes separated medially; telson with crenulated posterior margin .... *Aega symmetrica*
5. Frontal margin of head slightly produced, pointed at tip .... *Rocinela belliceps*
- Frontal margin of head produced into a long median process, truncate at tip .... *Rocinela angustata*

SUBORDER MICROCERBERIDEA

Only one species, *Microcerberus abbottii* (fig. 41), of this small suborder has been found in California, interstitially in wet sand at Monterey Bay. Being tiny (less than 2 mm) and cryptic, members of this group have been overlooked by collectors. Microcerberids resemble anthurid isopods in having an elongate body and subchelate first pereopods. In having terminal styliform uropods, however, they are more like asellote isopods. An asellote species, *Caecianiropsis psammophila,* also lives interstitially in intertidal sands of central California and shows the same adaptations to this habitat as microcerberids, i. e., elongation, small size, and loss of eyes and pigmentation.

SUBORDER ASELLOTA

Of the ten families known from California of this large and diverse suborder, five are included in the following keys.

KEY TO FAMILIES OF ASELLOTA

1. First 2 pleopods in male free, not coupled together; female with 1st pleopod greatly reduced, 2nd absent; 3rd pair in both sexes not fused medially, but with expanded exopods forming a compound operculum covering posterior pleopods; in fresh water (Group **Aselloidea,** fig. 8) ....ASELLIDAE

   Asellus tomalensis is the only representative of this freshwater family on the California coast.

– First 2 pleopods in male with elongate bases coupled and forming large operculum; female with 1st pair of pleopods fused to form single large operculum covering posterior pleopods; 2nd pair absent in female only; 3rd pleopods in both sexes not operculate; marine .... Group **Paraselloidea** 2

2. Eyes on lateral, pedunclelike projections; terminal joint (dactylus) of pereopods 2–7 with 2 claws 3

– Eyes dorsolateral on head, not pedunculate; dactylus of pereopods 2–7 with 2–3 claws 4

3. Pleotelson somewhat pear-shaped (figs. 42, 45, 48, 51); uropods greatly reduced, barely visible dorsally (figs. 44, 47, 50, 53) .... MUNNIDAE (p. 298)

– Pleotelson broad, shieldlike; uropods short but clearly visible in dorsal view; rami evident ....ANTIASIDAE

   Only one species of this family, *Antias hirsutus,* is known from California (Tomales Bay).

4. Both pairs of antennae small, flagellum lacking or rudimentary; articles of peduncles of 2nd pair dilated; uropods short, inserted in subterminal excavations of telson, not extending much beyond its posterior margin, if at all (fig. 56) ....JAEROPSIDAE (p. 300)

– Antennae 2, long with multiarticulate flagellum (Caution: often broken off), articles of peduncle not dilated; uropods well developed (figs. 54, 55, 57) ....JANIRIDAE (p. 300)

KEY TO MUNNIDAE

1. Uropods minute, leaflike, lacking spinelike protuberances; male 1st pleopods apically pointed (figs. 42–44) ....*Munna (Uromunna) ubiquita*

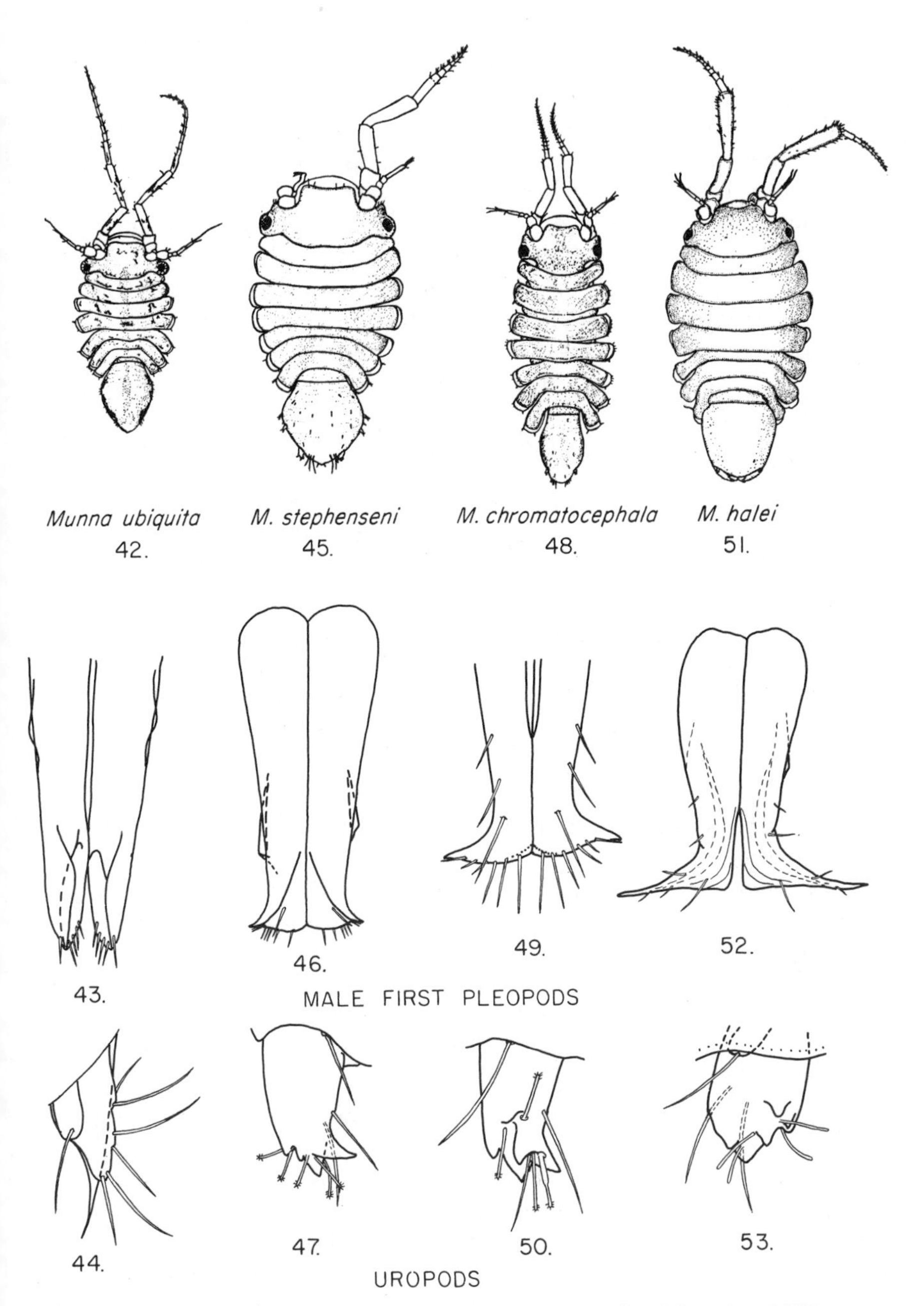

PLATE 68. **Isopoda** (6). **Asellota:** Munnidae: 42–53 after Menzies, 1952.

- – Uropods round in cross-section; male 1st pleopods with apices laterally expanded (figs. 46, 49, 52) 2
- 2. Uropods lacking spinelike protuberances (figs. 51–53) dentate shelf present below uropods .... *Munna (Munna) halei*
- – Uropods with at least one large, spinelike protuberance (figs. 47, 50); no dentate shelf below uropods 3
- 3. Lateral borders of pleotelson with 2 or 3 large, 2-pointed setae (spines); body broad (figs. 45–47) .... *Munna (Neomunna) stephenseni*
- – Lateral borders of pleotelson lacking large, 2-pointed setae; body elongate (figs. 48–50) .... *Munna (Neomunna) chromatocephala*

KEY TO JAEROPSIDAE

- 1. Pleotelson with spineless lateral borders .. *Jaeropsis lobata*
- – Pleotelson with spines on lateral borders 2
- 2. Pleotelson with 5–7 spines on each lateral border (fig. 56) .. .. *Jaeropsis dubia dubia*
- – Pleotelson with 3 spines on each lateral border .... *Jaeropsis dubia paucispinis*

KEY TO JANIRIDAE

(Intertidal and subtidal species)

- 1. Antenna 2 with squama (scale) attached at 3rd article of base . 2
- – Antenna 2 lacking squama 14
- 2. Propodus (next to last joint) of 1st leg with conspicuous serrated margin on proximal 3rd of inferior border; basal 3 articles of maxillipedal palp as wide as endite (see fig. 10) ... . *Janiralata* 3
- – Propodus of 1st leg with proximal 3rd of inferior border smooth; maxillipedal palp with 2nd and 3rd articles much wider than endite .... *Ianiropsis* 7
- 3. Pleotelson with distinct, medially recurved, spinelike posterolateral angles 4
- – Pleotelson with posterolateral angles evenly curved, lacking distinct angles or spinelike processes 6
- 4. Head with deep incision on lateral margins .... *Janiralata triangulata*
- – Head with lateral margins entire 5
- 5. Head with elongate, pointed rostrum; spinelike posterolateral angles of telson reaching or slightly exceeding posterior extent of medial posterior lobe .... *Janiralata solasteri*
- – Head with short, triangular rostrum; spinelike posterolateral

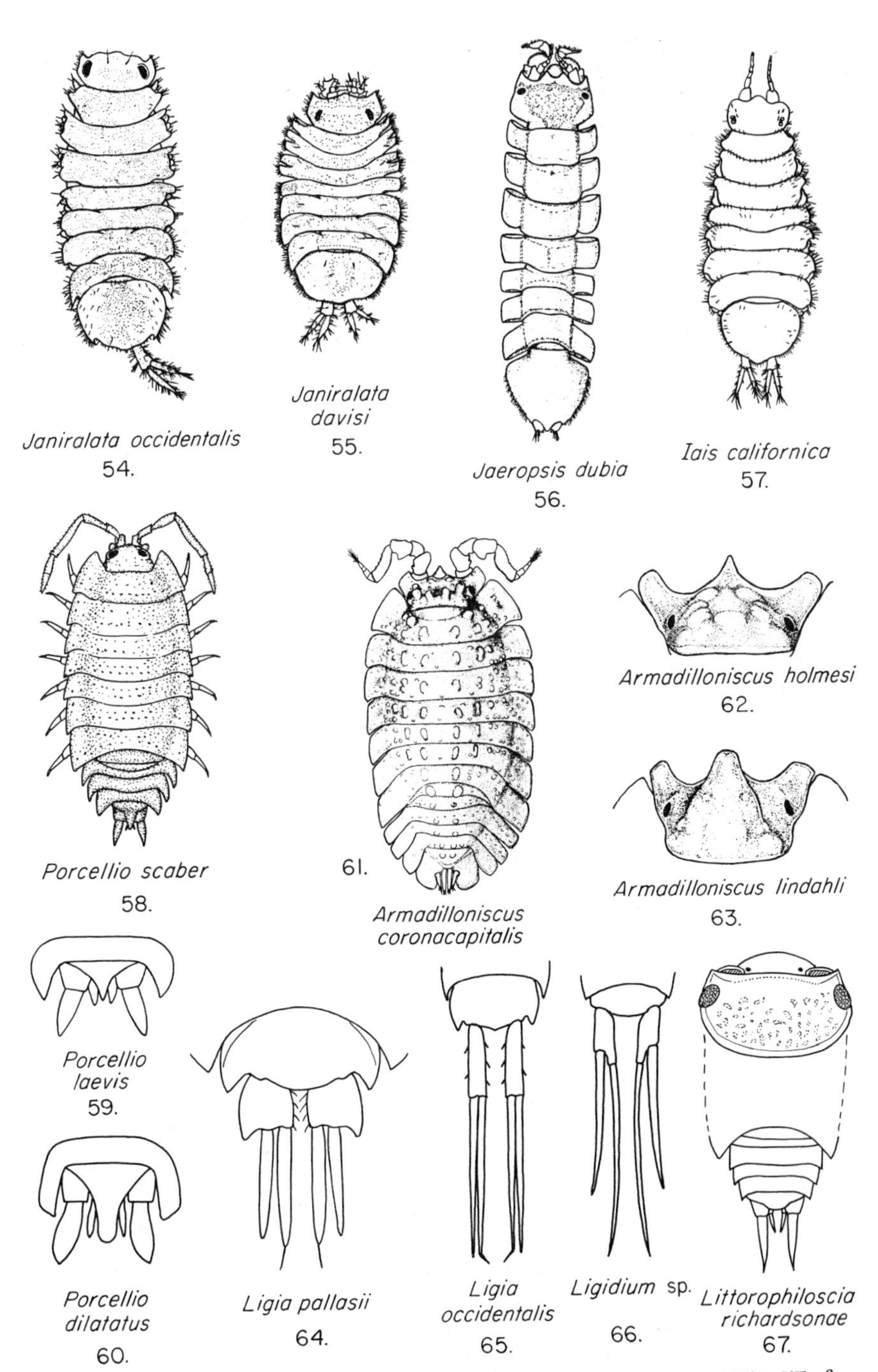

PLATE 69. **Isopoda** (7). **Asellota:** Janiridae: 54, 55 after Menzies, 1951; 57 after Menzies and Barnard, 1951; Jaeropsidae: 56 after Menzies, 1951; **Oniscoidea:** 58 after Van Name, 1936; 59, 60 after Miller, 1936; 61–63 after Menzies, 1950; 64–66 after Richardson, 1905; 67 after Lemos de Castro, 1961.

angles of telson not exceeding posterior extent of medial posterior lobe (fig. 54) .... *Janiralata occidentalis*

6. Head with expanded anterolateral angles (fig. 55) .... *Janiralata davisi*
– Head lacking expanded anterolateral angles *Janiralata rajata*

7. Lateral borders of pleotelson with spinelike serrations 8
– Lateral borders of telson spineless (fine setae may be present) 10

8. Pleotelson with 4–7 spinelike serrations on each side; lateral apices of 1st male pleopod not directed abruptly posteriorly .... *Ianiropsis analoga*
– Pleotelson with 2–3 spinelike serrations on each side; lateral apices of 1st male pleopod directed abruptly posteriorly ... 9

9. Pleotelson with 2 spinelike serrations on each side .... *Ianiropsis epilittoralis*
– Pleotelson with 3 spinelike serrations on each side .... *Ianiropsis tridens*

10. Uropods ½ or less length of pleotelson 11
– Uropods considerably exceeding ½ pleotelson length 13

11. Pleotelson with distinct posterolateral angles lateral to insertions of uropods; head lacking anteriorly projecting anterolateral angles .... *Ianiropsis kincaidi derjugini*
– Pleotelson lacking posterolateral angles lateral to uropod insertions; head with or without anterolateral projections 12

12. Head with distinct, spinelike anterolateral projection on each side of frontal margin .... *Ianiropsis magnocula*
– Head lacking anterolateral projections, with anterolateral margins evenly rounded .... *Ianiropsis minuta*

13. Uropods exceeding length of telson; lateral apices of 1st male pleopod bifurcate .... *Ianiropsis montereyensis*
– Uropods not exceeding pleotelson length; lateral apices of 1st male pleopod not bifurcate ..... *Ianiropsis kincaidi kincaidi*

14. Body greatly elongated (6 times longer than wide); eyes and pigmentation lacking; interstitial in sand .... *Caecianiropsis psammophila*
– Body not greatly elongated; eyes present; little or no pigmentation; associated with *Sphaeroma pentodon*, usually among pleopods or in its burrow (fig. 57) .... *Iais californica*

SUBORDER ONISCOIDEA

Only the maritime (coastal) terrestrial isopods are included in the key and list below. Several are widely distributed species associated with man (synanthropic).

KEY TO ONISCOIDEA (Terrestrial Isopods)

1. Flagellum of 2nd antennae (1st is rudimentary in Oniscoidea) with not more than 4 articles; head with anterolateral lobes 5
– Flagellum of 2nd antennae with more than 4 articles; head without anterolateral lobes .... LIGIIDAE 2
2. Uropods without process at inner distal margin of basal joint; telson with more or less pointed posterolateral projections (figs. 64, 65) .... *Ligia* 3
– Uropods with process at inner distal margin for articulation with endopod; telson without posterolateral projections (fig. 66) .... *Ligidium* 4
3. Eyes separated in front by distance equal to twice length of eye; basal article of uropod about as broad as long (fig. 64) .... *Ligia (Ligia) pallasii*
– Eyes separated in front by distance equal to length of eye; basal article of uropod several times longer than broad (fig. 65) .... *Ligia (Megaligia) occidentalis*
4. Surface of body smooth and shiny .... *Ligidium gracilis*
– Surface of body rough and scaly .... *Ligidium latum*
5. Uropods not reaching beyond posterior border of telson; body markedly convex and capable of rolling into a ball (pillbugs) . 6
– Uropods reaching beyond posterior border of telson; body not markedly convex and incapable of rolling into a ball (sowbugs), (except *Alloniscus perconvexus* and *Armadilloniscus lindahli*) 8
6. Exopod of uropods (ventral view) large, flattened, filling space between pleonite 5 and telson, inserted at apex of flattened basal article (ARMADILLIDIIDAE) .... *Armadillidium vulgare*
– Exopod of uropod (dorsal view) minute, inserted at or near inner lateral margin of flattened basal article that fills space between pleonite 5 and telson and extends to top of telson .... CUBARIDAE 7
7. Eyes small (about 4 ocelli) .... *Cubaris microphthalma*
– Eyes of normal size (with many more than 4 ocelli) .... *Cubaris* spp.
8. Flagellum of antenna 2 with usually at least 4 articles (sometimes seemingly less because of obscure articulation) 16
– Flagellum of antenna 2 composed of 2 or 3 articles .... ONISCIDAE 9
9. Flagellum of antenna 2 triarticulate; exopods of pleopods lack pseudotracheae or "white bodies" (respiratory invaginations) 10
– Flagellum of antenna 2 biarticulate; exopods of first 2 pairs of pleopods with pseudotracheae (seen as "white bodies" in

fresh specimens; absent in *"Porcellio" littorina*) 11

10. Pleon abruptly narrower than pereon (fig. 67) .... *Littorophiloscia richardsonae*
- – Pleon not abruptly narrower than pereon .... *Alloniscus perconvexus*

11. Pleon abruptly narrower than pereon; frontal lobe of head absent .... *Porcellionides pruinosus*
- – Pleon not abruptly narrower than pereon; frontal lobe of head present (fig. 58) .... *Porcellio* 12

12. Hinder angle of pereonite 1 not produced posteriorly 13
- – Hinder angle of pereonite 1 produced posteriorly 14

13. First article of flagellum of antenna 2 less than 1/2 length of 2nd (terminal) joint .... *"Porcellio" littorina*
- – First article of flagellum of antenna 2 generally longer than 2nd .... *Porcellio (Mesoporcellio) laevis*

14. Apex of telson spatulate (fig. 60) *Porcellio (Porcellio) dilatatus*
- – Apex of telson triangulate (fig. 58) 15

15. Inferior margin of ischium of 7th leg of male straight or slightly concave; pleopods unpigmented .... *Porcellio (Porcellio) scaber scaber*
- – Inferior margin of ischium of 7th leg of male markedly concave (bowed); posterior 3 pairs of pleopods pigmented .... *Porcellio (Porcellio) scaber americanus*

16. Pleon abruptly narrower than pereon; eyes lacking (TRICHONISCIDAE) .... *Protrichoniscus heroldi*
- – Pleon not abruptly narrower than pereon; eyes present (SCYPHACIDAE) .... *Armadilloniscus* 17

17. Rostrum broad, truncate (fig. 63); animal capable of rolling into ball .... *Armadilloniscus lindahli*
- – Rostrum pointed; incapable of rolling into a ball 18

18. Dorsum of head with large tubercles; pereon with at least 2 conspicuous rows of tubercles on dorsal side; posterior borders closely set with minute tubercles giving them beaded appearance (fig. 61) .... *Armadilloniscus coronacapitalis*
- – Dorsum of head with low, rounded tubercles (fig. 62); pereon relatively smooth; posterior borders not beaded .... *Armadilloniscus holmesi*

LIST OF ISOPODA

Suborder **Gnathiidea** (Adults ectoparasitic on fish; young free-living, benthic)

GNATHIIDAE

* *Gnathia crenulatifrons* Monod, 1926. Subtidal.

Suborder **Epicaridea** (parasitic on Crustacea)

BOPYRIDAE

Aporobopyrus muguensis Shiino, 1964. In branchial cavity of *Pachycheles rudis*.

Argeia pauperata Stimpson, 1857. In gill chamber of *Crangon franciscorum*.

Argeia pugettensis Dana, 1853. In gill chamber of crangonid shrimps.

Ione cornuta Bate, 1864 (=*Ione brevicauda* Bonnier, 1900). In gill chamber of *Callianassa californiensis*.

Phyllodurus abdominalis Stimpson, 1857. Under abdomen of *Upogebia pugettensis*.

Pseudione sp. On *Pagurus hirsutiusculus* at Bodega Head.

CRYPTONISCIDAE

Cryptothir balani (Bate, 1860). Has also been assigned to *Hemioniscus*, *Cryptothiria*, and *Liriope* by European authors who found it in *Balanus balanoides*. In central California, *Cryptothir* occurs in species of *Balanus* and *Chthamalus*.

ENTONISCIDAE

Portunion conformis Muscatine, 1956. In body cavity of *Hemigrapsus* spp. See Muscatine, 1956, J. Wash. Acad. Sci. 46: 122–126; Piltz, 1969, Bull. So. Calif. Acad. Sci. 68: 257–259.

Suborder **Valvifera**

ARCTURIDAE (ASTACILLIDAE)

Idarcturus hedgpethi Menzies, 1951.

IDOTEIDAE

Edotea sublittoralis Menzies and Barnard, 1959.

† *Idotea (Idotea) fewkesi* Richardson, 1905.

Idotea (Idotea) rufescens Fee, 1926.

Idotea (Idotea) urotoma Stimpson, 1864 (= *Cleantis heathi* Richardson, 1899).

* Not in key.
† Many authors, including Richardson (1905) and Menzies (1950), spell the generic name "*Idothea*" but the original spelling by Fabricius was "*Idotea*".

Idotea (Pentidotea) aculeata (Stafford, 1913).

Idotea (Pentidotea) kirchanskii Miller and Lee, 1970. See Miller and Lee, 1970, Proc. Biol. Soc. Wash. 82: 789–798.

Idotea (Pentidotea) montereyensis Maloney, 1933 (= *Idotea gracillima* (Dana) of Richardson, 1905; Schultz, 1969). On *Phyllospadix.* See Lee, 1966, Comp. Biochem. Physiol. 18: 17–36; 1966, Ecology 47: 930–941 (pigmentation, color change, ecology).

Idotea (Pentidotea) resecata Stimpson, 1857. On *Zostera* and *Macrocystis.* See Menzies and Waidzunas, 1948, Biol. Bull. 95: 107–113 (post-embryonic growth); Lee and Gilchrist, 1972, J. Mar. Biol. Ecol. 10: 1–27 (coloration and ecology).

Idotea (Pentidotea) schmitti Menzies, 1951 (= *Pentidotea whitei* Stimpson of Richardson, 1905).

Idotea (Pentidotea) stenops (Benedict, 1898).

Idotea (Pentidotea) wosnesenskii (Brandt, 1851). Common on kelp.

Synidotea berolzheimeri Menzies and Miller, 1972. Often on hydroid *Aglaophenia.*

Synidotea bicuspida (Owen, 1839).

Synidotea harfordi Benedict, 1897. Known from Morro Bay south.

Synidotea laticauda Benedict, 1897. San Francisco Bay only; on hydroids, fouled pilings, etc.

Synidotea pettiboneae Hatch, 1947 (= *S. consolidata* Benedict of Richardson, 1905; Menzies and Miller, 1954; Schultz, 1969).

Synidotea ritteri Richardson, 1904. Often on *Aglaophenia.*

Suborder **Anthuridea**

ANTHURIDAE

Colanthura squamosissima Menzies, 1951.

Cyathura munda Menzies, 1951.

Paranthura elegans Menzies, 1951.

Suborder **Flabellifera**

AEGIDAE (fish parasites)

Aega lecontii (Dana, 1854).

Aega microphthalma Dana, 1854.

Aega symmetrica Richardson, 1905.

Aega tenuipes Schioedte and Meinert, 1879–1880.

Rocinela angustata Richardson, 1898.

Rocinela belliceps (Stimpson, 1864).

CIROLANIDAE

Cirolana harfordi Lockington, 1877. Carnivorous scavenger, common in mussel beds.

Excirolana chiltoni (Richardson, 1905).

Excirolana kincaidi (Hatch, 1947). On sandy beaches.

Excirolana linguifrons (Richardson, 1899). On sandy beaches.

CYMOTHOIDAE (Parasites on fish)

Lironeca californica Schioedte and Meinert, 1883–1884 (*Livoneca* is a misspelling).

Lironeca vulgaris Stimpson, 1857.

Nerocila californica Schioedte and Meinert, 1881–1883.

LIMNORIIDAE (see Menzies, 1957)

Limnoria (Limnoria) lignorum Rathke, 1799. A widely distributed woodborer; northern.

Limnoria (Limnoria) quadripunctata Holthuis, 1949. Woodborer; in bays and harbors.

Limnoria (Limnoria) tripunctata Menzies, 1951. Widely distributed on Pacific coast; woodborer, as above. See Beckman and Menzies 1960, Biol. Bull. 118: 9–16 (reproductive temperature and geographic range); Johnson and Menzies 1956, Biol. Bull. 110: 54–68 (migratory habits).

Limnoria (Phycolimnoria) algarum Menzies, 1956. Borer in laminarian holdfasts.

SPHAEROMATIDAE

Bathycopea daltonae (Menzies and Barnard, 1959) (=*Ancinus daltonae*). Subtidal on sandy bottoms, camouflaged by color. Two species of the related genus *Ancinus* occur in similar habitats in southern California. For taxonomy see Loyola e Silva 1971, Arq. Museu Nacional 54: 209–223; Menzies and Barnard 1959, Pacific Nat. 1: 3–35; Trask 1970, Bull. So. Calif. Acad. Sci. 69: 145–149.

Dynamenella dilatata (Richardson, 1899).

"Dynamenella" benedicti (Richardson, 1899).

"Dynamenella" glabra (Richardson, 1899).

"Dynamenella" sheareri (Hatch, 1947).

Exosphaeroma amplicauda (Stimpson, 1857).

Exosphaeroma inornata Dow, 1958. See Dow 1958, Bull. So. Calif. Acad. Sci. 57: 95–97.

Exosphaeroma octoncum Richardson, 1899.

Exosphaeroma rhomburum (Richardson, 1899).

Gnorimosphaeroma lutea Menzies, 1954. Fresh and brackish water. See Eriksen, 1968, Crustaceana 14: 1–12 (ecology); Riegel, 1959, Biol. Bull. 116: 272–284 (osmoregulation); Hoestlandt, 1973.

Gnorimosphaeroma noblei Menzies, 1954.

Gnorimosphaeroma oregonense (Dana, 1854–1855). Marine, sometimes in brackish water. See Riegel, above (osmoregulation); Hoestlandt, 1973; Riegel, 1959.

Gnorimosphaeroma rayi Hoestlandt, 1969. Possibly introduced with oysters from Japan; in Tomales Bay. See Hoestlandt 1969, C.R. Acad. Sci. Paris 268: 325–327; Hoestlandt, 1973.

Paracerceis cordata (Richardson, 1899).

Sphaeroma pentodon Richardson, 1904 (= *S. quoyana* H. Milne Edwards, 1840, according to Rotramel, 1972). Burrows in driftwood, styrofoam floats, and mud banks in estuaries.

Tecticeps convexus Richardson, 1899. Subtidal, sand or gravel bottoms, with camouflage coloration and a strong odor.

Suborder **Microcerberidea**

MICROCERBERIDAE

Microcerberus abbottii Lang, 1961. Interstitial in sand. See Lang 1961, Ark. Zool. (2) 13: 493–510.

Suborder **Asellota**

(Group ASELLOIDEA)

ASELLIDAE

Asellus tomalensis Harford, 1877. Freshwater lakes and sluggish streams.

(Group PARASELLOIDEA)

ANTIASIDAE

Antias hirsutus Menzies, 1951.

JAEROPSIDAE

Jaeropsis dubia dubia Menzies, 1951.

Jaeropsis dubia paucispinis Menzies, 1951.

Jaeropsis lobata Richardson, 1899.

‡ JANIRIDAE

Caecianiropsis psammophila Menzies and Pettit, 1956. Interstitial in sand, sometimes with *Microcerberus abbottii.*

‡ *Iais californica* (Richardson, 1904). Symbiotic with *Sphaeroma pentodon,* intertidal. See Menzies and Barnard, 1951, Bull. So. Calif. Acad. Sci. 50: 136–151; Rotramel, 1972.

‡ *Ianiropsis analoga* Menzies, 1952. Intertidal.

Ianiropsis epilittoralis Menzies, 1952. Intertidal.

Ianiropsis kincaidi derjugini Gurjanova, 1933.

Ianiropsis kincaidi kincaidi Richardson, 1904. 0–66 meters depth.

Ianiropsis magnocula Menzies, 1952. 20–57 meters depth.

Ianiropsis minuta Menzies, 1952. Intertidal, rocks and sand.

Ianiropsis montereyensis Menzies, 1952. Intertidal, rocky beaches.

Ianiropsis tridens Menzies, 1952. Intertidal; common on algae, occasionally in sponges.

‡ *Janiralata davisi* Menzies, 1951. Lower intertidal.

Janiralata occidentalis (Walker, 1898). 18–69 meters depth.

Janiralata rajata Menzies, 1951. Monterey Bay; from ray at 36 meters.

Janiralata solasteri (Hatch, 1947). Shallow water to 218 meters.

Janiralata triangulata (Richardson, 1905).

‡ In the use of the initial I or J in certain names in this group of isopods, the list follows the original (taxonomically correct) spelling. In Roman usage the initial consonant J was written with the same sign as the vowel I, leading to some confusion among modern writers. Some of these have named genera using the initial I, others with the initial J, although basically the same consonant (J) is involved. These names are not pronounced with the vowel "I" sound, but with the "J" or "Y" sound. (Eds.)

MUNNIDAE

Munna (Munna) halei Menzies, 1952. Intertidal.

Munna (Neomunna) chromatocephala Menzies, 1952. Intertidal.

Munna (Neomunna) stephenseni Gurjanova, 1933. Intertidal to 18 meters.

Munna (Uromunna) ubiquita Menzies, 1952. Low intertidal to 37 meters.

Suborder **Oniscoidea**

ARMADILLIDIIDAE

Armadillidium vulgare (Latreille, 1804). Cosmopolitan; synanthropic (introduced by or associated with man).

CUBARIDAE

* *Cubaris californica* (Budde Lund, 1885) [?=*C. affinis* (Dana)].

Cubaris microphthalma (Arcangeli, 1932).

LIGIIDAE

Ligia (Ligia) pallasii Brandt, 1833. Sea cliffs and caves, especially near freshwater seeps.

Ligia (Megaligia) occidentalis Dana, 1853. Rocky shores.

Ligidium gracilis (Dana, 1856). Riparian.

Ligidium latum Jackson, 1932. Riparian.

ONISCIDAE

Alloniscus perconvexus Dana, 1854. Sandy beaches above high-tide marks, buried in sand and under driftwood.

Littorophiloscia richardsonae (Holmes and Gay, 1909). Maritime; Salicornia marshes. See Lemos de Castro, 1965, Arquivos do Museu Nacional 53: 85–98.

"Porcellio" littorina Miller, 1936. Not a true *Porcellio* because it lacks pseudotracheae, but proper atracheate genus undetermined.

Porcellio (Mesoporcellio) laevis Latreille, 1804. Cosmopolitan, synanthropic.

Porcellio (Porcellio) dilatatus Brandt and Ratzeburg, 1833 (= *Porcellio spinicornis occidentalis* Miller, 1936; Menzies and Miller, 1954). An introduced European species.

Porcellio (Porcellio) scaber americanus Arcangeli, 1932. Common along coast.

Porcellio (Porcellio) scaber scaber Latreille, 1804. Rare.

Porcellionides pruinosus (Brandt, 1833) (= *Metoponorthus pruinosus*). Cosmopolitan, synanthropic.

SCYPHACIDAE See Menzies, 1950, Proc. Calif. Acad. Sci. 26: 467–481; Schultz, 1972, Proc. Biol. Soc. Wash. 84: 477–487.

Armadilloniscus coronacapitalis Menzies, 1950. Near shore, under rocks, driftwood, etc.

Armadilloniscus holmesi Arcangeli, 1933 [= *A. tuberculatus* (Holmes and Gay)]. Maritime.

Armadilloniscus lindahli (Richardson, 1905). Maritime; unique among west-coast species of the genus in being able to roll into a ball.

TRICHONISCIDAE

Protrichoniscus heroldi Arcangeli, 1932. Muddy beaches, Muir Woods (Marin County).

REFERENCES ON ISOPODA

Barnard, K. H. 1925. A revision of the family Anthuridae (Crustacea, Isopoda) with remarks on certain morphological peculiarities. J. Linn. Soc. London 36: 109–166.

Bodle, J. E. 1969. Pleopod and penial structure in the isopod family Sphaeromatidae with special reference to its taxonomy. Unpublished M.A. Thesis, Univ. Calif., Davis. 71 pp.

Danforth, C. G. 1970. *Epicaridea (Crustacea: Isopoda) of North America.* University Microfilms: Ann Arbor, Michigan, 190 pp. (Published on demand.)

Filice, F. P. 1958. Invertebrates from the estuarine portion of San Francisco Bay and some factors influencing their distribution. Wasmann J. Biol. 16: 159–211.

Haderlie, E. C. 1969. Marine fouling and boring organisms in Monterey Harbor. II. Second year of investigation. Veliger 12: 182–192.

Hansen, H. J. 1905. On the propagation, structure and classification of the family Sphaeromidae. Quart. J. Micr. Sci. 49: 69–135.

Hatch, M. H. 1947. The Chelifera and Isopoda of Washington and adjacent regions. Univ. Wash. Publ. Biol. 10: 155–274.

Hoestlandt, H. 1973. Étude systématique de trois espèces Pacifiques nordaméricaines du genre *Gnorimosphaeroma* Menzies (Isopodes Flabellifères). I. Considérations générales et systématique. Arch. Zool. Expér. Gén. 114: 349–395.

Holmes, S. J. and M. E. Gay 1909. Four new species of isopods from the coast of California. Proc. U.S. Nat. Mus. 36: 375–379.

Iverson, E. W. 1974. Range extensions for some California marine isopod crustaceans. Bull. So. Calif. Acad. Sci. 73: 164–169.

Jackson, H. G. 1922. A revision of the isopod genus *Ligia* (Fab.). Proc. Zool. Soc. London. for 1922: 683–703.

Menzies, R. J. 1950. The taxonomy, ecology, and distribution of Northern California isopods of the genus *Idothea* with the description of a new species. Wasmann J. Biol. 8: 155–195.

Menzies, R. J. 1951. A new species of *Limnoria.* Bull. So. Calif. Acad. Sci. 50: 86–88.

Menzies, R. J. 1951. New marine isopods, chiefly from Northern California with notes on related forms. Proc. U.S. Nat. Mus. 101: 105–156.

Menzies, R. J. 1952. Some marine asellote isopods from Northern California, with descriptions of nine new species. Proc. U.S. Nat. Mus. 102: 117–159.

Menzies, R. J. 1954a. A review of the systematics and ecology of the genus "*Exosphaeroma*" with the description of a new genus, a new species, and a new subspecies (Crustacea, Isopoda, Sphaeromidae). Amer. Mus. Novitates 1683: 1–24.

Menzies, R. J. 1954b. The comparative biology of reproduction in the woodboring isopod crustacean *Limnoria.* Bull. Mus. Comp. Zool. Harvard 112: 364–388.

Menzies, R. J. 1957. The marine borer family Limnoriidae (Crustacea, Isopoda). Bull. Mar. Sci. Gulf and Caribbean 7: 101–200.

Menzies, R. J. and M. A. Miller 1972. Systematics and zoogeography of the genus *Synidotea* (Crustacea: Isopoda) with an account of Californian species. Smithsonian Contr. Zool. 102: 1–33.

Miller, M. A. 1936. California isopods of the genus *Porcellio* with descriptions of a new species and a new subspecies. Univ. Calif. Publ. Zool. 41: 165–172.

Miller, M. A. 1938. Comparative ecological studies on the terrestrial isopod Crustacea of the San Francisco Bay region. Univ. Calif. Publ. Zool. 43: 113–142.

Miller, M. A. 1968. Isopoda and Tanaidacea from buoys in coastal waters of the continental United States, Hawaii, and the Bahamas (Crustacea). Proc. U.S. Nat. Mus. 125: 1–53.

Richardson, H. 1905. Monograph on the isopods of North America. Bull. U.S. Nat. Mus. 54: 727 pp.

Richardson, H. 1909a. The isopod crustacean, *Ancinus depressus* (Say). Proc. U.S. Nat. Mus. 36: 173–177.

Richardson, H. 1909b. Isopods collected in the northwest Pacific by the U.S. Bureau of Fisheries Steamer "Albatross" in 1906. Proc. U.S. Nat. Mus. 37: 75–129.

Riegel, J. A. 1959. A revision in the sphaeromid genus *Gnorimosphaeroma* Menzies (Crustacea: Isopoda) on the basis of morphological, physiological and ecological studies on its "subspecies." Biol. Bull. 117: 154–162.

Rotramel, G. 1972. *Iais californica* and *Sphaeroma quoyanum,* two symbiotic isopods introduced to California (Isopoda, Janiridae and Sphaeromatidae). Crustaceana, Suppl. 3: 193–197.

Schultz, G. A. 1969. *How to Know the Marine Isopod Crustaceans.* Dubuque: W. C. Brown, vii + 359 pp.

Shiino, S. M. 1964. On three bopyrid isopods from California. Rept. Fac. Fish. Pref. Univ. Mie 5: 19–25.

Van Name, W. G. 1936. The American land and fresh-water isopod Crustacea. Bull. Amer. Mus. Nat. Hist. 71: 1–535, (and Supplements, 1940, 77: 109–142 and 1942, 80: 299–329).

Wilson, W. J. 1970. Osmoregulatory capabilities in isopods: *Ligia occidentalis* and *Ligia pallasii.* Biol. Bull. 138: 96–108.

PHYLUM ARTHROPODA: CRUSTACEA, AMPHIPODA: GAMMARIDEA

The large order **Amphipoda** is represented locally by three suborders: **Gammaridea, Caprellidea,** and **Hyperiidea.** The Gammaridea (p. 313) are by far the most abundant and familiar (Pl. 75). They occur in fresh and marine waters, and in the semiterrestrial supralittoral fringe of the seashore. Caprellidea (p. 367) are exclusively marine. They have a rudimentary abdomen and their bodies are very elongated (Pl. 86) in free-living families, and very flattened in the Cyamidae or whale-lice (Pl. 90, figs. 19–21). A third suborder, Hyperiidea, is pelagic and distinguished by a very large head and eyes. Hyperiids are not as a rule seen intertidally except for individuals commensal beneath the bells of stranded medusae.

SUBORDER GAMMARIDEA

Gammaridean amphipods, like copepods, are an intrinsically difficult group to identify because they are numerous and highly diverse, yet superficially look a good deal alike. Except for a few distinctive species, most gammarids are easily confused by the beginner. Reliable specific identification is not a matter to be undertaken casually; it requires careful, systematic examination and dissection. To make these keys usable to the practicing zoologist or ecologist who is willing to make the necessary effort to get a specific identification, Dr. Barnard has written a detailed dissection and identification procedure, followed by a basic key (p. 324) to families or groups of families (Keys A–N). This approach may deter all but the serious student—but no one except a serious student should tackle amphipod identification anyway. In view of the enormous numbers and ecological importance of amphipods in the intertidal biota, this effort to provide a reliable and workable set of keys is necessary.

A detailed key to the gammarids is not a good introduction for a beginner or a student in a short, general invertebrate course. Fortunately, examples of easily identifiable gammaridean amphipods are easy to find: beachhoppers of the genus *Orchestoidea* (Talitridae) are readily identified (Key O) and are large enough to be used for class study and dissection to make clear the basic morphology of the group. For the usual invertebrate zoology course, such an introduction will suffice. The detailed procedures of Dr. Barnard's account should then be followed if identifications are to be made.

At the risk of superficiality, and with no claim that it will suffice for more than the commoner and more obvious families, we also provide a simple (or oversimplified) key to some two dozen families of gammaridean amphipods (p. 324), which will help the beginner to become aware of the diversity among amphipods, and, hopefully, will encourage him to go further.

For detailed treatments of the morphology and systematics of gammaridean amphipods, the student should consult Barnard (1969, but note that the last literature cited is 1965) and Bousfield (1973).

We are grateful to John W. Chapman for helpful suggestions and for careful checking of the keys.

IDENTIFICATION OF GAMMARIDEAN AMPHIPODS

J. Laurens Barnard

National Museum of Natural History, Smithsonian Institution, Washington, D.C.

(PLATES 70–85)

The following section on identification procedures, and Keys A through N, apply to aquatic, intertidal Gammaridea. The familiar beach-hoppers can be placed directly into the family **Talitridae,** and identified by Key O (p. 352).

The body of a gammaridean (fig. 1), usually flattened from side to side, is composed of a clearly defined head, a thorax of seven freely articulated segments (**pereonites**) lacking a carapace, and an **abdomen** of six segments (**pleonites**). The abdomen is divided into two parts: the **pleosome** with three segments, each bearing a pair of **pleopods** (swimmerets); and a **urosome,** bearing three pairs of rigid **uropods** projecting posteriorly. The **telson** is a flap over the anus attached to pleonite 6. The thorax bears seven pairs of pereopods, the first two pairs of which are often modified for grasping and are then called **gnathopods.**

Numbering systems vary among specialists; here we designate the first two legs gnathopods 1 and 2 and number the remaining pereopods 3 to 7. In most of the literature on Californian amphipods, writers

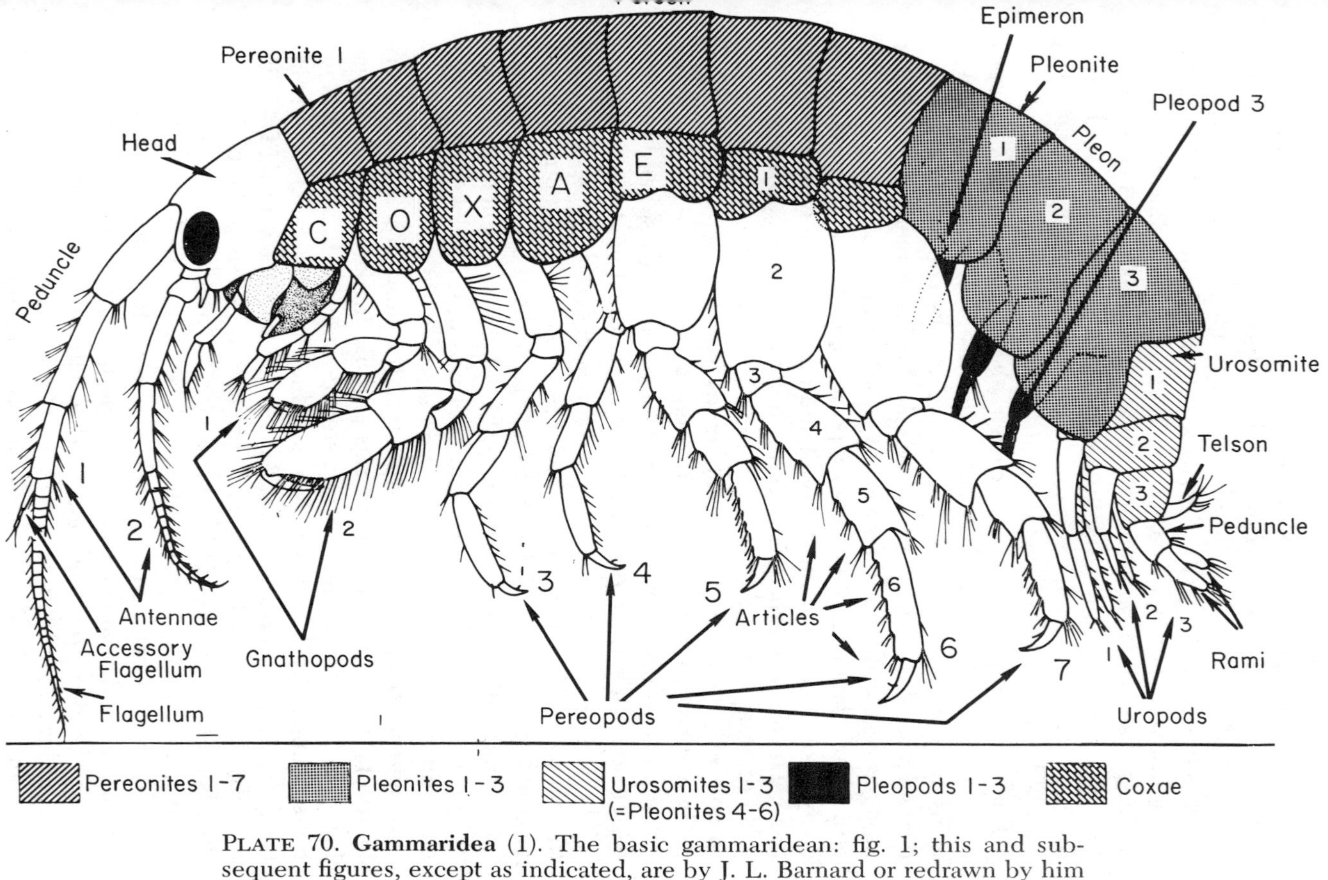

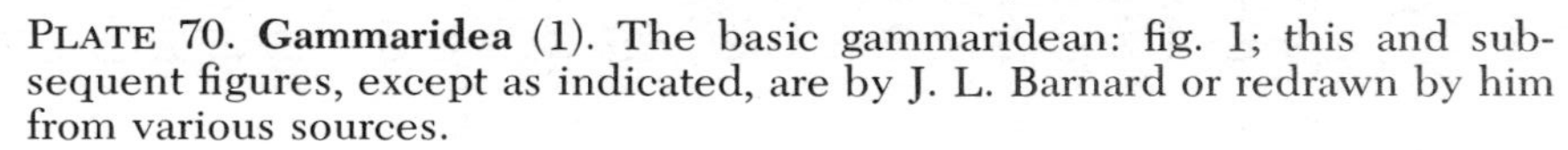

PLATE 70. **Gammaridea** (1). The basic gammaridean: fig. 1; this and subsequent figures, except as indicated, are by J. L. Barnard or redrawn by him from various sources.

have used the terms gnathopods 1 and 2, followed by a new numerical sequence of pereopods 1, 2, 3, 4, 5. But, since gnathopods are merely specialized pereopods, we follow the European custom of equating gnathopods 1 and 2 with pereopods 1 and 2, and continuing the sequence with pereopods 3, 4, 5, 6, 7, as in figure 1. Because gnathopods are used to grasp members of the opposite sex in mating and to transfer spermatophores, the gnathopods are as important in the evolution of diversity as are the sexual devices of plants, and therefore should be accorded distinction from other pereopods.

Each pereopod has seven **articles** (the word "joint" is inappropriate). The first article, the **coxa,** is usually widely expanded downward to cover part of the remaining leg. Article 6 on the gnathopods is termed a **hand,** and on all pereopods article 7 is termed a **dactyl.** Common names for other articles are not widely used in the literature. The term **subchelate** is applied to pereopods and gnathopods on which the dactyl closes tightly against an inflated margin of the hand called the **palm.** The terms **merochelate, carpochelate,** and **propodochelate** refer to an opposable thumb formed from article 4 (**merus**), 5 (**carpus**), or 6 (**propodus**), respectively, of gnathopods 1 and 2. Pereopods without prehensility are termed simple. Pleopods are rarely used in taxonomy.

Uropods normally are composed of a **peduncle** (base) and two **rami** (extensions). The rami of uropod 3 are occasionally reduced or absent.

The telson is often split into two lobes or fused into a solid piece (dorsal view). The distinction between a fleshy and a laminar telson is vital to identification. A **fleshy** telson, as restricted herein, is always entire (unsplit), thick, short, and puffy. A **laminar** telson is either split or entire and may have a ventral keel giving it a thickened appearance, but the edges and apex remain laminar (thin and flattened).

Besides a **rostrum** and a pair of eyes, the head (fig. 2g) bears nine kinds of attachments, all paired except those marked with an asterisk, listed in order from anterior to posterior: **antenna 1, antenna 2, epistome** *, **upper lip** * (figs. 2a, 3), **mandible** (figs. 2b, 4–12), **lower lip** * (figs. 2c, 15, 16), **maxilla 1** (figs. 2d, 19, 20), **maxilla 2** (figs. 2f, 17, 18), **maxilliped** (fig. 2e). The rostrum, projecting between the first antennae, is rarely elongate. The eyes are rarely fused together. The epistome is a cephalic sclerite seldom of taxonomic importance (herein mainly in Phoxocephalidae). When observation of the epistome is required, the first and second antennae must be spread apart widely and the front of the head observed in lateral view. If the epistome bears a significant tooth or **cusp,** it can be seen emerging forward from the mouthpart bundle. Care must be exercised not to confuse the **epistomal spike** with the **mandibular palps;** the latter are flexible and setose whereas the epistomal spike is solid, smooth, and fixed. An epistomal spike may also be confused with the **excretory spouts** projecting

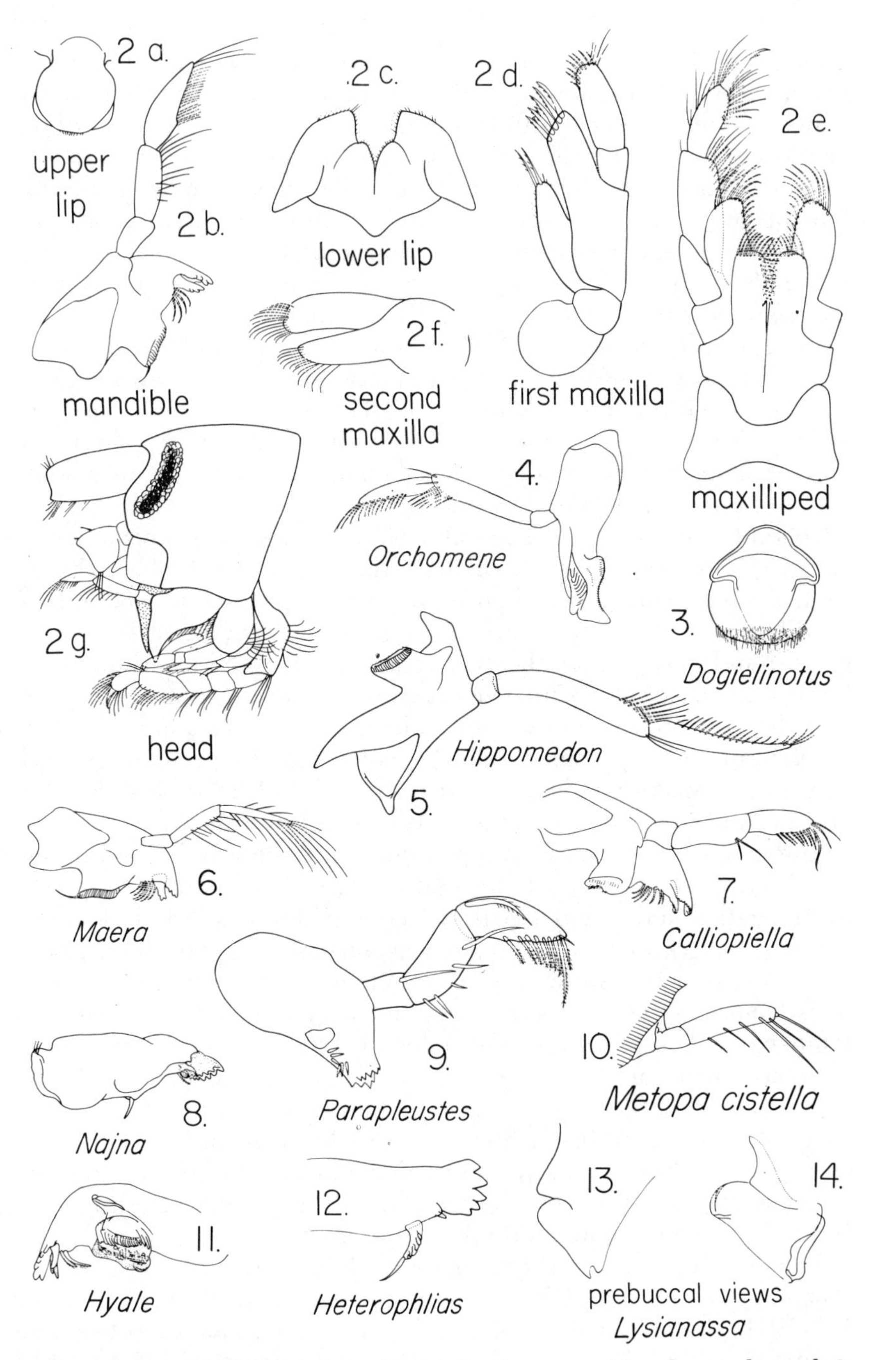

PLATE 71. **Gammaridea** (2). 2a–g, basic mouthparts and head, as indicated; 3, upper lip; 4–12, mandibles; 13, 14, lateral profile of prebuccal area of *Lysianassa*.

from article 2 of each second antenna. The upper lip is ignored in this presentation. Mandibles (figs. 2b, 4–12) each have a grinding molar, which is often reduced with loss of grinding surface. The apex of the mandible (incisor) usually projects as a series of teeth near which are several spines. A triarticulate **palp** is usually attached to the mandible; but the palp is often absent (Hyalidae, Talitridae) or sometimes reduced to only one or two articles. Care should be taken to observe the presence or absence of a palp before mandibles are removed as palps are easily lost by accident during dissection.

The lower lip (figs. 2c, 15, 16) lies behind the mandibles and in front of the first pair of maxillae; it is extremely difficult to remove without damage (see directions for dissection). The first pair of maxillae (figs. 2d, 19, 20) each have an inner plate, an outer plate, and a palp of two articles. The inner plate is not closely contiguous with the outer plate and is often overlooked or lost during dissection because it is partially attached to the lower lip. The outer plate bears very heavy spines (7, 9, or 11) and the palp is occasionally reduced to one article or is absent. Each maxilla 2 is composed of two simple, setose plates. The maxillipeds are usually so large as to cover most of the other mouthparts from below; hence they are removed first. The maxillipeds are fused at the base so as to appear as a single branched appendage. Each branch has an inner plate, an outer plate, and a palp composed of four articles; the fourth article is usually pointed and is termed a dactyl. The palp is occasionally reduced to three articles (or absent in hyperiids) and the plates are often severely reduced in size.

Antenna 1 is composed of a peduncle of three articles and a long flagellum of numerous articles. An accessory flagellum often is attached to the medial and apical surface of peduncular article 1; if it is composed of two or more articles it can usually be seen from lateral view as a projecting secondary ramus, but numerous species of amphipods have a minute accessory flagellum which must be searched for intensively. The presence or absence of the accessory flagellum and often its shape, are important taxonomic characters. Its observation requires mounting the antenna in glycerine on a slide under a thin cover slip. Some grains of fine sand should be scattered in the glycerine so that the cover slip can be moved easily to grip and roll the antenna over slowly during the search for the accessory flagellum (use 100X objective of compound microscope).

Sexual differentiation is often strong. Males often bear an enlarged, ornamented, and heavily prehensile gnathopod 2 (paired) (figs. 169–201); elongate and heavily setose antennae; heavily setose and enlarged uropod 3; powerful pleopods; and, occasionally, enlarged eyes. However, many species have no obvious external differences between the sexes, except that adult and breeding females bear four

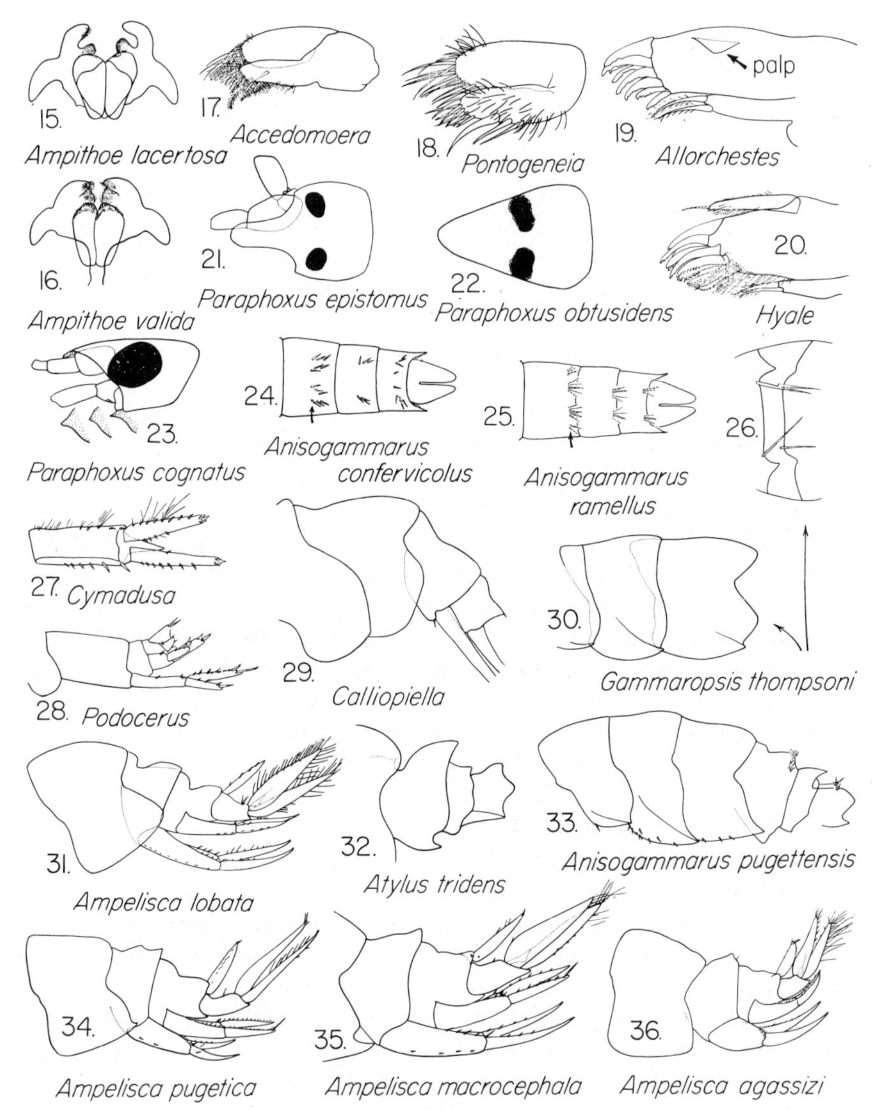

PLATE 72. **Gammaridea** (3). 15, 16, lower lip; 17, 18, second maxilla; 19, 20, first maxilla; 21, 22, head, dorsal; 23, head, lateral; 24–26, urosomites dorsal; 27, uropod 1; 28–36, lateral views of pleonal parts.

pairs of setose (or not) lamellae in the brood space between the bases of coxae 2 to 5; the lamellae are attached to the coxae but must not be confused with the fleshy gills also attached to these coxae. Some gills have secondary appendages so that confusion can occur in the identification of broodplates; no short sentence can fully explain the problems to be encountered, but fortunately the difficult cases in California amphipods are rare. Eggs are encased in the brood pouch formed by the interleaving setae of the brood lamellae. In adult and breeding males, a minute pair of **penial processes** hangs ventrally from sternite 7, between the coxae of the seventh thoracic segment. They are attached to the sternum and not to the coxae but can be confused with small gills attached to coxa 7. If the supposed penes break off easily, they are probably gills; penes often bear tiny spines; they are extremely difficult to see and can often be confused with broken ends of tendons after the seventh pereopods are removed.

DISSECTION

Choose an unbroken specimen bearing two pairs of antennae, a pair of third uropods and a telson, two pairs of gnathopods and at least articles 1 to 3 of all other legs. Examine urosome to determine whether it is composed of one, two, or three segments. Manipulate telson to determine its fleshy or laminar condition; a laminar telson usually is freely articulate at its base. Remove urosome and mount dorsal side up on depression slide filled with glycerine * and emplace cover slip. Examine coxa 1 behind head and determine whether it is significantly smaller than and largely hidden by coxa 2 or whether it is nearly as large as coxa 2 and freely visible. In this process tilt the amphipod so that dissecting light shines properly on these parts and carefully count all coxae to ensure a proper orientation. Remove right antenna 1 and mount on flat side with glycerine and a dozen minute sand grains; add cover slip. Sand grains facilitate rolling the antenna by movement of the cover slip during observation. Hold carcass with heavy dissecting needle or forceps; remove right and left pereopod 5, including coxa, by grasping deeply into basal musculature with jeweler's forceps; place both pereopods on glycerine puddle of slide, orienting both so as to appear from left side of animal. Mount cover slip. Repeat process for pereopods 4 and 3, gnathopod 2, and gnathopod 1, in this order, making a slide for each pair. Label each slide with grease pencil.

Carefully study slides of urosome (telson, uropod 3), possible fusion of urosomites and proportions of these segments, accessory flagellum, and the general condition of gnathopods 1 and 2. Proceed to key and go as far as possible until observation is required on some part yet undissected or unobserved. Make additional slides of parts as needed (or better, fully dissect the amphipod into its component parts, a slide for each kind, requiring about twenty slides). If observations on head and **epimera** (sides of pleonites; sing. **epimeron**) are difficult to make on

* Test amphipod for glycerine shrinkage before mounting any parts; if amphipod shrinks rapidly or develops frost and air bubbles, use a slow-drip method for an hour to replace alcohol preservative with glycerine. Hyalidae are especially sensitive to glycerine.

the dissecting microscope because of small size, place these sections on depression slide with cover slip and make observations on compound microscope. Remove pleopods for clearer view of epimera, but observe head before removing mouthparts and left antennae. Then remove left antenna 1 for view of rostrum (rarely necessary).

Upper lip and lower lip are rarely needed in this analysis. In removing mouthparts, spear head side to side with needle in dish of alcohol, with head on its left side (downward) and upside down (so that mouthparts project toward 12 o'clock). For lefthanded persons, the head should be right side downward. Use forceps to grab maxillipeds at base, snap off, place on mounting medium puddle; further remove any curved basal projections without separating branches of appendage, and add cover slip after orienting maxilliped so original curved base points upward. Remove and mount right and left maxilla 2. Identify mandibles as heavily sclerotized, solidly attached, and somewhat twisted appendages; search for palp; do not remove mandible but grab deeply with forceps all of remaining fleshy and flexible mouthpart bundle *behind* mandibles and remove to slide. This mass should include right and left maxilla 1 and all of lower lip attached together. Carefully tease away lower lip while leaving inner plates of maxillae attached to their outer plates, then separate maxillae if further necessary for observation. Mandibles are very difficult to remove; attempt partial excision of basal musculature before grabbing and pulling mandible free with forceps; use extreme care not to grab mandible at levels of molar, incisor, or palp as these may be easily shattered or broken away. If necessary, attempt rotation of mandible outward with slight pressure of forceps so as to identify medial molar before grasping heavily. Mount each mandible in glycerine on separate slide with several coarse sand grains so that mandible can be rolled over and properly oriented later; label left and right mandibles. In later rolling process, the palp may be removed to yet another slide but it should be analyzed before removal.

AIDS TO IDENTIFICATION

Many characters of the keys are briefly described and supplemented by illustrations. Plates 77 and 78 (figs. 75–138) show many kinds of uropod 3 and telson, characteristic of limited groups of amphipods, by family or genus. Attempt your first identification using a collection of specimens in good condition, free of debris, with largely unbroken appendages, and containing numerous specimens of gross similarity and both sexes identified. Use a male for the first dissection and pursue it through the key. Return to the supposed females where necessary and check the main key characters against the female. If uncertain of your ultimate identification, commence identification of another species, preferably one with distinct gross appearance. Practice with as many species as possible until experience is gained in the interpretive process. If literature is at hand, check specimens against any information available. California amphipods are not yet fully analyzed and many species are not described or adequately illustrated.

The keys are divided into sections A to O so as to group genera into similar clusters. Focus your attention on genera rather than on families. So many cases of convergent evolution occur in this order that familial concepts do not aid in rapid identification, as they do, for instance, in polychaetes. In our present state of inadequate knowledge, much time can be wasted debating the evolutionary descent of various families and even deciding taxonomically the families to which certain genera belong. This problem is most critical in the Gammaridae, Eusiridae (=Calliopiidae and Pontogeneiidae), Pleustidae, Haustoriidae, Phoxocephalidae and in the various families of the superfamily Corophioidea, such as the Corophiidae (now including Aoridae, Photidae, and Isaeidae), Ampithoidae, Ischyroceridae, Podoceridae, and Cheluridae (Barnard, 1973).

The inner ramus of uropod 3 is lost frequently during preservation; check the dorsal mount of the urosome carefully, counting all six uropods (three pairs) and identifying the terminal third pair. Then observe the presence of two rami per peduncle; if a long outer ramus remains but an inner ramus is absent, the latter may have been lost accidentally. This is especially true for species in Key K; a few species of Gammaridae (Key N) have an immensely long outer ramus and such a short inner ramus that the latter may be concealed by or confused with the telson. Remove uropod 3 if necessary to a separate slide for clear observation. If the inner ramus has been lost accidentally, a sclerotic socket usually remains to mark its presence.

Spines and setae are homologous but setae are highly flexible and can be bent in the middle without breaking; spines are simply thickened setae, so inflexible and brittle as to be shattered when bent in the middle.

In many taxa, specific differences are based on only one sex; the opposite sexes have not been studied intensively enough to find their specific distinctions. Certain taxa are keyed out twice in a key to facilitate their identification where judgments on certain characters may be difficult to make.

IDENTIFICATION BY MEANS OF TELSON AND UROPOD 3

Rapid identification of genera often can be facilitated by reference to shapes of uropod 3 and telson in combination (Plates 77, 78). This is a supplementary procedure to accelerate progress through the keys or to confirm various questionable endpoints. To some extent, keys A through M are arranged so as to progress from anomalous to average conditions of uropods and telson.

The ordinary uropod 3 (fig. 75) is composed of a short peduncle and elongate lanceolate rami of various taxa in keys F, K, L, M, N; but variants include foliaceous rami (fig. 87), shortened inner ramus (fig. 81), a biarticulate outer ramus (fig. 82), or short, apically spinose, truncate rami (fig. 86). The rami of certain corophiids (fig. 90) are evenly strap-shaped; in *Aoroides* and *Gammaropsis* (fig. 90), the peduncle may be slightly elongate.

Elongation of peduncle, with equally long rami, characterize *Amphilochus* (fig. 88), the rami usually being very plain or poorly setose. Thickened rami characterize lysianassids (figs. 84, 91). *Photis* (fig. 96) has a strongly shortened inner ramus, while *Ampithoe* (fig. 95) has two pad-shaped rami, the inner setose, the outer bearing two large hooks. In contrast, the Ischyroceridae have two plain rami, the inner lacking setae, the outer bearing some kind of microscopic apical hook or denticles (figs. 89, 92, 97, 98).

Loss of the inner ramus characterizes hyalids (fig. 94), but a small ramus remains on *Parallorchestes* (fig. 93) and *Chelura* (fig. 104), the latter with an immense, flabellate, outer ramus. *Corophium* bears a setose, padlike ramus (fig. 102). *Cerapus* (fig. 105) may have a small inner ramus but usually does not, uropod 3 being tiny and bearing a reduced outer ramus with a hook. Uropod 3 of *Ericthonius* (fig. 103) looks like that of ischyrocerids but without the inner ramus. Stenothoids (fig. 99) have two articles in the ramus so that uropod 3 appears to have three articles in tandem.

Finally, several genera lack rami altogether, having only a thick and poorly setose peduncle (figs. 101, 106).

The basic telson (figs. 107, 127), two short, cleft lobes with irregular setation, is found on *Gammarus* and *Anisogammarus*. In the ordinary California amphipod (fig. 109), the telson is slightly more elongate than in the basic kind. The telsons (figs. 108, 110) of two species of *Elasmopus* and species of *Melita* may be recognized by the presence of spine groups. *Hyale* (fig. 111) has a small telson divided into triangular flaps that appear fleshy and tent-shaped from lateral aspect, whereas *Allorchestes* (fig. 114) has a laminar, compressed-lobe telson with a short slit. The odd *Eohaustorius* has the lobes completely divorced from each other (fig. 130). Pontogeneiids (fig. 109) can be recognized by the nakedness, or presence of only thin setae (not thick spines), of the telson. *Oligochinus* (fig. 115) has scarcely any apical cleft.

Uncleft (entire) telsons are found in a variety of calliopiids, pleustids, lysianassids, and *Synchelidium* (figs. 120–122). The telson is slightly elongate and tends to be pointed in *Calliopiella* (fig. 118) and becomes fully elongate in amphilochids (fig. 126) Stenothoids have an ovate telson usually with large dorsal spines (fig. 117).

The fleshy telson (figs. 123, 131) characterizes corophioideans; it is very thick, poorly articulate, entire, and often bears tiny apical knobs and a dorsoposterior, cup-shaped hollow. In *Podocerus* (fig. 129), the telson is very large and dorsally setiferous, whereas in *Ericthonius* (fig. 124), the telson becomes extremely short and bears hook-shaped denticles (observe under high power). In *Corophium* and eophliantids (fig. 116), the fleshy telson is reduced in size and thickness and becomes confounded with such non-fleshy telsons as in *Synchelidium* (fig. 121), although manipulation will demonstrate a puffiness in the *Corophium* telson. Finally, two species of lysianassids have a thick, uncleft telson (fig. 125), elongate and narrow.

SIMPLE KEY TO SOME COMMON FAMILIES OF GAMMARIDEAN AMPHIPODS

(Anamixidae, Najnidae, and Acanthonotozomatidae are not keyed here)

1. Telson thick, puffy, uncleft 2
– Telson thin, laminar, cleft or uncleft (if slightly puffy then telson must be distinctly cleft) 9
2. Uropod 3 seemingly absent, hidden beneath telson 3
– Uropod 3 large and visible 5
3. Body tubular, like a tanaid (cheliferan); bores in kelp .... EOPHLIANTIDAE (Key A)
– Body broad, flat, especially noted by flexion of abdomen beneath thorax 4
4. Gnathopod 1 and especially gnathopod 2 large and subchelate .... PODOCERIDAE (Key A)
– Gnathopods 1–2 small and simple like pereopods 3–4 .... PHLIANTIDAE (Key A)
5. Gnathopod 2 with article 3 elongate (fig. 174) .... LYSIANASSIDAE (in part) (Key J)
– Gnathopod 2 with article 3 normally short 6
6. Outer ramus or only ramus of uropod 3 with apical hook or 2 large, hooked spines 7
– Outer ramus of uropod 3 unhooked 8
7. Outer ramus of uropod 3 stout, with 2 hooked spines; inner ramus flat and apically setose .....AMPITHOIDAE (Key A)
– Outer ramus of uropod 3 slender, hooked, and denticulate

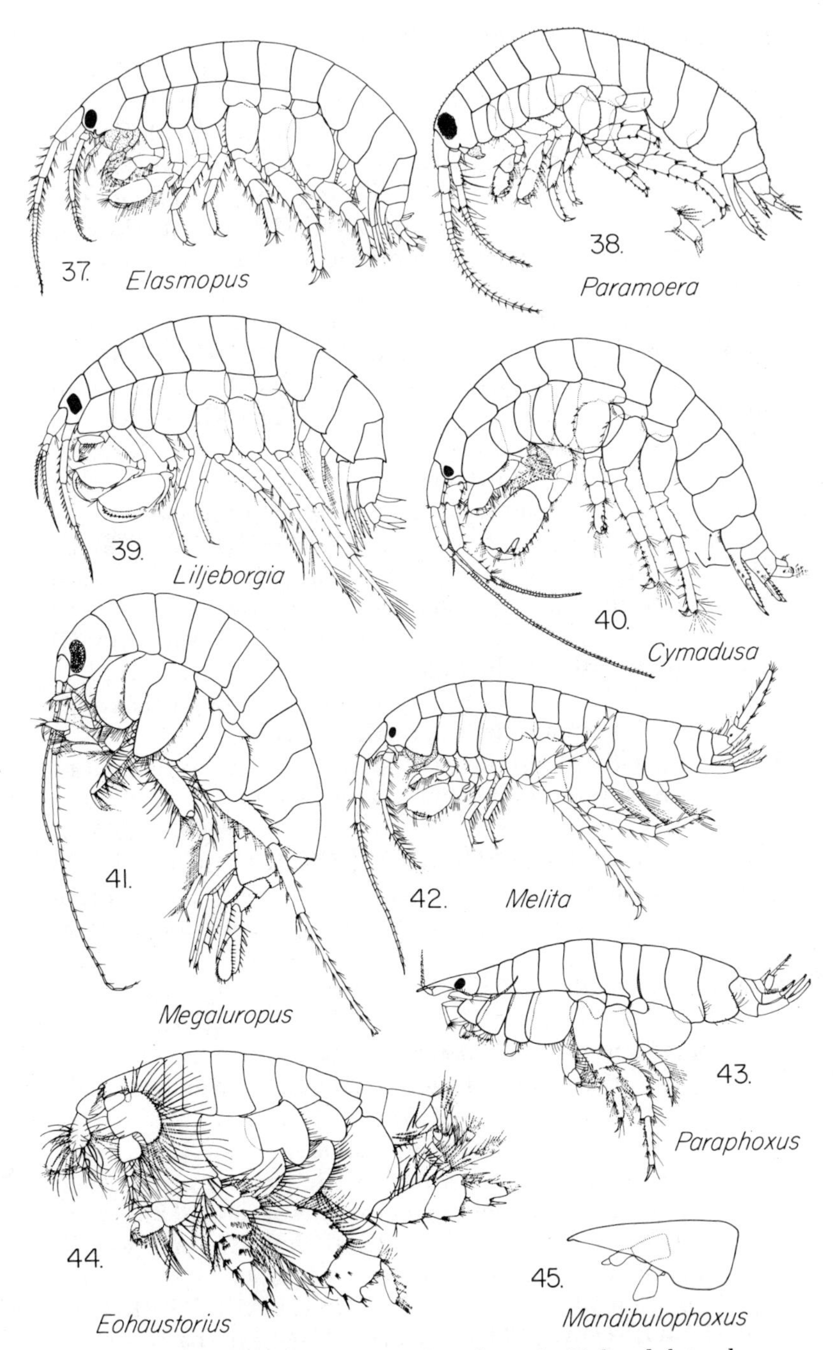

PLATE 73. **Gammaridea** (4). 37–44, bodies, lateral; **45**, head, lateral.

apically; inner ramus slender and naked ....ISCHYROCERIDAE (Key A)

8. Uropod 2 with thin peduncle and long rami (normal) ....COROPHIIDAE (includes AORIDAE, ISAEIDAE and PHOTIDAE) (Key A)
- Uropod 2 with flabellate peduncle, rami tiny, subquadrate; bores in wood ....CHELURIDAE (Key A)
9. Coxa 1 less than 3/8 as large in area as coxa 2; coxae 3–4 enlarged 10
- Coxa 1 at least 1/2 as large in area as coxa 2 11
10. Uropod 3 with 2 rami ........AMPHILOCHIDAE (Key I)
- Uropod 3 with one ramus (uropod 3 composed of 3 articles in tandem) ......................STENOTHOIDAE (Key G)
11. Urosomites 2–3 fused together 12
- Urosomites 2–3 freely articulate 13
12. Head short; eyes 2 and multifaceted, pigment in facets .... DEXAMINIDAE (Key E)
- Head elongate; eyes with pigment separated into blobs inside head, facets forming 4 external corneal lenses (as in spiders) ..............................AMPELISCIDAE (Key E)
13. Gnathopod 2 with article 3 elongate (fig. 174) ....LYSIANASSIDAE (in part) (Key J)
- Gnathopod 2 with article 3 normally short 14
14. Mandible lacking palp; uropod 3 with inner ramus absent or vestigial, remaining ramus not longer than peduncle ....**Talitroidea** (TALITRIDAE, HYALIDAE, DOGIELINOTIDAE) (Keys H, O)
- Mandible with palp; uropod 3 with visible inner ramus, often shortened but if so shortened then outer ramus longer than peduncle 15
15. Gnathopod 1 with special carpochelate form (fig. 157) .... LEUCOTHOIDAE (Key F)
- Gnathopod 1 not of special carpochelate form 16
16. Telson uncleft; accessory flagellum always 1-articulate or absent 17
- Telson cleft; accessory flagellum variable 18
17. Eyes coalesced into dorsocentral faceted mass; peduncle of uropod 3 elongate (as long as outer ramus of uropod 2); pereopod 7 very elongate ..........OEDICEROTIDAE (Key K)
- Eyes separated laterally; peduncle of uropod 3 and pereopod 7 not elongate ....PLEUSTIDAE, EUSIRIDAE (=CALLIOPIIDAE, PONTOGENEIIDAE) (in part) (Key K)
18. Gnathopods 1–2 both large and similar to each other (fig. 183) .... LILJEBORGIIDAE (Key M)

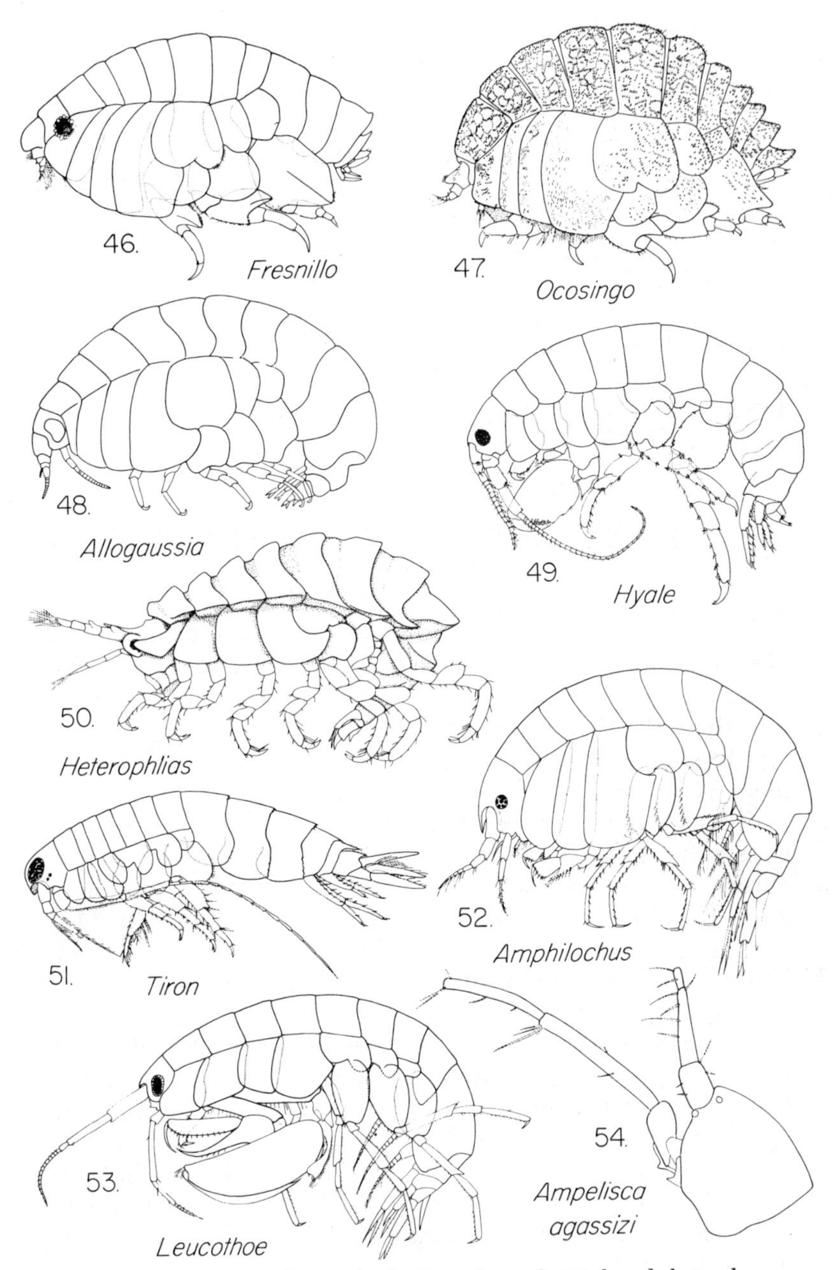

PLATE 74. **Gammaridea** (5). 46–53, bodies, lateral; 54, head, lateral.

– Gnathopods 1–2 either small or of strongly differing sizes 19
19. Eyes coalesced into dorsocentral faceted mass; head galeate (helmet-shaped) (fig. 68) ....SYNOPIIDAE (Key M)
– Eyes separated laterally; head not galeate 20
20. Accessory flagellum 1-articulate or absent .... EUSIRIDAE (in part) (Key K)
– Accessory flagellum 2+ articulate 21
21. Pereopod 7 shorter and of different form than pereopod 6 (figs. 202, 203) .... PHOXOCEPHALIDAE (Key L)
– Pereopod 7 as long as and similar in form to pereopod 6 ... 22
22. Lobes of telson connected to each other .... GAMMARIDAE (Key N)
– Lobes of telson disjunct .... HAUSTORIIDAE (Key L)

BASIC KEY TO GAMMARIDEA

(except beach-hoppers, Talitridae)

1. **Telson fleshy, puffy, uncleft; when manipulated moving as one piece, not easily articulate at base ....Key A (p. 330)**
– Telson with certain parts laminar, often with ventral keel (if slightly puffy, then telson must be distinctly cleft); when manipulated, generally freely articulate and easily moved in one piece or half at a time 2
2. Urosomites 2–3 coalesced (fig. 34) .... Key E (p. 340)
– Urosomites all free 3
3. Gnathopod 1 of both sexes of special leucothoid carpochelate form (fig. 157) ....Key F (p. 342)
– Gnathopod 1 not of leucothoid form 4
4. Uropod 3 of stenothoid form (fig. 99); peduncle with only 1 ramus, latter 2-articulate .... Key G (p. 342)
– Uropod 3 not of stenothoid form, generally with 2 rami, but if uniramous then ramus only 1-articulate 5
5. Uropod 3 with 1 ramus or rarely with extremely minute, inner ramus; mandible lacking palp; accessory flagellum absent . ...Key H (p. 343)
– Uropod 3 with 2 rami; if inner ramus shortened, confirm necessary presence of a mandibular palp and accessory flagellum (often minute) 6
6. Coxa 1 small and mostly concealed by large coxa 2 (count forward from largest coxa 4 to confirm) (figs. 210, 211) ....Key I (p. 344)
– Coxa 1 visible, with surface area at least ²/₃ that of coxa 2 7

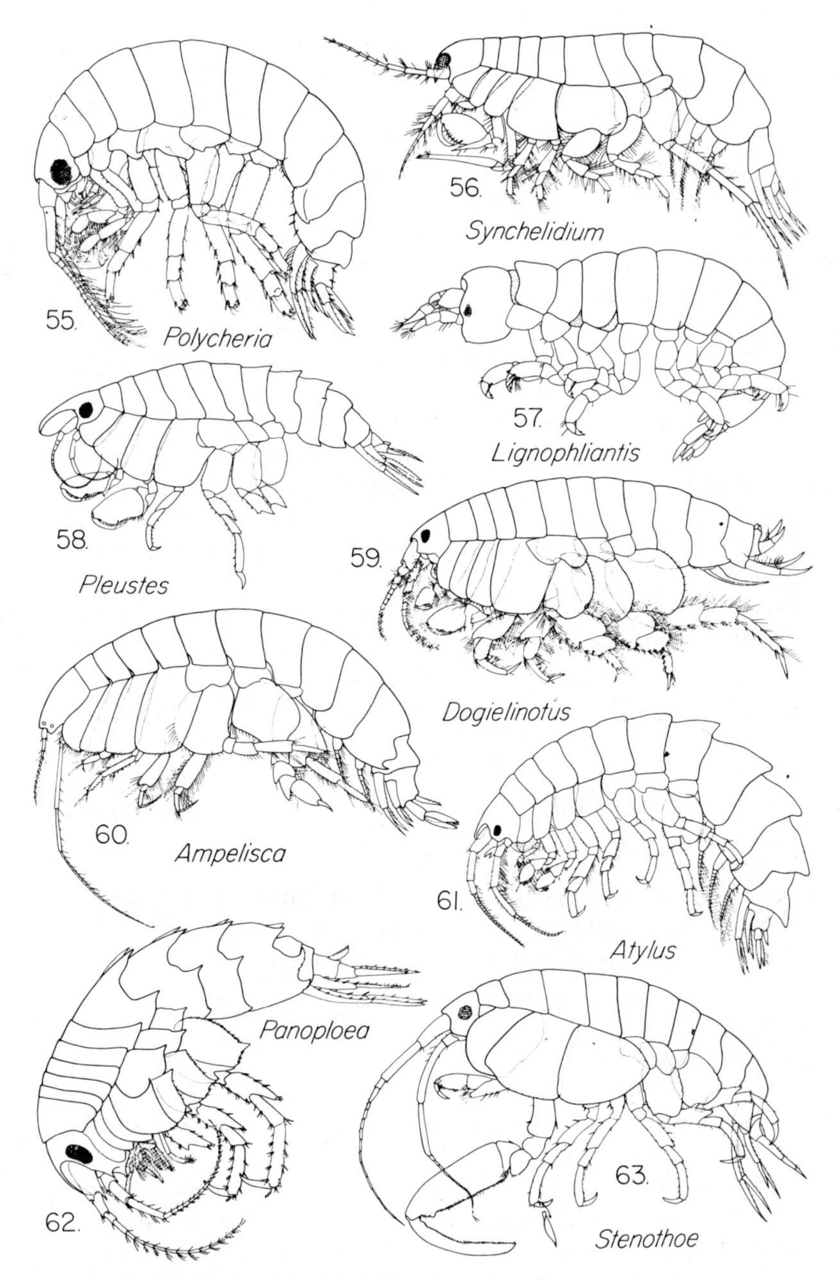

PLATE 75. **Gammaridea** (6). 55–63, bodies, lateral.

7. Gnathopod 2 of lysianassid mitten shape (fig. 174), note article 3 elongate; antenna 1 with large accessory flagellum; body usually very white, compact, shiny, slippery, hard ... . Key J (p. 344)
– Gnathopod 2 of ordinary subchelate form (fig. 1) 8
8. Accessory flagellum absent or with only 1 tiny article .... Key K (p. 345)
– Accessory flagellum with 2 or more articles 9
9. Telson formed of 2 setose lobes attached to pleonite 6, but fully separate at base by wide margin of pleonite (fig. 130) .... *Eohaustorius*, Key L (p. 348)
– Telson, if cleft, with lobes attached to each other basally; if entire, formed of one solid piece 10
10. Head galeate (helmet-shaped) (fig. 68); with 3 eyes, one main eye dorsally and one accessory eye on each side near base of antenna 2; accessory eyes with few ommatidia .... *Tiron*, Key M (p. 349)
– Head not galeate; rostrum essentially absent or visorlike; 2 eyes 11
11. Head with visorlike rostrum (fig. 23) .... Key L (p. 348)
– Head with vestigial rostrum (fig. 2g) 12
12. Dactyls of gnathopods 1–2 with large serrations on inner margin (fig. 183); **or** mandibular molar vestigial, formed as a small setose hump (fig. 9) .... Key M (p. 349)
– Dactyls of gnathopods 1–2 with small or no serrations on inner margin; **and** mandibular molar large, with rasping surface .... Key N (p. 349)

KEY A: (TELSON FLESHY) COROPHIIDAE, AMPITHOIDAE, ISCHYROCERIDAE, PODOCERIDAE, CHELURIDAE, PHLIANTIDAE, EOPHLIANTIDAE

1. Urosomite 1 elongate (fig. 28) or more than twice as long as urosomite 2 (urosomite 3 occasionally absent, therefore count carefully commencing with pleonites 1–3, then urosomite 1) 2
– Urosomite 1 not elongate 5
2. (*Note 3 choices*) Head with large, spatulate rostrum (observe dorsally) (fig. 72); mandible lacking palp; coxae 1–4 each as large in surface area as surface of pereonite 1 (observe dorsally); gnathopods 1–2 simple and thin (body, fig. 50) .... *Heterophlias seclusus*
– Head lacking rostrum; mandible lacking palp; coxae 1–4 very small; gnathopods 1–2 simple and thin (fig. 135); body cylin-

drical (fig. 57), about 1.5 mm long; boring in rhizomes of kelp .... *Lignophliantis pyrifera*

– Head lacking rostrum (or very small); mandible with palp; coxae 1–4 very small; gnathopods 1–2 subchelate, gnathopod 2 stout (figs. 173, 187; body, *Podocerus*, fig. 66) 3

3\. Pereonites 6–7 and pleonites 1–2 with sharp, dorsal keel-teeth .... *Podocerus cristatus*

– Pereonites 6–7 and pleonites 1–2 with low dorsal humps .. 4

4\. Article 4 of male gnathopod 2 not projecting distally along hand towards apex of dactyl (fig. 177; female, 187) .... *Podocerus brasiliensis*

– Article 4 of male gnathopod 2 projecting distally along hand toward dactyl (fig. 173) .... *Podocerus spongicolus*

5\. Telson fully divided into 2 bluntly triangular, somewhat fleshy lobes (fig. 111)............ *Hyale*, see Key H (p. 343)

– Telson uncleft 6

6\. Pleonite 3 with immense dorsal tooth projecting posteriorly; urosomites 1–3 together forming long, flat, wide box as long as pereonites 1–5 together; uropod 2 of odd form (fig. 69); uropod 3 with 1 ramus as long as pereonites 1–5 together; boring in wood .... *Chelura terebrans*

– Pleonite 3 lacking dorsal tooth; urosomites 1–3 short, often freely articulate, occasionally fused together, forming thin box or plate; uropod 2 normal; uropod 3 with ramus or rami not highly elongate 7

7\. Outer ramus of uropod 3 with 2 large hooked spines reverting laterally (dorsal view); inner ramus present, as long as outer, apically setose (fig. 95)..... AMPITHOIDAE Key B (p. 334)

– Outer ramus (or only ramus) of uropod 3 with only 1 hooked spine or none, tiny denticles or none; inner ramus present or absent, never as blunt and fleshy as in *Ampithoe* of figure 95 . 8

8\. Uropod 3 with 2 rami 9

– Uropod 3 with 1 ramus 20

9\. Uropod 3 elongate; peduncle as long as rami of uropods 1–2, but rami of uropod 3 only 3/4 or less as long as peduncle of uropod 3; outer ramus with apical hook or apicolateral denticles (high-power microscopy); inner ramus lacking setae (except minutely) (figs. 89, 92) .... ISCHYROCERIDAE Key C (p. 335)

– Uropod 3 with either ramus as long as peduncle or longer . 10

10\. Gnathopod 1 slightly to greatly larger than gnathopod 2 (longer or stouter from lateral view) 11

– Gnathopod 1 smaller than gnathopod 2 13

11\. Accessory flagellum absent; male gnathopod 1 immensely

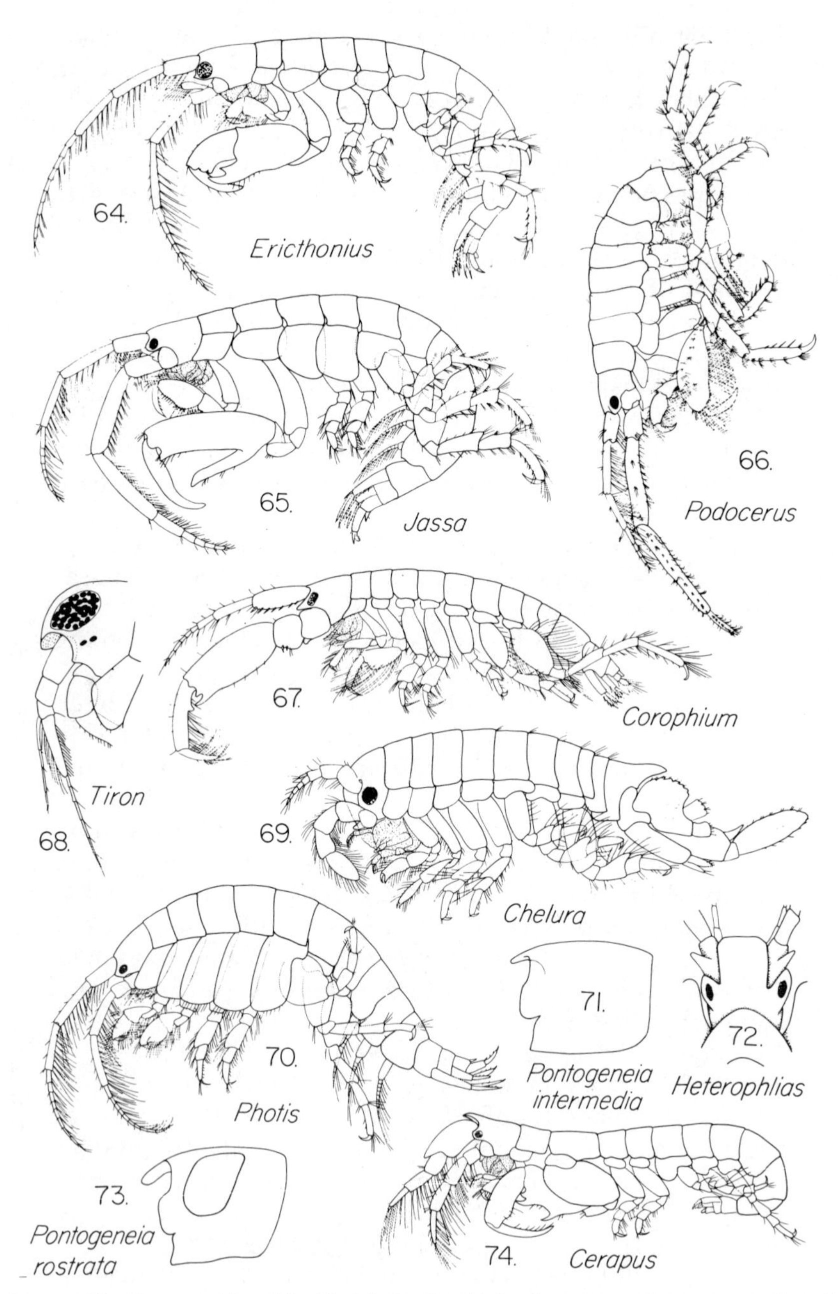

PLATE 76. **Gammaridea** (7). 64–67, 69, 70, 74, bodies, lateral; 68, 71, 73, heads, lateral; 72, head, dorsal.

merochelate (fig. 164) ....*Aoroides columbiae*
– Accessory flagellum present 12
12. Male gnathopod 1 large and regularly subchelate (fig. 163) ....*Lembos* spp.
– Male gnathopod 1 large and carpochelate (fig. 166) ....*Microdeutopus schmitti*
13. Inner ramus of uropod 3 less than 4/5 as long as outer ramus (fig. 96; body, *Photis,* fig. 70) 14
– Inner ramus of uropod 3 more than 4/5 as long as outer 17
14. Palm of male gnathopod 2 with bifurcate (cryptically) disjunct process (fig. 184) ....*Photis bifurcata*
– Palm of male gnathopod 2 lacking special process 15
15. Inner margin of dactyl on male gnathopod 2 with large hump (fig. 193) ....*Photis brevipes*
– Inner margin of dactyl on male gnathopod 2 evenly curved 16
16. Palm of male gnathopod 2 with large, shallow, hemispherical excavation, dactyl scarcely overlapping palm (fig. 194); coxa 3 about 1.6 times as broad as coxa 4 ....*Photis californica*
– Palm of male gnathopod 2 with narrow, slitlike, deep excavation, dactyl greatly overlapping palm (fig. 195); coxa 3 about 1.2 times as broad as coxa 4 ....*Photis conchicola*
17. Coxa 2 of male with large posterior projection (fig. 219) .... *Gammaropsis mamolus*
– Coxa 2 of male straight posteriorly 18
18. Hand of male gnathopod 2 tapering, thus simple (fig. 201) .... *Megamphopus effrenus*
– Hand of male gnathopod 2 strongly subchelate 19
19. Urosomites 1–2 each with pair of dorsal setae and cusps (fig. 26); epimera 2–3 each with faint lateral ridge (fig. 30) ....*Gammaropsis thompsoni*
– Urosomites 1–2 dorsally smooth; epimera 2–3 laterally smooth ....*Megamphopus martesia*
20. Ramus of uropod 3 with many long setae (fig. 102); gnathopod 2 of corophiid filtering form (fig. 198) ....*Corophium,* Key D (p. 338)
– Ramus of uropod 3 naked or with 1–2 tiny spines (or peduncle with short setae); gnathopod 2 either ordinary or carpochelate 21
21. Uropod 2 with 1 ramus (body, fig. 74; telson, fig. 128) ....*Cerapus tubularis*
– Uropod 2 with 2 rami (body, *Ericthonius,* fig. 64) 22
22. Gnathopod 1 of male enlarged and carpochelate; apical end of article 5 with 1 enlarged tooth and 2 smaller teeth (fig. 237) ....*Grandidierella japonica*
– Gnathopod 2 of male enlarged; male gnathopod 2 immensely

carpochelate (telson, fig. 124) .... *Ericthonius* 23

23. Article 5 of male gnathopod 2 with apical tooth bifid (fig. 169) .... *Ericthonius brasiliensis*
– Apical tooth simple (fig. 170) .... *Ericthonius hunteri*

KEY B: AMPITHOIDAE

(Body, fig. 40)

(Extirpate lower lip carefully, destroying maxilla 1 if necessary; mount on glycerine slide with cover slip elevated by sand grains.)

1. Pleonal epimeron 3 with small point at posteroventral corner 2
– Pleonal epimeron 3 with rounded posteroventral corner 5
2. Accessory flagellum present; uropod 1 with ventral tooth on peduncle between rami reaching nearly halfway along rami (fig. 27) .... *Cymadusa uncinata*
– Accessory flagellum absent; uropod 1 with interramal tooth vestigial or absent 3
3. Outer lobes of lower lip with gape between sublobes (fig. 15) 4
– Outer lobes of lower lip with sublobes compressed together (fig. 16) .... *Ampithoe valida*
4. Adult male gnathopod 2 with transverse palm (fig. 191) .... *Ampithoe lacertosa*
– Adult male gnathopod 2 with oblique and concave palm (fig. 179) .... *Ampithoe simulans*
5. Hand of gnathopod 1 thin, rectangular, palm transverse (fig. 153) 6
– Hand of gnathopod 1 thick, ovate, palm oblique (fig. 150) .. 7
6. Flagellum of antenna 2 with 5–7 articles; antenna 2 stout; gnathopod 2 of both sexes much stouter than and unlike gnathopod 1 .... *Ampithoe lindbergi* *
– Flagellum of antenna 2 with 15–25 articles; antenna 2 thin; gnathopod 2 of both sexes thin and like gnathopod 1 (fig. 153) .... *Ampithoe humeralis* *
7. Hand of adult male gnathopod 2 ordinary, palm either subtransverse, *or* oblique and weakly sinuous 8
– **Hand of adult male gnathopod 2 develops, by stages, a large** cleft or deep hollow perpendicular to palm (figs. 179, 192) . 10

* A third species, *Ampithoe* sp. (cf. *A. mea* Barnard, 1965) has a flagellum intermediate between that of *A. lindbergi* and *A. humeralis;* its antenna 2 has about 11 articles and female gnathopod 2 is like gnathopod 1, but male gnathopod 2 is enlarged.

8. Apex of telson with 2 enlarged and folded "rabbit ears" (fig. 119) .... *Ampithoe aptos*
– Apex of telson with 2 minute, lateral knobs (fig. 123) 9
9. Outer lobes of lower lip with gape between sublobes (fig. 15); article 5 of peduncle on antenna 2 of adults heavily setose (fig. 223) .... *Ampithoe plumulosa*
– Outer lobes of lower lip with sublobes compressed together (fig. 16), article 5 of peduncle on antenna 2 of adults scarcely setose .... *Ampithoe valida*
10. Outer lobes of lower lip with sublobes compressed (fig. 16); male gnathopod 2 with large thumb meeting dactyl (fig. 192) .... *Ampithoe pollex*
– Outer lobes of lower lip with gape between sublobes (fig. 15); male gnathopod 2 lacking extended thumb (fig. 179) .... *Ampithoe simulans*

KEY C: ISCHYROCERIDAE

(Study uropod 3 under high magnification, using oil immersion if possible.)

1. Peduncle of uropod 1 with lateral row of plumose setae (fig. 138) .... *Parajassa angularis*
– Peduncle of uropod 1 bearing only short, stout spines 2
2. Outer ramus of uropod 3 with 2–4 large, outer denticles proximal to apical spine hook (figs. 97, 98) 3
– Outer ramus of uropod 3 with 4 or more medium to tiny outer denticles proximal to apical spine (figs. 89, 92) 4
3. Outer ramus of uropod 3 with granules or serrations proximal to main denticles .... *Ischyrocerus anguipes*
– Outer ramus of uropod 3 smooth proximally (body, fig, 65) .. .. *Jassa falcata*
4. Coxa 1 about 2/3 as long as coxa 2 ....*Microjassa litotes*
– Coxa 1 as long as coxa 2 (reaching tangent of curve on coxa 2) .. 5
5. Outer ramus of uropod 3 with 2 parallel rows of main denticles; eyes small .... *Ischyrocerus* sp. A
– Outer ramus of uropod 3 with single row of main denticles; eyes large 6
6. Large denticles on outer ramus of uropod 3 (fig. 97) .... *Ischyrocerus anguipes*
– Tiny denticles on outer ramus of uropod 3 (fig. 92) .... *Ischyrocerus* sp. B.

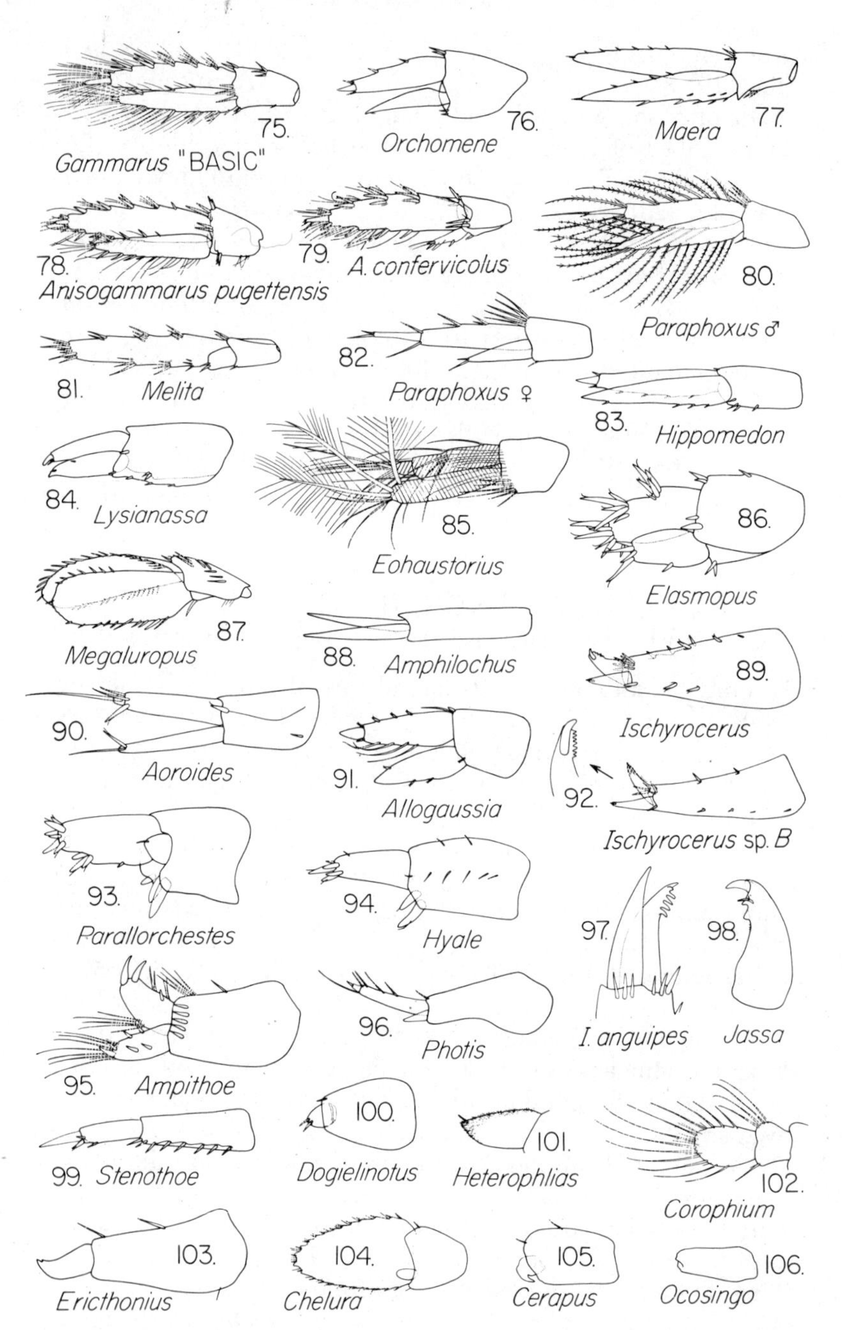

PLATE 77. **Gammaridea** (8). 75–106, uropod 3, left outer ramus toward top of figure, except 97 (outer ramus to right) and 98 (outer ramus only).

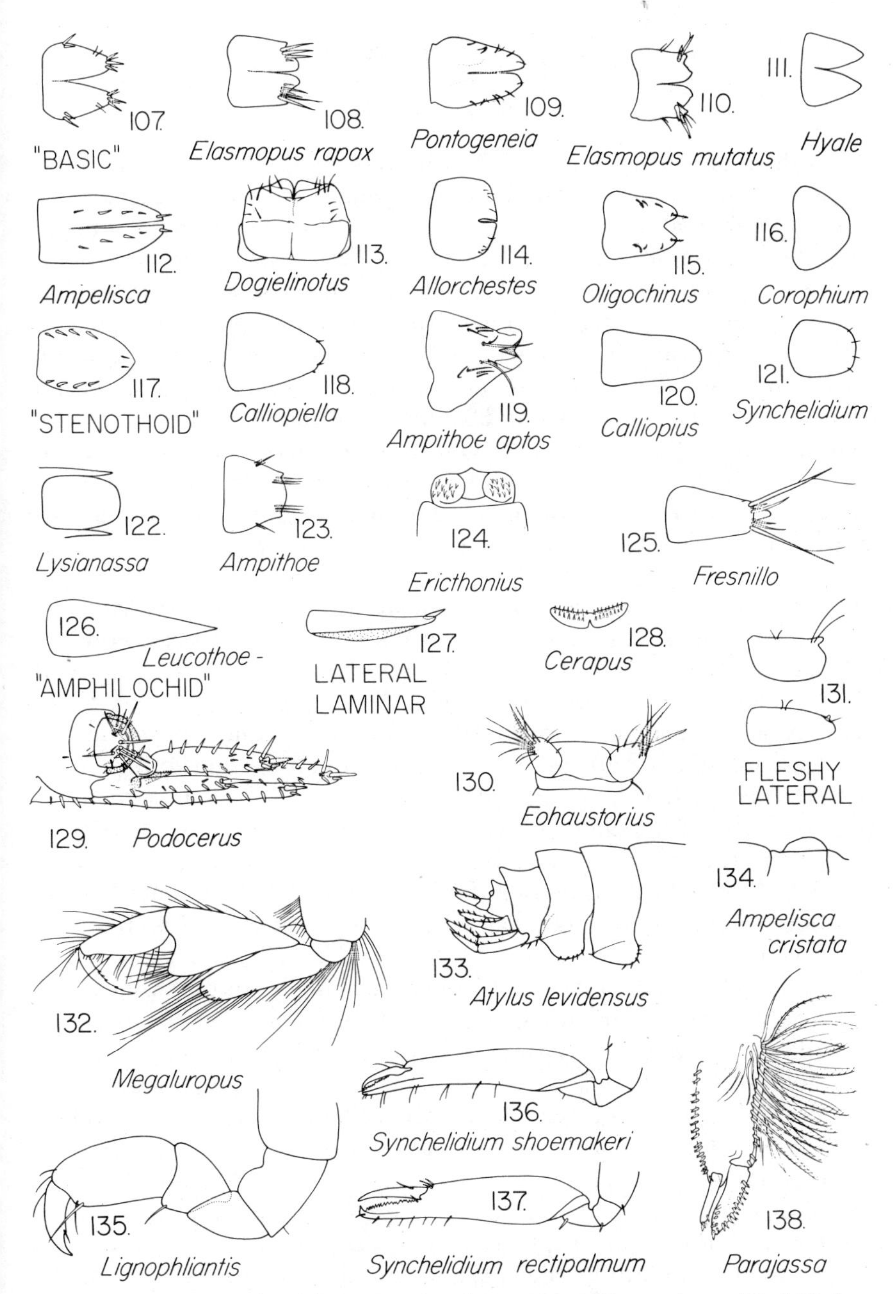

PLATE 78. **Gammaridea** (9). 107–131, telson; 132, 135, gnathopod 1; 133, lateral pleon; 134, pleonite 4, lateral; 136, 137, gnathopod 2; 138, uropod 1.

KEY D: COROPHIIDAE: *Corophium*

(Body, fig. 67)

Identification: because article 2 is hidden medially, antenna 2 often appears to have only four peduncular articles; the two longest articles are numbers 4 and 5. The two sexes of various species may be mixed together in single samples, one or the other sex being absent but paired with the opposite sex of another species. Great care is therefore required in identifying species of this genus. Males have a heavily thickened antenna 2 with one or more fixed teeth on the apex of article 4, whereas females have stout spines mixed among the setae on the ventral margins of articles 4 and 5 (antenna 2 of females must be rolled outward to see these spines). Several species have females with male-like antennae. Males of several species are indistinguishable; one must resort to females for final identification, but uncertainties are prevalent if more than one species occurs in a sample.

1. Urosomites all coalesced, dorsal surface of urosome smooth and free of articulation lines (test with needle scraped several ways across urosome) (figs. 144, 146) 2
– Urosomites freely articulate (fig. 148) 8
2. Uropods 1–2 attached ventrally on sides of urosome; posterolateral margins of urosome lacking invaginations and upturned slightly as sharp rim (fig. 146) ....*C. baconi*
– Uropods 1–2 attached at invaginations on sides of urosome, rim absent (fig. 144) 3
3. Article 4 of male antenna 2 (female like male) with ventral articulate spines (spines break when bent, setae flex when bent) (fig. 140) 4
– Article 4 of male antenna 2 lacking articulate ventral spines 5
4. Ventral spines on article 4 of male antenna 2 set as single spines in tandem; length of article 5 as great as breadth of article 4 .... *C. californianum*
– Ventral spines on article 4 of male antenna 2 set as pairs and triads in tandem; length of article 5 twice breadth of article 4 (fig. 140) .... *C. oaklandense*
5. Males 6
– Females (head, fig. 143) 7
6. Male rostrum minute (fig. 147), its length less than 2/3 diameter of eye ....*C. acherusicum*
– Male rostrum long and thin (fig. 145), its length greater than diameter of eye; males of these species inseparable, see females of couplet 7 .... *C. insidiosum*, *C. uenoi*

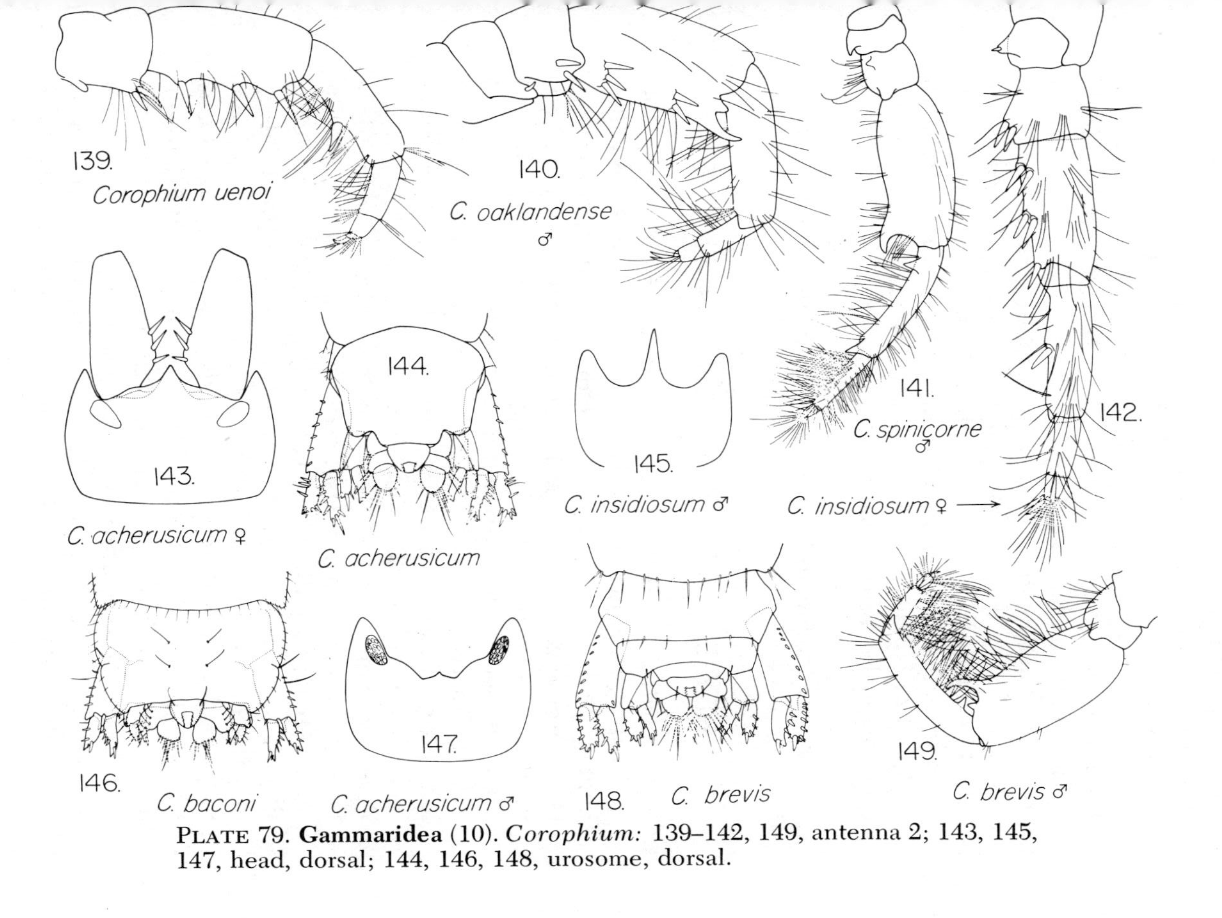

PLATE 79. **Gammaridea** (10). *Corophium:* 139–142, 149, antenna 2; 143, 145, 147, head, dorsal; 144, 146, 148, urosome, dorsal.

7. Ventral margin of article 4 on female antenna 2 with 2–3 pairs of spines in tandem, and a 4th single, apical spine (fig. 142); females of these species inseparable but see males of couplet 6 .... *C. acherusicum, C. insidiosum*
– Ventral margin of article 4 on female antenna 2 with 2–3 single spines set in tandem (fig. 139) .... *C. uenoi*
8. Article 4 of male antenna 2 with main tooth forming open halfmoon (fig. 141); females like males .... *C. spinicorne*
– Article 4 of male antenna 2 with main tooth plus accessory tooth in halfmoon (fig. 149); females with normal antenna 2 9
9. Article 4 of female antenna 2 with 2 single, ventral spines in tendem; male rostrum nearly as long as ocular lobe (dorsal view) .... *C. stimpsoni*
– Article 4 of female antenna 2 with 3 pairs of ventral spines in tandem; male rostrum minute .... *C. brevis*

KEY E: AMPELISCIDAE, DEXAMINIDAE

1. Eyes usually with red pigment arranged in 5 (appearing 3–4) internal cephalic masses; anterolateral margins of head with 4 faint corneal lenses, 2 on each side of head, appearing as contact lenses (fig. 54); internal pigment occasionally bleached; head as long as pereonites 1–3 together; pereopod 7 distinct from pereopod 6 (figs. 204, 205); body, *Ampelisca*, fig. 60 4
– Eyes ordinary, formed of compacted ommatidia in right and left masses; head shorter than pereonites 1–2 together; pereopod 7 like pereopod 6 2
2. Pereopods simple; mandible with palp (body, *Atylus*, fig. 61) . 3
– Pereopods cheliform (fig. 213); mandible lacking palp; inhabiting compound ascidians (body, fig. 55) ...*Polycheria osborni*
3. Pleonite 3 rounded dorsally; urosomite 1 with tooth and deep incision anterior to tooth (fig. 32); pereopod 7 with large, ventrally produced, posterodistal cusp (fig. 216) ..*Atylus tridens*
– Pleonite 3 with subacute dorsal tooth or squared-off extension; urosomite 1 with dorsal extension but no incision (fig. 133); pereopod 7 lacking gross ventroposterior tooth (minute cusp apparent) .... *Atylus levidensus*
4. Article 3 of pereopod 7 much longer than article 4 (fig. 205) .... *Ampelisca milleri*
– Article 3 of pereopod 7 shorter than article 4 (including posterior lobe of article 4) 5
5. Dorsal carina on urosomite 1 very thin, lamellar, dorsally rounded (fig. 134) .... *Ampelisca cristata*
– Dorsal carina on urosomite 1 thick, nonlamellar 6

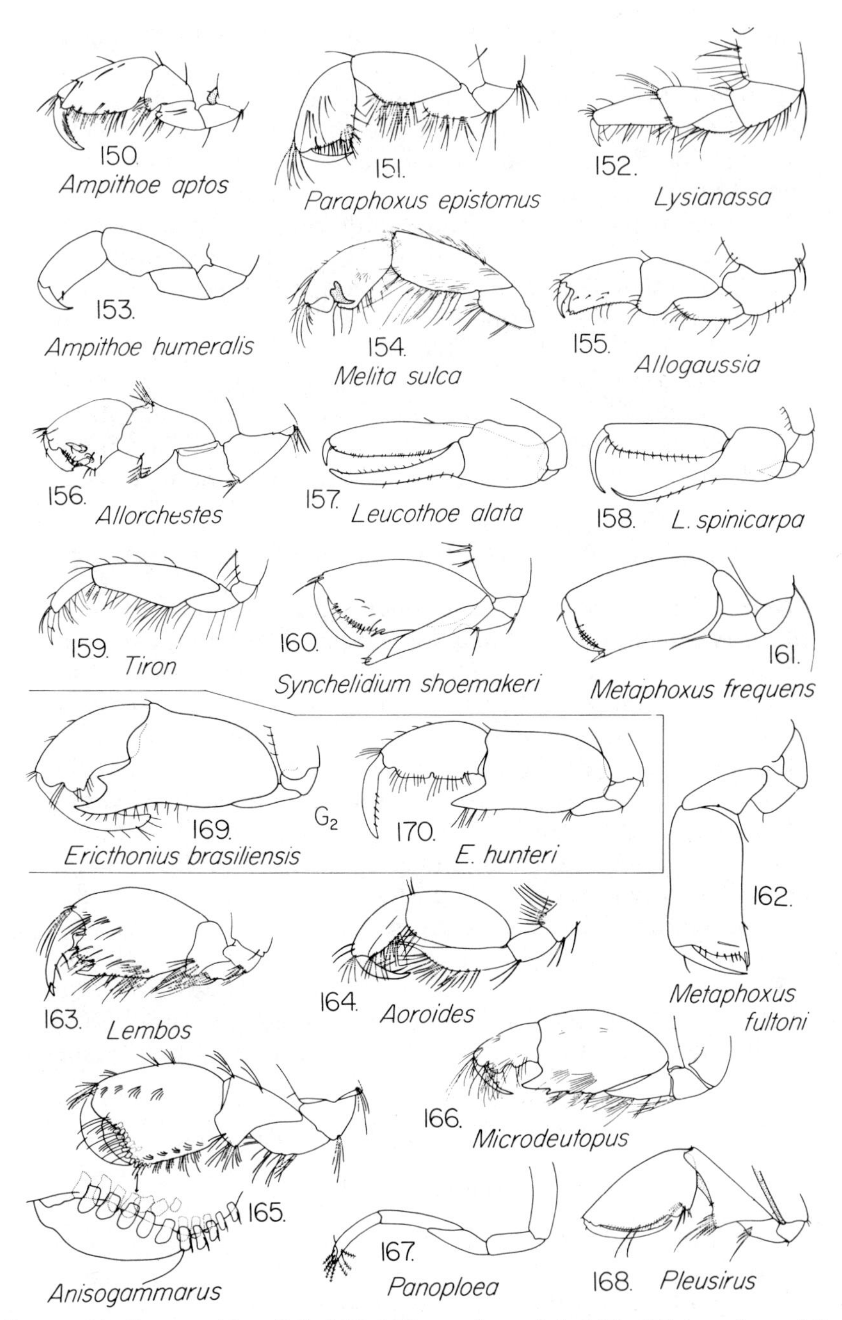

PLATE 80. **Gammaridea** (11). 150–168, gnathopod 1; 169, 170, gnathopod 2.

6. Epimeron 3 with distinct, sharp, medium to large tooth at posteroventral corner 7
– Epimeron 3 quadrate or slightly rounded posteroventrally .. 8
7. Dorsal carina on urosomite 1 saddle-shaped (fig. 34); article 4 of pereopod 7 with posterior lobe covering most of article 5 (fig. 206) .... *Ampelisca pugetica*
– Dorsal carina on urosomite 1 not saddle-shaped (fig. 35); article 4 of pereopod 7 without conspicuous lobe covering posterior margin of article 5 .... *Ampelisca macrocephala*
8. Base of peduncle on uropod 1 not stouter (lateral view) than peduncle of uropod 2, (fig. 31); article 2 of pereopod 7 with special posterior bevel (fig. 207) .... *Ampelisca lobata*
– Base of peduncle on uropod 1 much stouter (1.5 times) than peduncle of uropod 2 (fig. 36); article 2 of pereopod 7 ordinary (fig. 206) .... *Ampelisca agassizi*

KEY F: ANAMIXIDAE, LEUCOTHOIDAE

1. Dactyl of gnathopod 2 able to touch long process of article 5 (fig. 196); mandibles and maxillae replaced by ventral keel on head .... *Anamixis linsleyi*
– Dactyl of gnathopod 2 barely unable to touch long process of article 5 (fig. 185); mandibles and maxillae present and large 2
2. Coxa 1 small, mostly hidden by coxa 2 (fig. 211); palm of gnathopod 2 subtransverse, strongly distinct from posterior margin of hand (figs. 186, 197). .... *Leucothoides pacifica*
– Coxa 1 large, visible; palm of gnathopod 2 very oblique, long, curved, poorly defined from posterior margin of hand (body, *Leucothoe,* fig. 53) 3
3. Dactyl of gnathopod 1 shorter than article 3 of gnathopod 1 (fig. 157) .... *Leucothoe alata*
– Dactyl of gnathopod 1 about 1.5 times as long as article 3 of gnathopod 1 (fig. 158) .... *Leucothoe spinicarpa*

KEY G: STENOTHOIDAE

(Body, fig. 63)

1. Article 2 of pereopod 6 linear, thin; article 2 of pereopod 7 expanded (figs. 221, 222) 2
– Article 2 of pereopods 6 and 7 expanded 4
2. Mandibular palp 1-articulate or absent *Stenothoides burbanki*
– Mandibular palp 2–3 articulate 3
3. Article 5 of gnathopod 1 twice as long as article 6; mandibular

palp large and 3-articulate (fig. 7) ...... *Mesometopa esmarki*
- Article 5 of gnathopod 1 only as long as article 6; mandibular palp minute and 2-articulate (fig. 10) ... *Mesometopa sinuata*

4. Palp of maxilla 1 uniarticulate * 5
   Palp of maxilla 1 biarticulate * .... *Stenothoe valida*
5. Mandibular palp 1-articulate; telson lacking spines; dactyls of pereopods 3–4 with numerous denticles (fig. 220) .... *Stenula incola*
- Mandibular palp 2-articulate (fig. 10); telson with 4 large spines; dactyls of pereopods 3–4 smooth .... *Metopa cistella*

KEY H: HYALIDAE, DOGIELINOTIDAE AND NAJNIDAE

(For obvious beach-hoppers from sandy shores, see Key O, TALITRIDAE)

1. Semiterrestrial, on sandy beaches or in marshes; telson entire or only poorly cleft, and bearing more than 2 stiff spines ... .TALITRIDAE, Key O (p. 352)
- Aquatic; telson cleft or entire, but stiff spines 2 or fewer 2

2. Pereopods 3–4 bearing setae as long as article 5; prebuccal region with pendant epistome (fig. 3; body, fig. 59; uropod 3, fig. 100) .... *Dogielinotus loquax*
- Pereopods 3–4 bearing setae shorter than article 7 3

3. Mandibular molar composed of weak hump bearing spine (fig. 8) .... *Najna* sp.
- Mandibular molar large, with rasping surface (fig. 11) 4

4. Apex of uropod 3 with weak, thin, flexible setae (and 1 spine); palp of maxilla 1 minute (in fig. 19 arrow points to lobe, palp invisible); telson rectangular (dorsal view), cleft only halfway, lobes appressed (fig. 114; gnathopod 1, male, fig. 156) .... *Allorchestes angusta*
- Apex of uropod 3 with thick, inflexible spines; palp of maxilla 1 reaching base of spines on outer plate (fig. 20); telson deeply cleft, composed of 2 bluntly triangular lobes (fig. 111) 5

5. Uropod 3 with minute inner ramus (fig. 93) ....*Parallorchestes ochotensis*
- Uropod 3 lacking inner ramus (fig. 94; body, *Hyale*, fig. 49) 6

6. Ramus of uropod 3 with 1 apical spine widely disjunct from other and placed on dorsal margin ....*Hyale grandicornis californica*
- All spines on ramus of uropod 3 fully apical, occasionally 1 spine pointing dorsally 7

* This observation is very difficult; make slide carefully and observe under high power.

7. Coxae 1–4 with sharp and enlarged posterior cusp (fig. 218) .... *Hyale anceps*
– Coxae 1–4 with obsolescent posterior cusp or acclivity (fig. 215) 8
8. Setae on article 5 and flagellum of antenna 2 (adults) minute and insignificant .... *Hyale frequens*
– Setae on article 5 and flagellum of antenna 2 (adults) in dense clusters longer than width of antennal segments .... *Hyale plumulosa*

KEY I: AMPHILOCHIDAE

(Body, fig. 52)

1. Projecting lobe of article 5 on gnathopod 2 reaching about halfway or less along posterior margin of hand (fig. 188); mandibular molar a tiny setose hump ..... *Amphilochus litoralis*
– Projecting lobe of article 5 on gnathopod 2 reaching fully along posterior margin of hand (fig. 189) 2
2. Article 4 of gnathopod 2 with 2 marginal spines and 1 terminal (fig. 189); middle ommatidia of eyes visible; mandibular molar a tiny setose hump ....*Amphilochus neapolitanus*
– Article 4 of gnathopod 2 with only 1 terminal spine, (fig. 190); middle ommatidia of eyes concealed by dense core of pigment; mandibular molar a large box with several serrations—like head of electric hair clippers ....*Gitanopsis vilordes*

KEY J: LYSIANASSIDAE

1. Uropod 3 composed only of peduncle (fig. 106); inner rami of uropods 1–2 much shorter than outer rami (or inner rami vestigial); body and coxae covered with fuzz and scales 2
– Uropod 3 with 2 rami; inner rami of uropods 1–2 nearly as long as outer rami; body and coxae glass-smooth or pitted 3
2. Four posterior body segments forming erect dorsal peaks (body, fig. 47) .... *Ocosingo borlus*
– Two posterior body segments forming erect dorsal peaks (body, fig. 46) .... *Fresnillo fimbriatus*
3. Telson entire 4
– Telson cleft 5
4. Gnathopod 1 simple (fig. 152); lower lip projecting as flat plate forward from mouthparts (figs. 13, 14); telson with apical setae or naked .... *Lysianassa* spp.
– Gnathopod 1 subchelate (fig. 155); epistome projecting slightly

forward from upper lip; telson with 2 apical spines (body, fig. 48) .... *Allogaussia recondita*

5. Mandibular palp attached level with molar (fig. 5); epimeron 3 with large notch above posteroventral tooth (uropod 3, fig. 83) .... *Hippomedon denticulatus*
– Mandibular palp attached proximal to molar (fig. 4); epimeron 3 without notch above posteroventral tooth (uropod 3, fig. 76) .... *Orchomene pacifica*

KEY K: ACANTHONOTOZOMATIDAE, OEDICEROTIDAE, PLEUSTIDAE, EUSIRIDAE

1. Coxae 1–4 pointed (fig. 62); maxillipedal palp 3-articulate; gnathopod 1 thin and minutely propodochelate (fig. 167) .... *Panoploea* spp.
– Coxae 1–4 apically rounded or truncate 2
2. Eyes dorsally coalesced; uropod 3 elongate, peduncle nearly as long as rami of uropods 1–2, rami of uropod 3 as long as peduncle; pereopod 7 nearly 1.5 times as long as pereopod 6 (figs. 208, 209; body, fig. 56) 3
– Eyes 2 and lateral; pereopod 7 not immensely longer than pereopod 6 4
3. Palm of gnathopod 1 transverse (as in fig. 162); dactyl of gnathopod 2 at least 0.3 times length of article 6 (fig. 137) .... *Synchelidium rectipalmum*
– Palm of gnathopod 1 oblique (fig. 160); dactyl of gnathopod 2 about 0.2 times length of article 6 (fig. 136) .... *Synchelidium shoemakeri*
4. Telson cleft (occasionally weakly) 5
– Telson entire 10
5. Antenna 1 lacking accessory flagellum 6
– Antenna 1 with minute accessory flagellum (see maxilla 2, fig. 18) 8
6. Antenna 1 with tooth on article 3 (fig. 217) .... *Pontogeneia inermis*
– Antenna 1 lacking tooth on article 3 7
7. Rostrum as in fig. 73 .... *Pontogeneia rostrata*
– Rostrum as in fig. 71 .... *Pontogeneia intermedia*
8. Telson cleft 1/4 or 1/5 its length (fig. 115) *Oligochinus lighti*
– Telson cleft halfway or more 9
9. One ramus of uropod 2 longer than peduncle, rostrum extending beyond tangent of lateral cephalic lobe (see maxilla 2, fig. 17) .... *Accedomoera vagor*

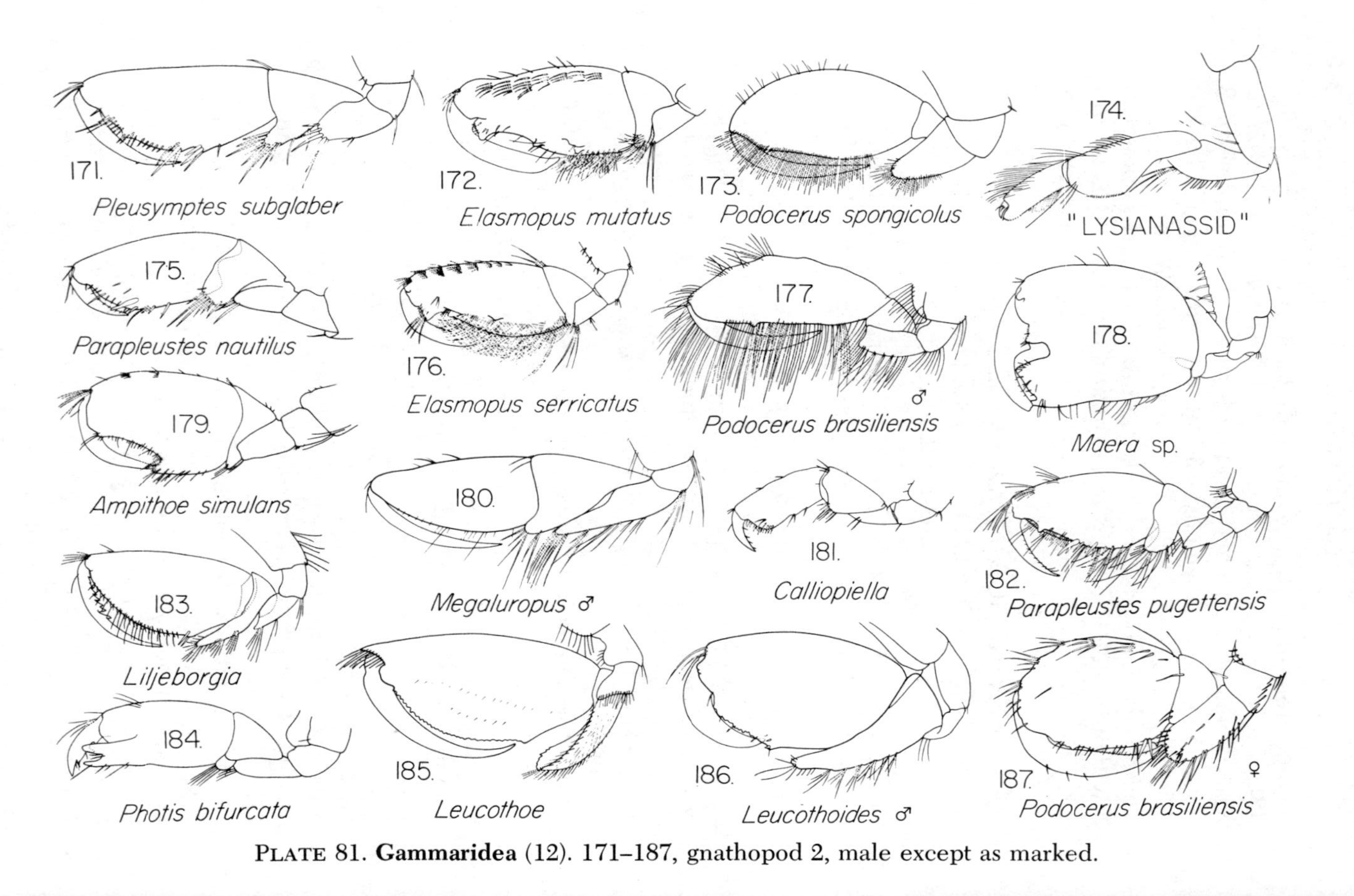

PLATE 81. **Gammaridea** (12). 171–187, gnathopod 2, male except as marked.

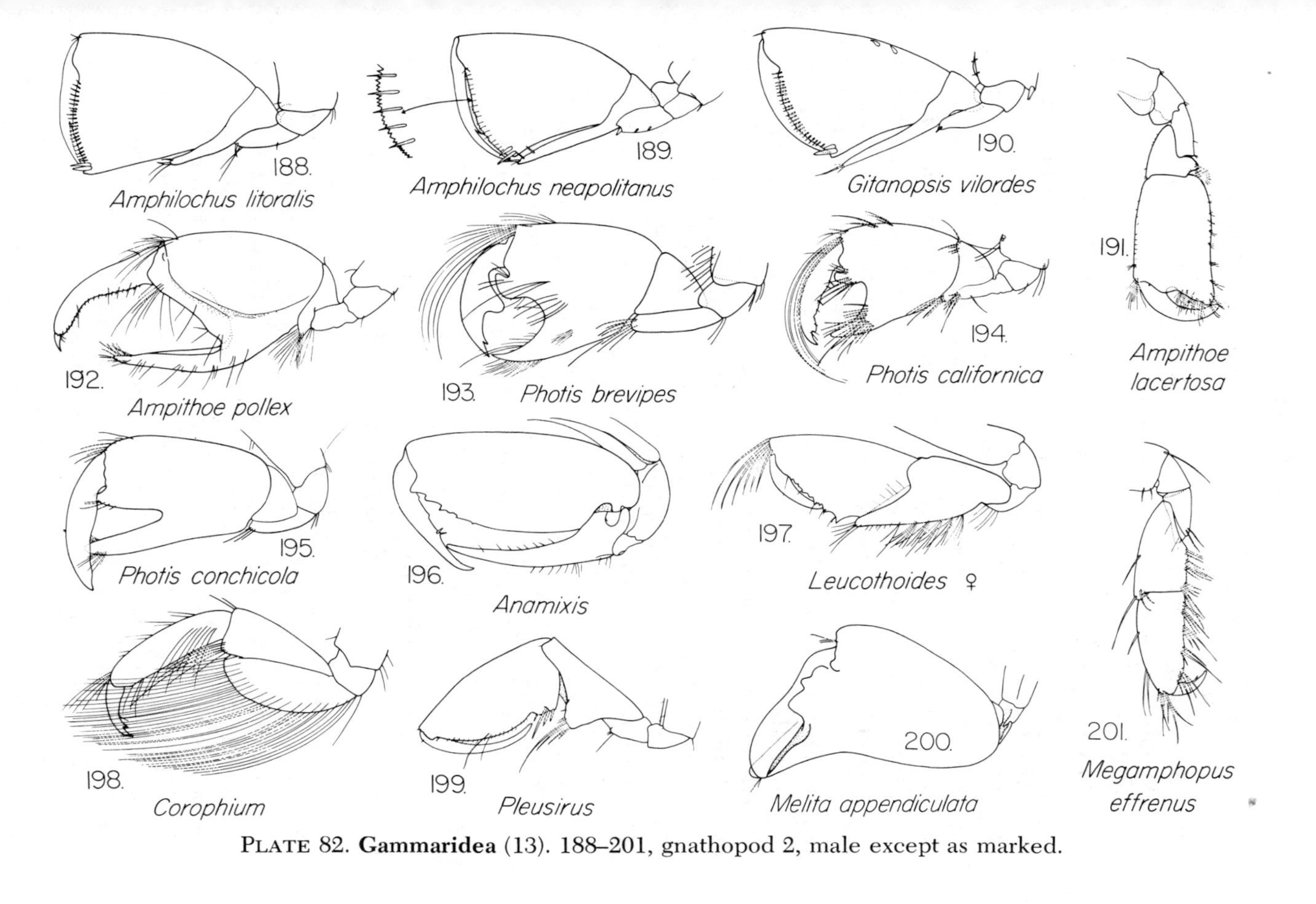

PLATE 82. **Gammaridea** (13). 188–201, gnathopod 2, male except as marked.

– Rami of uropod 2 shorter than peduncle, rostrum obsolete (body, fig. 38) .... *Paramoera mohri*
10. Mandibular molar with rasplike grinding surface (fig. 5) ... 11
– Mandibular molar forming small smooth hump, occasionally bearing articulate spine(s) (figs. 8, 9) 13
11. Article 3 of antenna 1 with distoventral tooth (fig. 217) .... *Calliopius* sp.
– Article 3 of antenna 1 lacking tooth 12
12. Hands of gnathopods rectangular, thin (fig. 181); epimeron 3 with rhombic posterior margin, no tooth (fig. 29) .... *Calliopiella pratti*
– Hands of gnathopods ovate to subovate (fig. 171); epimeron 3 with posterior hump and sharp posteroventral tooth .... *Pleusymptes subglaber*
13. Hands of gnathopods attached to thin apex of article 5 (figs. 168, 199) .... *Pleusirus secorrus*
– Hands of gnathopods attached to thick apex of article 5 14
14. Rostrum immense, tonguelike; some posterior body segments with weak teeth and rugosities (fig. 58) ... *Pleustes depressa*
– Rostrum small and inconspicuous; body segments smooth .. 15
15. Antenna 1 longer than first 7 body segments together; hands of gnathopods ovate (fig. 182); article 6 of pereopods 3–7 with setae only .... *Parapleustes pugettensis*
– Antenna 1 as long as first 3 body segments together; hands of gnathopods rectangular (fig. 175); article 6 of pereopods 3–7 with inflexible stout spines .... *Parapleustes nautilus*

KEY L: PHOXOCEPHALIDAE, HAUSTORIIDAE

1. Pereopod 7 like pereopod 6 in overall appearance and length; pereopod 4 reversed like pereopods 5–7 (body, fig. 44; uropod 3, fig. 85) .... *Eohaustorius* spp.
– Pereopod 6 much longer than pereopod 7 (fig. 43); latter with broadened article 2 of form in fig. 203; pereopod 4 normal .. 2
2. Head as long as first 4 pereonites together, blind (fig. 45); uropods 1–2 with freely articulate spines near apices of rami .. .. *Mandibulophoxus gilesi*
– Head as long as first 3 pereonites together, with eyes; uropods 1–2 apically naked 3
3. Palp of maxilla 1 uniarticulate; hand of gnathopod 2 about 3 times as long and 2–3 times as broad as article 5 (figs. 161, 162) 4
– Palp of maxilla 1 biarticulate; hand of gnathopod 2 shorter than and nearly as broad as article 5 5

4. Palm of gnathopod 1 oblique (fig. 161) *Metaphoxus frequens*
– Palm of gnathopod 1 transverse (fig. 162) *Metaphoxus fultoni*
5. Rostrum abruptly narrowed in front of eyes (dorsal view) (fig. 21; body, *Paraphoxus*, fig. 43; uropod 3, figs. 80, 82) 6
– Rostrum tapering evenly in front of eyes (fig. 22) 8
6. Gnathopods 1–2 stout, articles 5 and 6 subequal to each other in length (fig. 151); epistome with long, sharp tooth .... *Paraphoxus epistomus*
– Gnathopods 1–2 slender, article 5 much longer than article 6; epistome lacking long tooth 7
7. Article 2 of pereopod 7 with 3–4 medium-sized teeth .... *Paraphoxus tridentatus*
– Article 2 of pereopod 7 with 7 or more small teeth .... *Paraphoxus milleri*
8. Epistome with anterior cusp (fig. 23) ..*Paraphoxus cognatus*
– Epistome lacking cusp 9
9. Inner and outer dorsal margins of peduncle on uropod 1 with 2 or more spines each; ovigerous females 6.0 mm or longer (body length) ....*Paraphoxus obtusidens*
– At least 1 dorsal margin of peduncle on uropod 1 with only 1 or no spine; ovigerous females not exceeding 5 mm in length.. .... *Paraphoxus spinosus*

KEY M: SYNOPIIDAE, LILJEBORGIIDAE

1. Head galeate (fig. 68); eyes coalesced into 1 dorsal eye, each side of head with tiny accessory eye near base of antenna 2; gnathopods 1–2 elongate, thin, almost simple (fig. 159) (body, fig. 51) .... *Tiron biocellata*
– Head ordinary; eyes lateral, no accessory eyes; gnathopods 1–2 very large and subchelate 2
2. Article 5 of gnathopods 1–2 strongly produced, slender and elongate, dactyls with deep inner serrations (fig. 183); outer ramus of uropod 3 with 1 article (body, fig. 39) *Liljeborgia* spp.
– Article 5 of gnathopods 1–2 weakly produced, thick and blunt, dactyls with smooth inner margins; outer ramus of uropod 3 with 2 articles ....*Listriella* spp.

KEY N: GAMMARIDAE

(Check mandibular molar—if not heavily ridged, return to Key M)

1. Urosomite 1 (and usually 2 and 3) armed with dorsal bundles of articulate spines; gnathopod 1 slightly larger than 2; palms of gnathopods with peglike spines (fig. 165) 2

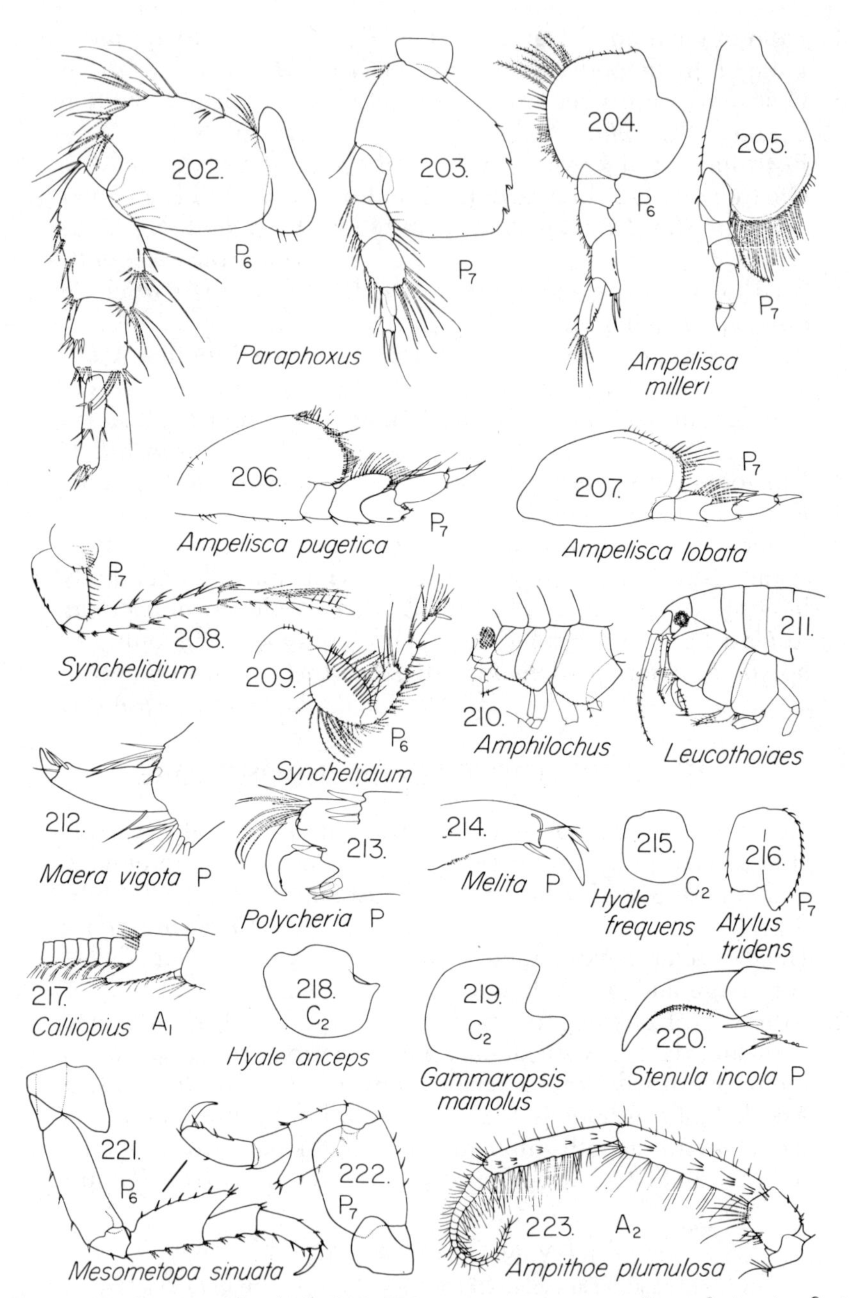

PLATE 83. **Gammaridea** (14). 202–223, various parts as indicated: A_1, A_2, first and second antennae; P, P_6, P_7, pereopods; C_2, second coxae; 202–209, 221, 222, pereopods; 210, 211, 215, 218, 219, coxae; 216, article 2 of pereopod 7; 217, article 3 of antenna 1; 223, antenna 2; 212–214, 220, dactyls of pereopods.

– Urosomite 1 occasionally serrate, usually smooth, articulate spines if present scattered and not grouped; gnathopod 2 larger than 1; palms of gnathopods with ordinary pointed spines 4

2. Urosomite 2 with large, fixed, dorsal tooth; no spines (fig. 33) (see uropod 3, fig. 78) .... *Anisogammarus pugettensis*

– Urosomite 2 with groups of articulate dorsal spines; no tooth 3

3. Urosomite 1 with 4 dorsal groups of 3–4 stout spines each; groups with spines aligned anterior to posterior (fig. 24) (see uropod 3, fig. 79) .... *Anisogammarus confervicolus*

– Urosomite 1 with 4 dorsal groups of 1–2 stout spines and 1–4 thin setae; groups with spines and setae aligned side to side (fig. 25) .... *Anisogammarus ramellus*

4. Coxa 3 shorter than coxae 2 or 4 (fig. 41) (very difficult to observe in this context; compare gnathopods carefully to figs. 132 and 180); uropod 3 often missing; rami foliate (fig. 87) .... *Megaluropus* spp.

– Coxa 3 as long as coxae 2 and 4; gnathopod 2 ordinary 5

5. Pleon dorsally smooth; inner plate of maxilla 2 lacking stout medial setae 6

– Pleon with dorsal teeth (often minute) or serrations on certain segments; maxilla 2 variable 11

6. Article 3 of mandibular palp falcate and with a comb of setae (fig. 2b; (body, *Elasmopus,* fig. 37) 7

– Article 3 of mandibular palp ordinary, thin, and without a comb of setae (fig. 6; uropod 3, fig. 77) 9

7. Epimeron 3 with small, sharp, posteroventral tooth; male gnathopod 2 palm toothless and fully setose .... *Elasmopus antennatus*

– Epimeron 3 rounded-quadrate at posteroventral corner; male gnathopod 2 palm with 1 or more teeth often hidden in dense clusters of setae 8

8. Each lobe of telson with 4–6 spines occupying most of apical breadth; adult male gnathopod 2 with setae on palm opposite dactyl (fig. 176) .... *Elasmopus serricatus*

– Each lobe of telson with 2–3 spines near lateral margin; medial parts of apices naked and pointed (fig. 110); adult male gnathopod 2 with palm opposite dactyl largely free of setae (fig. 172) .... *Elasmopus mutatus*

9. Dactyls of pereopods with tiny tooth on outer margin near apex (fig. 214) (telson must be much longer than broad) 10

– Dactyls of pereopods lacking accessory tooth (fig. 212) .... *Maera vigota*

10. Palm of adult gnathopod 2 oblique .... *Maera simile*

– Palm of adult gnathopod 2 transverse (fig. 178) ....*Maera* sp. (*"inaequipes"*)

11. Rami of uropod 3 long and extending equally; uropod 3 often lost, therefore check for dense *medial* setae on inner plate of maxilla 1 .... *Ceradocus spinicaudus*

– Outer ramus of uropod 3 very long, inner ramus very short and scalelike (fig. 81); uropod 3 often lost, therefore check for naked medial margin on inner plate of maxilla 1; setae all apical (body, *Melita*, fig. 42) 12

12. Dorsal segmental teeth occurring only on urosomites 1–2 13

– Dorsal segmental teeth occurring on pleonites 1–3 and on urosomites 14

13. Urosomite 1 with only 1 dorsal tooth; male gnathopod 1 with deformed dactyl (fig. 154) ....*Melita sulca*

– Urosomite 1 with dorsal teeth (side to side); male gnathopod 1 with normal dactyl .... *Melita californica*

14. Male gnathopod 2 immense, chelate (fig. 200); dactyls of pereopods 3–7 with accessory tooth (fig. 214); all spines of telson on proximal 3/4 of lobes .... *Melita appendiculata*

– Male gnathopod 2 of ordinary size, subchelate; dactyls of pereopods 3–7 lacking accessory tooth; all spines of telson on distal 1/4 of lobes ....*Melita dentata*

KEY O: TALITRIDAE

(PLATES 84 and 85)

Talitrids comprise mainly beach-hoppers, common in evening and early morning on the damp sands, where they feed upon seaweeds cast up at the previous high tides. Fresh beach-wrack may contain purely aquatic amphipods—their death is rapid in air, whereas beach-hoppers survive well in air. Since the patterns and colors by which they may be identified in life are lost in preservatives, a morphological key to the family Talitridae precedes a key to *Orchestoidea* based on color (Bowers, 1963).

MORPHOLOGICAL KEY TO TALITRIDAE

E. L. Bousfield
National Museum of Natural Sciences, Ottawa, Canada

1. Uropod 1, outer ramus marginally smooth, interramal spine prominent (fig. 231A); pleopod 3 minute, lacking distinct rami

(fig. 231B); animals not sexually dimorphic (i.e., gnathopod 1 simple, gnathopod 2 minutely chelate in both sexes and immatures) 2

– Uropod 1, outer ramus with marginal spines, interramal spine minute or lacking (fig. 234D); pleopod 3 about equal to 1 and 2, normally biramous (fig. 234F); animals dimorphic (i.e., gnathopod 2 always powerfully subchelate in mature males) 4

2. Gnathopod 1, segment 6 shorter than segment 5; pleopod 1 and 2 subequal, outer margin of peduncles with some plumose setae (fig. 231C); telson, dorsolateral margins with 4–5 stout spines lining each side; mature animals small (5–11 mm) .... *Talitroides* 3

– Gnathopod 1, segments 5 and 6 subequal; pleopod 2 distinctly smaller than 1, peduncular margins lacking plumose setae; telson with 2–3 spines on each side; mature animals larger (12–14 mm) .... *"Talitrus" sylvaticus*

3. Gill on pereopod 6 distinctly L-shaped, subtruncate distally; antenna 1 reaching (or nearly so) distal end of peduncle of antenna 2; uropod 1, interramal spine with complex tip (fig. 231A) ....*Talitroides topitotum*

– Gill on pereopod 6 curved, distally narrowing, rounded; antenna 1 reaching about midpoint of peduncular segment 5 of antenna 2; uropod 1, interramal spine with simple tip ....*Talitroides alluaudi*

4. Gnathopod 1 distinctly subchelate (more strongly so in male), dactyl slender (figs. 234A, B; 236); pereopod 7 longer than 6; uropod 3, ramus narrowing distally, shorter than peduncle (fig. 234C) .... *Orchestia* 5

– Gnathopod 1 simple (both sexes), dactyl strong, heavy (fossorial) (figs. 232A, 233A, 235A); pereopod 6 longer than 7; uropod 3 ramus broad distally, about as long as peduncle .... *Orchestoidea* 7

5. Pereopods 3 and 4 slender, segment 5 about equal to 6 (as in pereopod 2, fig. 234E); gnathopod 2 (mature male) palm with triangular tooth near hinge, dactyl sinuous; antenna 2 long, peduncle thick or inflated (male) ......*Orchestia chiliensis*

– Pereopods 3 and 4 short and stout, segment 5 distinctly shorter and thicker than segment 6; gnathopod 2 (mature male) palm smoothly convex; antenna 2 short, peduncle not heavily thickened (male) 6

6. Pleopods weak, rami 4–6 segmented; gnathopod 1 (male), segment 4 lacking posterior translucent process ("blister") . ... *Orchestia georgiana*

– Pleopods strong, rami 7–10 segmented; gnathopod 1 (male),

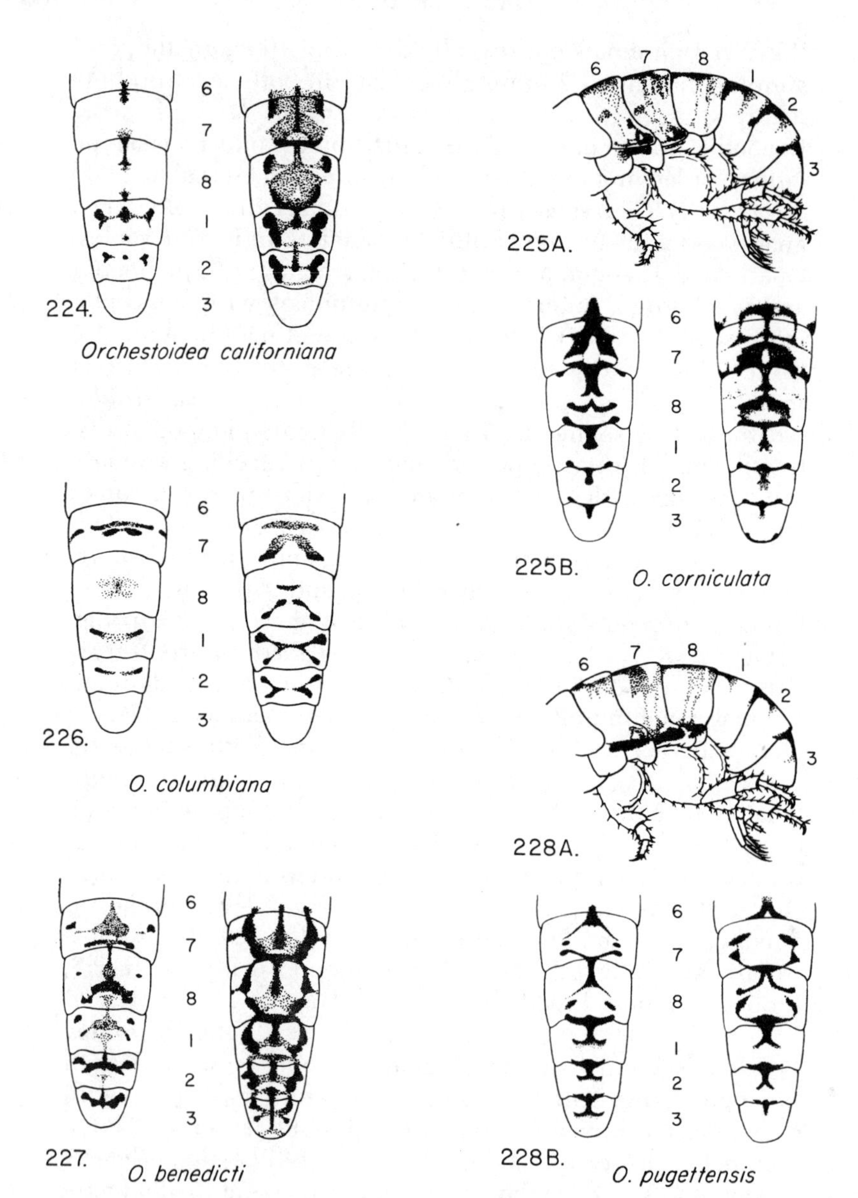

PLATE 84. **Gammaridea** (15). Talitridae, color patterns of *Orchestoidea:* 224–228B, species of *Orchestoidea*, as indicated, after Bowers, 1963. 225A and 228A, lateral; rest dorsal views; paired figures show extent of pattern variation.

segment 4 with small, posterior, translucent process (fig. 236, arrow) . . . . *Orchestia traskiana*

7. Uropod 2, inner margin of outer ramus bearing lateral spines (fig. 233B); flagellum of antenna 2 not shorter than peduncle, usually much longer, especially in males (fig. 229) 8
– Uropod 2, inner margin of outer ramus without lateral spines; flagellum of antenna 2 shorter than peduncle (fig. 230) 9
8. Margins of pleonites with numerous small spines; female gnathopod 1 with translucent process ("blister") on lower margin of article 5; rami of pleopods less than 1/2 length of peduncle . . . . *Orchestoidea californiana*
– Margins of pleonites smooth; female gnathopod 1 without translucent process on article 5 (fig. 233A); rami of pleopods ½ to ¾ length of peduncle . . . . *Orchestoidea columbiana*
9. Telson notched at tip (fig. 232B); anteroventral margin of pleonite 1 with 1–7 spines; male gnathopod 1 with conspicuous blister on lower margin of article 6, near base of claw (fig. 232A, arrow) . . . . *Orchestoidea pugettensis*
– Telson entire at tip; anteroventral margin of pleonite 1 smooth; male gnathopod 1 without conspicuous blister on article 6 (fig. 235A, arrow) 10
10. Posterior margin of pleonites with 10 or more small spines (fig. 235B); adult animals large (15–22 mm) . . . . *Orchestoidea corniculata*
– Posterior margin of pleonites with 1–5 spines; adult animals small (9–13 mm) . . . . *Orchestoidea benedicti*

A FIELD (COLOR-PATTERN) KEY TO *ORCHESTOIDEA*

Darl E. Bowers
Mills College, Oakland

1. Mature animals 2
– Immature animals and others not distinguishable by first 4 couplets 5
2. Antenna 2 when folded reaching back to or past middle of body; flagellum longer than peduncle (fig. 229) 3
– Antenna 2 when folded not reaching middle of body; flagellum shorter than peduncle (fig. 230) 4
3. Color of antennae 2 rosy red . . . . . *Orchestoidea californiana*
– Color of antennae 2 bluish white . . . . *Orchestoidea columbiana*

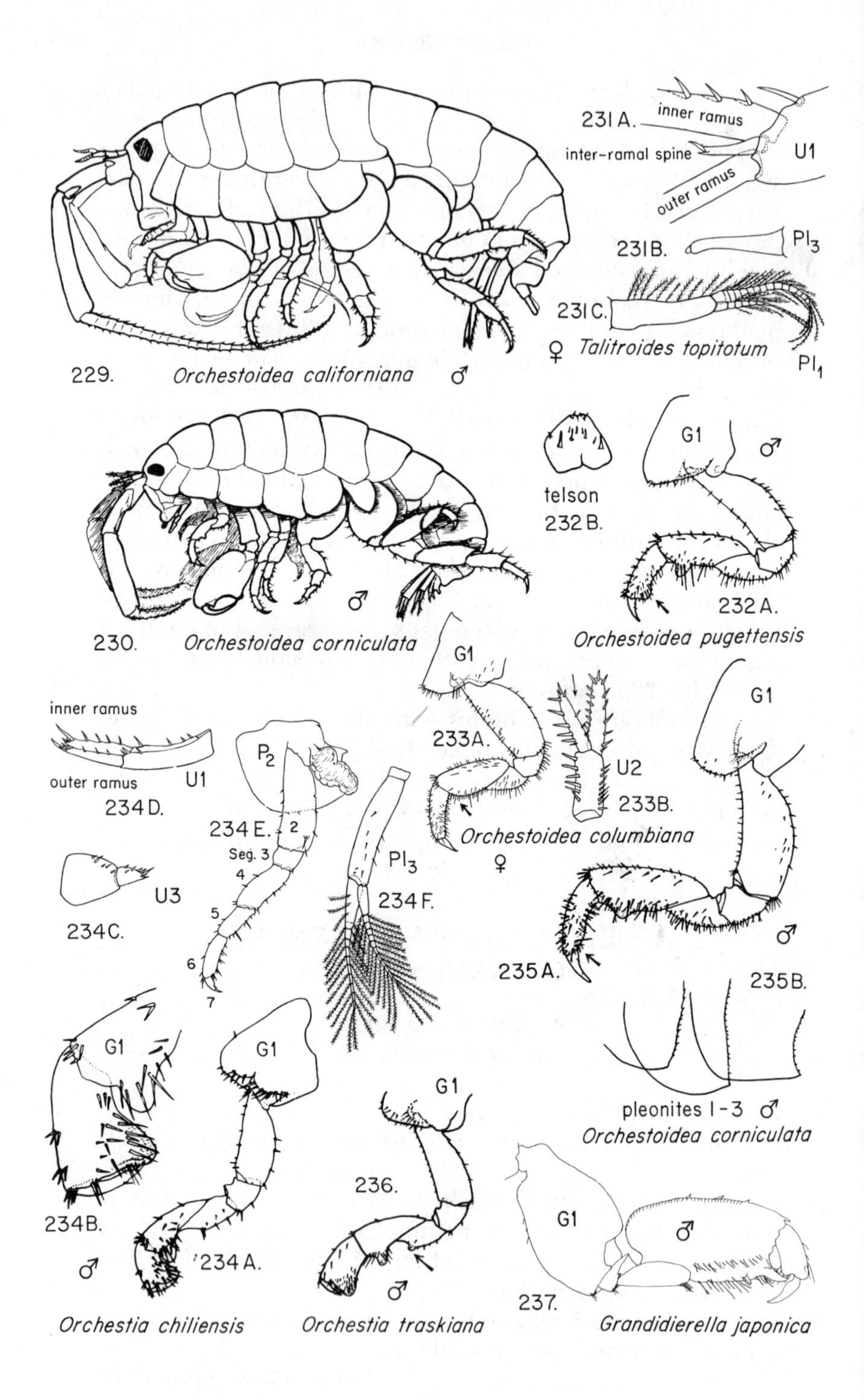
231 A.
inner ramus
inter-ramal spine
outer ramus
U1
231B.
Pl3
231C.
Pl1
♀ Talitroides topitotum
229. Orchestoidea californiana ♂
telson
232B.
G1
♂
232A.
Orchestoidea pugettensis
230. Orchestoidea corniculata
♂
G1
233A.
U2
233B.
Orchestoidea columbiana
♀
G1
inner ramus
outer ramus
U1
234D.
P2
234E.
Seg. 3
Pl3
234F.
U3
234C.
235A.
♂
235B.
pleonites I-3 ♂
Orchestoidea corniculata
G1
234B.
♂
234A.
G1
236.
♂
G1
♂
237.
Orchestia chiliensis
Orchestia traskiana
Grandidierella japonica

4. Color of antennae 2 usually salmon pink ....
.................................*Orchestoidea corniculata*
– Color of antennae 2 otherwise 5
5. Dorsal pigment pattern containing "butterfly" designs (figs. 224, 226, 227) 6
– Dorsal pigment pattern containing T-shaped figures; the lower limb of the "T" may be faint or missing (figs. 225B, 228B) .. 8
6. Mid-dorsal line absent; "butterfly" spots are flattened (fig. 226)
.... *Orchestoidea columbiana*
– Mid-dorsal line present 7
7. No markings on 3rd abdominal segment (fig. 224); sides of body relatively free of pigment marks ....
................................*Orchestoidea californiana*
– Markings on 3rd abdominal segment (fig. 227); sides of body blotched in checkerboard pattern ...*Orchestoidea benedicti*
8. Two diffuse spots on sides of body (fig. 225A) ....
.... *Orchestoidea corniculata*
– Three discrete spots on sides of body (fig. 228A) ....
................................*Orchestoidea pugettensis*

LIST OF GAMMARIDEA

(Except Talitridae, see p. 363)

SYMBOLS USED IN NOTES

A: nestling in algae or surfgrass.
B: burrower in sand or mud bottoms or sedimentary patches of intertidal rocky regions.
I: inquilinous, ectocommensal or ectoparasitic, usually with sucking or mucus-lapping mouthparts, hosts poorly known.
TA: tube builder attaching tube to algae.
TB: tube builder attaching to debris on mud bottom.
TR: tube partially inserted in mud bottom.

PLATE 85. **Gammaridea** (16). Talitridae (cont.); Corophiidae (237 only): 229, *Orchestoidea californiana;* 230, *O. corniculata;* 231A,B,C, *Talitroides topitotum;* 232A, 233A, 234A, 235A, 236, first gnathopods of *Orchestoidea* and *Orchestia;* other parts as indicated: U, uropod; P, pereopod; Pl, pleopod; 237, *Grandidierella japonica*, male first gnathopod. 234, Bousfield and Carlton, 1967; 237, Chapman and Dorman, 1975; 231A,B,C Shoemaker, 1936; 229, 230 after Bowers in Ricketts, Calvin, and Hedgpeth: *Between Pacific Tides*, 4th ed., 1968, used with permission of Stanford University Press; rest, Bousfield, 1959, 1961.

In the following list, pertinent references are placed in parentheses after the author and date of each species.

Accedomoera vagor Barnard, 1969. A.

Allogaussia recondita Stasek, 1958. I, in digestive cavity of *Anthopleura elegantissima.*

Allorchestes angusta Dana, 1854. (Barnard, 1952), A.

Ampelisca agassizi (Judd, 1896) (=*A. compressa* Holmes, and *A. vera* Barnard, 1954b). (Mills, 1967), TR.

Ampelisca cristata Holmes, 1908. (Barnard, 1954b), TR.

Ampelisca lobata Holmes, 1908. (Barnard, 1954a), TR, A.

Ampelisca macrocephala Liljeborg, 1852. (Barnard, 1954b), TR.

Ampelisca milleri Barnard, 1954. TR, occasionally TB.

Ampelisca pugetica Stimpson, 1864. (Barnard, 1954b), TR.

Amphilochus litoralis Stout, 1912. (Barnard, 1962c), I, among coralline algae.

Amphilochus neapolitanus Della Valle, 1893. (Barnard, 1962c), I, among coralline algae.

Ampithoe aptos (Barnard, 1969, as *Pleonexes*). TA, especially on *Macrocystis* stipes. *Ampithoe* was formerly also spelled *Amphithoe.*

Ampithoe humeralis Stimpson, 1864. (Barnard, 1965), TA.

Ampithoe lacertosa Bate, 1858. (Barnard, 1965), TA, on *Macrocystis.*

Ampithoe lindbergi Gurjanova, 1938. (Gurjanova, 1951; Barnard, 1965), TA.

Ampithoe plumulosa Shoemaker, 1938. (Barnard, 1965), TA, on red algae.

Ampithoe pollex Kunkel, 1910. (Barnard, 1965), TA, on red algae.

Ampithoe simulans Alderman, 1936 (=*A. dalli* Shoemaker, 1938). (Barnard, 1965), TA.

Ampithoe valida Smith, 1873. (Barnard, 1965), TA, especially in estuaries.

Ampithoe sp., *cf. A. mea* Barnard, 1965. TA.

Anamixis linsleyi Barnard, 1955. I, on sponges and tunicates, especially on pilings.

Anisogammarus confervicolus (Stimpson, 1857). (Shoemaker, 1964), A, marshes.

Anisogammarus pugettensis (Dana, 1853). (Barnard, 1954a), A, marshes.

Anisogammarus ramellus (Weckel, 1907). (Shoemaker, 1942b), A, marshes, brackish to fresh water.

Aoroides columbiae Walker, 1898 (=*A. californica* Alderman, 1936). (Barnard, 1954a), TA.

Atylus levidensus Barnard, 1956. (Mills, 1961), A.

Atylus tridens (Alderman, 1936, *Nototropis*). (Mills, 1961), A.

Calliopiella pratti Barnard, 1954. (Barnard, 1969a), A.

Calliopius sp., *cf. C. laeviusculus* of Barnard, 1954a (probably a new species). A.

Ceradocus spinicaudus (Holmes, 1908). (Barnard, 1962b), A, on *Macrocystis.*

Cerapus tubularis Say, 1817 (=*C. abditus* Templeton, 1836). (Barnard, 1962a), tube builder; its heavy, striped tube is dragged about in fashion of hermit crab.

Chelura terebrans Philippi, 1839. (Barnard, 1950), bores wood, associated with *Limnoria,* introduced.

Corophium acherusicum Costa, 1857. (Shoemaker, 1949), TB, TA, especially on harbor pilings; introduced.

Corophium baconi Shoemaker, 1934. (Shoemaker, 1949), TB, TA.

Corophium brevis Shoemaker, 1949. TB, estuaries.

Corophium californianum Shoemaker, 1934. TB.

Corophium insidiosum Crawford, 1937. (Shoemaker, 1949), TB, estuaries, harbor pilings; introduced.

Corophium oaklandense Shoemaker, 1949. Tube builder; Oakland Estuary.

Corophium spinicorne Stimpson, 1857. (Shoemaker, 1949) TB, estuaries.

Corophium stimpsoni Shoemaker, 1941. (Shoemaker, 1949), TB, estuaries.

Corophium uenoi Stephensen, 1932. (Barnard, 1952), TB, estuaries and lagoons; possibly introduced.

Cymadusa uncinata (Stout, 1912). (Barnard, 1965), TA, on *Macrocystis.*

Dogielinotus loquax Barnard, 1967. Shallow-water sand bottoms, seaward of surf zone.

Elasmopus antennatus (Stout, 1913). (Barnard, 1962b), A.

Elasmopus mutatus Barnard, 1962. A, here considered a species rather than a subspecies of *E. rapax.*

Elasmopus serricatus Barnard, 1969. A, here considered a species rather than a subspecies of *E. rapax*.

Eohaustorius spp. (Barnard, 1957a, 1962d), shallow-water sand bottoms, seaward of surf zone.

Ericthonius brasiliensis (Dana, 1853). (Sars, 1895), TB, TA.

Ericthonius hunteri (Bate, 1862). (Sars, 1895; Barnard, 1962a), TB, TA.

Fresnillo fimbriatus Barnard, 1969. Probably I.

Gammaropsis mamolus (Barnard, 1962). TA.

Gammaropsis thompsoni Walker, 1898 (=*G. tenuicornis* Holmes, 1904). (Shoemaker, 1955a), TA.

Gitanopsis vilordes Barnard, 1962. I in A.

Grandidierella japonica Stephensen, 1938. TR, abundant in estuaries; introduced from Japan; see Chapman and Dorman, 1975, Bull. So. Calif. Acad. Sci. 74.

Heterophlias seclusus Shoemaker, 1933. (Barnard, 1962b), nestler under rocks; scavenges dead rhizomes of algae trapped in sand.

Hippomedon denticulatus (Bate, 1857). (Sars, 1895; Gurjanova, 1962), intertidal of open coast.

Hyale anceps (Barnard, 1969). A.

Hyale frequens (Stout, 1913). (=*Hyale nigra* of Barnard, 1962c and *H. rubra* of authors). A, various Hyales also under edges of limpet shells.

Hyale grandicornis californica Barnard, 1969. A.

Hyale plumulosa (Stimpson, 1857). A, especially estuaries.

Ischyrocerus anguipes Krøyer, 1838 (=*I. minutus* Walker, 1898). (Sars, 1895), TA.

Ischyrocerus sp. A, Barnard, 1969. TA.

Ischyrocerus sp. B, Barnard, 1969. TA.

Jassa falcata (Montagu, 1808). (Sars, 1895, Pl. 212, as *Podocerus falcatus*), TA especially T on pilings.

Lembos spp., Barnard, 1962. TA.

Leucothoe alata Barnard, 1959. I, in tunicates and sponges.

Leucothoe spinicarpa (Abildgaard, 1789). (Sars, 1895; Barnard, 1962c), I, in tunicates and sponges.

Leucothoides pacifica Barnard, 1955. I, in sponges and tunicates.

Lignophliantis pyrifera Barnard, 1969. Bores into rhizomes of giant kelp.

Liljeborgia spp., Barnard, 1969. Several unidentified species in northern California, A.

Listriella spp., Barnard, 1959b. Species identified by pigmentary pattern in certain preservatives; phenotypic system not yet analyzed in central California; mud-bottom commensals of maldanid polychaetes.

Lysianassa spp. (=*Aruga* spp.). (Barnard, 1959b, 1969a) Species in northern California not yet analyzed; mud ingesters often found in sediment pockets of intertidal; facultative diatom feeders, scavengers.

Maera simile (Stout, 1913). (Barnard, 1959a), A.

Maera vigota Barnard, 1969. Nestler under rocks.

Maera sp. ("*inaequipes*" of Alderman, 1936 and Barnard, 1959a; a new species). A.

Mandibulophoxus gilesi Barnard, 1957. Shallow-water sand bottoms seaward of surf zone.

Megaluropus spp., (Barnard, 1962). Genus poorly studied in northern California; A, but primarily neritic, nektonic, or demersal.

Megamphopus effrenus Barnard, 1964. TA.

Megamphopus martesia Barnard, 1964. TA.

Melita appendiculata (Say, 1818) [=*M. fresneli* (Audouin, 1826)]. (Shoemaker, 1955b), A.

Melita californica Alderman, 1936. A.

Melita dentata (Krøyer, 1842). (Sars, 1895), A.

Melita sulca (Stout, 1913). (Barnard, 1969a), A.

* *Melita* sp. A, abundant in Lake Merritt, Oakland; possibly introduced.

Mesometopa esmarki (Boeck, 1872). I.

Mesometopa sinuata Shoemaker, 1964. I.

Metaphoxus frequens Barnard, 1960. B.

Metaphoxus fultoni (Scott, 1890). (Barnard, 1964b), B.

Metopa cistella Barnard, 1969. I.

* Not in key.

Microdeutopus schmitti Shoemaker, 1942. TA.

Microjassa litotes Barnard, 1954. (Barnard, 1962a), TA.

Najna sp. (=*N. ?consiliorum* of Barnard, 1962c; a new species). A, especially on kelp. See Barnard on Najnidae, 1972, New Zealand Inst. Oceanography, Mem. 62.

Ocosingo borlus Barnard, 1964. (Barnard, 1969a), probably I.

Oligochinus lighti Barnard, 1969. A.

Orchestia spp. See list of Talitridae (page 363).

Orchestoidea spp. See list of Talitridae (page 363).

Orchomene pacifica (Gurjanova, 1938). (Barnard, 1964b), see *Lysianassa* for ecology.

Panoploea spp., Barnard, 1969a. Genus poorly studied, I.

Parajassa angularis Shoemaker, 1942. TA.

Parallorchestes ochotensis (Brandt, 1851). (Barnard, 1962c), A.

Paramoera mohri Barnard, 1952. (Barnard, 1969a), A.

Paraphoxus cognatus Barnard, 1960. B.

Paraphoxus epistomus Shoemaker, 1938. (Barnard, 1960), B.

Paraphoxus milleri (Thorsteinson, 1941). (Barnard, 1960), B, estuaries.

Paraphoxus obtusidens (Alderman, 1936). (Barnard, 1960), B.

Paraphoxus spinosus Holmes, 1903. (Barnard, 1960), B.

Paraphoxus tridentatus (Barnard, 1954). (Barnard, 1960), B.

Parapleustes nautilus Barnard, 1969. A.

Parapleustes pugettensis (Dana, 1853). (Barnard and Given, 1960; Shoemaker, 1964), A.

Photis bifurcata Barnard, 1962. TA.

Photis brevipes Shoemaker, 1942. (Barnard, 1962a), TA.

Photis californica Stout, 1913. (Barnard, 1962a), TA.

Photis conchicola Alderman, 1936. (Barnard, 1962a), TA.

Pleusirus secorrus Barnard, 1969. A.

Pleustes depressa Alderman, 1936. A.

Pleusymptes subglaber (Boeck, 1861). (Sars, 1895, as *Parapleustes;* species unconfirmed recently), A.

Podocerus brasiliensis (Dana, 1853). (Barnard, 1959a), among hydroids, especially on harbor pilings.

Podocerus cristatus (Thomson, 1879). (Barnard, 1962a), among hydroids.

Podocerus spongicolus Alderman, 1936. Among hydroids.

Polycheria osborni Calman, 1898. (Skogsberg and Vansell, 1928; as *P. antarctica* in Alderman, 1936), I. Burrows in tests of the compound ascidian *Aplidium;* filters diatoms.

Pontogeneia inermis (Krøyer, 1838). (Sars, 1895), A.

Pontogeneia intermedia Gurjanova, 1938. (Gurjanova, 1951; Barnard, 1969a), A.

Pontogeneia rostrata Gurjanova, 1938. (Gurjanova, 1951; Barnard, 1962b, 1969a), A.

Stenothoe valida Dana, 1852. (Barnard, 1953, 1959a), I.

Stenothoides burbanki Barnard, 1969. I.

Stenula incola Barnard, 1969. I.

Synchelidium rectipalmum Mills, 1962. B.

Synchelidium shoemakeri Mills, 1962. B.

Talitroides, see list of Talitridae, below.

"*Talitrus,*" see list of Talitridae, below.

Tiron biocellata Barnard, 1962. ?B or A.

LIST OF TALITRIDAE

E. L. Bousfield

Orchestia chiliensis Milne-Edwards, 1840 (=*O. enigmatica* Bousfield and Carlton, 1967). Introduced; in Lake Merritt, Oakland, and possibly more widely distributed; under debris on sand beaches.

Orchestia georgiana Bousfield, 1958. Recorded from Washington, British Columbia, and San Diego; may occur in our area; see Bousfield 1958, 1961.

Orchestia traskiana Stimpson, 1857. On rocky beaches, occasionally on sandy beaches with algae; under debris and boards in salt marshes.

Orchestoidea benedicti Shoemaker, 1930. Common on fine-sand beaches with *O. californiana.*

Orchestoidea californiana (Brandt, 1851). Large and common, high on wide, exposed beaches of fine sand; digs burrows of elliptical cross-section.

Orchestoidea columbiana Bousfield, 1958. On coarse-sand beaches with little seaweed; see Bousfield, 1961.

Orchestoidea corniculata Stout, 1913. Large and common, on steep, protected beaches with coarse sand and considerable seaweed; burrow nearly circular in cross-section. See Craig, 1973, Mar. Biol. 23: 101–109 (ecology).

Orchestoidea pugettensis (Dana, 1853). Under debris on coarse-sand beaches with little seaweed.

Talitroides alluaudi Chevreux, 1898. A very small introduced species, in gardens, leaf litter, often common along shore immediately at and above high-water drift line, overlapping *Orchestia;* may occur in San Francisco area.

Talitroides topitotum Burt, 1934. An introduced Indian and Australian species, under damp leaf litter, gardens, greenhouses, in southern California, probably also in San Francisco area. First recorded in California as *Talitrus sylvaticus* (Shoemaker, 1936, J. Wash. Acad. Sci. 26: 60–64).

"Talitrus" sylvaticus Haswell, 1879. An introduced Australian terrestrial species, abundant under damp leaf litter and debris in Golden Gate and other parks in San Francisco.

REFERENCES ON GAMMARIDEA

Alderman, A. L. 1936. Some new and little known amphipods of California. Univ. Calif. Publ. Zool. 41: 53–74.

Barnard, J. L. 1950. The occurrence of *Chelura terebrans* Philippi in Los Angeles and San Francisco Harbors. Bull. So. Calif. Acad. Sci. 49: 90–97.

Barnard, J. L. 1952. Some Amphipoda from central California. Wasmann J. Biol. 10: 9–36.

Barnard, J. L. 1953. On two new amphipod records from Los Angeles Harbor. Bull. So. Calif. Acad. Sci. 52: 83–87.

Barnard, J. L. 1954a. Marine Amphipoda of Oregon. Oregon State Monog., Studies in Zoology 8: 1–103.

Barnard, J. L. 1954b. Amphipoda of the family Ampeliscidae collected in the eastern Pacific Ocean by the *Velero III* and *Velero IV*. Allan Hancock Pac. Expeds. 18: 1–137.

Barnard, J. L. 1954c. A new species of *Microjassa* (Amphipoda) from Los Angeles Harbor. Bull. So. Calif. Acad. Sci. 53: 127–130.

Barnard, J. L. 1955a. Two new spongicolous amphipods (Crustacea) from Newport Harbor, California. Pac. Sci. 9: 26–30.

Barnard, J. L. 1955b. Notes on the amphipod genus *Aruga* with the description of a new species. Bull. So. Calif. Acad. Sci. 54: 97–103.

Barnard, J. L. 1956. Two rare amphipods from California with notes on the genus *Atylus*. Bull. So. Calif. Acad. Sci. 55: 35–43.

Barnard, J. L. 1957a. A new genus of haustoriid amphipod from the northeastern Pacific Ocean and the southern distribution of *Urothoe varvarini* Gurjanova. Bull. So. Calif. Acad. Sci. 56: 81–84.

Barnard, J. L. 1957b. A new genus of phoxocephalid Amphipoda (Crustacea) from Africa, India, and California. Ann. Mag. Nat. Hist. ser. 12, 10: 432–438.

Barnard, J. L. 1959a. Estuarine Amphipoda. In Ecology of Amphipoda and Polychaeta of Newport Bay, California, by J. L. Barnard and D. J. Reish. Allan Hancock Found. Publs., Occ. Pap. 21: 13–69.

Barnard, J. L. 1959b. Liljeborgiid amphipods of southern California coastal bottoms, with a revision of the family. Pac. Nat. 1(4): 12–28.

Barnard, J. L. 1960. The amphipod family Phoxocephalidae in the eastern Pacific Ocean, with analyses of other species and notes for a revision of the family. Allan Hancock Pac. Expeds. 18: 175–368.

Barnard, J. L. 1962a. Benthic marine Amphipoda of southern California: Families Aoridae, Photidae, Ischyroceridae, Corophiidae, Podoceridae. Pac. Nat. 3: 1–72.

Barnard, J. L. 1962b. Benthic marine Amphipoda of southern California: Families Tironidae to Gammaridae. Pac. Nat. 3: 73–115.

Barnard, J. L. 1962c. Benthic marine Amphipoda of southern California: Families Amphilochidae, Leucothoidae, Stenothoidae, Argissidae, Hyalidae. Pac. Nat. 3: 116–163.

Barnard, J. L. 1962d. A new species of sand-burrowing marine Amphipoda from California. Bull. So. Calif. Acad. Sci. 61: 249–252.

Barnard, J. L. 1964a. Los anfipodos bentonicos marinos de la Costa occidental de Baja California. Rev. Soc. Mex. Hist. Nat. 24: 205–274.

Barnard, J. L. 1964b. Marine Amphipoda of Bahia de San Quintin, Baja California. Pac. Nat. 4: 55–139.

Barnard, J. L. 1965. Marine Amphipoda of the family Ampithoidae from southern California. Proc. U.S. Nat. Mus. 118 (3522): 1–46.

Barnard, J. L. 1967. New and old dogielinotid marine Amphipoda. Crustaceana 13: 281–291.

Barnard, J. L. 1969a. Gammaridean Amphipoda of the rocky intertidal of California: Monterey Bay to La Jolla. Bull. U.S. Nat. Mus. 258: 1–230.

Barnard, J. L. 1969b. The families and genera of marine gammaridean Amphipoda. Bull. U.S. Nat. Mus. 271: 1–535. (Note: last literature cited, December, 1965.)

Barnard, J. L. 1973. Revision of Corophiidae and related families (Amphipoda). Smithson. Contr. Zool. no. 151, iv + 27 pp.

Barnard, J. L. and R. R. Given 1960. Common pleustid amphipods of southern California, with a projected revision of the family. Pac. Nat. 1(17): 37–48.

Boeck, A. 1872. Bidrag til Californiens Amphipodefauna. (separate from) Forhandl. Vidensk.-Selsk. Christiana, 1872, 22 pp.

Bousfield, E. L. 1957. Notes on the amphipod genus *Orchestoidea* on the Pacific Coast of North America. Bull. So. Calif. Acad. Sci. 56: 119–129.

Bousfield, E. L. 1958. Distributional ecology of the terrestrial Talitridae (Crustacea: Amphipoda) of Canada. Proc. 10th Internat. Congr. Entomology 1: 883–898.

Bousfield, E. L. 1961. New records of beach hoppers (Crustacea: Amphipoda) from the coast of California. Nat. Mus. Canada, Contr. Zool., Bull. 172: 1–12.

Bousfield, E. L. 1973. *Shallow-water Gammaridean Amphipoda of New England.* Cornell University Press, 312 pp.

Bousfield, E. L. and J. T. Carlton 1967. New records of Talitridae (Crustacea: Amphipoda) from the central California coast. Bull. So. Calif. Acad. Sci. 66: 277–284.

Bowers, D. E. 1963. Field identification of five species of Californian beach hoppers (Crustacea: Amphipoda). Pac. Sci. 17: 315–320.

Bowers, D. E. 1964. Natural history of two beach hoppers of the genus *Orchestoidea* (Crustacea: Amphipoda) with reference to their complemental distribution. Ecology 45: 677–696.

Calman, W. T. 1898. On a collection of Crustacea from Puget Sound. Ann. N.Y. Acad. Sci. 11: 259–292.

Gurjanova, E. 1951. Bokoplavy morej SSSR i sopredel'nykh vod (Amphipoda-Gammaridea). Opred. po Faune SSSR, Akad. Nauk SSSR 41: 1–1029 (in Russian).

Gurjanova, E. 1962. Bokoplavy severnoi chasti Tixogo Okeana (Amphipoda-Gammaridea) chast' 1. Opred. po Faune SSSR, Akad. Nauk SSSR, 74: 1–440 (in Russian).

Holmes, S. J. 1904. Amphipod crustaceans of the Expedition. Harriman Alaska Exped. 10: 233–246.

Holmes, S. J. 1908. The Amphipoda collected by the U.S. Bureau of Fisheries Steamer, *Albatross,* off the west coast of North America in 1903 and 1904, with descriptions of a new family and several new genera and species. Proc. U.S. Nat. Mus. 35: 489–543.

Mills, E. L. 1961. Amphipod crustaceans of the Pacific coast of Canada, I. Family Atylidae. Bull. Nat. Mus. Canada 172: 13–33.

Mills, E. L. 1962. Amphipod Crustaceans of the Pacific Coast of Canada, II. Family Oedicerotidae. Nat. Hist. Papers, Nat. Mus. Canada 15: 1–21.

Mills, E. L. 1967. A reexamination of some species of *Ampelisca* (Crustacea: Amphipoda) from the east coast of North America. Canad. J. Zool. 45: 635–652.

Sars, G. O. 1895. *Amphipoda. An account of the Crustacea of Norway with short descriptions and figures of all the species.* 1: i–viii + 1–711.

Shoemaker, C. R. 1933. Two new genera and six new species of Amphipoda from Tortugas. Carnegie Inst. Wash. Paps. 28: 245–256 (Publication 435).

Shoemaker, C. R. 1934. Two new species of Corophium from the west coast of America. J. Wash. Acad. Sci. 24: 356–360.

Shoemaker, C. R. 1938. Three new species of the amphipod genus *Ampithoe* from the west coast of America. J. Wash. Acad. Sci. 28: 15–25.

Shoemaker, C. R. 1942a. Notes on some American fresh-water amphipod crustaceans and descriptions of a new genus and two new species. Smithson. Misc. Colls. 101(9): 1–31.

Shoemaker, C. R. 1942b. Amphipod crustaceans collected on the Presidential Cruise of 1938. Smithson. Misc. Colls. 101(11): 1–52.

Shoemaker, C. R. 1949. The amphipod genus *Corophium* on the west coast of America. J. Wash. Acad. Sci. 39: 66–82.

Shoemaker, C. R. 1955a. Notes on the amphipod crustacean *Maeroides thompsoni* Walker. J. Wash. Acad. Sci. 45: 1–59.

Shoemaker, C. R. 1955b. Amphipoda collected at the Arctic Laboratory, Office of Naval Research, Point Barrow, Alaska, by G. E. MacGinitie. Smithson. Misc. Colls. 128(1): 1–78.

Shoemaker, C. R. 1964. Seven new amphipods from the west coast of North America with notes on some unusual species. Proc. U.S. Nat. Mus. 115: 391–430.

Skogsberg, T. and G. H. Vansell 1928. Structure and behavior of the amphipod, *Polycheria osborni.* Proc. Calif. Acad. Sci. (4) 17: 267–295.

Stasek, C. R. 1958. A new species of *Allogaussia* (Amphipoda, Lysianassidae) found living within the gastrovascular cavity of the sea-anemone *Anthopleura elegantissima.* J. Wash. Acad. Sci. 48: 119–126.

Stimpson, W. 1857. On the Crustacea and Echinodermata of the Pacific shores of North America. Boston J. Nat. Hist. 6: 444–532.

Walker, A. O. 1898. Crustacea collected by W. A. Herdman, F.R.S., in Puget Sound, Pacific coast of North America, September, 1897. Trans. Liverpool Biol. Soc. 12: 268–287.

PHYLUM ARTHROPODA: CRUSTACEA, AMPHIPODA: CAPRELLIDEA

John C. McCain
Bernice P. Bishop Museum, Honolulu, Hawaii

(PLATES 86–90)

The **Caprellidea** consists of the abundant and familiar caprellids or "skeleton shrimps" (fig. 1) and the not so familiar cetacean parasites known as cyamids or "whale lice." This suborder of amphipods is characterized by reduction of the abdomen, sometimes to a tiny, unsegmented vestige. Generally the first pereonite is at least partially fused with the head to form a cephalon. Sexual dimorphism is marked; adult females bear a conspicuous marsupium. The pereopods vary from fully segmented with a coxal plate to completely absent, and this variation is used to a large extent for generic separation.

Historically, the Caprellidea have been considered a suborder quite distinct from the Gammaridea but, following the discovery of *Caprogammarus*, which is intermediate between caprellidean and gammaridean stocks, questions have been raised about this classical division. In the past, the Caprellidea have usually been divided into two families, **Cyamidae** and **Caprellidae,** but the latter has now been separated into four families. For a discussion of the relationship between the suborders of amphipods and revision of the families of Caprellidea, see McCain (1968, 1970).

The families **Phtisicidae, Aeginellidae,** and **Caprellidae** are represented in the intertidal zone of central California. Most caprellids are nonspecific as to the substrate upon which they live; however, most require something to which they can cling, and therefore are not found on bare sandy or muddy bottoms. Algae, sea grasses, sponges, hydroids, and bryozoans are common habitats. Some, such as the tiny *Caprella greenleyi* (fig. 18), are commonly associated with other invertebrates. Caprellids feed primarily upon diatoms but they also consume small invertebrates and perhaps detritus. These small amphipods are prey for many bottom-feeding fishes, such as cod, blennies, skates, and sea bass, as well as for grass shrimps, sea anemones, and other invertebrates.

The **Cyamidae** (figs. 19–21) are highly specialized ectoparasites of cetaceans, and are encountered intertidally only on stranded whales and dolphins. See Leung (1967) for a key.

The following key should permit identification of most central California intertidal caprellids. It should be borne in mind that the Caprellidea of southern California may have a subtropical component, and have not been studied thoroughly. The morphological characters used in the keys are illustrated in Figure 1. The advice of Diana R. Laubitz of the National Museum of Natural Sciences, Ottawa, Canada in the preparation of this section is gratefully acknowledged.

KEY TO FAMILIES OF INTERTIDAL CAPRELLIDEA

1. Gills on pereonites 2 to 4; mandible lacking molar (fig. 1) .... PHTISICIDAE (Key A)
– Gills on pereonites 3 and 4; mandible with molar (fig. 1) 2
2. Pereonites 3 and 4 with rudimentary appendages (examine with great care; these are often overlooked!); mandibular palp present (fig. 1) ............AEGINELLIDAE (Key B)
– Pereonites 3 and 4 lack appendages; mandibular palp absent ....CAPRELLIDAE (Key C)

KEYS TO SPECIES

A. PHTISICIDAE

The only California species is *Perotripus brevis* (fig. 6).

B. AEGINELLIDAE

1. Pereopods 3 and 4 of single article 2
– Pereopods 3 and 4 of 2 articles 4
2. Pereopod 5 inserted at midlength of pereonite 5 .... *Tritella tenuissima*
– Pereopod 5 inserted at posterior of pereonite 5 3
3. Lateral spines on pereonites 2–4 directed laterally; flagellum of antenna 2 slender, with long setae (fig. 3) .... *Tritella pilimana*
– Lateral spines on pereonites 2–4 directed anteriorly; flagellum of antenna 2 stout, with short setae (fig. 2) .... *Tritella laevis*
4. Pereopod 5 of 6 articles (fig. 8) ....*Deutella californica*
– Pereopod 5 of 3 articles (fig. 7) .... *Mayerella banksia*

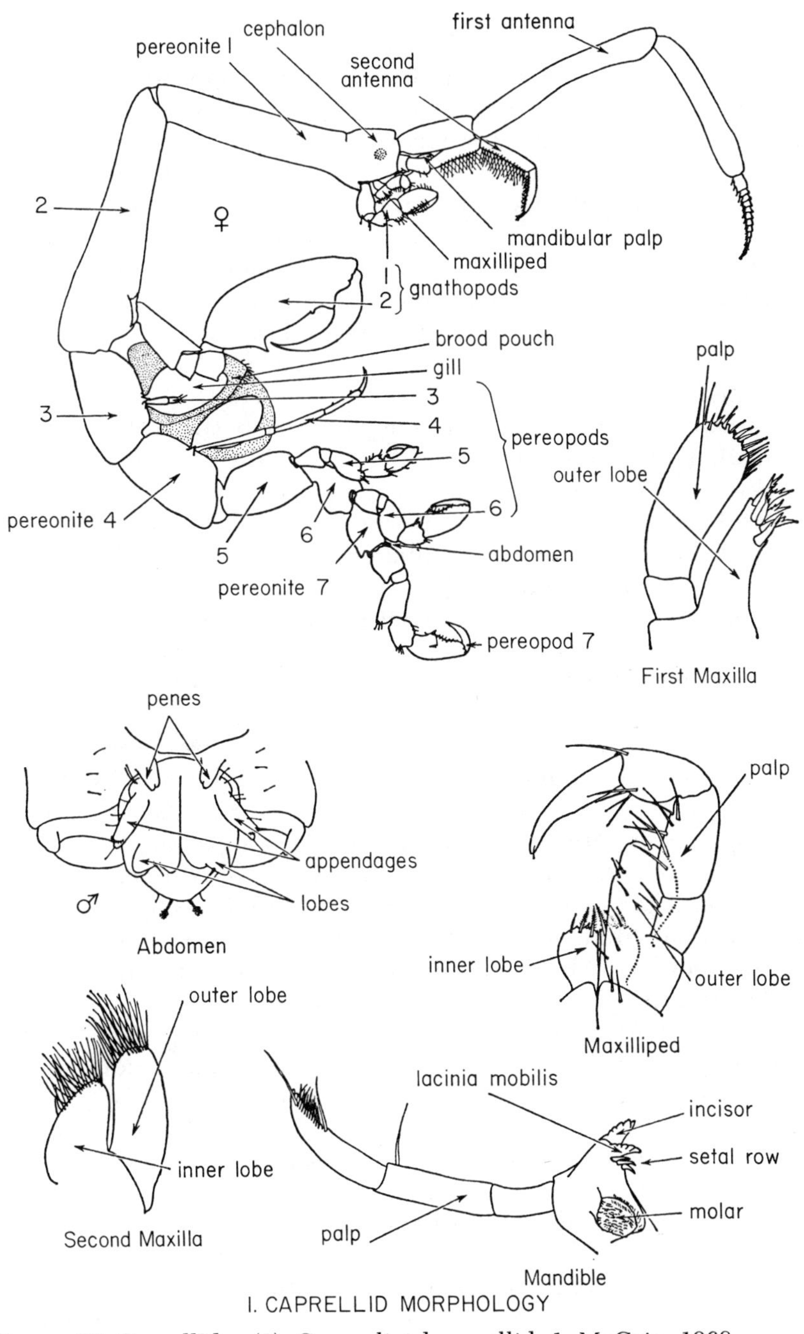

PLATE 86. **Caprellidea** (1). Generalized caprellid: 1, McCain, 1968.

C. CAPRELLIDAE

1. Female abdomen with pair of single article appendages ... 2
– Female abdomen with pair of lobes but no appendages 3
2. Flagellum of antenna 1 longer than peduncle (fig. 4) .... *Metacaprella anomala*
– Flagellum of antenna 1 shorter than peduncle (fig. 5) ....*Metacaprella kennerlyi*
3. Ventral spine present between the insertions of gnathopod 2 4
– Ventral spine lacking between the insertions of gnathopod 2 6
4. Cephalic spine long, anteriorly directed (fig. 17) ....*Caprella californica*
– Cephalic spine absent 5
5. Lateral projections on anterior of pereonite 5; spine at base of gnathopod 2 large (figs. 12A, B) ....*Caprella equilibra*
– Pereonite 5 without lateral projections; spine at base of gnathopod 2 small or absent (fig. 10) .... *Caprella mendax*
6. Cephalon without spine or tubercle 7
– Cephalon with spine or tubercle 9
7. Grasping spines on propodus of pereopods medial (fig. 16). ....*Caprella gracilior*
– Grasping spines on propodus of pereopods proximal 8
8. Flagellum of antenna 1 and 2 uniarticulate (fig. 18) .... *Caprella greenleyi*
– Flagellum of antenna 1 and 2 bi- or multiarticulate (fig. 13) .... *Caprella laeviuscula*
9. Propodus of pereopods lacking grasping spines ....*Caprella brevirostris*
– Propodus of pereopods with 2 or more grasping spines 10
10. Cephalon with anteriorly directed triangular projection 11
– Cephalon with tubercles or spines, but not an anteriorly directed triangular projection 15
11. Dorsal tuberculations on pereonites lacking or minute 12
– Dorsal tuberculations on pereonites present and obvious .. 14
12. Anteriorly directed projection on cephalon slight ....*Caprella uniforma*
– Anteriorly directed projection on cephalon well developed and distinct 13
13. Pereonite 5 usually shorter than pereonites 6 plus 7 ....*Caprella penantis*
– Pereonite 5 usually longer than pereonites 6 plus 7 (fig. 11) .... *Caprella natalensis*
14. Dorsal tuberculations on pereonites large; antenna 1 peduncle scarcely setose (fig. 9) .... *Caprella verrucosa*

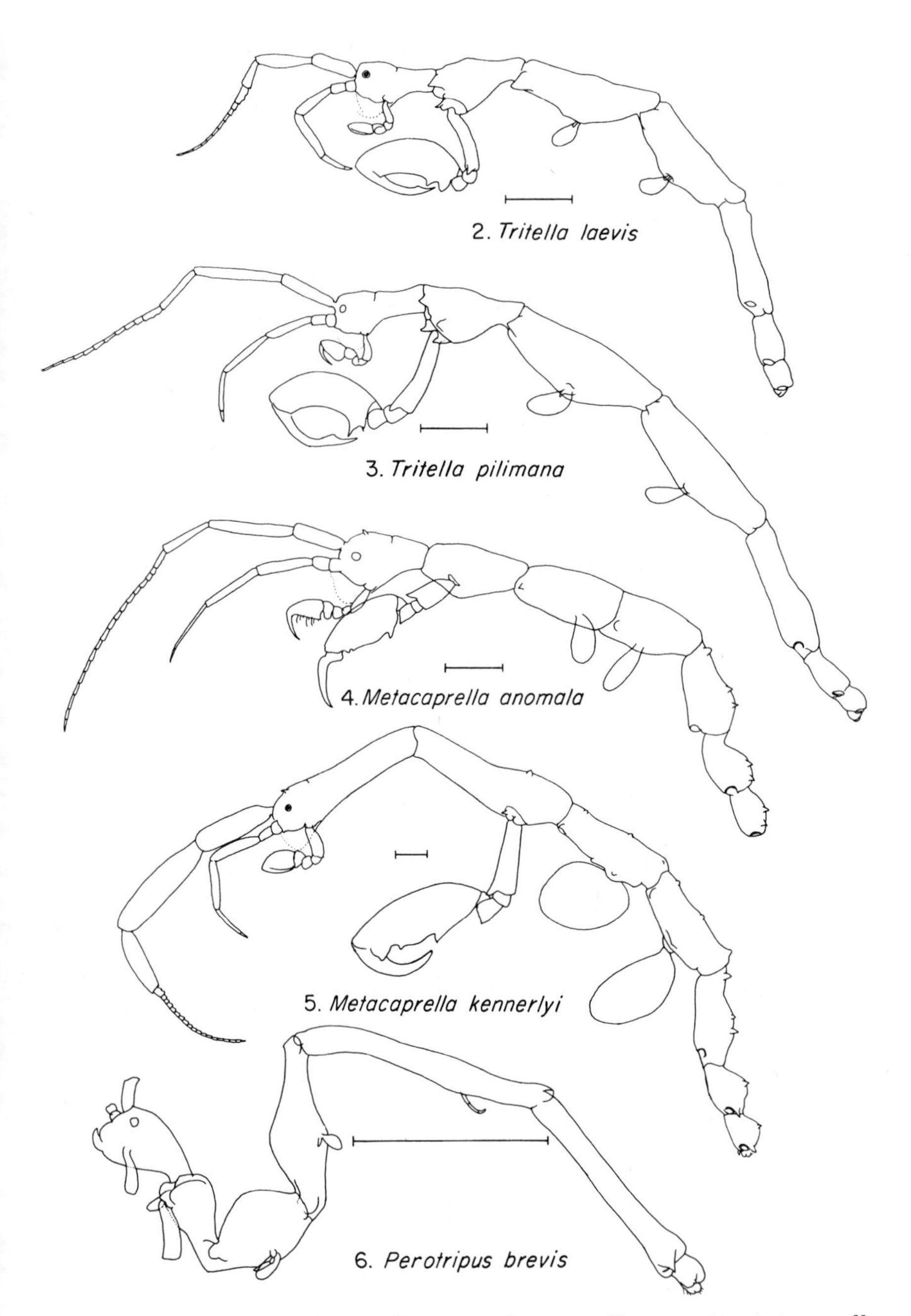

PLATE 87. **Caprellidea** (2). *Tritella* spp., *Metacaprella* spp., *Perotripus;* all males, from Laubitz, 1970. Scales = 1 mm.

- Dorsal tuberculations on pereonites small; antenna 1 peduncle finely setose (fig. 14) .... *Caprella incisa*

15. Cephalic spine dorsally directed, pointed *Caprella pilipalma*

- Cephalic projections are rounded knobs or blunt tubercles (fig. 15) .... *Caprella ferrea*

LIST OF CAPRELLIDEA

PHTISICIDAE

Perotripus brevis (LaFollette, 1915).

AEGINELLIDAE

Deutella californica Mayer, 1890.

Mayerella banksia Laubitz, 1970.

Tritella laevis Mayer, 1903.

Tritella pilimana Mayer, 1890.

Tritella tenuissima Dougherty and Steinberg, 1953.

CAPRELLIDAE

Caprella brevirostris Mayer, 1903.

Caprella californica Stimpson, 1857.

Caprella equilibra Say, 1818.

Caprella ferrea Mayer, 1903.

Caprella gracilior Mayer, 1903.

Caprella greenleyi McCain, 1969. Northern California and Oregon, on the starfish *Henricia leviuscula*, hydroids, algae, etc.

Caprella incisa Mayer, 1903.

Caprella laeviuscula Mayer, 1903.

Caprella mendax Mayer, 1903.

Caprella natalensis (Mayer, 1903). "*C. penantis* and *C. natalensis* are very similar in general appearance" (see Laubitz, 1972, p. 45 for details). Monterey Bay may be the northern limit of *C. penantis* and the southern of *C. natalensis;* further study of this complex is needed.

Caprella penantis Leach, 1814.

Caprella pilipalma Dougherty and Steinberg, 1953.

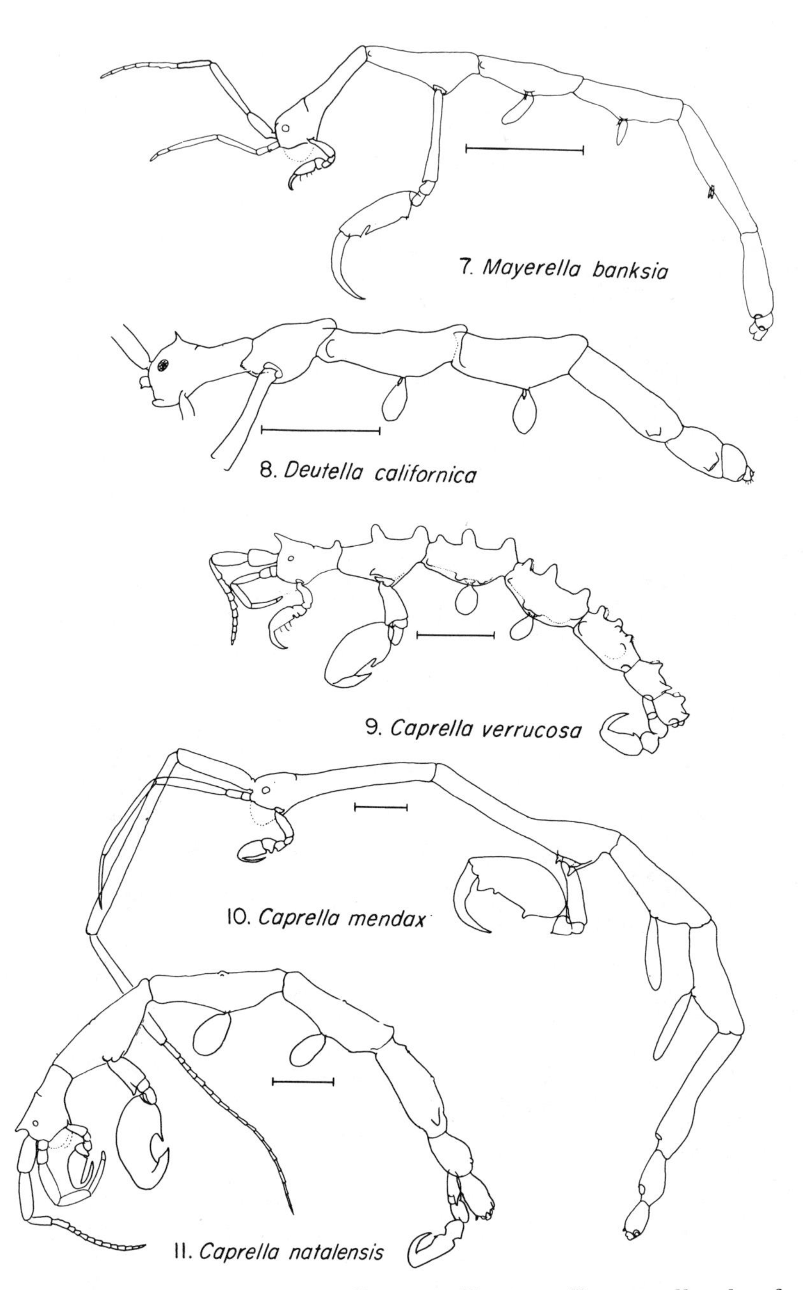

PLATE 88. **Caprellidea** (3). *Mayerella, Deutella, Caprella* spp.; all males, from Laubitz, 1970.

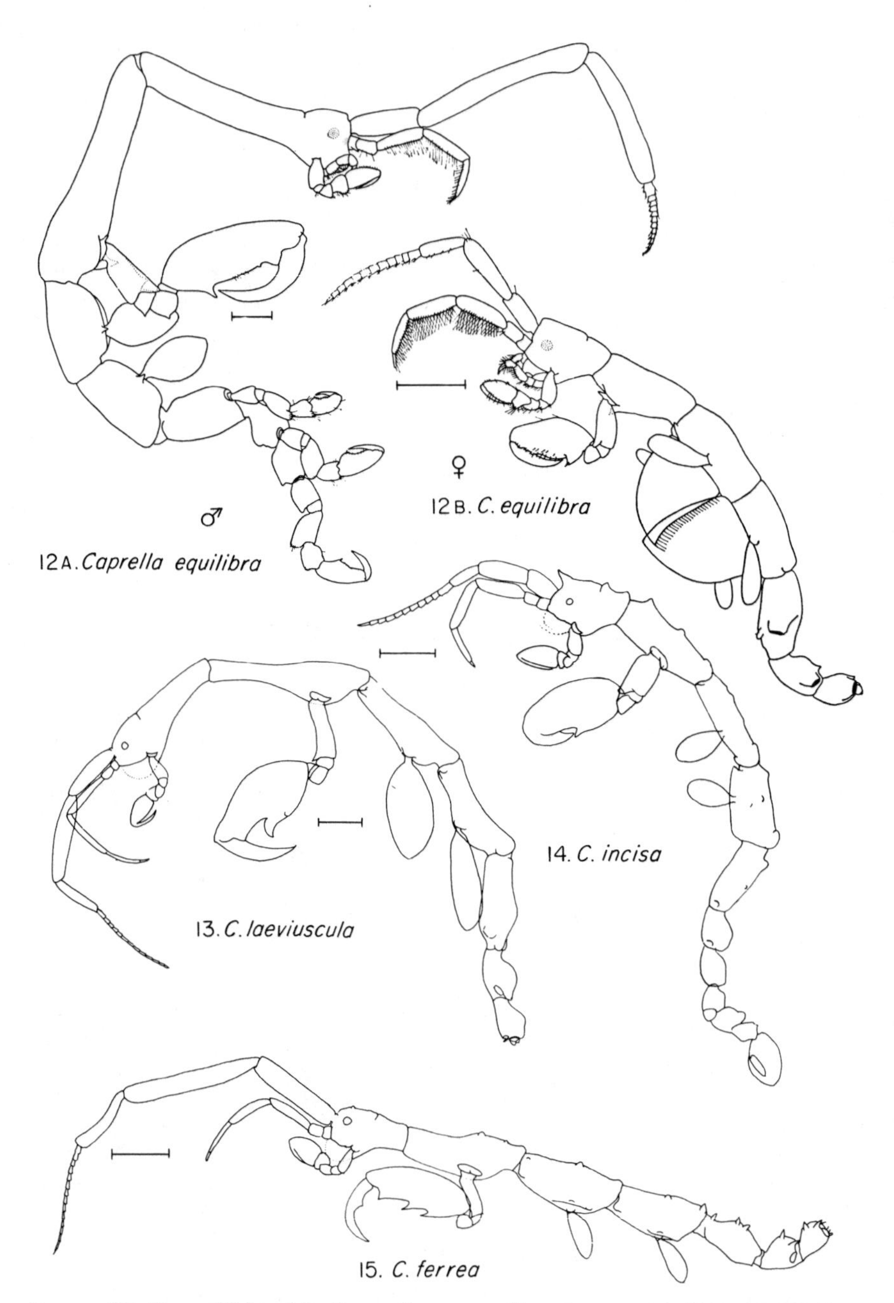

PLATE 89. **Caprellidea** (4). *Caprella* spp.; all males except 12B; 12A,B from McCain, 1968; rest from Laubitz, 1970. Scales = 1 mm.

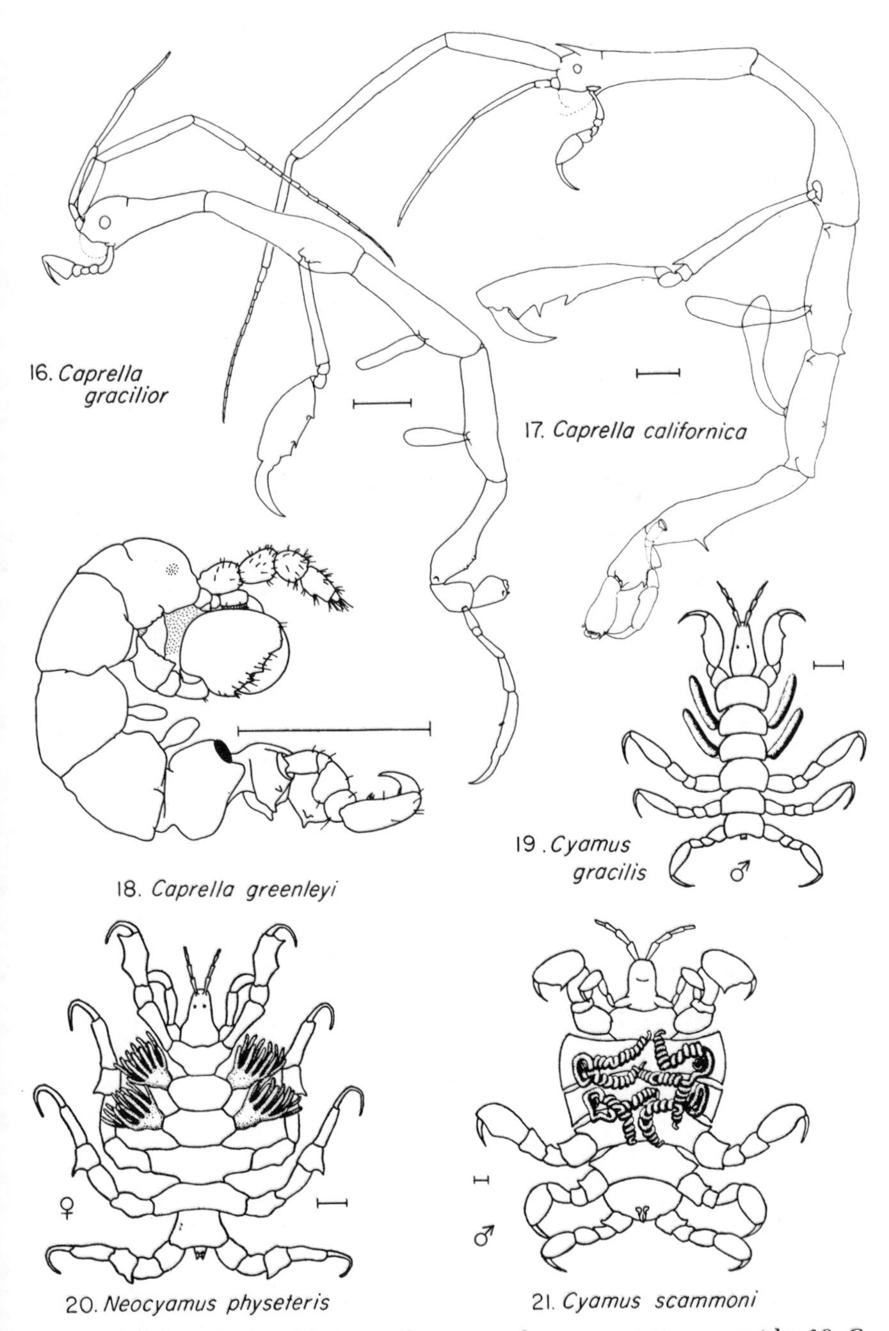

PLATE 90. **Caprellidea** (5). *Caprella* spp. and representative cyamids: 19, *C. gracilis*, from north Pacific right whale; 20, *N. physeteris*, from sperm whale; 21, *C. scammoni*, from gray whale; 16, 17, from Laubitz, 1970; 18 from McCain, 1969; cyamids from Leung, 1967. Scales = 1 mm.

Caprella uniforma LaFollette, 1915. Probably a junior synonym of *C. angusta* Mayer, 1903.

Caprella verrucosa Boeck, 1871.

Metacaprella anomala (Mayer, 1903).

Metacaprella kennerlyi (Stimpson, 1864).

REFERENCES ON CAPRELLIDEA

Dougherty, E. C. and J. E. Steinberg 1953. Notes on the skeleton shrimps (Crustacea: Caprellidae) of California. Proc. Biol. Soc. Wash. 66: 39–50.

Laubitz, D. R. 1970. Studies on the Caprellidae (Crustacea, Amphipoda) of the American North Pacific. Nat. Mus. Canada Publ. Biol. Oceanogr. no. 1: 89 pp.

Laubitz, D. R. 1972. The Caprellidae (Crustacea, Amphipoda) of Atlantic and Arctic Canada. Nat. Mus. Canada Publ. Biol. Oceanogr. no. 4: 82 pp.

Leung, Y-M. 1967. An illustrated key to the species of whale-lice (Amphipoda, Cyamidae), ectoparasites of Cetacea, with a guide to the literature. Crustaceana 12: 279–291.

McCain, J. C. 1968. The Caprellidae (Crustacea: Amphipoda) of the western North Atlantic. Bull. U.S. Nat. Mus. 278: vi + 147 pp.

McCain, J. C. 1969. A new species of caprellid (Crustacea: Amphipoda) from Oregon. Proc. Biol. Soc. Wash. 82: 507–510.

McCain, J. C. 1970. Familial taxa within the Caprellidea (Crustacea: Amphipoda). Proc. Biol. Soc. Wash. 82: 837–842.

McCain, J. C. and J. E. Steinberg 1970. Amphipoda I, Caprellidea I, Fam. Caprellidae. Crustaceorum Catalogus, part 2: 78 pp.

PHYLUM ARTHROPODA: CRUSTACEA, EUCARIDA

(PLATES 91–98)

A wide variety of forms are included in the **Eucarida.** Although many subdivisions of this large group have been proposed, Balss's (1957) scheme has received widespread support among crustacean specialists and in the main will be followed here.

ORDER EUPHAUSIACEA

(PLATE 54, fig. 6)

Euphausiids are small, pelagic, marine, shrimplike forms, differing from decapods in having biramous thoracic appendages and in other characters; most are luminescent and all have well-developed pleopods for swimming. Euphausiids or "krill," important as food of whalebone whales, are generally deep-living, but many perform diurnal vertical migrations, rising to the surface at night; locally, schools may be encountered swimming near shore (as in Monterey Bay) and occasionally species of *Thysanoessa* and *Euphausia* may be washed ashore. See Boden, Johnson, and Brinton (1955), Brinton (1962), and Mauchline and Fisher (1969).

ORDER DECAPODA

Members of this large and diverse order have five pairs of thoracic limbs, developed for walking or grasping, and three pairs of maxillipeds. Decapods include the familiar shrimps, prawns, lobsters, crayfishes, and crabs as well as others less familiar; the group is so diverse that it is divided into a number of sections and tribes in two suborders:

Suborder **Natantia**—swimming types
  Tribe Penaeidea—primitive prawns, no local species
  Tribe Caridea—shrimps and prawns
Suborder **Reptantia**—creeping or walking types
  Section Palinura—spiny lobsters and slipper lobsters, no local intertidal species
  Section Astacura—lobsters and crayfish, represented only by fresh-water species locally (see p. 386)
  Section Anomura—diverse forms, including:
    Tribe Thalassinidea—mud and ghost shrimps
    Tribe Paguridea—hermit crabs and stone crabs
    Tribe Galatheidea—porcelain crabs and pelagic galatheids ("red crabs")
    Tribe Hippidea—sand crabs
  Section Brachyura—true crabs

Of these, the **Caridea, Anomura,** and **Brachyura** are treated in detail below.

BIOLOGY OF DECAPOD CRUSTACEA

Armand M. Kuris
Bodega Marine Laboratory

Decapod crustaceans are relatively large and sturdy, well suited for field observations and study of living or freshly killed specimens. Recognition of life-history features will provide the student many clues for physiological, ecological, and behavioral problems. The following section may serve as a protocol for efficient observation of crustacean field biology. Details of life history have not been recorded for most of our local species. Thus the primary intention of this protocol is not to describe well-known systems, but to direct the attention of the observer to areas where interesting discoveries may be made.

Sexing. All local species of decapods attach their eggs to the pleopods of the female for embryonic development. Such **ovigerous females** provide the quickest means for sexual determination, but not all females breed at the same time or in all seasons.

The **Caridea** are only weakly sexually dimorphic; one must look for the presence (males) or absence (females) of an **appendix masculina** on the second pair of pleopods (figs. 6A, B).

Brachyura are markedly sexually dimorphic. Males have relatively large chelae and narrow, often triangular, abdomens (fig. 4C); first and second pleopods are specialized for copulation (figs. 8A, B); third and fourth pleopods are absent. Females have wide, flaplike abdomens (figs. 4A, B) and four pairs of pleopods with long setae for egg attachment.

Anomura have a variety of sexually dimorphic features. In general, males have fewer pairs of relatively small, slender or inflexible pleopods. Local female hermit crabs (*Pagurus*) have four pairs of pleopods, males have three (figs. 9A, B). Female porcelain crabs (*Petrolisthes*) have two pairs of pleopods, males have one (figs. 10A, B). Sand crab (*Emerita*) females have three pairs of pleopods, males have none. The striking differences in shape of the male and female pleopods should be noted. Correlation between structure and function of these appendages remains to be studied. Ghost and mud shrimp (*Callianassa* and *Upogebia*) males also have fewer and/or smaller and more slender pleopods than do females (figs. 11A, B; 12A, B). However, *Callianassa* is most easily sexed by noting the presence of the large major chela of the males, often twice as long as the minor chela; in females the major chela exceeds the minor by less than 50 percent in length. Stone crabs (*Hapalogaster*, *Oedignathus*) are quickly sexed, as only females show signs of segmentation on the left side of their soft, asymmetrical abdomens.

Size is another sexually dimorphic feature. Adult female sand crabs (*Emerita*) and most pea crabs (*Pinnixa*, *Fabia*) are much larger than males. Most female shrimp are also larger, but the size ranges of the sexes often overlap. Sizes are similar for both sexes of hermit crabs. In most of the remaining Brachyura and Anomura, males are larger than females, again with considerable overlap.

Upon reaching sexual maturity, most if not all decapods undergo a **molt of puberty** to attain their adult morphology. This is most easily seen in female brachyurans, in which the width of the abdomen relative to other parts of the crab greatly increases at the molt of puberty (figs. 4A, B). In all decapods, more subtle changes in the relative size of the chelae, abdominal width, presence of setae on pleopods and the abdominal margin, and shape of the genital opening can usually be detected in at least one sex. Relative growth techniques described by Teissier (1960) are very useful in maturation studies.

A related phenomenon is the acquisition of **breeding dress** prior to oviposition in many shrimp and possibly other decapods. Female shrimp acquire longer setae on their pleopods at the preceding molt. Shorter setae appear at molts preceding nonovigerous instars.

Reproduction. Ovaries of decapod Crustacea lie dorsal to the other organs in the carapace and anterior part of the abdomen (entirely abdominal for hermit crabs and ghost and mud shrimps). They can readily be seen through the transparent cuticle of most shrimps. Corre-

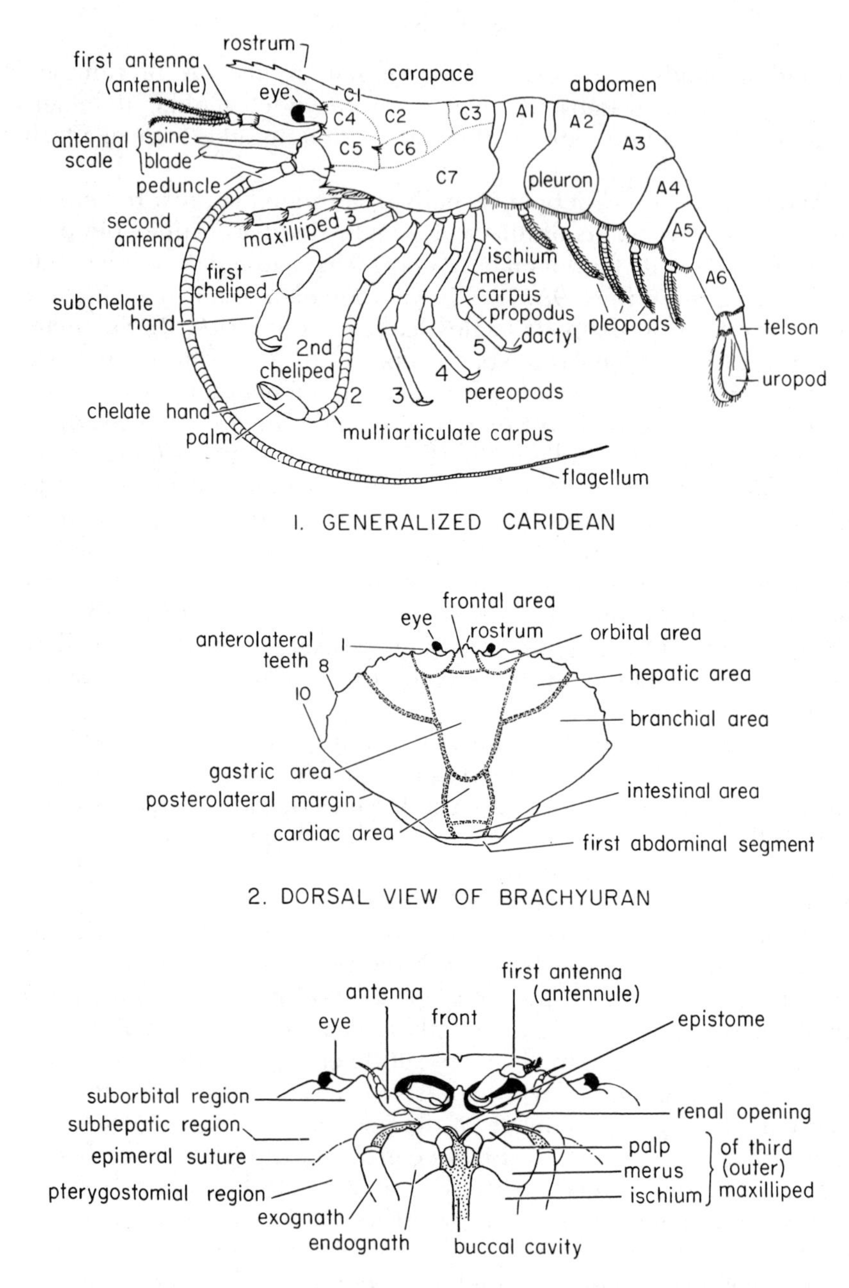

PLATE 91. **Crustacea, Decapoda** (1). General external morphology: 1, generalized caridean, parts as labeled, carapace region abbreviated: (C1) frontal, (C2) gastric, (C3) cardiac, (C4) orbital, (C5) antennal, (C6) hepatic, (C7) branchial; 2, dorsal view of brachyuran carapace (outline of *Cancer magister*); 3, anteroventral view of brachyuran. All modified after Schmitt.

sponding observations can also be made on typically opaque crabs by looking through the flexible **arthrodial membranes** connecting the ventral surfaces of thorax and abdomen. Ovarian color varies, being red (*Pugettia*), purple (*Pachygrapsus*), brown (*Fabia*), green (*Crangon*), orange (*Emerita*), or magenta (*Petrolisthes*); as the ovary matures and grows, pigmentation becomes more intense.

Relative size of the ripening ovary may be judged by its relation to certain morphological markers, such as sutures separating abdominal segments. Recently laid eggs lie in a gelatinous mass on the pleopods. Within a day an outer membrane forms around each egg and a thin strand attaches it to a pleopodal seta. Young eggs are evenly pigmented. As the yolk is gradually displaced by the growing embryo, a transparent area appears in the egg; in advanced embryos, eyespots and larval chromatophores may also be recognized. Broods about to hatch often have a grayish cast since all the yolk has been absorbed. Information on the reproductive state of male decapods is relatively difficult to obtain, dissection and microscopic examination often being required.

Molting. Discontinuity of growth is one of the most singular aspects of the lives of arthropods. **Ecdysis**—the actual shedding of the old skin, the **exuvium**—and the accompanying increase in size take only a few minutes. However, the entire interval between molts represents a dynamic, cyclical process. Following ecdysis, the animal is soft. In this **postmolt** period, the cuticle gradually hardens as different areas of the exoskeleton calcify sequentially. Compare a series of animals of a given species to detect this hardening sequence. Postmolt ends with the deposition of a thin, membranous layer. This may be detected by carefully cracking the carapace. If a shiny membrane holds the cracked pieces together, the animal has passed into the **intermolt** period.

The **premolt** period, or preparation for the coming molt, is signaled by the separation of the epidermis from the old cuticle and the deposition of the thin, pre-exuvial layers of the new cuticle directly under the old. These early premolt stages may be recognized in intact carideans and in ghost and mud shrimps by observing setal formation along the margins of the uropods, telson, or antennal scales. Dissection of the transparent mouthparts is usually necessary in more heavily calcified decapods. In late premolt, decalcification of the old exoskeleton may be detected if gentle pressure along the epimeral suture (fig. 3) causes it to crack. Molting is imminent if the epimeral suture of a crab, or the dorsal thoracic-abdominal suture of a shrimp, is visibly split. Passano (1960) gives a good summary of the molt cycle.

Exuvia may be distinguished from the empty remains of dead animals by the absence of pigment from the corneas of the eyestalks of the exuvia. Discovery of a fragile intact exuvium suggests that a very soft, recently molted animal may be hiding nearby. The previous owner of the exuvium can be certified by matching details of the pigment patterns of the exuvium and the soft animal. Comparison of the soft animal and its exuvium will demonstrate the growth increment per molt for animals of that size.

Regeneration. Decapods are able to cast off (**autotomize**) their limbs under duress and then regenerate the appendages in subsequent molts. Autotomy is readily demonstrated by squeezing basal segments of an appendage. The regenerating limb forms in a bud that protrudes from the stump of the autotomized limb. Limb buds are transparent in early regenerative stages. Only when the animal passes into premolt does the bud become pigmented (the new cuticle is being deposited on the regenerating appendage). Recently regenerated appendages are smaller than normal limbs, but this size discrepancy is no longer apparent after a second or third molt. In some species the regenerated appendage always has a distinctive appearance. Since spider crabs (e.g., *Pugettia*) cease molting after the molt of puberty, a calcified cap is secreted over stumps of limbs lost in their terminal instar. Presence of such a cap verifies that the animal is an adult. Frequency of missing appendages is high among porcelain crabs, true crabs, and *Betaeus* shrimps; it is low for hermit crabs and rare for most carideans.

Parasites. Decapods are hosts to certain large and conspicuous parasites. Asymmetrical swellings on the gill chambers of shrimps, porcelain crabs, ghost and mud shrimps, and hermit crabs usually contain bopyrid isopods (p. 284). Organisms adhering to the abdomens of hermit crabs, *Pugettia*, or *Upogebia* are either rhizocephalan cirripedes (p. 259), bopyrids, or (on *Upogebia*) the small clam *Orobitella* (= "*Pseudopythina*"). Rhizocephalans and bopyrids often cause **parasitic castration** by destroying the ovaries of female hosts or by modifying male secondary sexual characteristics toward the feminine condition (fig. 4D).

PLATE 92. **Crustacea, Decapoda** (2). 4, abdominal shape and chela length of *Hemigrapsus oregonensis* in various stages of maturation and parasitism by the entoniscid isopod parasitic castrator *Portunion conformis;* all crabs 10.3–10.8 mm carapace width; A–C as labeled, D., male feminized by *Portunion;* 5, *H. oregonensis*, parts as labeled (redrawn from Hedgpeth by Emily Reid); 6–12, sexual dimorphism of decapod pleopods, sex as indicated, numbering of pleopods starts with first visible pleopod; 6A–B, *Crangon:* both left, posterior face (body length 5 cm); 7A–C, *Pachygrapsus:* A, left, anterior face; B and C, left, posterior face (carapace width 2 cm); 8A–B, *Cancer:* both left, anterior face (carapace width 15 cm); 9A–B, *Pagurus:* both left (carapace length 8 mm); 10A–B *Petrolisthes:* both left, anterior face (carapace length 15 mm); 11A–B, *Callianassa:* A, posterior face; B anterior face (body length 10 cm); 12A–B *Upogebia:* A, posterior face; B anterior face (body length 10 cm).

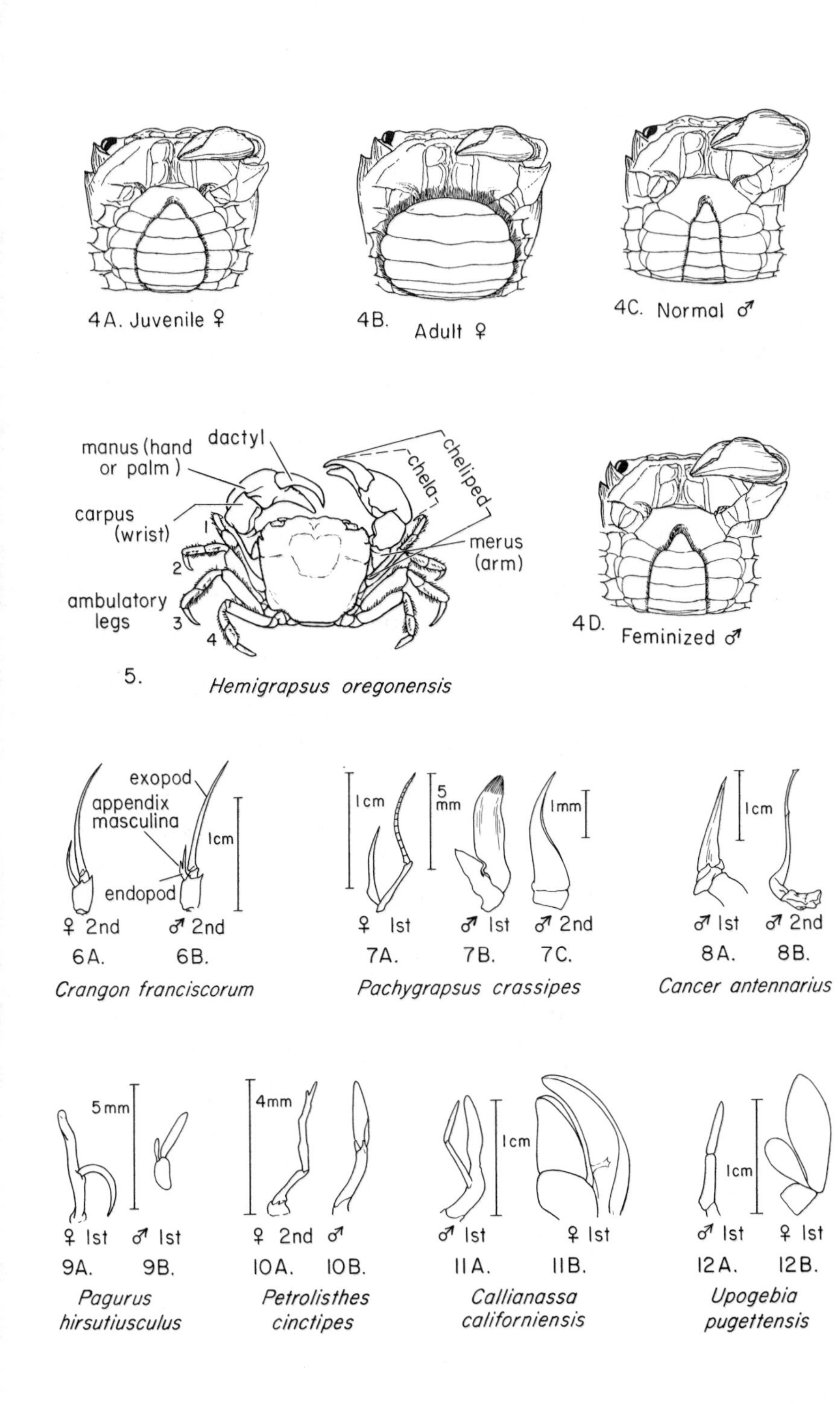

5. *Hemigrapsus oregonensis*

6A. 6B. *Crangon franciscorum*

7A. 7B. 7C. *Pachygrapsus crassipes*

8A. 8B. *Cancer antennarius*

9A. 9B. *Pagurus hirsutiusculus*

10A. 10B. *Petrolisthes cinctipes*

11A. 11B. *Callianassa californiensis*

12A. 12B. *Upogebia pugettensis*

Decapods also commonly serve as hosts for a variety of internal parasites including entoniscid isopods (p. 286), acanthocephalans (p. 125), nematodes, and larval trematodes and cestodes. Local shrimps and other decapods should also be investigated for larval nematomorphans (p. 125).

The exoskeletons of some decapods provide a substrate for the settlement and growth of encrusting organisms (barnacles, bryozoans, sponges, etc.). The size and number of these fouling organisms should be compared to the molt stage and size of the host. Barnacle orientation may indicate the pattern of water flow in the immediate vicinity of the host. Egg masses of ovigerous females should be inspected for the presence of egg-predator nemerteans (*Carcinonemertes,* p. 117) and other associated organisms.

Behavior. All copulating pairs should be carefully studied. Note molt stages, reproductive states, and relative sizes. Some species appear to form long-term pairs; *Alpheus, Betaeus,* and *Pachycheles* are possible examples (see MacGinitie, 1937). Note the kinds of snail shells that hermit crabs occupy, how they fit, and the relative sizes of the hermits and their shells.

In local decapods, burrowing ranges from the brief escape behavior of some caridean shrimps to the construction of deep, complex, permanent burrows by ghost and mud shrimps. *Crangon* often settles into shallow depressions leaving only its eyes and antennae exposed. Sand crabs burrow backward as waves recede on sandy beaches. Their long, setose, second antennae are then extended to feed in the moving sand. Ghost shrimps (*Callianassa*) build poorly defined burrows in muddy sand; mud shrimps (*Upogebia*) construct permanent U-shaped burrows with strong walls cemented by mucous secretions. *Upogebia* burrows may have enlarged sections, side chambers, and two or three openings.

Some decapods display strong tidal rhythms. *Emerita* moves up and down the beach with the incoming and outgoing tides. Diel rhythms are less obvious, but at least *Pachygrapsus* is distinctly nocturnal, foraging at night and tightly wedged into crevices by day. Many rhythmic behavior patterns remain to be observed and described.

KEYS TO DECAPOD CRUSTACEA

James T. Carlton
California Academy of Sciences, San Francisco
and
Armand M. Kuris
Bodega Marine Laboratory

(PLATES 91–98)

Schmitt's classic monograph, *Marine Decapod Crustacea of California* (1921), remains the single most useful reference for local decapods. Garth (1958) and Haig (1960) have revised the Majidae and Porcellanidae, respectively and treatments of smaller groups, among them Chace (1951) on *Hippolyte* and Hart's (1964) revision of *Betaeus*, are available. Holthuis's generic revision (1947) of *Spirontocaris* has been adopted here, but local species may require revision. The Pinnotheridae, or pea crabs, of central California remain poorly known systematically and biologically (see discussion in the species list). Compared to other groups, the number of introduced marine decapods in central California is very small; only the shrimp *Palaemon macrodactylus* (from the Orient, perhaps by way of ships) and the crab *Rhithropanopeus harrisii* (from the Atlantic coast with oysters) are recognized as introductions; there is, however, a possibility that some pea crabs in bays are introduced. Terms used in the following keys are illustrated in figures 1, 2, 3, and 5.

We acknowledge the aid of Dr. Rogene K. Thompson for notes on the thalassinids, and of F. G. Doherty and P. Jennings for notes on the pagurids and pinnotherids, respectively. The key to *Betaeus* is adapted from Hart (1964), and the key to *Petrolisthes* from Haig (1960).

KEY TO MAJOR DECAPOD CRUSTACEAN GROUPS

1. Abdomen shrimplike, with well-developed tail fan:
- Body generally laterally compressed, shrimplike in form; side plates (pleura) of 2nd abdominal segment overlap those of 1st; abdomen usually with a sharp bend **Caridea** (Key A, p. 386)
- Body dorsoventrally flattened; lateral margins of second abdominal segment not overlapping 1st segment; chelipeds very

large and strong; fresh water (in this area) .... **Astacura**

See Riegel, 1959. In central coastal California 2 introduced crayfish, *Pacifastacus leniusculus* (Dana, 1852) (formerly placed in the European *Astacus*) and *Procambarus clarkii* (Girard, 1852), occur in shallow muddy sloughs, irrigation ditches, lakes, streams, and at the heads of estuaries. (Riegel's synonymy of *P. trowbridgii* (Stimpson, 1859) with *P. leniusculus* has not been accepted by all workers.) An endemic species, *Pacifastacus nigrescens* (Stimpson, 1859) was present in the Bay Area in the 19th century, but may have been replaced by *P. leniusculus*. Epizoic branchiobdellid worms may be found on the body and gills.

2. Abdomen small, folded under thorax, symmetrical; uropods absent; last pair of legs not markedly reduced; antennae between eyes .... **Brachyura** (Key B, p. 393)
3. Abdomen usually asymmetrical and/or reduced; uropods present or absent; last (5th) pair of legs almost always reduced, folded up behind bases of preceding pair (except in *Upogebia* and *Callianassa*); posterior sternite of thorax not fused to others; antennae external to eyes ..... **Anomura** (Key C, p. 399)

KEY A: CARIDEA

1. Rostrum absent or very short, without dorsal teeth 2
– Rostrum present, distinct, usually well developed and spinose 15
2. Rostrum very short, dorsally flattened; eyes free (not covered by carapace); hands subchelate .... *Crangon* 3
– Rostrum absent or very small and spine-like; eyes free or covered by carapace; hands chelate 8
3. Carapace without a median spine in gastric region; rostrum narrow, tapering to an acute tip that curves strongly downward (fig. 45); telson distinctly shorter than uropods .... *Crangon stylirostris*
– With one median gastric spine; telson nearly equal to length of uropods 4
4. Hand of cheliped slender, elongate, with finger turned back nearly longitudinally when closed (figs. 18A, B, 41) .... *Crangon franciscorum*
– Hand of cheliped not slender and elongate; finger when closed more or less transverse, folding across palm 5
5. Prominent dark circular spot (blue center surrounded first by black and then by a yellow ring) on each side of 6th abdominal segment in living specimens; 5th segment of abdomen without median dorsal keel; antennal scale long, narrow, nearly 3x as long as

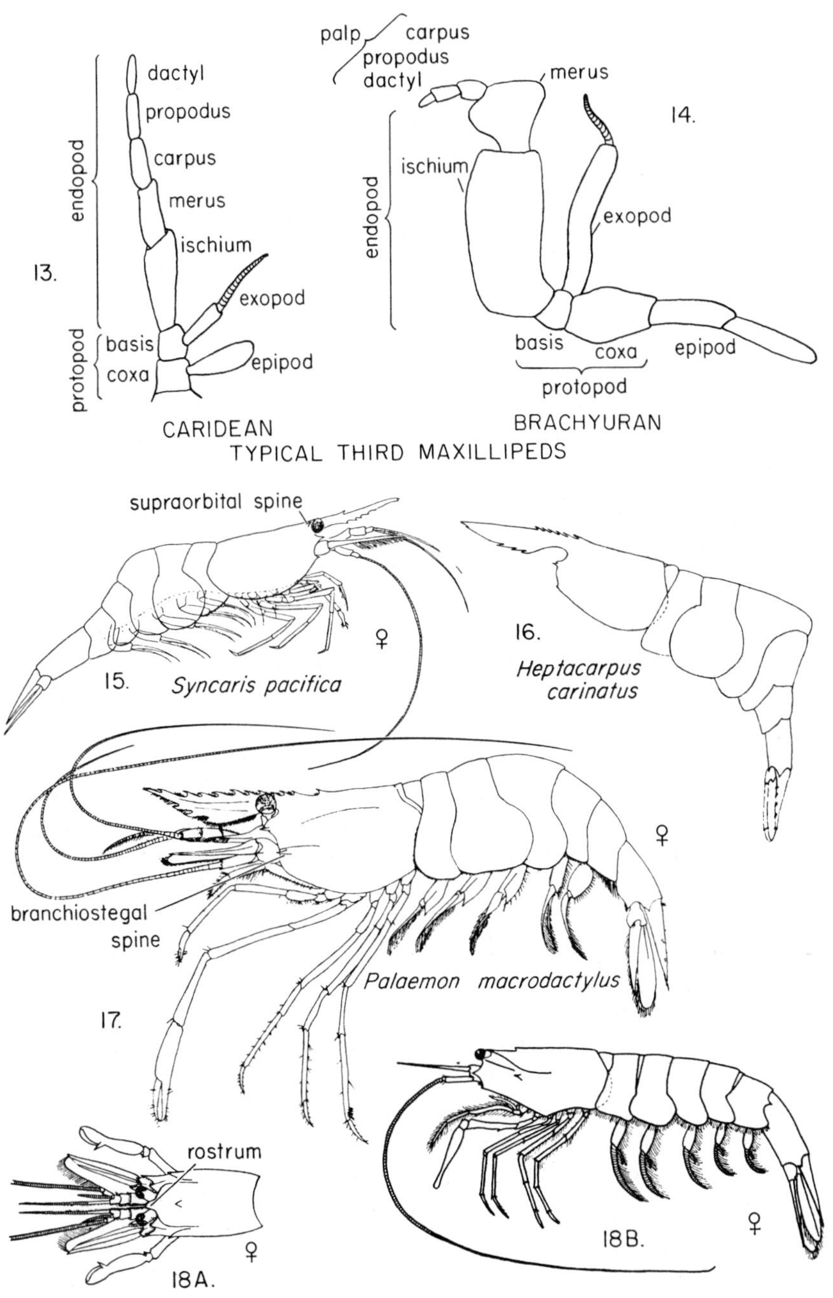

PLATE 93. **Crustacea, Decapoda** (3). Typical third maxillipeds: 13, caridean shrimp; 14, brachyuran crab; rest, as labeled; 15, after Hedgpeth, 1968; 16, after Holmes, 1900; 17, after Newman, 1963; 18, after Schmitt, 1921.

wide; finger folding nearly transversely across hand (fig. 44). .... *Crangon nigromaculata*

– Without dark spot on sides of 6th abdominal segment in living specimens; 5th segment of abdomen with a keel (carinate) or with slight medial protuberance 6

6. Antennal scale short, subquadrate, 1/2 length of carapace or less; spine short, not extending noticeably beyond anterior margin of blade, anterointernal angle of blade produced; rostrum short, rounded, not reaching cornea of eye .... *Crangon* sp.

– Antennal scale not short and stout, distinctly more than 1/2 length of carapace 7

7. Antennal scale nearly equal to length of carapace, spine produced, extending well beyond anterior margin of blade (fig. 43), anterointernal angle of blade not produced; rostrum slender, not markedly rounded, reaching beyond posterior edge of cornea; finger of hand when closed oblique, forming an angle of about 45° .... *Crangon alaskensis elongata*

– Antennal scale about 2/3 length of carapace, spine not extending as far beyond blade as above, anterointernal angle of blade produced (fig. 42A); rostrum short, rounded, reaching at least posterior corneal edge; finger of hand when closed more nearly transverse than longitudinal (fig. 42B) .... *Crangon nigricauda*

8. Eyes free; chelae not powerfully developed; rostrum very small, reduced to a small spine on frontal margin; 3 teeth on carapace behind rostral spine, the median the largest; one very prominent supraorbital spine extending beyond anterior margin of carapace; antennal scale broad, subrectangular .... *Lebbeus lagunae*

– Eyes covered by carapace; one or both chelae powerfully developed 9

9. Rostrum present; one chela greatly enlarged and complex, with dactyl above (figs. 38A, B) .... *Alpheus* spp.

– Rostrum absent; chelae usually similar and about equal, inverted so that dactyls are below (figs. 26–30); see note in species list .... *Betaeus* 10

10. Dactyls of ambulatory legs slender and simple 11

– Dactyls of legs stout and bifid 12

11. Chelae with fingers longer than palm (figs. 23, 28) .... *Betaeus longidactylus*

– Chelae with fingers not longer than palm (figs. 19, 22, 29, 30) .... *Betaeus harrimani*

12. Front of carapace rounded, not emarginate (notched) (fig. 24) .... *Betaeus macginitieae*

– Front emarginate (figs. 20, 21, 25) 13

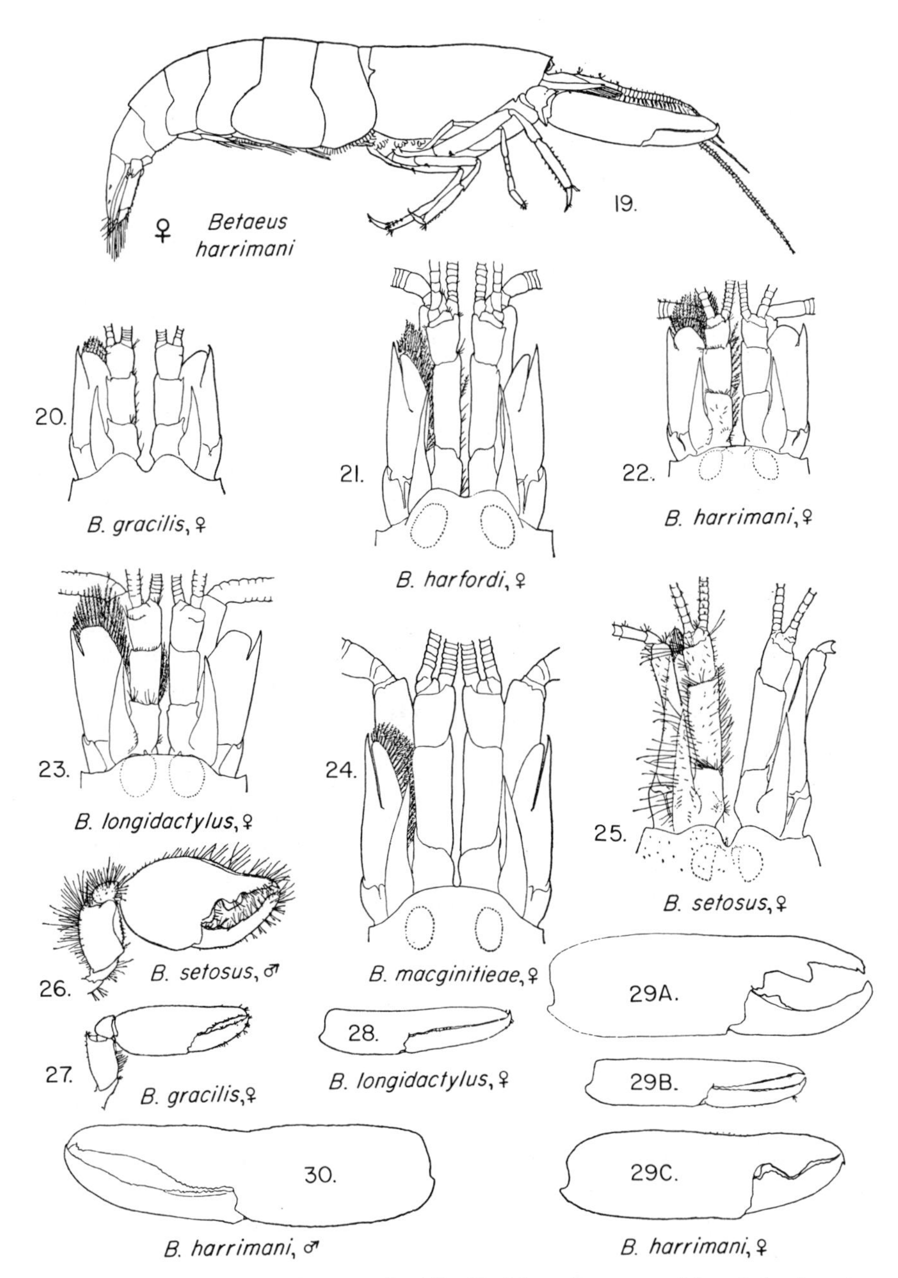

PLATE 94. **Crustacea, Decapoda** (4), **Caridea,** *Betaeus.* 19, adult female; 20–25, females, frontal region, dorsal; 26, male right cheliped; 27, female right cheliped; 28, female right chela; 29, variations in female right chela; 30, male left chela. All after Hart, 1964.

13. Emargination shallow (fig. 21); telson with posterolateral spines small or absent .... *Betaeus harfordi*
– Emargination deep (figs. 20, 25); telson with well-developed posterolateral spines 14
14. Peduncle of antennule less than 1/2 carapace length; lower inner ridge of merus of cheliped with long bristles, upper ridge ending in sharp tooth; chelae 3x as long as wide, with fingers subequal to palm length (figs. 20, 27) .... *Betaeus gracilis*
– Antennular peduncle approximately equal to carapace length; merus of cheliped with lower inner ridge usually tuberculate, upper ridge with tuft of hairs, not ending in sharp tooth; chela 2x as long as wide, with fingers longer than palm (figs. 25, 26) ....*Betaeus setosus*
15. Both legs of 1st pair simple; 2nd pair of legs very unequal, both with multiarticulate carpus; medium- to large-sized, to about 13 cm or more ....*Pandalus danae*
– Both legs of 1st pair chelate (chelae small); 2nd pair equal or nearly so 16
16. Carpus of 2nd legs not annulated 17
– Carpus of 2nd legs annulated 18
17. Rostrum with 1–2 dorsal teeth; supraorbital spine present (figs. 15, 40); in freshwater streams ....*Syncaris pacifica*
– Rostrum with at least 8 dorsal teeth; at least 3 teeth behind orbit; supraorbital spine absent; branchiostegal spine arising behind anterior margin of carapace (fig. 17); in brackish water of bays ....*Palaemon macrodactylus*
18. Carpus of 2nd legs with 3 annulations; color bright green (figs. 39A, B, C) ....*Hippolyte californiensis*
– Carpus of 2nd legs with 7 annulations; color green, red brown, mottled, or various 19
19. With 2–3 small, supraorbital spines in a longitudinal series; rostrum high, leaflike, with 3 dorsal teeth bearing serrate margins; 3rd maxilliped with small exopod; body opaque in life .... *Spirontocaris prionota*
– No supraorbital spines; rostral teeth various, not as above; 3rd maxilliped without exopod ....*Heptacarpus* 20
20. Rostrum (length measured from posterior margin of orbit to tip) generally as long as or longer than rest of carapace 21
– Rostrum generally shorter than rest of carapace 23
21. Anterior half of rostrum with some dorsal teeth; rostral teeth in mature specimens 4–8 dorsally and 1–5 below $(\frac{4\text{–}8}{1\text{–}5})$(fig. 37), subadults with fewer teeth; uniform green with broken, red brown stripes on carapace (do not confuse with *H. pictus*, the rostrum of which does not reach the end of the antennal scale) ....*Heptacarpus paludicola*

– Anterior half of rostrum lacking dorsal teeth 22

22. Sixth abdominal segment less than 2x as long as wide; rostrum deep, 1/4 as deep as long; rostral teeth $\frac{4\text{–}6}{4\text{–}6}$ (fig. 16); color highly variable, often matches algal substrate .... *Heptacarpus carinatus*

– Sixth abdominal segment elongate, more than 2x as long as wide; rostrum very narrow; rostral teeth $\frac{4}{4\text{–}5}$ (fig. 33); color bright green .... *Heptacarpus gracilis*

23. Rostrum elongate, generally reaching beyond middle of antennal scale, but not to end; rostral teeth in mature specimens 6–7 dorsally and 2–4 below ($\frac{6\text{–}7}{2\text{–}4}$) (figs. 36A, B), subadults with fewer teeth; greenish, translucent, with oblique red bands on carapace and crimson bars on legs (do not confuse with *H. paludicola,* the rostrum of which is larger and reaches to or beyond end of antennal scale) .... *Heptacarpus pictus*

– Rostrum short, generally not reaching middle of antennal scale 24

24. Rostrum not reaching as far as cornea of eye; rostral teeth $\frac{5\text{–}7}{0}$ (fig. 35); anteriormost rostral teeth often above and slightly behind, rather than well behind, tip; color highly variable, including red–brown, greenish with white carapace, or mottled colors .... *Heptacarpus taylori*

– Rostrum reaching as far as or farther than cornea 25

25. Rostrum reaching beyond 1st segment of peduncle of antennule; in male may only overlap 2nd antennular segment; rostral teeth $\frac{5\text{–}8}{1\text{–}3}$ (fig. 34A); dactyls of ambulatory legs long and slender, about 1/3 to 1/2 length of propodus (fig. 34B) ... . *Heptacarpus cristatus*

– Rostrum not reaching beyond 1st segment of antennular peduncle; dactyls of legs short and stout, not long and slender .. 26

26. Antennal scale equal to or shorter than telson; rostral teeth $\frac{5\text{–}6}{0}$ (figs. 32A, B); (do not confuse with *H. palpator,* which is smaller, has a longer rostrum and a longer and narrower antennal scale) .... *Heptacarpus brevirostris*

– Antennal scale distinctly longer than telson; rostral teeth $\frac{5\text{–}6}{0\text{–}1}$ (figs. 31A, B); (compare *H. brevirostris,* above) .... *Heptacarpus palpator*

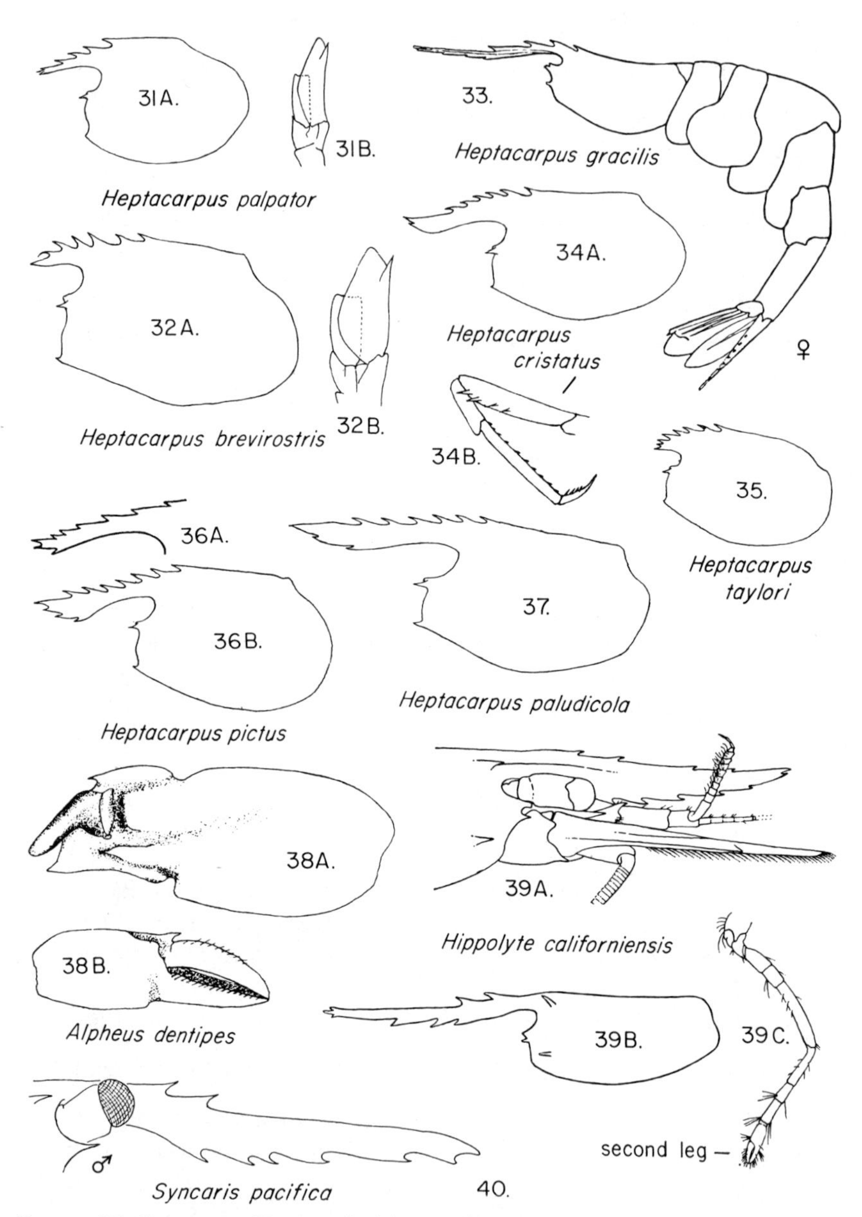

PLATE 95. **Crustacea, Decapoda** (5), **Caridea.** 31, 32, 34A, 35–37, 39B, outline view of carapace, as labeled; 31B, 32B, antennal scale; 34B, first pereopod; 36A–36B, rostral variations; 38A, large chela, 38B, small chela; 39A, frontal part of female, 39B, rostral variation; 39C, female second right leg; all, species as labeled. 31, 32, 34–38 after Holmes, 1900; 33, 36A, after Schmitt; 39A, 39C after Chace, 1951; 40 after Hedgpeth, 1968.

KEY B: BRACHYURA

1. Carapace round, with 2 prominent posterior spines, or ovate with large, straight, lateral spines and more than 12 teeth on anterolateral margin; mouth field triangular, narrow in front 2
– Carapace nearly square, triangular, ovate, or round; if round, without spines on posterior margin and, if bearing long lateral spines, then with not more than 10 anterolateral teeth; mouth field square 3
2. Carapace round with 2 short, prominent spines posteriorly (fig. 48); color white, often with purple patches .... *Randallia ornata*
– Carapace ovate with pronounced lateral spines and about 15 small, anterolateral teeth (fig. 49); reddish .... *Mursia gaudichaudii*
3. Carapace nearly square; sides approximately parallel; anterior edge nearly transverse; eyes at anterolateral corners 4
– Carapace triangular, oval, or nearly round, not square; sides not parallel 6
4. Carapace with transverse flat ridges, strongest laterally; 2 teeth on anterolateral margin; surface blackish green with numerous red or purple transverse lines .... *Pachygrapsus crassipes*
– Carapace smooth; 3 teeth on anterolateral margin; without transverse lines .... *Hemigrapsus* 5
5. Color red, purple, or whitish; no hair on legs; chelipeds red-spotted (fig. 54) .... *Hemigrapsus nudus*
– Color dull brownish green; legs hairy; chelipeds without reddish spots (figs. 4A–D, 5) (see note in species list on distinguishing young *Hemigrapsus*) ... *Hemigrapsus oregonensis*
6. Body narrow anteriorly; rostrum single or bifid 7
– Body broad anteriorly; rostrum usually reduced or absent 15
7. Rostrum single 8
– Rostrum bifid 9
8. Chelipeds short and stout; carapace broadly pyriform (pear-shaped), with tubercles and fine hairs; short, prominent, spinelike tubercle on 1st abdominal segment .... *Pyromaia tuberculata*
– Chelipeds much longer and heavier than ambulatory legs; carapace broadly triangular (fig. 47) .... *Heterocrypta occidentalis*
9. Carapace about as broad as long with lateral margins markedly flattened and produced, leaflike; surface smooth, often

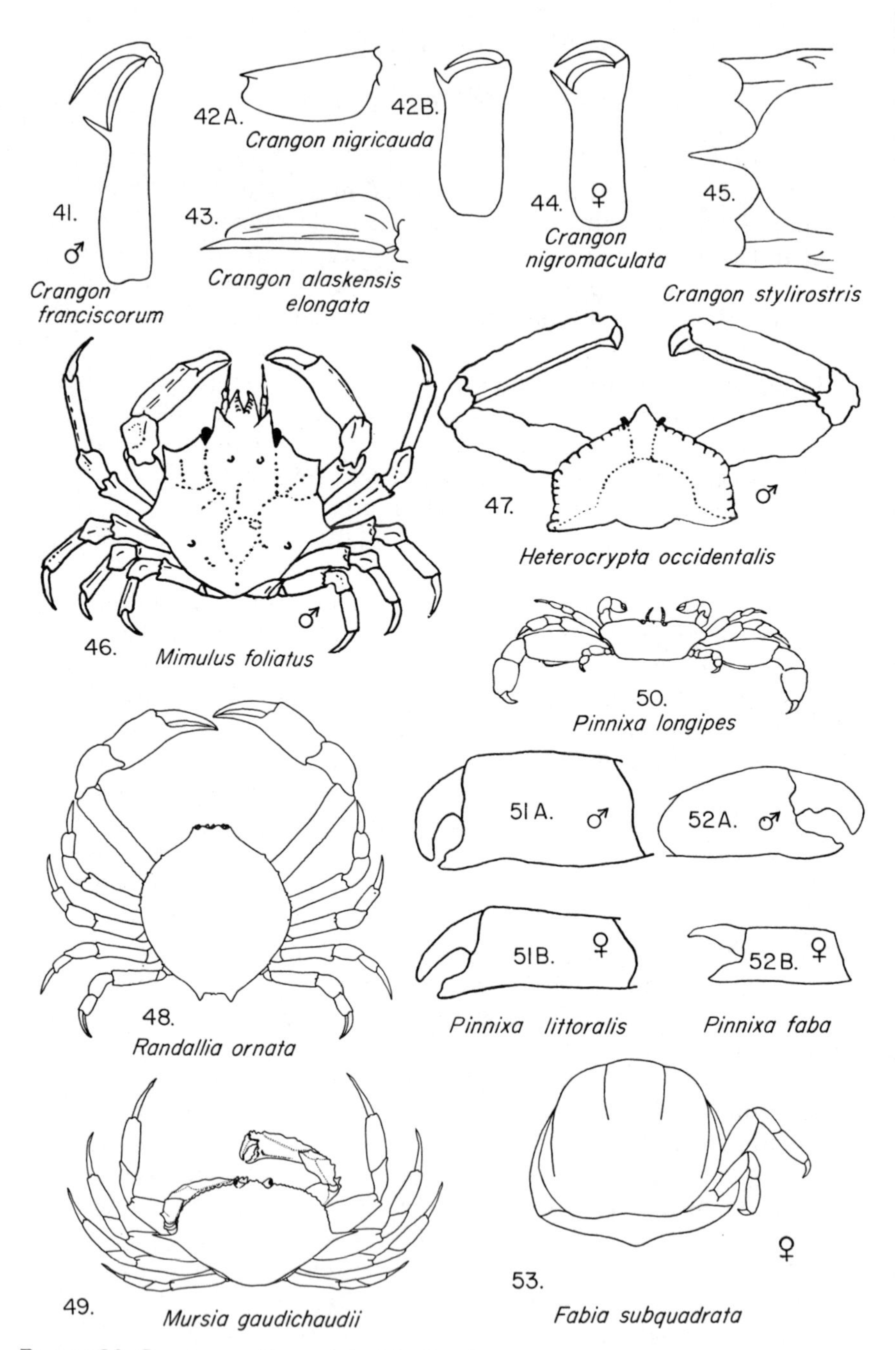

PLATE 96. **Crustacea, Decapoda** (6), **Caridea.** 41, 42B, 44, chela; 42A, 43, antennal scale; 45, dorsal anterior view of carapace; all, species as labeled; **Brachyura:** 51A, male left chela, 51B, female left chela; 52A, male right chela; 52B, female right chela; rest, as labeled. 41, 42B, 43, 44 after Rathbun, 1904, redrawn by Emily Reid; 42A, 45 after Holmes, 1900, redrawn by E. Reid; rest after Schmitt.

encrusted with sponges, bryozoans, etc. (fig. 46) .... *Mimulus foliatus*

– Carapace longer than broad; lateral margins not flattened and produced; surface smooth or rough, often encrusted, obscuring carapace 10

10. Posterolateral margin of carapace without spines 11

– Prominent posterolateral projections ....*Pugettia* 13

11. Rostrum consisting of 2 long, very slender spines; preorbital spine (internal to eye) absent; postorbital spine (external to eye) prominent, slender and acute, far from eye .... *Oregonia gracilis*

– Rostrum otherwise; preorbital spine present 12

12. Rostrum thin, flat, ovate-lanceolate; postorbital spine short, not hairy on anterior face, deeply cupped anterolaterally; carapace tuberculate ....*Scyra acutifrons*

– Rostrum thick, subacute, not lanceolate; postorbital spine hairy on anterior face, not deeply cupped anterolaterally (figs. 58A, B, C); carapace hairy ....*Loxorhynchus crispatus*

13. Surface of carapace smooth; distance between eyes less than 1/3 width of carapace in adult specimens (fig. 56) ....*Pugettia producta*

– Carapace tuberculate or spiny; distance between eyes about 1/2 greatest width of carapace 14

14. Carapace distinctly broader posteriorly; anterolateral teeth narrow, laterally directed; legs moderately long and slender; merus of cheliped with a few tubercles dorsally, not carinate ....*Pugettia richii*

– Carapace not expanded posteriorly; anterolateral teeth broad, anteriorly directed; legs relatively short; merus of cheliped with irregularly dentate keel dorsally ..... *Pugettia gracilis*

15. Frontal area (area between eyes) either 5-toothed or divided by median notch; carapace hard, anterolateral margin toothed; free-living 16

– Frontal area entire; carapace often membranous, frequently rounded or may be much wider than long; carapace margin without sharp teeth or spines; commensals in polychaete tubes, molluscs, sea cucumbers, echiuran and ghost- and mud-shrimp burrows 22

16. Frontal area 5-toothed; carapace broadly oval; antennules fold back longitudinally ....*Cancer* 17

– Frontal area divided by median notch; carapace broader anteriorly; antennules fold back transversely or obliquely 32

17. Frontal area markedly produced beyond outer orbital angles forming 5 nearly equal teeth; fingers of chelipeds black-

tipped (fig. 57); adults uniformly brick red above, young often brightly colored with spots or stripes ....*Cancer productus*

– Frontal area not markedly produced, with 5 unequal teeth 18

18. Carapace widest at 10th anterolateral tooth (1st anterolateral tooth is external to eye) (fig. 2); fingers of chelipeds not black-tipped ....*Cancer magister*

– Carapace widest at 8th or 9th tooth 19

19. Carapace widest at 8th tooth; 10th and 11th teeth distinct; teeth with entire edges, curving forward; red-spotted beneath; black on fingers of chelipeds (fig. 61) ....*Cancer antennarius*

– Carapace generally widest at 9th tooth; not red-spotted beneath 20

20. Upper surface of carapace hairy (pubescent); teeth sharp, curving, with entire edges .... *Cancer jordani*

– Carapace smooth, not hairy (glabrous); teeth blunt, with serrate posterior edges 21

21. Fingers of chelipeds white-tipped; merus of 3rd (outer) maxilliped rounded anteriorly .... *Cancer gracilis*

– Fingers black-tipped; merus of outer maxilliped truncate anteriorly ....*Cancer anthonyi*

22. Carapace distinctly wider than long; 3rd ambulatory legs the longest 23

– Carapace orbicular or quadrate, not markedly wider than long 30

23. Third ambulatory legs slightly longer than others; legs slender and somewhat rounded; carapace hard; anterolateral margin curving gradually into posterolateral margin ....*Scleroplax granulata*

– Third ambulatory legs distinctly longer and larger than others; legs flattened; carapace membranous; anterolateral margin forming an angle with posterolateral margin (a difficult genus, see note in species list) ....*Pinnixa* 24

24. Dactyl of 3rd ambulatory leg strongly hooked 25

– Dactyl of 3rd leg straight or moderately curved 26

25. Carapace oblong, about 1½x as wide as long; fingers of cheliped in female not gaping when closed (figs. 52A, B) ....*Pinnixa faba*

– Carapace pointed at sides, about 2x as wide as long; fingers of female gaping when closed (figs. 51A, B) ....*Pinnixa littoralis*

26. Fourth ambulatory leg when extended not reaching end of merus of 3rd leg; carapace very wide, nearly 3x as wide as long; 3rd leg relatively enormous (fig. 50); one of the smallest

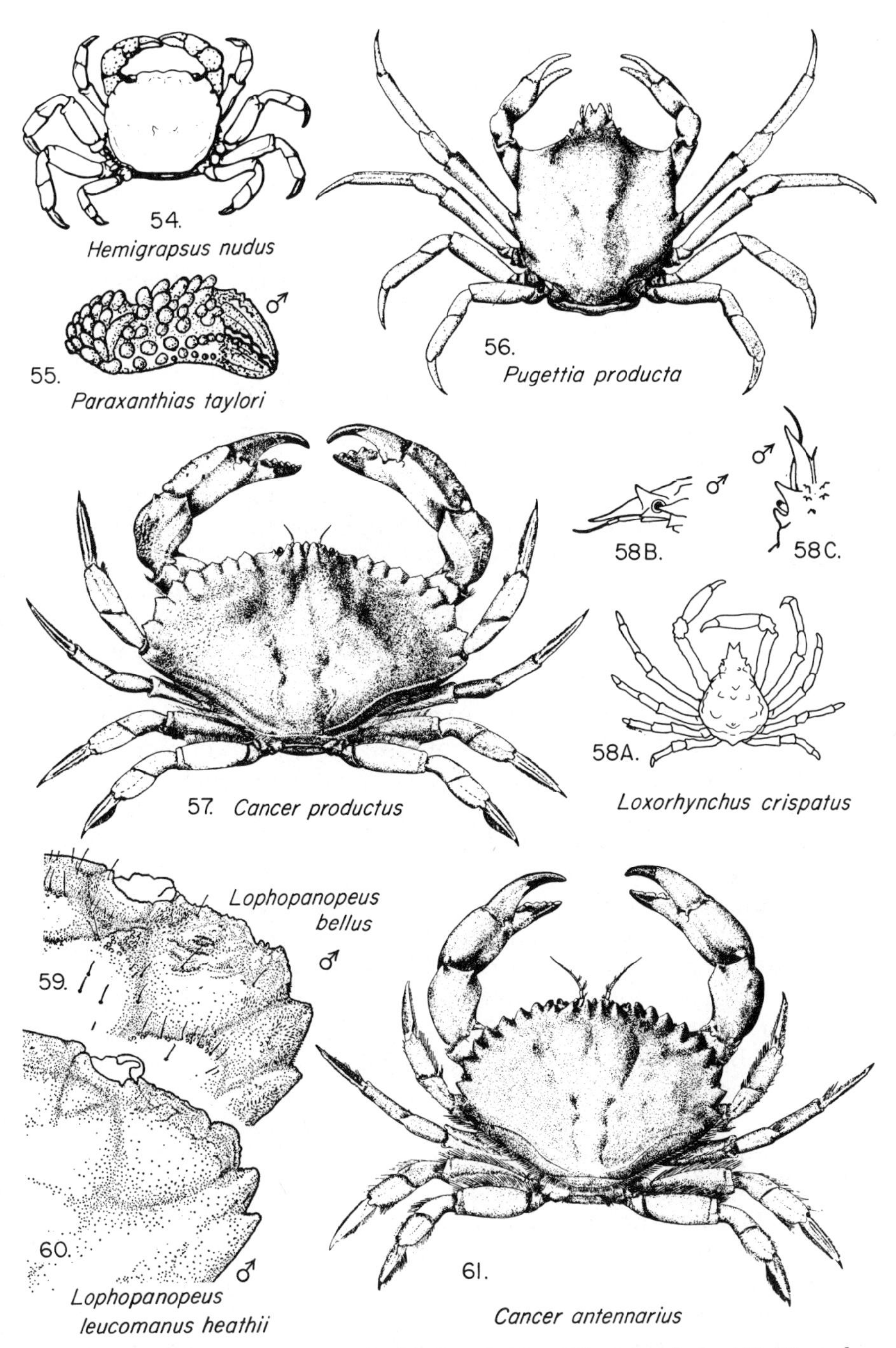

PLATE 97. **Crustacea, Decapoda** (7), **Brachyura.** 55, male chela; 59–60, male, dorsal right side of carapace; rest, as labeled. 54 after Hedgpeth, 1962; 58A after Rathbun, 1925; 59–60 after Menzies, 1954; rest after Schmitt.

of local crabs, and approached by none in its great relative width and development of 3rd legs ....*Pinnixa longipes*
- Fourth ambulatory leg, when extended, reaching end of or beyond merus of 3rd leg; 3rd leg not enormously large in proportion to body 27

27. Propodus of 3rd ambulatory leg about as long as wide 28
- Propodus of 3rd leg distinctly longer than wide 29

28. Dactyl of 3rd leg approximately equal to length of propodus; lower margin of hand nearly straight; width of carapace 20mm or more; sharp, transverse cardiac ridge ....(male) *Pinnixa franciscana*
- Dactyl of 3rd ambulatory leg markedly shorter than propodus; lower margin of hand convex, width of carapace to about 11mm; small cardiac ridge; light spots on carapace .... *Pinnixa tubicola*

29. Immovable finger of cheliped straight or nearly so; lower margin of hand may be slightly concave; carapace about 2x as wide as long ....(female) *Pinnixa franciscana*
- Immovable finger curved upward distally; lower margin of hand nearly straight or somewhat convex; carapace less than 2x as wide as long ....*Pinnixa* spp.

30. Carapace smooth, glossy, sharply flexed downwards, with 2 grooves leading back from orbits (fig. 53); palm of cheliped widened anteriorly, bearing 2 rows of hairs along lower margin ....(female) *Fabia subquadrata*
- Carapace without longitudinal grooves 31

31. Anterolateral and frontalorbital margins with a continuous margin of short, dense hairs (pubescence), hand of cheliped with dense pubescence on upper and lower margins; 4th pair of legs noticeably shorter than others; immovable finger with large, serrate lobe on upper margin; movable finger with 1 large tooth on lower margin ......(male) *Fabia subquadrata*
- Lacking dense pubescence on carapace margins; legs approximately equal in length; carapace red-spotted .... *Opisthopus transversus*

32. Chelipeds with numerous, prominent, rounded tubercles (fig. 55); legs hairy ....*Paraxanthias taylori*
- Chelipeds otherwise 33

33. Fingers whitish; in brackish water ....*Rhithropanopeus harrisii*
- Fingers black; not in brackish water 34

34. Anterolateral margin with 8–10 small, subequal, acute teeth; carapace broadly oval .... *Cycloxanthrops novemdentatus*
- Anterolateral margin with 3 prominent, subequal teeth (figs. 59, 60) .... *Lophopanopeus* 35

35. Carapace and distal segments of ambulatory legs hairy (fig. 59) .... *Lophopanopeus bellus*
– Carapace and ambulatory legs smooth, not pubescent (fig. 60) .... *Lophopanopeus leucomanus heathii*

KEY C: ANOMURA

1. Abdomen short, reflexed beneath thorax 2
– Abdomen elongate, not reflexed, may be twisted 12
2. Body egg-shaped; 2nd to 4th legs with last joint curved and flattened; on sandy beaches 3
– Body not egg-shaped; 2nd to 4th legs with last joint ending in sharp, pointed dactyl; in rocky areas 4
3. First pair of legs simple; carapace without sharp spines along anterolateral margin (fig. 65) ....*Emerita analoga*
– First pair of legs subchelate; carapace with sharp, anterolateral spines ....*Blepharipoda occidentalis*
4. Uropods absent; carapace wider posteriorly or completely covering appendages; abdomen not folded up against body 5
– Uropods present; carapace nearly round in outline; abdomen folded against body 7
5. Carapace wider than long, completely covering legs and body; abdomen small and flattened *Cryptolithodes sitchensis*
– Carapace as long as, or longer than, broad; abdomen thick and fleshy 6
6. Legs and carapace hairy, flattened *Hapalogaster cavicauda*
– Legs and carapace roughly tuberculate, not hairy; legs nearly cylindrical ....*Oedignathus inermis*
7. Body and chelae thick; chelae unequal and tuberculate or granular; carpus of chelipeds as long as broad *Pachycheles* 8
– Body and chelae flattened; chelae equal or nearly so, smooth; carpus of chelipeds longer than broad .... *Petrolisthes* 9
8. Telson with 5 plates, lacking small plate at anterior margin of each lateral plate (figs. 63A, B) .... *Pachycheles rudis*
– Telson with 7 plates, a small plate at anterior margin of each lateral plate (fig. 64) ....*Pachycheles pubescens*
9. Carpus of chelipeds over 2x as long as wide 10
– Carpus 2x as long as wide or less 11
10. Carapace covered with short, transverse, hairy striations and large, flattened tubercles; carpus about 2½x as long as wide; distal portion of maxillipeds bright orange red .... *Petrolisthes rathbunae*
– Carapace nearly smooth posteriorly, often granular anteriorly, never with hairy striations; carpus a little over 2x to nearly 3x as long as wide; outer edge of palp of maxilliped blue (see

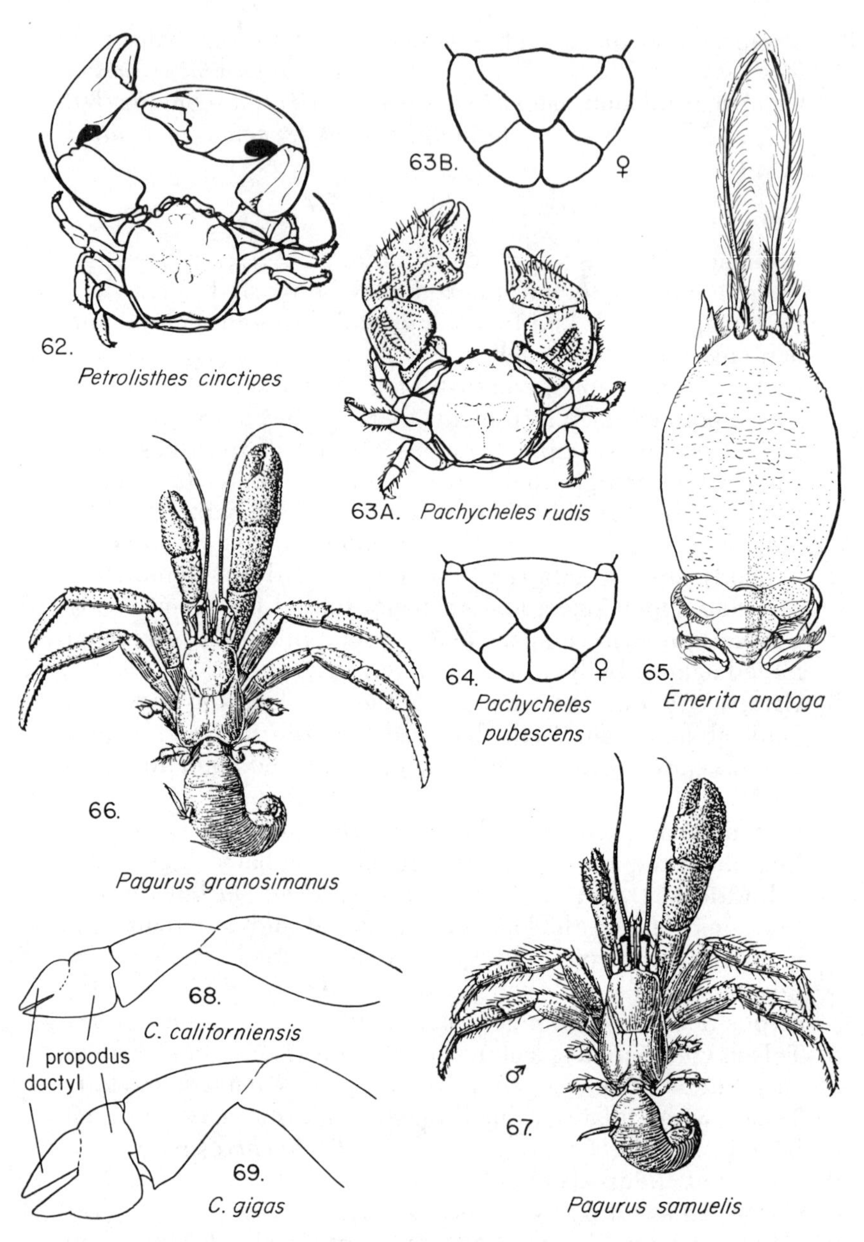

PLATE 98. **Crustacea, Decapoda** (8), **Anomura.** 63B, 64, female telsons; 67, young male; 68–69, second right pereopod; rest, as labeled. 62, 63A, 65 after Hedgpeth; 68, 69 Rogene Thompson, redrawn by Emily Reid; rest after Schmitt.

note in species list) .... *Petrolisthes manimaculis*

11. Anterior margin of carpus with a distinct proximal lobe, margins of carpus converging distally from their highest points; palp of maxilliped orange red (fig. 62) .... *Petrolisthes cinctipes*
- – Carpus without lobe on anterior margin, margins subparallel; outer edge of palp of maxilliped bright blue (see note in species list) .... *Petrolisthes eriomerus*

12. Burrowing in mud or sand; abdomen symmetrical, extended, externally segmented; ghost and mud shrimps 13
- – Living in snail shells; abdomen asymmetrical, soft and twisted, not externally segmented; hermit crabs 15

13. First pair of legs approximately equal and subchelate, others simple; eyestalks cylindrical, corneas terminal; 4 pairs of fanlike pleopods; body hairy, often bluish .... *Upogebia pugettensis*
- – First pair of legs very unequal and chelate, 2nd pair chelate; eyestalks flattened, corneas dorsal; 3 pairs of fanlike pleopods; body smooth, whitish to reddish .... *Callianassa* 14

14. Propodus and dactyl of 2nd pereopod approximately equal (fig. 68); inner edge of dactyl of large male cheliped serrate, with recurved hook at distal end; adult length under 90 mm .... *Callianassa californiensis*
- – Propodus of 2nd pereopod curved and wider than dactyl (fig. 69); inner edge of dactyl of large male cheliped serrate, with prominent proximal flattened tooth; adult length to 125 mm .... *Callianassa gigas*

15. Left cheliped equal to or larger than right; outer maxillipeds approximated at base; eyestalks closely set, with median, dorsal, brown stripe; body whitish; abdomen very hairy ....*Isocheles pilosus*
- – Right cheliped larger than left; outer maxillipeds widely separated at base .... *Pagurus* 16

16. Antennae red (for pickled specimens, see references in species list) 17
- – Antennae not red 19

17. Ambulatory legs with bands of blue or white on distal portion of propodus; rostrum (medial projection on anterior margin of carapace) triangular, acute (fig. 67) ....*Pagurus samuelis*
- – Legs not banded blue or white; rostrum acute or blunt 18

18. Eyes with yellow circles; rostrum triangular, acute; carpus of large cheliped with deep, flat, perpendicular inner face; dactyl of ambulatory legs with white spot at tip ....*Pagurus hemphilli*

- Eyes without yellow circles; rostrum rounded and blunt (fig. 66); carpus of large cheliped without very deep, flat, perpendicular inner face .... *Pagurus granosimanus*

19. Antennae dark, with fine white bands (when viewed at distance); body dark orange-brown; propodus of ambulatory legs generally prominently banded with white proximally and distally; rostrum triangular, acute ...... *Pagurus hirsutiusculus*
    Antennae light, translucent orange; body whitish; orange-red bands on ambulatory legs; inverted V (Λ) formed by larger spiny tubercles on dorsal surface of large chela, inner arm of Λ continuing on proximal inner edge of carpus; rostrum blunt, broadly rounded .... *Pagurus beringanus*

LIST OF INTERTIDAL DECAPOD CRUSTACEA

Suborder **NATANTIA**

Tribe **Caridea**

ATYIDAE

Syncaris pacifica (Holmes, 1895). Restricted to freshwater streams of Marin, Sonoma, and Napa counties; see Hedgpeth, 1968, Intern. Revue Ges. Hydrobiol. 53: 511–524.

PALAEMONIDAE

Palaemon macrodactylus Rathbun, 1902. Introduced accidentally from the Orient; in San Francisco Bay, especially in brackish water, where it may be abundant along wharf pilings and in algae; see Newman, 1963, Crustaceana, 5: 119–132; Little, 1969, Crustaceana 17: 69–87 (larval development).

PANDALIDAE

Pandalus danae Stimpson, 1857. The "coon-stripe" shrimp; sublittoral, but occasional in shallow water near harbor channels and over eelgrass (*Zostera*) beds.

HIPPOLYTIDAE The following species, with the exception of *Hippolyte,* were formerly placed in the broad genus *Spirontocaris;* Holthuis's generic revision (1947), which has received widespread acceptance, is followed here. See also Pike and Williamson, 1961.

Heptacarpus brevirostris (Dana, 1852). In low-intertidal pools in algae and under rocks; on floats, pilings; sublittoral on algae and on rocky bottoms.

Heptacarpus carinatus Holmes, 1900. In algae, often matching color of algae.

Heptacarpus cristatus (Stimpson, 1860). Generally sublittoral; occasional in low intertidal of rocky coasts, among eelgrass, or in algae on pilings.

* *Heptacarpus franciscanus* (Schmitt, 1921). Recorded from shallow water of San Francisco Bay over sandy and rocky bottoms; see Schmitt, 1921.

Heptacarpus gracilis (Stimpson, 1864).

Heptacarpus palpator (Owen, 1839). In eelgrass; generally sublittoral.

Heptacarpus paludicola Holmes, 1900. In *Zostera* beds and on algae, such as *Ulva*, in shallow pools on mudflats; on wharf pilings, floats; also in mid-intertidal pools of rocky coast; common to abundant.

Heptacarpus pictus (Stimpson, 1871). Abundant to common in middle and lower tidepools of rocky coasts; also in *Zostera* beds, and on floats.

Heptacarpus taylori (Stimpson, 1857). Mid to low intertidal under rocks and clumps of bryozoa, sponges; in algae; wharf pilings.

Hippolyte californiensis Holmes, 1895. Common locally in bays on *Zostera*, matching its color and oriented longitudinally along blade; see Chace (1951).

Lebbeus lagunae (Schmitt, 1921). Monterey; rare, in rocky pools; generally southern.

Spirontocaris prionota (Stimpson, 1864). In lower tidepools of rocky coast.

ALPHEIDAE

Alpheus spp. (formerly *Crangon;* see below under Crangonidae) The "pistol shrimps," producing a snapping noise by clicking dactyl of chela against palm (see Johnson, Everest, and Young, 1947, Biol. Bull. 93: 122–138); in very low rocky intertidal and in sponges and ectoprocts in kelp holdfasts. Several species, including *A. dentipes* Guérin, 1832 (figs. 38A, B) and *A. bellimanus* Lockington, 1877, have been recorded locally; see Schmitt, 1921.

Betaeus gracilis Hart, 1964. Intertidal on rocky shores; kelp holdfasts. See Hart (1964) for a detailed treatment of this genus. Size, shape, and dentition of chelae vary with age, sex, and extent of regeneration and are not reliable systematic characters. A 6th species, *B. ensenadensis* Glassell, 1938, occurs from San Diego (Mission Bay) south.

Betaeus harfordi (Kingsley, 1878). Commensal in mantle cavities of abalones (*Haliotis* spp.); leaves host readily and often missed when abalones are collected.

Betaeus harrimani Rathbun, 1904. In burrows of *Upogebia* and *Callianassa* on mudflats.

* Not in key.

Betaeus longidactylus Lockington, 1877. Intertidal on rocky shores; kelp, kelp holdfasts, and in eelgrass; also recorded in southern California from *Urechis* and *Upogebia* burrows.

Betaeus macginitieae Hart, 1964. Occur in pairs under sea urchins (*Strongylocentrotus* spp.).

Betaeus setosus Hart, 1964. Under rocks and in algae on semiprotected rocky coasts; in kelp holdfasts (especially *Laminaria*) and surfgrass (*Phyllospadix*) roots; on pilings.

* *Synalpheus lockingtoni* (Coutière, 1909). Has been taken at Elkhorn Slough, rare, largely southern. Distinguished from *Alpheus* by strong spination of tip of 3rd maxilliped, presence of pterygostomian spine (at anteroventral corner of carapace), and bifid dactyls.

CRANGONIDAE The generic name *Crago* was suppressed in 1955 by a ruling of the International Commission on Zoological Nomenclature, restoring *Crangon* for the common shrimps treated here and *Alpheus* for the snapping shrimps. See Zarenkov (1965, Zoologicheskii Zhurnal, 44: 1761–1775, in Russian) for review of group in general and aspects of evolution and biology. These shrimp are parasitized by bopyrid isopods, such as *Argeia* (see page 287), which may appear as large, asymmetrical swellings in the gill chambers.

Crangon alaskensis elongata Rathbun, 1902. In shallow water of bays on soft bottoms; common; should not be confused with *C. nigricauda*, a larger species.

Crangon franciscorum Stimpson, 1859. In bays on mud bottoms, common to abundant; also offshore in deeper waters. See Israel, 1936, Calif. Fish Game, Bull. 46 (life history, biology).

* *Crangon munitella* Walker, 1898. Assigned by Zarenkov questionably to *Mesocrangon;* recorded by Schmitt from shallow rocky bottoms in San Francisco Bay.

Crangon nigricauda Stimpson, 1856. Common to abundant in bays and offshore in deeper waters; among eelgrass, rocks, and on sand bottoms. See Israel, 1936, above.

Crangon nigromaculata Lockington, 1877. On mud and sand bottoms.

Crangon stylirostris Holmes, 1900. Common in surf zone of semiprotected sandy beaches; sublittoral on sandy-rocky bottoms.

Crangon sp. Coarse sand of coves along rocky outer coast; matching substrate in color; occasionally washed ashore in wrack.

Suborder **REPTANTIA**

Section **Brachyura**

LEUCOSIIDAE

Randallia ornata (Randall, 1839). Sublittoral; sandy substrates; occasionally washed inshore.

CALAPPIDAE

Mursia gaudichaudii (Milne-Edwards, 1837). Sublittoral, occasional specimens in harbors released from fishing boats.

PARTHENOPIDAE

Heterocrypta occidentalis (Dana, 1854). Sublittoral, sandy bottoms, rare in intertidal.

MAJIDAE (=INACHIDAE). The spider and decorator (masking) crabs. See monographic treatment by Garth (1958).

Loxorhynchus crispatus Stimpson, 1857. Sublittoral; low intertidal on semiprotected rocky coasts in crevices; often heavily encrusted with hydroids, sponges, algae, and so forth.

Mimulus foliatus Stimpson, 1860. Low intertidal of rocky coast, among algae, under rocks, often encrusted with sponges or bryozoans.

Oregonia gracilis Dana, 1851. Occasional on wharf pilings and in *Zostera;* usually sublittoral and generally northern; boreal, also occurring in Japan. Sexual dimorphism pronounced; see Garth, 1958.

Pugettia gracilis Dana, 1851. In low rocky intertidal; in bays among *Zostera.*

Pugettia producta (Randall, 1839) (=*Epialtus productus*). The kelp crab; low intertidal and sublittoral of protected and semiprotected rocky coasts, in kelp beds and other macro-algae and on jetties, wharf pilings; adults often encrusted with barnacles, bryozoans, and sponges; a lively and aggressive spider crab with a strong pinch. Occasionally with parasitic rhizocephalan cirripedes.

Pugettia richii Dana, 1851. Low intertidal among corallines and other algae; often encrusted with hydroids and coralline algae.

Pyromaia tuberculata (Lockington, 1877) (=*Inachoides tuberculatus*). Sublittoral on wharf pilings, often encrusted with sponges and algae; common in shallow dredge hauls in San Francisco Bay. Schmitt's figure is of a Gulf of California subspecies; see Garth, 1958.

Scyra acutifrons Dana, 1851. Uncommon in low intertidal of semiprotected rocky coasts, often encrusted. Schmitt's figure is of a young specimen.

CANCRIDAE

Cancer antennarius Stimpson, 1856. The rock crab; lower intertidal, partially imbedded in sand among rocks; protected and semiprotected coast as well as in bays; often encrusted; common. Iphitimid polychaetes recorded from branchial cavity in southern California specimens (Pilger, 1972, Bull. So. Calif. Acad. Sci. 70: 84–87).

Cancer anthonyi Rathbun, 1897. Under rocks, low intertidal; in bays.

* *Cancer gibbosulus* (De Haan, 1835). A small species (reaching about 35 mm in width) which may be mistaken for small *C. antennarius* (the granules on the carapace in *gibbosulus* are in scattered groups; in *antennarius* crowded); uncommon, in bays.

Cancer gracilis Dana, 1852. Intertidal to sublittoral in bays; megalops and post-larvae live on scyphozoan medusae.

Cancer jordani Rathbun, 1900. Intertidal in bays; uncommon.

Cancer magister Dana, 1852. The edible, market or Dungeness crab, the latter name from a village on Juan de Fuca Strait, Washington; generally offshore, on sandy bottoms; occasionally inshore.

* *Cancer oregonensis* (Dana, 1852). Lower intertidal of semiprotected rocky coast, under well-embedded rocks; rare south of Oregon.

Cancer productus Randall, 1839. Bays, under rocks or partly buried in sand and mud; also under rocks of semiprotected outer coast; active nocturnally; common.

XANTHIDAE The pebble crabs; see Knudsen, 1957, Bull. So. Calif. Acad. Sci. 56: 133–142 (molting); 1959, Wasmann J. Biol. 17: 95–104 (autotomy and regeneration); 1959, Ecology 40: 113–115 (shell formation and growth); 1960, Ecol. Monogr. 30: 165–185 (ecology) on *Cycloxanthops*, *Lophopanopeus* and *Paraxanthias*.

Cycloxanthops novemdentatus (Lockington, 1877). Low intertidal under rocks in gravel and shell substrate; rare north of Point Conception.

Lophopanopeus bellus (Stimpson, 1860). Intertidal under rocks, stones of protected and unprotected coast; see Menzies, 1948, Allan Hancock Found. Publs. Occ. Pap. 4, 45 pp. (taxonomy).

Lophopanopeus leucomanus heathii Rathbun, 1900. Intertidal in coarse sand under rocks and in surfgrass roots; see Menzies, 1948, above.

Paraxanthias taylori (Stimpson, 1860). Lower intertidal, protected outer coast, under well-impacted rocks; rare north of Point Conception. See Knudsen, 1959, Bull. So. Calif. Acad. Sci. 58: 138–145 (life history).

Rhithropanopeus harrisii (Gould, 1841). Introduced from Atlantic coast; common to abundant in sloughs, estuarine habitats with mud banks.

PINNOTHERIDAE The pea crabs, commensals with annelids, echiurans, molluscs, in crustacean burrows and sea cucumbers. See Pearce (below); Wells, 1928, Publ. Puget Sound Biol. Sta. 6: 283–314; 1940, Univ. Wash. Publ. Ocean. 2: 19–50; Rathbun, 1918; and Schmitt, 1921.

Fabia subquadrata Dana, 1851 [= *Pinnotheres concharum* (Rathbun, 1893), in part, the male; see Davidson, 1968, Bull. So. Calif. Acad. Sci. 67: 85–88]. In bivalve molluscs, especially *Mytilus californianus;* the large, soft-shelled females remain within valves of host, the small, hard-shelled males move between hosts and are able to swim; see Pearce, 1966, Pac. Sci. 20: 3–35 (biology).

Opisthopus transversus Rathbun, 1893. Recorded from various molluscan (*Megathura, Tresus, Cryptochiton*) and sea-cucumber hosts; see Hopkins and Scanland, 1964, Bull. So. Calif. Acad. Sci. 63: 175–180; Beondé, 1968, Veliger 10: 375–378 (review of hosts).

Pinnixa faba (Dana, 1851). In gaper clam *Tresus* and other bivalves; see Pearce, 1966, in H. Barnes, ed., *Some Contemporary Studies in Marine Science,* pp. 565–589, Allen & Unwin, (biology).

Pinnixa franciscana Rathbun, 1918. Recorded from burrows of *Urechis, Upogebia, Callianassa,* and terebellid polychaetes.

Pinnixa littoralis Holmes, 1895. In *Tresus, Protothaca,* and other bivalves; see Pearce, 1966, reference under *P. faba.*

Pinnixa longipes (Lockington, 1877). Common in tubes of *Axiothella* and occasional with *Pectinaria* and *Pista.*

Pinnixa tubicola Holmes, 1895. In terebellid worm tubes, such as *Eupolymnia.*

Pinnixa spp. The taxonomy of the small pea crabs in central California remains to be fully worked out; males or females of some species have gone unrecognized or under other names; juveniles differ from adults in proportions, cheliped dentition, and other characters. Introduced species in bays and range extensions should be expected. Additional local species include *P. schmitti* Rathbun, 1918 and *P. weymouthi* Rathbun, 1918; *P. barnharti* Rathbun, 1918, has been recorded from southern California and Puget Sound and may occur in our area; *P. occidentalis* Rathbun, 1893, from shallow water, may occur inshore.

Scleroplax granulata Rathbun, 1893. Common in *Urechis* burrows; also found with *Upogebia* and *Callianassa.*

GRAPSIDAE

Hemigrapsus nudus (Dana, 1851). The purple shore crab; mid-intertidal of semiprotected and protected rocky coasts and bays, locally abundant; prefers coarse sand to gravel substrates overlain with large rock cover; sluggish.

Hemigrapsus oregonensis (Dana, 1851). Mid and low intertidal of bays under rock cover overlying mud or muddy sand, abundant; sometimes exposed and active over large areas of mudflats; sublittoral populations in shallow water of bays with profuse *Ulva* cover; small populations along protected outer coast under rocks over mud; also in burrows in *Salicornia* marshes; moderately active. The entoniscid isopod *Portunion conformis* occurs in both species of *Hemigrapsus*. Very small *H. nudus* may be distinguished from very small *H. oregonensis* by a combination of the following characters: in *oregonensis* there is a marked frontal notch, in *nudus* a shallow depression; in *oregonensis* the lateral spines are sharp and clearly set out, in *nudus* they are not sharp and not as clearly separated from the side; the dactyls of ambulatory legs 1–3 are long in *oregonensis*, shorter in *nudus;* the dactyl of leg 4 is quite flat in *nudus,* less so in *oregonensis*.

Pachygrapsus crassipes Randall, 1839. The lined shore crab; ubiquitous in upper intertidal of rocky areas of bays as well as rocky outer coast, abundant; in burrows in *Salicornia* marshes; active, aggressive, nocturnal; see Hiatt, 1948, Pac. Sci. 2: 135–213 (biology); Bovbjerg, 1960, Ecology 41: 668–672 (behavioral ecology).

* *Planes cyaneus* Dana, 1852 (=*P. minutus* of authors). Pelagic; occasionally washed ashore on drift logs with goose barnacle *Lepas;* see Chace, 1951, Proc. U.S. Nat. Mus. 101: 65–103. *P. marinus* Rathbun, 1914 has been washed ashore in Oregon and north (see Chace, 1951; 1966, Proc. U.S. Nat. Mus. 118: 623–661).

Section **Anomura**

Tribe **Thalassinidea**

CALLIANASSIDAE See Stevens, 1928, Publ. Puget Sound Biol. Sta. 6: 315–369.

Callianassa californiensis Dana, 1854. The pink or ghost shrimp; burrowing in mud or sand of upper to mid-intertidal in bays; locally abundant; burrows with poorly defined walls; commensals include shrimp *Betaeus,* polynoid worm *Hesperonoe,* various pinnotherid crabs and copepods (*Hemicyclops* and *Clausidium*); bopyrid isopod *Ione* may occur in gill chamber (see p. 286). See MacGinitie, 1934, Amer. Midl. Nat. 15: 166–177 (natural history); Gooding, 1960, Proc. U.S. Nat. Mus. 112: 159–195 (commensal copepods); Powell, 1974, Univ. Calif. Publ. Zool. 102 (functional morphology of stomach).

Callianassa gigas Dana, 1852 (=*C. longimana* Stimpson, 1857). Low intertidal to sublittoral, rare, burrowing in sand; builds deep burrows.

Upogebia pugettensis (Dana, 1852). The blue mud shrimp; mid to lower intertidal of bays; locally common; builds U- or Y-shaped, firm-walled burrows in mud or muddy sand; commensals include *Betaeus, Hesperonoe,* pinnotherids, and copepods (see Gooding, 1960, above); the isopod *Phyllodurus abdominalis* (p. 286) and the clam *Orobitella rugifera* (p. 559) may both occur on the abdomen; see MacGinitie, 1930, Amer. Midl. Nat. (10) 6: 36–44 (natural history); Thompson, 1972, unpublished Ph.D. Thesis (Zoology), Univ. Calif. Berkeley (burrows); Powell, 1974, Univ. Calif. Publ. Zool. 102 (gut morphology).

Tribe **Paguridea**

DIOGENIDAE

Isocheles pilosus (Holmes, 1900) (= *Holopagurus pilosus*). A hermit crab of the low intertidal in sand on semiprotected beaches, often in *Polinices* shells; more common sublittorally.

PAGURIDAE

PAGURIDAE Hermit crabs; see McLaughlin, 1974; Schmitt, 1921; Stevens, 1925, Publ. Puget Sound Biol. Sta. 3: 273–309; papers by Reese (1962, Anim. Behav. 10: 347–360; 1963, Behaviour 21: 78–126); Orians and King, 1964, Pac. Sci. 18: 297–306; Bollay, 1964, Veliger 6, Suppl., 71–76 on shell use and selection. Various species of the spionid worm *Polydora* occur with hermit crabs; see Blake and Evans, 1973, Veliger 15: 235–249.

Pagurus beringanus (Benedict, 1892). Low intertidal on rock jetties; sublittoral.

Pagurus granosimanus (Stimpson, 1859). Exposed and semiprotected outer coast, lower intertidal pools; common.

Pagurus hemphilli (Benedict, 1892). Mid to low intertidal.

Pagurus hirsutiusculus (Dana, 1851). Mid intertidal of rocky coast, common; in bays, under rock cover, and in tidepools, coarse sand to gravel substrates.

* *Pagurus quaylei* Hart, 1971. In gravelly areas, shallow water; see Hart, 1971, J. Fish. Res. Bd. Canada 28: 1527–1544.

Pagurus samuelis (Stimpson, 1857). Rocky coasts, mid to lower tidepools, abundant, in *Tegula* and other shells; occasionally in coarse substrates in bays.

LITHODIDAE (Stone crabs)

Cryptolithodes sitchensis Brandt, 1853. The umbrella crab, low intertidal to sublittoral; in crevices, on sponges; algae.

Hapalogaster cavicauda Stimpson, 1859. Low intertidal of protected rocky coast; uncommon.

Oedignathus inermis (Stimpson, 1860). Exposed and semiprotected rocky coasts, among coralline algae, under mussel beds or in rock crevices; uncommon.

Tribe **Galatheidea**

GALATHEIDAE

* *Pleuroncodes planipes* Stimpson, 1860. Pelagic "red crab," sometimes beached in vast swarms from Monterey south; for occurrence and biology see Glynn, 1961, Calif. Fish Game 47: 97–101; Longhurst and Seibert, 1971, Pac. Sci. 25: 426–428; Boyd, 1967, Pac. Sci. 21: 394–403.

PORCELLANIDAE Porcelain crabs; see Haig (1960) for detailed descriptions and systematic review; Gonor and Gonor, 1973, Fishery Bull. 71: 189–234 on larval morphology and larval biology of *Petrolisthes cinctipes, P. eriomerus, Pachycheles pubescens, P. rudis.*

Pachycheles pubescens Holmes, 1900. Low intertidal, rocky areas.

Pachycheles rudis Stimpson, 1859. Low intertidal, semiprotected rocky coast; in *Laminaria* and *Egregia* holdfasts, under rocks, on wharf pilings, and other habitats; bopyrid isopod (*Aporobopyrus*) may occur in branchial cavity (p. 286); see MacMillan, 1972, Biol. Bull. 142: 57–70, on larval development of *P. pubescens* and *P. rudis*.

* *Petrolisthes cabrilloa* Glassell, 1945. Morro Bay south; see Haig, 1960.

Petrolisthes cinctipes (Randall, 1839). Mid and upper intertidal of protected, semiprotected rocky coast, under rocks; common; see Wicksten, 1973, Bull. So. Calif. Acad. Sci. 72: 161–163 (feeding).

Petrolisthes eriomerus Stimpson, 1871. Mid intertidal of protected rocky coasts; bays, under rocks over gravel substrates; also in eelgrass and kelp holdfasts; the rhizocephalan *Lernaeodiscus* has been reported from specimens from southern California.

Petrolisthes manimaculis Glassell, 1945 (=*P. gracilis* of Schmitt, 1921). Low intertidal under rocks. Females and juveniles often closely resemble *P. eriomerus*, but the 2 species can be distinguished by the length-width ratio of the carpus: in *P. manimaculis* the length of the carpus is not less than 2.4x the width, in *P. eriomerus* it seldom exceeds 2x, and is never as much as 2.4x the width (Haig, 1960, p. 76).

Petrolisthes rathbunae Schmitt, 1921. Under stones, low intertidal.

Tribe **Hippidea**

ALBUNEIDAE

Blepharipoda occidentalis Randall, 1839. Very low intertidal, more common sublittorally; exposed sandy beaches; see Knight, 1968, Proc. Calif. Acad. Sci. (4) 35: 337–370 (larval development, distribution, ecology).

HIPPIDAE

Emerita analoga (Stimpson, 1857). Sand crab; intertidal of exposed sandy beaches, abundant but distribution patchy; moves up and down beach with tidal cycle, burrowing to depth of several cm when tide is out; see Efford 1966–1971, Crustaceana 10: 167–182; 13: 81–93; 16: 15–26; 18: 293–308; 21: 316–317 (biology, feeding, distribution); Cubit, 1969, Ecology 50: 118–123 (migration, aggregation); Dillery and Dudley, 1968, Ecology 49: 746–751 (seasonal reproduction); Gilchrist and Lee, 1972, Comp. Biochem. Physiol. 42(B): 263–294 (carotenoid pigments; reproduction).

REFERENCES ON EUPHAUSIACEAN AND DECAPOD CRUSTACEA

Balss, H. 1957. Decapoda 13: 1673–1770. In *Klassen und Ordnungen das Tierreichs*, H. G. Bronn, 5, I, abt. 7.

Boden, P., M. W. Johnson, and E. Brinton 1955. The Euphausiacea (Crustacea) of the North Pacific. Bull. Scripps Inst. Oceanogr. 6: 287–400.

Boolootian, R. A., *et al.* 1959. Reproductive cycles of five west coast crabs. Physiol. Zool. 32: 213–220. (*Pachygrapsus, Emerita, Hemigrapsus nudus, Pugettia producta, Petrolisthes cinctipes*)

Brinton, E. 1962. The distribution of Pacific euphausiids. Bull. Scripps Inst. Oceanogr. 8: 51–270.

Chace, F. A. 1951. The grass shrimps of the genus *Hippolyte* from the west coast of North America. J. Wash. Acad. Sci. 41: 35–39.

Garth, J. S. 1958. Brachyura of the Pacific coast of America: Oxyrhyncha. Allan Hancock Pac. Exped. 21, Pts. 1 (text) and 2 (tables and plates).

Gurney, R. 1939. *Bibliography of the larvae of decapod Crustacea.* London: Ray Society, 123 pp.

Gurney, R. 1942. *Larvae of decapod Crustacea.* London: Ray Society, 306 pp.

Haig, J. 1960. The Porcellanidae (Crustacea, Anomura) of the eastern Pacific. Allan Hancock Pac. Exped. 24, 440 pp.

Hart, J. F. L. 1964. Shrimps of the genus *Betaeus* on the Pacific coast of North America with descriptions of three new species. Proc. U.S. Nat. Mus. 115: 431–466.

Hartnoll, R. G. 1969. Mating in the Brachyura. Crustaceana 16: 161–181.

Holmes, S. J. 1900. Synopsis of the California stalk-eyed Crustacea. Occ. Pap. Calif. Acad. Sci. 7: 262 pp.

Holthuis, L. B. 1947. The Decapoda of the Siboga Expedition. Part IX. Siboga Exped. Monogr. 39a[8], 100 pp.

Knudsen, J. W. 1963. Observations of the reproductive cycles and ecology of the common Brachyura and crablike Anomura of Puget Sound, Washington. Pac. Sci. 18: 3–33.

MacGinitie, G. E. 1937. Notes on the natural history of several marine Crustacea. Amer. Midl. Nat. 18: 1031–1037.

Mauchline, J. and L. R. Fisher 1969. The biology of euphausiids. Adv. Mar. Biol. 7, 454 pp.

McLaughlin, P. A. 1974. The hermit crabs (Crustacea: Decapoda: Paguridea) of northwestern North America. Zool. Verhandel. no. 130, 396 pp.

Passano, L. M. 1960. Molting and its control. In *The Physiology of Crustacea*, T. H. Waterman, ed., Vol. 1, Chap. 15, pp. 473–536. Academic Press.

Pike, R. B. and D. I. Williamson 1961. The larvae of *Spirontocaris* and related genera (Decapoda, Hippolytidae). Crustaceana 2: 187–208.

Rathbun, M. J. 1904. Decapod crustaceans of the northwest coast of North America. Harriman Alaska Exped. 10: 190 pp.

Rathbun, M. J. 1918. The grapsoid crabs of America. U.S. Nat. Mus. Bull. 97: 461 pp. (Includes Pinnotheridae, Grapsidae.)

Rathbun, M. J. 1925. The spider crabs of America. U.S. Nat. Mus. Bull. 129: 613 pp. (Includes Majidae, Parthenopidae.)

Rathbun, M. J. 1930. The cancroid crabs of America of the families Euryalidae, Portunidae, Atelecyclidae, Cancridae, and Xanthidae. U.S. Nat. Mus. Bull. 152: 609 pp.

Rathbun, M. J. 1937. The oxystomatous and allied crabs of America. U.S. Nat. Mus. Bull. 166: 278 pp. (Includes Leucosidae, Calappidae.)

Riegel, J. A. 1959. The systematics and distribution of crayfishes in California. Calif. Fish Game 45: 29–50.

Schmitt, W. L. 1921. The marine decapod Crustacea of California. Univ. Calif. Publ. Zool. 23: 470 pp.

Teissier, G. 1960. Relative growth. In *The Physiology of Crustacea,* ed., T. H. Waterman, Academic Press. Vol. 1, Chap. 16, pp. 537–560.

Weymouth, F. W. 1910. Synopsis of the true crabs (Brachyura) of Monterey Bay, California. Stanford Univ. Publs., Univ. Series no. 4: 64 pp.

PHYLUM ARTHROPODA: ARACHNIDS, TARDIGRADES, AND INSECTS

Although insects and arachnids have achieved their striking evolutionary success in numbers and diversity on land, representatives of several families of the subclass **Insecta** have re-entered the intertidal, and among marine "arachnids" (subphylum **Chelicerata**) are the curious and numerous sea spiders (**Pycnogonida**), the large family of marine mites (Halacaridae), and the ancient horseshoe crabs (*Limulus,* order **Xiphosura,** not found on this coast). At or above high-tide mark occur tiny pseudoscorpions (order **Chelonethida,** p. 431).

Hedgpeth, who has contributed much to our knowledge of the pycnogonids, has revised his key for this edition (p. 413). Marine mites, although common, are usually passed over in despair by the average invertebrate zoologist. The revised discussion and key by Newell (p. 425), representing otherwise inaccessible information, should be valuable to students of marine communities and populations. The minute **Tardigrada** or water bears (p. 431) are included in this chapter with insects and arachnids because they have arthropod affinities and are most likely to be encountered by students seeking mites and pycnogonids. The keys to intertidal insects by Daly, Schlinger, and Doyen (pp. 432, 436, 446) should aid and encourage studies of this often-overlooked component of our shore fauna.

PYCNOGONIDA

Joel W. Hedgpeth
Oregon State University, Marine Science Center, Newport, Oregon

(Plates 99–102)

Pycnogonids or "sea spiders" are exclusively marine arthropods found from intertidal to abyssal depths. Occasionally they occur in large swarms but usually are sparse and not often observed in temperate intertidal regions because of their cryptic coloration and resemblance to hydroid stalks and other fragments of stems and twisted bits of stuff under rocks and in crevices. They are often considered to be related to arachnids and are conceded to be chelicerates; I prefer to regard them as a subphylum or major branch of arthropods because of their multiple gonopores and unique protonymphon larva (fig. 25); the only other free larval stage in the arthropods is the crustacean nauplius. All shallow-water and intertidal species of which we have any knowledge spend at least part of their lives as parasites, usually of some cnidarian, and many of the adults persist as succivorous predators of cnidarians and other soft-bodied invertebrates. Some species are parasitic, at least for a time, in molluscs. Usually the protonymphon larva encysts in a cnidarian and breaks free at an advanced stage. However, we know little of the sex life and early life history of most coastal species. The most conspicuous and easily recognized intertidal species are those of the genus *Pycnogonum* (figs. 2, 3), from which the entire group takes its name, although their stout, knobby aspect is the least characteristic shape among the more than 500 species of Pycnogonida.

Naturally I profess a great fondness for these obscure animals; this attraction was shared by the late Professor William A. Hilton of Pomona, who published many papers on California species. Unfortunately for the unwary, Dr. Hilton, despite his great personal charm and talents as a teacher, was not a very critical systematist, and many of his publications lack the detail necessary to be certain of what he was talking about; hence his papers on Pycnogonida are useful only to those who may have access to his specimens. Most of his material is in the U.S. National Museum, the remainder at the Allan Hancock Foundation. The definitive monograph on the species of the Pacific coast of North America remains to be written; in the meanwhile Cole's Harriman Alaska Report (1904) is the most useful single reference, although incomplete.

There are no discrete orders in the Pycnogonida, but about a dozen families are recognized, based primarily on permutations of the various appendages and their stage of development. Several of the families are not represented in the California intertidal zone. Examples of Nymphonidae, the largest family in number of species although restricted for the most part to the single genus *Nymphon* (fig. 1), occur near shore in subtidal waters and may be brought ashore by divers. Some southern California species included in the previous version of this key have been omitted, both to simplify the key and to discourage students from relying on this key as a definitive source for identifications in that region. On the other hand, this does not imply that it is considered definitive for the central California coast; surprises are always turning up.

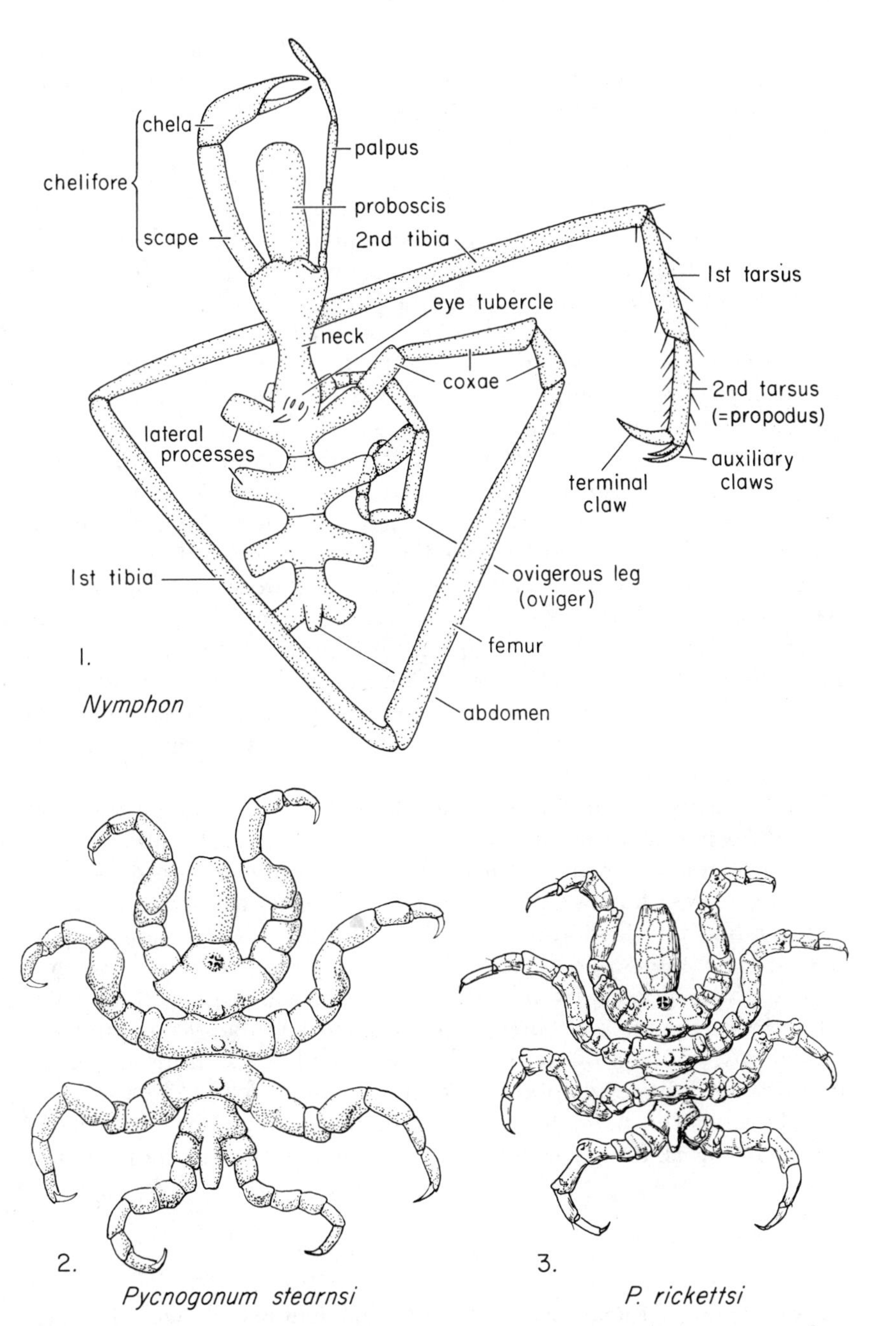

PLATE 99. **Pycnogonida** (1). 1, *Nymphon,* to illustrate characters used in key; 2, *Pycnogonum stearnsi;* 3, *P. rickettsi.* All, Hedgpeth, original.

The characters used in the key should be visible under the medium power of a dissecting microscope, although for many of the smaller spinose species it is difficult to make out details because of the debris adhering to the surface of the animal. Much of this may be removed with a camel's hair brush or picked off carefully with fine needles. Specimens are best killed for study by placing them in fresh water for a few minutes so that they die in an extended condition. Individuals should be handled separately since pycnogonids tend to cling to any available object, including each other.

Although the terms used in the key are illustrated in figure 1, experience has demonstrated that students are often unable to decide which appendage is which, perhaps because the form illustrated is one in which all possible appendages occur. This is not the case with many local forms; the major basis for classification in these animals is the presence or absence or various combinations of anterior appendages. The situation is further confused by sexual dimorphism; the females of some groups either lack the special egg carrying structures (**ovigers**) or have much smaller or less completely developed ones than the males. Quite often, however, females may be identified by the swollen appearance of the longest leg joints (**femora**). It is hoped that a preliminary inspection of the figures will assist the student in determining whether he is in the right part of the key, even if it is impossible to discern the number of segments in the smaller anterior appendages or the ovigers.

The key is useful only for fully adult specimens; juveniles are often difficult to assign to genus or even family. If several similar specimens are found and some of these bear clusters of eggs ventrally, they are adult males; females of similar or larger size are probably adult. The definitive indication of sex is the condition of the gonopores; on most species these pores occur on the ventral surface of the second coxae of the last two pairs of legs (on all legs in some species). In many ammotheids the male gonopores are situated on tubercles or spurlike projections, while in the female they are on a low swelling. At sexual maturity the female gonopores are large enough for the extrusion of eggs, but those of the male are much smaller. In immature or sexually quiescent forms, however, the size difference of the gonopores is not apparent.

KEY TO PYCNOGONIDA

1. Legs short, thick, shorter or not conspicuously longer than combined length of proboscis and trunk, giving the total animal an oval or elliptical appearance at rest; without anterior dorsal appendages .... *Pycnogonum* 2

– Legs usually much longer than combined length of proboscis and trunk, spindly or angular, producing a circular outline at rest; awkward and gangly animals; with anterior appendages, either chelifores or palpi or both 3

2. Body apparently smooth but minutely granular; even-toned in color (ivory to light pink); dorsal tubercles when present not much higher than their dorsal diameter (fig. 2) .... *Pycnogonum stearnsi*

– Body of uneven tone, usually light brown with dark lines or clear areas between darker patches ("reticulated"); dorsal tubercles higher than basal diameter (fig. 3) .... *Pycnogonum rickettsi*

In both species, males are the smaller, often with masses of eggs ventrally; the larger females lack the small, anterior, ventral ovigers for holding the egg masses.

3. With both chelifores and palpi; chelifores chelate or achelate, reduced to stubs, or chelae reduced to terminal knobs, although in many juveniles there are chelae which are lost in the adult stage; palpi conspicuously multijointed; ovigers usually present in both sexes and conspicuous 4

– One or the other pairs of anterior appendages lacking; when the only pair of appendages is the chelifores, they are usually chelate and functional in the adult; ovigers reduced or absent in the females 19

4. Chelifore with 2 basal segments; sometimes small chelae present, or reduced to nonfunctional knobs; total number of joints is 3 (including the chela); the entire appendage shorter than proboscis 5

– Chelifore with a single basal segment and chela or knoblike terminal joint or subchela; "2-jointed," or chelifore reduced to a stub or an elongate, tuberclelike structure 8

5. With prominent, spine-bearing tubercles in a double row on longer joints of legs and single row along median line of trunk; a conspicuous species (fig. 9) .... *Nymphopsis spinosissima*

– Less robust, usually small (less than 1 cm total extent); without prominent, multiple, spine-bearing tubercles on legs ... 6

6. Trunk rather delicate, with lateral processes narrowly separated and long, pointed, dorsal tubercles along midline; chelifores with blunt, spurlike processes in addition to spines ... 7

– Trunk compact, circular; chelifores rather thick, without spurs or processes; dorsal trunk tubercles rounded (fig. 6) .. .. *Ammothella tuberculata*

7. Propodus about 4 times as long as its dorsoventral width,

without large basal spines at "heel," but with regular series of large spines along "sole;" auxiliary claw 3/4 as long as terminal claw (fig. 11) .... *Ammothella setosa*
– Propodus about 3 times as long as wide, heel well developed, with 3 large, basal spines; auxiliary claw 1/2 as long as terminal claw (fig. 14) .... *Ammothella menziesi*
8. Chelifores reduced to a single knob or tubercle 9
– Chelifores consisting of a basal segment ("scape") and terminal chela reduced in most species to a knoblike structure but in some, especially subadults, functionally chelate 14
9. Extended or elliptical in outline circumscribed by legs, lateral processes well separated; palpi well developed 10
– Trunk compact, lateral processes contiguous or nearly so, circular in outline; palpi short, usually less than 6-jointed .... *Tanystylum* 11
10. Chelifores easily recognized, papillate processes; eye tubercle and eyes present; legs long, slender; a medium-sized (18–20 mm extent) form with conspicuous brownish purple bands on the legs (fig. 12) .... *Lecythorhynchus hilgendorfi*
– Chelifores reduced to almost imperceptible stumps; eye tubercle and eyes absent; legs short (fig. 15); a minute, interstitial form (2–3 mm extent) ...*Rhynchothorax philopsammum*
11. Proboscis tapered to subconical point 12
– Proboscis rounded; trunk very compact .... *Tanystylum occidentalis*
12. With prominent basal spines on heel of propodus 13
– Without prominent basal spines on heel of propodus; small delicate form .... *Tanystylum intermedium*
13. Three straight or slightly curved basal spines on propodus 2; abdomen about as long as last pair of lateral processes (fig. 16); conspicuous, brown, knobby form with white articulations .... *Tanystylum californicum*
– Basal spines of propodus 2 angular; abdomen longer than last pair of lateral processes (fig. 17) ... *Tanystylum duospinum*
14. Never chelate in adult 15
– Chelate in adult (fig. 10) .... *Achelia chelata*
15. Trunk and legs conspicuously spinose, with small spinose

PLATE 100. **Pycnogonida** (2). 4, *Achelia gracilipes;* 5, *A. nudiuscula;* 6, *Ammothella tuberculata;* 7, *Achelia spinoseta* with palp and claw at right; 8, *A. simplissima* with palp and claw, left; 9, *Nymphopsis spinosissima;* 10, *Achelia chelata*, A from front and B from above; 11, *Ammothella setosa* with claw, from holotype; 12, *Lecythorhynchus hilgendorfi;* 13, *Achelia echinata* with claw; 14, *Ammothella menziesi* with claw, from holotype. 5, from Hall, 1913; 10, from Hedgpeth, 1950; rest, Hedgpeth, original.

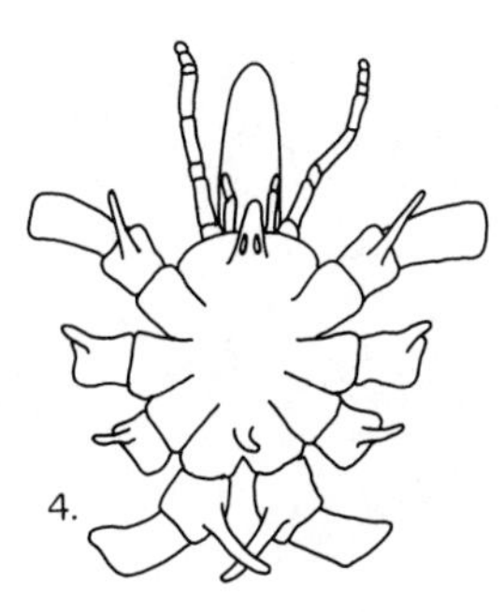

4. *Achelia gracilipes*

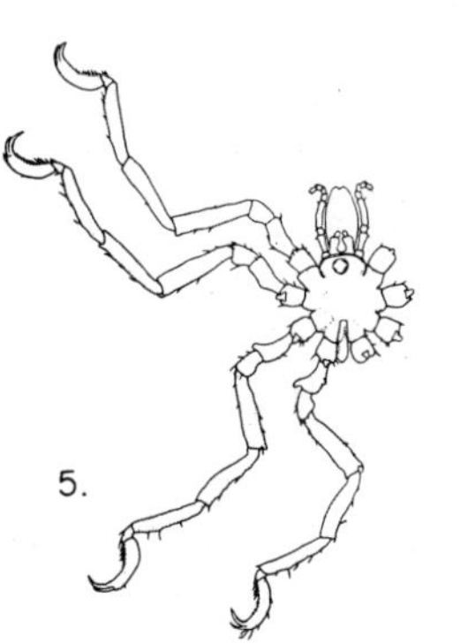

5. *Achelia nudiuscula*

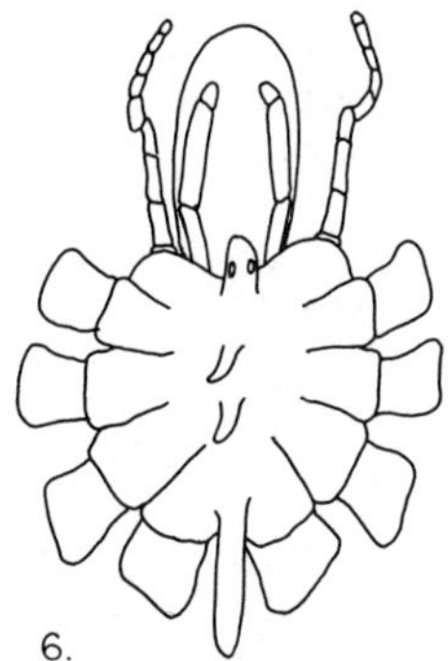

6. *Ammothella tuberculata*

7. *Achelia spinoseta*

8. *Achelia simplissima*

9. *Nymphopsis spinosissima*

10. *Achelia chelata*

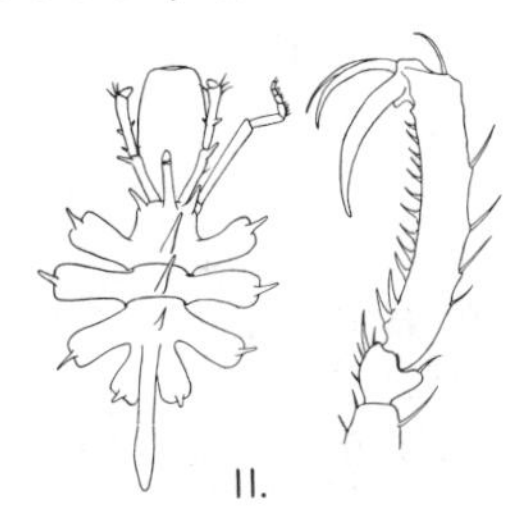

11. *Ammothella setosa*

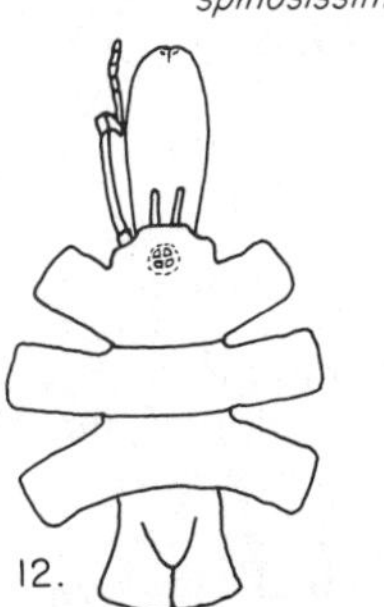

12. *Lecythorhynchus hilgendorfi*

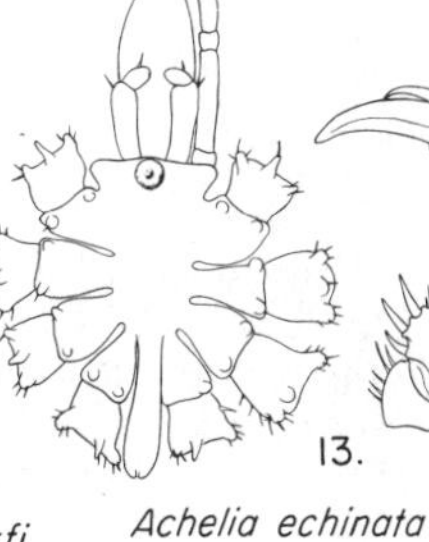

13. *Achelia echinata*

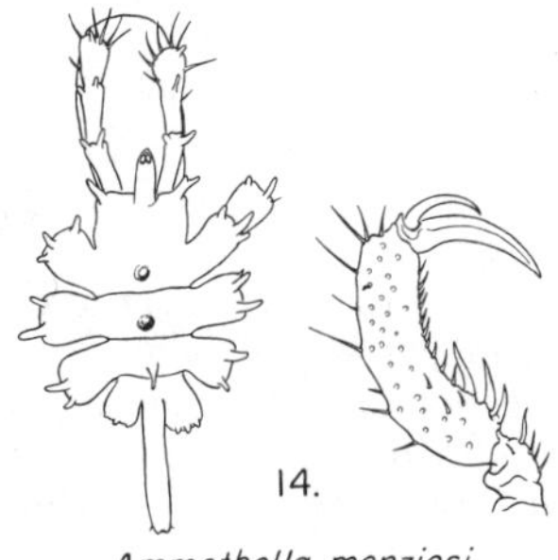

14. *Ammothella menziesi*

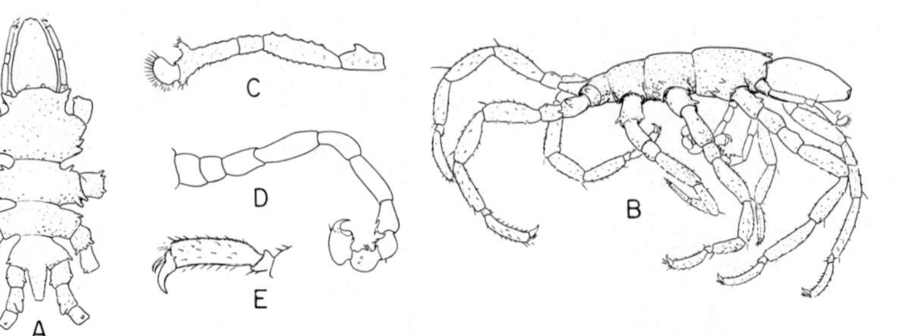

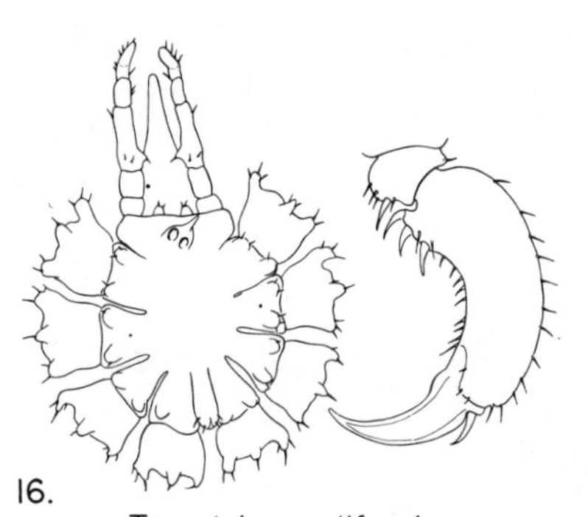

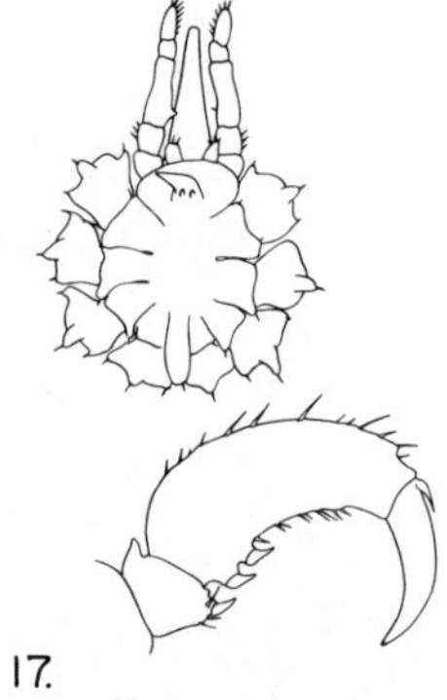

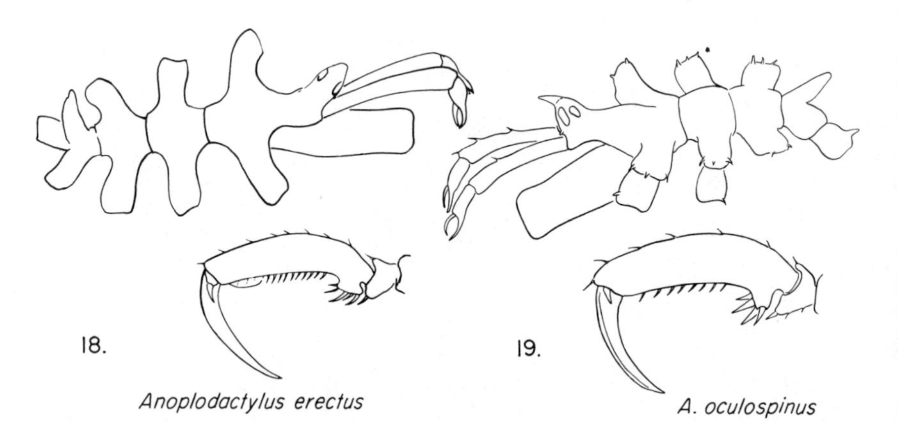

PLATE 101. **Pycnogonida** (3). 15, *Rhynchothorax philopsammum* (female); A, trunk, dorsal; B, whole animal from right; C, palp; D, ovigerous leg; E, tarsus, propodus, and terminal claw; 16, *Tanystylum californicum* and claw; 17, *T. duospinum* and claw; 18, *Anoplodactylus erectus* and claw; 19, *A. oculospinus* and claw. 15, from Hedgpeth, 1951; rest, Hedgpeth, original.

tubercles on 1st coxae 16
- Not so spiny, without spiny processes or knobs on 1st coxae, but some species with fingerlike spurs 17
16. Terminal segments of palpi with ventral lobes ("pectinate") (fig. 7) ....*Achelia spinoseta*
- Palpi without ventral lobes, but subcylindrical (fig. 13) ....*Achelia echinata*
17. First coxae with fingerlike dorsal processes; propodus with basal spines; palpi 8-jointed 18
- First coxae without such processes; no large basal spines on propodus; palpi 7-jointed (fig. 8) ....*Achelia simplissima*
18. Processes on coxae 3/4 as long as the joint; proboscis narrowly elliptical (fig. 4) .... *Achelia gracilipes*
- Processes on coxae less then 1/2 as long as the joint; proboscis broadly elliptical (fig. 5) ....*Achelia nudiuscula*
19. Chelifores conspicuous, with large functional chelae extending over the proboscis; eye tubercle well developed, with eyes 20
- Chelifores apparently absent; palpus 5- or 6-jointed; eye tubercle and eyes lacking; interstitial (see no. 10) ....*Rhynchothorax philopsammum*
20. Trunk compact, lateral processes contiguous; legs comparatively short and stout 21
- Trunk extended, lateral processes separated; legs long and slender 22
21. Trunk and legs not conspicuously spinose; propodus with large central spine opposable to terminal spine, forming a subchelate process (fig. 24) ....*Decachela discata*
- Trunk and legs spinose; propodus not subchelate ....*Pseudopallene* sp.
22. Cephalic segment extended forward as a conspicuous neck, overhanging insertion of proboscis; auxiliary claws minute, seta-like 23
- Cephalic segment not extended forward; auxiliary claws usually present, small but well developed 24
23. Without conspicuous spines on trunk or legs; distal 1/3 of base of sole with a blade or lamella (fig. 18) .... *Anoplodactylus erectus*
- Trunk and leg moderately spinose; spines of sole of propodus separated (fig. 19) ....*Anoplodactylus oculospinus*
24. Lateral processes separated by less than 1/2 their own diameter (trunk usually circular); auxiliary claws minute, inconspicuous ....*Halosoma* 25
- Lateral processes separated by at least 1/2 their own diameter;

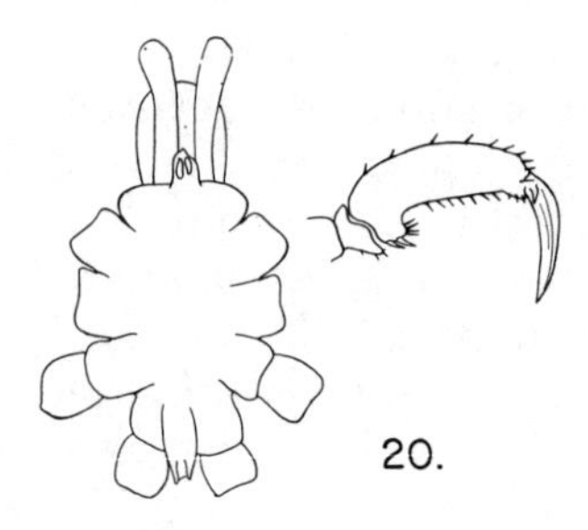

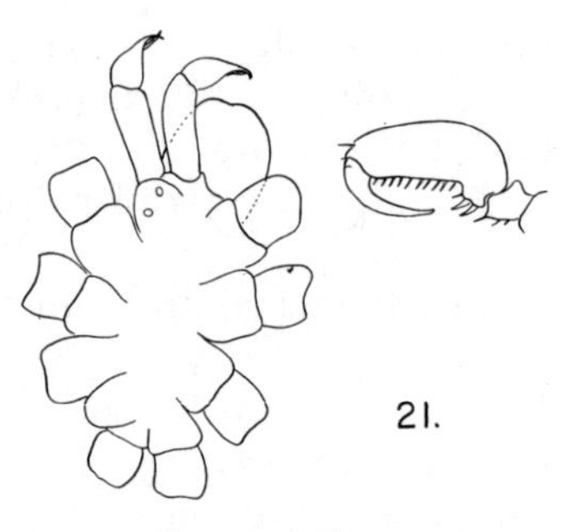

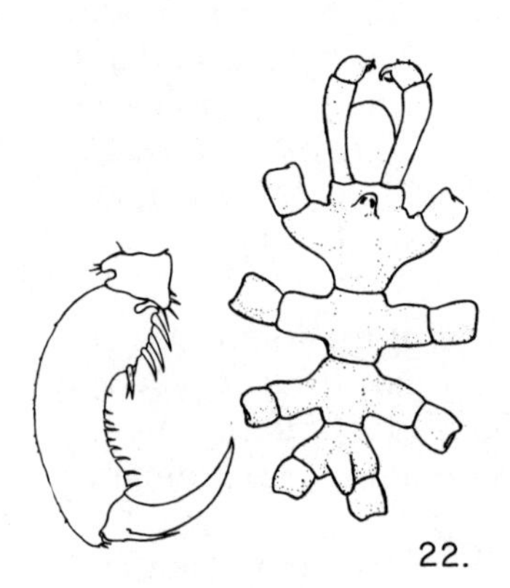

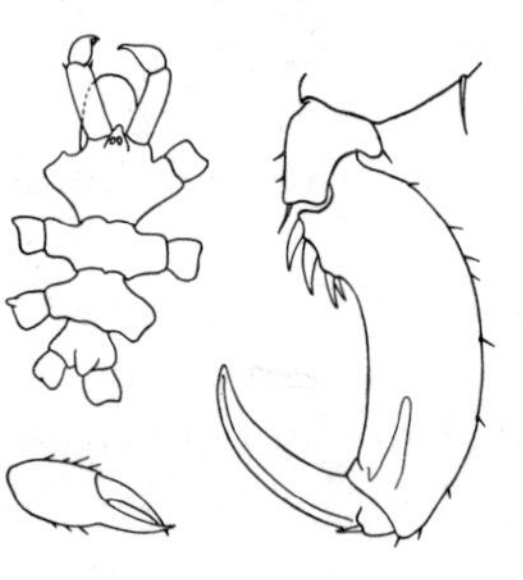

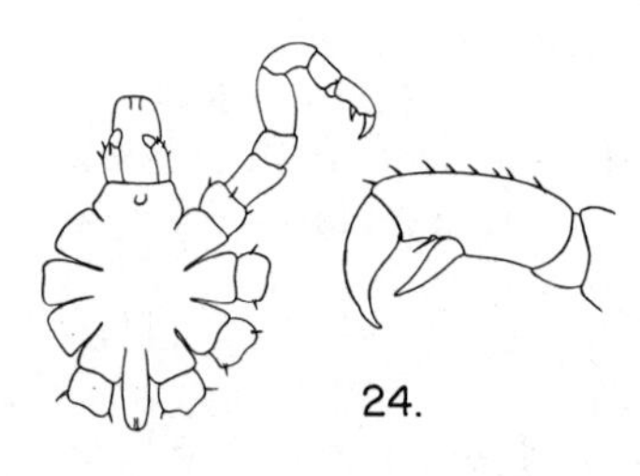

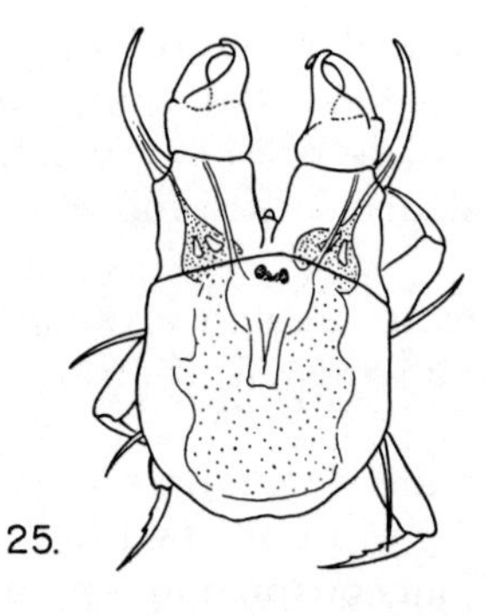

PLATE 102. **Pycnogonida** (4). 20, *Halosoma viridintestinale* and claw; 21, *H. compactum* and claw, from holotype; 22, *Phoxichilidium femoratum* (after Hedgpeth, 1948) and claw (after Lebour); 23, *P. quadridentatum* and claw; 24, *Decachela discata* and claw, from holotype; 25, a protonymphon larva (from Hedgpeth, 1955). Except as noted, Hedgpeth, original.

auxiliary claws small but well developed (fig. 22) ....
.... *Phoxichilidium femoratum*

25. Small, delicate, with a bright green intestine branching out into the legs; lateral processes distinct, close but not touching (fig. 20) .... *Halosoma viridintestinale*
– Similar to above, but with broader and blunter eye tubercle; lateral processes touching (fig. 21) ...*Halosoma compactum*

LIST OF PYCNOGONIDA

Achelia chelata (Hilton, 1939). Often in mussel beds and, in winter, parasitic on the ctenidia of *Mytilus californianus*.

Achelia echinata Hodge, 1864. Reported from Mile Rock, San Francisco. This, or a closely related species, is also reported from Europe and the northwest Pacific.

Achelia gracilipes (Cole, 1904).

Achelia nudiuscula (Hall, 1913). A bay form; one of the few pycnogonids tolerating reduced salinity.

Achelia simplissima (Hilton, 1939).

Achelia spinoseta (Hilton, 1939).

Ammothella menziesi Hedgpeth, 1951.

Ammothella setosa Hilton, 1942.

Ammothella tuberculata Cole, 1904. The most common pycnogonid of *Phyllospadix* holdfasts, etc.

Anoplodactylus erectus Cole, 1904.

Anoplodactylus oculospinus Hilton, 1942. Both species of *Anoplodactylus* are reported as far north as Monterey Bay; the occurrence of juvenile *Anoplodactylus* at Duxbury Reef suggests that one or both of these species may occur intertidally in central California, possibly during warm-water years.

Decachela discata Hilton, 1939. Known also from Japan; type locality and only California record of this curious little species is Monterey.

Halosoma compactum (Hilton, 1939). Known only from the type; possibly not distinct from the following.

Halosoma viridintestinale Cole, 1904. Common on hydroids and *Zostera;* in Tomales Bay it is the most abundant and characteristic pycnogonid of shallow, sheltered waters.

Lecythorhynchus hilgendorfi (Böhm, 1879) (= *L. marginatus* Cole, 1904). This common north Pacific species, living among hydroids and in sheltered crevices, is one of the characteristic species of the central California intertidal zone as well as Japanese shores.

Nymphopsis spinosissima (Hall, 1912).

Phoxichilidium femoratum (Rathke, 1799). A widely distributed boreal species, this is the only truly intertidal *Phoxichilidium* found in central California.

* *Phoxichilidium quadridentatum* Hilton, 1942. (fig. 23), similar to *P. femoratum;* occurs in tremendous numbers on buoys off the Golden Gate, but has not yet been found on the shore. In *P. quadridentatum* there are 4 spines at the heel of the propodus, 2 large ones side by side and a pair of smaller ones; in *P. femoratum* the 4–5 spines are placed in a single row.

Pseudopallene sp. A boreal-arctic genus, not yet reported in this area.

Pycnogonum rickettsi Schmitt, 1934. Usually associated with *Anthopleura, Metridium,* or *Aglaophenia.*

Pycnogonum stearnsi Ives, 1892. Same hosts as above; *P. stearnsi* occurs from southern California to Oregon; *P. rickettsi* from Monterey to Friday Harbor. The two species occur sympatrically at Duxbury Reef on the same species of sea anemones, but so far have not been found on the same individual host.

Rhynchothorax philopsammum Hedgpeth, 1951. So far known only from the inner side of Tomales Point, several inches beneath the surface of coarse sand in sheltered coves.

Tanystylum californicum Hilton, 1939. A characteristic species of the central California intertidal, especially on *Aglaophenia.*

Tanystylum duospinum Hilton, 1939.

Tanystylum intermedium Cole, 1904. Characteristically a southern species.

Tanystylum occidentalis (Cole, 1904).

REFERENCES ON PYCNOGONIDA

Benson, P. H. and D. D. Chivers 1960. A pycnogonid infestation of *Mytilus californianus.* Veliger 3: 16–18.

Cole, L. J. 1904. Pycnogonida of the west coast of North America. Harriman Alaska Expedition 10: 249–298.

Hedgpeth, J. W. 1941. A key to the Pycnogonida of the Pacific Coast of North America. Trans. San Diego Soc. Nat. Hist. 9: 253–264.

Hedgpeth, J. W. 1951. Pycnogonids from Dillon Beach and vicinity, California, with descriptions of two new species. Wasmann J. Biol. 9: 105–117.

Ziegler, A. C. 1960. Annotated list of Pycnogonida collected near Bolinas, California. Veliger 3: 19–22.

* Not in key.

MARINE MITES (HALACARIDAE)

Irwin M. Newell
University of California, Riverside

(PLATES 103–104)

Intertidal mites include representatives of the three major orders of **Acari**, but most marine mites are in a single family, the Halacaridae (order **Trombidiformes**). This family has been unusually successful in its evolutionary adaptation to numerous marine niches with the result that more than thirty generically distinct groups now exist, including hundreds of species. Some are phytophagous, others predaceous, and several have developed truly parasitic habits. Halacaridae probably evolved in the sea from several terrestrial lines, and should be regarded as polyphyletic. Subsequently they have invaded fresh water, where there are twelve more genera or subgenera.

The other two major orders of Acari have few intertidal representatives. One common species of the order **Parasitiformes** is *Gammaridacarus brevisternalis* Canaris, 1962, which attaches to the undersides of the well-known amphipod beach hoppers, *Orchestoidea* spp. Other genera of Parasitiformes occur intertidally, but are not well studied. The **Sarcoptiformes** are represented by the marine genus *Hyadesia* (suborder Acaridiae) and several genera of the suborder Oribatei that are not well known in marine habitats.

Of the marine genera of Halacaridae, at least fourteen are known from the north Pacific and one, *Thalassacarus*, is known only from this region. Halacaridae occupy a great number of marine habitats, even to depths of more than 5000 meters (Newell, 1971). Nevertheless, the ecological distribution of any given species is probably fairly restricted. For example, *Isobactrus* spp. are usually confined to brackish tide pools or estuaries; *Rhombognathus* spp. are rarely encountered subtidally, and never below the euphotic zone; *Scaptognathus*, *Anomalohalacarus*, and *Actacarus* are interstitial in coarse sand.

Halacaridae range from 0.18 to 2 mm in body length. They are usually abundant: a liter of coralline algae may contain hundreds of individuals and up to fifteen species. Despite their small size they are a conspicuous and omnipresent element, well worth a few hours of the student's time. Biological studies are of particular importance, since little is known of the actual niches occupied by the various species.

Halacaridae are easily collected by placing algae, gravel, barnacles, mussels, and other substrates in sea water, anesthetizing for about 10 minutes with chloroform, and washing the substrate vigorously with either salt or fresh water. Washings can be preserved in 65% alcohol. If the mites are wanted alive, the chloroform treatment should be greatly reduced or eliminated, and a vigorous jet of tap water should be used to separate the mites from the substrate. They should then be returned to sea water for further study. Intertidal mites tolerate immersion in fresh water for 1 to 3 hours, but longer periods are usually fatal.

Temporary mounts of Halacaridae can be made in Berlese fluid or Hoyer's modification of it. They may also be cleared with lactic acid and transferred to 15% glycerine in water, which is then allowed to evaporate slowly while the mites are being examined. For permanent mounts the mites should be cleared with enzymes and mounted in Hyrax or glycerine, following procedures outlined by Newell (1947).

The following key is based on adults only. Males are distinguishable by the **phorotype** (fig. 5), the organ which produces the spermatophore (the phorotype is erroneously called "penis" in most writings). Probably all species of Halacaridae utilize spermatophores. There is a six-legged larva in the life cycle, followed by one, two, or three nymphal instars, depending on the genus. Protonymphs have one, deutonymphs two, and tritonymphs (known in *Isobactrus*) have three pairs of provisory genital acetabula, but no genitalia. Protonymphs also have the femur of leg IV undivided.

KEY TO GENERA OF MARINE HALACARIDAE OF THE EASTERN NORTH PACIFIC

1. Insertions of palpi lateral to rostrum (fig. 1); trochanters separated by an interval appreciably greater than their width, so that they are largely or fully visible in ventral view; abundant to rare, on various substrates, occasionally interstitial in coarse sand 2
– Insertions of palpi dorsal to rostrum (figs. 2–4); trochanters separated by an interval less than their width, so that they are largely concealed in ventral view; palpi very long to short; generally rare, often interstitial in coarse sand 12
2. Middle piece of claw articulating directly with tip of the tarsus (fig. 20) (NOTE: oil immersion may be necessary at first to interpret this important character); color in life yellow, brown, or rarely green or greenish black; predaceous or parasitic 3
– Middle piece of claw articulating with an intermediate sclerite, the **carpite,** and this in turn articulates freely with (fig. 18), or is a flexible extension of (fig. 19), the tip of the

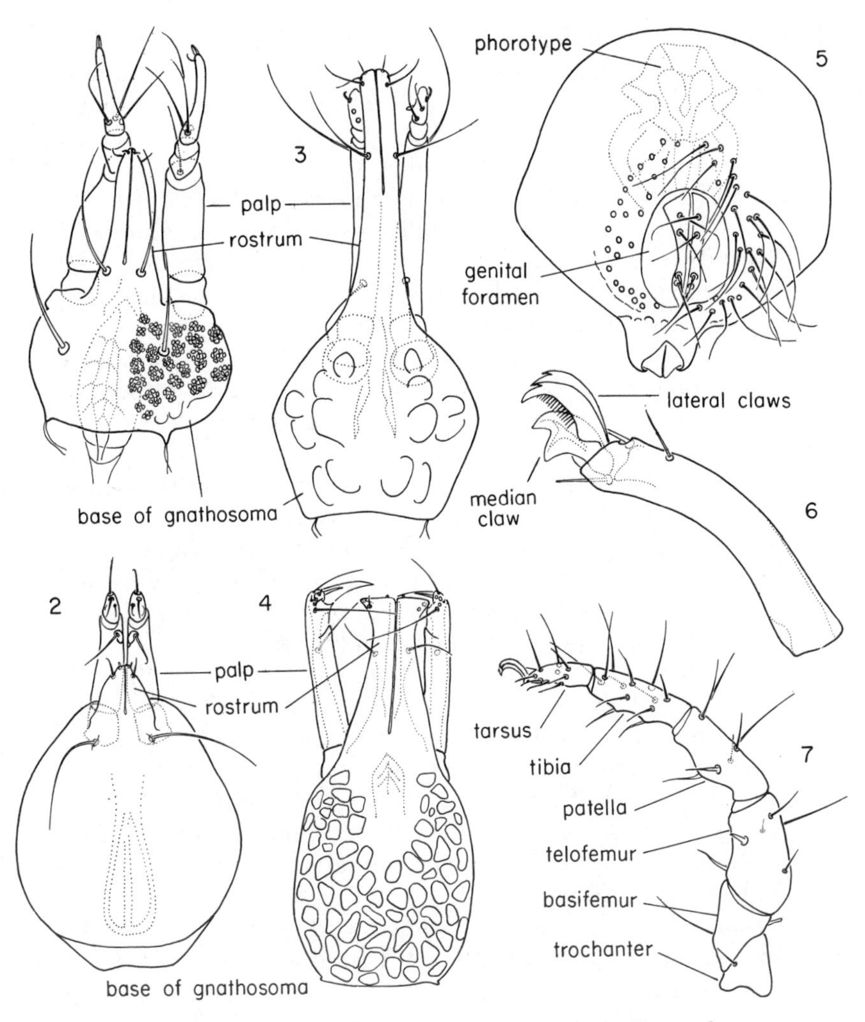

PLATE 103. **Halacaridae** (1). 1, *Copidognathus curtus* Hall, gnathosoma, ventral; 2, *Simognathus* sp., gnathosoma, ventral; 3, *Lohmannella falcata* (Hodge), gnathosoma, ventral; 4, *Scaptognathus* sp., gnathosoma, ventral; 5, *Copidognathus curtus* Hall, genitoanal plate, male, phorotype in dotted line; 6, *Bradyagaue bradypus* Newell, tarsus 1 of leg III, showing curvature and massive median claw; 7, *Halacarus frontiporus* Newell, leg I, showing segmentation. Newell, original.

tarsus (NOTE: some species of *Agauopsis* have a minute, carpitelike structure at the tip of the tarsus, but it appears to be a rigid, rodlike extension of the tarsus); color in life green to black; phytophagous; not living under conditions precluding algal growth 11

3. Patella of palp with a seta, variable in position and form (figs. 12, 13); NOTE: In *Copidognathus pseudosetosus* and related species, there is a sharp spine here, but there is no alveolus (socket) and it is not a seta (fig. 14) 5
– Patella of palp lacking a seta (figs. 1, 14) 4
4. Tarsi of legs bowed; median claw massive, thicker than lateral claws (fig. 6); slow-moving forms, adapted for clinging to hydroids or bryozoans ....*Bradyagaue* Newell, 1971
– Tarsi of legs not bowed, but straight (fig. 7); median claw minute, not as thick as lateral claws ....*Copidognathus* Trouessart, 1888
5. Patella of legs relatively long, nearly as long as either the telofemur or the tibia (fig. 7); femur of palp with 2 setae (fig. 12); usually rare 6
– Patella of legs distinctly shorter than telofemur or tibia; femur of palp with only 1 seta 7
6. Idiosoma (body, exclusive of gnathosoma or "capitulum") slender, very flexible in life, modified for moving freely and quickly through interstices in sand, posterior dorsal plate of female divided into right and left halves ....*Anomalohalacarus* Newell, 1949
– Idiosoma neither slender nor flexible, not modified for interstitial life; posterior dorsal plate either absent, or (if present) not divided into right and left halves ....*Halacarus* Gosse, 1855
7. Ocular plates large, readily visible on dorsal surface (figs. 8–10); habitats variable, but not normally interstitial in coarse sand 8
– Ocular plates very small, at sides of idiosoma (body), often easily overlooked; normally interstitial in coarse sand and under boulders ....*Actacarus* Schulz, 1936

PLATE 104. **Halacaridae** (2). 8, *Rhombognathus* sp., dorsum, right side; 9, *Isobactrus* sp., dorsum, right side; 10, *Thalassacarus commatops* Newell, dorsum, right side; 11, *Agauopsis productus* Newell, basifemur-tibia I, right side, ventral view; 12, *Halacarus frontiporus* Newell, left palp, anterior (= "medial") view; 13, *Agauopsis* sp., right palp, dorsal view; 14, *Copidognathus pseudosetosus* Newell, left palp, dorsal view, showing spine (not a seta) on patella of palp; 15, *Thalassacarus commatops* Newell, tarsus of chelicera, side view; 16, *Thalassarachna capuzinus* (Lohmann), tarsus of chelicera, side view; 17, *Agaue longiseta* Newell, tarsus of chelicera, side view; 18, *Rhombognathus* sp., ambulacrum and tip of tarsus III, left side, ventral view, showing rod-like carpite; 19, *Isobactrus* sp., ambulacrum III, showing moniliform carpite; 20, *Copidognathus curtus* Hall, ambulacrum II, ventral view, (carpite absent). Note: In figs. 18–20, middle piece (where present) and carpite are shown as dotted outlines, and are surrounded by thin, membranous cuticle. Newell, original.

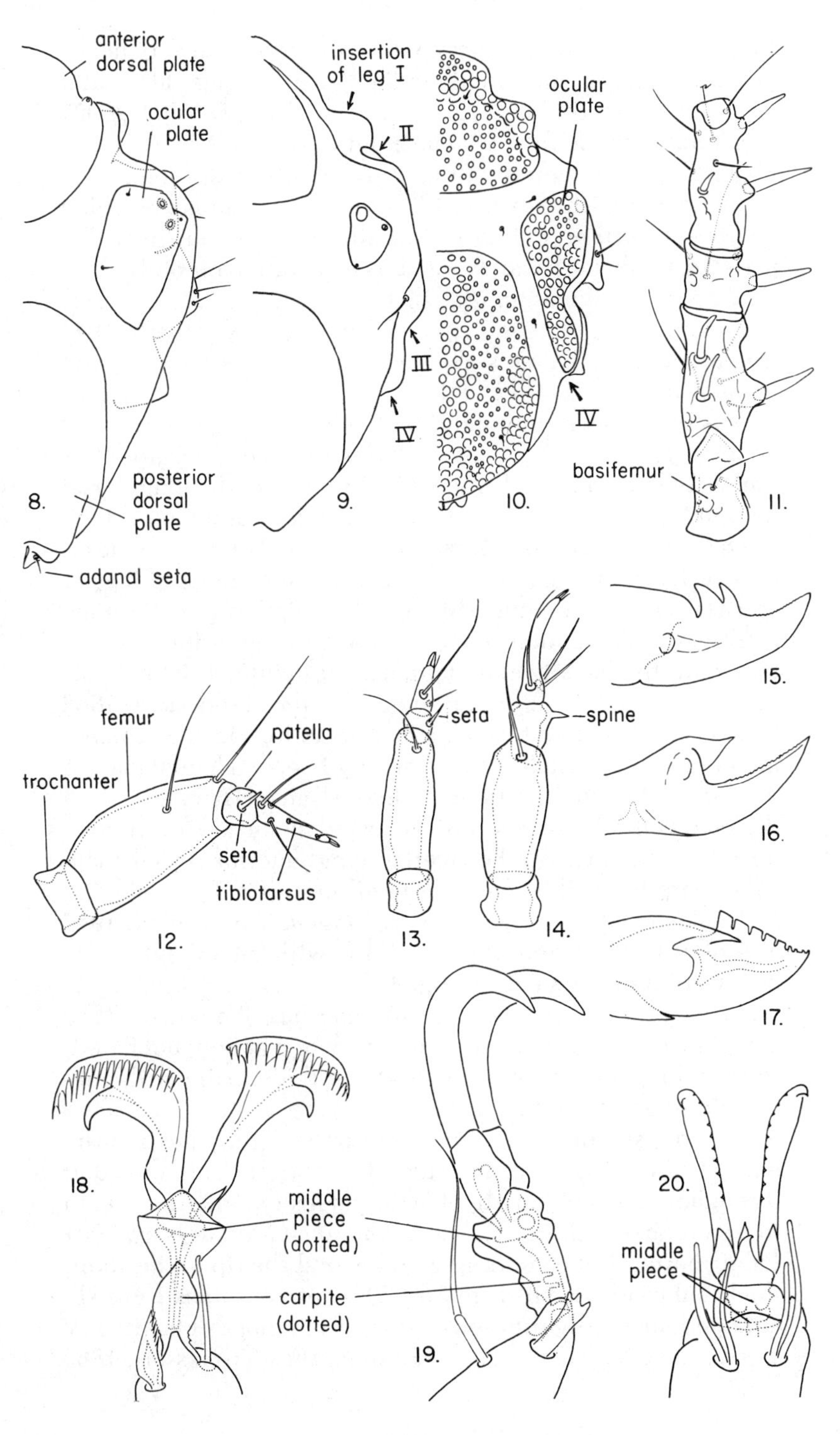
anterior dorsal plate
ocular plate
posterior dorsal plate
adanal seta
8.
insertion of leg I
II
III
IV
9.
ocular plate
IV
10.
basifemur
11.
femur
trochanter
patella
seta
tibiotarsus
12.
seta
13.
spine
14.
15.
16.
17.
18.
middle piece (dotted)
carpite (dotted)
19.
20.
middle piece

8. Leg I rakelike in appearance, with a row of several very heavy peg setae along anterior ("medial") margin (fig. 11); palpi very short, straight .... *Agauopsis* Viets, 1927
– Leg I not rakelike in appearance, although some heavy setae may be present ventrally or anteroventrally; palpi longer .. 9
9. Ocular plates with a thick, tail-like extension (cauda), reaching nearly to insertions of legs IV (fig. 10); cheliceral tarsus with 2 massive teeth on basal 1/2 of dorsal margin, minutely denticulate in distal margin (fig. 15) .... *Thalassacarus* Newell, 1949
– Ocular plates without such a cauda, not reaching beyond level of insertions of legs III 10
10. Tarsus of chelicera minutely denticulate throughout (fig. 16) .... *Thalassarachna* Packard, 1871
– Tarsus of chelicera with a few (5–7) coarse teeth along dorsal margin (fig. 17) .... *Agaue* Lohmann, 1889
11. Each ocular plate with 2 setae (fig. 8); with 3 or more setae on or near lateral margin of body, between insertions of legs II and III; carpite straight, stiff, rod-like (fig. 18); gnathosoma readily visible in dorsal view, projecting anteriorly or anteroventrally; usually abundant, except in interstitial habitats .... *Rhombognathus* Trouessart, 1888
– Ocular plates without setae, a few setae free in the striated, membranous cuticle (fig. 9); with only 1 seta on lateral margin of body between insertions of legs II and III; gnathosoma directed ventrally so that it is concealed in dorsal view (undistorted specimens) by the overhanging anterior dorsal plate (AD); carpite flexible, curved, moniliform (fig. 19); generally in brackish water .... *Isobactrus* Newell, 1947
12. Tip of rostrum flared at end; palpi with an exceptionally heavy spiniform seta at tip (fig. 4) .... *Scaptognathus* Trouessart, 1889
– Tip of rostrum narrowly or bluntly rounded at end, not flared; rostrum long and slender, or short and thick; palpi with only minute setae at tip (figs. 2, 3) 13
13. Palpi long, slender, extending to or only slightly beyond the end of the long rostrum (fig. 3); rostrum parallel-sided throughout most of length; claws of tarsus I similar in form to those on tarsi II–IV .... *Lohmannella* Trouessart, 1901
– Palpi shorter, but extending well beyond the tip of the short, thick, subtriangular rostrum (fig. 2); claws of tarsus I grossly enlarged, markedly different in form from those of tarsi II–IV .... *Simognathus* Trouessart, 1889

REFERENCES ON MARINE MITES

Canaris, A. G. 1962. A new genus and species of mite (Laelaptidae) from *Orchestoidea californiana* (Gammaridea). J. Parasitol. 48: 467–469.

Newell, I. M. 1947. A systematic and ecological study of the Halacaridae of eastern North America. Bull. Bingham Oceanogr. Coll. 10 (3): 1–232. Methods of handling marine mites and keys to subfamilies and genera are included.

Newell, I. M. 1952. Further studies on Alaskan Halacaridae. Amer. Museum Novitates 1536: 1–56. Contains references to earlier Alaskan studies.

Newell, I. M. 1953. The natural classification of the Rhombognathinae (Acari, Halacaridae). Syst. Zool. 2: 119–135.

Newell, I. M. 1971. Halacaridae (Acari) collected during Cruise 17 of the R/V *ANTON BRUNN* in the southeastern Pacific Ocean. *ANTON BRUNN* Report no. 8: 1–58. Important bibliographical citations since Newell, 1947.

ORDER PSEUDOSCORPIONIDA (CHELONETHIDA)

(PLATE 105)

Pseudoscorpions are minute (2–4 mm) animals resembling tiny scorpions, but with a flattened, disc-shaped posterior body and relatively enormous pincers. They are active hunters, and in woodlands are found in leaf mold and under bark. Two species often encountered along the shore are *Garypus californicus* Banks, 1909 (fig. 2), under driftwood and beach wrack on stony beaches above high tide mark from Baja California to northern California, and *Halobisium occidentale* Beier, 1931 (fig. 3), under *Salicornia* and logs from central California to Alaska. For keys to genera, see Chamberlin (1931).

REFERENCES ON PSEUDOSCORPIONIDA

Chamberlin, J. C. 1931. The arachnid order Chelonethida. Stanford Univ. Publ. Biol. Sci. 7: 1–284.

Weygoldt, P. 1969. *The Biology of Pseudoscorpions.* Harvard University Press, 145 pp.

MARINE TARDIGRADES

(PLATE 105)

The curious tardigrades ("water bears") are tiny animals, rarely as much as 1 mm in length, mostly found in freshwater and terrestrial habitats. They are generally placed near the arthropods, but have some features suggesting aschelminth affinities (Renaud-Mornant and Pollock, 1971). There are four pairs of legs with toes or claws, a coelomic (or pseudocoelomic) space, and simple organ systems, as would be expected in such very small animals. Several marine genera are found in the interstitial habitat or in debris; some are reported commensal with other invertebrates. In the local intertidal, the cosmopolitan *Echiniscoides sigismundi* (Schultze, 1865) is apparently common in fine, filamentous algae growing on the plates of barnacles and on rocks in exposed situations (fig. 1).

REFERENCES ON MARINE TARDIGRADES

Crisp, D. J. and J. Hobart 1954. A note on the habitat of the marine tardigrade *Echiniscoides sigismundi* (Schultze). Ann. Mag. Nat. Hist. (ser. 12) 7: 554–560.

Renaud-Mornant, J. and L. W. Pollock 1971. A review of the systematics and ecology of marine Tardigrada. Smithson. Contr. Zool. 76: 109–117.

Schuster, R. O. and A. A. Grigarick 1965. Tardigrades from western North America. Univ. Calif. Publ. Zool. 76: 1–67.

ORDERS OF INTERTIDAL INSECTS; COLLEMBOLA, HEMIPTERA

Howell V. Daly
University of California, Berkeley

(Plate 105)

Among many contrasts between marine and terrestrial biotas, few are as striking as that involving insects and other terrestrial arthropods. Nearly three-fourths of the earth's animal species are insects, yet only 3 percent of insect species are aquatic and only a fraction of these are marine or intertidal. Stated in another way: the largest group of animals is virtually absent from the largest habitat. This does not mean that insects as organisms are scarce on the coasts and ocean; on the contrary, some are exceedingly abundant. Taken individually, each physical barrier has been overcome by specializations in certain species: permanent subaquatic respiration (the midge *Pontomyia natans* in Samoa); tidal submergence (species in several orders including the bug *Aepophilus bonairei* in England); wave action (especially the Clunione midges which occur in California); salinity (especially salt-marsh mosquitoes, ephydrid flies, and the water-boatmen *Trichocorixa*); depth (the midge *Chironomus oceanicus* from twenty fathoms); and ocean surface (the pelagic water striders, *Halobates* spp.). Collectively, however, the physical barriers plus intense biological competition have excluded the insects from a major adaptive radiation in the marine environment. A more detailed review and an annotated bibliography are provided by Usinger (1957).

The insects discussed in this manual are limited to species regularly living part or all of their lives in the intertidal zone of the outer coast. Only the water strider, *Halobates sericeus* Eschscholtz (Gerridae, order Hemiptera) inhabits the open ocean. It occurs in warm, offshore waters fifty miles or more from the coast as far north as San Francisco. *Halobates* eggs have been found glued to floating material at sea. Our species is one of the true pelagic forms which are taken near shore only after severe storms. In Hawaii the same species has been observed feeding on small coelenterates at the water surface (Herring, 1961).

The rich faunas of the beaches above the tide, coastal dunes, salt marshes, and estuaries are beyond the scope of this treatment. Occasional insects may be found in the intertidal zone as transients from these areas and the adjacent land. Unless a given specimen is clearly an intertidal resident, readers are advised to identify order and family with the aid of general entomology textbooks or keys (Borror and DeLong, 1971; Essig, 1958). Included in this manual are insects in brackish tidal pools, and intertidal rocks and beaches.

KEY TO ORDERS OF ADULT INTERTIDAL INSECTS

1. Wings absent; abdomen with only 6 segments (fig. 4); small (1.5 mm); grayish blue ....
marine springtails, PODURIDAE, order **Collembola** (p. 434)
– Wings present, though sometimes concealed by leathery front wings (elytra), or reduced to small, articulated lobes; abdomen with 8–10 visible segments; size and color variable 2
2. Front wings overlapping apically when at rest, the basal part divided by converging sutures which form a triangle on the dorsum (fig. 6); brackish pools ....
.... water boatmen, CORIXIDAE, order **Hemiptera** (p. 434)
– Front wings not as above: either entirely free and membranous, reduced to vestigial stumps, or covering the dorsum without overlapping 3
3. Front wings free and membranous or reduced to vestigial stumps; hind wings highly modified as halteres ....
.... flies and midges, order **Diptera** (p. 436)

- Front wings leathery, completely or sometimes only partly covering dorsum, concealing hind wings when at rest (fig. 5) ....beetles, order **Coleoptera** (p. 446)

PRIMITIVE WINGLESS INSECTS

The marine springtail *Anurida maritima* (Guérin, 1836) (fig. 4) is an important intertidal scavenger on both coasts of North America. Dexter (1943) found up to one hundred per square inch in favored areas: on and under rocks and hard mud banks in the tidal zone. They cluster in air-filled pockets under rocks during submergence. Salmon (1964: 257) provides an extensive list of references.

A "silver-fish", *Neomachilis halophila* Silvestri 1911, is common in the highest intertidal zone of beaches, on the seaward edge of the last terrestrial vegetation, in areas covered with rocks too high on the beach to be disturbed by most high tides, and especially among rocks piled so as to leave small spaces between rocks and sand (Benedetti, 1973).

ORDER HEMIPTERA

Water boatmen, *Trichocorixa*, are characteristic of brackish pools throughout the world and can tolerate salinities considerably above that of the sea (Hutchinson, 1931). They are also found with the brine shrimp *Artemia salina* (Pl. 54, fig. 1) and brine flies, *Ephydra gracilis*, in the brine pools of southern San Francisco Bay. Our species are distinguished as follows: *T. reticulata* (Guérin-Meneville, 1857) (fig. 6) has the width of the interocular space at narrowest point distinctly exceeding width of eye along hind margin as seen from above; *T. verticalis* (Fieber, 1851) has the interocular space subequal to or less than the width of the hind margin of the eye.

REFERENCES ON COLLEMBOLA, HEMIPTERA, AND MARINE INSECTS IN GENERAL

Benedetti, R. 1973. Notes on the biology of *Neomachilis halophila* on a California sandy beach. Pan-Pac. Ent. 49: 246–249.

Borror, D. J., and D. M. DeLong 1971. *An Introduction to the Study of Insects.* 3rd ed. Holt, Rinehart and Winston, 812 pp.

Dexter, R. W. 1943. *Anurida maritima:* an important sea-shore scavenger. J. Econ. Entomology 36: 797.

Essig, E. O. 1958. *Insects and Mites of Western North America.* The Macmillan Co., 1050 pp.

Herring, J. L. 1961. The genus *Halobates* (Hemiptera: Gerridae). Pacific Insects 3(2–3): 223–305.

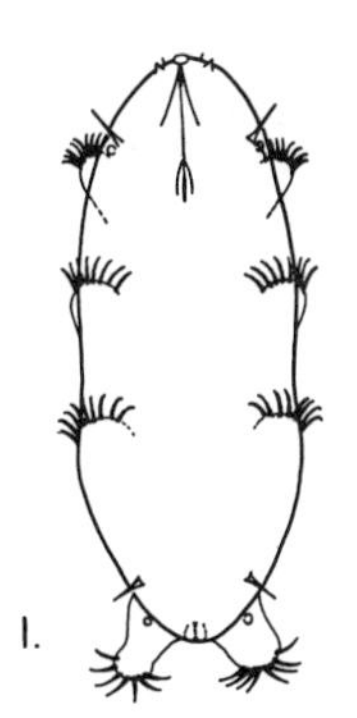

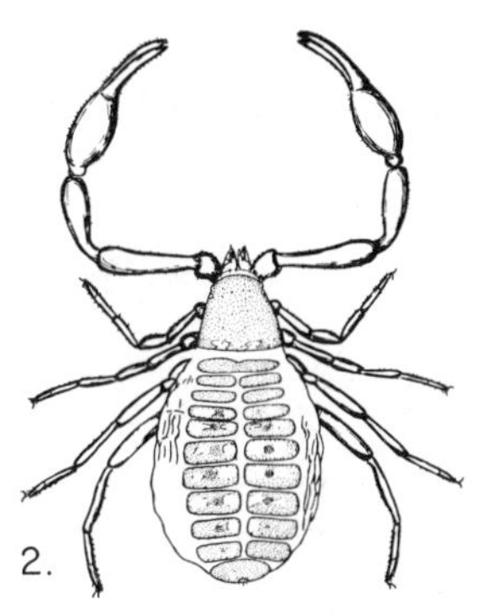

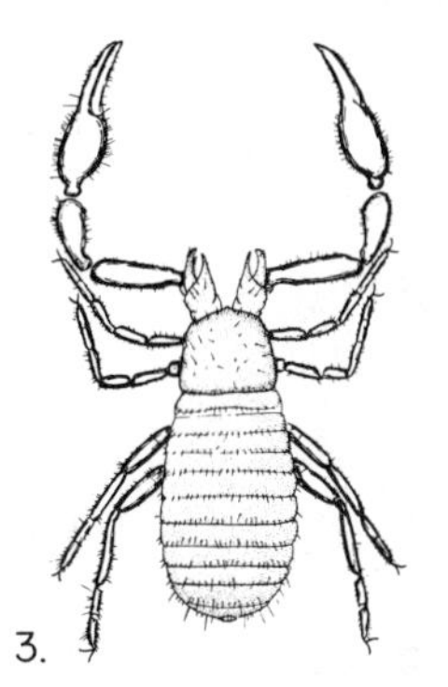

Echiniscoides sigismundi — *Garypus californicus* — *Halobisium occidentale*

TARDIGRADA — CHELONETHIDA (PSEUDOSCORPIONIDA)

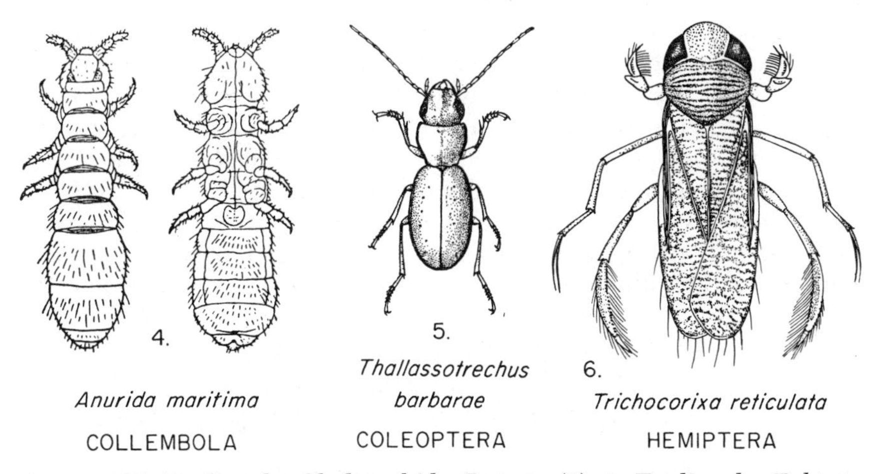

PLATE 105. **Tardigrada, Chelonethida, Insecta** (1). 1, **Tardigrada:** *Echiniscoides sigismundi;* 2, 3, **Chelonethida:** 2, *Garypus californicus;* 3, *Halobisium occidentale;* 4–6, **Insecta:** 4, *Anurida maritima* (Collembola), dorsal (L) and ventral (R); 5. *Thallasotrechus barbarae* (Coleoptera); note: a southern California species is shown; 6, *Trichocorixa reticulata* (Hemiptera). 1, after Schuster and Grigarick, 1965; 2, 3, Schuster, original; 4, Imms, 1957; 5, Schaeffer, 1901; 6, Usinger, 1963, redrawn by Emily Reid.

Hutchinson, G. E. 1931. On the occurrence of *Trichocorixa* Kirkaldy (Corixidae, Hemiptera-Heteroptera) in salt water and its zoogeographical significance. Amer. Nat. 65: 573–574.

Salmon, J. T. 1964. An index to the Collembola. Royal Society of New Zealand, Bull. No. 7, in two volumes, 644 pp.

Usinger, R. L. 1957. Marine insects. *In* Hedgpeth, J. W., ed., *Treatise on Marine Ecology and Paleoecology,* Vol. 1, *Ecology,* Geol. Soc. Amer. Mem. 67: 1177–82.

INTERTIDAL INSECTS: ORDER DIPTERA

Evert I. Schlinger
University of California, Berkeley

(PLATES 106, 107)

There are nineteen known species representing five families of marine flies in the central California intertidal. These species are briefly discussed below and keyed under each family. In addition, about thirty more species representing nine other families are frequently found in or near the intertidal zone, and these are listed at the end of this section. By comparison, Saunders (1928) found six species in three families of flies in his detailed studies of the marine shore insects of the Pacific coast of Canada.

Some of these marine flies are specialized members of taxa which are well adapted to fresh water, for example, Chironomidae and Dolichopodidae; others are primarily terrestrial, for example Tipulidae and Dryomyzidae; the family Canaceidae seems to have more worldwide saline species than freshwater species. The reader is referred to the *Catalogue of the Diptera of North America* (Stone, *et al.*, 1965) and to the *Flies of Western North America* (Cole, 1969) for references and synonomy of all species discussed in this text. For the special terminology of the following keys, consult a standard entomological text.

KEY TO SUBORDERS AND FAMILIES OF DIPTERA COMMON IN THE COASTAL INTERTIDAL ZONE

1. Antenna of 6 or more segments; palpus often elongated, 4- or 5-segmented; calypters (squamae) absent; abdomen often without much hair and rarely with bristles .... Suborder **Nematocera** 2
– Antenna of 3 segments, the terminal segment often annulated or bearing a style or arista; palpus short, 1- or 2-segmented; calypters present or absent; abdomen often with both hair and bristles .... Suborder **Brachycera** 3
2. Mesonotum with V-shaped suture, starting on each side in front of wing base and pointed in middle part at scutellum; at least 9 wing veins reach wing margin; discal cell present .. .. TIPULIDAE (p. 439)
– Mesonotum with transverse suture; less than 9 wing veins

reach wing margins; discal cell absent ....
.... CHIRONOMIDAE (p. 439)

3. Anal cell much longer than 2nd basal cell and narrowed or closed at wing margins 4
– Anal cell as long as or shorter than 2nd basal cell, or absent; when present it ends or is closed some distance from wing margin 5
4. Head with depressed region at vertex; proboscis stiff and horny .... ASILIDAE *
– Head without depression at vertex; proboscis fleshy-lobed
.... THEREVIDAE *
5. Antennae not enclosed above with a crescent-shaped frontal suture; antennal arista usually terminal 6
– Antennae enclosed above with a crescent-shaped frontal suture; antennal arista usually dorsal 7
6. Discal cell usually separate from 2nd basal cell; anterior crossvein located beyond basal 4th of wing ..EMPIDIDAE *
– Discal cell confluent with 2nd basal cell; anterior crossvein located within basal 5th of wing ....
.... DOLICHOPODIDAE (p. 440)
7. Second antennal segment with longitudinal seam along upper outer edge; lower calypter large 8
– Second antennal segment without such seam; lower calypter undeveloped or vestigial 9
8. Hypopleura with one or more rows of vertical bristles; anal vein usually does not reach margin of wing ....
.... SARCOPHAGIDAE *
– Hypopleura without bristles; anal vein reaches wing margin
.... ANTHOMYIIDAE *
9. Costa not fractured near end of subcosta or near end of humeral crossvein 10
– Costa fractured at end of subcosta, or at end of humeral crossvein or both 12
10. Thorax flattened; last tarsal segment flat and enlarged .....
.... COELOPIDAE *
– Thorax convex; last tarsal segment not flattened 11
11. Mouth rim raised; palpus with apical bristle; antennae close together at their bases .... DRYOMYZIDAE (p. 442)
– Mouth rim not raised; palpus without apical bristle; antennae separated by upper ridge of face ...... HELCOMYZIDAE *

* These families include species which may be found intertidally but are not considered of primary significance in that zone. Common genera and species in these families are given in list at end of Diptera section.

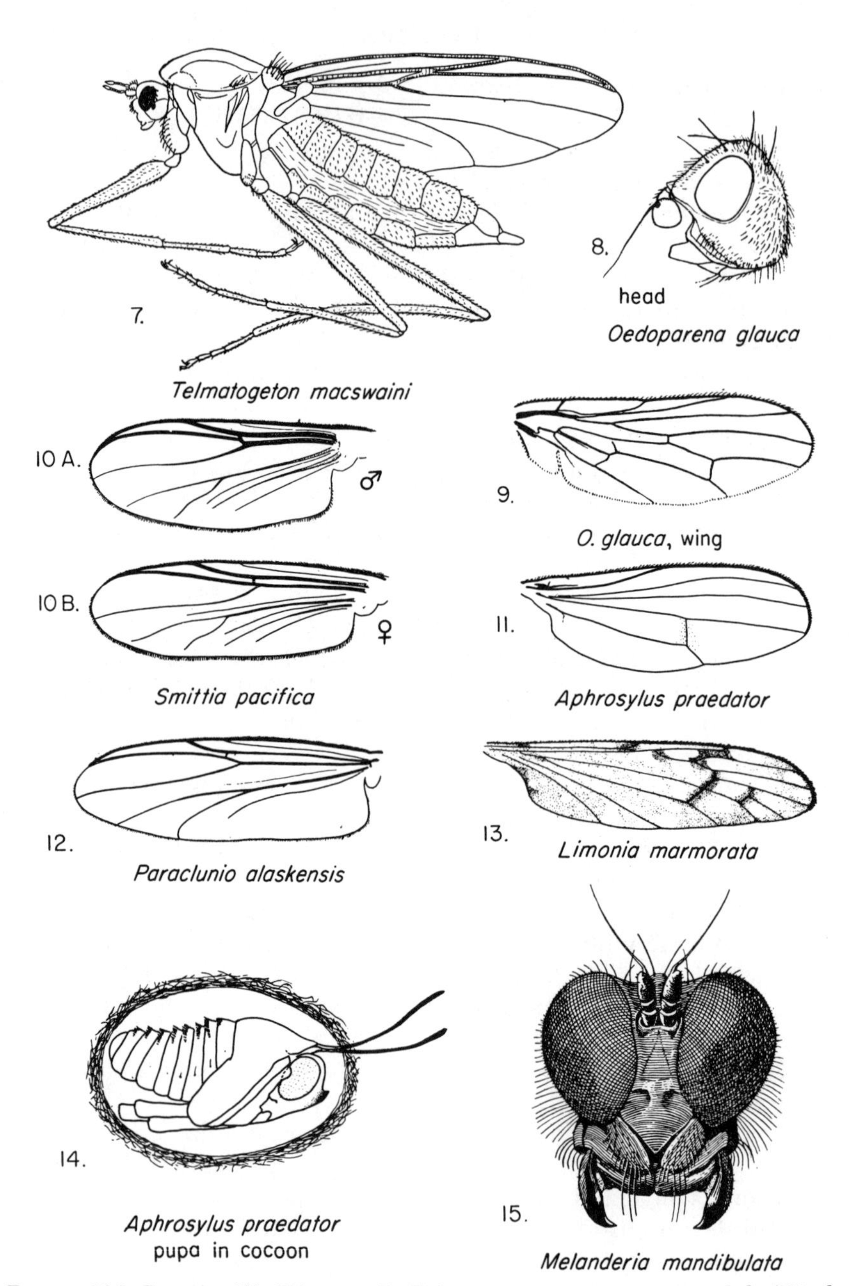

PLATE 106. **Insecta** (2). **Diptera:** 7, *Telmatogeton macswaini,* adult (Wirth, 1949); 8, *Oedoparena glauca,* head; 9, *O. glauca,* wing (Curran, 1934); 10, *Smittia pacifica,* wing: A, male; B, female; 11, *Aphrosylus praedator,* wing; 12, *Paraclunio alaskensis,* wing; 13, *Limonia marmorata,* wing; 14, *A. praedator,* pupa in cocoon (10–14 after Saunders, 1928); 15, *Melanderia mandibulata,* head, note that apparent "mandible" is a lobe of the labellum (Snodgrass, 1922).

12. Costa broken only at end of the subcosta 13
– Costa broken both at end of subcosta and humeral crossvein .... EPHYDRIDAE *
13. Oral vibrissae present at vibrissal angle; preapical tibial bristles present .... HELEOMYZIDAE *
– Oral vibrissae absent although one or more pairs of genal bristles may be present; tibiae without preapical bristles ... . CANACEIDAE (p. 442)

TIPULIDAE

A truly marine tipulid fly, *Limonia (Idioglochina) marmorata* (fig. 13), occurs on the California coast (see Alexander, 1967). The immature stages were first found by Saunders (1928) on Vancouver Island, living in filamentous algae under *Fucus* and among barnacles, small mussels, and other marine animals. Most commonly the larvae were found in the bright green *Enteromorpha,* which covered rocks a foot or two below high-water mark. The larvae are most often observed feeding on the vegetation from the end of their loosely woven silken tubes. The larvae overwinter in these tubes and adults emerge from May to August.

CHIRONOMIDAE

This family of midge flies appears to be the most adaptive group of dipterans inhabiting the coastal marine area, where no less than eight species representing five genera are found. These species may be wingless and small (1 mm), or fully winged and up to 7 mm. Larvae of most species are known and have been found feeding in algal mats of species of *Enteromorpha* and *Ulva* (see Saunders, 1928; Wirth, 1949). Some species are rather abundant in areas of sewage outfalls, and *Paraclunio alaskensis* (figs. 12, 16–24) may be an important species in consuming algae which develop massively in such areas (see Cheesemen and Preissler, 1972; Morley and Ring. 1972).

KEY TO SPECIES OF INTERTIDAL MIDGES

1. Pronotal lobes widely separated; antennae not plumose 2
– Pronotum not, or only slightly, notched anteriorly on median line; male antennae usually plumose 6
2. Hind tarsus with 2nd segment not longer than 3rd 3
– Hind tarsus with 2nd segment longer than 3rd 4
3. Wings straplike, reaching to 4th segment of abdomen, halters present ....*Eretmoptera browni*

- Wings vestigial, not reaching to abdomen; halters vestigial (figs. 25–27) .... *Tethymyia aptena*
4. Hairs of legs strong and sometimes with scales; front legs of male modified; femora swollen 5
- Hairs of legs weak; legs unmodified (fig. 7) .... *Telmatogeton macswaini*
5. Tarsal claws slender, those of male deeply cleft; tibiae with rows of strong bristly hairs (figs. 12, 16–24) .... *Paraclunio alaskensis*
- Tarsal claws flattened and broadened, those of male shallowly cleft; tibiae with rows of hairs mostly replaced by scales .... *Paraclunio trilobatus*
6. Wings milky white with yellowish basal area; male antenna with 8 or 13 flagellar segments 7
- Wings milky white throughout; male antenna with 12 flagellar segments .... *Smittia marina*
7. Maxillary palpus 4-segmented; male antenna with 8 flagellar segments .... *Smittia clavicornis*
- Maxillary palpus 5-segmented; male antenna with 13 flagellar segments (figs. 10A, B) .... *Smittia pacifica*

DOLICHOPODIDAE

There are two genera of marine dolichopodid flies in California. Biological information on both is scant, but immature stages and their habits are known for *Aphrosylus*. Adults of *A. praedator* (fig. 11) are predaceous upon both larvae and adult midges (*Smittia*) on Vancouver Island, according to Saunders (1928). The larvae are likely also predaceous, but their true feeding preferences remain unknown. They inhabit the same mats of filamentous algae as do the midges. The pupae are formed in rounded, elliptical cocoons loosely spread in the algae (fig. 14).

Melanderia mandibulata and *M. crepuscula* are both found in our area (Arnaud, 1958). The adults are remarkable in having the lobes of the labellum developed into "mandible-like" structures (fig. 15) possibly used in predatory activities.

KEY TO SPECIES OF INTERTIDAL DOLICHOPODIDAE

1. Mouthparts appearing "mandible-like" (fig. 15) 2
- Mouthparts without "mandibles" 3
2. Front femur of male on inner side near base with dense tuft of

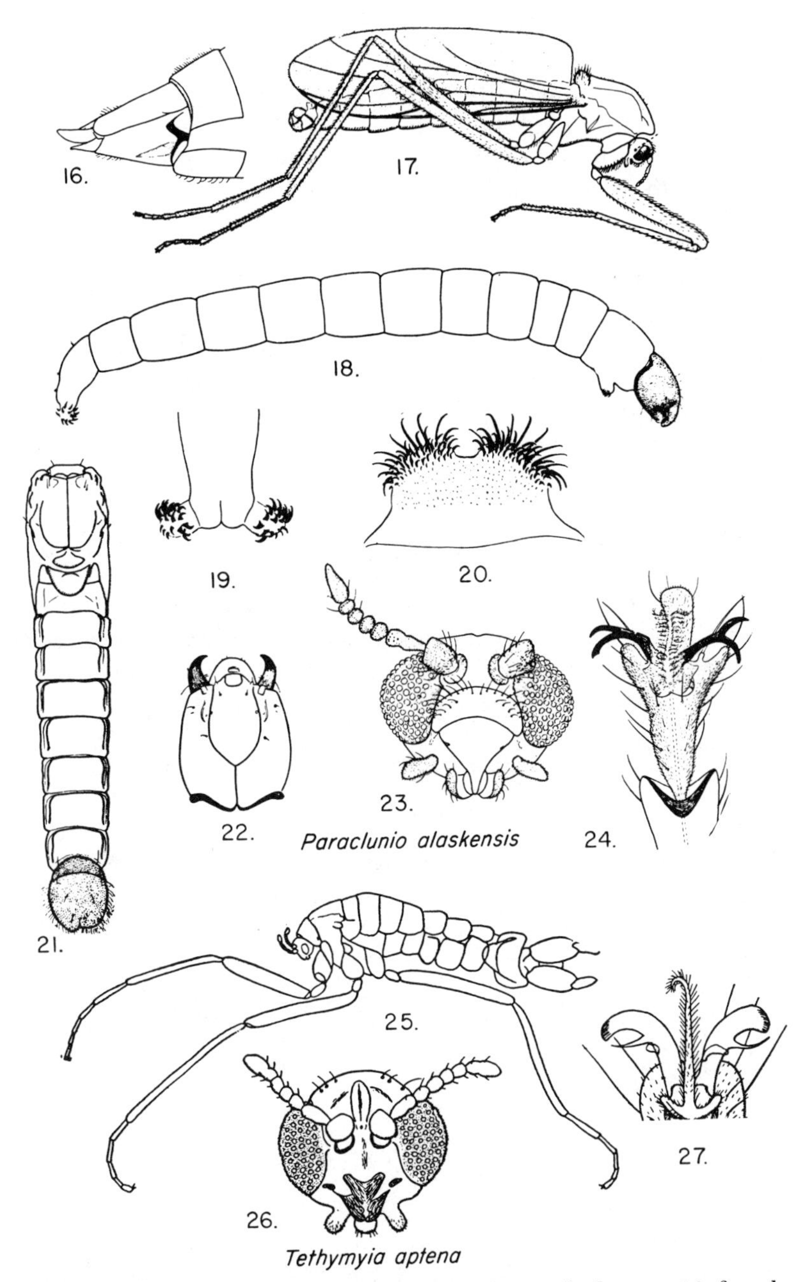

PLATE 107. **Insecta** (3). **Diptera:** 16–24, *Paraclunio alaskensis:* 16, female, tip of abdomen; 17, male; 18, larva; 19, anal pseudopods of larva; 20, prothoracic pseudopods of larva; 21, pupa; 22, larval head, dorsum; 23, female, head; 24, male, last tarsal segment (16–24, Saunders, 1928); 25–27, *Tethymyia aptena:* 25, adult; 26, head; 27, male, fore-tarsal claws (25–27, Wirth, 1949).

8–10 bristles on a very slight protuberance .... *Melanderia mandibulata*
- Front femur of male without tuft of bristles on inner side near base, but with long, black hair .... *Melanderia crepuscula*

3. Wing clear without distinct black spot over hind crossvein .. 4
- Wing with distinct black spot over hind crossvein .... *Aphrosylus direptor*

4. Antennal arista bare 5
- Arista pubescent .... *Aphrosylus grassator*

5. Wing with 2nd, 3rd, and 4th longitudinal veins greatly broadened; fore and middle tibiae with delicate black hairs on lower surface .... *Aphrosylus wirthi*
- Wing without such broadened veins; fore and middle tibiae without such hairs (fig. 11) .... *Aphrosylus praedator*

CANACEIDAE

Little is known biologically about these small shoreflies, but Wirth (1951) indicates that some species breed preferably on tide-covered rocks which are clothed with such green algae as *Enteromorpha* and *Ulva*. Three species are known from this region.

KEY TO SPECIES OF INTERTIDAL CANACEIDAE

1. Mesofrons with one pair of bristles, just outside the ocellars 2
- Mesofrons with several pairs of long, interfrontal bristles .... *Canace aldrichi*

2. Four genal bristles; anterior notopleural bristles present but small .... *Canaceoides nudatus*
- Three genal bristles; anterior notopleural bristles absent .... *Nocticanace arnaudi*

DRYOMYZIDAE

According to Knudsen (1968) one species, *Oedoparena glauca* (figs. 8, 9), of this small family is a predator during its larval stages on the barnacles *Balanus glandula* and species of *Chthamalus*. The adults apparently feed upon diatoms and are never observed away from the *Endocladia-Balanus* association. After mating, males often remain on the back of the female while she lays eggs in and around the mouth of a large barnacle. Large larvae move from barnacle to barnacle and take several months to become adults. Death of the barnacle may be due indirectly to starvation, since the larval feeding action seems to restrict the barnacles' cirri from operating properly and thus impairs their feeding. The adult fly is lead-colored with grayish hyaline wings and yellow halteres.

LIST OF INTERTIDAL DIPTERA INCLUDED IN KEYS

TIPULIDAE

Limonia (Idioglochina) marmorata (Osten Sacken, 1861) (= *Dicranomyia signipennis* Coquilett, 1905).

CHIRONOMIDAE, Orthocladiinae (Clunioninae of authors)

Metriochemini

Eretmoptera browni (Kellogg, 1900).

Smittia clavicornis (Saunders, 1928).

Smittia marina (Saunders, 1928).

Smittia pacifica (Saunders, 1928).

Tethymyia aptena Wirth, 1949.

Telmatogetonini

Paraclunio alaskensis (Coquillett, 1900).

Paraclunio trilobatus Kieffer, 1911.

Telmatogeton macswaini Wirth, 1949.

DOLICHOPODIDAE

Aphrosylus direptor Wheeler, 1897.

Aphrosylus grassator Wheeler, 1897.

Aphrosylus praedator Wheeler, 1897.

Aphrosylus wirthi Harmston, 1951.

Melanderia crepuscula Arnaud, 1958.

Melanderia mandibulata Aldrich, 1922.

CANACEIDAE

Canace aldrichi Cresson, 1936.

Canaceoides nudatus (Cresson, 1926).

Nocticanace arnaudi Wirth, 1954.

DRYOMYZIDAE

Oedoparena glauca (Coquillett, 1900). Predatory as larvae upon barnacles (see Knudsen, 1968)

LIST AND DIAGNOSES OF REPRESENTATIVE INTERTIDAL DIPTERA NOT INCLUDED IN KEYS

EMPIDIDAE (small-winged or brachypterous, predaceous flies found in shoreline sandy areas)

Chersodromia cana Melander, 1945 (*Coloboneura* and *Thinodromia* of authors).

Chersodromia inchoata (Melander, 1906).

Chersodromia insignita Melander, 1945.

Chersodromia magacetes Melander, 1945.

Parathalassius aldrichi Melander, 1906.

Parathalassius melanderi Cole, 1912.

ASILIDAE (large, predaceous "robber flies," found on sand and occasionally close to ocean water line).

Lasiopogon actius Melander, 1923.

Stichopogon coquilletti (Bezzi, 1910).

THEREVIDAE (small to large flies similar to Asilidae, but adults are not predaceous. Larvae are predaceous on several sand-dwelling insects, particularly in dune areas).

Chromolepida bella Cole, 1923.

Dialineura melanophleba (Loew, 1876).

Pherocera sp.

Thereva hirticeps Loew, 1874.

Thereva pacifica Cole, 1923.

EPHYDRIDAE (mostly small flies with large "mouths," often called brine flies or shore flies).

No particular species are noted, but taxa likely to be found intertidally close to freshwater areas may be in such genera as *Ephydra*, *Parydra*, *Scatella*, *Dimecoenia*, *Scatophila*, and *Lipochaeta*.

COELOPIDAE (small, flattened "kelp flies," sometimes common on drying seaweed; their larvae feed on decaying kelp).

Coelopa (Neocoelopa) vanduzeei Cresson, 1914.

HELCOMYZIDAE (large, sea-beach flies which develop in rotting seaweed. Although not yet known from California, they occur from Oregon north and may be found in California).

HELEOMYZIDAE (small to large flies rarely encountered in the intertidal zone; several species of one sand-colored genus are an exception).

Anorostoma grande Darlington, 1908.

Anorostoma maculatum Darlington, 1908.

Anorostoma wilcoxi Curran, 1933.

ANTHOMYIIDAE (medium-sized flies whose larvae are entomophagous on kelp-fly larvae; see Huckett, 1971).

Fucellia antennata Stein, 1910.

Fucellia assimilis Malloch, 1918.

Fucellia costalis Stein, 1910.

Fucellia fucorum (Fallen, 1819).

Fucellia pacifica Malloch, 1923.

Fucellia rufitibia Stein, 1910.

Fucellia separata Stein, 1910.

Fucellia thinobia (Thomson, 1869).

SARCOPHAGIDAE (large scavenger flies which may occasionally enter the intertidal zone)

REFERENCES ON DIPTERA

Alexander, C. P. 1967. The crane flies of California. Bull. Calif. Ins. Survey, 8: 1–269.

Arnaud, P. H. 1958. A synopsis of the genus *Melanderia* Aldrich (Diptera: Dolichopodidae). Proc. Ent. Soc. Wash. 60: 179–186.

Cheesemen, D. T., and P. Preissler 1972. Larval distribution of *Paraclunio alaskensis* at Point Pinos sewage outfall, Monterey County, California. Pan-Pacific Ent. 48: 204–207.

Cole, F. R. 1969. *The Flies of Western North America* (with collaboration of E. I. Schlinger). University of California Press, 693 pp.

Huckett, H. C. 1971. The Anthomyiidae of California, exclusive of the subfamily Scatophaginae (Diptera). Bull. Calif. Ins. Survey 12: 1–121.

Knudsen, M. 1968. The biology and life history of *Oedoparena glauca* (Diptera: Dromyzidae), a predator of barnacles. (Unpublished master's thesis in Parasitology, University of California, Berkeley.)

Morley, R. L. and R. A. Ring 1972. The Intertidal Chironomidae (Diptera) of British Columbia I. Keys to their Life Stages (pp. 1093–1098), II. Life History and Population Dynamics (pp. 1099–1121). Canad. Ent. 104: 1093–1121.

Saunders, L. G. 1928. Some marine insects of the Pacific Coast of Canada. Ann. Ent. Soc. Amer. 21: 521–545.

Stone, *et al.* 1965. *A Catalog of the Diptera of America north of Mexico.* Washington, D.C.: Agricultural Handbook 276, USDA, 1969 pp.

Wirth, W. W. 1949. A revision of the Clunionine midges with descriptions of a new genus and four new species (Diptera: Tendipedidae). Univ. Calif. Publ. in Entomol. 8: 151–182.

Wirth, W. W. 1951. A revision of the dipterous family Canaceidae. Occas. Papers B. P. Bishop Museum, Honolulu 20: 245–275.

INTERTIDAL INSECTS: ORDER COLEOPTERA

John T. Doyen
University of California, Berkeley

(PLATES 105, 108)

Most intertidal beetles are specialized members of typically terrestrial families. With few exceptions, families adapted to fresh water do not enter the intertidal zone. For example, only two hydrophilid beetles (*Cercyon*) (fig. 28) are intertidal, where they invade piles of decaying marine algae along the line of highest tides. However, they are much more abundant in algae which has been washed high onto beaches by storms. The Dytiscidae, Elmidae, and Dryopidae, common and widespread in fresh water, have no marine representatives in California.

Decaying algae also harbor the weevil *Emphyastes*, which occasionally enters rock crevices in the high intertidal. Melyrids (*Endeodes*) are most frequently seen crawling on dry beach sand, but also occur in rock crevices below the line of highest tides. The staphylinids *Bryobiota*, *Emplenota*, *Pontomalota*, and *Thinusa* are locally abundant beneath rotting algae, and are probably temporarily submerged by very high tides. Although not known to occur in the intertidal zone, the staphylinids *Cafius*, *Hadrotes* and *Thinopinus* should be mentioned because of their abundance on seashores. *Cafius* and *Hadrotes* are characteristic inhabitants of decaying algae and also frequently fly or crawl about beaches. *Thinopinus* burrows into sand beneath piles of algae; adults wander about on the surface at night; all are confined to seashores. Their large size (at least 5 mm) distinguishes them from other maritime Staphylinidae.

Thallasotrechus (Carabidae) (fig. 5), *Ochthebius* (Hydraenidae), *Aegialites* (Salpingidae) (figs. 34A, B), and the staphylinids *Amblopusa, Diaulota* (fig. 30), and *Liparocephalus* (figs. 29A, B) apparently never leave the intertidal zone. They spend much of their lives, particularly during high tides, deep in cracks in rocks and consequently are rarely encountered. The Staphylinidae are the largest group of strictly intertidal Coleoptera, with representatives in all major oceans. Most species are wingless and probably disperse with ocean currents.

The biology of western American intertidal Coleoptera is very poorly known. It has been suggested that some species are predaceous on mites and other organisms (Saunders, 1928; Leech and Chandler, 1956; Spilman, 1967), while others probably feed on algae (Sugihara, 1938). The immature stages of *Aegialites californicus* (figs. 34A, B; despite name, occurs from British Columbia to Alaska) are described by Wickham (1904) and available biological information is summarized by Spilman (1967). Most intertidal Staphylinidae of the Pacific coast are keyed and described by Moore (1956). For taxonomic treatments of other families, see Blackwelder (1931, 1932), Moore (1954), and Van Dyke (1918). Original descriptions of species are available through the Leng Catalogue. General accounts of the families appear in Arnett (1960) and Hatch (1957, 1965).

KEY TO ADULT INTERTIDAL COLEOPTERA

1. First visible abdominal sternite completely divided by hind coxal cavities (suborder **Adephaga;** CARABIDAE) (fig. 5) . . . . *Thallasotrechus nigripennis*
– First visible abdominal segment transversely complete . . . . suborder **Polyphaga** 2
2. Antennae about twice as long as maxillary palpi; last 3–5 segments abruptly enlarged as a pubescent club (fig. 28) 3
– Antennae at least 4 times as long as maxillary palpi; filiform or with last 2 segments enlarged as club 5
3. Antennal club with 5 pubescent segments; abdomen with 6–7 visible sternites (HYDRAENIDAE) . . *Ochthebius vandykei*
– Antennal club with 3 pubescent segments; 5 visible abdominal sternites (fig. 28) . . . . HYDROPHILIDAE 4
4. Elytra deeply striate, punctate only laterally . . . . *Cercyon fimbriatus*
– Elytra faintly striate, finely punctate over entire surface . . . . *Cercyon luniger*
5. Elytra truncate, abbreviated, exposing at least 4 abdominal tergites 6

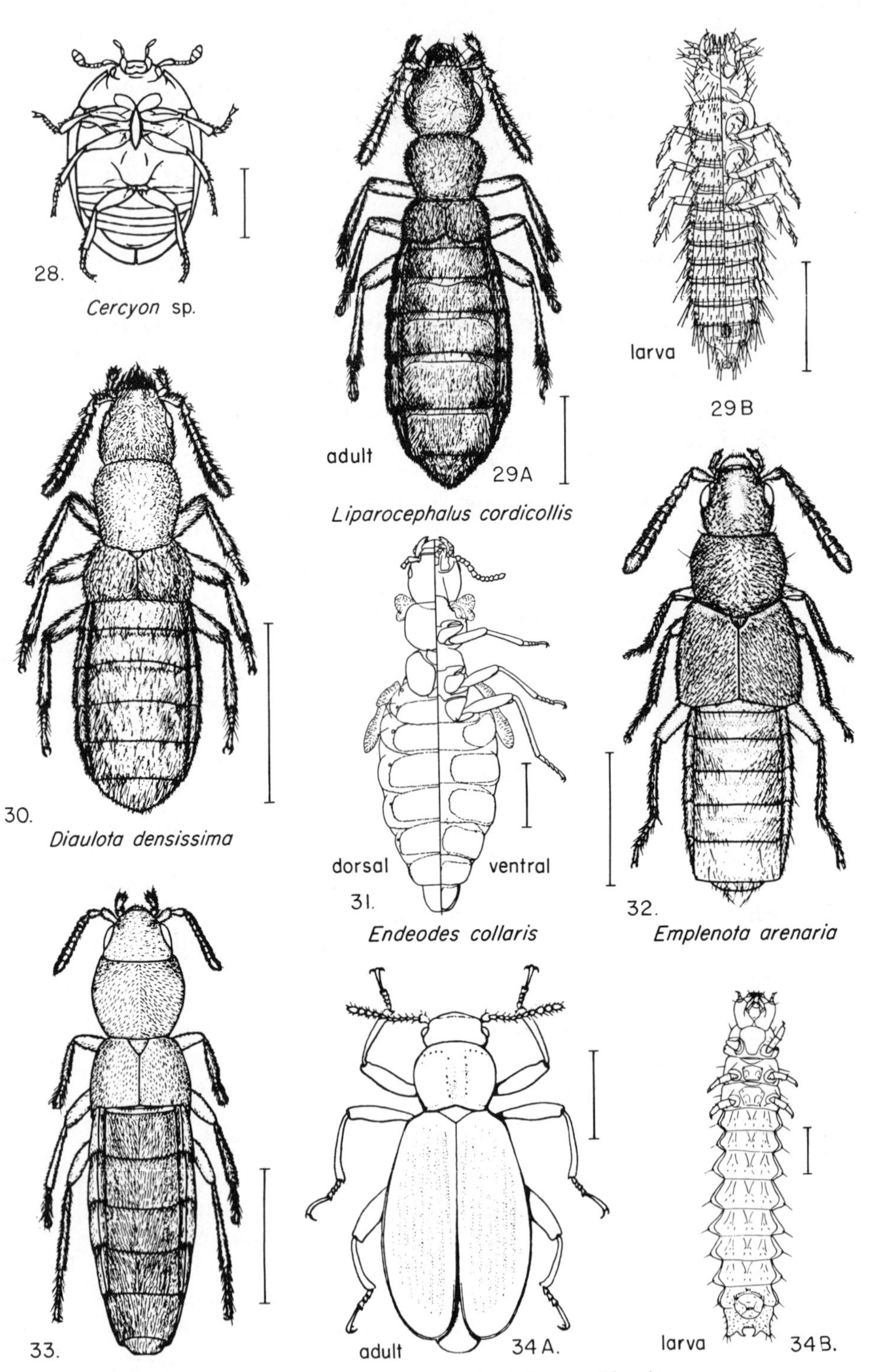
28.
Cercyon sp.
adult
29A
larva
29B
Liparocephalus cordicollis
30.
Diaulota densissima
dorsal
ventral
31.
Endeodes collaris
32.
Emplenota arenaria
33.
Pontomalota luctuosa
adult
34 A.
larva
34 B.
Aegialites californicus

- – Elytra covering entire abdomen 17
6. Prothorax and abdomen with yellow or orange protrusible vesicles ....MELYRIDAE 16
- – Prothorax and abdomen without protrusible vesicles .... STAPHYLINIDAE 7
7. Anterior and middle tibiae coarsely spinose .... *Thinusa maritima*
- – Anterior and middle tibiae pubescent, no trace of spinules 8
8. Middle tarsi with 4 segments 9
- – Middle tarsi with 5 segments 14
9. Abdominal tergites not impressed at base (fig. 29A, B)...... .... *Liparocephalus cordicollis*
- – At least 2nd and 3rd abdominal tergites transversely impressed at base 10
10. Fifth abdominal tergite transversely impressed at base ....*Amblopusa borealis*
- – Fifth abdominal tergite not impressed at base 11
11. Base of head not longitudinally impressed 12
- – Base of head longitudinally impressed ... *Bryobiota bicolor*
12. Head black 13
- – Head reddish brown ....*Diaulota vandykei*
13. Each elytron as wide as lo g (fig. 30)..*Diaulota densissima*
- – Each elytron wider than long ....*Diaulota fulviventris*
14. Front tarsi with 5 segments (fig. 32) ...*Emplenota arenaria*
- – Front tarsi with 4 segments .... *Pontomalota* 15
15. Abdomen entirely black (fig. 33)...*Pontomalota californica*
- – Apex of 3rd and 4th and base of 5th abdominal tergites clouded with blackish brown ...... *Pontomalota nigriceps*
16. Head black (fig. 31) ....*Endeodes collaris*
- – Head reddish ....*Endeodes rugiceps*
17. Head prolonged anteriorly as a rostrum; antennae with last 2 segments abruptly enlarged as club (CURCULIONIDAE).. ....*Emphyastes fucicola*
- – Head not prolonged anteriorly; antennae filiform (figs. 34A, B) ....SALPINGIDAE 18

PLATE 108. **Insecta** (4). **Coleoptera:** 28, *Cercyon* sp., ventral (Mulsant, 1844); 29A, *Liparocephalus cordicollis,* adult (Hatch, 1957); 29B, larva, dorsal surface on left, ventral on right (Chamberlin and Ferris, 1929); 30, *Diaulota densissima,* adult, dorsal (Hatch, 1957); 31, *Endeodes collaris,* adult, dorsal surface on left, ventral on right (Blackwelder, 1932); 32, *Emplenota arenaria,* adult, dorsal (Hatch, 1957); 33, *Pontomalota luctuosa,* adult, dorsal (Hatch, 1957); 34A, *Aegialites californicus,* adult, dorsal (Spilman, 1967); 34B, larva, ventral (Wickham, 1904). Caution: figs. 33 and 34 are Pacific Northwest species. Scale bars = 1 mm.

18. Elytra deeply striate; integument shining, metallic .... *Aegialites fuchsii*
– Elytra without striae; integument dull, opaque .... *Aegialites subopacus*

LIST OF INTERTIDAL COLEOPTERA

CARABIDAE

Thallasotrechus nigripennis Van Dyke, 1918. Locally abundant in rock crevices just below high-tide line.

HYDRAENIDAE

Ochthebius vandykei Knisch, 1924.

HYDROPHILIDAE

Cercyon fimbriatus Mannerheim, 1852. Both species of *Cercyon* are nearly always found in piles of decaying algae.

Cercyon luniger Mannerheim, 1853.

STAPHYLINIDAE

Amblopusa borealis Casey, 1906.

Bryobiota bicolor Casey, 1885.

* *Cafius canescens* Maklin, 1852. *Cafius* spp. are the most obvious beetles of decaying algae; probably predaceous on dipteran larvae.

* *Cafius seminitens* Horn, 1884.

Diaulota densissima Casey, 1893.

Diaulota fulviventris Moore, 1956.

Diaulota vandykei Moore, 1956.

Emplenota arenaria Casey, 1893.

* *Hadrotes crassus* Mannerheim, 1846. Commonly found in decaying algae; probably predaceous on dipteran larvae.

Liparocephalus cordicollis LeConte, 1880. Larvae, pupae, and adults all occur in the mid-tidal; pupae enclosed in silken cocoons in rock crevices.

Pontomalota californica Casey, 1885.

* Not in key.

Pontomalota nigriceps Casey, 1885.

* *Thinopinus pictus* LeConte, 1852. Common nocturnal predators on small arthropods throughout maritime California.

Thinusa maritima Casey, 1885.

MELYRIDAE

Endeodes collaris LeConte, 1852.

Endeodes rugiceps Blackwelder, 1932. Locally abundant under driftwood near high-tide line.

SALPINGIDAE

Aegialites fuchsii Horn, 1893.

Aegialites subopacus Van Dyke, 1918. Active throughout year in rock crevices of mid-intertidal; larvae and pupae occur during summer months.

CURCULIONIDAE

Emphyastes fucicola Mannerheim, 1852. Locally abundant beneath beached algae.

REFERENCES ON INTERTIDAL COLEOPTERA

Arnett, R. A. 1960. *The Beetles of the United States.* Washington, D.C.: Catholic University of America Press, xi + 1048 pp.

Blackwelder, R. E. 1931. The Sphaeridiinae of the Pacific coast (Coleoptera, Hydrophilidae). Pan-Pac. Entomol. 8: 19–32.

Blackwelder, R. E. 1932. The genus *Endeodes* LeConte (Coleoptera, Melyridae). Pan-Pac. Entomol. 8: 128–136.

Hatch, M. H. 1957. *The Beetles of the Pacific Northwest.* Part II, Staphyliniformia. University of Washington Press, ix + 384 pp.

Hatch, M. H. 1965. Part IV, Macrodactyles, Palpicornes, and Heteromera. University of Washington Press, viii + 268 pp.

Leech, H. B. and H. P. Chandler 1956. Aquatic Coleoptera. *In* R. L. Usinger, ed., *Aquatic Insects of California,* University of California Press, ix + 508 pp.

Leng Catalogue (see below)

Leng, C. W. 1920. *Catalogue of the Coleoptera of America, north of Mexico.* Mt. Vernon, N.Y., 470 pp.

Leng, C. W. and A. J. Mutchler 1927. Suppl. (1919–1924), 78 pp.

Leng, C. W. and A. J. Mutchler 1933. 2nd and 3rd Suppl. (1925–1932), 112 pp.

Blackwelder, R. E. 1939. 4th Suppl. (1933–1938), 146 pp.

Blackwelder, R. E. and R. M. 1948. 5th Suppl. (1939–1947), 87 pp.

Moore, I. 1954. Notes on *Endeodes* LeConte with a description of a new species from Baja California. Pan-Pac. Entomol. 30: 195–198.

Moore, I. 1956. A revision of the Pacific Coast Phytosi with a review of the foreign genera (Coleoptera: Staphylinidae). Trans. San Diego Soc. Nat. Hist. 12: 103–152.

Moore, I. 1956. Notes on some intertidal Coleoptera with descriptions of the early stages (Carabidae, Staphylinidae, Malachiidae). Trans. San Diego Soc. Nat. Hist. 12: 207–230.

Saunders, L. G. 1928. Some marine insects of the Pacific coast of Canada. Ann. Entomol. Soc. Amer. 21: 521–545.

Spilman, T. J. 1967. The heteromerous intertidal beetles. Pac. Insects 9: 1–21.

Sugihara, Y. 1938. An observation on the intertidal rock-dwelling beetle, *Aegialites stejnegeri sugiharai* Kono, in the Kuriles. Entomol. World 6: 6–12.

Van Dyke, E. C. 1918. New inter-tidal rock-dwelling Coleoptera from California. Entomol. News 24: 303–308.

Wickham, H. F. 1904. The metamorphosis of *Aegialites*. Canad. Entomol. 36: 57–60.

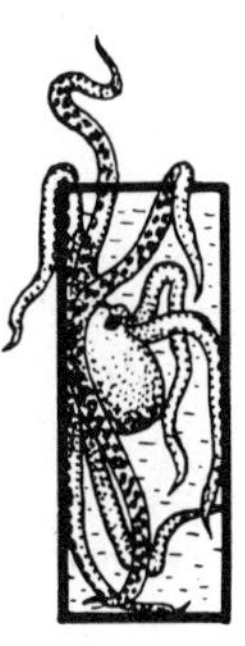

PHYLUM MOLLUSCA: INTRODUCTION AND SMALLER GROUPS

Mollusca rank second only to the Arthropoda as an interesting and diverse group. They have adapted to almost all habitats, marine, freshwater, and terrestrial, from the 35,000 foot depths of the Philippine Trench to mountain masses 15,000 feet above the sea, from the tropics to the Arctic, from the surface of the open oceans to barren deserts. Since earliest times, man has used them for food, as witness the great prehistoric shell heaps (kitchen middens) in all parts of the world, for tools and weapons, for money, and for jewelry. Their shells have appealed to lovers of beauty as much as to scientists and to the great and indefatigable fraternity of conchologists. References to molluscs pervade our legends, our literature, and our everyday speech; they provide such symbols as the snail for slowness, the clam for silence, the pearl for virtue, and the octopus for greed; they have been the livelihood of the poor, and have dyed the robes of kings.

Mollusca apparently arose from a stem that also produced the Annelida, as evidenced by the extremely similar "spiral cleavage" in the eggs of the more primitive members of these phyla. In both groups a larva of the trochophore type is produced, after which the development takes its characteristic course. The segmented annelids develop an elongated body, an extensive coelom, and a muscular body wall provided with setae. But the molluscs reduce the coelom to a small pericardium and associated structures and develop a compact body, divisible into head, foot, and visceral mass. The latter is covered by a **mantle,** which often secretes a shell and overhangs the sides and rear of the body to form a **pallial chamber** or **mantle cavity.** In this space are found respiratory organs of unique form, known as **ctenidia** (sing. **ctenidium).** The primitive aspect is best shown in the lower gastropods (order Archeogastropoda); here each ctenidium has the form of a long axis bearing two lateral rows of flat plates (filaments), set closely together. On the neighboring faces of these filaments, cilia move water to create a respiratory current. In different groups of molluscs we find ctenidia variously modified, especially for feeding in the lamellibranchs, or lost altogether as in nudibranchs and pulmonates.

The study of ctenidia and associated organs provides one of the most illuminating means of appreciating the adaptations of molluscs to diverse ecological situations. The most important work in this field is that of Yonge, whose paper on the pallial organs of the gastropods and other molluscs (1947) should be read by any serious student of Mollusca.

Molluscs apparently arose as creeping types, probably living on hard surfaces and scraping up their food by means of a unique organ, the **radula,** which is found in all classes except the lamellibranchs. Lamellibranchs have extensively modified their ctenidia for the filtration of microscopic food material from the water and have lost all semblance of a head. Further details are discussed in relation to the various classes and orders.

GENERAL REFERENCES ON MOLLUSCA

Arnold, R. 1903. *The paleontology and stratigraphy of the marine Pliocene and Pleistocene of San Pedro, California.* Calif. Acad. Sci. Mem. 3, 420 pp.

Dall, W. H. 1921. Summary of the marine shellbearing mollusks of the northwest coast of America, from San Diego, California, to the Polar Sea . . . Bull. U.S. Nat. Mus. 112: 217 pp. (See also additions and emendations, Proc. U.S. Nat. Mus. 63 (2478): 4 pp.; Bulletin 112 was updated by the checklist and bibliography of Keen 1937, Stanford University Press, 87 pp.)

Fretter, V. ed. 1968. *Studies in the Structure, Physiology and Ecology of Molluscs.* Symp. Zool. Soc. London and Malacological Soc. London, no. 22, Academic Press 377 pp.

Grant, U. S. IV and H. R. Gale 1931. *Catalogue of the marine Pliocene and Pleistocene Mollusca of California.* Mem. San Diego Soc. Nat. Hist. 1: 1036 pp.

Hyman, L. H. 1967. *The Invertebrates: Mollusca I,* Vol. VI. McGraw-Hill, 792 pp. (Covers Polyplacophora, Gastropoda, Aplacophora, and Monoplacophora.)

Keen, M. 1971. *Sea Shells of Tropical West America.* Stanford University Press, 1064 pp. (Valuable on a supraspecific level for the central California coast.)

Keen, M. and E. Coan 1974. *Marine Molluscan Genera of Western North America: an illustrated key.* 2nd Ed. Stanford University Press, 208 pp.

Keep, J. 1935. *West Coast Shells.* Rev. by J. L. Baily Jr. Stanford University Press, 350 pp.

McLean, J. H. 1969. *Marine Shells of Southern California.* Los Angeles County Mus. Nat. Hist. Science Series 24, Zoology 11, 104 pp.

Morton, J. E. 1967. *Molluscs.* 4th Ed. Hutchinson University Library, 244 pp.

Oldroyd, I. S. 1925–1927. *The Marine Shells of the West Coast of North America.* Stanford Univ. Publs. Geol. Sci. (2 vols. in 4 parts.)

Packard, E. L. 1918. Molluscan fauna from San Francisco Bay. Univ. Calif. Publ. Zool. 14: 199–452.

Purchon, R. D. 1968. *The Biology of the Mollusca.* Pergamon Press, 560 pp.

Thiele, J. 1929–1935. *Handbuch der Systematischen Weichtierkunde.* Gustav Fischer, Jena, 1–2, 1154 pp. (A standard systematic reference.)

Wilbur, K. M. and C. M. Yonge, eds. 1964, 1966. *Physiology of Mollusca.* Vols. I and II, Academic Press.

Yonge, C. M. 1947. The pallial organs in the aspidobranch Gastropoda and their evolution throughout the Mollusca. Phil. Trans. Roy. Soc. London (B) 232: 443–518.

SMALLER MOLLUSCAN GROUPS: SCAPHOPODA, CEPHALOPODA, POLYPLACOPHORA

Allyn G. Smith
California Academy of Sciences, San Francisco

CLASS SCAPHOPODA

(PLATE 120)

The small class Scaphopoda, often called tooth or tusk shells, have tubular, tapering, curved shells, open at both ends, with an extensible lobed foot adapted for burrowing in soft mud and a group of slender oral tentacles (**captacula**) used in feeding (fig. 106). There is a small but powerful radula capable of cracking the tests of Foraminifera, and bands of cilia within the mantle cavity for maintaining respiration and water circulation. Three genera occur off the central California coast, but none lives intertidally. *Dentalium hexagonum* Pilsbry & Sharp, 1897, a ribbed species 30 mm long, is sometimes taken in depths as shallow as 7 m; *D. pretiosum berryi* Smith & Gordon, 1948, a smooth species about 45 mm long, occurs at 35 m and deeper; *Siphonodentalium quadrifisatum* (Pilsbry & Sharp, 1898), a small species with a four-lobed apex, may be found occasionally as shallow as 4 m; and *Cadulus fusiformis* Pilsbry & Sharp, 1898, a 10 mm, smooth, shiny species with slightly constricted aperture, occurs at 4 m and below. Other species occur in deeper water. Indians of the Pacific Northwest collected *Dentalium pretiosum* Sowerby, 1860 with long-handled rakes, strung them with deer sinew, and used the strings as wampum, or as ornaments on clothing and papoose carriers (Clark, 1963).

REFERENCES ON SCAPHOPODA

Clark, R. B. 1963. The economics of *Dentalium*. Veliger 6: 9–19.

Emerson, W. K. 1962. A classification of the scaphopod mollusks. J. Paleontol. 36: 461–482.

Pilsbry, H. A. and Sharp, B. 1897–98. Class Scaphopoda. Manual of Conchology 17: i–xxxii + 280 pp.

CLASS CEPHALOPODA

Cephalopods are the most highly developed molluscs. They are free-swimming or crawl on the bottom by flexible arms. There are several orders in the subclass Coleoidea: **Octopoda,** the eight-armed octopuses, which include the argonaut or "paper nautilus;" **Teuthoidea,** the squids, with an additional pair of longer tentacular arms used in grasping prey; and **Sepioidea,** the cuttle-fishes.

On the central California coast, only the Pacific giant octopus, probably *Octopus dofleini martini* Pickford, 1964, is common in the lowest intertidal zone, where small individuals occur under rocks and in crevices. These local intertidal octopuses were formerly called *O. apollyon* (Berry, 1912), but this is a northern species. A smaller, offshore species, *O. rubescens* Berry, 1953 of southern California, has been reported as far north as Monterey Bay. In southern California the commonest species is the two-spotted octopus, *O. bimaculatus* (Verrill, 1883), which is generally found in the lower intertidal *Laminaria* zone and in deeper rocky areas. A sibling species, *O. bimaculoides* Pickford & McConnaughey, 1949, is found somewhat higher in intertidal rocky flats, shallow bays, and on mud flats, and hence has been called the mudflat octopus. Care should be exercised in handling live octopuses, especially small ones, as they have a venom gland and can inflict a painful bite which may result in lesions and swelling (see Halstead, 1949; Berry and Halstead, 1954).

Our common squid, *Loligo opalescens* Berry, 1911, is abundant offshore, although not seen living in the intertidal zone (see Fields, 1965). It is fished commercially in Monterey Bay and off southern California and is common in fish markets. The white, cylindrical, gelatinous egg masses are occasionally washed ashore in vast numbers.

Identification of octopuses and squids requires careful measurement of body parts, hence they should be preserved so as to minimize contraction or distortion. Preferably, specimens should be relaxed under anesthesia and then fixed with arms extended for greater ease of measurement. For anesthetization, magnesium sulfate (Epsom Salts) can be added in progressively increasing strengths until the animal is immobilized, followed by preservation in 4% formaldehyde (10% of full strength formalin). The fixing fluid should be flushed into the mantle chamber with a syringe and, in larger animals, it is essential that full strength formalin be injected into the viscera.

Minute endoparasitic Mesozoa commonly occur in the nephridial cavity of octopuses (see p. 121).

REFERENCES ON CEPHALOPODA

Berry, S. S. 1912. A review of the cephalopods of Western North America. Bull. U.S. Bur. Fisheries 1910. 30: 269–336.

Berry, S. S. 1953. Preliminary diagnoses of six West American species of *Octopus*. *In* S. S. Berry, ed., Leaflets in Malacology 1 (10): 51–58; see also 1 (11): 66.

Berry, S. S. and B. W. Halstead 1954. Octopus bites—A second report. *Ibid.* 1 (11): 59–65.
Fields, W. G. 1965. The structure, development, food relations, reproduction, and life history of the squid *Loligo opalescens* Berry. Calif. Fish and Game, Fish Bull. 131: 1–108.
Halstead, B. W. 1949. Octopus bites in human beings. *In* S. S. Berry, ed., Leaflets in Malacology, 1 (5): 17–22.
Pearcy, W. G. 1965. Species composition and distribution of pelagic cephalopods from the Pacific Ocean off Oregon. Pacific Science 19: 261–266.
Pickford, G. E. 1964. *Octopus dofleini* (Wülker), the giant octopus of the North Pacific. Bull. Bingham Oceanogr. Coll., Peabody Mus. Nat. Hist., Yale Univ. 19: 1–70.
Pickford, G. E. and B. H. McConnaughey 1949. The *Octopus bimaculatus* problem: A study in sibling species. *Ibid:* 12: 1–66.
Robson, G. C. 1929. A monograph of the Recent Cephalopoda. Pt. I, 236 pp.; 1932, Pt. II, 359 pp. British Mus. Nat. Hist.
Roper, C. F. E., R. E. Young, and G. L. Voss 1969. An illustrated key to the families of the order Teuthoidea (Cephalopoda). Smithson. Contr. Zool. 13: 1–32.
Young, R. E. 1972. The systematics and areal distribution of pelagic cephalopods from the seas off southern California. Smithson. Contr. Zool. 97: 1–159.

CLASS POLYPLACOPHORA (CHITONS)

(PLATE 109)

The familiar chitons furnish excellent material for studies of intertidal zonation and behavior. All chitons are marine and live on hard surfaces, clinging by the well-adapted broad foot. The low-arched form of the body and shells makes them capable of withstanding wave shock, and the flexible edge of the mantle, or **girdle,** allows them to fit closely the contours of rough surfaces to reduce desiccation when exposed during low tides. The overlapping valves permit chitons to roll up for protection if dislodged, but if left undisturbed they flatten out and move toward a protected spot away from the light. Most chitons are not well adapted to cope with silt or suspended matter, which clogs their **ctenidia** (gills). This fact, together with their adaptation to hard surfaces, suggest why they live most successfully in clear water on rocky coasts. Chitons are most abundant in the middle to lower intertidal zones or in moderate depths.

The characters used in identification are mainly the length of the expanded adult animal in comparison with its width and height, the configuration and sculpture of the eight overlapping **valves,** and the width of the girdle as well as its ornamentation (figs. 1, 2). For some species color is a reliable identifying character, but in others the color may not only vary from specimen to specimen but may also mask important types of sculpture, especially the dorsal microsculpture of the valves.

Chiton valves are of three forms: an anterior or **head** valve, six median or **intermediate** valves, and a posterior or **tail** valve. Each valve has two principal shell layers: an upper outer layer consisting of the exposed part **(tegmentum)**, and a lower inner layer **(articulamentum)**, covered by the girdle at the edges of the valves. When freed of the girdle by soaking in warm water for some time, the six intermediate valves and the tail valve are seen to have pairs of somewhat semicircular extensions of the articulamentum, called **sutural laminae**, that project under the valve ahead of each. In addition, narrow extensions of the articulamentum, called **insertion plates**, occur at the sides of the intermediate valves, the anterior margin of the head valve, and the posterior margin of the tail valve; these plates are buried in the girdle. In most chiton species the insertion plates are cut into teeth by **slits**, the number and character of which are often important in identification. A chiton with an average of 10 slits in the insertion plate of the head valve, 2 on each side of the intermediate valves and 8 in the tail valve insertion plate, is said to have a **slit formula** of 10/2/8. To confirm identification, it is often desirable to free the valves from the girdle, either entirely or partially, with care to avoid damaging them.

Configuration of the valves may range from a low, round-backed aspect to a high-arched form with straight or gothic-arched side slopes. When the apices of the valves have pointed posterior extensions, they are said to be **mucronate** (beaked). The **jugal ridge** or **jugum** is the median longitudinal line at the top of a valve and may be rounded or sharply angled. Dorsal sculpture usually is divided into valve areas, generally quite distinct. In an intermediate valve the pair of raised triangular segments extending diagonally from the apex to the valve's side margins are called **lateral areas**; they generally are set off from the valve's **central areas** on either side of the jugum, by well-defined diagonal ribs or **cords**. Sometimes the area along the jugal ridge is sculptured differently from that on the central areas and is termed the **jugal area**. A chiton tail valve has a more or less prominent projection, a **mucro** (beak); its shape and position can assist in identification. A chiton head valve usually has sculpture like that on the lateral areas of the intermediate valves; the area behind the mucro on the tail valve generally has similar sculpture; the area in front of it is sculptured like that on the central areas.

The girdle varies markedly in width and thickness. It may be entirely smooth, or rough and variously decorated with minute, sandlike **granules**; glassy **spicules**, singly or in bunches; short rodlike processes or **spines**; or small, imbricating (overlapping), smooth or striated **scales**. Girdles of some species have **bristles** that may be simple or branched, stiff or flexible, coarse or fine. In the giant *Cryptochiton stelleri* the girdle covers the valves completely except when extremely young; in the "whale-back" *Katharina tunicata* only a small portion of each valve is exposed.

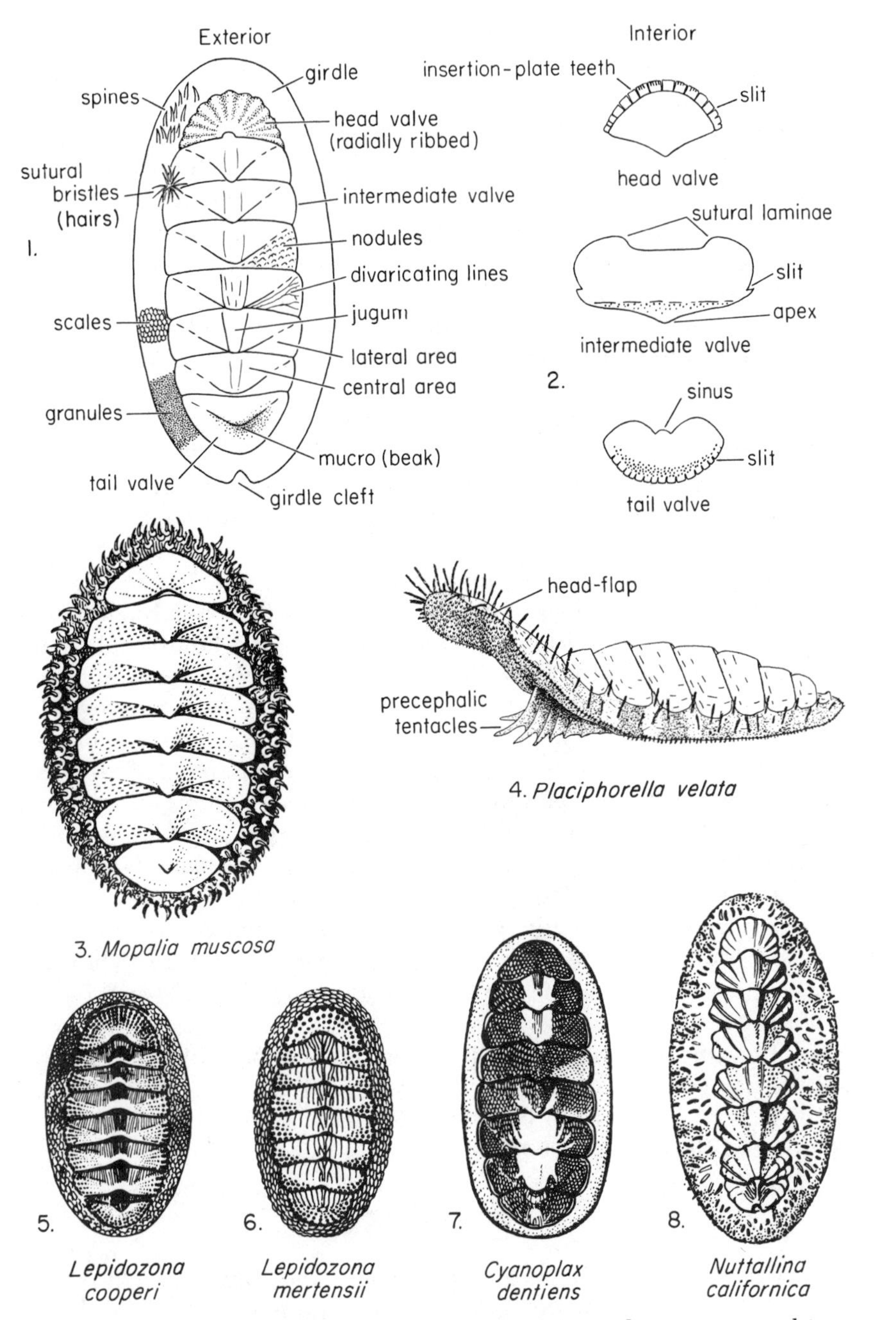

PLATE 109. **Mollusca. Polyplacophora (Chitons)**: 1, diagrammatic chiton showing girdle and shell ornamentation; 2, terminology of valves; 3, *Mopalia muscosa;* 4, *Placiphorella velata;* 5–8, other representative chitons. 1 and 2 redrawn by Emily Reid after Yonge, 1960; 4, McLean, 1962.

The head; foot, and rows of ctenidia are concealed under the girdle. The head bears simple sensory lappets, no definite tentacles, and no eyes; it cannot be protruded from beneath the girdle. Light-sensitive species have tiny eyelike receptors or **aesthetes** set in microscopic pits in the valve tegmentum layer. In central California chitons, aesthetes generally are not visible except under a microscope and then only in young individuals. The ctenidia, consisting of triangular plumes, lie in the space surrounding the edge of the foot; there may be several to many pairs. Both nephridial and genital openings are paired, the nephridial located in the gill-groove at the sides and in front of the anus, the genital in front of the former. All chitons have a well-developed radula with a simple tooth at its center, flanked by a combination of eight variously formed teeth on either side. The sexes are separate.

"Chiton sticks" are recommended for collecting chitons. These are strips of wood or of hard plastic, in several widths and lengths to accommodate chitons of various sizes, with one or two feet of selvage tape one-half or three-quarter inches wide attached to one end. As each chiton is removed from the substrate with a thin-bladed knife or spatula, it is placed foot down on the stick and bound lightly but firmly by several turns of wetted tape, which is then secured to prevent unwinding. Animals that tend to roll up should not be forced to flatten, as this may break the shells, but, if held down with light pressure on the stick under water, they will usually flatten out of their own accord. Both sides of a chiton board can be used. Chitons so tied down may be killed by immersion in hot (not boiling) water for a few minutes and will remain in an extended condition, after which they may be preserved in 70–75% ethyl alcohol. Formalin corrodes the calcareous shells and girdle spicules and should not be used for final preservation. Ten percent formalin may be used for a 24 to 48 hour fixing period, with subsequent transfer to alcohol. Glycerol is not recommended as a preservative, but 1–2% added to alcohol prevents drying out. Identification of dried chitons is often impossible because the girdle hairs or spicules break off or are lost. Methods for collecting and preserving chitons have been published by Berry (1961) and Burghardt and Burghardt (1969).

KEY TO CHITONS

1. Valves completely covered by a thick, brick-red or brown-red girdle, with crowded fascicles of minute spinelets; adults very large (length 15–33 cm) .... *Cryptochiton stelleri*
– Valves exposed 2
2. Relatively small portion of valve area exposed; girdle black,

wide, and smooth; 7–12 cm .... *Katharina tunicata*
– Most of valve areas exposed 3
3. Pattern of red and white wavy lines, or red with brown or black spotting; girdle smooth, leathery; valve surfaces smooth, lateral areas not conspicuous; 2–5 cm .... *Tonicella lineata*
– Color otherwise; girdle not smooth, variously decorated with bristles (hairs), spines, spicules, or scales 4
4. Girdle without scales, but with bristles, spines, or spicules 5
– Girdle with small, closely-packed, overlapping scales (microscopic in *Stenoplax fallax*) 17
5. Girdle with bristles, not prominently spiculose or spiny ... 6
– Girdle without bristles, with or without prominent spicules or spines 13
6. Girdle extended in front to form a prominent head-flap (fig. 4); girdle bristles sparse, unbranched; valves short and broad, without prominent nodulose sculpture, often overgrown; juveniles usually with brightly colored spots; 2.5–5 cm .... *Placiphorella velata*
– Girdle not forming a head flap, of about equal width all around 7
7. Girdle bristles conspicuous, somewhat fleshy, occurring singly at sutures between valves and in series around the margins of the end valves (caution, bristles often broken off) 8
– Girdle bristles not occurring singly at valve sutures 9
8. Girdle cleft posteriorly; valves with lateral areas coarsely nodulose and central areas with longitudinal granulose ribs; color gray (sculpture and color similar to *Mopalia muscosa*); 2.5–5 cm .... *Mopalia porifera*
– Girdle not cleft; valves microscopically granular; color dark or bright green, red, crimson, or a combination of these; 2–3 cm .... *Basiliochiton heathii*
9. Girdle bristles prominently branching, sparsely set, flexible, straplike, with short, white, sharply pointed, branching spines and bundles of short, transparent spicules at their bases; in sculpture, color, and size resembles *Mopalia lowei*, but more common .... *Mopalia ciliata*
– Girdle bristles not branching 10
10. Girdle bristles coarse and stiff, straplike, closely packed; valves low-arched and carinate, olive-black or gray; valves often overgrown and eroded; lateral areas coarsely pustulate and central areas with longitudinal, granulose ribs (fig. 3); valve interiors blue green, stained with lilac toward centers; adults large (5–9 cm) .... *Mopalia muscosa*
– Girdle bristles flexible and less coarse; sculpture beaded, of rows of rounded pustules 11
11. Girdle bristles closely spaced, especially near the girdle margin, the central cores surrounded by short, appressed spines that are

slightly recurved and free only at their tips; beaded sculpture strong and well developed; color usually green but often with areas of crimson, white, dark brown, and sometimes yellow; valve interiors blue green, often stained with lilac at centers, or whitish stained with rose; 3–6 cm .... *Mopalia lowei*

– Girdle bristles or hairs sparsely set 12

12\. Beaded sculpture well developed (but not as strong as in *Mopalia muscosa*); girdle with flexible, threadlike hairs; posterior girdle cleft prominent; color brown or green brown; valve interiors not chalky, white; 8–10 cm .... *Mopalia hindsii*

– Beaded sculpture fine, nearly obsolete; girdle without threadlike hairs; bristles thick, recurved; underside of girdle orange in life; color green-gray with brown markings radiating from the valve apices forming a branching pattern; valve interiors chalky, blue and white; 5–7 cm .... *Mopalia lignosa*

13\. Girdle with short, rigid, brown spines with a few white spines intermingled, closely packed; valves narrow, nearly as long as wide, commonly much eroded; underside of girdle orange in life; sculpture of strong ribs over a basic granular pattern (fig. 8); body elongate; 2.5–5 cm; slit formula 10–11/2/8–9 .... *Nuttallina californica*

– Girdle without strong spines, appearing smooth or velvety, clothed with fine, closely packed, sandlike grains, with or without short, glassy spicules 14

14\. Girdle with sparsely scattered, short, transparent, glassy spicules; valves with rows of prominent, round, projecting tubercles, generally smaller and more numerous on the central areas; color red or orange, with black tail valve (Monterey area), or spotted with green or brown; 1–2 cm; slit formula 9–12/1/7–8 .... *Chaetopleura gemma*

– Girdle with fine, sandlike grains, without spicules 15

15\. Insertion plates without slits; valves round-backed; sculpture granular, in finely beaded rows or chains; foot liver-colored; 1–1.5 cm .... *Leptochiton rugatus*

– Insertion plates with slits; valves short and broad, low-arched or mucronate; sculpture microscopically granulate, with or without low, warty protuberances 16

16\. Lateral areas with warty sculpture, especially in adults; valves low-arched, not carinate; body broadly oval; color green or gray; valve interiors an intense blue green; 2–3 cm; slit formula 11/1/11 .... *Cyanoplax hartwegii*

– Lateral areas without warty protuberances (fig. 7); valves mucronate; body more elongate; color of adults green or red brown, with juveniles often brightly marked with white, crimson, or black; 1–2 cm; slit formula 11/1/10–12 .... *Cyanoplax dentiens*

17. With strong, radiating, rounded ribs on end valves and on lateral areas of intermediate valves; body elongate 18
– Valves not strongly ribbed 19
18. Head valve with 7–9 strong radiating ribs, lateral areas with single, strong, undivided ribs, all ribs crenulating the valve margins; color brown, green, or cream; 1–3.5 cm; slit formula 9/1/13–20 .... *Callistochiton crassicostatus*
– Head valve with about 11 strong radiating ribs, ribs on lateral areas of intermediate valves divided by median groove; tail valve excessively prominent and high-arched in subspecies *mirabilis;* color tan with darker markings; slit formula 9–11/1/22–26 .....*Callistochiton palmulatus*
19. Lateral areas raised, sculptured with well-developed radial rows of rounded pustules, the end valves similarly pustulate 20
– Lateral areas without radial rows of pustules 21
20. Central areas latticed, the longitudinal riblets stronger (fig. 6); valves with sharp jugal ridge; color red or speckled with cream and dark brown; girdle sometimes with 2 pairs of opposing patches of white scales; adults 2.5–5 cm; slit formula 10–11/1/10 .... *Lepidozona mertensii*
– Ribbing and nodulose sculpture weaker (fig. 5); color invariably a uniform olive or slate gray; slit formula 11/1/11 .... *Lepidozona cooperi*
21. Lateral areas with 3–4 granulose riblets; sculpture sharp, central area with raised ribs, and intermediate lattice network, color green, often with white and gray lines or spots; body ornate; 1–2 cm; slit formula 10/1/9 ....*Lepidozona sinudentata*
– Lateral areas without 3–4 granulose riblets 22
22. Sculpture weak, with many fine, closely placed riblets, the lateral areas concentrically grooved with fine striations; valves carinate; body elongate; color uniform slate gray, blue gray, or olive brown (a rare variant is brilliant cobalt blue); 3–5 cm; slit formula 14–16/2–3/22 .... *Ischnochiton regularis*
– Sculpture and color otherwise 23
23. Body form oval; sculpture of fine riblets on minutely granulated surface, appearing smooth; lateral areas slightly elevated, not prominent; valves low-arched, not especially carinate; color extremely variable, olivaceous with brown spots, speckled, occasionally pure white or with end valves bright blue; 1–2.5 cm; slit formula 10–11/1/9–10 .....*Ischnochiton radians*
– Body very elongate, twice as long as wide; end valves large (tail valve especially long); lateral areas prominent; sculpture rough, somewhat vermiculate, with concentric grooving; adults large (5–11 cm) 24
24. Valves often worn smooth, central areas irregularly, longitudinally lined, without marked pitting; girdle with small, finely

striated scales; color white with green or beige markings; valve interiors white, sometimes suffused with light blue, tinged with rose at centers; slit formula 10–14/2–4/9–12 .... *Stenoplax heathiana*

– Valves with marked pitting in central areas; girdle scales microscopic, the girdle appearing smooth, almost leathery; color brighter than *Stenoplax heathiana*, with tones of rose red, dark red, or beige with white markings; valve interiors highly colored with rose pink and blue; slit formula 8–9/1/9–11 .... *Stenoplax fallax*

LIST OF CHITONS

Unless otherwise indicated, habitats range from middle and low intertidal into the subtidal at moderate depths.

ACANTHOCHITONIDAE

Cryptochiton stelleri (Middendorff, 1846). Scarce in the Monterey area; more common further north. See Heath, 1897, Proc. Phil. Acad. Sci. 299–302, pl. 8 (juvenile morphology); MacGinitie and MacGinitie, 1968, Veliger 11: 59–61 (food, growth, age, external cleaning); Webster, 1968, Veliger 11: 121–125 (commensals).

CALLISTOPLACIDAE

Callistochiton crassicostatus Pilsbry, 1893.

Callistochiton palmulatus mirabilis Pilsbry, 1893. More common than the following subspecies.

Callistochiton palmulatus palmulatus Carpenter in Pilsbry, 1893.

Nuttallina californica (Reeve, 1847). One of the commonest intertidal chitons; upper mid-tidal on exposed rocks.

* *Nuttallina thomasi* Pilsbry, 1898. Low intertidal on holdfasts of the sea palm *Postelsia;* Monterey area; rare; smaller than the preceding species.

CHAETOPLEURIDAE

Chaetopleura gemma Dall, 1879.

ISCHNOCHITONIDAE

Basiliochiton heathii (Pilsbry, 1898). Low intertidal and subtidal; scarce; see Berry, 1925, Proc. Acad. Nat. Sci. Phil. 77: 23–29 (taxonomy).

Cyanoplax dentiens (Gould, 1846) (= *C. raymondi*).

* *Cyanoplax fackenthallae* Berry, 1919. Low intertidal, concealed in kelp holdfasts, which it mimics in color; scarce.

* Not in key.

Cyanoplax hartwegii (Carpenter, 1855). Common in middle to low intertidal on exposed rocks and under coralline algae.

Ischnochiton radians Carpenter in Pilsbry, 1892. Low intertidal and subtidal; fairly common. May be the same as *I. interstinctus* (Gould, 1846), a less colorful species occurring from Puget Sound north.

Ischnochiton regularis (Carpenter, 1855).

Lepidozona cooperi (Pilsbry, 1892).

Lepidozona mertensii (Middendorff, 1846).

Lepidozona sinudentata (Carpenter in Pilsbry, 1892). Low intertidal and subtidal; scarce.

Stenoplax fallax (Pilsbry, 1892) (= *Ischnochiton fallax*). With the following, but much less common.

Stenoplax heathiana Berry, 1946 (=*Ischnochiton heathiana*). Low intertidal on smooth boulders imbedded in clean sand; numbers of the minute commensal snail *Vitrinella oldroydi* are occasionally found under the mantle.

Tonicella lineata (Wood, 1815). Common on coralline-encrusted, shaded rocks, well camouflaged by its color pattern.

LEPIDOPLEURIDAE

Leptochiton rugatus (Pilsbry, 1892).

MOPALIIDAE

* *Dendrochiton thamnoporus* (Berry, 1911). Low intertidal and subtidal; rare; small (0.5–1 cm); valves often tinged with lilac or rose; central areas cut by 9 or 10 curved grooves; girdle with bristle-like hairs at sutures.

Katharina tunicata (Wood, 1815). Occurs with *Nuttallina californica* among corallines and mussels on exposed rocks.

Mopalia ciliata (Sowerby, 1840).

Mopalia hindsii (Reeve, 1847) (Includes *M. hindsii recurvans* Barnawell, 1960). On rocks and pilings in bays and estuaries as well as along the open coast; tolerates silt better than most chitons; the small triclad *Nexilis epichitonius* occurs in the mantle cavity.

* *Mopalia imporcata* Carpenter, 1864. Low intertidal and subtidal; scarce; sculpture like *M. muscosa* but smaller (2.5–5 cm); girdle with single, fleshy bristles at sutures and around end valves.

Mopalia lignosa (Gould, 1846). Distinguished from other Mopalias by its dendritic color pattern.

Mopalia lowei Pilsbry, 1918.

Mopalia muscosa (Gould, 1846). One of the commoner intertidal chitons; see Monroe and Boolootian, 1965, Bull. So. Calif. Acad. Sci. 64: 223–228 (reproductive biology).

Mopalia porifera Pilsbry, 1893. Low intertidal, generally under rocks imbedded in sand or sandy mud; scarce.

Placiphorella velata Dall, 1879. Can entrap small prey beneath anterior girdle flap; see McLean, 1962, Proc. Malac. Soc. London 35: 23–26 (feeding).

REFERENCES ON CHITONS

Barnawell, E. B. 1960. The carnivorous habit among the Polyplacophora. Veliger 2: 85–88.

Berry, S. S. 1917, 1919. Notes on West American chitons—I and II. Proc. Calif. Acad. Sci. (4) 7: 229–248, 4 figs., and 9: 1–36, pls. 1–8.

Berry, S. S. 1961. Chitons, their collection and preservation. In *How to Collect Shells* (2nd ed., pp. 44–49). American Malacological Union.

Burghardt, G. E. and L. E. Burghardt 1969. *A Collector's Guide to West Coast Chitons.* San Francisco Aquarium Soc., Spec. Publ. 4: 45 pp.

Giese, A. C., J. S. Tucker and R. A. Boolootian 1959. Annual reproductive cycles of the chitons *Katharina tunicata* and *Mopalia hindsii.* Biol. Bull. 117: 81–88.

Heath, H. 1899. The development of *Ischnochiton.* Zool. Jahrb. 12: 567–656. (Based on *Stenoplax heathiana.*)

Hyman, L. H. 1967. *The Invertebrates:* Mollusca I, Vol. VI. McGraw-Hill, pp. 70–142. (Polyplacophora.)

Lowenstam, H. A. 1962. Magnetite in denticle capping in Recent chitons (Polyplacophora). Bull. Geol. Soc. Amer. 73: 435–438.

Omelich, P. 1967. The behavioral role and structure of the aesthetes of chitons. Veliger 10: 77–82.

Pilsbry, H. A. 1892–1894. Polyplacophora (Chitons). In *Manual of Conchology* 14: xxiv + 350 pp.; 15: 133 pp.

Pilsbry, H. A. 1898. Chitons collected by Dr. Harold Heath at Pacific Grove, near Monterey, California. Proc. Acad. Nat. Sci. Phil. 1898: 287–290.

Smith, A. G. 1960. Amphineura. *In* R. C. Moore, ed., *Treatise on Invertebrate Paleontology,* Part I, Mollusca 1, pp. 41–76. Univ. Kansas Press and Geol. Soc. Amer.

Smith, A. G. 1966. The larval development of chitons (Amphineura). Proc. Calif. Acad. Sci. (4) 32: 433–446.

Thorpe, S. R., Jr. 1962. A preliminary report on spawning and related phenomena in California chitons. Veliger, 4: 202–210.

Tucker, J. S. and A. C. Giese 1959. Shell repair in chitons. Biol. Bull. 116: 318–322. (Concerns *Cryptochiton stelleri, Katharina tunicata* and *Mopalia hindsii.*)

Yonge, C. M. 1939. On the mantle cavity and its contained organs in the Loricata (Placophora). Quart. J. Micro. Sci. 81: 367–390.

PHYLUM MOLLUSCA: SHELLED GASTROPODS

James T. Carlton and Barry Roth
California Academy of Sciences, San Francisco

(PLATES 110–120)

The gastropods are the largest class of molluscs and exhibit enormous diversity in form and habitat. Garden snails and slugs are familiar to all; limpets, top shells, periwinkles, abalones, slipper shells, and whelks are known to most who visit the seashore. The beauty of many gastropod shells, especially from tropical regions, has long made them favorite objects for collectors, and our relatively advanced knowledge of the taxonomy of the gastropods is in large part due to the interest of amateur shell collectors.

This chapter deals with those gastropods with external shells which occur between Monterey Bay and Bodega Bay, including the prosobranchs, the marine pulmonates, and most of the shelled opisthobranchs.

Gastropods possess a muscular foot for creeping or burrowing, a head with mouth, characteristic scraping radula (absent in some) and, usually, sensory tentacles and eyes. A mantle secretes the shell and provides, in the pallial cavity, a shelter for the respiratory organs **(ctenidia)**. A hallmark of the gastropods, setting them apart from other molluscs, is the phenomenon of **torsion,** which occurs early in development. Torsion consists of a 180° counterclockwise rotation of the visceral mass upon the head and foot; the result is that the mantle cavity, ctenidia, and anus, which were originally at the rear, come to lie just behind the head. Torsion in its fullest expression characterizes the subclass **Prosobranchia,** where the ctenidia lie anteriorly and the nervous system is twisted into a crude figure 8 (the **streptoneurous** condition). Other groups of gastropods have tended to modify the extreme effects of torsion, one change being a straightening out of the nervous system to the **euthyneurous** condition. Euthyneury has been attained in two ways: in the subclass **Opisthobranchia,** the body has "unwound" itself in detorsion; in the subclass **Pulmonata,** the body has retained much of its torsion, but the central nervous system has straightened out by condensation into a ring of ganglia around the esophagus (as in the garden snail *Helix*).

Torsion is not the same thing as the coiling of the visceral hump and shell of many gastropods. Coiling is lost in limpetlike gastropods and in land slugs and nudibranchs, which as adults have reduced or lost the shell and flattened the visceral hump. Coiling is not unique to the gastropods, but is also found in the cephalopod *Nautilus* and was characteristic of many ancient, shelled cephalopods.

CLASSIFICATION

The classification used here is outlined below, with the more common synonymous terms being indicated. In an alternate system adopted by some workers, **Prosobranchia** are referred to as **Streptoneura** ("twisted nerves") with a second main group, **Euthyneura** ("straight nerves"), encompassing **Opisthobranchia** and **Pulmonata.** Although the shell is useful in identification, the classification of gastropods is based largely on structure of the ctenidia, heart, nervous system, and radula. The radula is extremely useful in classification and an understanding of its basic morphology and terminology (see Hyman, 1967; Fretter and Graham, 1962) and of preparation techniques (see Fritchman 1960, Veliger 3: 52–53; Radwin 1969, Veliger 12: 143–144) is useful.

Subclass **PROSOBRANCHIA** ("gills forward") (=**STREPTONEURA,** "twisted nerves")

Ctenidia anteriorly located; nervous system streptoneurous; sexes separate. The great majority of local marine gastropods are prosobranchs. There are three orders:

Order **Archaeogastropoda** (= **Aspidobranchia; Diotocardia**)

Ctenidia paired (abalones, keyhole limpets) or single (limpets, top and turban shells, and a few smaller families), **bipectinate** (with a double set of flattened, cilia-bearing filaments along axis). Those with two ctenidia have the shell slit or pierced with one or more holes for discharge of water. Radula typically with numerous teeth, reduced in the limpets. Most are grazing herbivores, some are carnivorous on sponges.

Order **Mesogastropoda** (=**Pectinibranchia** [**Taenioglossa**] or **Ctenobranchia; Monotocardia**)

Ctenidia single, **monopectinate** (a single row of filaments), fused with roof of mantle cavity; radula broad, typically with seven teeth in each row, but reduced or lost in parasitic members; generally without proboscis or long siphon. A diverse group, largely herbivores, but also including mucus- and filter-feeders (vermetids, slipper shells), ectoparasites (epitoniids, eulimids), and carnivores (cerithiopsids, *Triphora*, the moon snail *Polinices*, *Lamellaria*, *Velutina*, *Erato*, *Trivia*).

Order **Neogastropoda** (= **Pectinibranchia** [**Stenoglossa**])

Ctenidia single, monopectinate; radula narrow, with one to three teeth in each row; generally with a proboscis and a long siphon formed by mantle edge; **osphradium** (sensory organ testing the water passing over the ctenidia) well developed, probably for locating prey. Most are active, vagile carnivores although some, such as *Nassarius*, are also scavengers.

Subclass **OPISTHOBRANCHIA** ("gills behind")

The opisthobranchs include diverse hermaphroditic forms; the best known are the colorful shell-less nudibranchs or sea-slugs, considered in the next chapter. Several groups retain a shell in the adult; these include two orders which are keyed in this chapter. Detorsion has taken place in the opisthobranchs. Ctenidia are located in the rear and the nervous system is generally euthyneurous although a primitive, twisted (streptoneurous) condition is found in some opisthobranchs, such as Acteonidae.

Order **Cephalaspidea**

Shell generally present (absent in *Runcina*, p. 518); with a broad head-shield; radula present. Includes carnivores, herbivores, and detritus-feeders.

Order **Pyramidellida**

Shell small, whitish, nuclear whorls **heterostrophic** (coiled in a different plane to succeeding whorls); radula absent; gut modified with stylet, long proboscis, and buccal pump for suctorial feeding. Ectoparasites or commensals with other invertebrates.

Subclass PULMONATA ("possessing a lung")

The vascularized lining of the mantle cavity serves as a lung in the pulmonates; ctenidia are lost; detorsion has taken place and the nervous system is euthyneurous through condensation; hermaphroditic. Includes mostly freshwater and terrestrial species, but also some common marine or brackish-water species.

Order Basommatophora

Eyes at the base of the cephalic tentacles. Herbivores and filter-feeders, locally including the small, salt-marsh snail *Ovatella myosotis* and the limpetlike, rocky intertidal species, *Trimusculus reticulatus* and *Williamia peltoides*.

Order Stylommatophora

Eyes at the tips of the tentacles. Includes terrestrial snails and slugs. The small, shell-less marine *Onchidella* is placed here in the pulmonates by many workers; others consider it to comprise an order of the Opisthobranchia, or to be in the order Gymnophila (Soleolifera) of the Euthyneura.

TERMINOLOGY

Most of the terms used in the key are illustrated in figure 1. The gastropod shell is usually a spirally coiled tube of progressively increasing diameter; each turn of the spiral is called a **whorl.** Coiling may take place in a single plane (**planorboid** coiling) or, more frequently, in descending stages along an axis. The nuclear or embryonic whorls, smallest and first-formed, are found at the **apex.** These and subsequent whorls, except the last, form the **spire.** The last whorl is the **body whorl** and terminates at the **aperture.** Many, but not all, gastropods possess a separate **operculum,** which occludes part or all of the aperture when the animal withdraws into its shell. The spiral trace of each whorl's juncture with its adjacent whorl is the **suture.** Direction of coiling is usually to the right (**dextral**) or rarely to the left (**sinistral**). To determine the direction of coiling, hold the shell with the apex uppermost and the aperture facing the observer. If in this position the aperture lies to the right of the central axis, the shell is dextral; if to the left, it is sinistral. The top of a shell in this orientation is roughly posterior, its base anterior; this may not, however, coincide with the anteroposterior axis of the living animal. In certain types of gastropods, such as limpets, spiral structure is not evident in the adult shell, which is openly conical or cap-shaped. The term apex, when applied to these shells, refers to the peak of the cone.

The surface of a shell may be smooth or variously sculptured. Common sculptural features are **ribs, ridges, nodes,** and **threads;** incised lines are known as **striae; lamellae** are thin, raised, platelike structures. Surface features are described as **axial** if oriented along the axis of coiling, **spiral** if following the direction of the growing whorl. On cap-shaped shells, sculpture which extends from the center of the shell outward toward the margin is termed **radial;** sculpture which parallels the margin is termed **concentric.** Netlike sculpture composed of equally strong radial and concentric ribs is **reticulate** or **cancellate.** Many gastropods, notably certain Muricidae, periodically develop prominent axial thickenings which may be ornamented with scales or spines; these thickenings are called **varices** (sing. **varix**) and represent periods when the animal was not actively enlarging its shell. Fine **growth lines** may traverse the whorls; they represent traces of the former position of the aperture. A thin, exterior layer of organic material, at times textured or embellished, forms the **periostracum,** which may be removed by wear, exposing the calcareous shell layer beneath.

Height and length of a coiled gastropod shell are equivalent terms and signify measurement along the axis of coiling. Width or breadth is measured at right angles to height. The area of greatest breadth is

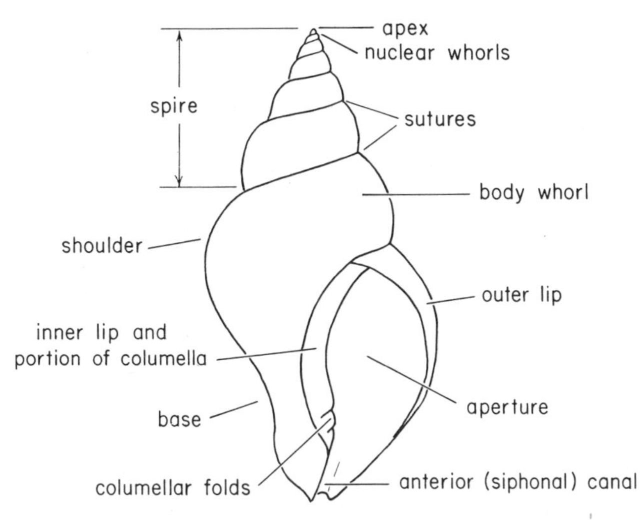

PLATE 110. **Prosobranchs** (1). 1, diagrammatic prosobranch labeled to show parts, drawn by Emily Reid.

termed the shell's **periphery.** The central pillar of the shell is the **columella,** which may bear teeth or folds. A hole or indentation in the base of the columella is called an **umbilicus.**

COLLECTION AND PRESERVATION

Examination of certain habitats will reveal many species which would otherwise be overlooked in normal collecting from rocks or algae. Dark overhanging ledges or deep crevices hide nocturnal, negatively phototactic species; *Lamellaria* and other snails match exactly the color and shape of ascidians and sponges they prey upon; apertures of hermit-crab shells may be examined for *Crepidula,* and the base and column of sea anemones are often fruitful hunting grounds. Numerous sedentary polychaetes, molluscs, echinoderms, and other invertebrates may be examined for ectoparasitic pyramidellids.

Microgastropods may be recovered by vigorous shaking or washing of algae, eel- or surfgrass and algal holdfasts, and rock undersurfaces. Sand, gravel, or detritus under rocks, or algae, may be dried and the microgastropods later sorted out. Screened and dried sediment samples placed in carbon tetrachloride (use adequate ventilation) will sink and small molluscs will float at the surface. Minimum damage to the habitat should always be the rule in collecting; overturned rocks should be replaced, and only small amounts of algae should be taken from any one area. Large collections of a single species from one area should be avoided.

Snails may be relaxed in a 7% solution of $MgCl_2 \cdot 6H_2O$, or by sprinkling menthol crystals on the surface of a shallow dish of water, or by one or two drops of propylene phenoxetol in 200–400 ml of water. Animals may be preserved in 75% ethyl or isopropyl alcohol; formalin will decalcify the shells. Some shells of each species should be dried and kept away from sunlight (in order to preserve the color) after the animal has been removed.

THE FAUNA AND THE KEYS

The intertidal shelled gastropod fauna of central California is generally well known. The majority of species (approximately 130) were described before 1875, with P. P. Carpenter, W. H. Dall, and A. A. Gould naming more than 70 of the approximately 160 species treated below. Since 1920 few new species have been found in our intertidal fauna. Range extensions of southern species during warm-water years, new records of microgastropods, and newly introduced species are to be expected.

The following key includes most of the intertidal shelled gastropods occurring between Monterey Bay and Bodega Bay. Certain species common in southern California or in extreme northern California have not been included, and users of the key are cautioned against attempting to apply it outside its intended range. The keys to shelled opisthobranchs in the next chapter, which is somewhat more extensive geographically, include two species not keyed below: *Acteocina inculta* and *Tylodina fungina,* both of which are found only south of Monterey Bay. Worn, eroded, and juvenile specimens will not key out or will key out incorrectly; only fresh adult specimens should be used. Sizes given in the key are average sizes. Some species key out in more than one place.

REFERENCES

For detailed treatments of the morphology, biology, and ecology of gastropods, Cox (1960), Fretter and Graham (1962), Wilbur and Yonge (1964, 1966), Hyman (1967), Purchon (1968), and Franc (1968) are recommended. Standard systematic works include Thiele (1929–1935), Wenz (1938–1944) and Zilch (1959–1960). For local genera and species, Dall (1921), Keep and Baily (1935), McLean (1969) and Keen and Coan (1974) are excellent guides; Palmer's (1958) treatment of species described by P. P. Carpenter is an extremely useful compilation; Boss, Rosewater, and Ruhoff (1968) and Johnson (1964) have provided summaries of the taxa of W. H. Dall and A. A. Gould respectively.

ACKNOWLEDGMENTS

We thank Eugene V. Coan, Anthony D'Attilio, Meredith L. Jones, A. Myra Keen, James H. McLean, George E. Radwin, Allyn G. Smith, and Rudolf Stohler for critical commentary on the manuscript. A portion of the basic design of the limpet key is from Dr. McLean's thesis; Dr. McLean, in addition, advised on numerous taxonomic questions. Dr. Stohler graciously helped in many ways, providing many new figures, permitting use of *Veliger* figures, and aiding in preparation of the half-tone plates of limpets.

KEY TO SHELLED GASTROPODS

1. Shell external; tubular, cap-shaped, or spirally coiled; if partially covered by mantle, shell with hole at apex, or mantle fully retractable into shell 2
– Shell internal, fragile, globose, without hole at apex, wholly covered by mantle or exposed by small pore when animal disturbed; animal oval to subcircular, resembling a dorid nudibranch, but without a dorsal rosette of gills or dorsal anal opening; with an anterior siphon and tentacles (fig. 24); characteristically on compound ascidians .... *Lamellaria* spp.
2. Shell spirally coiled 3
– Shell tubular or cap-shaped, not spirally coiled 7
3. Shell uncoiling with growth, early whorls irregular; solitary or forming irregular clumps of wormlike tubes; in life attached to substrate 4
– Shell not uncoiling with growth; not affixed permanently to substrate 40
4. Solitary; shell coiled, flat, with a strong dorsal ridge and scaly longitudinal ribs (figs. 32A, B); tube diameter 1.5–3mm; often on abalones, corroding a channel into shell surface .... *Dendropoma lituella*
– Principally gregarious, generally occurring in clumps or masses (except *Serpulorbis squamigerus* in central California; see species list) 5
5. Tube diameter about 2mm; sculpture of diagonal wrinkles on early portion of tube, projecting tubes smooth; with internal spiral thread on columella in medial whorls (fig. 33) occasional isolated individuals tightly coiled .... *Petaloconchus montereyensis*
– Tube diameter greater than 2mm 6
6. Lacking operculum; shell relatively large, diameter 9mm or more; tube scaly to wrinkled, longitudinally ribbed, not

embedded in substrate (fig. 31) *Serpulorbis squamigerus*

– With operculum; tube similar to above in size and sculpture, but initial whorls embedded in substrate (if on a shell) and later whorls somewhat corroded where one tube crosses another (this distinction difficult, see note in species list) .... *Dendropoma rastrum*

7. Shell tubular, slightly curved, minute, with 30–40 closely set rings; a plug at one end (fig. 21) *Caecum californicum*

– Shell cap-shaped 8

8. With hole at apex 9

– Apex entire, without hole 12

9. Apical hole large, widely oval; animal much larger than shell, the mantle nearly covering shell 10

– Apical hole small, either circular or elongate, mantle not extending over shell 11

10. Shell small (to 16mm); margin set off internally by a broad, shallow, encircling groove; ends slightly elevated (figs. 20A–C), shell buff color with radiating brown or gray bands; mantle variously colored (red, orange, lemon yellow, gray, brown) .... *Megatebennus bimaculatus*

– Shell large (to 13 cm); inner margin crenulate, lacking groove (figs. 23A–C); shell buff color, without radiating bands; mantle black (fig. 23D) ......*Megathura crenulata*

11. Internal apical callus truncate posteriorly; cancellate, coarse, radial sculpture with every 4th rib larger; weaker concentric sculpture (fig. 18); shell color, gray-brown radiating bands on grayish white background ..*Diodora aspera*

– Callus rounded posteriorly, not sharply truncate; radiating ribs of varying sizes, or faint radial striae; no concentric sculpture (fig. 19); shell pink with red-brown or black rays; foot yellow, mantle red striped ....*Fissurella volcano*

12. Interior with a deck or platform at posterior end 13

– No internal deck 18

13. Deck attached only on one side of shell and with a raised medial fold; shell nearly circular (fig. 22); mottled or radially striped brown and white ... *Crepipatella lingulata*

– Deck attached on both sides of shell (figs. 36, 38, 40) .... *Crepidula* 14

14. Shell brown or white with brown markings; generally deeply convex; periostracum thin or imperceptible 15

– Shell white; generally flat or concavely curved, at times markedly so; periostracum evident 17

15. Apex strongly turned to one side and united with margin of shell; deck margin waved (fig. 36); shell dirty white with

brown blotches or interrupted, wavy, chestnut-colored markings .... *Crepidula fornicata*

– Apex not turned markedly to one side, overhanging posterior margin of shell, but not united with margin 16

16. Shell dark brown; deck corners produced, extending forward on both sides; apex distant from shell margin (fig. 37) .... *Crepidula adunca*

– Shell light red brown, exterior often encrusted or eroded, occasionally spotted and striped reddish brown; deck corners not extending forward; apex close to shell margin (figs. 38A, B) .... *Crepidula convexa*

17. Thick, shaggy, golden brown periostracum; shell relatively thick, often broadly oval .... *Crepidula nummaria*

– Thin, shiny, brown periostracum; shell relatively thin; shape variable, may occur as foliated, elongate shells in pholad holes (compare *C. nummaria*); smooth, very thin, concave specimens in hermit-crab shells (compare *C. plana* in bays; see species list), or as flat, white shells on rocks .... *Crepidula perforans*

18. Sculpture reticulate; shell white, nearly circular; apex central (figs. 42A–C) without periostracum .... *Trimusculus reticulatus*

– Sculpture, if present, not reticulate; shell white or variously colored; apex central to marginal 19

19. Shell smooth, thin, waxy, orange or red-brown with translucent, lighter-colored rays; apex hooked, 1/3 distance from posterior end; thin, transparent periostracum .... *Williamia peltoides*

– Shell sculptured, or, if smooth, color and texture not as above 20

20. Shell white under brown periostracum (only traces may be present); sculpture of concentric lamellae or faint or prominent radial ribbing .... *Hipponix* 21

– No obvious periostracum; sculpture various.............. 22

21. Sculpture of flat, concentric lamellae bearing fine radial striae under periostracum; apex low, subcentral or near margin .... *Hipponix cranioides*

– Sculpture of strong radial ridges with weaker concentric sculpture; periostracum with fine hairs; apex elevated, overhanging posterior margin (fig. 35) .. *Hipponix tumens*

22. Ends of muscle scar joined by a sinuous line; shell long-oval, low, large (length to about 100 mm) and heavy; apex near anterior end; maculated brown and white; shell often eroded; inner margin dark brown, with prominent, owl-

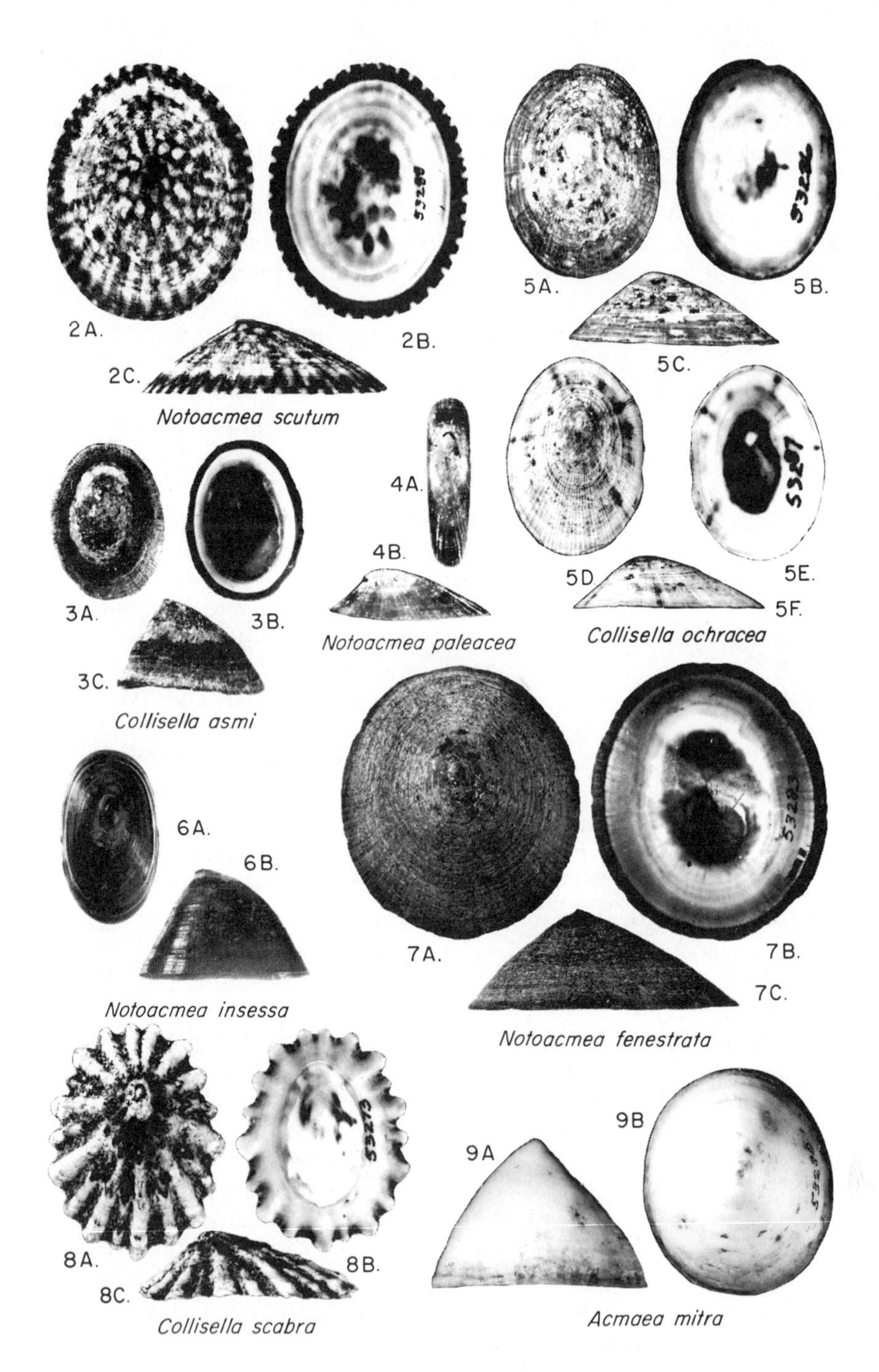

PLATES 111 and 112. **Prosobranchs** (2 and 3). Acmeidae: 2–17, all specimens from the California Academy of Sciences collection, photographed by Maurice Giles.

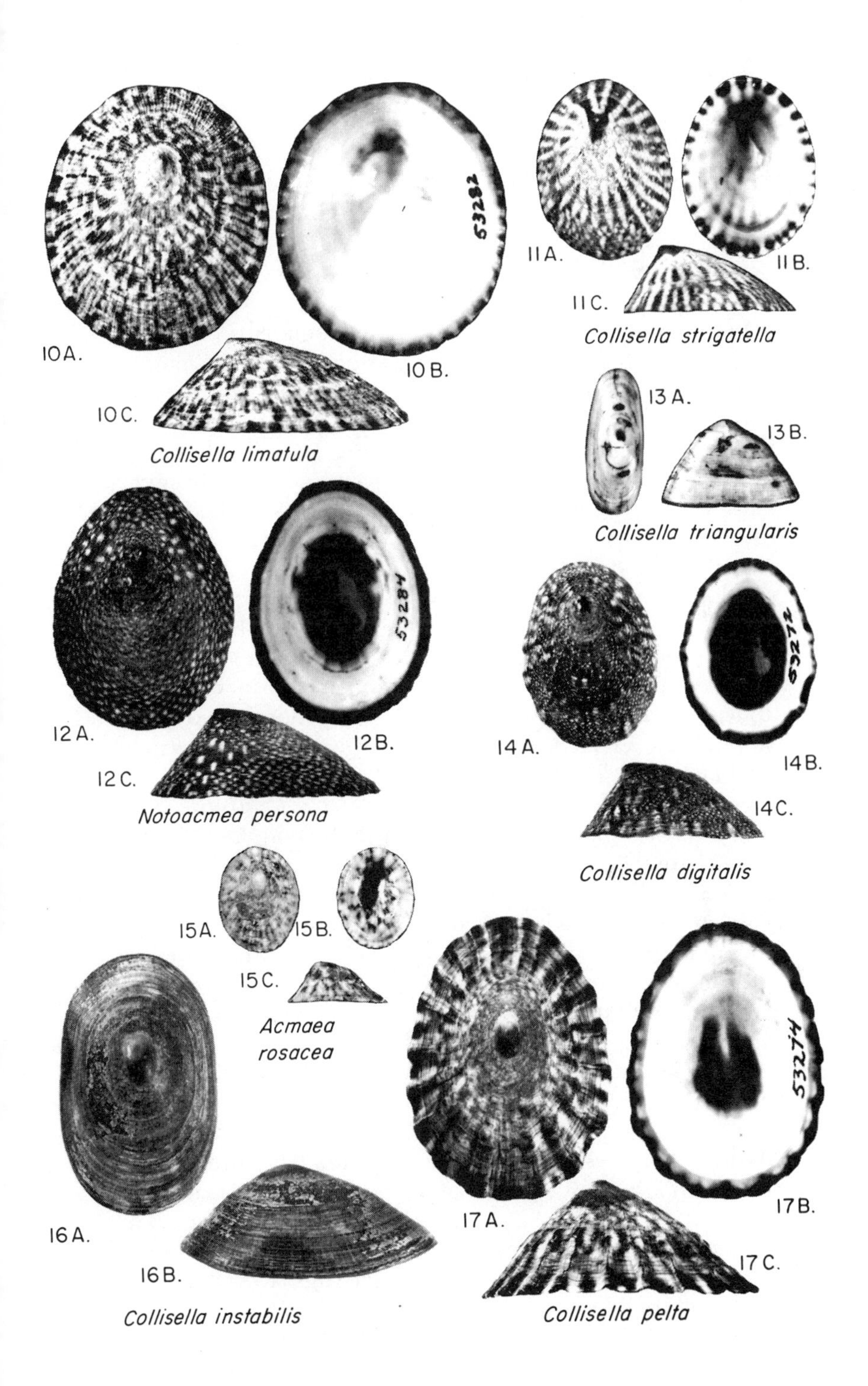
10A.
10B.
53282
10C.
Collisella limatula
11A.
11B.
11C.
Collisella strigatella
13A.
13B.
Collisella triangularis
12A.
12B.
53284
12C.
Notoacmea persona
14A.
14B.
53272
14C.
Collisella digitalis
15A.
15B.
15C.
Acmaea
rosacea
16A.
16B.
Collisella instabilis
17A.
17B.
53274
17C.
Collisella pelta

shaped muscle scar at center; (fig. 34) ....*Lottia gigantea*
– Ends of muscle scar joined by a faint, thin, simply curved line (fig. 30); shell length to about 60 mm 23

23. With heavy radial ribs; may be eroded, with ribs visible only at margin of shell 24
– Smooth, or with fine radial ribs or striations 26

24. Rib surfaces usually light-colored (often with darker-colored spines) and darker spaces between; ribs projecting strongly in all directions, forming strong scalloped margin; apex about 1/3 of distance from anterior end, but position variable; anterior slope convex (figs. 8A–C) animal with black spots on head and sides of foot ....*Collisella scabra*
– Rib surfaces usually not lighter than intermediate areas; margin not strongly scalloped; animal lacks dark spots on head and sides of foot 25

25. Anterior slope generally concave, posterior convex; apex generally above or, at times, overhanging the anterior margin; ribs strongest on posterior slope, may be absent at anterior end; (figs. 14A–C) ....*Collisella digitalis*
– Anterior slope not concave, apex subcentral; smoother than above 2 species; ribs generally equally developed on all slopes (figs. 17A–C); color various, brown, green, or greenish black, checkered with white tessellations or peripheral rays and bands of white ....*Collisella pelta*

26. Sides of shell parallel or nearly so; shell compressed, aperture narrow, oblong 27
– Sides not parallel, aperture oval 31

27. Ends of shell curved upward; nearly smooth with obscure, low, radial ribbing at margin (figs. 16A, B); exterior brown, interior bluish, stained with brown apically; on *Laminaria* stipes in low intertidal ....*Collisella instabilis*
– Ends of shell not curved upward 28

28. Mature shell under 4 mm wide; exterior white with brown rays or light brown, with fine striae or small, broad, rounded ribs 29
– Mature shell over 6 mm wide; exterior dark brown, lustrous, smooth, with fine radial sculpture; occurring on *Egregia* 30

29. Generally high-conic; apex subcentral; anterior and posterior slopes convex; fine radiating striae (figs. 13A, B); white, often bearing brown rays, with brown spot posterior to apex; on coralline algae ....*Collisella triangularis*
– Low; apex near or overhanging anterior end; anterior slope straight, posterior convex; with small, rounded, radial ribs;

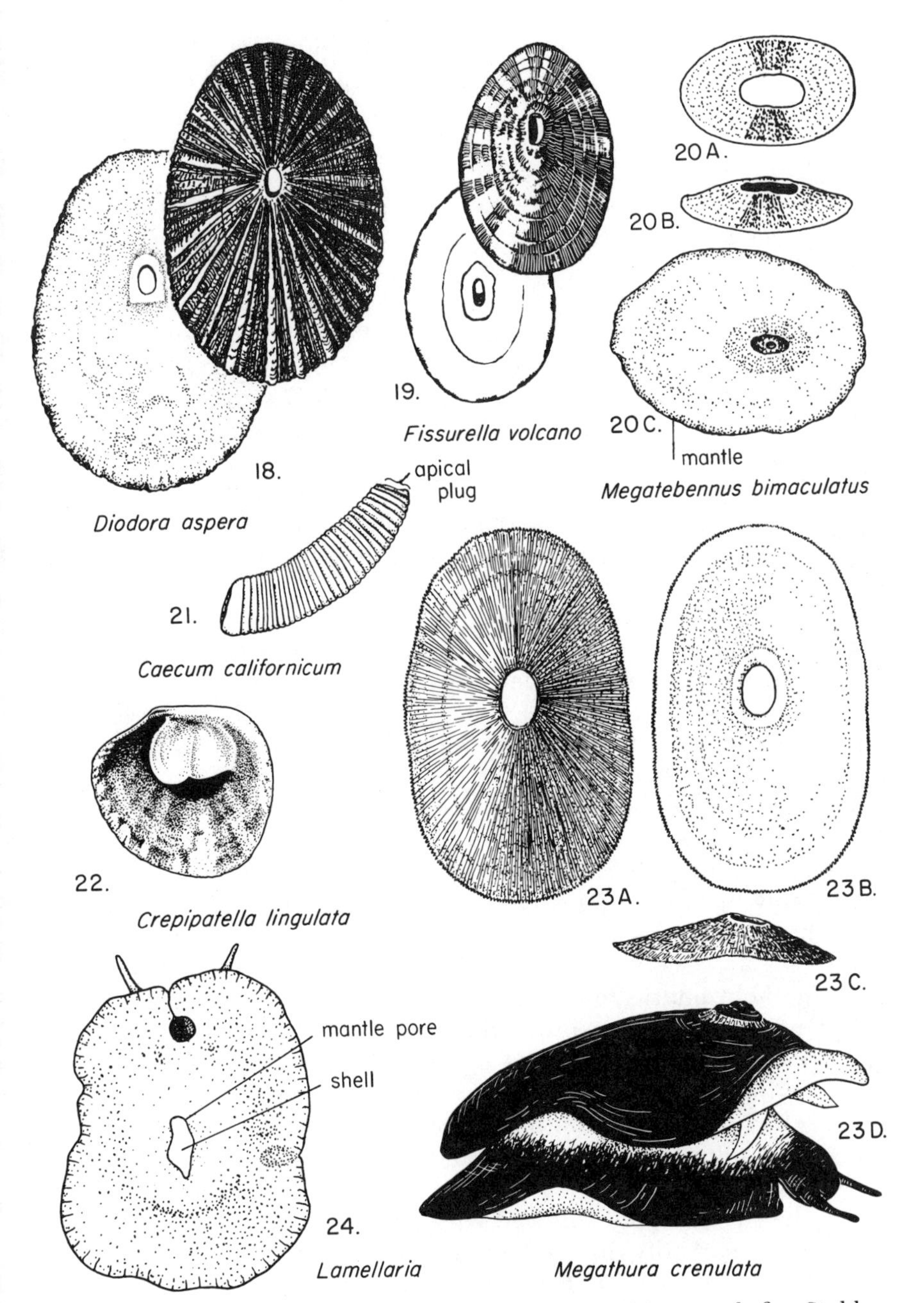

PLATE 113. **Prosobranchs** (4). 18, dorsal after Keep, 1904, ventral after Stohler (by Emily Reid); 19, after Keen and Pearson, 1952; 20A–C and 23A–D, Stohler (by E. Reid); 21, Abbott, 1954; 22 and 24, by E. Reid.

right anterior margin with pallial notch (figs. 4A, B); generally light brown, darker near margin, apex eroding to white; on surfgrass *Phyllospadix* .... *Notoacmea paleacea*

30. Interior dark brown (figs. 6A, B) ...... *Notoacmea insessa*
– Interior gray to light brown, usually with internal post-apical brown spot; externally resembling *N. insessa* (see species list) .... subadult *Collisella pelta*

31. Black to dark gray brown inside and out, occasionally bluish white between internal shell margin and apex; small (to about 11 mm), elevated, exterior often eroded, generally with fine radial striae visible at least at margin (figs. 3A–C); generally on turban snail *Tegula funebralis* (but not the only limpet on this *Tegula*) .... *Collisella asmi*
– Not black-brown inside and out 32

32. Entirely white; high-conic; slopes straight; with fine concentric growth lines and fine radial striae (figs. 9A, B); sculpture often completely concealed by pink coralline algae .... *Acmaea mitra*
– Shell not uniformly white 33

33. Pink, mottled with white streaks and white and yellow brown dots; thin, elevated, small (to about 8 mm); smooth, or with fine radial ribs; anterior slope straight, posterior slope convex (figs. 15A–C) .... *Acmaea rosacea*
– Color otherwise 34

34. With fine, imbricate (scaly) radial ribbing; margin serrate (with sawlike notching); shell low or elevated (figs. 10A–C); color buff yellow, with fine darker mottlings, or green brown with white tessellations or bands; sides of foot and head black to gray contrasting with white undersurface .... *Collisella limatula*
– Without imbricate radial ribbing 35

35. Internal area between shell margin and apex suffused with brown; low to moderately elevated; shell with weak radial sculpture (figs. 7A–C); olive to gray with small, white tessellations; apex often eroded to brown (contrast *Collisella strigatella*, which is smaller and has apex eroding to white) .... *Notoacmea fenestrata*
– Shell interior not suffused with brown 36

36. Apex markedly directed anteriorly 37
– Apex subcentral, not directed anteriorly 38

37. Shell large, to 53 mm in length; anterior slope straight, other slopes convex; surface with fine, regular striae (figs. 12A–C); usually olivaceous green with scattered white tessellations or rays; apex eroding to brown with 2 lateral

streaks of white dots .... *Notoacmea persona*

– Shell to about 20 mm; anterior slope slightly concave, posterior convex; surface with few fine, raised, radial lines (sometimes obsolete) (figs. 11A–C); color variable, often with fine bluish white spots drawn out as white streaks at margin; apex eroding to white (contrast young *Collisella pelta*, which may be ribbed, has slopes generally straight, and with internal postapical brown spot) .... *Collisella strigatella*

38. Shell elevated, height usually greater than 1/3 of length; surface with fine, irregular, radial ribbing; apex slightly hooked, eroding to bluish white; anterior slope straight to somewhat concave; shiny, dark brown to black, with or without scattered white dots and rays; internal postapical brown spot; in mussel beds and associated with alga *Pelvetia* .... *Collisella pelta*

– Shell depressed, height usually less than 1/3 of length 39

39. Adult shell thin, small to medium (to about 30 mm); apex 1/3 of distance from anterior end; sculpture of threadlike radial riblets; color variable, either tessellate pattern of oval white spots or solid color (white to buff) (figs. 5A–F) .... *Collisella ochracea*

– Adult shell thick, large (to about 63 mm); apex subcentral; sculpture coarser, of flat-topped ridges, lacking fine radial riblets; color pattern of variable spotting, rarely a solid color (figs. 2A–C) .... *Notoacmea scutum*

40. Aperture without an anterior siphonal canal or notch 41

– Aperture with an anterior canal or notch 98

41. Shell broad, ear-shaped, with a row of holes along one side .... *Haliotis* 42

– Without a row of holes 45

42. Shell greenish black to dark blue; holes round and flush with surface, which is nearly smooth; muscle scar may be present in older specimens .... *Haliotis cracherodii*

– Shell not black or dark blue; holes oval, raised above shell surface 43

43. Sculpture of low, rounded, spiral ridges crossed by closely spaced, raised striations; shallow, indistinct groove along shell margin; shell flat, elongated; brick red with white, blue, and green mottling; no muscle scar .... *Haliotis walallensis*

– Sculpture irregularly lumpy, undulating; shell moderately deep, not flat and elongated 44

44. Shallow, broad channel parallel to shell edge; shell thin,

mottled green-brown or red-brown with scattered blue and white areas; no muscle scar ...... *Haliotis kamtschatkana*

– Lacking broad channel; shell dull brick red, may have light color bands (pink, white, green); muscle scar prominent; usually covered with algae, barnacles, and other organisms .... *Haliotis rufescens*

45. Shell minute (0.8 mm height), with an elongate slit (foramen) in outer lip near aperture; white, turbinate; 3 rapidly enlarging whorls; sculpture of axial folds .... *Sinezona rimuloides*

– Shell without elongate slit in outer lip 46

46. Minute (about 1 mm diameter), planorboid, of about $2^1/_2$ regularly increasing whorls; spire depressed and base broadly umbilicate; shell smooth, lacking axial sculpture, translucent, with thin, brown periostracum and white "quarter lines" radially traversing each whorl at about 90° increments .... *Omalogyra* sp.

– Shell otherwise 47

47. Shell bulloid (figs. 27, 29) 48

– Shell not bulloid 51

48. Shell small (to about 4 mm in height), umbilicate, thin, translucent, smooth, with weak axial growth lines; 3 whorls, with globular nucleus .... *Diaphana californica*

– Shell medium to large, lacking an umbilicus 49

49. Aperture broadly rounded anteriorly, not flaring (fig. 27); shell pink-gray to brown with cloudy maculations bordered by white on their left edges; to 8 cm ... *Bulla gouldiana*

– Aperture flaring anteriorly; without brown maculations; shell white, green, or yellowish, to 25 mm ..... *Haminoea* 50

50. Upper portion of body whorl tapered, with a shallow constriction; ratio of dimensions A and B (fig. 29) about 1:2; usually less than 15 mm in length, but may be larger .... *Haminoea virescens*

– Upper portion of body whorl relatively broad, somewhat flattened, not constricted; ratio of dimensions A and B about 1:1; usually more than 15 mm in length .... *Haminoea vesicula*

51. Shell cylindrical (figs. 43, 44) with a small conic spire ... . *Acteocina* 52

– Shell not cylindrical 53

52. With strongly keeled shoulder (fig. 43); axial striations on upper $^1/_2$ of whorl; small, to about 6 mm *Acteocina harpa*

– Shoulder rounded (fig. 44); sculpture of numerous, spiral striations, reflected by brown periostracum; larger, to 22

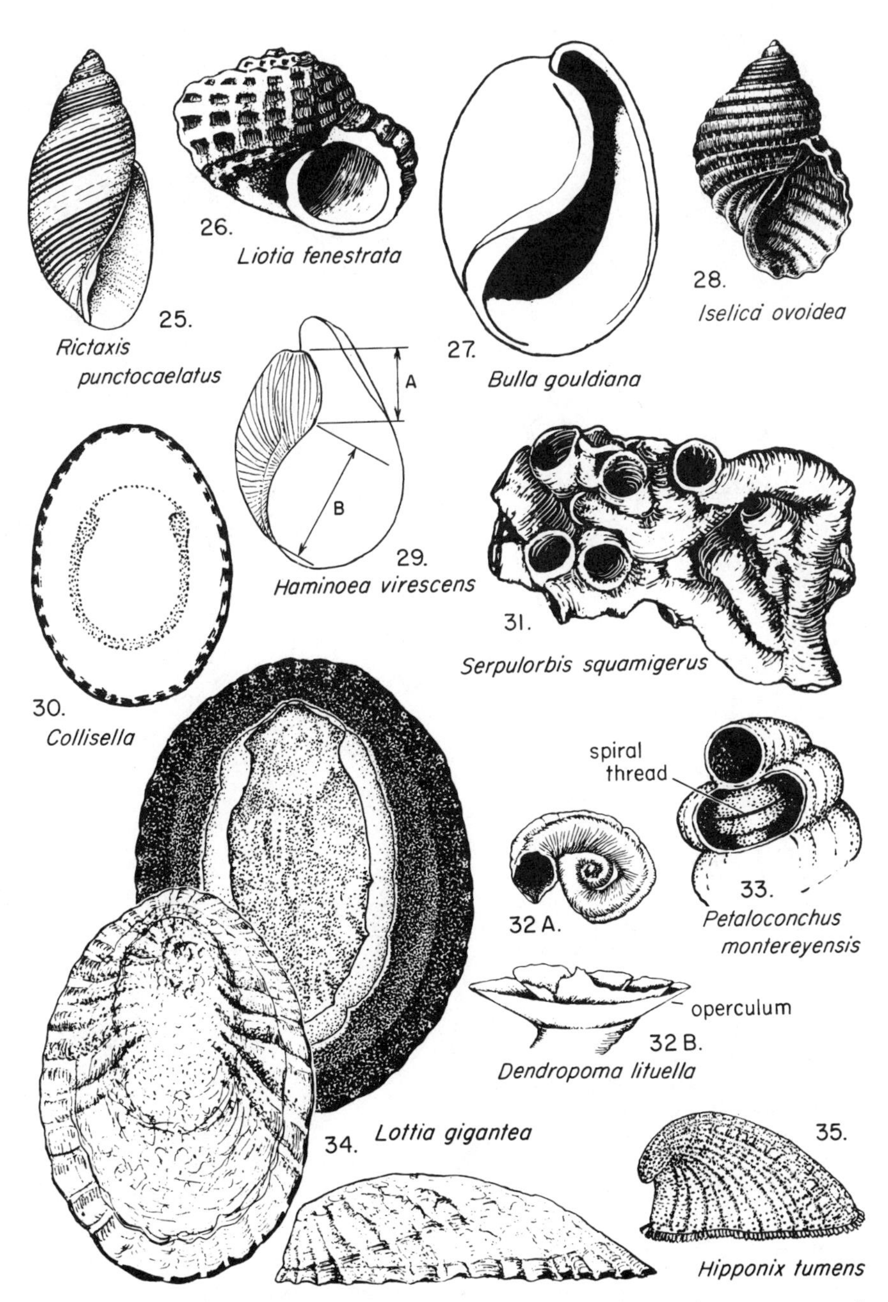

PLATE 114. **Prosobranchs** (5). 25, Stohler (by E. Reid); 26, 27, 30, Keen and Pearson, 1952; 28, Dall, 1921; 29, Marcus and Marcus, 1967; 31, 32A–B, Keen, 1963; 33, E. Reid; 34, dorsal and lateral, Hedgpeth, 1962, ventral, Stohler (by E. Reid); 35, original.

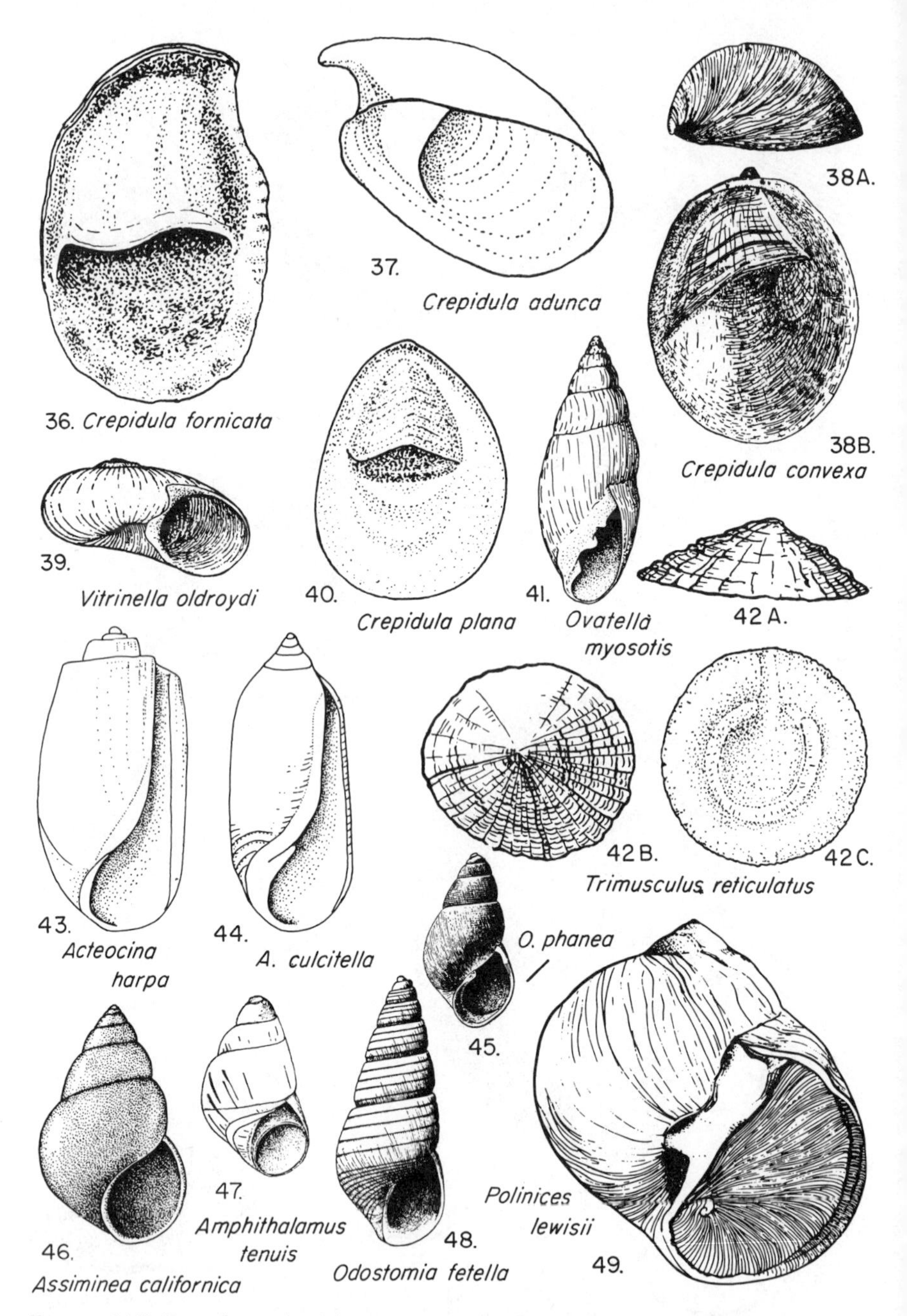

PLATE 115. **Prosobranchs** (6). 36, 40, Jacobson and Emerson, 1961; 37, original; 38A–B, Franz and Hendler, 1970; 39, Bartsch, 1907; 41, Stohler (by E. Reid); 42A–C, lateral and dorsal, Keen and Pearson, 1952, ventral, Stohler (by E. Reid); 43, 44, after Keen and Pearson, 1952, redrawn by B. Roth; 45, 48, Dall and Bartsch, 1909; 46, E. Reid from Bartsch, 1920; 47, Bartsch, 1911; 49, Hedgpeth, 1962.

mm .... *Acteocina culcitella*

53. Interior of shell pearly, showing rainbow colors 54
– Interior of shell porcelainlike, not pearly 70
54. Base of columella with denticles or small nodes 55
– Base of columella without nodes 62
55. Columellar nodes strong; operculum not calcareous, with numerous spiral lines; shell to 50 mm or more in height .... *Tegula* 56
– Nodes weak; operculum calcareous, with few spiral lines; shell less than 10 mm .... *Homalopoma* 59
56. Umbilicus covered by a callus, nearly always closed 57
– Umbilicus open 58
57. Shell purplish black to black; scaly band below suture; mature specimens with 2 teeth on columella (lower tooth occasionally worn) (fig. 57) .... *Tegula funebralis*
– Shell brown or orange brown; no scaly subsutural band; 1 tooth on columella (fig. 55) .... *Tegula brunnea*
58. Top of inner lip receding into aperture (figs. 52A, B); umbilicus defined by a strong spiral cord; brown .... *Tegula montereyi*
– Top of inner lip produced into flange on apertural side of umbilicus (figs. 51A, B); no strong spiral cord defining umbilicus; brown or gray, at times with orange, white, or brown spots on periphery .... *Tegula pulligo*
59. Globose; nearly smooth with faint, incised, spiral line (fig. 59); gray to reddish gray or brown *Homalopoma baculum*
– With numerous rounded spiral cords, or with a few prominent spiral ribs 60
60. Numerous rounded spiral cords over body whorl and base (fig. 58); often purple or red; juveniles grayish; highly variable in color .... *Homalopoma luridum*
– With a few prominent ribs 61
61. Spiral and axial ribbing; white with pink dots on spiral ribs .... *Homalopoma radiatum*
– With prominent spiral ribs only; color white, pink, or red .... *Homalopoma paucicostatum*
62. Shell medium to large (2.5–7.5 cm in height), conical; brick red; sculpture of diagonal folds and small, rounded nodes; periphery angulate or rounded base with strong spiral cords (fig. 68) .... *Astraea gibberosa*
– Shell small to medium-sized (not more than 40 mm in height) 63
63. Sculpture cancellate, of deep square pits formed by strong spiral and axial ribs; minute (height to 3 mm), white, spire

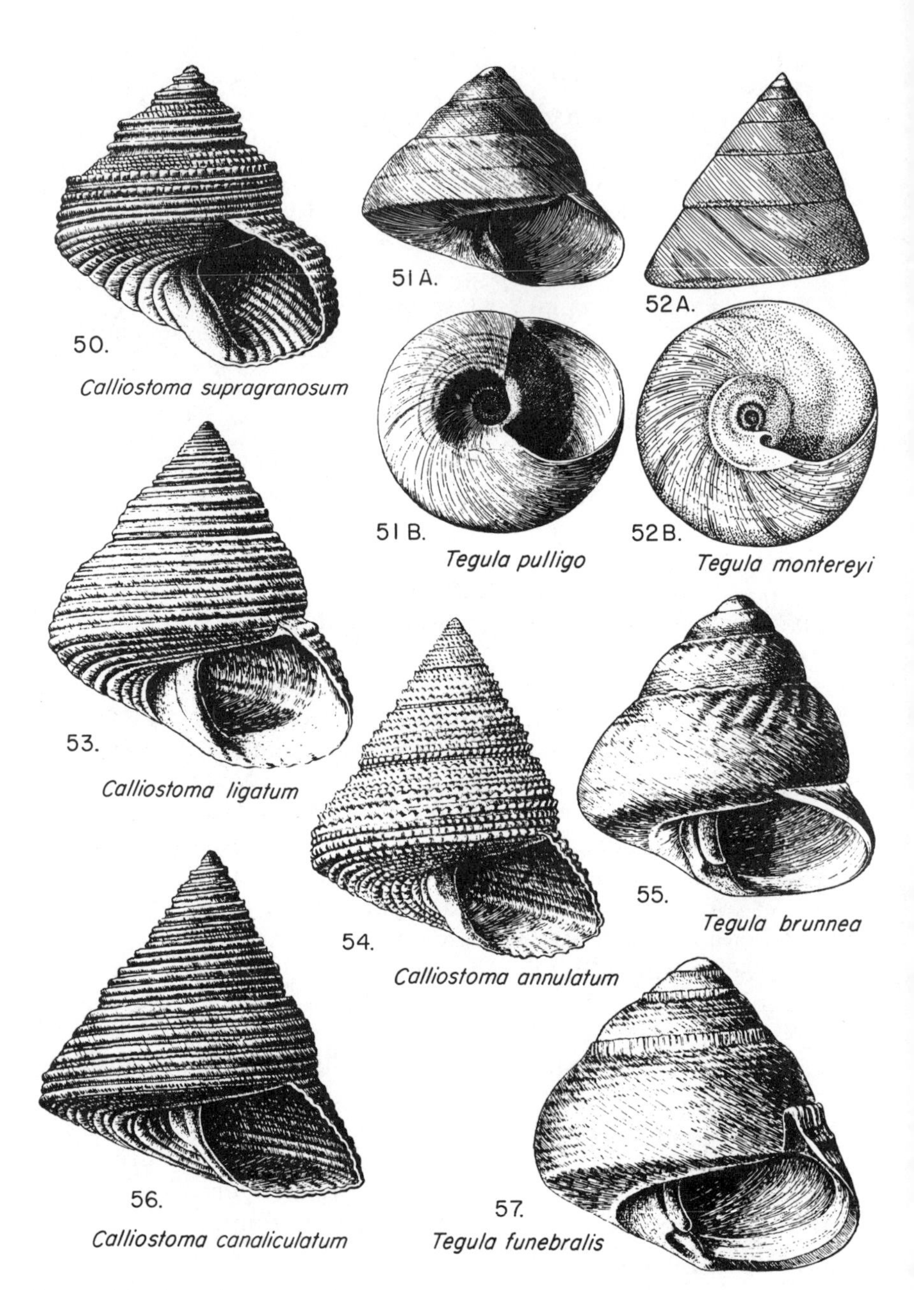

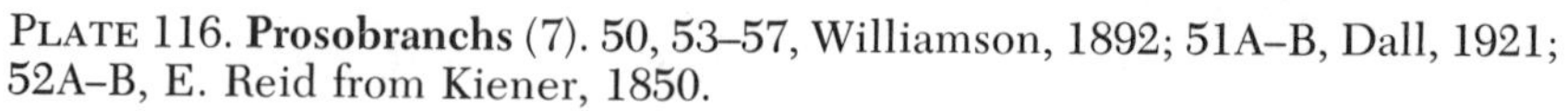

PLATE 116. **Prosobranchs** (7). 50, 53–57, Williamson, 1892; 51A–B, Dall, 1921; 52A–B, E. Reid from Kiener, 1850.

flattened, umbilicus deep, aperture circular (fig. 26) .... *Liotia fenestrata*
- Not cancellate, but with numerous spiral ribs 64
64. Without an umbilicus .... *Calliostoma* 65
- With an umbilicus 68
65. With beaded spiral cords (fig. 54); shell golden yellow to yellowish brown, with purple band adjacent to columella .... *Calliostoma annulatum*
- Color otherwise 66
66. Whorls flat-sided, yellowish tan to white or buff, with prominent revolving ridges paler in color than interspaces (fig. 56); blue stain next to columella .... *Calliostoma canaliculatum*
- Sides of whorls rounded, not flat 67
67. Shell chocolate brown, with narrow, light tan spiral ridges (fig. 53); nacre blue; height to 33 mm *Calliostoma ligatum*
- Shell yellow-brown with irregular mottling; fine revolving ridges strongly beaded on early whorls (fig. 50); height to 14 mm .... *Calliostoma supragranosum*
68. Uniformly red-brown to orange; subconical (fig. 76); height to 10 mm or more .... *Margarites salmoneus*
- Color usually variegated, grayish with bands of purple-brown, but may be solid color; to about 5 mm .. *Lirularia* 69
69. Base inflated, with a shallow, spiral channel; basal cords more numerous than *L. funiculata*, but not as strong; in open rocky low intertidal (fig. 77) .... *Lirularia succincta*
- Base without channel; basal cords strong, few; periphery rounded; on mudflats .... *Lirularia funiculata*
70. Diameter of shell greater than height 71
- Diameter of shell equal to or less than height 74
71. Lenticular (fig. 39), minute (about 2 mm diameter), white; whorls and aperture rounded; umbilicus wide, open; often eroded, with nucleus missing; under mantle of chiton *Stenoplax heathiana* and in sand .... *Vitrinella oldroydi*
- Globose; small to medium-sized (diameter generally of 6 mm or more), fragile, thin; aperture large, flaring 72
72. Lacking a distinct periostracum; generally whitish, translucent; surface smooth, or with fine spiral striae, or malleated (like hammered metal) and with transverse growth lines .... *Lamellaria* spp.
- With a distinct brown or yellow periostracum .. *Velutina* 73
73. Brown, with small spiral rows of bristles; shell globose .... *Velutina* cf. *V. velutina*
- Yellow, smooth, with minute spiral striae crossed by axial

ridges; shell depressed .... *Velutina prolongata*
74. Columella with 1 or more prominent folds 75
– Columella lacking strong conspicuous folds 78
75. Columella with 1 fold; shell with incised, pitted, spiral striations or grooves; white, with 2 spiral, gray-black bands on body whorl (fig. 25) .... *Rictaxis punctocaelatus*
– Columella with 3 or 4 folds 76
76. Three columellar folds (3rd may be weakly expressed) (fig. 41); spire elevated; color variable, brown or brown-purple to yellow; juveniles with small hairs around spire .... *Ovatella myosotis*
– Four columellar folds; spire low or concealed; white 77
77. Spire concealed; outer lip toothed within (fig. 60) .... *Granulina margaritula*
– Spire low but not concealed; outer lip smooth within .... *Cystiscus jewettii*
78. Nuclear whorls heterostrophic (fig. 66A) 79
– Nuclear whorls normal 81
79. Shell slender, with many whorls; axial and spiral sculpture various, often with low, axial ribbing (figs. 66, 66A) .... *Turbonilla* spp.
– Shell broadly ovate to conic, generally of few whorls 80
80. Shell broadly ovate, with strong spiral ribs and thin axial lamellae; outer lip lobed (fig. 28); white with brown periostracum .... *Iselica ovoidea*
– Shell sculpture otherwise; ovate to conic, may be smooth or highly ornamented; white (figs. 45, 48); see species list .... *Odostomia* spp.
81. Shell slender, elongate; numerous whorls (figs. 61, 71–75) 82
– Shell otherwise; with few whorls 86
82. Shell small, unsculptured, glassy white; aperture oval, shell moderately slender, whorls flat-sided or slightly inflated; spire often tilted slightly to one side (fig. 61) *Balcis* spp.
– With distinct axial sculpture 83
83. Shell dark brown; axial ribs low; a few rounded varices on lower whorls (fig. 72) .... *Cerithidea californica*
– Shell white; with many prominent, regular, vertical lamellae or ribs over entire shell 84
84. Base not set off by a spiral keel; axial sculpture of thin, sharp lamellae, continuous from whorl to whorl (fig. 75); fresh specimens with a characteristic purplish or brown line below sutures .... *Epitonium tinctum*
– Base set off by a spiral keel .... *Opalia* 85
85. Axial ribs acute; spiral keel strong, projecting; ribs not

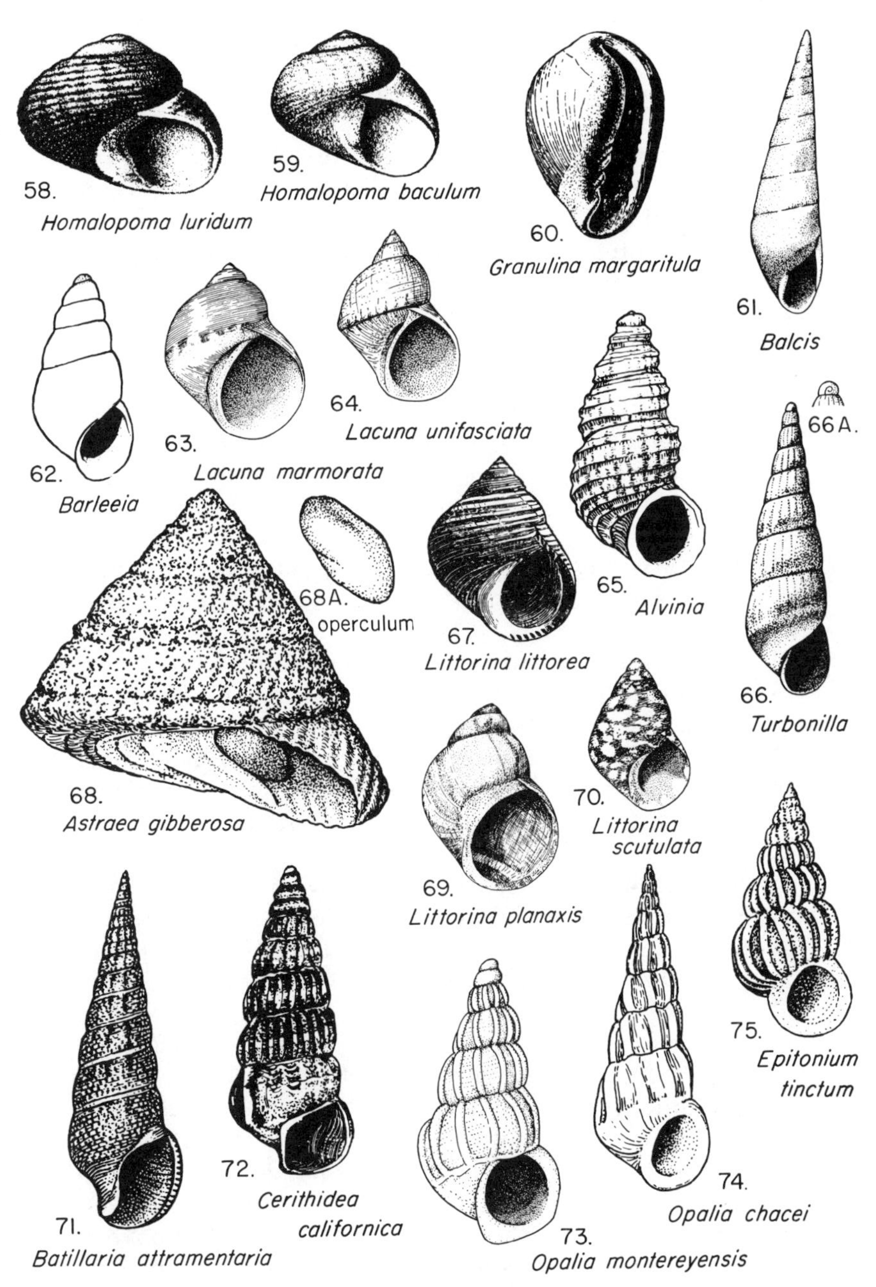

PLATE 117. **Prosobranchs** (8). 58, 59, Tryon, 1888; 60, Williamson, 1892; 61, E. Reid from Bartsch, 1917; 62, 65, Keen and Pearson, 1952; 63, 64, E. Reid from Tryon, 1887; 66, E. Reid from Bartsch, 1917; 67, Gould, 1870; 68, Stohler (by E. Reid); 70, E. Reid; 71, Keen, 1963; 72, Hedgpeth, 1962; 73, E. Reid from Berry, 1948; 74, Stohler (by E. Reid); 75, E. Reid from Tryon, 1887, and Strong, 1930; 69, from Hedgpeth in Ricketts, Calvin, and Hedgpeth: *Between Pacific Tides*, 4th ed., 1968; used with permission of Stanford University Press.

continuing over onto shell base (fig. 73) .... *Opalia montereyensis*
- Axial ribs broadly rounded; spiral keel low to obscure; about every third rib stronger, continuing over keel onto base of shell (fig. 74) .... *Opalia chacei*

86\. Adult shell minute (height generally less than 4 mm) 87
- Adult shell small to large (generally greater than 4 mm) 92

87\. Inner lip offset from columella by a shelf; (fig. 47); slightly over 1 mm in height .... *Amphithalamus tenuis*
- Inner lip continuous with columella 88

88\. Inner lip a small, thickened callus; whorls rounded, convex; shell smooth, stoutly conical (fig. 46); brownish; abundant in *Salicornia* marshes .... *Assiminea californica*
- Inner lip without expanded parietal callus; not occurring in salt marshes 89

89\. Shell with strong to weak cancellate sculpture (fig. 65); white or brown .... *Alvinia* spp.
- Shell smooth 90

90\. Shell mottled red and white, occasionally with brown blotches on periphery; thin, green periostracum; sea-green calcareous operculum; shell ovate ......*Tricolia pulloides*
- Color otherwise; operculum not calcareous, with peglike, internal projection; shell ovate-conic (fig. 62) ....*Barleeia* 91

91\. Sharp basal keel; whorls flat-sided; shell shiny, dark red brown; to about 4 mm .... *Barleeia acuta*
- Slight basal angulation; whorls somewhat rounded; color variable, translucent yellow-white to red or dark brown; to about 3 mm .... *Barleeia haliotiphila*

92\. Shell mottled red and white, occasionally with brown blotches on periphery; thin, green periostracum; sea-green calcareous operculum .... *Tricolia pulloides*
- Color pattern, if present, not markedly irregular 93

93\. Large (adult 12.5–15 cm), heavy, globose shell with a large, broadly rounded body whorl; umbilicus deep, upper part covered by a wide columellar callus; shoulder tabulate with shallow sulcus below (fig. 49) .... *Polinices lewisii*
- Shell less than 1 cm in height; umbilicus a narrow groove or lacking entirely; columella shelf-like, without a thickened callus 94

94\. Columellar groove present (figs. 63, 64); shell thin; animal with a pair of posterior metapodial tentacles ......*Lacuna* 95
- No columellar groove; shell solid; animal without metapodial tentacles .... *Littorina* 97

95\. Shell high-spired, conical; brown line (at times broken) on periphery (defined by a raised ridge); columellar groove narrow (fig. 64) .... *Lacuna unifasciata*

– Shell low-spired, not conical; periphery rounded to carinate, lacking brown line 96

96. Shell varying from depressed to elliptical, surface smooth with fine spiral striae; generally brownish, marbled with white, especially at periphery; often with white stripe inside base of aperture; columellar groove generally narrow, broad in elliptical shells (fig. 63); height to 6–7mm .... *Lacuna marmorata*

– Generally globose; surface wrinkled with fine, wavy, spiral striae; solid brown color, generally lacking a white band in aperture; columellar groove broad; to 12 mm .... *Lacuna porrecta*

97. Shell stout, robust, somewhat globose; columella broad, flat, polished; narrow white band inside aperture (fig. 69); color gray brown, often with white maculations .... *Littorina planaxis*

– Shell conical; columella thin; aperture without white band (though external color bands may show through shell); color pattern more or less dotted or checkered (fig. 70) .... *Littorina scutulata*

98. Aperture slotlike, running full length of shell; shell dark purple brown; about the shape and size of a coffee bean; numerous transverse ribs extending from a dorsal longitudinal furrow to aperture .... *Trivia californiana*

– Length of aperture less than total length of shell 99

99. Shell obconic (inversely conical), spire dome-shaped; dull gray, tan, or gray brown with fine spiral markings of brown under a heavy, dark brown periostracum .... *Conus californicus*

– Shell not obconic 100

100. Aperture with a posterior siphonal notch ("turrid notch") at or near suture (fig. 87) 101

– Aperture entire at posterior end 103

101. With light-colored spiral nodes; rusty brown or yellow brown to blackish .... *Pseudomelatoma torosa*

– Without light-colored axial nodes 102

102. Posterior notch deep, at suture; with about 15 strong axial ridges per whorl, crossed by spiral ribs (fig. 87) .... *Clathurella canfieldi*

– Posterior notch shallow, near suture; axial and spiral sculpture producing rectangular cancellations .... *Clathromangelia interfossa*

103. Aperture long, nearly 7/8 length of shell 104
– Aperture less than 3/4 length of shell, not including siphonal canal 106
104. Shell white, ovate; mature shells less than 4 mm; aperture with a distinct anterior notch visible in dorsal (back) view .... *Granula subtrigona*
– Shell red to gray dorsally, glossy, inverted pear-shaped; mature shells to 16 mm .... *Erato* 105
105. Outer lip with 7–10 denticles; 5–8 denticles on inner lip; color purple red dorsally; to about 16 mm *Erato vitellina*
– Outer lip with about 12 denticles; only traces of denticles on inner lip; gray to orange-brown or reddish brown dorsally; to about 8 mm .... *Erato columbella*
106. **Shell turreted (with 8 or more whorls, figs. 71, 80, 88) ... 107**
– **Shell not turreted (6 or fewer whorls, figs. 82, 91, 99) 115**
107. With irregularly spaced, axial varices fading out on lower whorls; generally lacking prominent varices at base (fig. 71) .... *Batillaria attramentaria*
– Sculpture beaded or of spiral ribs 108
108. Anterior canal short, poorly differentiated .... *Bittium* 109
– Canal well developed 112
109. Shell with spiral ribs separated by prominent grooves (fig. 88); color dirty white with axial or spiral brown bands .... *Bittium eschrichtii*
– With distinct axial and spiral sculpture 110
110. Sculpture of square pits formed by strong axial sculpture; 2 strong spiral cords per whorl, 3 rounded cords at base .... *Bittium interfossa*
– Sculpture not of prominent square pits 111
111. Early whorls cancellate, later whorls with axial sculpture subdued and spiral cords flattened .. *Bittium attenuatum*
– Shell with 3 spiral cords per whorl on early whorls, smaller accessory cords on final whorl; base with spiral cords .... *Bittium purpureum*
112. Shell sinistral, small (to about 6 mm), slightly convex, with beaded sculpture; canal closed and turned upward (fig. 81) .... *Triphora* spp.
– Shell dextral 113
113. Sculpture of raised spiral cords with minute axial threads in interspaces; brown, flat-sided ..... *Seila montereyensis*
– Sculpture cancellate, whorls somewhat inflated 114
114. Whorls convex, sutures deeply impressed; with weak axial ribs, 4 spiral ribs per whorl bearing elongate beads; base concave .... *Metaxia convexa*

- Whorls nearly flat, sutures not deeply impressed; sculpture beaded with strong axial and spiral ribbing (fig. 80) .... *Cerithiopsis* spp.

115. With a distinct revolving furrow around base 116
- Lacking distinct revolving furrow 117

116. With orange callus spreading over front of body whorl; shell relatively large (to 47 mm), broad, periphery rounded to carinate; axial sculpture of widely spaced folds, limited to upper part of body whorl (fig. 95) .. *Nassarius fossatus*
- Without orange callus; shell small (to 20 mm), generally slender, periphery rounded; axial sculpture of rounded folds (varying from prominent to poorly developed or absent) continuous over body whorl (fig. 90) ....*Nassarius mendicus*

117. Columella with 1 or more folds 118
- Columella without folds 127

118. With incised spiral grooves restricted to base of shell; columellar fold may be weak ....*Mitrella* 119
- Spiral sculpture not restricted to base of shell 121

119. Shell small, slender, with a chevron pattern of thin revolving, brown lines on a yellow-brown background (not an irregular pattern of wavy longitudinal lines) .... *Mitrella aurantiaca*
- Color tan to yellow-brown or variously mottled 120

120. Shoulder of body whorl varying from smooth to strongly keeled (keel usually lighter in color than rest of shell); whorls somewhat inflated; outer lip sinuous (figs. 82A, B); periostracum smooth; color usually yellow-brown to dark brown, at times with white and darker brown mottling ....*Mitrella carinata*
- Slender, whorls nearly flat-sided (fig. 79); periostracum forming thin, projecting, axial blades in living animals; color usually tan, sometimes darker, may show fine white dots ....*Mitrella tuberosa*

121. Folds present on entire length of columella (fig. 84) .... *Amphissa columbiana*
- Folds limited to upper portion or to base of columella 122

122. Folds on upper portion of columella; sculpture of prominent spiral cords and axial ribs which are strongest in early whorls 123
- Columellar folds at base 124

123. Axial and spiral ribs equally spaced, strongly beaded at intersections, producing squarish cancellations (fig. 83) .... *Cymakra aspera*

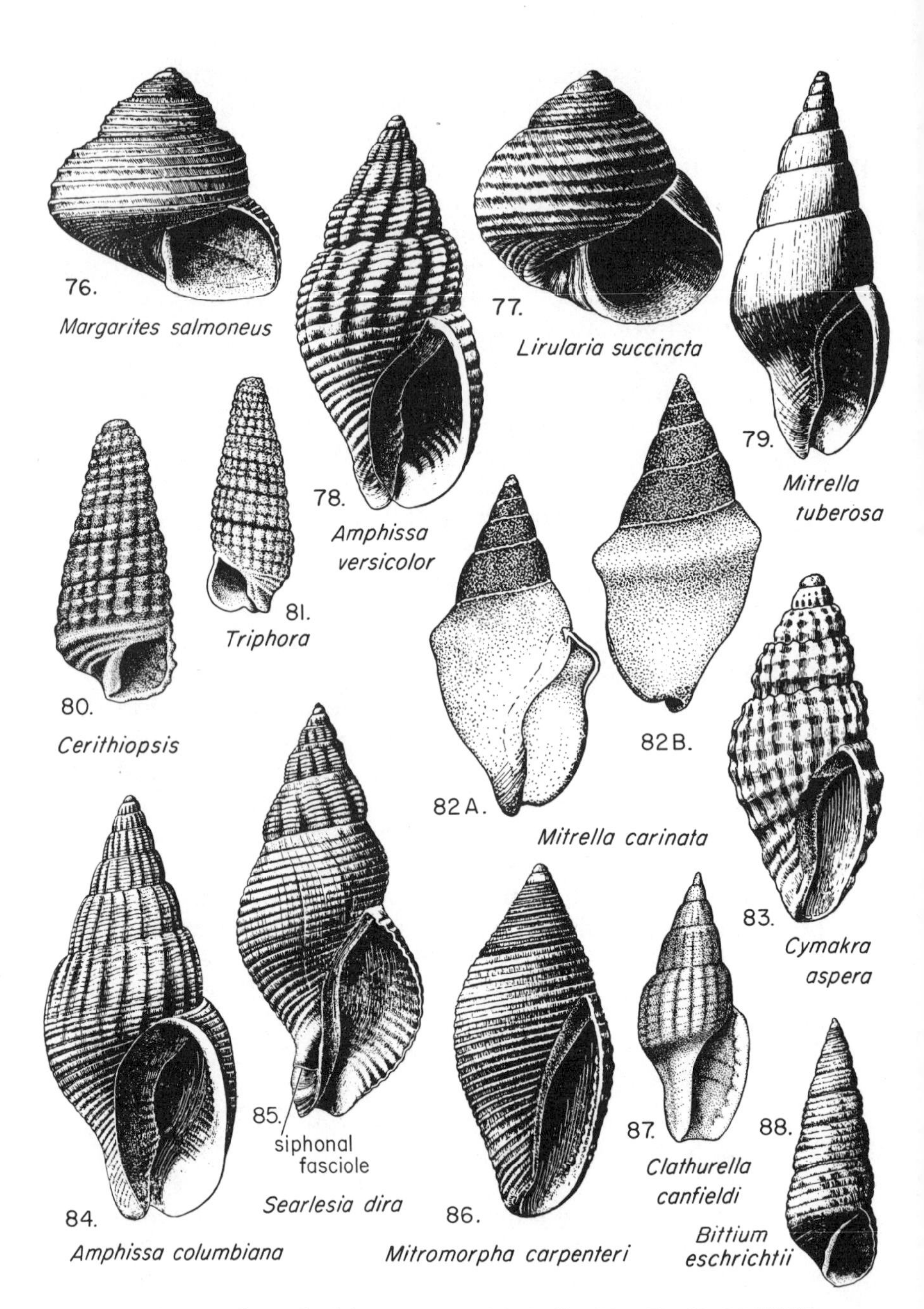

PLATE 118. **Prosobranchs** (9). 76–78, 84–85, Dall, 1921; 79, 83, 86, Williamson, 1892; 80, E. Reid from Bartsch, 1911; 81, E. Reid from Bartsch, 1907; 82A,B, Stohler (by E. Reid); 87, E. Reid from Dall, 1871; 88, E. Reid from Bartsch, 1911.

- Spiral sculpture strong, crossed by axial furrows; purple brown mottled with white .... *Cymakra gracilior*

124. Shell sculpture of revolving, weakly beaded lines crossed by growth lines and oblique folds; apex often eroded; aperture black-glazed (fig. 91); shell dark brown to black, often with adherent detritus and algae .... *Nassarius obsoletus*

- Shell unsculptured, polished .... *Olivella* 125

125. Shell to about 30 mm in length, broad, and robust; variously colored, from almost all white to a black-gray, often violet at base, offset with a dark line; columellar callus relatively strong; fold at base of columella often with several incised spiral lines (fig. 99) .... *Olivella biplicata*

- Shell smaller (to 19 mm), either slender or stout; with or without longitudinal, zigzag, brownish stripes 126

126. Shell stout and chunky (fig. 92); often with brown, longitudinal, zigzag lines on a brownish buff, gray, or olive gray background; occasionally with a red-brown spot beside fold at base of columella .... *Olivella pycna*

- Shell oblong and slender (fig. 98); may have brown longitudinal lines, color generally gray-brown to tan with faint purplish brown maculations near suture .. *Olivella baetica*

127. Anterior canal short to almost obsolete; outer lip generally rounding smoothly to anterior end of shell 128

- Anterior canal moderate to long; outer lip sinuous (curving inward at anterior end) 136

128. Aperture long, narrow, more than 1/2 length of shell; sculpture of even, flat-topped, narrow spiral cords; anterior end truncate; outer lip denticulate (fig. 86); shell dark brown .... *Mitromorpha carpenteri*

- Aperture (not including anterior canal) 1/2 length of shell or less 129

129. Outer lip with a projecting tooth near base (young specimens may lack tooth); revolving, interrupted, brown bands .... *Acanthina* 130

- Outer lip smooth or evenly toothed 131

130. Prominent keel at shoulder; spire produced (fig. 97) .... *Acanthina spirata*

- Shoulder rounded or weakly angulate; spire low .... *Acanthina punctulata*

131. Base and spire with similar sculpture (spiral or with scaly axial lamellae) .... *Nucella* 132

- Base of shell and spire differently sculptured 134

132. Shell with smooth, often closely set, spiral ridges; interspaces with minute axial scales; somewhat elongate, nar-

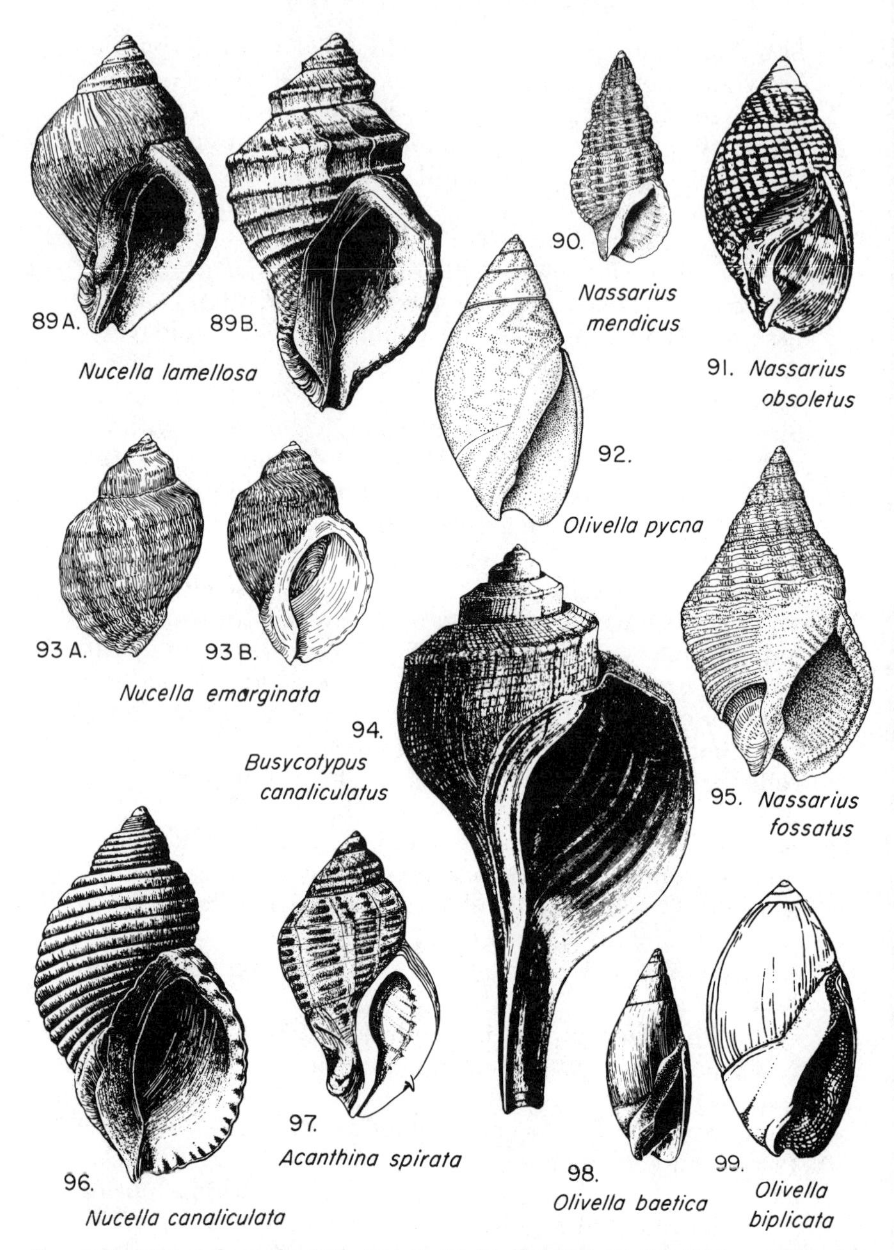

PLATE 119. **Prosobranchs** (10). 89A,B, 96, Dall, 1915; 91, 94, Pilsbry, 1891; 90, E. Reid; 92, E. Reid from Berry, 1935; 93A,B, E. Reid; 95, E. Reid, based on Stohler, *Veliger;* 98, Williamson, 1892; 99, Hedgpeth, 1962; 97 from Hedgpeth in Ricketts, Calvin, and Hedgpeth: *Between Pacific Tides*, 4th ed., 1968; used with permission of Stanford University Press.

row umbilical chink (fig. 96) .... *Nucella canaliculata*

– Shell smooth, or with nodulose spiral cords, or with prominent axial lamellae 133

133. Shell with irregularly nodulose, often well-separated, spiral cords (some populations with shells nearly smooth); columella excavated; anterior canal short; umbilicus closed (figs. 93A, B); common in high intertidal .... *Nucella emarginata*

– Sculpture various: nearly smooth, or with prominent axial lamellae, or with spiral cords and weaker, irregular, axial swellings; anterior canal moderately long; umbilicus small, sometimes closed (figs. 89A, B); low intertidal .... *Nucella lamellosa*

134. Body whorl with distinct spiral sculpture only; low, rounded axial ribs on spire; columella arched, glossy; canal short, twisted (fig. 85); color dull gray or brownish purple; sculpture may be obscured by growths of purple coralline algae .... *Searlesia dira*

– Body whorl with both spiral and axial sculpture above periphery; lower 1/2 of body whorl with only spiral lines; columellar area with a glossy shield, which may bear a few denticles (figs. 78, 84) .... *Amphissa* 135

135. Body whorl with about 12 sinuous axial ribs (fig. 78); shell solid; color variable, mottled gray, red brown, yellow, or white with darker spiral banding; height to 17 mm .... *Amphissa versicolor*

– Axial ribs finer and more numerous, nearly vertical; thin-shelled (fig. 84); yellow-brown with mauve mottling; larger, may be 25 mm or more in height . . *Amphissa columbiana*

136. Three or more varices per whorl; canal restricted; often with projecting tooth on outer lip near base. .*Ceratostoma* 137

– Shell nearly smooth or with axial ribs, not true varices ... 138

137. Three prominent varices per whorl; shell surface smooth, lustrous; in low intertidal of outer coast or semiprotected situations .... *Ceratostoma foliatum*

– Number of varices inconstant; shell surface dull, texture chalky; sculpture of alternating large and small spiral cords; on Japanese oyster beds in bays . .*Ceratostoma inornatum*

138. No axial ribs; outer lip smooth; shell relatively thin; large, height to 17.5 cm; sutures strongly channeled; shoulders keeled; long, slightly curved canal; with a yellow-brown, felt-like, hairy periostracum, often partly worn off; (fig. 94) .... *Busycotypus canaliculatus*

– With axial ribs; outer lip toothed or striate within 139

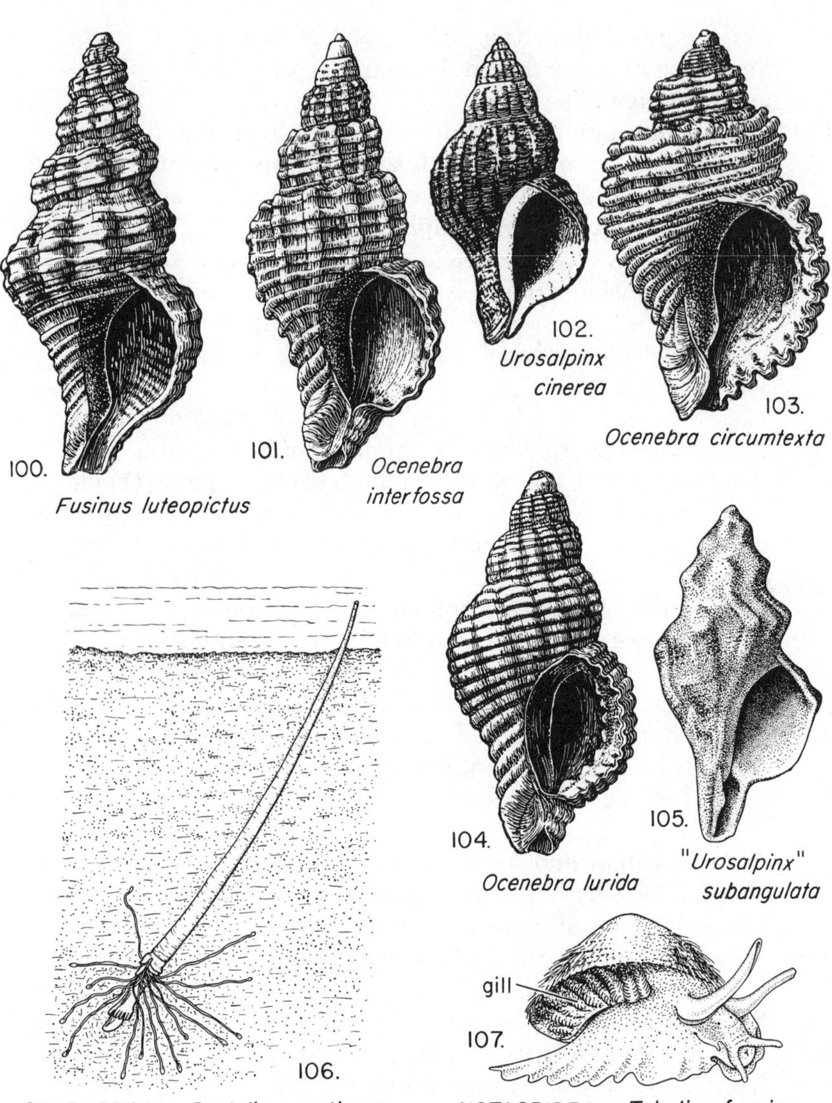

PLATE 120. **Prosobranchs** (11), **Scaphopoda, Notaspidea:** 100, 103, 104, Williamson, 1892; 101, Dall, 1921; 102, Keen and Pearson, 1952; 105, E. Reid from Stearns, 1873; 106, **Scaphopoda,** *Dentalium pretiosum,* feeding, captacula (majority not shown) extending actively into mud, sketched by R. I. Smith at Friday Harbor; 107, **Notaspidea,** *Tylodina fungina,* E. Reid from MacFarland, 1966.

139. Siphonal fasciole (a raised, spiral band adjacent to, and formed by, successive margins of the anterior canal) weak or absent 140
- Siphonal fasciole strong (figs. 101, 103, 104) 141

140. Numerous fine spiral cords, in older specimens most pronounced between axial ribs; canal short, constricted (fig. 102); shell color variable, gray, yellow-brown, aperture purplish; in bays, on oysters and barnacles *Urosalpinx cinerea*
- Strong spiral cords, not particularly numerous, continuous over axial ribs; canal well developed, broad (fig. 100); shell dark brown, white spiral band at periphery; open coast, low intertidal, under algae .... *Fusinus luteopictus*

141. Axial sculpture of 7–9 strong ribs, sharply angled at shoulder of whorl; spiral sculpture reduced to fine threads (fig. 105); whitish, often with brown spiral line at periphery .... *"Urosalpinx" subangulata*
- Spiral sculpture as strong as or stronger than axial sculpture .... *Ocenebra* 142

142. Sculpture cancellate 143
- Spiral sculpture much more prominent than axial sculpture 144

143. Spiral ribs all of nearly the same size; brown, often with yellow rib at periphery .... *Ocenebra atropurpurea*
- Spiral ribs alternately large and small (fig. 101); grayish brown .... *Ocenebra interfossa*

144. Color dull white with interrupted, broad, brown band at periphery (fig. 103) .... *Ocenebra circumtexta*
- Color otherwise, variously white, brown, yellow, or purplish, lacking peripheral brown band 145

145. With varix just behind outer lip .... *Ocenebra gracillima*
- Without varix behind outer lip (fig. 104) *Ocenebra lurida*

LIST OF SHELLED GASTROPODS

Subclass **PROSOBRANCHIA**

Order **Archaeogastropoda**

HALIOTIDAE See Cox, 1962, Calif. Dept. Fish Game, Fish. Bull. 118 (review); Owen, McLean, and Meyer, 1971, Bull. Los Angeles Co. Mus. no. 9 (hybridization).

Haliotis cracherodii Leach, 1814. Black abalone; mid to low intertidal; in crevices or under large rocks.

Haliotis kamtschatkana Jonas, 1845. Specimens from south of Marin County may show characters approaching the subspecies *H. k. assimilis* Dall, 1878 found south of Point Conception, which has a higher, more rounded, less corrugated shell. Sublittoral, occasionally washed inshore.

Haliotis rufescens Swainson, 1822. Red abalone; primarily sublittoral, intertidal in areas of considerable wave action. See Olsen, 1968, Biol. Bull. 134: 139–147 (banding patterns).

Haliotis walallensis Stearns, 1899.

SCISSURELLIDAE

Sinezona rimuloides (Carpenter, 1865). Minute (0.8 mm height); interstitial in sand and gravel; recovered by floating finely screened, dry, bottom samples in heavy liquids. See McLean, 1967, Veliger 9: 404–419 (taxonomy, distribution).

FISSURELLIDAE

* *Diodora arnoldi* McLean, 1966. Sublittoral, occasionally washed ashore; differs from *D. aspera* in smaller size, nearly parallel sides, and oval rather than round apical hole.

Diodora aspera (Rathke, 1833). Low intertidal zone under rocks, in crevices; diet includes encrusting bryozoans (Gonor 1968, Veliger 11: 134); commensal polychaete *Arctonoe vittata* often in mantle cavity.

Fissurella volcano Reeve, 1849. Rocky intertidal zone, on and under coralline-encrusted rocks.

Megatebennus bimaculatus (Dall, 1871). Low intertidal zone on compound ascidians.

Megathura crenulata (Sowerby, 1825). Giant keyhole limpet; low intertidal zone to sublittoral, in rocky areas; pea-crab *Opisthopus transversus* among commensals.

ACMAEIDAE

The following species, except *Lottia*, were formerly placed in the broad genus *Acmaea;* however, widely differing anatomical and shell characters require that most of these be placed in other genera (see McLean, 1969; Golikov and Kussakin 1972, Malacologia 11: 287–294). See series of papers in Veliger 11, Suppl., 112 pp. (1968); Fritchman 1961–1962, Veliger 3–4 (reproductive cycles); Wolcott 1973, Biol. Bull. 145: 389–422 (physiological ecology, zonation).

Acmaea mitra Rathke, 1833. Low intertidal zone to largely sublittoral; often encrusted with pink coralline algae *Lithothamnium* and *Lithophyllum*, upon which limpet feeds; frequently in abandoned sea-urchin holes; shells common in beach drift.

Acmaea rosacea Carpenter, 1864. Not an *Acmaea* in the strict sense (J. McLean). Largely sublittoral or in low intertidal zone, under small, smooth stones or on loose clam shells; feeds on coralline algae.

* Not in key.

Collisella asmi (Middendorff, 1847). In central California occurring on *Tegula funebralis*, feeding on microscopic epizoic algae; not the only and rarely the most common limpet on *Tegula*. See Alleman 1968, Veliger 11, Suppl., 61–63.

Collisella digitalis (Rathke, 1833). High intertidal, often on vertical rock faces, secreting mucous sheet between shell margin and surface; also on stalked barnacle *Pollicipes*. See Giesel, 1969, Ecology 50: 1084–1087 (growth) and 1970, Evolution 24: 98–119; Haven, 1973 (below, under *C. scabra*); also various papers in Veliger 13–15, on ecology and biology.

Collisella instabilis (Gould, 1846). In central California occurring on and feeding on stipes of alga *Laminaria dentigera*.

Collisella limatula (Carpenter, 1864). Elevated form in bays formerly called subspecies *moerchii* Dall, 1878; feeds on coralline and other encrusting red algae; see Seapy 1966, Veliger 8: 300–310 (reproduction and growth).

Collisella ochracea (Dall, 1871). At lowest tides under small, smooth stones and rocks.

Collisella pelta (Rathke, 1833). A eurytopic, widely varying species associated with various algae (*Egregia, Postelsia, Laminaria, Endocladia*), mussel beds, *Pollicipes*, under rocks, and in other habitats. Specimens occurring on *Egregia* and resembling *N. insessa* may move onto rocks when half grown and change shape, sculpture, and color pattern (J. McLean). See Jobe, 1968, Veliger 11 Suppl. 69–72 which, however, does not cover all variations.

Collisella scabra (Gould, 1846). High intertidal, often on horizontal rock surfaces; exhibits homing behavior; ecology and biology: see Haven, 1971, Veliger 13: 231–248, 1973, Ecol. 54: 143–151 and Sutherland, 1970, Ecol. Monogr. 40: 169–188.

Collisella strigatella (Carpenter, 1864) (=*Acmaea paradigitalis* Fritchman, 1960). This is the *A. pelta* x *A. digitalis* "hybrid" of previous editions of this manual, and of Stephenson and Stephenson (1972, see general bibliography). Mid to upper intertidal on partially protected, sloping surfaces; also in some areas on *Tegula funebralis* (J. McLean). See Fritchman, 1960, Veliger 2: 53–57.

Collisella triangularis (Carpenter, 1864). Radula with uncini (rudimentary marginal teeth) and thus assignable to *Collisella* (M. L. Jones, J. McLean). Low intertidal to sublittoral, on coralline algae (which often covers shell) and occasionally on coralline-encrusted *Tegula brunnea*.

Lottia gigantea Sowerby, 1834. The owl limpet; mid-intertidal in areas of heavy wave action. See Abbott, 1956, Nautilus 69: 79–87 (water circulation); Stimson, 1970, Ecology 51: 113–118 (territorial behavior), 1973, 54: 1020–1030.

Notoacmea fenestrata (Reeve, 1855) (=*N. f. cribraria* Carpenter, 1866). *Notoacmea* is the correct spelling. On smooth rocks and boulders set in loose sand, gravel, or mud, moving to base of rocks at low tide.

Notoacmea insessa (Hinds, 1842). On stipes of *Egregia menziesii* (in central California); feeds on the kelp, forming a conspicuous scar. See Proctor, 1968, Diss. Abst. 29B: 2305–2306 and Black, 1972, Diss. Abst. Intern. 33B: 2026.

Notoacmea paleacea (Gould, 1853). Occurs on surfgrass *Phyllospadix torreyi* in northern California; see Yonge, 1962, Veliger 4: 119–123.

Notoacmea persona (Rathke, 1833). High intertidal, nocturnal, found in sheltered, shaded areas, such as cave roofs and deep crevices; see Kenny, 1969, Veliger 11: 336–339 (growth).

Notoacmea scutum (Rathke, 1833) (=*Acmaea testudinalis scutum*). Rocky shores, frequently with small growths of the algae *Ulva* and *Enteromorpha* on shell.

TROCHIDAE

Calliostoma annulatum (Lightfoot, 1786). With *C. canaliculatum* and *C. ligatum* on offshore *Macrocystis* stands and in low rocky intertidal zone; little is known of the ecology and biology of the Calliostomas.

Calliostoma canaliculatum (Lightfoot, 1786).

Calliostoma ligatum (Gould, 1849) (=*C. costatum* Martyn, 1784).

Calliostoma supragranosum Carpenter, 1864 (=*C. splendens* Carpenter, 1864). Not uncommon, generally sublittoral, but shells occasionally washed ashore and occupied by hermit crabs.

Lirularia funiculata (Carpenter, 1864). Mud flats, on hard substrates, algae, *Zostera*.

* *Lirularia parcipicta* (Carpenter, 1864) (=*Margarites parcipictus*). Uncommon intertidally; sublittoral among rocks, in gravel, among algae.

Lirularia succincta (Carpenter, 1864) (=*Margarites succinctus*). Low intertidal zone; abundant on gravel, under loose rocks.

Margarites salmoneus (Carpenter, 1864). On and under surfaces of rocks, low intertidal zone; may be a southern subspecies of *M. pupillus* (Gould, 1849) (J. McLean).

Tegula brunnea (Philippi, 1848). Brown turban; occurs lower than *T. funebralis* and on offshore kelp beds near surface.

Tegula funebralis (A. Adams, 1855). Black turban; midtide levels, avoiding exposed outer-coast situations; occasional specimens are umbilicate; see Veliger 6, Suppl. 82 pp., (1964) for papers on ecology, biology; Frank 1965, Growth 29: 395–403 (growth); Paine 1971, Limnol. Oceanogr. 16: 86–98 (population, energy flow).

Tegula montereyi (Kiener, 1850). Low intertidal zone and on offshore kelp beds.

Tegula pulligo (Gmelin, 1791). Uncommon in low intertidal zone and on offshore kelp beds. The biology of the Tegulas, with the exception of *T. funebralis*, has received little attention.

LIOTIIDAE

Liotia fenestrata Carpenter, 1864. In gravel under rocks.

TURBINIDAE

Astraea gibberosa (Dillwyn, 1817) [=*A. inaequalis* (Martyn, 1784)]. Low intertidal zone to offshore depths on rocky surfaces.

Homalopoma baculum (Carpenter, 1864).

Homalopoma luridum (Dall, 1885) [=*H. carpenteri* (Pilsbry, 1888)].

Homalopoma paucicostatum (Dall, 1871). Under rocks; offshore under kelp in gravel and shell bottoms.

Homalopoma radiatum (Dall, 1918) [=*H. fenestratum* (Dall, 1919), according to P. LaFollette and J. H. McLean]. Found with *H. paucicostatum* from low intertidal to sublittoral, under rocks; see note in McLean (1969, p. 24).

PHASIANELLIDAE

Tricolia pulloides (Carpenter, 1865). In gravel, under rocks, or associated with surfgrass or algae.

Order **Mesogastropoda**

LACUNIDAE Local species have been identified as *Lacuna vincta* (Montagu, 1803) (=*L. solidula* Lovèn and *L. carinata* Gould) which occurs from Puget Sound north.

Lacuna marmorata Dall, 1919. Common; intertidal on rocks, algae, on surfgrass *Phyllospadix*.

Lacuna porrecta Carpenter, 1864. On algae and eelgrass.

Lacuna unifasciata Carpenter, 1857. Common, generally Monterey Bay south, in kelp beds, eelgrass, algae.

LITTORINIDAE See North, 1954, Biol. Bull. 106: 185–197 (biology, natural history); Bock and Johnson, 1967, Veliger 10: 42–54 (zonation), and Veliger 7(2) for several papers.

* *Littorina littorea* (Linnaeus, 1758). Globose, solid, brown-gray to black shell (fig. 67), introduced from Atlantic into California bays.

Littorina planaxis Philippi, 1847. In splash zone, higher than *L. scutulata;* often attached to rock surface by a mucous film.

Littorina scutulata Gould, 1849. In upper rocky intertidal, lower than *L. planaxis;* also in somewhat reduced salinities of bays.

HYDROBIIDAE

* *Tryonia imitator* (Pilsbry, 1899). Small hydrobiid now restricted to only a few brackish-water localities in central California; a victim of extensive estuarine modification and destruction. See Taylor, 1966, Malacologia 4: 53, and Veliger 9: 197.

RISSOIDAE

Alvinia spp. In older literature as *Alvania.* A number of species are common locally in gravel, sand, and under rocks. See McLean, 1969; Bartsch, 1911, Proc. U.S. Nat. Mus. 41: 333–362 (taxonomy).

Amphithalamus tenuis Bartsch, 1911. In sand and gravel under algae.

Barleeia acuta (Carpenter, 1864) (=*Diala acuta* and *B. dalli* Bartsch, 1920).

Barleeia haliotiphila Carpenter, 1864 (=*B. oldroydi* Bartsch, 1920). Common among algae, rocks, gravel, sand; also in kelp holdfasts, and reported from high intertidal *Endocladia-Balanus* zone.

* *Barleeia marmorea* (Carpenter, 1864).

* *Barleeia subtenuis* Carpenter, 1864 (=*B. sanjuanensis* Bartsch, 1920). Broader and with a larger aperture than *B. haliotiphila.*

ASSIMINEIDAE

Assiminea californica (Tryon, 1865) [=*A. translucens* (Carpenter, 1866)]. Abundant in *Salicornia* marshes on mud, under debris, boards, etc.

RISSOINIDAE

* *Rissoina* spp. Generally sublittoral; minute, white, axially ribbed with flaring aperture and thickened outer lip. See Bartsch, 1915, Proc. U.S. Nat. Mus. 49: 33–62 (taxonomy).

VITRINELLIDAE

Vitrinella oldroydi Bartsch, 1907. Commensal in mantle cavity of the chiton *Stenoplax heathiana* (which occurs on smooth boulders set in sand, low intertidal); also in sand, gravel.

OMALOGYRIDAE

Omalogyra sp. Minute, in low rocky intertidal on algae on which egg capsules are deposited; see Fretter, 1948. J. Mar. Biol. Assoc. U.K. 27: 597–632 (on *O. atomus*).

CAECIDAE

Caecum californicum Dall, 1885. Interstitial in sand, gravel; on eelgrass.

VERMETIDAE

The tubes of vermetid snails are distinguished from the tubes of serpulid worms by several characteristics: vermetids have a three-layered shell, are glossy and white (often tinged with brown) within, and begin with a spirally coiled embryonic shell; tube worms have a two-layered shell, are dull and lusterless within, and begin with a single noncoiled tubular chamber. See Keen, 1961, Bull. Brit. Mus. (Nat. Hist.) 7: 181–213 (taxonomy).

Dendropoma lituella (Mörch, 1861) (=*Spiroglyphus lituellus*). Often found embedded in abalone shells; also on other shells and rocks.

Dendropoma rastrum (Mörch, 1861). May occur in clusters on soft rock or on shells, such as abalones. Has been confused with *Serpulorbis squamigerus* with which it may occur in the same cluster (M. Keen). The sculpture is similar to that of *S. squamigerus*, but *D. rastrum* possesses an operculum and shows slight corrosion where one tube crosses another; this is difficult to see without practice.

Petaloconchus montereyensis Dall, 1919. Under rocks in low intertidal in areas of heavy but broken wave action; possibly unique among gastropods in periodic production of a new, and molting of old, operculum. See Hadfield, 1970, Veliger 12: 301–309 (anatomy, ecology, biology).

Serpulorbis squamigerus (Carpenter, 1857) (= *Aletes squamigerus*). Twisted masses found south of Point Conception, generally found only as individuals in central California; on rocks, shells, pilings. See Hadfield, above.

POTAMIDIDAE

Batillaria attramentaria (Sowerby, 1855) [=*B. cumingi* (Crosse, 1862)]. *B. zonalis* (Bruguière, 1792) a name previously used on our coast for this introduced Japanese snail, is a different species which has not been introduced. *B. multiformis* (Lischke, 1869) (the southern Japanese form of *B. attramentaria*, regarded by some as synonyms, by others as distinct species), while found on incoming Japanese oyster seed, has apparently not become established on the Pacific coast. *B. attramentaria* is very abundant in Tomales Bay and Elkhorn Slough, in dense aggregations in soft mud. See Whitlatch 1974, Veliger 17: 47–55.

Cerithidea californica (Haldeman, 1840). Horn snail; in bays, estuaries, on mud, in aggregations under debris, boards, etc.

CERITHIIDAE

Bittium attenuatum Carpenter, 1864. In sand, gravel, under rocks and in surfgrass holdfasts. Little is known of the ecology or biology of the Bittiums.

Bittium eschrichtii (Middendorff, 1849) (= *B. e. montereyense* Bartsch, 1907). In clean, coarse sand among rocks.

Bittium interfossa (Carpenter, 1864). In gravel, under algae; uncommon intertidally.

Bittium purpureum (Carpenter, 1864). Algae, surfgrass holdfasts, in sand.

CERITHIOPSIDAE

Cerithiopsis spp. About 7 intertidal species have been recorded locally associated with sponges, or on abalones, pilings, rocks. See Bartsch 1911, Proc. U.S. Nat. Mus. 40: 327–367 (taxonomy); also McLean, 1969 and Smith and Gordon, 1948. Fig. 80 is *C. montereyensis* Bartsch, 1911.

Metaxia convexa (Carpenter, 1857) (= *M. diadema* Bartsch, 1907). Low tide to sublittoral, gravel, rocks.

Seila montereyensis Bartsch, 1907.

TRIPHORIDAE

Triphora spp. In sand, gravel, rubble; may be associated with sponges. Fig. 81 is *T. pedroana* Bartsch, 1907; *T. montereyensis* Bartsch, 1907 is known from Monterey Bay.

EPITONIIDAE

* *Epitonium cooperi* Strong, 1930. Very deep sutures; on sandy or muddy bottoms.

Epitonium tinctum (Carpenter, 1864). In sand at base of sea anemones *Anthopleura elegantissima* and *A. xanthogrammica;* feeds at high tide upon anemone tentacles, the tips of which are torn off after injection of a salivary toxin (Hochberg 1971, Echo 4 (West. Soc. Malacologists), pp. 22–23).

Opalia chacei Strong, 1937. Under rocks; may be associated with sea anemones; perhaps a southern subspecies of *O. wroblewskyi* (Mörch, 1875) (J. McLean).

Opalia montereyensis (Dall, 1907). Largely sublittoral; both Opalias occur occasionally as hermit-crab shells intertidally.

JANTHINIDAE

* *Janthina* spp. One or more species of this pelagic, purple, subconical snail may occasionally wash ashore. See Berry 1957, Amer. Malacol. Union, Ann. Rept., p. 27 and Laursen 1953, Dana Report no. 38.

EULIMIDAE

Balcis spp. Several species may be encountered, possibly ectoparasitic on echinoderms, as well as in sand, gravel, holdfasts. See Bartsch 1917, Proc. U.S. Nat. Mus. 53: 295–356 (taxonomy); *B. rutila* (Carpenter, 1864) occurs on sublittoral starfish; *B. thersites* (Carpenter, 1864) is intertidal from Monterey south.

HIPPONICIDAE

Hipponix cranioides Carpenter, 1864 (=*H. antiquatus* of authors, not of Linnaeus, 1767). In colonies under narrow overhanging crevices, low tide, secreting a ventral calcareous plate. See Yonge 1953, Proc. Calif. Acad. Sci. (4) 28: 1–24 and 1960, 31: 111–119 (anatomy, biology, ecology); Cowan 1974, Veliger 16: 377–380 (taxonomy).

Hipponix tumens Carpenter, 1864. Low intertidal, in rock crevices; see Cowan 1974 (above).

CALYPTRAEIDAE (slipper shells)

Crepidula adunca Sowerby, 1825. Common on larger snails such as *Tegula* See Putnam 1964, Veliger 6: Suppl., 63–66 (dispersal of young).

Crepidula convexa Say, 1822 (=*C. glauca* Say, 1822). Introduced with Atlantic oysters; in bays, often on shells such as hermit-crab-occupied *Nassarius obsoletus* in San Francisco Bay. See Franz and Hendler 1970, Univ. Conn. Occ. Pap. (Biol. Sci. Ser.) 1: 281–289 (taxonomy); Hendler and Franz 1971, Biol. Bull. 141: 514–526 (reproductive biology and population dynamics).

Crepidula fornicata Linnaeus, 1758. Introduced with Atlantic oysters; may occur in Tomales Bay; shape highly variable; occurs in characteristic "stacks" (see Ricketts, Calvin, and Hedgpeth, 4th ed., pp. 50–51, 379).

Crepidula nummaria Gould, 1846 (=*C. nivea* of authors). Also see *C. plana* and *C. perforans*. Low intertidal zone of outer coast, under rocks, occasionally in abandoned pholad holes.

Crepidula perforans (Valenciennes, 1846). Compare with *C. nummaria* (which possesses a shaggy golden brown periostracum) and *C. plana* (see below); in abandoned pholad holes, hermit-crab shells, and under rocks along the open rocky coast.

* *Crepidula plana* Say, 1822. Thin, white shell with thin periostracum (fig. 40); introduced with Atlantic oysters; in bays on rocks and often (as concave specimens) in hermit-crab shells. This is the *C. nivea* of Packard's (1918) north and south San Francisco Bay stations; *C. perforans* (above) occurs on the open coast, and morphological distinctions between the two species require investigation.

Crepipatella lingulata (Gould, 1846). On rocks, shells, intertidal to offshore depths.

NATICIDAE

Polinices lewisii (Gould, 1847). Moon snail; common on mud and sand flats in bays, lagoons, *Zostera* flats, also offshore bottoms. Feeds on bivalves such as *Macoma;* forms characteristic "sand collars" in which egg capsules are embedded (the veligers settling on sublittoral algae). See Bernard, 1967, Fish. Res. Board Canada, Tech. Rept. 42 (MS).

LAMELLARIIDAE

Lamellaria spp. A poorly known group, with perhaps 5 species in central California, some considered assignable to *Marsenina.* Intertidal to sublittoral, under rocks, on pilings; feed upon compound ascidians and sponges; often overlooked because of their remarkable similarity in color and texture to the host (see Ghiselin, 1964, Veliger 6: 123–124: *L. stearnsi* Dall, 1871 on *Trididemnum opacum*). The animal illustrated in figure 24 is diagrammatic.

VELUTINIDAE

Velutina prolongata Carpenter, 1865. Generally sublittoral; rare in intertidal.

Velutina cf. *V. velutina* (Müller, 1776) (= *V. laevigata* of authors). Intertidal; may be common under rocks, in crevices, associated with ascidians; California specimens may represent an undescribed species (see MacGinitie, 1959, Proc. U.S. Nat. Mus. 109: 95–96).

TRIVIIDAE

Erato columbella Menke, 1847. Under rocks, low intertidal to offshore, associated with and feeding upon ascidians.

Erato vitellina Hinds, 1844. As above, may be encountered in beach drift. Northern limit is recorded as Bodega Bay, a record established by R. E. C. Stearns on a visit in June, 1867.

Trivia californiana (Gray, 1827). The coffee-bean shell, most common sublittorally, associated with ascidians.

Order **Neogastropoda**

MURICIDAE

Ceratostoma foliatum (Gmelin, 1791) (= *Purpura foliata*). Among rocks under algae; in sea-urchin beds. A large species which may develop prominent frilling, especially when living sublittorally; see Spight and Lyons 1974, Mar. Biol. 24: 77–83 (development and function of shell sculpture).

Ceratostoma inornatum (Récluz, 1851). [= *Ocenebra japonica* (Dunker, 1869) according to G. Radwin and A. D'Attilio, San Diego Natural History Museum]. Introduced from Japan; on rocks and on active oyster beds in Tomales Bay.

Ocenebra atropurpurea Carpenter, 1865 [= *O. clathrata* (Dall,

1919)]. Often as hermit-crab shells intertidally; may be a synonym of *O. interfossa* (G. Radwin and A. D'Attilio).

Ocenebra circumtexta Stearns, 1871. Common mid-intertidal species.

Ocenebra gracillima Stearns, 1871. Under rocks; more common in south.

Ocenebra interfossa Carpenter, 1864. Common under alga-covered rocks.

Ocenebra lurida (Middendorff, 1848). *O. munda* Carpenter, 1864 and *O. sclera* (Dall, 1919) are regarded by some workers as synonyms of *O. lurida*. Common on and under rocks.

Urosalpinx cinerea (Say, 1822). The oyster drill, introduced with Atlantic oysters; may be common among oysters and barnacles in estuaries; buries in mud in winter. See Franz 1971, Biol. Bull. 140: 63–72 (aspects of biology); Carriker 1969, Amer. Zool. 9: 917–933 (boring).

"Urosalpinx" subangulata (Stearns, 1873). Questionably in this genus; formerly in *Ocenebra* (Vokes 1971, Bull. Amer. Paleo. 61: 102).

THAIDIDAE See Paris 1960, Veliger 2: 41–47; Connell 1970, Ecol. Monogr. 40: 49–78 on predation by *Nucella*.

Acanthina punctulata (Sowerby, 1825). Upper intertidal zone on rocks; moving downward during breeding season.

Acanthina spirata (Blainville, 1832). Where the unicorn snails *A. punctulata* and *A. spirata* occur together, as at Monterey, *spirata* generally is in the lower, and *punctulata* in the upper intertidal. In bays and other protected situations, feeding on barnacles and molluscs.

Nucella canaliculata (Duclos, 1832) [= *N. c. compressa* (Dall, 1915)]. This and following *Nucella* were formerly placed in *Thais*, a subtropical and tropical genus. Common on rocks at low tide. See Houston 1971, Veliger 13: 348–457 (reproductive biology), but see also note by Lyons and Spight 1973, Veliger 16: 193.

Nucella emarginata (Deshayes, 1839). Common to abundant at upper tide levels. See Houston, 1971 (above).

Nucella lamellosa (Gmelin, 1791). Low tide on rocks, often in protected bays; highly variable sculpture, smoother specimens occurring in more exposed situations, delicate and prominent lamellae developing in protected areas; see Spight 1973, J. Expt. Mar. Ecol. 13: 215–228; 1974, Ecology 55: 712–729.

MELONGENIDAE

Busycotypus canaliculatus (Linnaeus, 1758) (= *Busycon canaliculatum*). Channeled whelk, introduced from Atlantic; sublittorally and in mud at low tide in San Francisco Bay; strings of large egg capsules are commonly washed ashore. See Stohler, 1962, Veliger 4: 211–212.

BUCCINIDAE

Searlesia dira (Reeve, 1846). Dire whelk; on coralline-encrusted rocks and among gravel and rocks in crevices; abundant further north, but of spotty occurrence in central California. See Lloyd, 1971, Diss. Abst. 32B: 4301 (biology, feeding).

COLUMBELLIDAE

Amphissa columbiana Dall, 1916. On algae-covered rocks; under rocks in sand and gravel.

Amphissa versicolor Dall, 1871. Under rocks loosely set in gravel; several color varieties have been given subspecific names.

Mitrella aurantiaca (Dall, 1871). Chiefly sublittoral; uncommon in low intertidal zone among rocks and algae.

Mitrella carinata (Hinds, 1844) (= *M. gausapata* of authors). Common on algae and on rocks.

Mitrella tuberosa (Carpenter, 1864). In sand and gravel at low tide and in beach drift; more common sublittorally.

NASSARIIDAE See Demond, 1952, Pac. Sci. 6: 300–317 (review).

Nassarius fossatus (Gould, 1850). Common on mud and sand in bays, estuaries; see MacGinitie, 1931, Ann. Mag. Nat. Hist. (10) 8: 258–261 (egg-laying).

Nassarius mendicus (Gould, 1850) (= *N. cooperi* Forbes, 1850, based on fewer-ribbed specimens). Common in sand, mud, on rocks, of open coast and in bays.

Nassarius obsoletus (Say, 1822). Introduced from Atlantic coast; an omnivore and deposit feeder very abundant on San Francisco Bay mudflats. Only American nassariid having a crystalline style. See Scheltema, 1964, Ches. Sci. 5: 161–166 (feeding and growth); Crisp, 1969, Biol. Bull. 136: 355–373; an extensive literature exists on Atlantic coast populations. Once called *Nassa*, and probably assignable to the genus *Ilyanassa*, since it lacks the deep basal groove, caudal cirri, bifurcated foot, and other features characteristic of *Nassarius;* while such variation in name is frustrating to many experimentalists, it is to be hoped that the name will become stabilized in the future.

* *Nassarius perpinguis* (Hinds, 1844). Generally sublittoral on sandy bottoms; with a narrow shelf below suture and fine cancellate sculpture with beaded axial ridges.

FASCIOLARIIDAE

Fusinus luteopictus (Dall, 1877). Low intertidal, on and under rocks.

MARGINELLIDAE

See Coan and Roth, 1966, Veliger 8: 276–299 (taxonomy); marginellids are found in gravel, among coralline algae, and in surfgrass holdfasts in the low intertidal.

Cystiscus jewettii (Carpenter, 1857).

Granula subtrigona (Carpenter, 1864).

Granulina margaritula (Carpenter, 1857) [=*Cypraeolina pyriformis* (Carpenter, 1864)].

OLIVIDAE

See Olsson, 1956, Proc. Acad. Nat. Sci. Phil. 108: 155–225 (taxonomy); all are found on sandy bottoms.

Olivella baetica Carpenter, 1864.

Olivella biplicata (Sowerby, 1825). The purple olive; common intertidally, burrowing in clean sand of sloping, protected beaches and offshore of more exposed beaches. See Edwards, 1968, Veliger 10: 297–304 (reproduction) and 1969, Veliger 11: 326–333 (predators); Stohler, 1969, Veliger 11: 259–267 (growth).

Olivella pycna Berry, 1935.

TURRIDAE

Most are found in rocky areas intertidally to offshore bottoms.

Clathromangelia interfossa (Carpenter, 1864) (=*Mangelia interlirata* Stearns, 1871).

Clathurella canfieldi Dall, 1871. In sand among surfgrass roots.

Cymakra aspera (Carpenter, 1864) (=*Mitromorpha aspera*). In possessing columellar plicae, cancellate sculpture, and in other aspects is allied to, but not strictly, *Cymakra* (J. McLean).

Cymakra gracilior (Tryon, 1884). Common, low intertidal.

Mitromorpha carpenteri Glibert, 1954 [=*M. filosa* (Carpenter, 1864), a preoccupied name].

Pseudomelatoma torosa (Carpenter, 1865).

CONIDAE

Conus californicus Reeve, 1844. Low intertidal in rock crevices or sand pockets; offshore on sand and rock bottoms; diverse diet of worms, molluscs, crustaceans, fish. See Saunders and Wolfson, 1961, Veliger 3: 73–76 and Kohn, 1966, Ecology 47: 1041–1043 (feeding).

Subclass **OPISTHOBRANCHIA**

Order **Cephalaspidea**

ACTEONIDAE

Rictaxis punctocaelatus (Carpenter, 1864) (= *Acteon punctocaelatus*). Of sporadic occurrence on mud and sand flats in bays.

BULLIDAE

Bulla gouldiana Pilsbry, 1893. The bubble shell; a southern species occasional in central California; on mudflats in lagoons, bays, estuaries.

ATYIDAE

Haminoea vesicula (Gould, 1855). Sporadically abundant among algae *Enteromorpha* and *Polysiphonia,* in sloughs, lagoons, bay mudflats.

Haminoea virescens (Sowerby, 1833). In higher tidepools of open-coast rocky areas.

DIAPHANIDAE

Diaphana californica Dall, 1919. Uncommon intertidally on algae; sublittorally in sand and kelp holdfasts.

ACTEOCINIDAE See Roller 1971, West. Soc. Malacologists, Echo 3, pp. 31–32 (taxonomy).

Acteocina culcitella (Gould, 1853). Sporadically common in bays and lagoons, on sand and mud.

Acteocina harpa (Dall, 1871) (= *Retusa harpa*). In sand, gravel, and mud, low intertidal to offshore.

* *Acteocina inculta* (Gould, 1855). Southern (Morro Bay south); common in mud of marsh channels, bays, lagoons. Keyed in Opisthobranchia Key A, p. 518.

Order **Notaspidea**

UMBRACULIDAE

* *Tylodina fungina* Gabb, 1865. (fig. 107) Southern (Cayucos, south); found on yellow sponges, which it closely resembles. Keyed in Opisthobranchia Key C, p. 522.

Order **Pyramidellida**

PYRAMIDELLIDAE

Iselica ovoidea (Gould, 1853) [= *I. fenestrata* (Carpenter, 1864), according to J. H. McLean]. Among *Mytilus* beds in bays; also associated with other invertebrates; in mud, gravel.

Odostomia spp. Ectoparasites extracting body fluids from host by means of a modified gut with long proboscis and stylet; no radula (see Fretter and Graham, 1949, J. Mar. Biol. Assoc. U.K. 28: 493–532). Numerous species recorded from central California, but most are poorly known, earlier workers failing to recognize intraspecific variability. Several species have been recorded from red-abalone shells, such as *Odostomia (Evalea) phanea* Dall and Bartsch, 1907 (fig. 45) and *O. (E.) tenuisculpta* Carpenter, 1864; others, such as *Odostomia (Menestho) fetella* Dall and Bartsch, 1909 (fig. 48), may be found in mud in shallow waters of bays. *Odostomia* have also been observed locally associated with the vermetid snail *Petaloconchus*, and the sabellid polychaete *Pseudopotamilla;* many similar associations will be found by careful observers. Degree of host specificity is unknown for most species. Backs of fresh abalone shells often provide a prolific hunting ground; the *Odostomia* may be feeding on vermetids, serpulids or other invertebrates rather than the abalone.

The group was monographed by Dall and Bartsch (1909, Bull. U.S. Nat. Mus. 68) with subsequent papers by Bartsch in USNM Proceedings, but many species were established on slight differences or single decollated specimens. Robertson (1967, Ann. Rept. Amer. Malacol. Union, pp. 12–13) discusses pertinent systematic characters. In estuaries one or more species of introduced *Odostomia* may be present (see p. 17). See Robertson and Orr, 1961, Nautilus 74: 85–91 (review of hosts); Bullock and Boss, 1971, Breviora no. 363 (host selection by *O. bisuturalis*); Clark, 1971, Veliger 14: 54–56 (Puget Sound, *O. columbiana* on snail *Trichotropis*).

Turbonilla spp. See Dall and Bartsch, above. Numerous species named, but little is known of most of them, and few recorded from central California intertidal zone. The species illustrated (fig. 66) is *T. franciscana* Bartsch, 1917, from San Francisco Bay.

Subclass **PULMONATA**

Order **Basommatophora**

MELAMPIDAE (formerly ELLOBIIDAE)

Ovatella myosotis (Draparnaud, 1801) [= *Phytia setifer* (Cooper, 1872)]. See Hedgpeth's *Seashore Life* (p. 107) for figure of juvenile. In *Salicornia* marshes, often very abundant on mud, under debris, and in crevices of old docks and pilings; possibly introduced.

SIPHONARIIDAE

Williamia peltoides (Carpenter, 1864) [= *W. vernalis* (Dall, 1870)]. In protected low intertidal to largely sublittoral, under rocks, on coralline algae and on coralline-covered shells of *Tegula, Astraea*, etc. See Yonge (1960) below.

TRIMUSCULIDAE (formerly GADINIDAE)

Trimusculus reticulatus (Sowerby, 1835) (= *Gadinia reticulata*). In groups on roofs of caves and under overhanging ledges in low intertidal; also in abandoned pholad holes. See Yonge, 1958, Proc. Malacol. Soc. London 33: 31–37, and 1960, Proc. Calif. Acad. Sci. (4) 31: 111–119, on aspects of biology, ecology; Walsby, 1975. Veliger 18: 139–145 (feeding).

REFERENCES ON SHELLED GASTROPODS

For references not cited here, see the introductory section to the Mollusca.

Boss, K. J., J. Rosewater, and F. A. Ruhoff 1968. The zoological taxa of William Healey Dall. Bull. U.S. Nat. Mus. 287: 427 pp.

Cox, L. R. 1960. Gastropoda, General Characteristics. *In* R. C. Moore, ed., *Treatise on Invertebrate Paleontology,* Univ. Kansas Press and Geol. Soc. Amer., Part I, Mollusca 1, pp. I84–I168.

Franc, A. 1968. Classe des Gastéropodes. *In* P.-P. Grassé, ed., *Traité de Zoologie,* 5 (3).

Fretter, V. and A. Graham 1962. *British Prosobranch Molluscs. Their Functional Anatomy and Ecology.* London: Ray Society, 755 pp.

Johnson, R. I. 1964. The Recent Mollusca of Augustus Addison Gould. Bull. U.S. Nat. Mus. 239: 182 pp.

Palmer, K. V. W. 1958. Type specimens of marine Mollusca described by P. P. Carpenter from the west coast (San Diego to British Columbia). Geol. Soc. Amer. Mem. 76: 376 pp.

Wenz, W. 1938–1944. *Gastropoda.* Allgemeiner Teil und Prosobranchia. In *Handbuch der Paläozoologie,* O. H. Schinderwolf, ed., Berlin, Gebrüder Borntraeger, 6(1) pts. 1–7, 1639 pp. (Offset reprint, 1961–1962)

Zilch, A. M. 1959–1960. *Gastropoda. Euthyneura.* (Continuation of Wenz). In *Handbuch der Paläozoologie,* O. H. Schinderwolf, ed., 6(2) pts. 1–4, 834 pp. Berlin: Gebrüder Borntraeger.

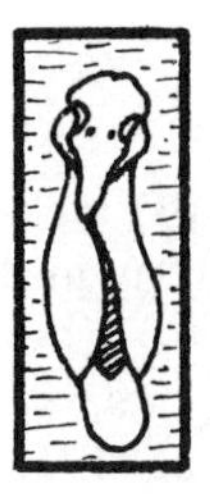

PHYLUM MOLLUSCA: GASTROPODA, OPISTHOBRANCHIA

James W. Nybakken
Moss Landing Marine Laboratories of the California State Universities and Colleges

(PLATES 121–123, and others as noted)

The subclass **Opisthobranchia** is a complex of forms, mostly shell-less or with reduced shells. It includes the well-known order **Nudibranchia** as well as several other orders formerly lumped as "tectibranchs." The abundant nudibranchs are primarily predators of sedentary or sessile invertebrates, such as hydroids, sea anemones, sponges, and ascidians. The more primitive cephalaspideans, anaspideans, and sacoglossans are primarily herbivores or detritus feeders.

It is often difficult for the beginner to decide whether or not a given specimen is an opisthobranch since both shelled and shell-less forms are represented in this subclass. Indeed, primitive shelled opisthobranchs are often externally indistinguishable from prosobranchs. Equally confusing is the fact that certain prosobranchs and pulmonates are or appear shell-less, resembling opisthobranchs. Accordingly, before the treatment of the opisthobranchs, we include a note on such troublesome forms:

A. Apparently shell-less non-opisthobranchs:

1) Order **Onchidiacea:** This order is represented by only one species in California, *Onchidella borealis* Dall, 1871 (fig. 4). It resembles a small, shell-less limpet, and is commonly regarded as a marine *pulmonate;* however, some workers continue to place it in the order Gymnophila of the Opisthobranchia. It is common in the exposed, upper-middle rocky intertidal zone. Body color varies from deep reddish brown to dark brown mottled with white and black; the eyes are borne on stalks; the body margin bears 20 to 24 papillae.

2) *Lamellaria* spp: Species of this genus of *prosobranch* snails have a reduced, flattened, spiral shell completely or nearly hidden in the mantle. These animals are oval to circular in outline, resembling some dorid nudibranchs, but lack a dorsal rosette of gills and dorsal anal opening (Pl. 113, fig. 24). They have a single anterior siphon and characteristically are found upon compound ascidians, which they match

in color and texture. They are keyed out under shelled gastropods (page 473).

B. **Shelled opisthobranchs:**

1) Order **Pyramidellida:** These are tiny, coiled, white snails, lacking a radula but with the proboscis adapted for suctorial feeding, and characteristically in commensal or ectoparasitic association with other invertebrates. They are keyed out with other shelled gastropods.

2) Certain **Cephalaspidea** have a strong, spiraled, external shell, into which the body can be completely withdrawn. The common *Rictaxis* (formerly *Acteon*) *punctocaelatus* is easily recognized by its black and white banded shell (Pl. 114, fig. 25). *Acteocina* is of comparable shell form but lacks an operculum, and is less common intertidally. Other cephalaspideans have globose spiral shells with wide apertures, partly hidden by the mantle. Such genera as *Bulla*, *Diaphana*, and *Haminoea* have a general resemblance to shell-less cephalaspideans, and so may be recognized as opisthobranchs without much difficulty.

3) Certain **Notaspidea** have a limpetlike external shell, but differ from proper limpets (**Patellacea**) in having the gill on the right-hand side of the body. *Tylodina fungina* often has a distinctive covering of bristles on its shell (Pl. 120, fig. 107).

Since most externally shelled opisthobranchs are covered in the key to shelled gastropods as well as in the opisthobranch keys, the student may find it simpler to use the former for gastropods with shells. But if it is known that the specimen is an opisthobranch, the following keys may provide a quicker identification.

The following keys do not include opisthobranchs which are normally subtidal or pelagic. However, we include species which often appear in the intertidal zone on floating objects or debris cast ashore. In this category are such species as *Fiona pinnata*, *Doridella steinbergae*, and *Corambe pacifica*. Similarly, *Diaphana californica* may be found in kelp holdfasts.

The pelagic opisthobranch pteropods are not keyed here, but the student may often find transparent, gelatinous, slipperlike structures on the beach which appear like parts of jellyfish (Scyphozoa). These are the pseudoconchs of the pteropod family **Cymbulidae.**

The sizes given in the keys are average sizes. The student should keep in mind that specimens could be much smaller or larger and still be that species. In some instances there is a difference in color or pattern between the young and the adult. This has been taken into account in the key so that either a juvenile or an adult should key out to the same species. For example, juvenile *Triopha maculata* lack the color pattern of the adult, but will still key out to *T. maculata*.

The student should also be aware that the systematics of the opisthobranchs of the California coast are in a state of flux. New species are still being discovered and named, while other species are reduced to synonymy as more is learned of the range of variation within a species. The following keys represent what we believe are most of the valid intertidal species.

Most opisthobranchs, and all nudibranchs, are best observed alive. Preservation of opisthobranchs, unless done very carefully, usually results in a colorless blob. Color preservation is often difficult and, in some nudibranchs such as *Hermissenda*, it is impossible to preserve color. Color can usually be preserved rather well in dorids by the method of Robilliard (1969). To effect good preservation, the specimen must be suitably narcotized or relaxed prior to preservation. A number of systems of relaxation have been suggested (Beeman, 1968a, Runham *et al.*, 1965) but perhaps the best, at least for nudibranchs, is propylene phenoxetol (900:1 with sea water) slowly dripped, conveniently by a string, into dishes of sea water in which the nudibranchs have been allowed to expand. This should be continued over several hours for dorids, but eolids may have the dilute propylene phenoxetol added directly in a single application. When the animals fail to respond to touch, they may be transferred to 75% ethyl alcohol.

The following keys do not require dissection of specimens. The characters used are mainly color, pattern, and external morphology of the animals. Technical terms used in the keys are illustrated in the figures. Sacoglossans and nudibranchs are included in a single key since they appear similar externally. This key is preceded by shorter keys to the much smaller orders Cephalaspidea, Anaspidea, and Notaspidea.

Many people have aided us in the preparation of this chapter. I would especially like to thank Gordon A. Robilliard, Richard A. Roller, James T. Carlton, Hans W. Bertsch, David H. Montgomery, Robert D. Beeman, and Virginia L. Waters, all of whom took great pains in critically reading the keys, and offered many constructive changes. I am indebted to Milos Radakovich for drawing most of the illustrations.

KEY TO THE ORDERS OF OPISTHOBRANCHIA

Terrence M. Gosliner and Gary C. Williams

1. Shell present, wholly or at least partially visible externally, spirally coiled 2
– Shell internal or absent or, if external, cap-shaped as in a limpet 3
2. Head shield present (figs. 1–3); aperture more than half the shell length ....(in part) **Cephalaspidea** (Key A)

– Head shield absent; aperture less than half the length of shell (see key to shelled gastropods, page 512) ..... **Pyramidellida**
3. Dorsal surface with projections, which may be either finger-like or a circlet of gills around the anus (figs. 6, 8, 9–11, 13, 18, 20) 4
– Dorsal surface without projections, although lateral flaps (parapodia) of the foot may project dorsally (fig. 5) 5
4. Rhinophores rolled (fig. 23) .....(in part) **Sacoglossa** (Key D)
– Rhinophores not rolled, but solid with various elaborations (figs. 24–27) ....(in part) **Nudibranchia** (except *Stiliger*)
5. Head shield present (figs. 1–3); rhinophores absent ....(in part) **Cephalaspidea** (Key A)
– Head shield absent; rhinophores present 6
6. Gill enclosed or partially hidden in a slit on the dorsal surface .... **Anaspidea** (Key B)
– Gill not enclosed in a slit on the dorsal surface 7
7. Gills, if present, located posteroventrally on both sides of the body between dorsum and foot margin .... (in part) **Nudibranchia** (Key D)
– Gill absent or located exclusively on right side of the body . 8
8. Gill on right side of animal, under overlapping dorsum between dorsum and foot; if external shell is present, it is cap-shaped, resembling a limpet .... **Notaspidea** (Key C)
– No gill present, rhinophores rolled (fig. 23) .... (in part) **Sacoglossa** (Key D)

The following keys to Cephalaspidea, Anaspidea, and Notaspidea are by Terrence M. Gosliner, Gary R. McDonald, James W. Nybakken, and Gary C. Williams.

KEY A: CEPHALASPIDEA

(PLATE 121 unless otherwise noted)

1. Shell absent or internal, not externally visible (figs. 1–3) 2
– Shell external and visible; may be thin and transparent or thick and calcified 5
2. Shell absent; body color orange, animal minute (less than 5 mm); possesses 4 gizzard plates .... *Runcina* sp.
– Shell internal; body tan, cream white, to dark brown or black, may be mottled or striped; animal not minute 3
3. Body color tan to dark brown with yellow longitudinal stripes and yellow and blue spotting (fig. 3) .. *Chelidonura inermis*

– Body color brownish to black or cream white without yellow striping or blue spotting 4

4. Head simply rounded, blunt without lateral projections; tail short, blunt, inconspicuous, the lobes equal in length; body color cream with brown mottling to dark brown or black (fig. 1) .... *Aglaja diomedea*

– Head with short, subacute lateral lobes; tail lobes elongate, pointed, and conspicuous, equal or subequal in length; body color brownish black with yellow to white spots (fig. 2) .... *Aglaja ocelligera*

5. Shell with prominent apical spire; aperture narrow, elongate 6

– Shell without apical spire; aperture wide anteriorly 9

6. Animal with an operculum; spirally sculptured shell with 2 revolving bands of white and black, separated by wide white bands; body white (Pl. 114, fig. 25) *Rictaxis punctocaelatus*

– Animal without operculum 7

7. Shell with strongly carinate (keeled) shoulder (Pl. 115, fig. 43) .... *Acteocina harpa*

– Shell with rounded shoulder 8

8. Shell without transverse striping; shell color cream or brown to pure white; animal small, usually less than 10 mm .... *Acteocina inculta*

– Shell with alternating narrow brown and white stripes; animal large, usually greater than 10 mm (Pl. 115, fig. 44) .... *Acteocina culcitella*

9. Shell with a pit at the apex; shell reddish brown to brownish gray with dark and light mottling; body orange to yellow brown; shell often quite large and heavy; 2–8 cm long (Pl. 114, fig. 27) .... *Bulla gouldiana*

– Shell lacking pit at apex; shell thin, whitish, translucent to opaque white 10

10. Shell usually less than 5 mm in length; shell cylindrical, apex flat; head composed of 2 rounded, cephalic lobes; body white .... *Diaphana californica*

– Shell usually larger than 10 mm in length, not cylindrical; globose head triangular in shape; body yellowish, brownish, or orange 11

11. Aperture 1/2 diameter of shell; shell usually more than 15 mm long, thin, translucent white .... *Haminoea vesicula*

– Aperture greater than 1/2 diameter of shell; shell usually less than 15 mm long, somewhat thickened, opaque white (Pl. 114, fig. 29) .... *Haminoea virescens*

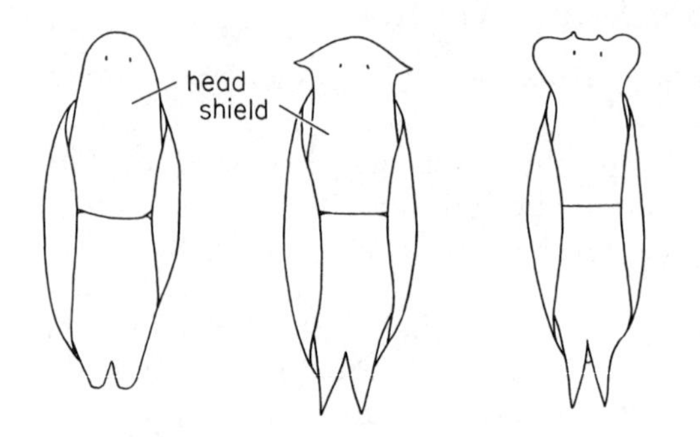

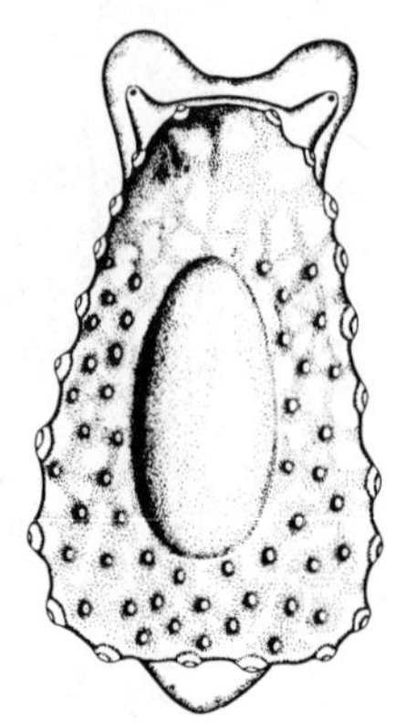

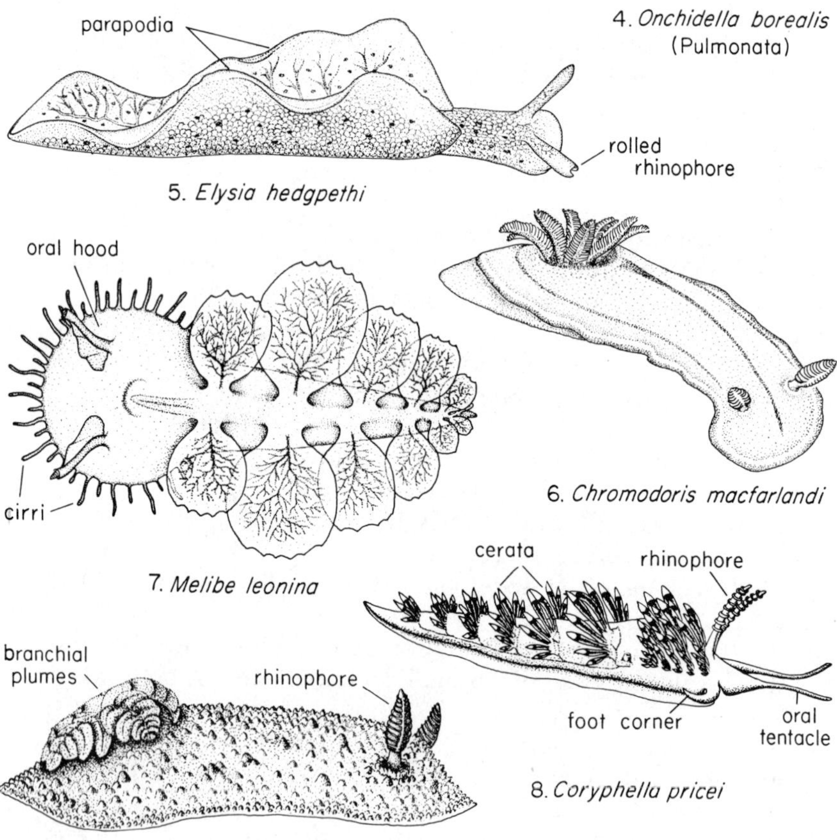

PLATE 121. **Opisthobranchs** (1). 1, *Aglaja diomedea;* 2, *A. ocelligera;* 3, *Chelidonura inermis;* 4, *Onchidella borealis*, an intertidal pulmonate; 5, *Elysia hedgpethi;* 6, *Chromodoris macfarlandi*, showing typical dorid nudibranch form; 7, *Melibe leonina;* 8, *Coryphella pricei*, showing typical eolid nudibranch form; 9, *Archidoris odhneri.* Figures 1–4 by Gary C. Williams; 5–9 redrawn from MacFarland (1966) by Milos Radakovich.

LIST OF CEPHALASPIDEA

Acteocina culcitella (Gould, 1853).

Acteocina harpa (Dall, 1871) (= *Retusa harpa*).

Acteocina inculta (Gould, 1855).

Aglaja diomedea (Bergh, 1894) (= *A. nana* Steinberg and Jones, 1960). On mudflats.

Aglaja ocelligera (Bergh, 1894). On mudflats.

† *Bulla gouldiana* Pilsbry, 1893. Largely a southern species; on mudflats.

Chelidonura inermis (Cooper, 1862) (= *Navanax inermis*). On mudflats.

† *Diaphana californica* Dall, 1919. In rocky intertidal and kelp holdfasts.

Haminoea vesicula (Gould, 1855). On boat landings or mudflats of bays.

Haminoea virescens (Sowerby, 1833). In higher pools of rocky intertidal.

Rictaxis punctocaelatus (Carpenter, 1864) (= *Acteon punctocaelatus*). Fairly common on mud and sand flats.

Runcina sp. High intertidal among alga *Endocladia;* around tubes of polychaete *Phragmatopoma.*

KEY B: ANASPIDEA

1. Body dorsoventrally flattened; length under 10 cm; parapodia reduced; body color green with black and white striping . . . . *Phyllaplysia taylori*
– Body laterally compressed, often exceeding 10 cm in length; parapodia highly developed; body color not green, unstriped 2
2. Parapodia joined posteriorly; body color uniform dark brown or black . . . . *Aplysia vaccaria*
– Parapodia not obviously joined posteriorly; body color mottled tan or brown . . . . *Aplysia californica*

LIST OF ANASPIDEA

(see Beeman, 1968b)

Aplysia californica Cooper, 1863.

† *Aplysia vaccaria* Winkler, 1955.

† Rare in our area.

Phyllaplysia taylori Dall, 1900 (= *Phyllaplysia zostericola* McCauley, 1960). Common on eelgrass *Zostera* in waters and on mudflats of bays.

KEY C: NOTASPIDEA

1. With limpetlike external shell often covered with bristles, body yellowish in color (Pl. 120, fig. 107) . . .*Tylodina fungina*
– External shell absent 2
2. Color cream to white ....*Berthella californica*
– Color pale yellow, finely punctate with darker yellow .... *Pleurobranchus strongi*

LIST OF NOTASPIDEA

Berthella californica (Dall, 1900) (= *Pleurobranchus californicus*). In open, low-tidal pools.

† *Pleurobranchus strongi* MacFarland, 1966. Occasionally found in rocky tide pools at low tide upon or near compound ascidians (MacFarland).

† *Tylodina fungina* Gabb, 1865. Feeds upon sponges; southern, not reported from central California.

KEY D: SACOGLOSSA AND NUDIBRANCHIA

Gary R. McDonald

Moss Landing Marine Laboratories of the California State Universities and Colleges

(PLATES 121–123)

1. Rhinophores in the form of a longitudinally rolled plate (except in *Stiliger*) (figs. 5, 23); oral tentacles absent; usually found on algae .... Order **Sacoglossa** 2
– Rhinophores not in form of a longitudinally rolled plate; oral tentacles usually present 7
2. With 2 parapodia carried folded together in a vertical position over back (fig. 5); ground color rich green to yellowish tan, with small spots of yellow, red, and blue; 10 mm ....*Elysia hedgpethi*
– Lacking parapodia; with dorsal cerata 3
3. Rhinophores quite reduced; anus borne on a long tube originating on median line among posterior cerata and resembling cerata; ground color greenish to yellowish tan with small

black spots on body; 8 mm .... *Alderia modesta*
– Rhinophores prominent; anus not borne on a long tube 4
4. Body, cerata, and proximal 1/2 of rhinophores with a branching system of olive to dark green lines; tail long and obvious; cerata rather long; ground color pale yellow; 8 mm .... *Placida dendritica*
– Branching system of lines not continuing up onto rhinophores 5
5. With a triangular patch of dark brown spots extending from base of rhinophores forward to the midline; neck with a band of brown forming a collar; cerata cores chocolate brown; 5 mm .... *Hermaea vancouverensis*
– Lacking triangular brown patch and brown cerata cores 6
6. Body covered with irregular patches of greenish to brownish black; head uniformly pigmented with dark green to brownish black; ground color yellowish white; 22 mm .... *Aplysiopsis smithi*
– With a Y-shaped mahogany line running posteriorly from bases of rhinophores to midline on top of head; body with irregular lines of mahogany; a bright pink spot below and behind eyes; ground color pale yellow; 10 mm .... *Aplysiopsis oliviae*
7. Ground color translucent grayish white; body with reddish brown spots and irregular lines; 2 parallel, reddish brown lines running from rhinophores posteriorly; rhinophores simple; 3 mm .... (Order **Sacoglossa**) *Stiliger fuscovittatus*
– Not as above .... Order **Nudibranchia** 8
8. Gills (branchial plumes) located posteriorly on dorsum (figs. 6, 9) or in 3 groups directed posteriorly 9
– Gills (branchial plumes) not located posteriorly on dorsum; respiratory surface otherwise 55
9. With extra-branchial appendages (fig. 13) 10
– Without extra-branchial appendages 16
10. With extra-rhinophoral appendages (fig. 13); 3 branchial plumes 11
– Without extra-rhinophoral appendages; usually with more than 3 branchial plumes 13
11. Each rhinophore shaft with 1 extra-rhinophoral appendage (fig. 13); ground color whitish; 5 longitudinal brownish lines running most of the length of body; distal tips of rhinophores, oral tentacles, tips of branchial plumes, extra-branchial, and extra-rhinophoral appendages orange yellow; 10 mm .... *Trapania velox*
– Each rhinophore shaft with 2 extra-rhinophoral appendages 12
12. With 2 extra-branchial appendages; ground color translucent

tan to whitish; head, tail, and sides of body with irregular patches of dark reddish brown; 5 mm . . *Ancula lentiginosa*

– With 5 or more extra-branchial appendages; ground color translucent yellowish white; 3 longitudinal yellow lines running most of the length of the body; rhinophores and extra-branchial appendages with orange yellow band near tip; 8 mm . . . . *Ancula pacifica*

13. Body covered with numerous tubercles; yellow spots on body and on tips of low tubercles; body color dark brownish; 3–5 branchial plumes; 10 mm . . . . *Polycera zosterae*

– Body not covered with numerous tubercles 14

14. Overall body color translucent grayish white; velar appendages (fig. 22) and extra-branchial appendages (fig. 13) with black on proximal 1/3 and a yellow ring immediately distal to black pigment; branchial plumes 5–6, blackish with yellow tips; 15 mm . . . . *Polycera tricolor*

– Overall body color blackish 15

15. With oblong orange spots occurring between longitudinal black lines running the length of the body; 8–11 branchial plumes; 12 mm . . . . *Polycera atra*

– With closely set, small black dots on body: rhinophores, corners of foot, velar (fig. 22), and extra-branchial appendages (fig. 13) with yellowish orange pigment; caudal crest and upper edge of foot with steaks of yellow orange; 7–9 branchial plumes; 12 mm . . . . *Polycera hedgpethi*

16. Rhinophores simple (fig. 24); ground color opaque white . . 17

– Rhinophores not simple; ground color variable 18

17. Dorsum with large cylindrical tubercles; usually with many irregularly scattered, dark brown to black spots; 3 branchial plumes; rhinophore sheaths with 5–6 high, rounded tubercles; 12 mm . . . . *Aegires albopunctatus*

– Dorsum nearly smooth, tubercles minute; lacking any brown or black pigment; about 8 branchial plumes; rhinophores long and tapering (fig. 24); lateral edge of dorsum with opaque white glands; 12 mm . . . . *Conualevia alba*

18. With dorsal processes other than tubercles or papillae on dorsum or sides of body 19

– Dorsum without dorsal processes, but with tubercles (fig. 9) or papillae or entirely smooth (fig. 6) 26

19. Dorsal processes simple 20

– Dorsal processes bifurcated, branched, or warty (tuberculate) near tip 23

20. Body color entirely rose pink; with many long dorsal processes covering entire dorsum; 7–14 branchial plumes which

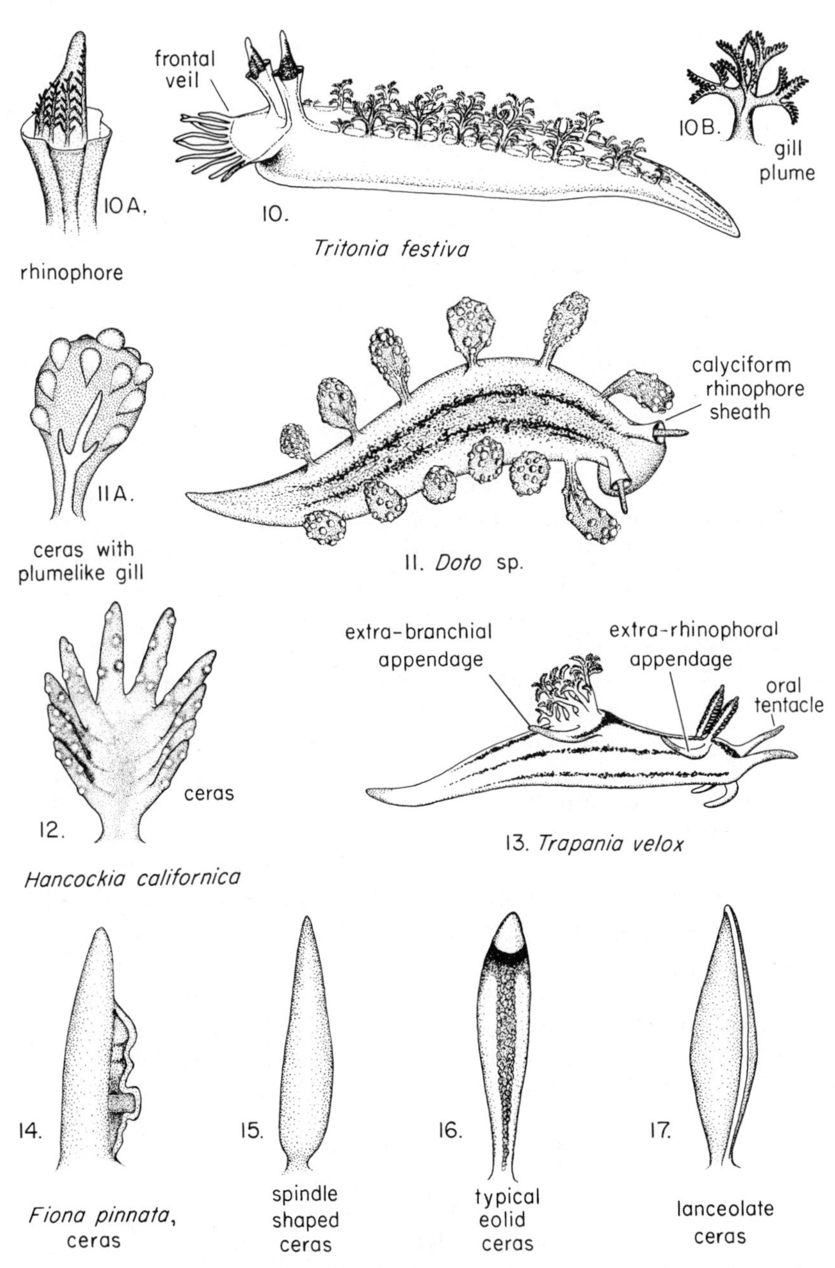

PLATE 122. **Opisthobranchs** (2). 10, *Tritonia festiva;* 10A, rhinophore; 10B, gill plume; 11, *Doto sp.;* 11A, *Doto,* ceras showing plumelike gill; 12, *Hancockia,* single ceras showing palmate shape; 13, *Trapania velox;* 14, *Fiona pinnata,* ceras showing sail-like expansion; 15, typical spindle-shaped ceras; 16, typical eolid ceras; 17, lanceolate ceras. Figures 10–13 from MacFarland (1966) and 14–17 from 35 mm slides, all redrawn by Milos Radakovich.

are usually a darker pink than dorsal processes; 15 mm .... *Hopkinsia rosacea*

– Body color not pink 21

21. Dorsal processes club-shaped and tipped with orange red; 5 branchial plumes; ground color white to pale yellow; white or orange low tubercles medially on dorsum; rhinophores tipped with orange red; 12 mm .... *Laila cockerelli*

– Dorsal processes not club-shaped, not tipped with orange red 22

22. Dorsum with a single median process; 8–11 branchial plumes; body flattened dorsoventrally; ground color whitish with minute brownish spots; 5 marginal processes on each side; 8 mm .... *Okenia plana*

– Dorsum with 5–7 median processes; 5–7 branchial plumes; body limaciform (sluglike); ground color translucent white with small yellow flecks and greenish gray patches; with numerous rather long papillae; 5 mm .... *Okenia angelensis*

23. Dorsum with tubercles or processes medially, in addition to dorsolateral and frontal marginal process; ground color whitish 24

– Dorsum smooth medially; with dorsolateral and frontal marginal processes; ground color yellow or orange 25

24. Dorsal processes and frontal marginal processes black, tipped with a subapical orange ring; 3 branchial plumes; 10 mm .... *Crimora coneja*

– Dorsal process and frontal marginal processes orange-tipped; 5 branchial plumes; 40 mm .... *Triopha carpenteri*

25. Branchial plumes whitish with reddish tips; ground color pale yellow to yellowish brown; dorsal processes tipped with reddish orange; powder blue spots in dorsum; 40 mm .... *Triopha grandis*

– Branchial plumes yellowish with orangish tips; ground color orange to orange brown, usually quite dark; dorsal processes usually tipped with dark orange; dorsum with numerous small blue dots in larger specimens; 15 mm .... *Triopha maculata*

26. Branchial plumes in 2 semicircles (⌒⌒) or in 3 groups directed posteriorly 27

– Branchial plumes in a single circlet 28

27. Branchial plumes 16–32, in form of 2 semicircles united anteriorly (⌒⌒); ground color brownish white, usually with 2–3 irregular, longitudinal, brownish stripes; many large, conspicuous tubercles on dorsum; 15 mm ....*Onchidoris bilamellata*

– Branchial plumes 3, directed posteriorly from beneath 3 thick, bluntly triangular lobes; ground color raw umber with

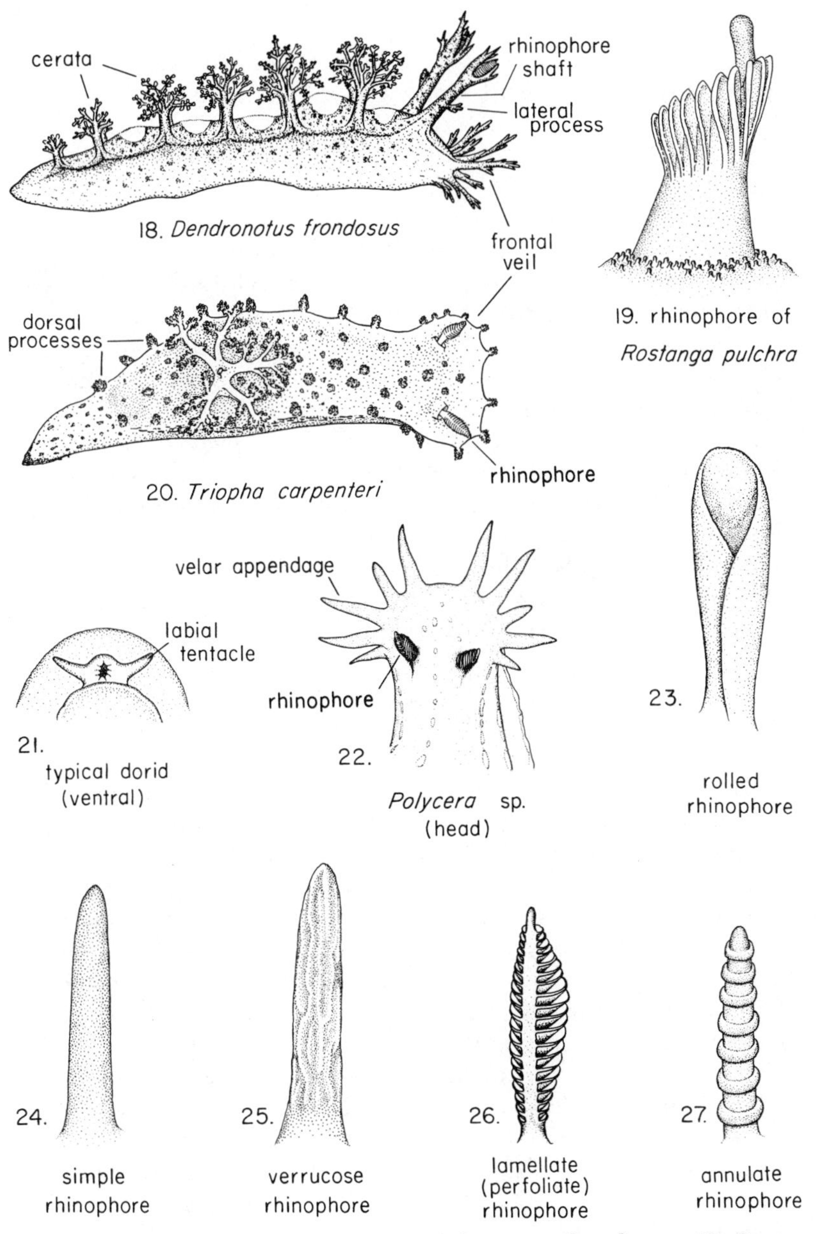

PLATE 123. **Opisthobranchs** (3). 18, *Dendronotus frondosus;* 19, *Rostanga pulchra,* detail of rhinophore; 20, *Triopha carpenteri;* 21, ventral view of typical dorid nudibranch showing labial tentacles; 22, *Polycera,* detail of head showing velar appendages; 23, rolled rhinophore; 24, simple rhinophore; 25, verrucose or warty rhinophore; 26, lamellate or perfoliate rhinophore; 27, annulate rhinophore. 18, 19, 21, 24–27 from MacFarland (1966); 22, 23 from 35 mm slides, redrawn by Milos Radakovich; 20 redrawn from MacFarland (1966) by Emily Reid.

many small dark brown to black spots on dorsum; dorsum with many velvety papillae; dorsum with a prominent, irregular ridge extending along midline; 12 mm .... *Atagema quadrimaculata*

28. Labial tentacles (fig. 21) absent or minute and rudimentary 29
– Labial tentacles present and obvious 30
29. Ground color yellow to reddish brown; dorsum with many low tubercles, each with a small, central white dot; 5 branchial plumes, white to pale yellowish; 25 mm .... *Doriopsilla albopunctata*
– Ground color white; dorsum with numerous brown to black blotches; dorsum smooth; mantle margin undulating; body rather long and narrow; 20 mm .... *Dendrodoris* sp.
30. Body color red; or if orange, then dorsum velvety (with minute papillae); 8 or more branchial plumes, same color as body 31
– Body color not red; if orange, then dorsum with large distinct tubercles or papillae; less than 8 branchial plumes; white.. 33
31. Rhinophores with a blunt, cylindrical process above the lamellae (fig. 19), nearly 1/4 the length of the entire rhinophore; frequently with minute brownish black dots on dorsum; 10 mm .... *Rostanga pulchra*
– Rhinophores perfoliate (fig. 26); lacking blunt, cylindrical process above lamellae 32
32. With 1 or 2 black spots on midline of dorsum, one just posterior to rhinophores and the other just anterior to the branchial plumes (rarely one or both may be lacking); 20 mm .... *Aldisa sanguinea*
– Lacking black spots; dorsum deep dark red in color, rather velvety smooth; body greatly flattened dorsoventrally; 30 mm .... *Platydoris macfarlandi*
33. Ground color blue or deep violet; dorsum smooth, lacking any obvious tubercles or papillae 34
– Ground color not blue or deep violet; dorsum with tubercles or papillae 36
34. Ground color deep violet; dorsum with 3 bright yellow, longitudinal lines; edge of dorsum with a submarginal white band; 25 mm .... *Chromodoris macfarlandi*
– Ground color blue 35
35. Dorsum with 2 longitudinal series of about 7 yellow to golden spots; anterior edge of dorsum and sides of foot also with yellow spots; ventral surface of posterior edge of dorsum with a series of about 8 small white dots; 40 mm .... *Hypselodoris californiensis*
– Dorsum with 2 bright yellow, longitudinal stripes, ending an-

teriorly at the rhinophores, and a single bright yellow stripe anterior to rhinophores; a median light blue line extending from between the rhinophores to the branchial plumes; margins of dorsum with narrow white band; 15 mm .... *Chromodoris porterae*

36. Labial tentacles large and auriculate (ear-shaped), reaching nearly to edge of dorsum; dorsum thickly set with long papillae 37
– Labial tentacles small and fingerlike or triangular, reaching less than halfway to edge of dorsum; dorsum with or without papillae 42
37. Ground color orange; dorsum with numerous flecks of yellow; 9 branchial plumes, whitish; usually with a pungent odor of cedar or sandalwood; 20 mm .... *Acanthodoris lutea*
– Body color not orange 38
38. Ground color brownish, flecked with irregular blotches of black; 7 branchial plumes, tipped with lemon yellow, usually with a pungent odor of cedar; 15 mm *Acanthodoris brunnea*
– Ground color white to gray white 39
39. Dorsal papillae tipped with black 40
– Dorsal papillae tipped with yellow 41
40. Edge of dorsum with an outer yellow and inner black line; 5–7 branchial plumes; rhinophores and branchial plumes tipped with reddish brown to black; 15 mm .... *Acanthodoris rhodoceras*
– Edge of dorsum without an outer yellow and inner black line; 9 branchial plumes; 15 mm .... *Acanthodoris pilosa*
41. Rhinophores and branchial plumes tipped with brown or maroon; 9 branchial plumes; 15 mm .... *Acanthodoris nanaimoensis*
– Rhinophores and branchial plumes tipped with yellow; 5 branchial plumes; edge of dorsum with yellow lines; 15 mm .... *Acanthodoris hudsoni*
42. Dorsum with black or brown pigment, not including rhinophores or branchial plumes 43
– Dorsum lacking black or brown pigment 49
43. Dorsum with a lateral series of small brown to black dots with yellow centers; ground color salmon to yellowish pink; 12 branchial plumes; 15 mm .... *Cadlina sparsa*
– Dorsum lacking black dots with yellow centers 44
44. Rhinophores tipped with maroon to brownish black; dorsum with small tubercles, largest of which are tipped with brown; ground color yellowish; 15 mm .... *Hallaxa* sp.
– Rhinophores not tipped with maroon to brownish black ... 45

45. Blackish pigment in form of many very small dots, with major concentration of pigment forming a blotch just anterior to branchiae; ground color yellow to yellowish brown; 8–10 branchial plumes, tripinnate; rhinophores perfoliate with 10–15 lamellae; tubercles small; 20 mm . . . *Discodoris heathi*
 – Blackish pigment not in form of many very small dots; lacking a major concentration of pigment just anterior to branchiae; branchial plumes less than 8 46
46. Dorsum with small tubercles giving dorsum a velvety appearance; dorsum with a few brown to black markings, generally as conspicuous rings, but occasionally as blotches; ground color whitish yellow to very pale brown; 6 branchial plumes, tripinnate; 20 mm . . . . *Diaulula sandiegensis*
 – Dorsum with large, conspicuous tubercles 47
47. Dorsum with 2 brown to blackish spots on midline, ground color whitish to yellow; dorsum with many pointed papillae; branchial plumes 6; 25 mm . . . . *Thordisa bimaculata*
 – Dorsum lacking 2 brown to blackish spots on midline 48
48. Blackish pigment in blotches on both dorsum and tubercles; 7 branchial plumes, yellowish; ground color light yellow; 25 mm . . . . *Archidoris montereyensis*
 – Blackish blotches on dorsum only, not on tubercles; 6 branchial plumes, whitish; ground color light yellow to yellow orange; 25 mm . . . . *Anisodoris nobilis*
49. Rhinophores brown to black 50
 – Rhinophores white to yellow 51
50. Branchial plumes brown to black; ground color whitish; dorsal and ventral surfaces of foot and sides of body with minute black spots; 15 mm . . . . *Cadlina limbaughi*
 – Branchial plumes white to yellow; ground color whitish to yellow; lateral edges of dorsum with 7–10 yellow dots on each side, anteriormost spot usually posterior to rhinophores; 15 mm . . . . *Cadlina flavomaculata*
51. Dorsum with yellow line around edge and with yellow-tipped tubercles; ground color whitish to pale yellow; rhinophores perfoliate with 16–18 lamellae; 6 branchial plumes, tipped with yellow; 25 mm . . . . *Cadlina luteomarginata*
 – Dorsum lacking yellow line around edge 52
52. Dorsum with many long, slender papillae giving animal a fuzzy appearance; 10–12 branchial plumes, simply pinnate; ground color white to yellowish white; dorsum with many opaque, white flecks; 12 mm . . . . *Onchidoris hystricina*
 – Dorsum lacking long, slender papillae 53
53. Dorsum with many tubercles, bulbous at tip, appearing

nearly spherical; ground color yellowish whitish to yellow orange; rhinophores orange yellow, with yellow tips; tubercles tipped with orange, 5 mm .... *Onchidoris muricata*
– Tubercles, if present, not bulbous, not appearing spherical 54
54. Ground color pure white with white tubercles of various sizes on dorsum; 7 branchial plumes, white, rather fluffy in appearance; rhinophores with 20–24 lamellae; 60 mm (fig. 9) .... *Archidoris odhneri*
– Ground color pale yellow; 10–12 branchial plumes, yellowish white; rhinophores with 10–12 lamellae; dorsum with several small yellow dots along lateral edges, anteriormost spots extending in front of rhinophores; 25 mm .... *Cadlina modesta*
55. With cerata (figs. 8, 18), or branched branchial plumes arranged laterally on dorsum (figs. 10, 10B) 56
– Without cerata 98
56. Cerata or branchial plumes branched (figs. 10, 12, 18) 57
– Cerata not branched (figs. 7, 8, 11) 65
57. Cerata palmately branched, each with 4–16 digitiform projections (fig. 12); cerata in 4–7 pairs; head with a broad, palmate, velar lobe on either side, each with 6–10 or more unequal digitiform processes; ground color reddish brown in mature individuals, younger individuals translucent greenish brown; 15 mm .... *Hancockia californica*
– Cerata not palmately branched 58
58. Processes on frontal veil simple (fig. 10) or lacking 59
– Processes on frontal veil branched (fig. 18) 62
59. Dorsum distinctly tuberculate, each tubercle tipped with white; ground color deep orange yellow; body margins with an irregular series of low, white branchial plumes; foot light salmon pink or yellow and edged with a narrow white band; 20 cm .... *Tochuina tetraquetra*
– Dorsum smooth, not distinctly tuberculate 60
60. Ground color white to yellowish white; frontal veil (fig. 10) with 7–12 digitiform processes; body margins with 11–15 branchial plumes (fig. 10B) on each side; fine, reticulate, opaque white lines on dorsum; foot edged with a narrow band of opaque white; 20 mm .... *Tritonia festiva*
– Ground color rose pink; frontal veil with 10–30 low, white tubercles; body margins with about 30 branchial plumes; margins of foot, margins of dorsum, and the edge of rhinophoral sheath bear a narrow, white line 61
61. Penis elongated, with terminal bulb and subterminal ridge; 15 cm .... *Tritonia diomedea*

– Penis squat and conical; 15 cm .... *Tritonia exsulans*
62. Rhinophore shaft with a lateral process (fig. 18) 63
– Rhinophore shaft lacking a lateral process; frontal veil with 4–6 stout, branched processes; margin of rhinophore sheath with 5–7 short, blunt processes which are shorter than clavus; dorsum with 4 distinct, longitudinal, light brown lines running from posterior of rhinophore shaft to tip of tail; ground color extremely variable, may be yellow, brown, orange, greenish, or white; about 6 pairs of cerata; 15 mm .... *Dendronotus subramosus*
63. Posterior face of rhinophore shaft with vertical row of 3–6 small, slightly branched processes; lateral process arising near base of rhinophore shaft; usually 4 pairs of branched velar processes; dorsal edge of foot usually with an opaque white line; ground color may be white, gray, brownish, orange red, or maroon; 4–7 pairs of cerata which may be tipped with orange, yellow, or purple; 60 mm .... *Dendronotus iris*
– Posterior face of rhinophore shaft lacking vertical row of small, slightly branched processes; margin of rhinophore sheath usually with 5 long processes 64
64. Ground color white; cerata in 4–5 pairs, usually tipped with orange red; frontal veil with 2 pairs of branched processes; 15 mm .... *Dendronotus albus*
– Ground color variable, translucent grayish white to greenish or even brownish; cerata in 5–8 pairs; frontal veil usually with 3 pairs of branched processes (fig. 18); 15 mm .... *Dendronotus frondosus*
65. With large oral hood (fig. 7), having 2 rows of cirri on margin, outer series long, inner series shorter; about 5–6 pairs of petal-like cerata; ground color yellowish brown to greenish brown; rhinophore shaft with a thin, triangular, sail-like expansion on inner margin; 35 mm .... *Melibe leonina*
– Lacking large oral hood with cirri; lacking sail-like expansion on inner margin of rhinophore shaft 66
66. Rhinophores perfoliate (fig. 26) or annulate (fig. 27), not smooth or warty 67
– Rhinophores smooth (simple) (fig. 24) or warty (fig. 25), not perfoliate or annulate 78
67. Cerata extending well in front of rhinophores and same color as rhinophores 68
– Cerata not extending well in front of rhinophores 70
68. Cerata spindle-shaped (fig. 15); lacking frontal veil; oral tentacles present; cerata with a subapical band of yellow orange

and tips blue or white; an orange red crest between rhinophores; ground color translucent grayish white; 20 mm .... *Antiopella barbarensis*

– Cerata lanceolate (fig. 17); with a distinct frontal veil; oral tentacles absent 69

69\. Cerata usually with irregular bumps; ground color light brown to greenish gray with fine yellowish white, olive green, and pink dots; with a pale red spot on the outer side of each ceras; 20 mm .... *Dirona picta*

– Cerata smooth, the lateral margin of each with an opaque white line; anterior edge of frontal veil and median crest of tail with opaque white line; ground color translucent grayish white; 25 mm .... *Dirona albolineata*

70\. Rhinophores arising from a single median stalk, Y-shaped; cerata pinkish red on proximal $1/2$, followed distally by a wide band of opaque white, then a narrow band of yellow orange below translucent tip; ground color whitish to pinkish, 15 mm .... *Babakina festiva*

– Rhinophores not on a single, common shaft 71

71\. With distinct white or pigmented lines on dorsum, edge of foot, or between oral tentacles 72

– Lacking distinct pigmented lines on dorsum, edge of foot, or between oral tentacles 74

72\. With a narrow orange red line between and on the proximal $1/2$ of the oral tentacles; cerata with pink tinge throughout; 30 mm .... *Phidiana pugnax*

– Lacking narrow, orange red line between oral tentacles; one or more pigmented lines running the length of the body ... 73

73\. With an opalescent blue line along edge of foot; double, opalescent, blue median line running the length of the body and enclosing a bright orange yellow elongate spot between and posterior to rhinophores and another in cardiac region; cerata with a subapical band of orange yellow, cores extremely variable in color; 30 mm .... *Hermissenda crassicornis*

– Lacking opalescent blue lines; an opaque white line dorsomedially and similar lines dorsolaterally; tips of cerata whitish, cores usually reddish to orange; 20 mm .... *Coryphella trilineata*

74\. Ground color deep purple; rhinophores perfoliate (fig. 26), lamellae deep maroon; cerata violet near base, grading to flaming scarlet or orange distally; 30 mm .... *Flabellinopsis iodinea*

– Ground color not deep purple 75

75\. Rhinophores annulate (fig. 27) 76

– Rhinophores perfoliate, lamellae nearly vertical; corners of foot produced 77

76. Oral tentacles and rhinophores brilliant vermillion above, with white tips; ground color light pinkish; 20 mm .... *Facelina stearnsi*

– Oral tentacles grayish white, encrusted with opaque white dots; rhinophores pale yellowish green distally; ground color grayish white; 20 mm (fig. 8) ....*Coryphella pricei*

77. Cerata orange red with white tips; ground color grayish white; dorsum along midline free of cerata; head usually with a pale orange spot; 20 mm ....*Spurilla oliviae*

– Cerata cores greenish brown, cerata with pink tinge, tipped with white; ground color dull orange; midline of dorsum with irregular patches of opaque white; oral tentacles very stout; 20 mm ....*Spurilla chromosoma*

78. With oral tentacles (fig. 13) 79

– Without oral tentacles; ground color white to yellowish white; a few black specks on dorsum; cerata usually in groups of 2; 3 mm ....*Tenellia adspersa*

79. Cerata each with thin, sail-like expansion on posterior edge from base upward for at least 3/4 of length of ceras (fig. 14); ground color translucent gray to brownish; 20 mm .... *Fiona pinnata*

– Cerata lacking posterior, sail-like membrane 80

80. Rhinophore sheath calyciform (fig. 11); large, bulbous cerata with a small, plumelike gill on inner surface (fig. 11A) 81

– Rhinophore sheaths lacking; cerata without plumelike gills 83

81. With brown to black pigment on body or cerata 82

– Lacking brown to black pigment on body or cerata; cerata cores orange to pinkish, cerata rather bulbous; ground color pale yellowish white; tubercles on cerata elongate; anal papilla high; 10 mm ....*Doto amyra*

82. Cerata with black rings at base of tubercles; upper border of rhinophore sheaths smooth; ground color whitish with brown pigment on head, back, and sides; 10 mm *Doto columbiana*

– Cerata lacking black rings at base of tubercles; upper border of rhinophore sheaths somewhat scalloped; ground color white to yellowish; cerata long with yellowish to brown cores; 10 mm ....*Doto kya*

83. Dorsum with distinct, median, white line; a greenish blue patch anterior and to the left of the cardiac region; cerata greenish at base with white tips, cores reddish brown; ground color white with pinkish tinge; 20 mm *Coryphella cooperi*

– Dorsum lacking distinct median white line 84

84. Rhinophores with various amounts of orange to red pigment 85
– Rhinophores lacking orange to red pigment 88
85. Dorsum with distinct opaque white blotches medially; foot corners distinct; vermillion on oral tentacles, rhinophores, head, and back; cerata tipped with white, and narrow white band below tip, cores brownish; 30 mm .... *Aeolidiella takanosimensis*
– Dorsum lacking distinct opaque white blotches medially; foot corners indistinct 86
86. Oral tentacles with white pigment on at least distal 1/2; tail with a median, opaque white line 87
– Oral tentacles orange red with white tips; tail lacking median, opaque white line; cerata recumbent with broad, opaque white band extending from base to tip of each ceras, covering distal 1/3, cores brownish; 8 mm .... *Catriona alpha*
87. Rhinophores white distally; light orange pigment covering front of head, extending 1/4 of the way up the rhinophores and on bases of oral tentacles; cerata with small, opaque white spots and with cores brownish to dark green; 8 mm .... *Trinchesia flavovulta*
– Rhinophores orange red for nearly entire length; large, orange red spot between oral tentacles and anterior to rhinophores; cerata with orange red tips and with blackish cores; 10 mm .... *Trinchesia lagunae*
88. Cerata irregularly inflated, warty in appearance 89
– Cerata not irregularly inflated, not warty 90
89. Rhinophores and oral tentacles with subapical band of brownish to light gray or greenish; sides of body with small yellow dots; 8 mm .... *Eubranchus rustyus*
– Rhinophores and oral tentacles lacking subapical bands; ground color yellowish white; numerous distinct brown dots on dorsum extending up onto cerata; each ceras with a subapical yellow ring; 20 mm .... *Eubranchus misakiensis*
90. With 2 median, light orange, oval spots, one spot anterior and one posterior to base of rhinophores; opaque white dots on distal 1/3 of rhinophores and oral tentacles; an irregular series of white blotches on median line of dorsum; cerata with 3 dark brownish black bands, core ochre; 10 mm .... *Emarcusia morroensis*
– Lacking median, light orange spots anterior and posterior to rhinophores 91
91. With band of purple midway on oral tentacles and rhinophores; cerata cores yellowish at base, olive green to brown in middle, and yellow at tip; 8 mm ...... *Trinchesia abronia*

– Lacking band of purple midway on rhinophores and oral tentacles 92

92. Cerata frosted with opaque white on distal 2/3; dorsum encrusted with opaque white; cerata cores pale green to raw umber; distal 1/3 of rhinophores encrusted with white; 8 mm .... *Trinchesia albocrusta*

– Cerata not frosted with opaque white on distal 2/3 93

93. With yellow bands near tips and bases of cerata, or with yellow on frontal margin between tentacles 94

– Lacking yellow bands on cerata or between oral tentacles 95

94. With yellow on frontal margin between oral tentacles; cerata with orange flecks, cores green with some brown; rhinophores and oral tentacles with white dots on distal 1/2; 8 mm .... *Trinchesia virens*

– Lacking yellow on frontal margin; cerata with yellow band near tip and another near base, cores yellow-brown to dark brown, body surface with white dots; 8 mm .... *Trinchesia fulgens*

95. With band of opaque white below tip of each ceras; cerata core greenish brown to brown; dorsum and cerata with opaque white dots; 8 mm ....*Trinchesia* sp.

– Lacking band of opaque white below tip of each ceras 96

96. Cerata flattened, broad below and pointed at tip; middle of dorsum free of cerata and encrusted with opaque white; ground color white, grayish, or pinkish; 25 mm .... *Aeolidia papillosa*

– Cerata not flattened, but round in cross-section 97

97. Cerata very long, longest about equal to body length; cerata cores brownish yellow; rhinophores united at base; foot forming pointed angles at the anterior sides; ground color grayish green to brownish; 6 mm ....*Cumanotus beaumonti*

– Cerata less than 1/3 body length; cerata cores reddish brown to pink; anterior margin of foot rounded; cerata tips white; ground color cream to pink; 15 mm ....*Precuthona divae*

98. Dorsum with undulating, longitudinal ridges of white on a pinkish brown background; gills located laterally below dorsum edge on either side; edge of dorsum with an anteromedian notch through which the rhinophores project; clavus of rhinophores with many longitudinal grooves; 25 mm ....*Armina californica*

– Dorsum without undulating longitudinal ridges; gills located posteroventrally; ground color pale gray, with small, yellowish brown blotches and reticulate lines 99

99. Dorsum with posterior notch through which gills may be

seen; rhinophore shaft with platelike expansion; 5 mm .... *Corambe pacifica*

– Dorsum without posterior notch; rhinophores smooth and tapering (fig. 24); 5 mm .... *Doridella steinbergae*

LIST OF SACOGLOSSA AND NUDIBRANCHIA

Sacoglossa

Alderia modesta (Lovén, 1844). Common on mats of the alga *Vaucheria* at edges of *Salicornia* marshes.

Aplysiopsis oliviae MacFarland, 1966 (=*Hermaeina oliviae*).

Aplysiopsis smithi Marcus, 1961 (=*Hermaeina smithi; Phyllobranchopsis enteromorphae* Cockerell and Eliot, 1905 in MacFarland, 1966). Usually on *Chaetomorpha, Rhizoclonium,* or *Enteromorpha* spp. See Gonor, 1961, Veliger 4: 85–98 (biology).

Elysia hedgpethi Marcus, 1961 (= *E. bedeckta* MacFarland, 1966). Usually on *Bryopsis* spp. or *Codium fragile.*

† *Hermaea vancouverensis* O'Donoghue, 1924. See Williams and Gosliner, 1973, Veliger 16: 114.

Placida dendritica (Alder and Hancock, 1843) (= *Hermaea ornata* MacFarland, 1966). Usually on *Bryopsis* spp. or *Codium fragile.*

Stiliger fuscovittatus Lance, 1962. Found on *Polysiphonia* and other algae; see Lance, 1962, Veliger 5: 33–38.

Nudibranchia

Acanthodoris brunnea MacFarland, 1905.

† *Acanthodoris hudsoni* MacFarland, 1905.

Acanthodoris lutea MacFarland, 1925.

Acanthodoris nanaimoensis O'Donoghue, 1921.

Acanthodoris pilosa (Abildgaard, 1789).

Acanthodoris rhodoceras Cockerell and Eliot, 1905.

Aegires albopunctatus MacFarland, 1905.

Aeolidia papillosa (Linnaeus, 1761). Common; frequently found with sea anemones, on which it feeds; see Waters, 1973, Veliger 15: 174–192 (food preferences).

† Rare in this area.

† *Aeolidiella takanosimensis* Baba, 1930. In bays and harbors; possibly introduced.

Aldisa sanguinea (Cooper, 1862).

Ancula lentiginosa Farmer, in Farmer and Sloan, 1964. See Farmer and Sloan, 1964, Veliger 6: 148–150.

Ancula pacifica MacFarland, 1905.

Anisodoris nobilis (MacFarland, 1905).

Antiopella barbarensis (Cooper, 1863) (= *A. aureocincta* MacFarland, 1966).

Archidoris montereyensis (Cooper, 1862). Commonly feeding on midtidal sponge *Halichondria panicea;* see Cook, 1962, Veliger 4: 194–196.

Archidoris odhneri (MacFarland, 1966) (=*Austrodoris odhneri*).

Armina californica (Cooper, 1862). Usually subtidal, frequently with sea pens or sea pansies (*Renilla*).

† *Atagema quadrimaculata* Collier, 1963 (= *Petelodoris spongicola* MacFarland, 1966).

† *Babakina festiva* (Roller, 1972) [=*Babaina festiva* (see Roller, 1973, Veliger 16: 117–118)].

Cadlina flavomaculata MacFarland, 1905.

† *Cadlina limbaughi* Lance, 1962. See Lance, 1962, Veliger 9: 155–159.

Cadlina luteomarginata MacFarland, 1966.

Cadlina modesta MacFarland, 1966.

Cadlina sparsa (Odhner, 1921).

Catriona alpha (Baba and Hamatani, 1963) (= *Cratena spadix* MacFarland, 1966). See Baba and Hamatani, 1963, Publ. Seto Mar. Biol. Lab. 11: 339–344.

Chromodoris macfarlandi Cockerell, 1902 (=*Glossodoris macfarlandi*).

Chromodoris porterae Cockerell, 1902 (= *Hypselodoris porterae*).

† *Conualevia alba* Collier and Farmer, 1964.

Corambe pacifica MacFarland and O'Donoghue, 1929. On *Membranipora* growing on *Macrocystis*, closely resembling the bryozoan.

† *Coryphella cooperi* Cockerell, 1901. Frequently found on the pink hydroid *Tubularia crocea.*

† *Coryphella pricei* MacFarland, 1966.

Coryphella trilineata O'Donoghue, 1921 (=*C. fisheri* MacFarland, 1966).

† *Crimora coneja* Marcus, 1961.

Cumanotus beaumonti (Eliot, 1906). See Hurst, 1967, for account of egg masses and veligers of this and other opisthobranchs.

†* *Dendrodoris nigromaculata* (Cooper, 1901).

Dendrodoris sp.

Dendronotus albus MacFarland, 1966.

Dendronotus frondosus (Ascanius, 1774) (= *Dendronotus venustus* MacFarland, 1966).

Dendronotus iris Cooper, 1863. Usually found subtidally with *Pachycerianthus fimbriatus* upon which it feeds; see Wobber, 1970, Veliger 12: 383–387.

Dendronotus subramosus MacFarland, 1966.

Diaulula sandiegensis (Cooper, 1862).

Dirona albolineata Cockerell and Eliot, 1905.

Dirona picta MacFarland in Cockerell and Eliot, 1905.

Discodoris heathi MacFarland, 1905.

Doridella steinbergae (Lance, 1962) (= *Corambella bolini* MacFarland, 1966). Usually on *Membranipora* growing on *Macrocystis;* see Lance, 1962, Veliger 5: 33–38.

Doriopsilla albopunctata (Cooper, 1863) [=*Dendrodoris albopunctata* and *D. fulva* (MacFarland, 1905)].

†* *Doris tanya* Marcus, 1971.

†* *Doris* sp. (sensu lato) MacFarland, 1966.

Doto amyra Marcus, 1961 [= *D. varians* MacFarland, 1966 (in part)]. Among hydroids on boat docks.

Doto columbiana O'Donoghue, 1921.

Doto kya Marcus, 1961. [=*D. varians* MacFarland, 1966 (in part)].

Emarcusia morroensis Roller, 1972.

Eubranchus misakiensis Baba, 1960. See Baba, 1960, Publ. Seto Mar. Biol. Lab. 8: 299–302.

* Not in key.

* *Eubranchus olivaceus* (O'Donoghue, 1922).

Eubranchus rustyus (Marcus, 1961) (=*E. occidentalis* MacFarland, 1966 and *Capellinia rustya*).

† *Facelina stearnsi* Cockerell, 1901.

Fiona pinnata (Eschscholtz, 1831). Usually with goose barnacles (*Lepas*) on floating objects.

Flabellinopsis iodinea (Cooper, 1862) (=*Flabellina iodinea*).

Hallaxa sp. An undescribed dorid-like species, superficially resembling but not related to *Cadlina*.

Hancockia californica MacFarland, 1923.

Hermissenda crassicornis (Eschscholtz, 1831). Ubiquitous and very common; see Bürgin, 1965, Veliger 7: 205–215 (color patterns).

Hopkinsia rosacea MacFarland, 1905.

Hypselodoris californiensis (Bergh, 1879) (=*Chromodoris californiensis*).

Laila cockerelli MacFarland, 1905.

Melibe leonina (Gould, 1852) (=*Chioraera leonina*). Usually found on *Macrocystis;* catches small crustaceans with oral hood.

Okenia angelensis Lance, 1966. Usually on pilings in bays.

Okenia plana Baba, 1960. See Baba, 1960, Publ. Seto Mar. Biol. Lab. 8: 79–83.

Onchidoris bilamellata (Linnaeus, 1767). Usually found with *Balanus* spp.

Onchidoris hystricina (Bergh, 1878).

† *Onchidoris muricata* (Müller, 1776) [=*O. aspera* (Alder and Hancock, 1842)].

Phidiana pugnax Lance, 1962 (=*P. nigra* MacFarland, 1966). See Lance, 1962, Veliger 4: 155–159.

† *Platydoris macfarlandi* Hanna, 1951.

Polycera atra MacFarland, 1905. Usually on the ectoproct *Bugula pacifica.*

Polycera hedgpethi Marcus, 1964. Usually on *Bugula pacifica;* see Marcus, 1964, Nautilus 77: 128–131.

Polycera tricolor Robilliard, 1971. See Robilliard, 1971, Syesis 4: 235–243.

† *Polycera zosterae* O'Donoghue, 1924. Often on the ectoproct *Bowerbankia gracilis.*

Precuthona divae Marcus, 1961 (= *Cuthona rosea* MacFarland, 1966). Preys on the hydrozoan *Hydractinia*.

Rostanga pulchra MacFarland, 1905. Commonly on red sponges, which it and its eggs match in color; see Cook, 1962, Veliger 4: 194–196 (feeding).

Spurilla chromosoma Cockerell and Eliot, 1905.

Spurilla oliviae (MacFarland, 1966) (=*Aeolidiella oliviae*).

† *Tenellia adspersa* (Nordmann, 1845) (= *T. pallida* Alder and Hancock, 1855). In bays and harbors; feeds on *Tubularia crocea*.

Thordisa bimaculata Lance, 1966.

Tochuina tetraquetra (Pallas, 1788).

† *Trapania velox* (Cockerell, 1901).

Trinchesia abronia (MacFarland, 1966). This and the following named species of *Trinchesia* were formerly in *Cratena*.

Trinchesia albocrusta (MacFarland, 1966).

Trinchesia flavovulta (MacFarland, 1966).

Trinchesia fulgens (MacFarland, 1966).

Trinchesia lagunae (O'Donoghue, 1926) (=*Cratena rutila* MacFarland, 1966).

† *Trinchesia virens* (MacFarland, 1966).

Trinchesia sp. In brackish waters of San Francisco Bay; feeds on the sea anemone *Haliplanella luciae;* probably introduced.

Triopha carpenteri (Sterns, 1873).

Triopha grandis MacFarland, 1905.

Triopha maculata MacFarland, 1905 (= *T. aurantiaca* Cockerell, 1908).

Tritonia diomedea Bergh, 1894 (=*Duvaucelia gilberti* MacFarland, 1966). This and the following species of *Tritonia* were formerly in *Duvaucelia*.

Tritonia exsulans Bergh, 1894.

Tritonia festiva (Stearns, 1873). See Gomez, 1973, Veliger 16: 163 (feeding).

REFERENCES ON OPISTHOBRANCHIA

Beeman, R. D. 1968a. The use of succinylcholine and other drugs for anesthetizing or narcotizing gastropod mollusks. Publ. Staz. Zool. Napoli 36: 267–270.

Beeman, R. D. 1968b. The order Anaspidea. Veliger 3 (Suppl.): 87–102.

Collier, L. and W. M. Farmer 1964. Additions to the nudibranch fauna of the east Pacific and the Gulf of California. Trans. San Diego Soc. Nat. Hist. 13: 377–396, pls. 1–6.

Hurst, A. 1967. The egg masses and veligers of thirty northeast Pacific opisthobranchs. Veliger 9: 255–288.

Lance, J. R. 1966. New distributional records of some northeastern Pacific Opisthobranchiata (Mollusca-Gastropoda) with descriptions of two new species. Veliger 9: 69–81.

MacFarland, F. M. 1966. *Studies of Opisthobranchiate Mollusks of the Pacific Coast of North America.* Mem. Calif. Acad. Sci. 6: 546 pp.

Marcus, Ernst 1961. Opisthobranch mollusks from California. Veliger 3: (Suppl.): 1–85.

Robilliard, G. A. 1969. A method of color preservation in opisthobranch mollusks. Veliger 11: 289–291.

Robilliard, G. A. 1970. The systematics and some aspects of the ecology of the genus *Dendronotus.* Veliger 12: 433–479.

Roller, R. A. 1970. A list of recommended nomenclatural changes for MacFarland's "Studies of Opisthobranchiate Mollusks of the Pacific Coast of North America." Veliger 12: 371–374.

Roller, R. A. 1972. Three new species of eolid nudibranchs from the West Coast of North America (Gastropoda-Opisthobranchia). Veliger 14: 416–423.

Runham, N. W., K. Isarankura, and B. J. Smith 1965. Methods for narcotizing and anaesthetizing gastropods. Malacologia 2: 231–238.

Russell, H. D. 1971. *Index Nudibranchia, a Catalog of the Literature 1554–1965.* Greenville: Delaware Mus. Nat. Hist., 141 pp.

PHYLUM MOLLUSCA: BIVALVIA

Eugene V. Coan and James T. Carlton
California Academy of Sciences, San Francisco

(PLATES 124–132)

The **Bivalvia,** which include clams, cockles, scallops, oysters, mussels, piddocks, and shipworms, are fundamentally bilaterally symmetrical molluscs in which the mantle encloses the head, foot, and visceral mass, and secretes a shell in the form of two lateral valves, hinged dorsally. With the retreat of the body from direct contact with the substrate, a unique mode of feeding using the ctenidia has developed.

Although the exact mode of feeding of ancestral bivalves is not known, it is likely (Stasek, 1961) that suspension feeding by palps and ctenidia was developed with the enclosure of the body by the mantle and shell. Modern representatives of the **Nuculoida** retain several primitive morphological characters, and also show a primitive form of suspension feeding. However, such forms today also possess specialized, elongate, palp appendages which can be extended from within the shell to sweep up detritus with cilia and to convey it to the mouth (Yonge, 1939). The radula and other structures of the head were lost as an indirect mode of feeding developed. The "protobranch" ctenidia of the primitive bivalve, presumably similar to those of *Nucula* (not represented in the local intertidal zone) or of *Acila* (common subtidally at Friday Harbor, Washington) consist of a central axis bearing, on either side, a series of flattened, ciliated filaments.

The evolution of "lamellibranch" ctenidia, which characterize the vast majority of modern bivalves, was accomplished by the elongation of the filaments, their folding back on themselves so that each ctenidium commonly resembles a tall, narrow W in cross-section, and the binding of adjacent filaments into extensive **lamellae,** as diagrammed in textbooks (although there are specialized departures from the W form). Complex feeding and rejection tracts of cilia on the ctenidia transport food and rejecta (pseudofeces) respectively, and the palps further sort the food before passing it into the mouth. Thus equipped, bivalves have become nearly complete introverts, using their ciliated ctenidia both for respiration and for filtering food from the water. But they are by no means completely out of touch with their environments; they have developed sensory tentacles on mantle edges and at siphonal apertures, and may even possess distinctive eyes along the edges of the mantle, as in *Hinnites* and other scallops.

Ctenidial food collecting, successful as it is, has imposed certain limitations; no bivalve can lead a terrestrial existence, and enclosure of the body within the mantle precludes a really active life. The ability to collect food from the water has ensured a steady source of food, and this has made possible the retreat of many bivalves into protected crevices or burrows.

Bivalves furnish splendid examples of evolutionary diversification and adaptive radiation. Easily recognizable adaptive modifications create features of taxonomic importance. This is of great value to the student, since closely related groups such as families, superfamilies, and orders show a general uniformity in way of life. On the other hand, similar ways of life have also produced parallelisms in structure and adaptation, such as we see among distantly related genera which cement themselves to hard substrates (e.g., *Chama*, *Ostrea*, and *Hinnites*). Certain structures, such as the hinge, often permit the recognition of affinities despite outward dissimilarity. Thus the ligament and cardinal teeth in *Tivela, Spisula,* and *Tresus* readily demonstrate the taxonomic affinity of the latter two genera, while in outward form and way of life *Spisula* and *Tivela* are most similar.

Modern classification of bivalves is based on a wide spectrum of characters, the most important of which are: (1) the structure of the ctenidia, including their relationship to the palps and the types of cilia on them; (2) the mode of life, such as burrowing, boring, attaching with a **byssus,** cementing to a substrate, or free-living; (3) the morphology of the shell, particularly the hinge teeth and the ligament, and the relative sizes and degree of gape of the two valves; (4) the surface sculpture of the valves; (5) the size and position of the adductor muscles pulling the shells closed, which create distinctive scars on the inside of the valves; (6) the degree of fusion of the mantle edges and the presence and nature of siphons; (7) the microstructure and mineralogy of the shell; (8) the morphology of the stomach; and (9) the form of the foot and the presence on it of attachment threads forming the byssus.

A long and abundant bivalve fossil record has enabled systematists to establish relationships and rankings which would be difficult to discern using only living forms. The number of characters now employed in bivalve classification and the seemingly great degree of parallel evolution in the expression of these characters has led workers to adopt "neutral" names for orders, not based on any one set of characters. Bivalves commonly have been, and still are in many texts, divided into the orders **Protobranchia, Filibranchia, Eulamellibranchia,** and **Septibranchia.** This arrangement, although simple and descriptively convenient, breaks down because some groups, such as the Mytiloida, have some members with filibranch and some with eulamellibranch ctenidia, a problem not solved by such attempts as the creation of an order "Pseudolamellibranchia."

Subclass **PALAEOTAXODONTA.** A compact and probably natural group of one order.

Order **Nuculoida.** These have protobranch ctenidia, a primitive taxodont hinge with a row of similar teeth, and they feed in large part by the palps. They are subtidal and are not included in the key.

Subclass **CRYPTODONTA.** This group includes only one modern order.

Order **Solemyoida.** The genus *Solemya* has primitive, protobranch ctenidia, which are the main organs of feeding, and, like the Nuculoida, a flattened foot. Solemyoida are also subtidal.

Subclass **PTERIOMORPHA.** Most systematists agree that this subclass is a natural group, a conclusion based both on fossil evidence and on the overall similarity of living representatives. Most members are epifaunal, attached to surfaces by a byssus or by cementation and, as a result, the foot is reduced or entirely absent. The mantle margins are less fused than in the subclass Heterodonta.

Order **Arcoida.** This order has the filter-feeding, filibranch ctenidium in which the elongate filaments are reflected so that each gill appears as a tall, narrow W in cross-section; adjacent filaments are united by patches of interlocking cilia to form lamellae. This is chiefly but not exclusively a tropical order, almost always with a taxodont hinge. The unique family Philobryidae, represented locally by *Philobrya setosa* (fig. 67), is tentatively placed here.

Order **Mytiloida.** Members of this order have either filibranch or eulamellibranch ctenidia, the latter with actual bridges of tissue between adjacent filaments. The adductor muscles are unequal in size **(heteromyarian).** The order includes the well-known mussels (Mytilidae), which lack true hinge teeth. Most are found attached to rocks or pilings by a byssus.

Order **Pterioida.** This group includes three families in the central California intertidal zone: Pectinidae (scallops), Ostreidae (true oysters), and Anomiidae (rock jingles or rock oysters). Only the much-enlarged posterior adductor muscle is present, a condition termed **monomyarian.** Both filibranch and eulamellibranch ctenidia are present in different members of the order. Many attach to the substrate, others are free living.

Subclass **HETERODONTA.** This group includes most of the familiar clams. Ctenidia are eulamellibranchiate; the mantle margins are well fused, and elongate siphons are present in most. Distinctive patterns of hinge teeth and ligament characterize the different families.

Order **Veneroida.** Most heterodonts are members of this order, in which the hinge teeth are well developed. Most veneroids are shallow to deep burrowers. Local families include Cardiidae, Carditidae, Chamidae, Corbiculidae, Ungulinidae, Lucinidae, Mactridae, Petricolidae, Solenidae, Veneridae, Erycinidae, Montacutidae, Kelliidae, Tellinidae, Psammobiidae, Semelidae, and Pisidiidae.

Order **Myoida.** In this group burrowing and boring are characteristic ways of life; most have long siphons, and the hinge has few teeth. Local families include Myidae, Hiatellidae, Pholadidae (piddocks), and Teredinidae (shipworms).

Subclass **ANOMALODESMATA.** Members of this group have siphons and many burrow into the substrate; the shells are generally thin and nacreous within; hinge teeth are inconspicuous or absent. There is only one order.

Order **Pholadamyoida.** In the central California intertidal zone there are a few eulamellibranchiate members of this order, which also includes the aberrant, deep-water **Septibranchia.** The only local intertidal family is the Lyonsiidae.

Most local bivalves are free-living infaunal burrowers (fig. 2) or nestlers, or they are epifaunal and attach to the substrate by cementation or a byssus. A number of species often occupy empty pholad holes; the external shape in these and other nestlers may vary considerably and, while such "situs" forms have sometimes been given subspecific or varietal names, these are of little significance. Only a few of our local species are commensal: *Cryptomya californica* lives in association with *Urechis* (Pl. 18, fig. 1) and burrowing anomuran shrimps, tapping their burrows with its short siphons (fig. 2); *Mytilimeria nuttallii* (fig. 46) lives buried in compound ascidian tests; *Orobitella rugifera* occurs byssally attached beneath the abdomen of *Upogebia* (fig. 70); and at least one species of *Mysella* is found on the legs and gills of the large sand crab *Blepharipoda occidentalis*. Other galeommataceans, such as *Lasaea* and *Kellia*, are nestlers among the byssal threads of mussels, in crevices, even in marine-laboratory seawater systems. Best represented in our fauna are the mytilids (with about 14 species), the venerids (about 12 species), the tellinids (about 11 species), and the pholads (about 8 species). It may be noted that some of the most abundant bivalves in our bays and lagoons are introduced;

these include *Ischadium demissum*, *Gemma gemma*, *Mya arenaria*, *Petricola pholadiformis*, *Lyrodus pedicellatus*, and *Teredo navalis* from the Atlantic, and *Musculus senhousia*, *Tapes japonica*, and *Lasaea* sp. from Japan. Others, such as the Atlantic quohog *Mercenaria mercenaria*, are not yet established locally but specimens may be encountered; the Japanese oyster *Crassostrea gigas*, while abundant in Tomales Bay, Drake's Estero, and other areas through direct introduction, does not reproduce here. Further introductions may be expected (see p. 17).

Figure 1 illustrates most of the basic terminology used in the keys to the bivalves. The first-formed part of the shell is the **beak.** The highest or most prominent point of each valve, at or near the beak, is called the **umbo** (pl. **umbones**). The outer surface of the valves may be covered with a fibrous or horny layer, the **periostracum.** Beneath this is the calcareous shell which may be variously sculptured with radial and/or concentric ridges. The **valves** are joined dorsally at the **hinge,** where there is a horny, elastic **ligament.** The ligament may be partly or entirely internal; if within, it is called a **resilium,** and if a calcareous shelf is built up for it, this structure is called a **chondrophore** (fig. 47B). In most bivalves the hinge is strengthened by interlocking teeth. In bivalves with a few strong teeth, those radiating directly from the beaks are **cardinals,** while those lying posterior or anterior to the beaks are **laterals.** In certain boring heterodonts, there may be a calcareous projection called a **myophore** or **apophysis** for the attachment of muscles in each valve below the hinge (fig. 56).

The inner surface of the valves bears the scars of muscle attachments, the most prominent being those of the **adductors.** The **pallial line** marks the attachment of the mantle. Posteriorly, this line may be indented to form a **pallial sinus,** marking the position of muscles for the siphons if any are present.

The following keys attempt to achieve a balance between demonstrating adaptive, phylogenetically related groups and providing somewhat artifical keys for easy identification. The keys will be most useful for fresh, mature, unworn specimens, and are designed for species occurring between Monterey Bay and Bodega Bay.

Bivalves may be relaxed in a 7.5% solution of $MgCl_2 \cdot 6H_2O$, or by menthol crystals or dilute propylene phenoxetol. Preservation is best in 75% ethyl alcohol, but some shells can be kept dry.

In addition to the References on Bivalvia (p. 577), the General References on Mollusca (p. 454) include valuable sources of information on the ecology, biology, physiology, and systematics of bivalves; see especially Keen and Coan (1974), Morton (1967), Purchon (1968), and Wilbur and Yonge (1964, 1966).

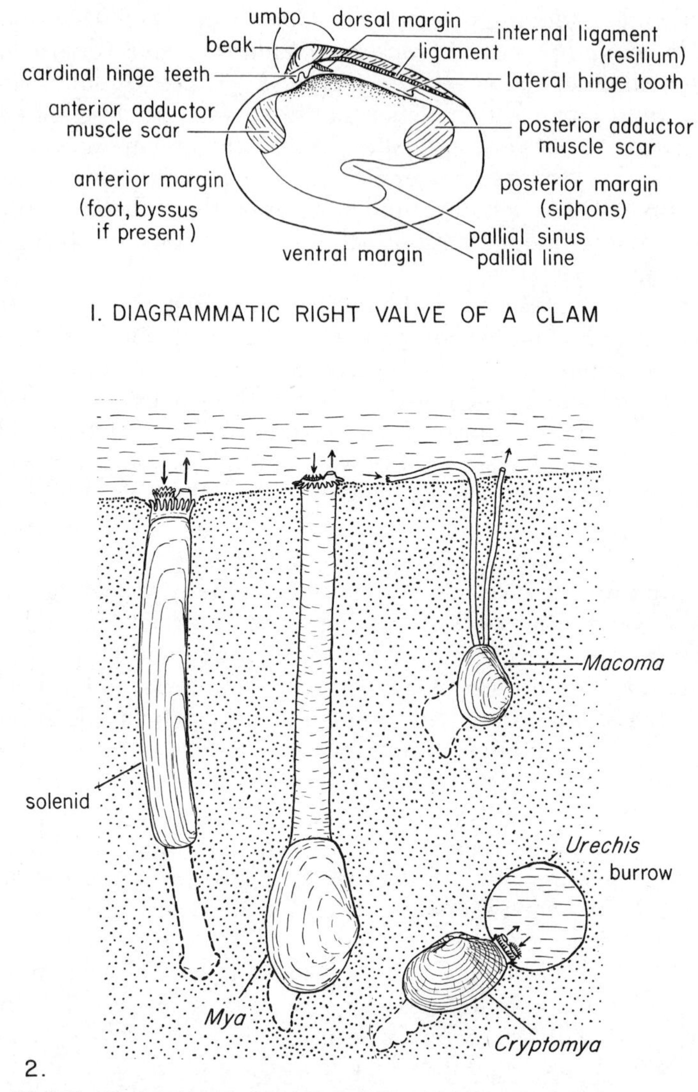

I. DIAGRAMMATIC RIGHT VALVE OF A CLAM

FORMS OF BIVALVE SIPHONS AND MODES OF BURROWING

PLATE 124. **Bivalvia** (1). 1, original (Emily Reid): 2, based on Yonge, 1949.

We wish to thank Peter D'Eliscu (University of Arizona) for advice on Pisidiidae and Vida Kenk (California State University, San Jose) for advice on Mytilidae. Rudolf Stohler (University of California, Berkeley), Editor of *The Veliger,* generously made available unpublished drawings.

KEY TO BIVALVE FAMILIES

1. Shell with filelike denticulations anteriorly; internally with a myophore (apophysis, fig. 56) (except in *Netastoma*); boring into heavy mud, clay, shale, wood, or shell 2
– Shell various, not as above 3
2. Boring into wood; pallets at siphonal (posterior) end (figs. 61–64); anterior end of shell indented with an angular notch (fig. 65) (shipworms) ....TEREDINIDAE (p. 575)
– Boring into a variety of substrates, but rarely wood; no pallets on siphon tips; anterior end arcuate (bow-shaped) (fig. 59) or evenly curved (fig. 57A) ....PHOLADIDAE (p. 572)
3. Shell fanlike, with radiating ribs and with dorsal margin produced into triangular "ears" at least in young (fig. 12A) ....PECTINIDAE

Represented locally by the rock scallop *Hinnites giganteus* (Gray, 1825) [= *H. multirugosus* (Gale, 1928), an unnecessary replacement name] (fig. 12); free-living when young but attaching as an adult to rocks and pilings where growth becomes irregular; see Yonge, 1951c; Grau, 1959, Allan Hancock Pac. Exped. 23: 134–137; Hertlein and Grant, 1972, pp. 211–212.

– Shell not fanlike 4
4. Shell firmly cemented to substrate by one valve (except in *Crassostrea gigas*); valves irregular and/or distinctly different from one another 5
– Shell not cemented to substrate; valves more or less regular, similar 7
5. Adductor muscles coalesced, resulting in one large, sometimes complex, muscle scar near center of shell 6
– Adductor muscles at opposite ends of shell, not coalesced . ... CHAMIDAE (p. 556)
6. One valve with a hole ventral to beaks; adductor muscle scar complex, with central area showing 2 superimposed secondary scars; on open coast, often attached to rocks ANOMIIDAE

The jingle shell or rock oyster *Pododesmus cepio* (Gray, 1850) (= *P. macrochisma* of authors, not of Deshayes, 1839, the latter perhaps a separable Alaskan species although subspecific ranking is possible, and *P. macroschisma*, an invalid emendation) is green to whitish interiorly and is common on rocks, abalone shells, and in dead shells (see Leonard, 1969, Veliger 11: 382–390). A southern species, *Anomia peruviana* Orbigny, 1836, with 3 smaller muscle scars in the central area of upper valve, has been reported as far north as Monterey, although not in recent years.

– Both valves entire; adductor muscle scar simple; in bays, attached to rocks and other shells by left valve, or free in mud (oysters) .... OSTREIDAE (p. 556)

7. Without a chondrophore (fig. 47B) or true, projecting, interlocking teeth on hinge (irregular denticles may be present) 8

– Hinge plate with a chondrophore or true teeth 11

8. Shell minute, triangular, with a single central adductor muscle scar (fig. 67A) .... PHILOBRYIDAE

Represented in the central California low intertidal only by the uncommon *Philobrya setosa* (Carpenter, 1864), which resembles a small mussel (height to about 5 mm) and lives attached to rocks or algae by a byssus.

– Shell with 2 adductor muscle scars 9

9. Adductor muscles and their scars not equal in size, anterior muscle smaller, located near beaks; shell brown or black, cylindrical or tapering anteriorly .... MYTILIDAE (p. 552)

– Adductor muscles and their scars approximately equal in size (but not necessarily in shape) 10

10. Ligament external; shell porcelainlike within; pallial line generally broken into patches (fig. 52) .... HIATELLIDAE (p. 570)

– Ligament internal; shell nacreous within; pallial line not broken into patches .... LYONSIIDAE (p. 575)

11. Shell cylindrical (length about 2.5 times height) to oval-ovate; gaping at both ends 12

– Shell otherwise 13

12. Beaks at or near anterior end; pallial sinus relatively shallow; some species with a prominent, internal, radial rib (razor clams) .... SOLENIDAE (p. 569)

– Beaks nearly central; pallial sinus relatively deep; never with prominent, radial, strengthening rib .... PSAMMOBIIDAE (p. 564)

13. Hinge with an internal ligament (entirely or in part) 14

– Hinge with ligament entirely on dorsal surface 17

14. Hinge with a spoon-shaped chondrophore (fig. 47B) in left valve .... MYIDAE (p. 570)

– Hinge without spoon-shaped chondrophore 15

15. Adult shells mostly small (under 25 mm), thin; young brooded in epibranchial chamber; some commensal; pallial line without a sinus .... ERYCINIDAE, MONTACUTIDAE, KELLIIDAE (Superfamily GALEOMMATACEA) (p. 558)

– Adult shells mostly large (over 25 mm); pallial line with a distinct sinus 16

16. Nestling; cardinal teeth weak, not Λ-shaped; shell with distinct radial or concentric sculpture ..SEMELIDAE (p. 564)

– Burrowing in sand or mud; cardinal teeth Λ-shaped; shell relatively smooth, with growth striae only MACTRIDAE (p. 563)

17. Sculpture predominately of radial ribs corrugating all or part of surface 18

– Sculpture various, including cancellate, but not predominately of radial ribs 21

18. Hinge without lateral teeth 19

– Hinge with lateral teeth 20

19. Three cardinal teeth in both valves . . VENERIDAE (p. 559)

– Three cardinal teeth in one valve, 2 in the other .... PETRICOLIDAE (p. 563)

20. Shell ovate, inflated; beaks central (figs. 17A, B) CARDIIDAE

One local species, the basket or heart cockle *Clinocardium nuttallii* (Conrad, 1837) [= *Cardium corbis* (Martyn, 1788) an unavailable name] (figs. 17A, B), occurs from mid-intertidal to offshore in sandy areas of bays. See Weymouth and Thompson, 1931, Bull. Bur. Fish. 46: 633–641 (age, growth); Fraser, 1931, Trans. Roy. Soc. Can. Sec. 5, 25: 59–72 (ecology).

– Shell quadrate (figs. 15A, B), not particularly inflated; beaks near anterior end ....CARDITIDAE (p. 563)

21. Anterior adductor muscle scar narrower than posterior, its lower end detached and bent inward (fig. 48) LUCINIDAE

Represented locally by *Epilucina californica* (Conrad, 1837) (fig. 48); nestling in gravel and crevices on rocky coast.

– Adductor muscle scars approximately equal in shape 22

22. Adult small to minute, less than 15 mm in length 23

– Adult more than 15 mm in length 24

23. Shell thin, ligament partly sunken below dorsal surface; hinge with 2 cardinal and 2 lateral teeth in both valves .... PISIDIIDAE (= SPHAERIIDAE)

The small, freshwater fingernail clams are occasionally encountered in deltas or at river mouths. *Pisidium casertanum* (Poli, 1791), with an anal siphon and anterior end much longer than posterior, is common in springs, ponds, and slow creeks, while *Sphaerium patella* (Gould, 1850), with the posterior end longer than the anterior and with anal and branchial siphons, is common in such streams as Salmon Creek (Sonoma County). See Herrington, 1962, Univ. Mich. Mus. Zool. Misc. Publ. 118 (systematic review).

– Shell relatively thick and heavy, ligament completely external; hinge with 3 cardinal teeth in both valves and, at most, one lateral tooth in either valve .... VENERIDAE (in part) (p. 559)

24. Three cardinal teeth in one or both valves 25

– Less than 3 cardinal teeth in both valves 27

25. Hinge with very elongate, serrate, lateral teeth .... CORBICULIDAE

The introduced Asiatic clam *Corbicula manilensis* (Philippi, 1844) (= *C. fluminea* of authors) has a thick, trigonal shell with low concentric ridges covered by a heavy black brown periostracum, and may be widely encountered on bay and ocean beaches as discarded fish bait; it is abundant in freshwater canals and irrigation channels, and large aggregations have locally clogged canal systems and water pipes. See Hanna (1966) for figures, discussion.

– Lateral teeth, if present, not very elongate or serrate 26

26. Three cardinal teeth in both valves . . VENERIDAE (p. 559)

– Three cardinal teeth in one valve, 2 in other .... PETRICOLIDAE (p. 563)

27. Pallial sinus long and well developed .... PSAMMOBIIDAE (in part), TELLINIDAE (Superfamily TELLINACEA) (p. 564)

– Pallial sinus inconspicuous or absent 28

28. Shell oval (fig. 24); pallial line simple, thin .... UNGULINIDAE (= DIPLODONTIDAE)

Represented in local rocky intertidal by *Diplodonta orbella* (Gould, 1851) (fig. 24) in holes in rocks and forming a "nest" of agglutinated detritus and sand under rocks; see note by Haas, 1943, Zool. Ser. Field Mus. Natl. Hist. 29: 9–12.

– Shell elongate (fig. 52); pallial line entire, thick, often in patches .... HIATELLIDAE (p. 570)

MYTILIDAE

1. Adult shell minute (5 mm or less), stubby, inflated, with a forwardly directed, prominent, anteroventral protuberance; oblique ridge extending from posteroventral margin to mid-shell region (fig. 4) .... *Musculus pygmaeus*

– Adult shell not minute 2

2. Beaks at end of shell 3

– Beaks near anterior end, but not terminal 5

3. Anterior end bridged by a shelly septum internally (fig. 6 B); shell with prominent radiating ribs (fig. 6A); black externally, purplish internally .... *Septifer bifurcatus*

– Anterior end open, lacking internal septum ....*Mytilus* 4

4. Shell generally smooth (fig. 5A); anterior adductor muscle scar on anteroventral margin (fig. 5B) .... *Mytilus edulis*

– Shell generally with irregular radial ribs especially on posterior end; anterior adductor scar situated more anteriorly (fig. 7) .... *Mytilus californianus*

5. Shell cylindrical, with dorsal and ventral margins more or less parallel, though at times irregular 6

– Shell not cylindrical 9

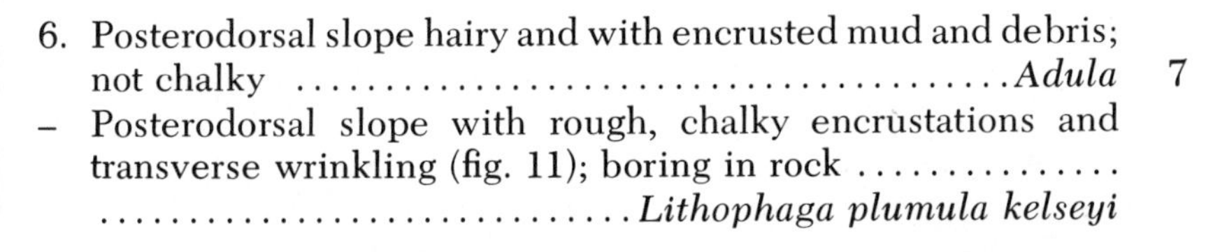

6. Posterodorsal slope hairy and with encrusted mud and debris; not chalky .... *Adula* 7
– Posterodorsal slope with rough, chalky encrustations and transverse wrinkling (fig. 11); boring in rock .... *Lithophaga plumula kelseyi*

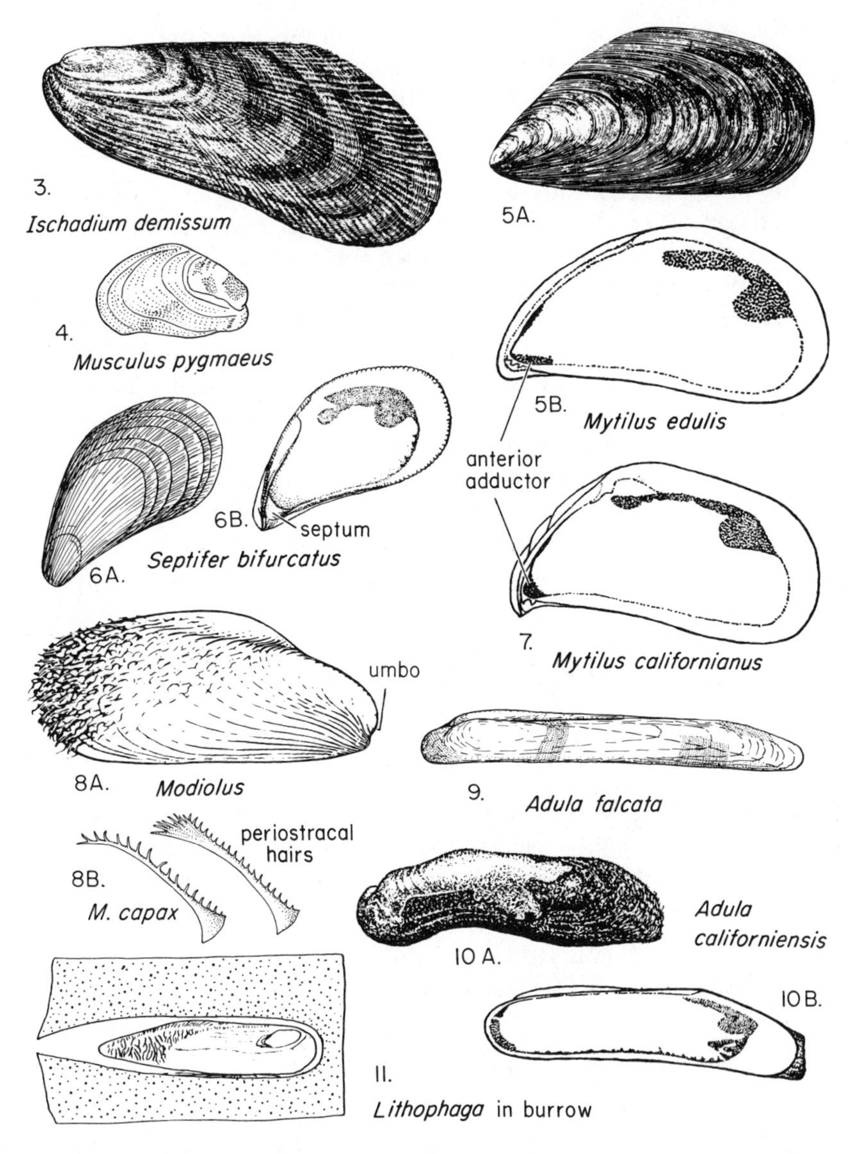

PLATE 125. **Bivalvia** (2). Mytilidae: 3, 5A, Pilsbry, 1891: 4, Glynn, 1964; 5B, 6B, 7, 8B, 10B, Soot-Ryen, 1955; 6A, Emily Reid; 8A, Keen, 1963; 9, Light *et al.*, 1954; 10A, Yonge, 1955; 11, Hodgkin, 1962.

7. With irregularly wrinkled, filelike, vertical striae (fig. 9); boring in soft shale .... *Adula falcata*
– Smooth or with a few radiating striations anteriorly 8
8. Shell elongate, generally tapering posteriorly, posterior not higher than anterior (fig. 10); generally boring in soft shale .... *Adula californiensis*
– Shell stouter, generally distinctly higher posteriorly; free-living (see note in species list) .... *Adula diegensis*
9. Shell with prominent radial ribbing (fig. 3); dark brown to blackish .... *Ischadium demissum*
– Shell without ribs 10
10. Shell smooth, without hairs; thin; often with wavy brown bands on greenish background .... *Musculus senhousia*
– Shell with periostracum bearing serrate or smooth hairs; may have posterior encrustations (fig. 8A) .... *Modiolus* 11
11. Periostracum with serrate hairs (fig. 8B); left valve generally more inflated than right valve .... *Modiolus capax*
– Periostracum with smooth, not serrate, hairs; usually equivalve 12
12. Shell short, stout, somewhat globose; umbones inflated, twisted downward and outward, protruding beyond anterior margin; adult to about 40 mm .... *Modiolus carpenteri*
– Shell elongate; umbones set well back from anterior margin; adult to 120 mm or more .... *Modiolus rectus*

LIST OF MYTILIDAE

See Soot-Ryen 1955, Allan Hancock Pac. Expeds. 20: 175 pp. for systematic review.

Adula californiensis (Philippi, 1847) (= *Botula californiensis*). Boring mechanically in soft shale, mudstone, but occasionally free-living (Soot-Ryen); intertidal to sublittoral. See Lough and Gonor 1971, Mar. Biol. 8: 118–125 (embryology, developmental rate); Yonge, 1955 (under *A. falcata*, boring).

Adula diegensis (Dall, 1911) (=*Botula diegensis*). Free-living on mud flats, pilings, with *Mytilus;* some specimens distinguished only with difficulty from *A. californiensis*.

Adula falcata (Gould, 1851) (= *Botula falcata*). Boring mechanically in soft shale or clay; see Yonge 1955, Quart. J. Micr. Sci. 96: 383–410 (boring).

Ischadium demissum (Dillwyn, 1817). Formerly in *Volsella, Modiolus, Arcuatula,* or *Geukensia,* the species is properly placed in *Ischadium* (V. Kenk). An introduced Atlantic mussel now abundant in mud in sloughs, bays, in cracks on pilings. Pierce 1973, Malacologia 12: 283–293, has argued for its placement in *Arcuatula.*

Lithophaga plumula kelseyi Hertlein and Strong, 1946. In Soot-Ryen as *L. subula* (Reeve, 1857). Boring in calcareous shale, shells; inner mantle fold may secrete acid mucus; see Yonge, 1955 (above); Hodgkin 1962, Veliger 4: 123–129 (boring).

Modiolus capax (Conrad, 1837) (= *Volsella capax*). Intertidal, on rocks, pilings; southern; central California records require confirmation.

Modiolus carpenteri Soot-Ryen, 1963 (=*M. fornicatus* Carpenter, 1865, pre-occupied). Occasional in low intertidal zone, largely sublittoral, and commonly washed ashore; among rocks, shells, gravel.

Modiolus rectus (Conrad, 1837) [= *Volsella recta*, = *M. flabellatus* (Gould, 1850)]. Largely sublittoral, rare in low intertidal zone; in mud, anterior end embedded in substrate.

* *Modiolus sacculifer* (Berry, 1953). Generally offshore in holdfasts, shell not elongate, umbones set back from produced anterior end, not overhanging anterior margin as in *M. carpenteri.*

Musculus pygmaeus Glynn, 1964. High intertidal, attached to blades or holdfast of alga *Endocladia muricata;* see Glynn, 1964, Veliger 7: 121–128.

Musculus senhousia (Benson, 1842) (Formerly in *Modiolus* and *Volsella; M. senhousei* is an invalid emendation). Abundant in mud, forming extensive mats; in fouling on pilings, among algae; introduced from Japan.

Mytilus californianus Conrad, 1837. California mussel; abundant in exposed rocky intertidal; also in bays, on pilings. See Coe and Fox, 1942, J. Exp. Zool. 90: 1–30; Fox and Coe, 1943, J. Exp. Zool. 93: 205–249; Coe and Fox, 1944, Biol. Bull. 87: 59–72; Coe, 1948, on aspects of biology; papers by Harger, and Harger and Landenberger, in Veliger 11–14 (1968–1971) on ecology, biology (*M. edulis* and *M. californianus*).

Mytilus edulis Linnaeus, 1758. Bay mussel; cosmopolitan; abundant on wharf pilings, floats, docks, rocks; occasional on outer coast. See Field, 1922, Bull. U.S. Bur. Fish. 38: 127–259; White, 1937, *Mytilus*, Liverpool Mar. Biol. Comm. Mem. 31, 117 pp.; Coe, 1945, J. Exp. Zool. 49: 1–14; 1946, J. Morph. 78: 85–104; Seed, 1968, J. Mar. Biol. Assoc. U.K. 48: 561–584 (variation in shape); 1969, Oecologia 3: 317–350 (growth, mortality); Reish, 1964, Veliger 6: 124–131, 202–207; Reish *et al.*, 1968, Veliger 10: 384–388; 1969, Veliger 11: 250–255 (*M. edulis* community).

Septifer bifurcatus (Conrad, 1837). Low intertidal, under rocks. See Yonge and Campbell, 1968, Trans. Roy. Soc. Edinburgh 68: 21–43.

* Not in key.

OSTREIDAE

1. Small (generally 50 mm or less), relatively thin; inner margin with fine crenulations near hinge; adductor muscle scar not pigmented, interior of shell greenish yellow; shell shape variable .... *Ostrea lurida*
– Larger (to 30 cm), thick; inner margin smooth near hinge; shell varies from oval to very long and narrow (fig. 18) *Crassostrea* 2
2. Shell generally with prominent, projecting fluting; radial grooves deep; muscle scar violet or whitish; external color gray-purple; attached or free-living in mud ..... *Crassostrea gigas*
– Shell without fluting, but with clear, concentric sculpture (fig. 18); muscle scar deeply impressed, dark blue-purple or brown; external color yellow brown; attached to rocks .... *Crassostrea virginica*

LIST OF OSTREIDAE

See Yonge, 1960, *Oysters,* London: Collins, 209 pp. and monograph by Stenzel, 1971, *Treatise on Invertebrate Paleontology,* N (Mollusca) (6) 3. Introduced species do not generally reproduce on central California coast.

* *Ostrea edulis* Linnaeus, 1758. European flat oyster; plantings have been made in Tomales Bay and Drake's Estero; see Hanna, 1966; Leonard, 1969, Veliger 11: 382–390.

Ostrea lurida Carpenter, 1864. Olympia or native oyster; common in mud, on rocks, pilings, in bays, often in clumps; see Hopkins, 1936, Ecology 17: 551–556; 1937, Bull. Bur. Fish. 48: 439–503 (reproduction, larval development); and Barrett, below.

Crassostrea gigas (Thunberg, 1795). Japanese or Pacific oyster; introduced; large oyster farms in Drake's Estero and Tomales Bay in central California; shells are common. See Barrett, 1963, Calif. Fish Game Fish Bull. 123; Quayle, 1969, Bull. Fish. Res. Board Canada 169; Berg, 1969, Veliger 12: 27–36; 1971, Calif. Fish Game 57: 69–75 (causes of mortality).

Crassostrea virginica (Gmelin, 1791). Virginia, Eastern, or Atlantic oyster; introduced, now uncommon in central California bays. See Galtsoff, 1964, Fish. Bull. 64 for monographic review; also Barrett, Quayle, and Berg, above.

CHAMIDAE

1. Attached by left valve; growth counter-clockwise, markedly foliose .... *Chama "pellucida"*
– Attached by right valve; growth clockwise, valves scaly, but not markedly foliose (figs. 16A, B) .... *Pseudochama exogyra*

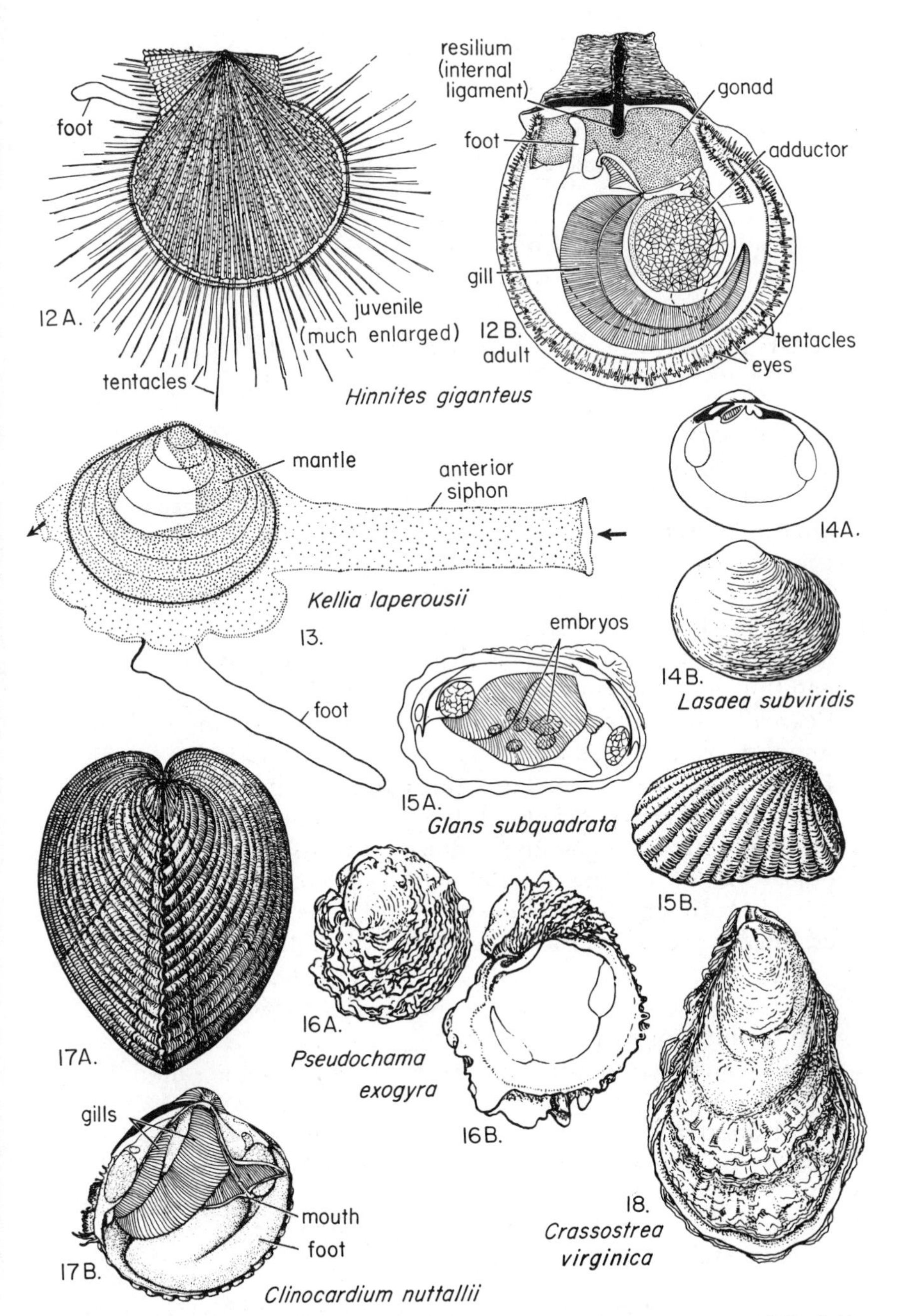

PLATE 126. **Bivalvia** (3). Various: 12A,B, Yonge, 1951; 13, Yonge, 1952; 14A, B, 15B, 16A, B, Keen, 1963; 15A, Yonge, 1969; 17A, Keep, 1904; 17B, Stasek, 1963; 18, Jacobson and Emerson, 1961.

LIST OF CHAMIDAE

See Yonge, 1967, Phil. Trans. Roy. Soc. London (B) 252: 49–105.

Chama "pellucida Broderip, 1835." Long identified as *C. pellucida*, a South American species; ours differs in several characters and requires a new name. Cemented to protected surfaces of mid-intertidal rocks; on pilings.

Pseudochama exogyra (Conrad, 1837). Cemented to algae-covered rocks on open coast from mid to low intertidal.

SUPERFAMILY GALEOMMATACEA (ERYCINIDAE, MONTACUTIDAE, KELLIIDAE)

1. Beaks nearly at anterior end .... *Mysella* spp.
– Beaks subcentral 2
2. Hinge with cardinal teeth only; no lateral teeth; shell quadrangular, yellowish; on abdomen of *Upogebia* (fig. 70) .... *Orobitella rugifera*
– Hinge with lateral teeth 3
3. Hinge with posterior lateral teeth, no anterior laterals; shell nearly round to elliptical; with fine, concentric, growth lines and thin, smooth, yellowish periostracum (fig. 13); in life mantle may extend to cover shell; prominent anterior siphon; length to 25 mm .... *Kellia laperousii*
– Hinge with posterior and anterior lateral teeth; shell reddish, periostracum wavy, wrinkled; length to 3 mm .... *Lasaea* 4
4. Shell flattened, not swollen, anterior end relatively elongate; dorsal margin joining anterior end in a round, smooth curve (figs. 14A, B); green or yellow gray, with pink at beaks, pink occasionally extending down ventral 1/2 of shell .... *Lasaea subviridis*
– Shell swollen; dorsal margin subparallel to ventral margin, joining anterior margin at an angle 5
5. Shell quadrate in outline, anterior end not elongate; medium to dark red, occasionally whitish .... *Lasaea cistula*
– Shell oblique, with elongate anterior end; brown, with suffusion of pink; stained dark red brown internally *Lasaea* sp.

LISTS OF GALEOMMATACEA

See Popham, 1940.

LIST OF ERYCINIDAE (= LASAEIDAE)

Lasaea cistula Keen, 1938 (=*L. rubra* of authors, not of Montagu, 1803). In algal holdfasts and among byssal threads of mussels, abundant, on open and semiprotected rocky coast. See Keen, 1938, Proc. Malac. Soc. London 23: 18–32; Ballentine and Morton, 1956, J. Mar. Biol. Assoc. U.K. 35: 241–274 (*L. rubra*, England, filtering, feeding, digestion).

Lasaea subviridis Dall, 1899. On open coast, among mussels, in kelp holdfasts. See Keen, 1938, above.

Lasaea sp. Abundant under rocks in mud, Tomales Bay; probably introduced.

LIST OF MONTACUTIDAE

Mysella spp. (= *Rochefortia* spp.). *Mysella tumida* (Carpenter, 1864) may occur in bays and offshore (see Mauer, 1967); a *Mysella*, which has been referred to both *M. pedroana* (Dall, 1899) and *M. golischi* (Dall, 1916), occurs on the legs and gills of the sand crab *Blepharipoda occidentalis* (see Hertlein and Grant, 1972, pp. 239–240).

Orobitella rugifera (Carpenter, 1864) (= *Pseudopythina rugifera*). Intertidal, attached by byssus to abdomen of *Upogebia;* see Narchi, 1969, Veliger 12: 43–52 (morphology).

* *Pristes oblongus* (Carpenter, 1864). Commensal with chiton *Stenoplax heathiana,* Monterey south. Like *Mysella,* the beaks are at the anterior end, but *Pristes* has two serrate cardinal teeth in both valves, whereas *Mysella* lacks cardinals in one valve.

LIST OF KELLIIDAE

Kellia laperousii (Deshayes, 1839). Often abundant, nestling on pilings, rocks, in musssel beds and empty barnacles, in bottles and pholad holes; a common fouler in seawater systems of marine laboratories. Small, round specimens have been reported as *Kellia suborbicularis* (Montagu, 1803); two species may be involved, but further study is necessary. See Yonge, 1952c; Howard, 1953, Wasmann J. Biol. 11: 233–240 (larvae); Oldfield, 1961, Proc. Malac. Soc. London 34: 255–295 (functional morphology of *K. suborbicularis,* Atlantic).

VENERIDAE

1. Adult shell small, less than 10 mm in length 2
– Adult shell more than 10 mm in length 4
2. Shell no longer than high, triangular; hinge without an anterior lateral tooth; inner ventral margin finely crenulate; pallial sinus bent sharply upward (fig. 22) .....*Gemma gemma*
– Shell elongate to oval; hinge with conspicuous, anterior lateral teeth in both valves; inner ventral margin smooth, with

obscure, oblique striae only; pallial sinus bent anteriorly .... *Transennella* 3

3. Shell uniform white to cream; siphons fused nearly to tips; lateral teeth relatively weak (fig. 21) .... *Transennella* sp.
– Shell light to deep purple; siphons with prominent cleft; lateral teeth strong .... *Transennella tantilla*
4. Shell smooth, with a shiny, adherent periostracum; large, trigonal (fig. 19) .... *Tivela stultorum*
– Shell sculptured; without a shiny periostracum 5
5. Sculpture entirely or almost entirely concentric, sometimes raised into lamellae 6
– Sculpture radial and concentric, the concentric sculpture not predominating and occasionally obsolete 10
6. Shell short, relatively small; nestling among rocks or in borer holes; pronounced, widely spaced, concentric lamellae (see note in species list) (fig. 25) .... *Irus lamellifer*
– Shell elongate or oval, without prominent concentric lamellae; in sand or mud 7
7. Hinge without anterior lateral teeth; valves not gaping 8
– Hinge with anterior lateral teeth; valves with a pronounced gape 9
8. Shell round-oval; valves convex, somewhat inflated; concentric sculpture smooth; beaks prominent; lunule distinct (fig. 20) .... *Mercenaria mercenaria*
– Shell elongate-oval; valves flattened; concentric sculpture of thin, sharp ridges; beaks not prominent; lunule absent .... *Protothaca tenerrima*
9. Concentric ribs heavy, conspicuous, well spaced (fig. 23); shell elongate; interior marked with purple posteriorly .... *Saxidomus nuttalli*
– Concentric ribs thin, low, and more closely spaced, giving shell a relatively smooth appearance; shell more rounded, posterior end not marked with purple internally .... *Saxidomus giganteus*
10. Shell elongate-oval; inner ventral margin smooth; hinge teeth weak; ligament prominent, elevated above dorsal margin; siphons separate at tips .... *Tapes japonica*
– Shell generally subcircular; inner ventral margin crenulate; hinge teeth strong; ligament sunken, not elevated above dorsal margin; siphons fused for entire length 11
11. Radiating ribs numerous, fine; concentric ridges faint to lacking in some specimens (see note in species list); color pattern extremely variable .... *Protothaca staminea*
– Radiating ribs and concentric ribs both predominant, forming

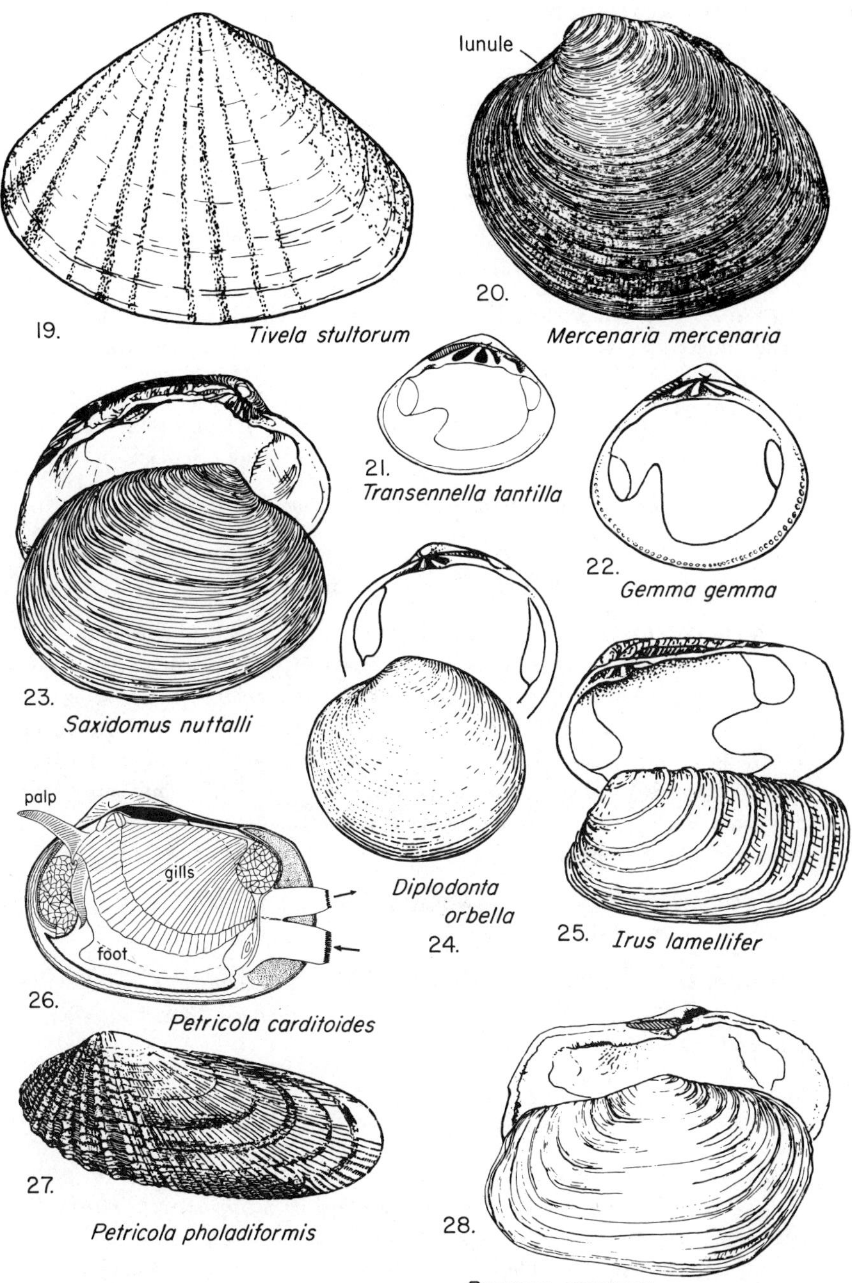

PLATE 127. **Bivalvia** (4). Various: 19, 25, R. Stohler, drawn by Emily Reid; 20, 27, Pilsbry, 1891; 21, 22, 24, Keen, 1963; 23, 28, Hedgpeth, 1962; 26, Yonge, 1958.

a sharp, coarse network; color generally drab .... *Protothaca laciniata*

LIST OF VENERIDAE

Gemma gemma (Totten, 1834). Gem clam; introduced from Atlantic; common to abundant in mud of bays. See Sellmer, 1967, Malacologia 5: 137–223 (functional morphology, ecology, life history); Narchi, 1971, Bull. Mar. Sci. 21: 866–885 (*G. gemma*, *Transennella tantilla*, anatomy).

Irus lamellifer (Conrad, 1837). Nestling among rocks and in pholad burrows along outer coast; on pilings, in fouling, in bays. See note under *Protothaca staminea.*

Mercenaria mercenaria (Linnaeus, 1758) (= *Venus mercenaria*). The quohog; introduced from Atlantic; in mud in bays; not common.

Protothaca laciniata (Carpenter, 1864). Uncommon, in sandy mud of bays.

* *Protothaca restorationensis* (Frizzell, 1930). Recorded as far south as Half Moon Bay; most likely a hybrid of *P. staminea* and *P. tenerrima.*

Protothaca staminea (Conrad, 1837) [= *Paphia staminea* and *Venerupis staminea;* = *P. s. orbella* (Carpenter, 1864), and other forms]. Rock cockle or common littleneck; common both in sand of bays and nestling among rocks (often in empty pholad holes) on outer coast; see Fraser and Smith, 1928, Trans. Roy. Soc. Canada (3) 22: 249–269 (notes on ecology); Schmidt and Warme 1969, Veliger 12: 193–199 (population characteristics). A nestling form of this species may have raised, concentric lamellae like *Irus,* but can be distinguished by its more prominent radial sculpture.

Protothaca tenerrima (Carpenter, 1857). Uncommon, in semiprotected areas of bays and offshore in sandy mud.

Saxidomus giganteus (Deshayes, 1839). Uncommon, a more northern species, in same habitat as *S. nuttalli;* see Fraser and Smith, 1928, Trans. Roy. Soc. Canada (3) 22: 271–277 (ecology).

Saxidomus nuttalli Conrad, 1837. Washington clam; common in bays and lagoons in mud or sand; also on outer coast in sand among rocks.

Tapes japonica Deshayes, 1853 (= *T. semidecussata* and *T. philippinarum* of authors, and variously placed in *Paphia, Venerupis, Protothaca,* and *Ruditapes,* the last now used as a subgenus). Manila, Philippine, or Japanese cockle, introduced from Japan; common in mud of bays.

Tivela stultorum (Mawe, 1823). Pismo clam; common on exposed sandy beaches from Half Moon Bay south; see Weymouth, 1923, Calif. Fish Game Fish Bull. 7, 120 pp.; Herrington, 1930, *Ibid.* 18, 67 pp.; Coe and Fitch, 1950, J. Mar. Res. 9: 188–210; Fitch, 1950, Calif. Fish Game 36: 285–312.

Transennella tantilla (Gould, 1853). Common in sand or sandy mud in semiprotected situations in bays as well as offshore; see Hansen, 1953, Vidensk. Medd. Dansk. Naturh. Foren. København. 115: 313–324 (brood protection, sex ratio); Narchi, 1971 (anatomy; above, under *G. gemma*); Mauer, 1967.

Transennella sp. In muddy sand of bays.

CARDITIDAE

1. Ventral margin of shell folded inward (in female) .... *Milneria minima*
– Ventral margin evenly arched, not bent inward (Figs. 15A, B) .... *Glans subquadrata*

LIST OF CARDITIDAE

Glans subquadrata (Carpenter, 1864) [= *G. carpenteri* (Lamy, 1922), an unnecessary replacement name]. Attains only about 10 mm in length and attaches by a byssus to the undersurfaces of rocks on open coast; broods young. See Yonge, 1969, Proc. Malac. Soc. London 38: 493–527.

Milneria minima (Dall, 1871). Reaches about 7 mm and also attaches with a byssus to rocks from the low intertidal to 50 m. Ventral margin is folded inward in the female to form a brood pouch.

PETRICOLIDAE

1. With fine radial sculpture over entire shell surface; shell shape extremely variable (internal, fig. 26); nestling in rock crevices .... *Petricola carditoides*
– With heavy radial sculpture, especially on anterior end; shell elongate (fig. 27); burrowing in mud *Petricola pholadiformis*

LIST OF PETRICOLIDAE

Petricola carditoides (Conrad, 1837). Common in rocky intertidal, nestling in rock crevices and in pholad holes, which it can enlarge by limited boring; see Yonge, 1958, Proc. Malac. Soc. London 33: 25–31.

Petricola pholadiformis Lamarck, 1818. Burrowing in mud; introduced from Atlantic coast.

MACTRIDAE

1. Shell broadly gaping posteriorly .... *Tresus* 2
– Shell narrowly gaping or closed 3
2. With 2 prominent, leathery plates on siphon tips; shell qua-

drate, distinctly elongated posteriorly (fig. 69); siphonal periostracum retained as tough, wrinkled membrane (see note in species list) . . . . *Tresus nuttallii*
– Siphonal plates indistinct; shell rounded, not markedly extended posteriorly; siphonal periostracum shed (see note in species list) . . . . *Tresus capax*
3. External ligament separated from internal ligament (resilium) by a thin, shelly plate (lamina) . . . . *Mactra* spp.
– External ligament separated from internal resilium by flat space, not a lamina . . . . *Spisula* spp.

LIST OF MACTRIDAE

See Packard, 1916, Univ. Calif. Publ. Bull. Dept. Geol. 9: 261–360 on the surf clams *Mactra* and *Spisula*.

Mactra spp. Mostly offshore in sand or mud; *M. dolabriformis* (Conrad, 1867) and *M. californica* Conrad, 1837 are rarely found in bays.

Spisula spp. Mostly offshore in sand or mud; *S. catilliformis* Conrad, 1868 is commonly washed ashore; see Hertlein and Grant, 1972, pp. 312 ff (key to species, discussion).

Tresus capax (Gould, 1850) (= *Schizothaerus capax*). The horseneck or gaper clam; a more northern species, uncommon in central California; see Swan and Finucane, 1952, Nautilus 66: 19–26 (distinctions between *T. capax* and *T. nuttallii*); Pearce, 1965, Veliger 7: 166–170 (ecology); Machell and DeMartini, 1971, Calif. Fish Game 57: 274–282 (reproductive cycle).

Tresus nuttallii (Conrad, 1837) (= *Schizothaerus nuttallii*). In sand in bays; see Pearce, 1965, above; Pohlo, 1964, Malacologia 1: 321–330 (ontogeny, ecology); Illg, 1949, Proc. U.S. Nat. Mus. 99: 391–428 (parasitic copepods); Stout, 1970, Veliger 13: 67–70 (epizoics on siphonal plates). The two species of *Tresus* are further distinguished by the presence of a "visceral skirt" (a prolongation of the inner palp lamellae which forms a curtainlike structure hanging from the dorsal extremities of and covering much of posterior of the visceral mass) in *T. capax* and its absence in *T. nuttallii* (see Pearce 1965, above, pl. 27); moreover, *T. capax* is generally higher dorsoventrally than *T. nuttallii*.

SUPERFAMILY TELLINACEA (PSAMMOBIIDAE, SEMELIDAE, TELLINIDAE)

1. Part of ligament internal, seated in a cup below beaks (a resilifer) or in a deep, posteriorly directed furrow; nestling in rocky intertidal . . . . SEMELIDAE 2
– Ligament external; generally not nestling 3

2. Internal portion of ligament seated in a resilifer projecting below beaks (fig. 29); conspicuous concentric sculpture only; shell white internally .... *Cumingia californica*
– Internal ligament seated in a posteriorly directed furrow (fig. 30); fine radial threads; shell usually reddish within .... *Semele rupicola*

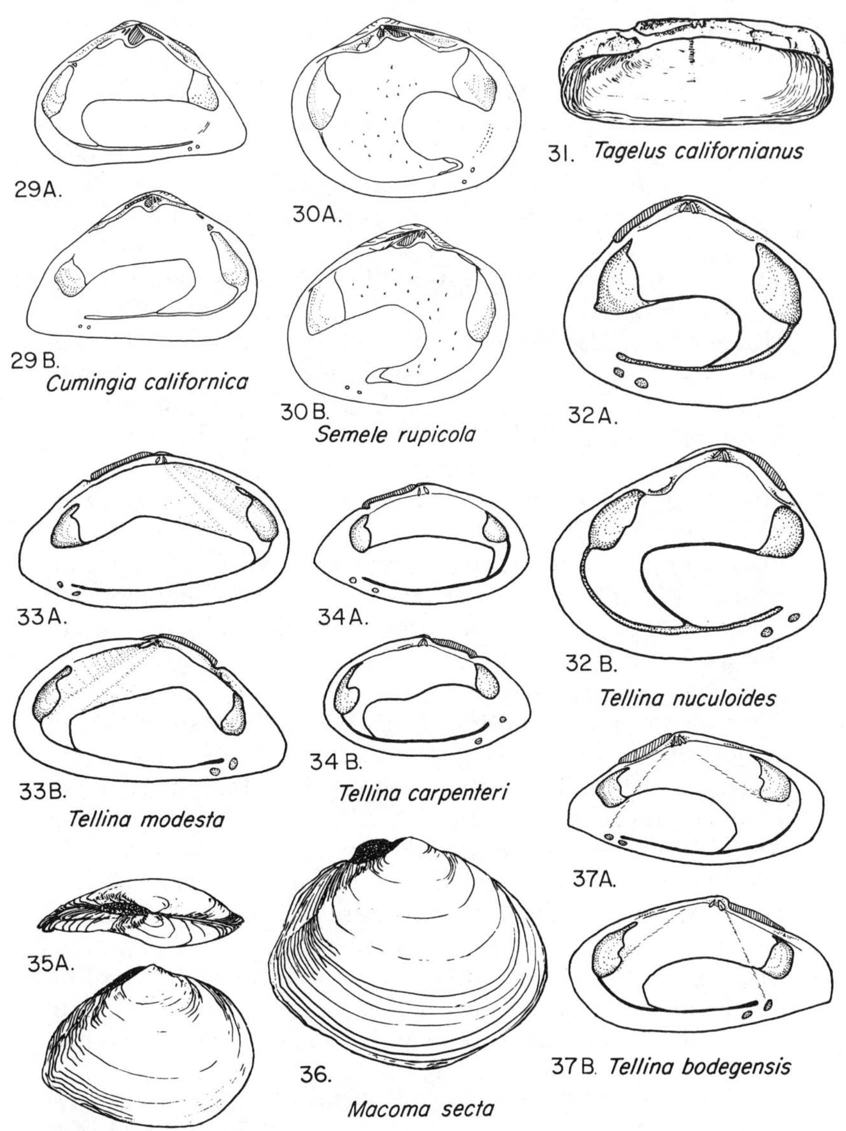

PLATE 128. **Bivalvia** (5). Tellinacea: 31, 33A,B, 35A,B, 36, Hedgpeth, 1962; 29A,B, 30A,B, Coan, 1973; 32A,B, 34A,B, 37A,B, Coan, 1971.

3. External ligament seated on a nymph (an internal shell buttress for the external ligament); elongate and cylindrical to ovate; some conspicuously colored with pink rays; no lateral teeth .... PSAMMOBIIDAE 4
– External ligament not seated on a nymph; shell never cylindrical; some with reddish tinge (not rays); some with lateral teeth .... TELLINIDAE 6
4. Shell elongate, cylindrical (fig. 31) .. *Tagelus californianus*
– Not cylindrical 5
5. Shell ovate, thin, generally purple internally; externally with shiny brown periostracum; left valve convex, right valve flatter .... *Nuttallia nuttallii*
– Shell oval-ovate, heavy, white internally; externally without conspicuous periostracum; with pink radial rays (fig. 68) ... . *Gari californica*
6. Hinge with lateral teeth, especially evident in right valve .. .. *Tellina* 7
– Hinge without lateral teeth .... *Macoma* 9
7. Shell with regular, concentric sculpture; to 60 mm in length (fig. 37) .... *Tellina bodegensis*
– Shell smooth externally or with occasional growth lines only; to 20 mm 8
8. Shell oval, heavy, often with greenish periostracum and reddish tinge (fig. 32) .... *Tellina nuculoides*
– Shell elongate, thin, white; never with a conspicuous periostracum (fig. 33) .... *Tellina modesta*
9. With a tendency to produce a posterior dorsal flange (posterior to end of ligament) 10
– Without a posterior dorsal flange 11
10. With a conspicuous posterior flexure; shell often produced posteriorly (figs. 39A, B) .... *Macoma indentata*
– Quadrate posteriorly, with a posterior fold, but not flexed or produced (figs. 36, 38A, B) .... *Macoma secta*
11. Anterior ventral edge of pallial sinus detached for at least 1/4 distance to posterior adductor muscle scar and more or less paralleling pallial line (true of both valves); to 30 mm in length 12
– Anterior ventral end of pallial sinus not detached from pallial line for a substantial distance, although it may overlap slightly near point of juncture (usually only in one valve); to 100 mm or more 13
12. Produced and slightly expanded posteriorly; longer anteriorly (figs. 41A, B) .... *Macoma yoldiformis*
– Pointed posteriorly, not produced or expanded; form longer

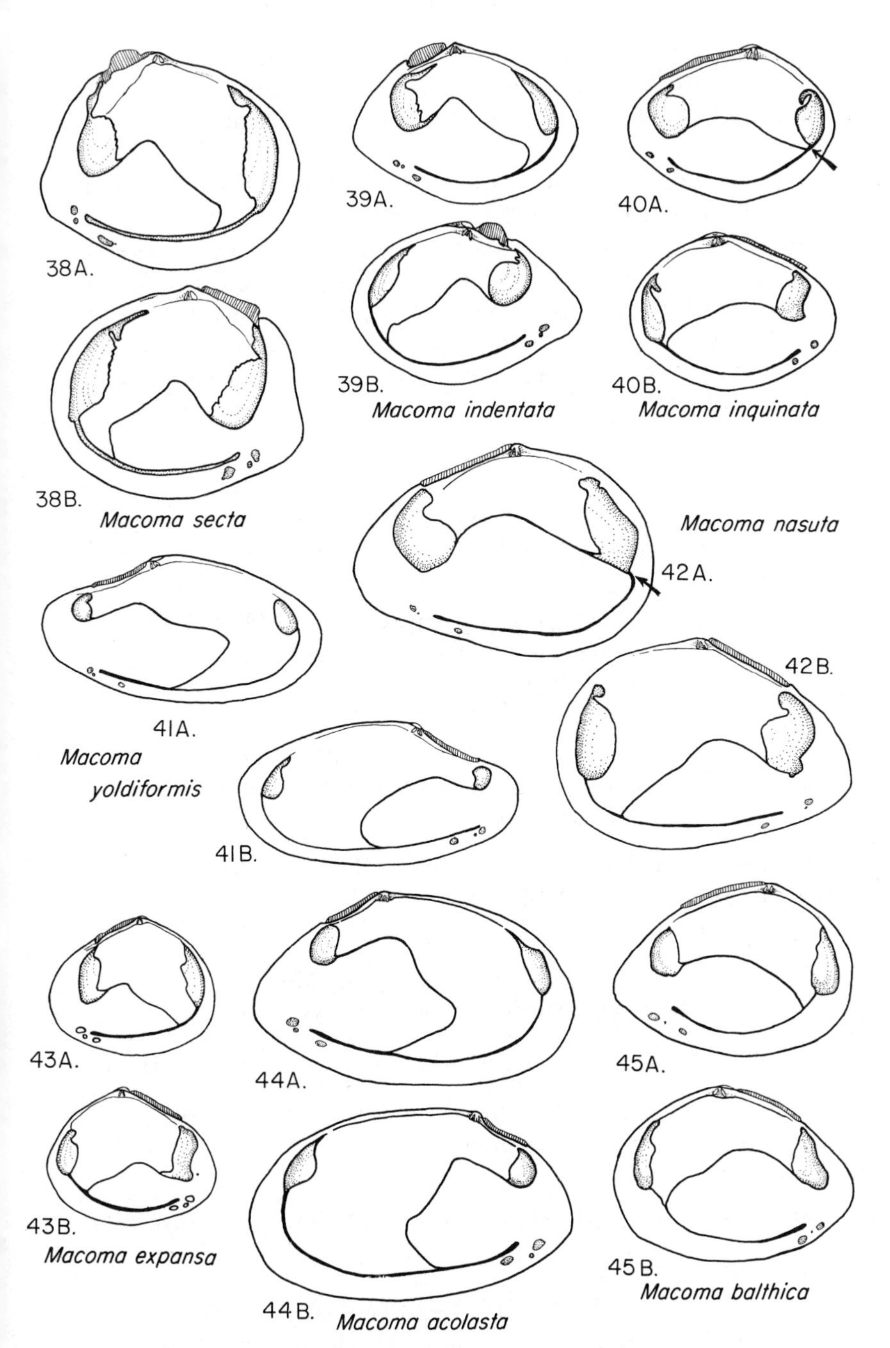

PLATE 129. **Bivalvia** (6). Tellinidae (*Macoma*): 38–45, Coan, 1971.

anteriorly to equilateral (figs. 44A, B) .... *Macoma acolasta*

13. Pallial sinus terminating only 3/4 of way to anterior adductor muscle scar (figs. 45A, B); often with pinkish tinge .... *Macoma balthica*
– Pallial sinus reaching to or almost to anterior adductor muscle scar in one or both valves; whitish 14
14. Pallial sinus in left valve nearly always reaching anterior adductor muscle scar and fusing with it (figs. 42A, B); shell bent to right posteriorly (figs. 35A, B) ....*Macoma nasuta*
– Pallial sinus not quite reaching or reaching only to ventral end of anterior adductor muscle scar (figs. 40A, B); equivalve, not bent posteriorly ....*Macoma inquinata*

LISTS OF TELLINACEA

See Yonge, 1949; Pohlo 1969, Proc. Malacological Soc. London 38: 361–364.

LIST OF PSAMMOBIIDAE (=GARIDAE)

See Coan, 1973, Veliger 16: 40–57.

Gari californica (Conrad, 1837). Among rubble, low intertidal.

Nuttallia nuttallii (Conrad, 1837) (= *Sanguinolaria nuttallii*). In sand and sandy mud of low intertidal, protected bays. See Pohlo, 1972, Veliger 14: 298–301 (feeding, associated morphology).

Tagelus californianus (Conrad, 1837). Low intertidal in sand and mud of protected bays. See Pohlo, 1966, Veliger 8: 225 (feeding).

LIST OF SEMELIDAE

See Coan, 1973, Veliger 15: 314–329 for systematic review.

Cumingia californica Conrad, 1837. Common nestler in rocky intertidal.

* *Semele rubropicta* Dall, 1871. Uncommon, offshore species; valves occasionally wash ashore; characterized by conspicuous radial and concentric sculpture and reddish radial rays.

Semele rupicola Dall, 1915. Nestler in rocky intertidal; uncommon.

LIST OF TELLINIDAE

See Coan, 1971, Veliger 14 Suppl. 63 pp. for systematic review; Reid and Reid, 1969, Can. J. Zool. 47: 649–657 (*Macoma* feeding).

Macoma acolasta Dall, 1921. Protected bays in low intertidal in sand; rare.

Macoma balthica (Linnaeus, 1758) [= *M. inconspicua* (Broderip and Sowerby, 1829)]. Common in mud in upper intertidal of bays, especially in brackish water; see Vassallo 1969, Veliger 11: 223–234 and 1971, 13: 279–284 on aspects of ecology in San Francisco Bay, its generally southern limit, where it may have been introduced from northern areas.

* *Macoma expansa* Carpenter, 1864. Of the same group as *M. secta* and *M. indentata*, but thinner, more inflated, with an only slightly developed posterior dorsal flange: rare, in sand offshore, in exposed areas: valves occasionally washing ashore.

Macoma indentata Carpenter, 1864. In silt to sand of bays, uncommon.

Macoma inquinata (Deshayes, 1855) (=*M. irus* of authors, not of Hanley, 1845). Common, in silt and mud in protected areas, most common in bays, but also below surf zone offshore.

Macoma nasuta (Conrad, 1837). Bent-nosed clam; common, in mud and muddy sand in protected areas, most common in bays at mid-tide, also below surf zone offshore. See Yonge, 1949.

Macoma secta (Conrad, 1837). Sand clam; common, intertidal in sand in semiprotected areas of bays and offshore of sandy beaches. See Yonge, 1949.

Macoma yoldiformis Carpenter, 1864. In silt to sand, in protected areas in low intertidal of bays; rare.

Tellina bodegensis Hinds, 1845. Low intertidal in sand of exposed beaches; in bays.

* *Tellina carpenteri* Dall, 1900. Similar to *T. modesta*, but light pink in color; occurs below low tide on various bottoms.

Tellina modesta (Carpenter, 1864) (=*T. buttoni* Dall, 1900). In sand to silty sand of bays, to well offshore. See Mauer, 1967 (aspects of biology).

Tellina nuculoides (Reeve, 1854) [=*T. salmonea* (Carpenter, 1864)]. On various bottoms in protected areas of bays. See Mauer, 1967 (aspects of biology).

SOLENIDAE

1. Beaks at anterior end of shell ....*Solen sicarius*
– Beaks central 2
2. Posterior end round; internal radial rib sloping anteriorly, relatively wide; not conspicuously colored externally; adult large, over 40 mm in length (figs. 49A, B) .... *Siliqua patula*
– Posterior end truncate; internal radial rib vertical, narrow; externally with concentric brown bands; adult not over 40 mm .... *Siliqua lucida*

LIST OF SOLENIDAE

See Hertlein, 1961, Bull. So. Calif. Acad. Sci. 60: 12–18 (*Siliqua*, taxonomy)

Siliqua lucida (Conrad, 1837). Protected sandy areas of bays.

Siliqua patula (Dixon, 1789). Razor clam; semiprotected, clean-sand beaches; see Weymouth, McMillin, and Holmes, 1925, Bull. U.S. Bur. Fish. 41: 201–236 and Weymouth and McMillin, 1930, *Ibid.*, 46: 543–567; Weymouth, McMillin, and Rich, 1931, J. Exp. Biol. 8: 228–249; Yonge, 1952a (aspects of morphology, reproduction, growth, biology, ecology); Pohlo, 1963, Veliger 6: 98–104 (morphology, burrowing).

Solen sicarius Gould, 1850. Protected areas of bays in mud or muddy sand; forming permanent burrows in which it moves freely up and down; see Pohlo, 1963, above, on morphology and burrowing in *S. rosaceus* Carpenter, 1864, a southern California species.

MYIDAE

1. Shell heavy, with concentric sculpture; anterior round, posterior truncate, gaping; periostracum thick at posterior end (fig. 53); boring into rock, hard clay ....*Platyodon cancellatus*
– Shell thin; in mud or sand 2
2. Pallial sinus deep; shell to 120 mm or more in length (fig. 47A) ....*Mya arenaria*
– Pallial line entire, pallial sinus inconspicuous; to 30 mm in length (fig. 2) ....*Cryptomya californica*

LIST OF MYIDAE

Cryptomya californica (Conrad, 1837). In sand or mud in bays; using its very short siphons, *Cryptomya* "taps" the burrows of other invertebrates, particularly of *Urechis* and *Upogebia* (fig. 2); also on open coast in gravel, among rocks; see Yonge, 1951a.

Mya arenaria Linnaeus, 1758. Soft-shelled or long-necked clam; in mud and sand of bays, burrowing to 30 cm deep; introduced from Atlantic coast. See MacNeil, 1965, U.S. Geol. Survey Prof. Paper 483-G, pp. 33–35.

Platyodon cancellatus (Conrad, 1837). A rock borer, common in shale, also in sandstone and hard clay; see Yonge, 1951b.

HIATELLIDAE

1. Large, quadrate in outline; pallial line entire, broad; hinge with 1 cardinal tooth in either valve; posterior end broadly

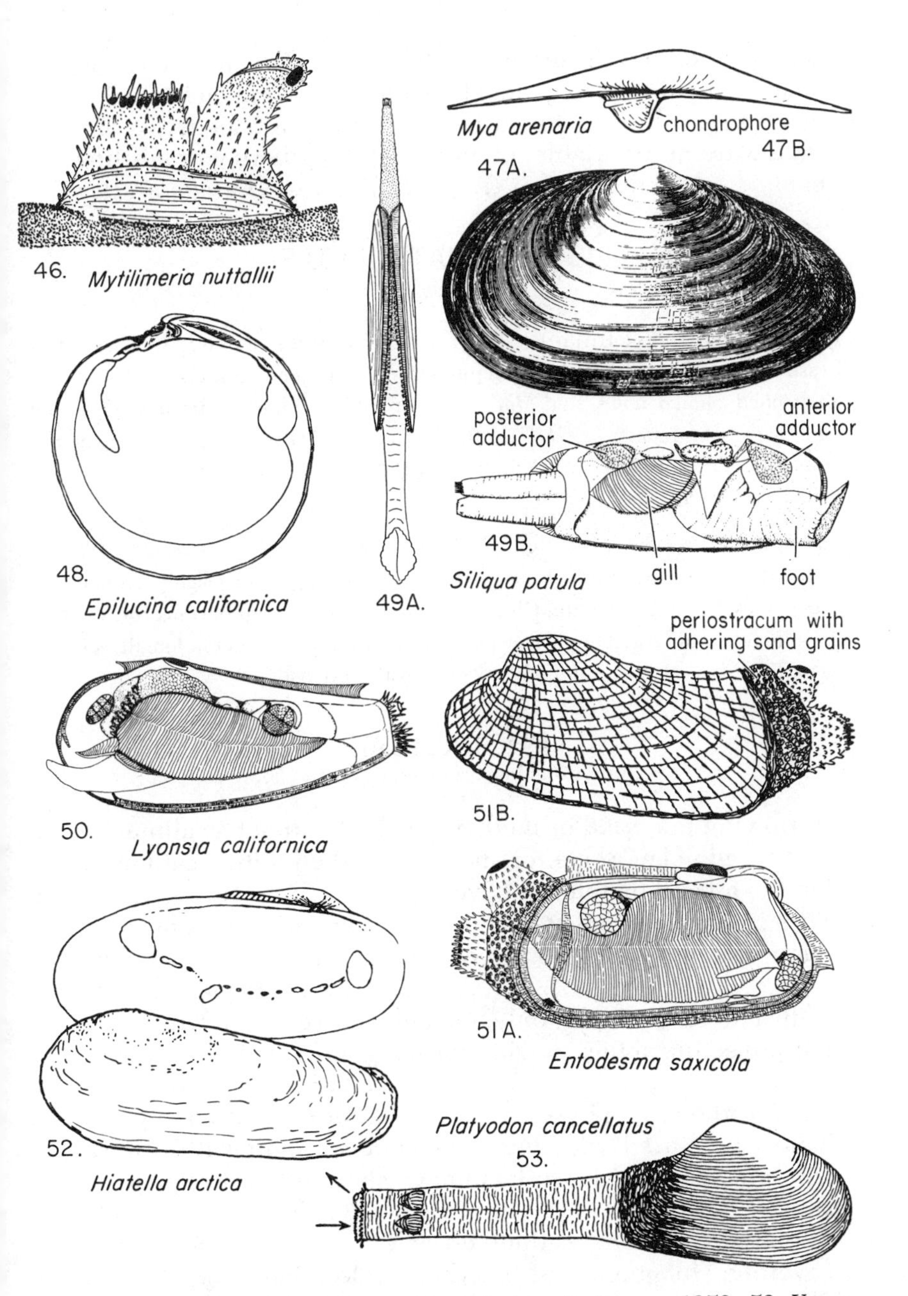

PLATE 130. **Bivalvia** (7). Various: 46, 49A, 51A,B, Yonge, 1952; 53, Yonge, 1951; 47A, Pilsbry, 1891; 48, *Treatise on Invertebrate Paleontology;* 49B, Pohlo, 1963; 50, Narchi, 1968; 47B, 52 (top) Keen, 1963; 52 (bottom) R. Stohler, drawn by E. Reid.

gaping; shell not distorted (fig. 28) ...... *Panopea generosa*
– Small to medium size; pallial line faint, often broken into irregular patches; hinge without teeth or with 1 or 2 teeth (fig. 52); shell extremely variable, often distorted by boring or nestling habit .... *Hiatella arctica*

LIST OF HIATELLIDAE

See Yonge, 1971.

Hiatella arctica (Linnaeus, 1767) (=*Saxicava arctica*). Attaches with a byssus, but can also bore; in bays, on pilings, in fouling, open coast in algal holdfasts, abandoned pholad holes, and *Mytilus* beds; elongate, boring specimens without teeth have been reported as *H. pholadis* (Linnaeus, 1771), a probable synonym; see Hunter, 1949, Proc. Roy. Soc. Edinburgh (B) 43: 271–289 (morphology, biology of British specimens).

Panopea generosa (Gould, 1850) (= *Panope generosa*, an invalid emendation). The geoduck (several variant spellings, but usually pronounced gooey-duck, possibly from the Chinook *gweduc*); a very deep burrower in soft bottoms from low intertidal to offshore; siphons may be several feet in length; see Illg, 1949, Proc. U.S. Nat. Mus. 99: 391–428 (parasitic copepods).

PHOLADIDAE

For terminology, see figures 54 and 56.

1. Burrowing into sand or mud; adult shell without a callum .. 2
– Boring into clay, shale, or shell; adult shell with a callum in form of a band or anterior covering 3
2. With a dorsal mesoplax only; umbonal-ventral sulcus present; often with projecting spines on anterior shell margin (fig. 59) .... *Zirfaea pilsbryi*
– With a dorsal protoplax only; no umbonal-ventral sulcus; without projecting spines on anterior margin (fig. 56) .... *Barnea subtruncata*
3. Shell without apophyses; callum present only as an anterior band (sculptured with high, thin flutes); adult with a long, tapering, calcareous siphonoplax; shell often irregular in shape (fig. 66) .... *Netastoma rostrata*
– Apophyses present; callum present in adult as an anterior covering; siphonoplax never wholly calcareous 4
4. Siphons with conspicuous, orange, chitinous patches and warty tips which cannot be retracted into shell; callum in adult not completely covering anterior aperture; shell oval (fig. 60) .... *Chaceia ovoidea* –
– Siphons retractable into shell, may be pustulose but never

with orange patches or warty tips; callum in adult completely covering anterior aperture; shell elongate 5

5. Shell with 2 dorsal plates, a mesoplax and a metaplax; posterior end of shell with overlapping chitinous plates and set off from central area; hypoplax present; siphonoplax absent (fig. 55) .... *Parapholas californica*

– Shell with only one dorsal plate, a mesoplax; posterior end of shell not set off; hypoplax absent; siphonoplax present or absent 6

6. Umbonal reflection free anteriorly; siphonoplax absent; siphons pustulose .... *Penitella gabbii*

– Umbonal reflection closely appressed for its entire length; siphonoplax present; siphons smooth 7

7. Siphonoplax composed of heavy, flexible, chitinous flaps which are not lined with calcareous granules; mesoplax pointed posteriorly, truncate anteriorly, and with lateral wings (figs. 57B, C); often in shale (shell, fig. 57A) *Penitella penita*

– Siphonoplax not flexible, composed of a heavy, chitinous, outer layer lined with coarse calcareous granules; mesoplax truncate posteriorly, pointed anteriorly, without lateral wings (figs. 58A, C); most often in *Haliotis* or other shells .... *Penitella conradi*

LIST OF PHOLADIDAE

See Turner, 1954–1955, Johnsonia 3: 1–160; Kennedy, 1974, Mem. San Diego Soc. Natl. Hist. 8: 128 pp. for monographic treatment.

Barnea subtruncata (Sowerby, 1834) (= *B. pacifica* Stearns, 1871). Burrowing in mud or clay of well-protected bays.

Chaceia ovoidea (Gould, 1851) (= *Pholadidea ovoidea*). Boring into shale.

Netastoma rostrata (Valenciennes, 1846) (= *Nettastomella rostrata*). Boring in shale at low tide; dredged offshore.

Parapholas californica (Conrad, 1837). Boring into a variety of substrates from clay to rock; siphonoplax absent but, instead, a thick tube ("chimney") of finely cemented particles is formed by the siphons as a result of boring activity.

Penitella conradi Valenciennes, 1846. Often found boring into shells such as *Mytilus* or *Haliotis* (forming "blister pearls" inside abalone shells); occasionally in shale or soft rock; see Smith, 1969, Amer. Zool. 9: 869–880 (functional morphology); Hansen, 1970, Veliger 13: 90–94.

Penitella gabbii (Tryon, 1863). Found along with *P. penita*, but much less common.

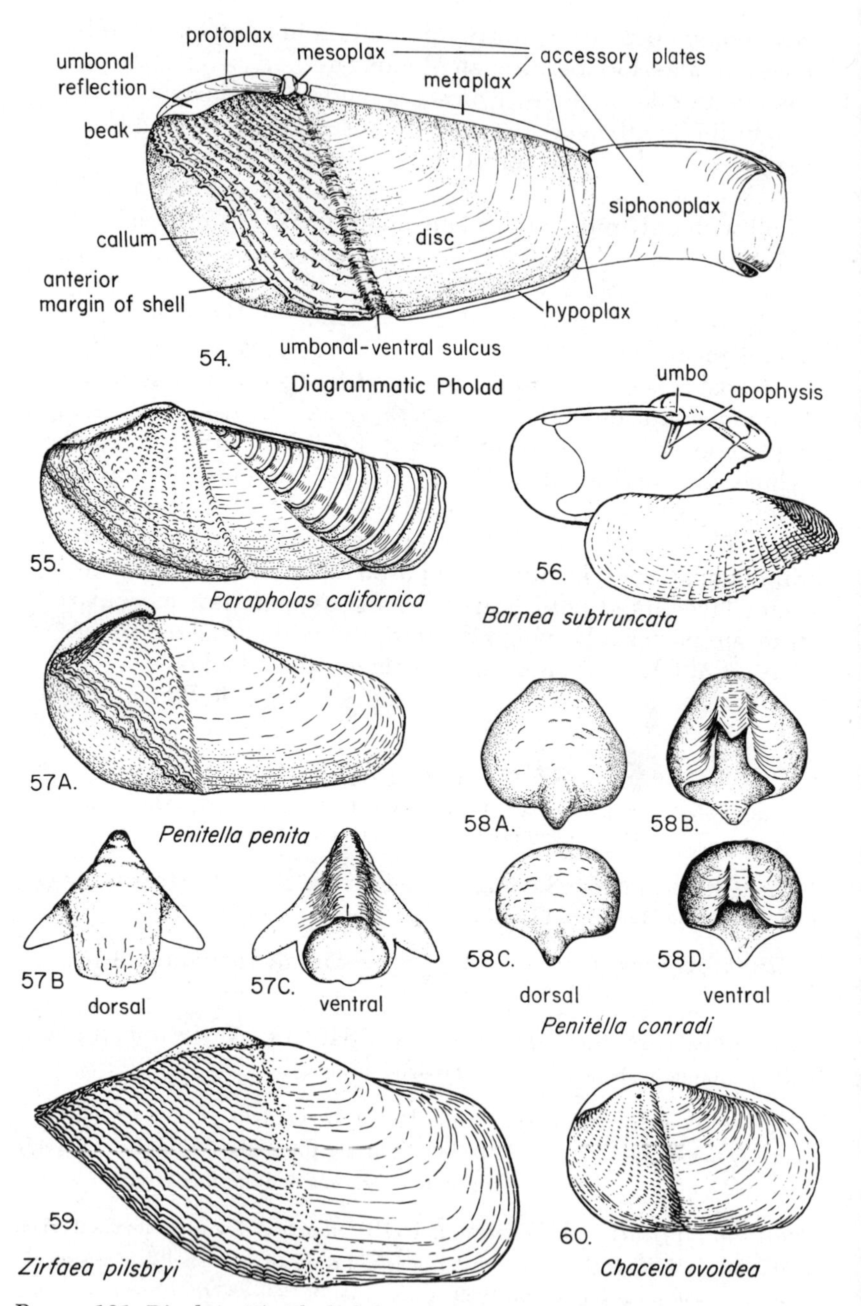

PLATE 131. **Bivalvia** (8). Pholadidea: 54, 57A,B,C, 58A,B,C,D, Turner, 1954; 55, original; 56, 60, Keen, 1963; 57A, 59, R. Stohler, drawn by E. Reid.

Penitella penita (Conrad, 1837) (=*Pholadidea penita*). Boring into a variety of substrates from stiff clay to sandstone and cement; common: see Evans, 1968, Ecology 49: 619–628 (growth rate) and earlier papers; Evans and LeMessurier, 1972, Can. J. Zool. 50: 1251–1258 (functional morphology).

* *Penitella turnerae* Evans and Fisher, 1966. See Kennedy, 1974, above; at Duxbury Reef.

Zirfaea pilsbryi Lowe, 1931. In bays, burrowing into mud and clay; see MacGinitie, 1935, Amer. Midl. Nat. 16: 731–735 (burrowing).

TEREDINIDAE

For terms, see figure 61

1. Pallets with elongate blade, composed of distinct, cone-shaped segments (fig. 64) .... *Bankia setacea*
– Pallets variable in shape, not segmented 2
2. Distal 1/2 of blade a dark colored periostracal cap overlapping calcareous basal portion (fig. 63) .... *Lyrodus pedicellatus*
– Blade almost entirely calcareous, slightly to deeply cupped; periostracum on distal 1/2 of blade not extending beyond calcareous portion (fig. 62) .... *Teredo navalis*

LIST OF TEREDINIDAE

All local species of shipworms occur in wood, such as wharf pilings, in bays and harbors; see monograph by R. D. Turner, 1966, *A Survey and Illustrated Catalogue of the Teredinidae (Mollusca: Bivalvia)*, Mus. Comp. Zool., Harvard, 265 pp., and C. L. Hill and C. A. Kofoid, eds., 1927, *Marine Borers and their Relation to Marine Construction on the Pacific Coast*, Final Report, San Francisco Bay Marine Piling Comm., San Francisco, 357 pp.

Bankia setacea (Tyron, 1863). Possesses external fertilization and planktonic larval stage; see Haderlie and Mellor, 1973, Veliger 15: 265–286 (aspects of biology).

Lyrodus pedicellatus (Quatrefages, 1849) (= *Teredo diegensis* Bartsch, 1916). Introduced; young are retained to late veliger stage; see Eckelbarger and Reish, 1972, Bull. So. Calif. Acad. Sci. 71: 48–50 (self-fertilization).

Teredo navalis Linnaeus, 1758. Introduced; young are retained until veliger stage; see papers by Miller, Blum, Dore and Lazier, 1922–1924, Univ. Calif. Publ. Zool. 22 and 26: 41–80 (biology, morphology, boring mechanism).

LYONSIIDAE

1. Outline circular or ovate; beaks twisted; living in compound ascidians (fig. 46) .... *Mytilimeria nuttallii*

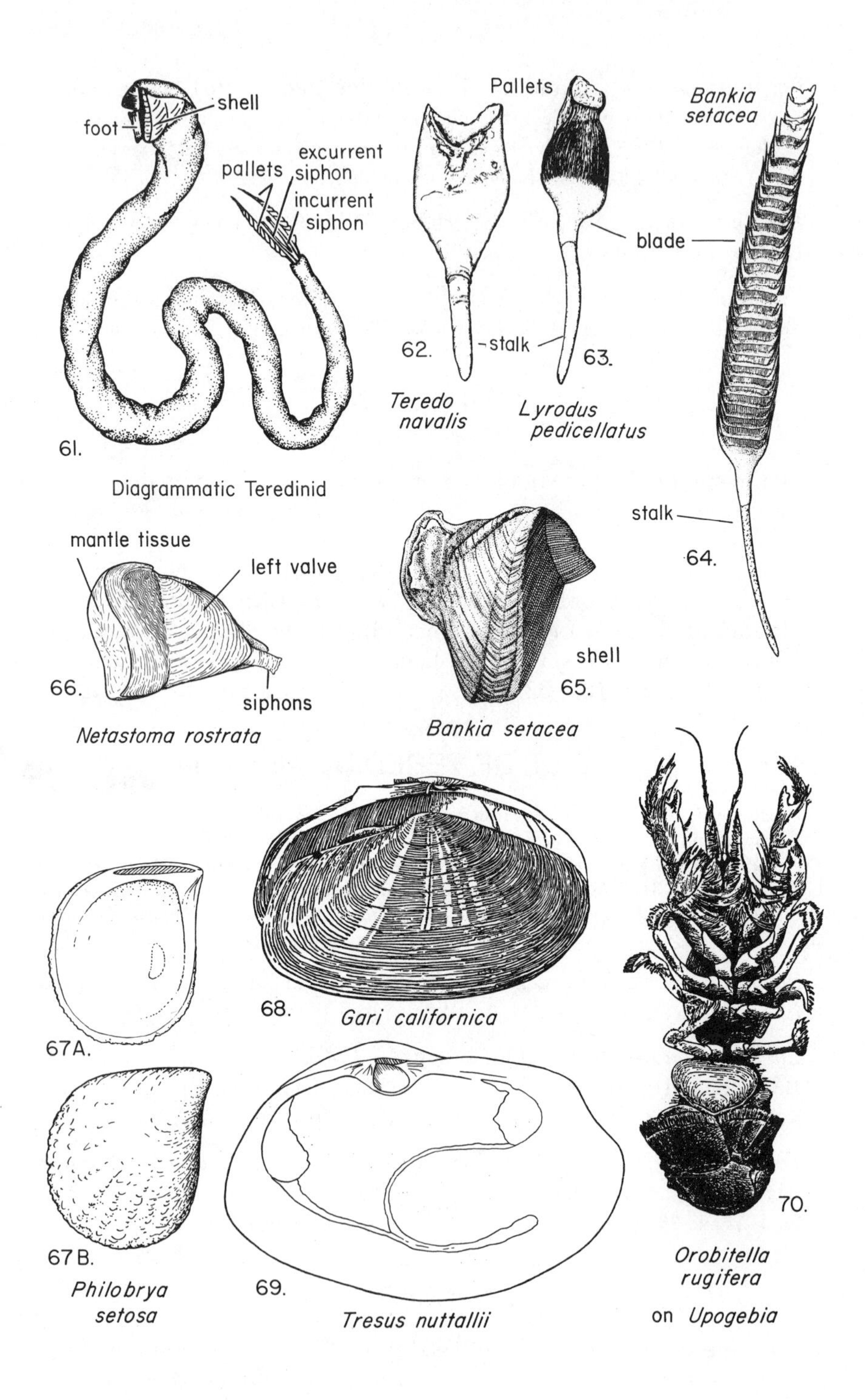
shell
foot
excurrent
pallets
siphon
incurrent
siphon
61.
Diagrammatic Teredinid
Pallets
62.
stalk
63.
Teredo
navalis
Lyrodus
pedicellatus
Bankia
setacea
blade
stalk
64.
mantle tissue
left valve
66.
siphons
Netastoma rostrata
shell
65.
Bankia setacea
67A.
68.
Gari californica
67B.
Philobrya
setosa
69.
Tresus nuttallii
70.
Orobitella
rugifera
on Upogebia

- Posterior end produced; beaks not twisted; free-living or byssally attached 2
2. Periostracum thin, striate; shell thin, elongate, pearly (internal, fig. 50); in muddy substrates .... *Lyonsia californica*
- Periostracum thin in adult; shell thicker, shorter, irregular in shape (figs. 51A, B); byssally attached in rocky areas .... *Entodesma saxicola*

LIST OF LYONSIIDAE

* *Entodesma inflatum* (Conrad, 1837). Small, lighter in color and more regular in shape than *E. saxicola;* reported in compound ascidians.

Entodesma saxicola (Baird, 1863). Byssally attached under rocks in low intertidal, in fouling on wharf pilings; see Yonge, 1952b.

Lyonsia californica Conrad, 1837. In muddy substrates in protected areas of bays; see Narchi, 1968, Veliger 10: 305–313 (functional morphology); Mauer, 1967 (aspects of biology).

Mytilimeria nuttallii Conrad, 1837. In compound ascidians *Eudistoma* and *Distaplia* in rocky intertidal; commonly washed ashore embedded in ascidians; see Yonge, 1952b.

REFERENCES ON BIVALVIA

(see also general molluscan references)

Coe, W. R. 1948. Nutrition, environmental conditions, and growth of marine bivalve mollusks. Jour. Mar. Res. 7: 586–601.

Fitch, J. E. 1953. Common marine bivalves of California. Calif. Dept. Fish Game Fish Bull. 90, 102 pp.

Hanna, G D. 1966. Introduced mollusks of western North America. Occ. Pap. Calif. Acad. Sci. 48, 108 pp.

Hertlein, L. G. and U. S. Grant IV 1972. The geology and paleontology of the marine Pliocene of San Diego, California (Paleontology: Pelecypoda). San Diego Soc. Natl. Hist. Mem. 2, pt. 2B, 409 pp. Excellent for extensive references to many Recent species.

Keen, M. and E. Coan 1974. *Marine Molluscan Genera of Western North America: an Illustrated Key.* Second edition. Stanford University Press, 208 pp.

Kellogg, J. L. 1915. Ciliary mechanisms of lamellibranchs with descriptions of anatomy. J. Morph. 26: 625–701. Many Pacific coast species illustrated.

Loosanoff, V. L. and H. C. Davis 1963. Rearing of bivalve mollusks. Adv. Mar. Biol. 1: 138 pp.

PLATE 132. **Bivalvia** (9). Teredinidae, others: 61–65, Turner, 1966; 66, drawn by Emily Reid; 67A,B, Keen, 1963, redrawn by E. Reid; 68, Keep and Baily, 1935; 69, Weymouth, 1921, redrawn by E. Reid; 70, Keep and Baily, 1935.

Loosanoff, V. L., H. C. Davis and P. E. Chanley 1966. Dimensions and shapes of larvae of some marine bivalve mollusks. Malacologia 4: 351–435.

Mauer, D. 1967. Mode of feeding and diet and synthesis of studies on marine pelecypods from Tomales Bay, California. Veliger 10: 72–76. *Tellina modesta* (as *T. buttoni*), *Tellina nuculoides* (as *T. salmonea*), *Mysella tumida, Transennella tantilla, Lyonsia californica.*

Moore, R. C. ed., *et al.* 1969, 1971. Mollusca, Bivalvia, Part N, of *Treatise on Invertebrate Paleontology,* Univ. Kansas Press and Geol. Soc. America.

Popham, M. L. 1940. The mantle cavity of some of the Erycinidae, Montacutidae and Galeommatidae with special reference to the ciliary mechanisms. J. Mar. Biol. Assoc. U.K. 24: 549–587.

Stasek, C. R. 1961. The ciliation and function of the labial palps of *Acila castrensis* (Protobranchia, Nuculidae), with an evaluation of the role of the protobranch organs of feeding in the evolution of the bivalvia. Proc. Zool. Soc. London 137: 511–538.

Yonge, C. M. 1939. The protobranchiate Mollusca: a functional interpretation of their structure and evolution. Phil. Trans. Roy. Soc. London (B) 230: 79–147.

Yonge, C. M. 1949. On the structure and adaptations of the Tellinacea, deposit-feeding Eulamellibranchia. Phil. Trans. Roy. Soc. London (B) 234: 29–76.

Yonge, C. M. 1951a. Studies on Pacific coast mollusks. I. On the structure and adaptations of *Cryptomya californica* (Conrad). Univ. Calif. Publ. Zool. 55: 395–400.

Yonge, C. M. 1951b. ———. II. Structure and adaptations for rock boring in *Platyodon cancellatus* (Conrad). Univ. Calif. Publ. Zool. 55: 401–407.

Yonge, C. M. 1951c. ———. III. Observations on *Hinnites multirugosus* (Gale). Univ. Calif. Publ. Zool. 55: 409–420.

Yonge, C. M. 1952a. ———. IV. Observations on *Siliqua patula* Dixon and on evolution within the Solenidae. Univ. Calif. Publ. Zool. 55: 421–438.

Yonge, C. M. 1952b. ———. V. Structure and adaptation in *Entodesma saxicola* (Baird) and *Mytilimeria nuttalli* Conrad, with a discussion on evolution within the family Lyonsiidae (Eulamellibranchia). Univ. Calif. Publ. Zool. 55: 439–450.

Yonge, C. M. 1952c. ———. VI. A note on *Kellia laperousii* (Deshayes). Univ. Calif. Publ. Zool. 55: 451–454.

Yonge, C. M. 1971. On functional morphology and adaptive radiation in the bivalve superfamily Saxicavacea (*Hiatella* (=*Saxicava*), *Saxicavella, Panomya, Panope, Crytodaria*). Malacologia 11: 1–44.

PHYLUM ECTOPROCTA (BRYOZOA)

John D. Soule, Dorothy F. Soule, and Penny A. Pinter
Allan Hancock Foundation, University of Southern California

(Plates 133–142 and Figure 84)

Bryozoans or "moss animals" are minute, sessile animals, mostly marine and colonial. Their oceanic distribution is worldwide and, although they are found from the intertidal zone to 6000 meters and from tropic to polar seas, they are most abundant in the littoral and neritic zones down to about 200 meters.

In its older meaning, the phylum **Bryozoa** (or **Polyzoa**) included polypoid lophophorate animals (but not phoronids), possessing a U-shaped gut with separate mouth and anal openings. The assemblage was divided into two classes, **Entoprocta** and **Ectoprocta,** according to whether the anus opened within or outside the circlet of ciliated tentacles. Later research showed other fundamental differences between these groups, particularly in the method of coelom formation: entoprocts are pseudocoelomate, but ectoprocts are coelomate. Separation of bryozoans into two phyla was suggested in the late nineteenth century, and this concept has been generally accepted by zoologists, although the groups are still treated together in many systematic works as a matter of convenience, and some zoologists (e.g., Nielsen, 1971) still support the older concept of the phylum Bryozoa. In this manual, Entoprocta are treated separately (p. 609) and, in this chapter, the term Bryozoa refers to Ectoprocta.

Individuals (**zooids**) of the ectoproct colony are boxlike and small, (figs. 1, 2), often less than 1 mm in length; the colonies (**zoaria**) vary greatly in size. The ciliated tentacles of each zooid are borne on a circular or crescentic ridge called the **lophophore;** the mouth lies within the tentacle circle, and the anus is outside. There is a complete digestive tract which, in some species, includes a gizzard. The body cavity is a true coelom which extends into the tentacles. A reduced nervous system is present, but there is no excretory system. The term "**polypide**" includes the tentacles, lophophore, and gut.

Each zooid is permanently enclosed in an exoskeleton of its own secretion (a **zooecium**) which is usually contiguous with the exoskeletons of adjacent zooids in the colony. The body wall lines the exoskeleton. The lophophore can be extended from the zooecium for feeding. Some ectoprocts have modified zooids called **avicularia,** pincerlike structures with a mandible, either borne on a stalk (*Bugula,* figs. 55–59) or sessile and attached directly to the zooecium (*Callopora,* figs. 1, 2). Globular structures called **ovicells,** for brooding larvae, may

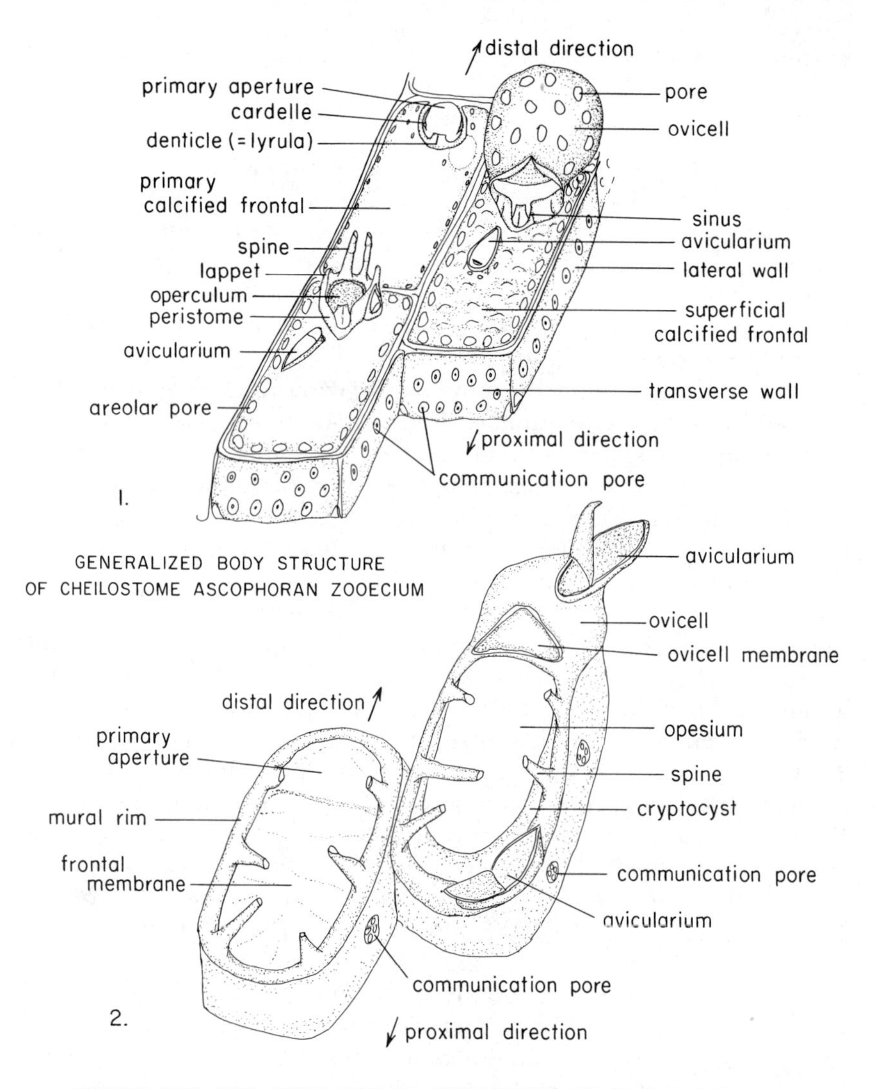

PLATE 133. **Ectoprocta** (1). Generalized cheilostome structure (grossly enlarged): 1, 2. These and following figures of ectoprocts drawn by Susan Soule Harrison under direction of the authors. Unless otherwise noted, all following figures are printed at a magnification of 25x.

be present at the distal end of some zooecia in cheilostomes, but are usually absent in young colonies. In some cyclostomes, the ovicell is a community chamber for the colony, with a separate opening (**ooeciostome**) which may be difficult to locate (figs. 12, 14, 15, 17B). In others the ovicell is a modified individual, a **gonozooecium** (figs. 16, 18, 19–21).

Early bryozoan classifications were based primarily on colony form because detailed studies of zooid structure had not yet been made. Later it was found that colony form was not adequate for classification. Some species vary in form according to the environment; one may cling closely to the substratum in turbulent waters, but rise in fronds when living in calmer waters. External features of the zooecia now largely form the basis of classification, but knowledge of typical colony forms is of considerable help to the student, and especially aids the nonspecialist in identifying the orders.

Bryozoan literature contains many growth-form terms based on the name of the genus most typical of that form; identification depends on a familiarity with generic characteristics. Regardless of the names chosen for these forms, the following basic growth patterns can be seen:

I. RECUMBENT OR INCRUSTING FORMS

a. **Soft-bodied species**

1. **stolonate:** forms with creeping branches extended over the substratum. Certain ctenostomes have true **stolons,** which are individual zooids lacking polyps and modified to form the extensions (*Bowerbankia,* fig. 7). Some Cheilostomata are **stoloniform,** the extensions being formed by the basal portions of zooids (*Aetea,* fig. 23). **Penetrating** forms construct stolonate colonies by burrowing into molluscan shells (Cheilostomata: *Penetrantia;* Ctenostomata: *Terebripora, Immergentia;* see Soule and Soule, 1969). Some species may be entirely recumbent, or contiguous with the substrate; this condition is spoken of as **adnate.**

2. **fleshy (carnose):** species that form gelatinous colonial masses, sometimes of considerable size (Ctenostomata: *Flustrellidra, Alcyonidium,* figs. 3–5).

b. **Incrusting,** more or less calcified, species:

1. **flexiform:** species that form a lightly calcified, flat crust over a soft, flexible substratum such as algal blades (*Membranipora,* figs. 24–30).

2. **rigid:** heavily calcified species that usually form unilaminar or multilaminar crusts over solid, inflexible surfaces such as shell and stone. Most cheilostomes are of this type. A few species form heavy, knobby incrustations over originally flexible surfaces, such as twigs, turning them into solid structures.

c. **Tubular species:** some cyclostomes form colonies of recumbent tubules with terminal apertures; some are erect (*Diaperoecia*, fig. 13A, B).

II. ERECT LEAFY, BRANCHING, OR BUSHLIKE FORMS

a. **Branching:** species with various modes of branching, some jointed, attached to the substratum loosely by rootlets (radicles). Anascan cheilostomes such as *Bugula* (figs. 54–59) and *Scrupocellaria* (figs. 49, 50) are typical. Most are not resistant to turbulent waters; easily confused with certain hydroids.

b. **Foliaceous (flustraform** or **frondose):** species that form leaflike colonies which resemble small cabbage plants or lichens (Cheilostomata, Anasca: *Thalamoporella*, fig. 40).

c. **Fenestrate:** erect species, with a "chicken-wire" appearance, which form large colonies; usually typical of deep-water forms. "Retepores" such as the beautiful cheilostome (Ascophora) *Phidolopora* (figs. 74; 84, page 608) are typical.

d. **Tubular:** certain Cyclostomata form colonies in which tubules extend upward from the basal disc of the colony (*Lichenopora*, fig. 12).

PRESERVATION METHODS AND MICROSCOPY

Bryozoans may be fixed and preserved in 75% alcohol; 10% formalin in sea water (1 part : 9 parts SW) may be used only if reagent grade (not commercial) formaldehyde is available, otherwise an acid pH may damage the calcareous exoskeleton. Specimens so preserved should be transferred to 70% alcohol for storage as soon as practicable.

Preserved material must be examined under a dissecting microscope for noncalcareous Entoprocta and ctenostome bryozoans; these are stained, dehydrated, and mounted under cover slips (see Humason, 1972, ch. 26). Remaining material may be stored wet, or dried and mounted uncovered on slides using casein "white glue," which is nonacid. Calcareous colonies mounted with balsam or "Duco cement" will disintegrate after a few years because of the acid reaction.

Field identifications of the more distinctive genera can be made with a hand lens, but species identification in general requires higher magnification, since individual zooids are usually less than 1 mm in length, their apertures about 0.1 mm, and other characters smaller still. The low power of a dissecting microscope is used for preliminary sorting and mounting, while magnifications of 75X to 100X are often needed for final identification.

Skeletal characteristics are useful in identification and can be made visible by a brief immersion in a bleaching solution, such as Clorox, which removes organic material. A damp, fine-grade, camel's hair brush is useful in handling delicate specimens. Colonies incrusted on large shells may be cut from the shells with a small abrasive disc on an electric drill. It is difficult to remove a colony from its substratum without damage. Occasionally it is desirable to tint a colony if the diagnostic characteristics are not readily visible; borax-carmine stain may be used.

Ectoprocts are not easily recognized by the beginning student. Some species resemble coralline algae, others resemble colonial tunicates or sponges, while bushy colonial species are often confused with hydroids. Examination with a hand lens will, however, show the patterns of multiple zooecia with their individual apertures. The polypides inside the chitinous cover of soft, bushy species of translucent ctenostomes distinguish them from hydroids. The white, "crunchy," bushy cyclostomes and many of the bushy cheilostomes are stiffer and more brittle than hydroids. Living ectoprocts, as seen under a dissecting microscope, will extend their delicate crowns of ciliated tentacles from the apertures. When disturbed, the lophophore is retracted base-first into the zooecium; the tentacles do not shorten as they do in hydroids, nor curl inwards onto the surface of the calyx as they do in entoprocts.

KEY TO ORDERS OF ECTOPROCTA

1. Colonies soft, gelatinous, leathery, or furry; zooids never calcified, either fleshy and contiguous or separate and linked by stolons; aperture simple, closed by constriction; no avicularia or external ovicells (figs. 3–9) .... order **Ctenostomata,** Key A (p. 584)
– Colonies extremely varied in form, with zooids encased in zooecia which are obviously calcified or chitinized, and usually provided with pores, spines, or other structures 2
2. Zooecia in form of whitish tubes with terminal aperture; surface may be finely perforate; edges of aperture only rarely bearing spines; no opercula or avicularia; colonies may be branching and arborescent (figs. 16, 18–22), or incrusting (figs. 10–12), or raised into stiff, flattened lobes (figs. 13, 17) ....order **Cyclostomata,** Key B (p. 584)
– Zooecia almost never simple tubes; of diverse form with aperture near the upper end of the "front"; aperture with an operculum; avicularia, pores, spines, etc., usual; colonies arborescent, foliaceous, or incrusting ....order **Cheilostomata** 3
3. Front basically uncalcified; no compensation sac .... suborder **Anasca,** Key C (p. 589)

The name is based on a negative character, the absence of an **ascus** or compensation sac beneath the usually uncalcified frontal membrane. However, there may be a calcified layer (**cryptocyst**) beneath the frontal membrane, or overarching or fused spines may make a calcified wall (**pericyst**) above the frontal membrane. Most conspicuously bushy, arborescent cheilostomes are anascans.

– Front rigidly calcified; internal compensation sac (ascus) present .... suborder **Ascophora,** Key D (p. 598)

The name refers to an internal compensation sac (**ascus**) which permits tentacle extrusion in those ectoprocts in which the frontal membrane is rigidly calcified. The sac may open by a distinctive, easily seen **ascopore** on the calcified frontal (figs. 60–63), or it may open into the aperture out of sight beneath the hinged base of the operculum. The frontals of the zooecia are heavily calcified; colony forms in the Ascophora are much varied.

KEY A: ORDER CTENOSTOMATA

1. Colonies fleshy or leathery, with squat, contiguous zooecia 2
– Colonies stolonate, with separate, erect zooecia 4
2. With smooth, membranous frontals 3
– With frontals bearing chitinous spines having 4–6 prongs (fig. 3) .... *Flustrellidra corniculata*
3. Zooecia small, irregularly circular, papillate; cuticle gelatinous, covered with silt (fig. 5) ... *Alcyonidium parasiticum*
– Zooecia large, hexagonal, elongate (fig. 4) .... *Alcyonidium polyoum*
4. Zooecia cylindrical (tubular), arising directly from stolon ... 5
– Zooecia with thin stalk (pedicel) 6
5. Zooecia with expanded basal area, may have secondary zooecia arising from primary zooecium (fig. 6) *Victorella pavida*
– Zooecia in clusters, often with spurlike, proximal (caudal) extension (fig. 7) .... *Bowerbankia gracilis*
6. Distal (oral) termination of zooecia square or irregularly rounded (fig. 8) .... *Triticella elongata*
– Distal termination of zooecia bilobate (liplike) (fig. 9) .... *Farrella elongata*

KEY B: ORDER CYCLOSTOMATA

Reproduction in specialized individuals (gonozooecia), or in interzooecial colonial brood chambers; colony forms much varied depending on substrate and environment; species difficult to determine in young, which lack the distinctive reproductive chambers. A comprehensive revision of Pacific Cyclostomata is needed, hence some identifications in the key are only to generic level.

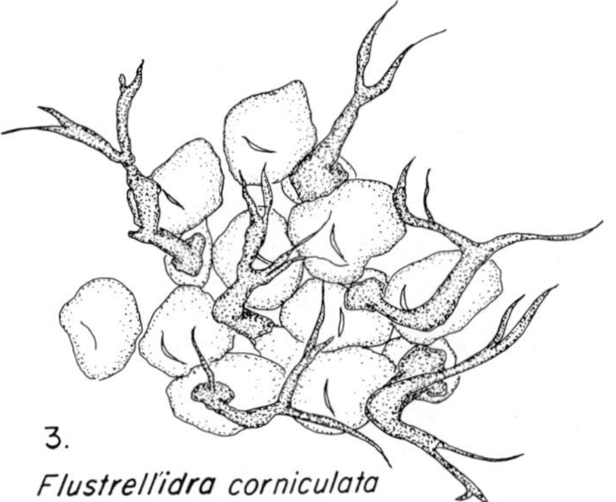

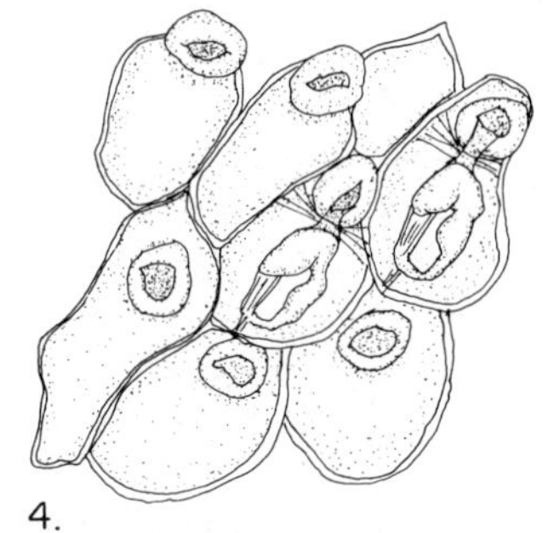

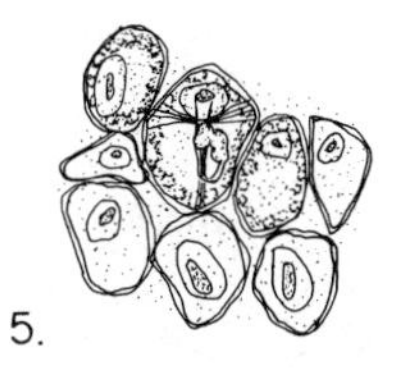

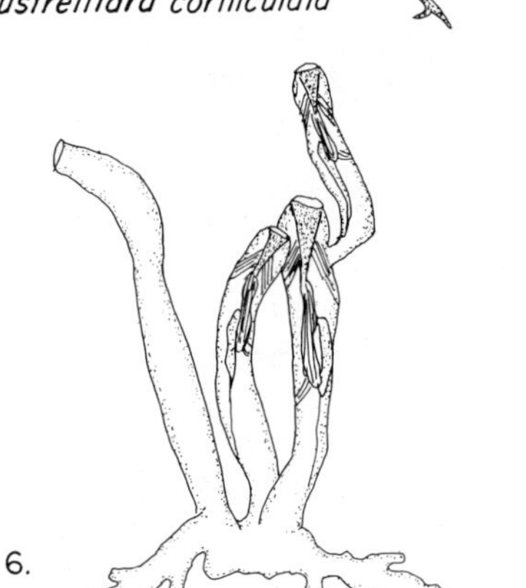

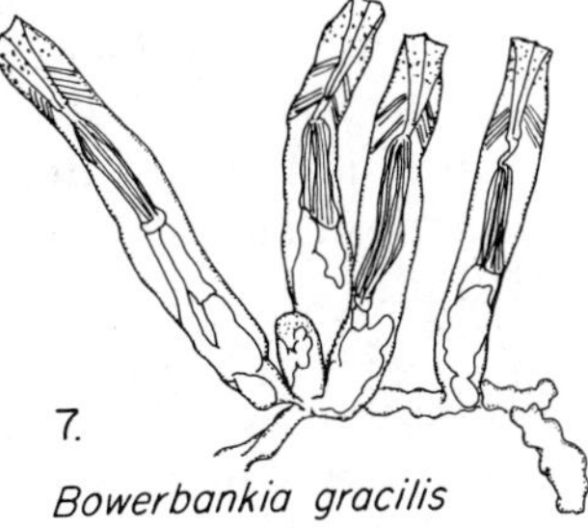

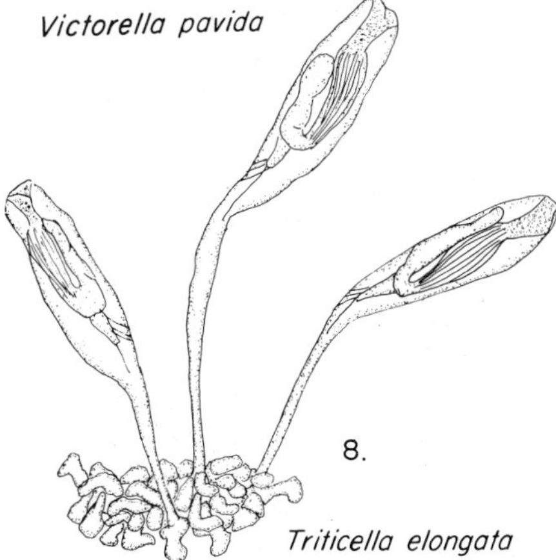

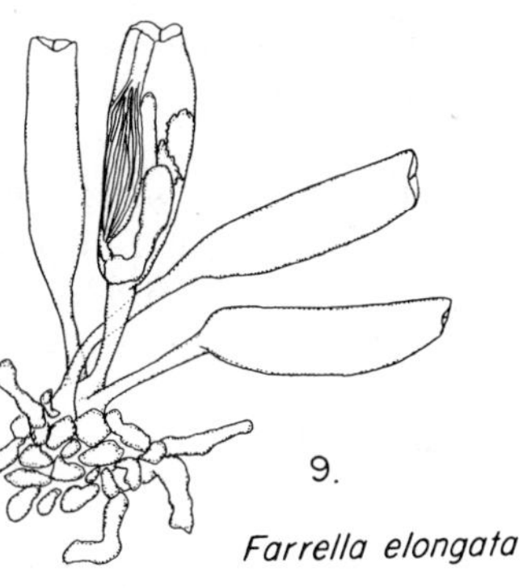

PLATE 134. **Ectoprocta** (2). Ctenostomata: 3–9.

1. Colonies incrusting or recumbent 2
– Colonies erect or partially erect 7
2. Colonies forming incrusting discs 3
– Colonies branching singly, or fanning into lobes 4
3. Colonies forming symmetrical, circular discs with zooecial tubules in regular, radiate, uniserial or multiserial rows; brood chamber beneath central area, with low, flaring opening (ooeciostome); new colonies budded as discrete discs, upward or laterally beyond edge of primary colony (fig. 12) .... *Lichenopora* spp.
– Colonies are irregular, circular discs with tubules in radiate rows in young, clustered in older colonies; brood chambers between radial rows; new colonies budded between rays or at edges in series of overlapping tubules (fig. 10) *Disporella* spp.
4. Colonies uniserial, adnate, branching, except multiserial around brood chamber; chamber with terminal opening (fig. 11) .... *Stomatopora granulata*
– Colonies mostly multiserial, irregular 5
5. Colonies coarse, irregular, whitish or lavender; tubules in bundles (fascicles); brood chamber lobed, opening through a tall, vertical, slitlike opening 6
– Colonies small, delicate, white, fan-shaped; tubules single in center, in bundles in outer areas (fig. 14) *Tubulipora pacifica*
6. Tubules in fascicles of 6–20 zooecia (fig. 15) *Tubulipora tuba*
– Tubules in fascicles of up to 6 zooecia .... *Tubulipora tuba* var. *fasciculifera*
7. Colonies partially erect, unjointed, rising from incrusting base, branching, sometimes forming stiff, whitish or brownish, fist-sized colonies inhabited by worms, crustaceans and by other more delicate bryozoans; zooecia multiserial, strong and large, in fascicles of 4–5 tubules on each side of midline; tubules usually fused, but sometimes raised, separate; brood chamber large, spread across branch below bifurcations; ooeciostome short, flared (figs. 13A, B) *Diaperoecia californica*
– Colonies erect, jointed, sometimes bushy 8
8. Long, jointed, filiform spines present; sterile internodes composed of 2 zooecia, fertile internodes 3–5; gonozooecium attached only at base, opening on dorsal (back) side (fig. 16) .... *Bicrisia edwardsiana*
– Filiform spines absent 9
9. Branching regular, in 2 alternating series; sterile internodes 1 to many zooecia; gonozooecium simple, adnate, with terminal opening 10
– Branching irregular, colony tangled, attached by radicles,

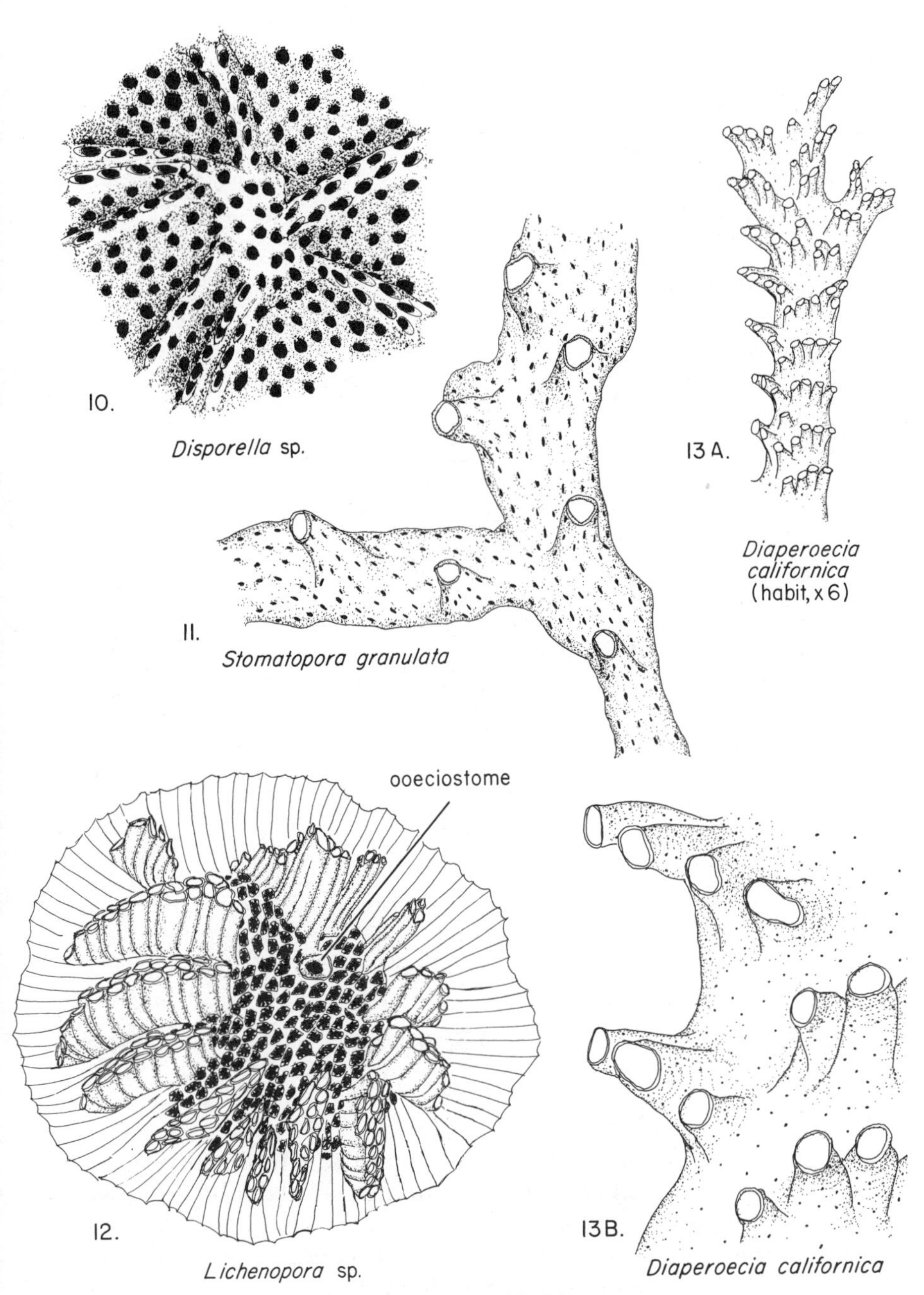

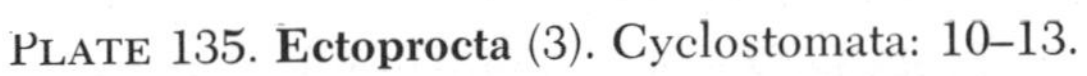

PLATE 135. **Ectoprocta** (3). Cyclostomata: 10–13.

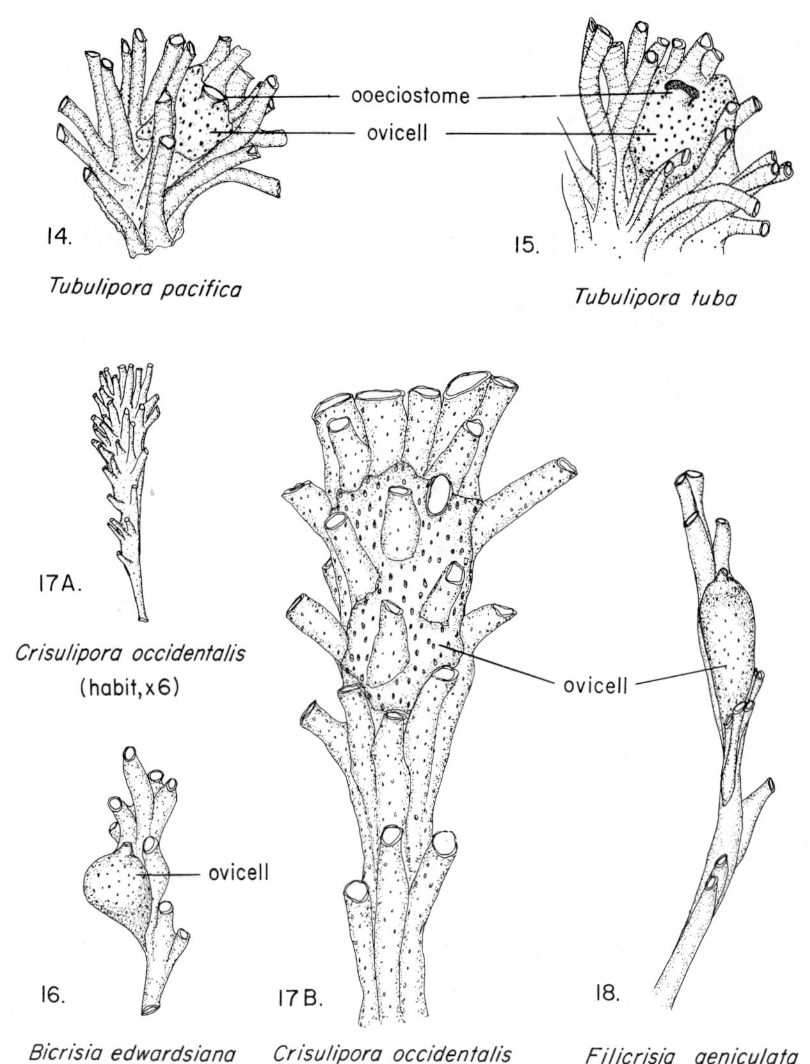

PLATE 136. **Ectoprocta** (4). Cyclostomata: 14–18.

sterile internodes 1–5 near base, up to 40 + terminally; gonozooecium spread between and around zooecial tubules, opening not terminal (figs. 17A, B) .....*Crisulipora occidentalis*

10. Internodes 3–5 straight, tubular zooecia with black joints; no radicles; gonozooecia simple, slender ....*Filicrisia* 11
– Internodes 5–30 curved, tubular zooecia; joints not black, radicles present; gonozooecia simple, inflated ....*Crisia* 12
11. Zooecia small; gonozooecium slender, opening on dorsal

(back) edge with tube bent forward (fig. 18) .... *Filicrisia geniculata*

– Zooecia slightly larger; gonozooecium a little inflated, opening on the ventral (front) edge with tube bent backward (fig. 19) .... *Filicrisia franciscana*

12. Internodes 3–5 zooecia near base, 7–12 terminal; zooecial tubules fused to tips, apertures turned forward bearing a point or spine distally; joints white to yellow; gonozooecium pear-shaped, partly adnate, opening circular, tube short, straight, directed upward, or with tip curved (fig. 20) .... *Crisia occidentalis*

– Internodes 12–30 or 40 + zooecia 13

13. Internodes 12–30+ zooecia; tubules fused almost to tips and turned forward, sometimes a distal point on tubules; joints yellow to brownish, gonozooecium large, a little flattened, fully adnate; opening elliptical, tube short, opening terminal or ventral (fig. 21) .... *Crisia serrulata*

– Internodes 12–20 to 40 +; coarser and larger than *C. serrulata*, zooecia without points or spines; joints dark brown; gonozooecium large, frontal inflated, truncate distally, opening round, tube directed forward (fig. 22) .... *Crisia maxima*

KEY C: ORDER CHEILOSTOMATA, SUBORDER ANASCA

1. Cryptocyst absent; no ovicells or avicularia; knobs on corner of each rectangular-shaped zooid, occasionally a tall, membranous tube (tower cell) on frontal membrane (fig. 25) .... *Membranipora membranacea*

– Cryptocyst present, calcified to varying degrees; or front is a pericyst (fused spines above frontal membrane) 2

2. Ovicells and avicularia absent, or a few vestigial avicularia on incrusting base; all frontals membranous; cryptocyst small 3

– Ovicells and/or avicularia present; frontals and cryptocysts various 10

3. Colony erect, forming bilaminar frills with layers back to back; vestigial avicularia rarely present on incrusting base (fig. 24) .... *Membranipora perfragilis*

– Colony incrusting 4

4. Zooecia with triangular, membranous, proximal corner spaces, or with separate or fused calcified tubercles at distal corners 5

– Zooecia with erect spines or smaller, chitinous spinules on frontal membrane 6

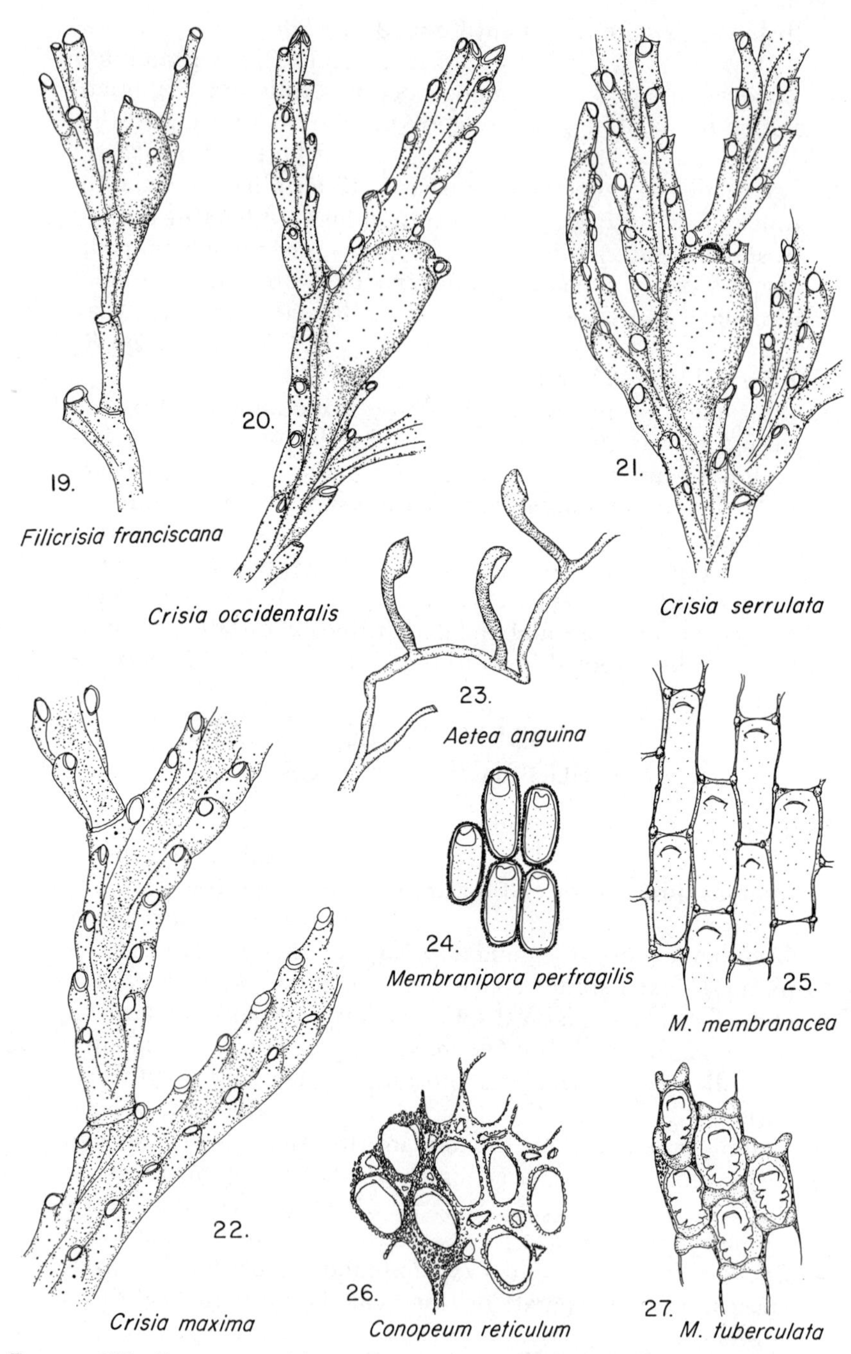

PLATE 137. **Ectoprocta** (5). Cyclostomata: 19–22; Cheilostomata, Anasca: 23–27.

5. Zooecia with small, triangular, membranous spaces at the proximal corners; frontal membranous area oval (triangular spaces sometimes form small knobs with age) (fig. 26) .... *Conopeum reticulum*
– Zooecia with calcified tubercles at distal corners 7
6. Zooecia rectangular; short, knoblike structures or erect, chitinous spinules present 8
– Zooecia oval, separated by grooves; 0–12 spines around margins; operculum often white, distinctive 9
7. Heavy, white tubercles at distal corners, sometimes merging along distal wall; cryptocyst developed into a proximal shelf with horizontal spinules, sometimes forked, directed medially; frontal membranes light tan (fig. 27) .... *Membranipora tuberculata*
– Tubercles at distal corners never merging along distal wall; cryptocyst oval, rim crenulated, without proximal shelf or spinules; zooecia separated by a dark line; frontal membrane dark yellow to black; operculum with dark border (fig. 28) .... *Membranipora fusca*
8. Cryptocyst a marginal shelf (fig. 2), with small horizontal spinules directed medially; distal corners of zooecium bear short, knoblike, calcified, hollow spines, or spines may be extended to pointed, chitinous tip; erect, chitinous spinules sometimes on frontal membrane (fig. 29) .. *Membranipora serrilamella*
– Cryptocyst narrow; small horizontal spinules; chitinous spinules on frontal membrane, larger erect spinules along lateral margins and at proximal corners (fig. 30) .... *Membranipora villosa*
9. Zooecia oval; up to 12 spines along edges of frontal membrane (fig. 31) .... *Electra crustulenta*
– Zooecia oval; single blunt spine at center of proximal margin; proximal portion of frontal membrane usually calcified (fig. 32) .... *Electra crustulenta arctica*
10. Zooecia with frontal mostly uncalcified (large opesium); cryptocyst small 11
– Zooecia with cryptocyst or with a pericyst 18
11. No ovicells; avicularia on slender stalks at margins, resembling spines; aperture bordered by heavy spines; smaller spines arch over opesium (fig. 33) *Cauloramphus spiniferum*
– Ovicells present, avicularia thicker, do not resemble spines 12
12. Spines absent; interzooecial avicularia present 13
– Spines present; avicularia not interzooecial 14
13. Ovicells small, caplike; interzooecial avicularia short, triangular, with one side curved (fig. 34) ...... *Hincksina velata*

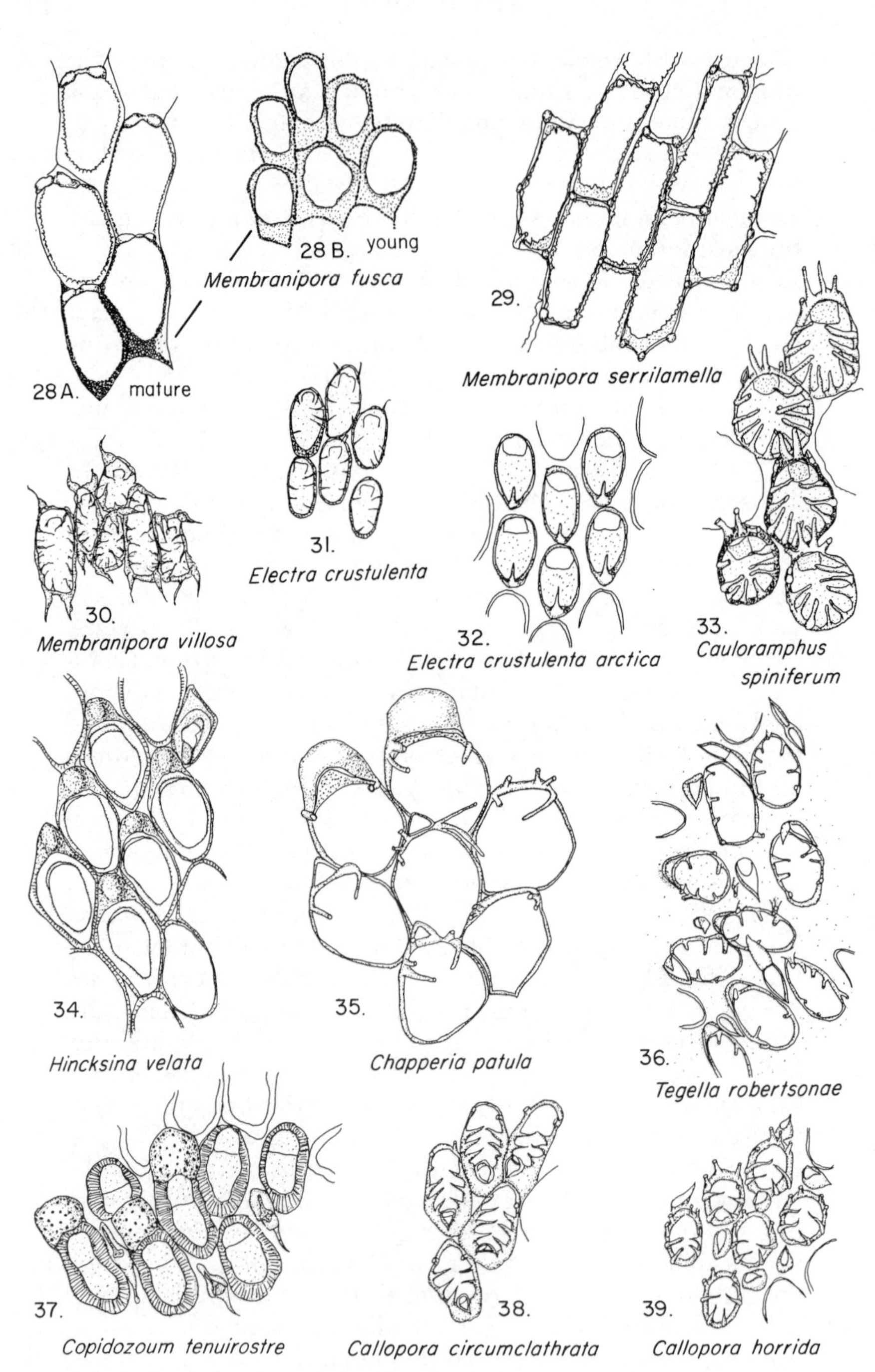

PLATE 138. **Ectoprocta** (6). Cheilostomata, Anasca: 28–39.

– Ovicell prominent, with tiny pores and/or knobs, granulated; interzooecial avicularia long, acute, slender, (fig. 37) .... *Copidozoum tenuirostre*

14. Colonies incrusting 15
– Colonies frondose or erect 24
15. Zooecia rounded, reddish; operculum fills much of frontal; distal margins raised; ovicell raised, hoodlike; 4–6 tall, heavy spines around aperture distally, none proximally; avicularia small, located on the median distal margin on zooecia lacking ovicells (fig. 35) .... *Chapperia patula*
– Zooecia oval, 1–6 spines around opesium and aperture; ovicells small, associated with large proximal avicularium of next zooecium 16
16. Large avicularium proximally on zooecium; close to ovicell of next proximal zooecium but not on top of it 17
– Ovicell small and mostly covered by large avicularium of next distal zooecium; transverse rib on ovicell near aperture; a pair of erect, hollow spines on either side of aperture, 1–3 smaller spines curving over opesia (fig. 36) .... *Tegella robertsonae*
17. Zooecia separated, calcified areas with pores between; 1–2 pairs of distal spines, 3–5 pairs arched over opesium; proximal avicularium usually triangular, sometimes rounded, not covering imperforate ovicell (resembles *Tegella*) (fig. 38) .... *Callopora circumclathrata*
– Zooecia contiguous, not separated by pores; 2–3 pairs of stout spines beside aperture, 2–3 pairs of fine, lateral spines arched over opesium (fig. 39) .... *Callopora horrida*
18. Calcified cryptocyst extensive below frontal membrane; interzooecial avicularia present 19
– Frontal composed of spines fused above frontal membrane (pericyst); avicularia present or absent 21
19. Zooecia resemble jug with two handles; circular openings (opesiules) on either side of perforated cryptocyst; interzooecial avicularia long with rounded tip; internal spicules seen under high magnification when zooecia are crushed; ovicells very large with a longitudinal keel; colonies incrusting or with incrusting base and erect branches with dark joints (fig. 40) .... *Thalamoporella californica*
– No opesiules, ovicells hard to see, imbedded, opening by small pore distal to apertures; colonies erect, cylindrical with black joints 20
20. Interzooecial avicularia larger than zooecia and with semicircular, brown mandibles (fig. 41) ...... *Cellaria mandibulata*
– Interzooecial avicularia smaller than zooecia, mandibles col-

orless (fig. 42) .... *Cellaria diffusa*

21. Frontal composed of 3–7 pairs of fused spines; tear-shaped spaces separating spines; ovicell small, shallow; no avicularia (fig. 43) .... *Figularia hilli*

– Fused frontal spines separated by rows of pores 22

22. No ovicells or spines around aperture; large, blunt, interzooecial avicularia (may be missing); pores large and conspicuous (fig. 44) .... *Lyrula hippocrepis*

– Ovicells present 23

23. Ovicells with a longitudinal keel; no avicularia or spines (fig. 45) .... *Reginella nitida*

– Ovicell with radiate grooves or a prominent, calcareous umbo; oral spines present, interzooecial avicularia slim, acute (fig. 46) .... *Colletosia radiata*

24. Colonies frondose, recumbent, and loosely attached by radicles; no avicularia .... *Dendrobeania* 25

– Colonies erect; avicularia present 26

25. Fronds varying in width; zooecia loosely connected to each other; 4–7 spines with distal spines stronger; spines may be absent where zooecia are contiguous (fig. 47) .... *Dendrobeania laxa*

– Fronds wide, lichenlike; zooecia closely connected; 2–3 spines distally (may be absent) (fig. 48) .... *Dendrobeania lichenoides*

26. Zooecia with scutum (modified spine arched over opesium) (figs. 49–52) 27

– Zooecia without scutum 31

27. Vibracula (whiplike structures, fig. 49) present on dorsal side of branches; bushy colonies .... *Scrupocellaria* 28

– No dorsal vibracula; branch internodes usually composed of 3 zooecia .... *Tricellaria* 30

28. Single dorsal vibraculum; vibracular chamber short, thin; ovicells without pores; scutum small 29

– Vibraculum large, chamber 2/3 length of zooecium, ovicell perforate; scutum large and oval (fig. 49) .... *Scrupocellaria diegensis*

29. Zooecia small; scutum small, narrow or paddle-shaped; frontal avicularia small; small, lateral avicularia or replaced by giant, triangular, hooked avicularium; vibraculum small, with transverse groove, sometimes absent .... *Scrupocellaria californica*

– Zooecia long, slender; scutum small and variable, usually 3-pronged; frontal avicularia small; lateral avicularia triangular or replaced by giant avicularium with long, rounded or serrate

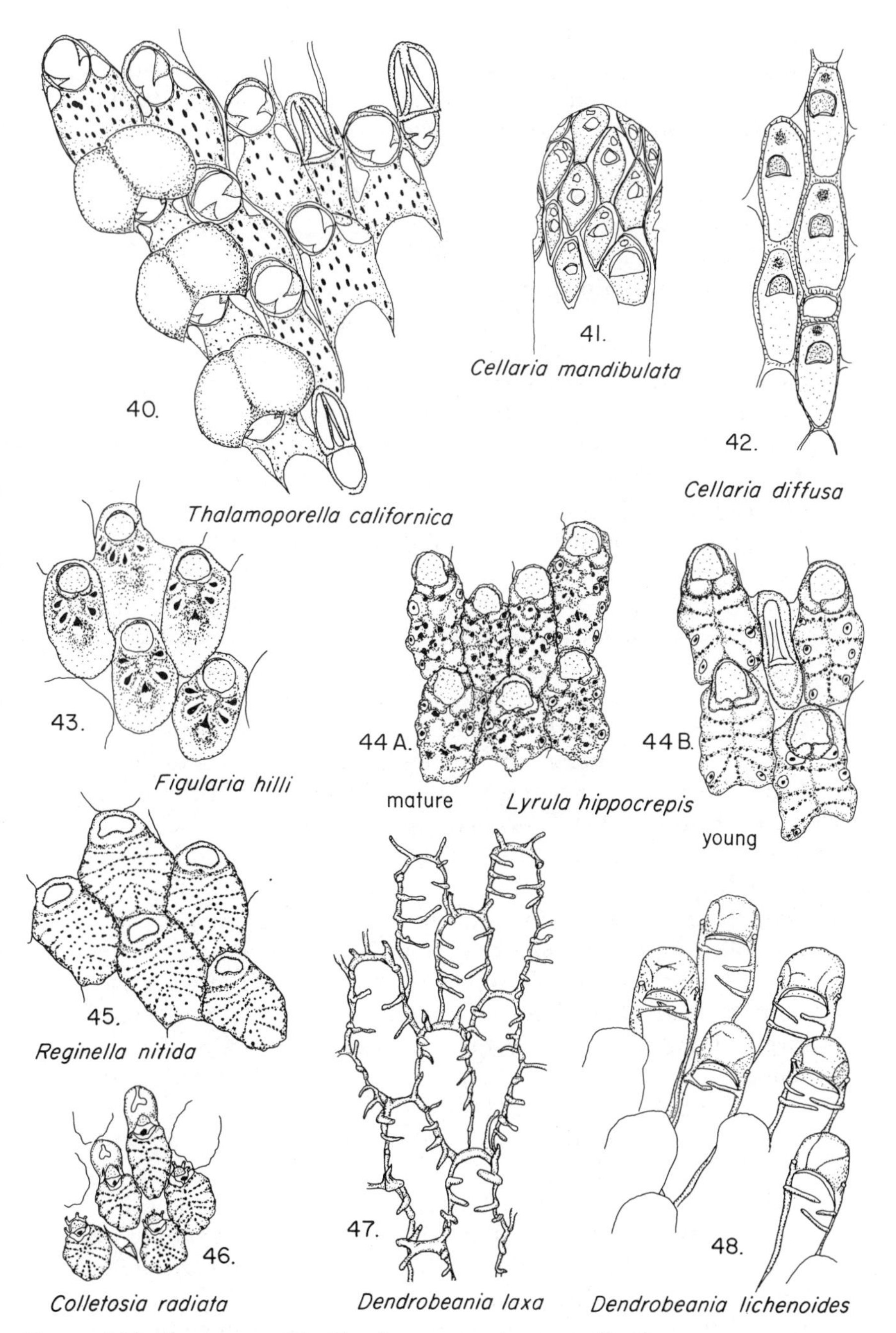

PLATE 139. **Ectoprocta** (7). Cheilostomata, Anasca: 40–48.

beak; vibraculum chamber triangular (fig. 50) .... *Scrupocellaria varians*

30. Scutum flattened, narrow, consisting of 1–3 points, attached below middle of opesium; ovicell with small pores (fig. 51) ....*Tricellaria occidentalis*

– Scutum narrow or with rounded tip, attached above middle of opesium; ovicell without pores (fig. 52) *Tricellaria ternata*

31. Colony with some zooecia modified as segmented stalk; ovicell a shallow hood attached at an angle on inner distal corner of zooecia; opesium reduced; zooecial proximal wall placed diagonally, not forked .... *Caulibugula* 32

– Colony without segmented stalked base; ovicells varying from a hood to completely enclosed; opesium large; zooecial proximal wall forked .... *Bugula* 33

32. Aperture tilted upward; 4–5 long spines at outer distal corner of zooecium (fig. 53) ....*Caulibugula ciliata*

– Aperture in same plane as frontal; 2–3 spines at outer corner ....*Caulibugula occidentalis*

33. Colony reddish purple, bushy; ovicells large and set at an angle on the distal wall; no avicularia or spines (fig. 54) ... .*Bugula neritina*

– Colony not reddish purple; avicularia present 34

34. Ovicells absent; small knob in normal ovicell position in the middle of the distal end; avicularia large, raised "bird's heads" attached to lateral wall proximal to middle; spines short, sturdy, at distal corners; zooecia may be multiserial (fig. 56) .... *Bugula pugeti*

– Ovicells present 35

35. Ovicell reduced, shallow, not covering entire larva 36

– Ovicell completely enclosed, globular, calcified; small avicularia attached at middle of inside lateral wall; large, raised, "bird's head" avicularia on outside edge; 3 spines; colony large, whitish to yellow brown, characteristically in spiral whorls (fig. 55) ....*Bugula californica*

36. Ovicell a calcified ledge with membranous sides, on a pedicel; avicularia small, near middle of outer lateral wall; colonies large, flaccid (fig. 57) ....*Bugula mollis*

– Ovicell a shallow hood, attached to distal end of zooecium 37

37. Avicularium on pedicel attached near base of zooecium; ovicell a very shallow hood (fig. 59) ....*Bugula pacifica*

– Avicularium very long, slim, hooked, stalk attached near middle or distal end of outer lateral wall; small spines on outer distal corner; ovicell like an overturned shallow bowl (fig. 58) ....*Bugula longirostrata*

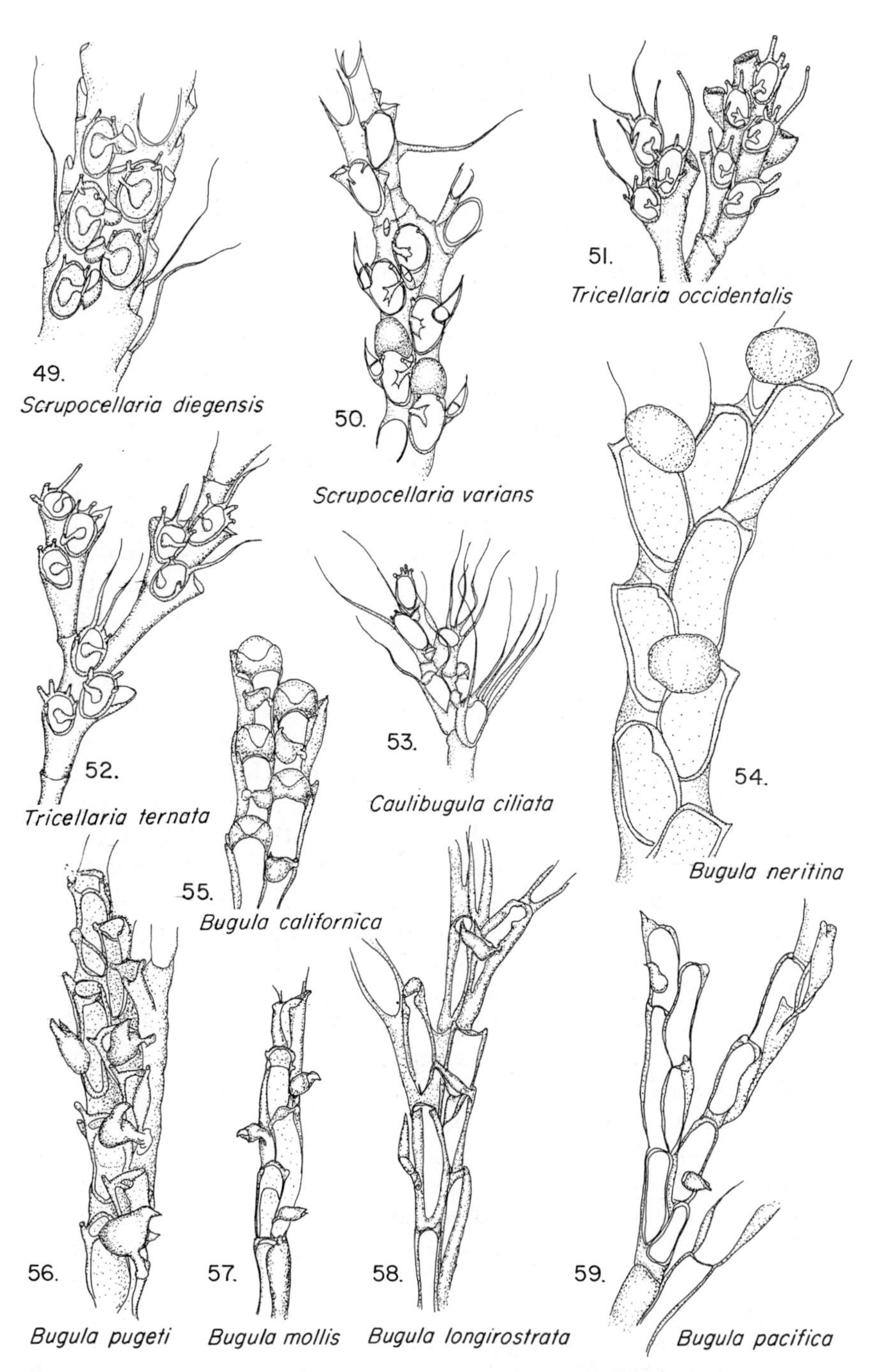

PLATE 140. **Ectoprocta** (8). Cheilostomata, Ascophora: 49–59.

KEY D: ORDER CHEILOSTOMATA, SUBORDER ASCOPHORA

1. Ascopore visible, opening separate from and proximal to the aperture (figs. 60–63) 2
– Ascopore inside aperture rim, not visible on frontal surface (figs. 64–83) 6
2. With avicularia; no frontal pores between ascopore and aperture 3
– Without avicularia; frontal pores between ascopore and aperture 5
3. Ascopore large and covered with sieve plate (figs. 60A, B); aperture with 5–7 spines; avicularia generally paired, occasionally single and located on either side of ascopore .... *Microporella cribrosa*
– Ascopore not covered by sieve plate 4
4. Avicularia paired, large, slim, acute, beside ascopore; aperture large; 5–7 heavy dark spines; frontal pores large; umbo often present proximal to ascopore; ovicells small, perforate, sometimes ribbed and with an umbo (fig. 61) .... *Microporella californica*
– Avicularium single, long, triangular, occasionally paired, placed a little proximal to ascopore, crescentic; sometimes an umbo proximally; aperture semicircular with 5–7 spines; frontal with tiny pores; ovicells globose and smooth or umbonate on top and ribbed around base (fig. 62) *Microporella ciliata*
5. Tremopores (tiny pores scattered over entire frontal) between ascopore and semicircular aperture (fig. 63); sometimes 4–5 small spines distal to aperture; ovicell large, perforate, with larger pores (areolae) around base .... *Fenestrulina malusii*
– As above, but conspicuous umbo proximal to ascopore and zooecia larger .... *Fenestrulina malusii* var. *umbonata*
6. Calcified frontal with pores uniformly distributed (figs. 64, 65) 7
– Calcified frontal smooth or granular, imperforate, at least in central portion (a pleurocyst) 13
7. No ovicells or oral spines; frontal pores and apertures large; rarely, a small, median, suboral avicularium on an umbo (fig. 64) .... *Cryptosula pallasiana*
– Ovicells present 8
8. Zooecial aperture rounded with a rolled or tubular collar .. 9
– Zooecial aperture with a curved or V-shaped proximal sinus, margins not raised 10
9. Zooecia very large, yellowish; aperture with raised, rolled

marginal collar; no avicularia; ovicell large with curved band across center (fig. 65) .... *Coleopora gigantea*

– Zooecia raised, vase-shaped; collar around aperture tall, tubular, with a pair of tiny, lateral, oral avicularia on the collar rim; ovicell hemispherical with central, finely perforate area; frontals perforate; resembles *Costazia,* which has an imperforate frontal (figs. 66, 67) .... *Lagenipora* spp.

10. Aperture with a narrow, slitlike, U-shaped proximal sinus; operculum with a straight proximal border except for tongue to fit into sinus; ovicell closed by operculum; no spines or avicularia; frontals perforate (fig. 68) ..... *Arthropoma cecili*

– Aperture with a wide, U- or V-shaped sinus 11

11. Aperture with a wide, U-shaped sinus 12

– Aperture with a deep, V-shaped sinus; avicularia beside sinus on raised mounds, mostly directed distally; frontals perforate, ovicells imperforate, with ridges; covered distally by secondary calcification; not closed by operculum; colonies incrusting in single layer; resembles *Stephanosella biaperta,* which has imperforate frontals (fig. 69) ..... *Schizoporella cornuta*

12. Aperture with a wide, proximal sinus and with an arched, calcified, distal rim; ovicell closed by operculum; no avicularia (fig. 70) .... *Hippodiplosia insculpta*

– Aperture with a wide, U-shaped sinus; paired or single avicularia usually placed slightly proximal to sinus, directed distolaterally; sometimes an umbo proximal to sinus; zooecia rectangular; frontal perforate; ovicells raised with small pores and ridges, not closed by operculum; colonies golden yellow to orange, multilaminar, sometimes raised in tubular branches (fig. 71) .... *Schizoporella unicornis*

13. No avicularia or spines; without true areolar (marginal frontal) pores 14

– With avicularia and/or with spines; with true areolar pores (fig. 76) penetrating frontal around margins near lateral walls 15

14. Frontal thin, with transverse lines, shiny, imperforate; aperture with proximal sinus; zooecia oval to subrectangular; ovicells large, with or without pores (fig. 73) .. *Hippothoa* spp.

– Frontals thick, rose color, imperforate, sometimes with an umbo; aperture shaped like derby hat, rounded distally with wider proximal border (fig. 72); operculum brown, margin reinforced by dark rim; ovicell small, with membranous area on top .... *Eurystomella bilabiata*

15. Zooecia contiguous, but forming large, erect, fenestrate (meshlike) colonies with branches; color orange to orange pink (figs. 74; 84, p. 608) .... *Phidolopora pacifica*

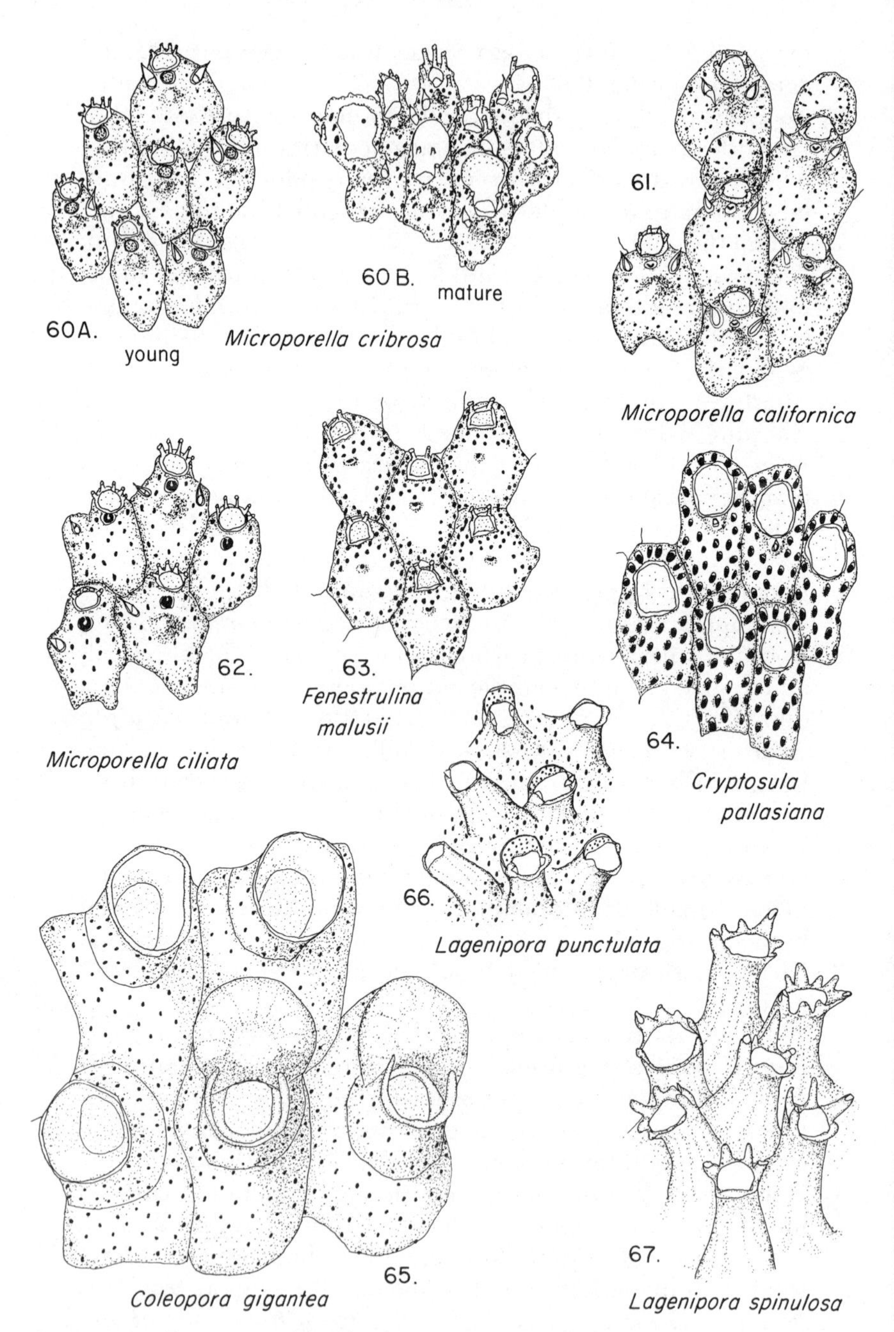

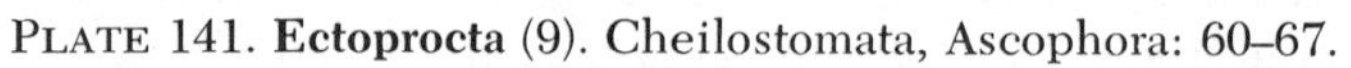

PLATE 141. **Ectoprocta** (9). Cheilostomata, Ascophora: 60–67.

– Zooecia contiguous, erect or recumbent, but not forming openings or fenestrae 16

16. No avicularia; aperture raised, tubular, ringed with 8–10 spines; a broad, median, proximal tooth (lyrula) inside aperture (fig. 75); ovicell imperforate; tiny tubular pores extend medially from lateral areolar pores ..... *Mucronella major*

– Avicularia present, variously located 17

17. Ovicells with pores; median tooth (lyrula) present in proximal part of aperture (fig. 77) 18

– Ovicells mostly without pores; with or without median tooth 20

18. Avicularium median and suboral, contiguous with median aperture sinus, mandible semicircular; no other frontal avicularia; ovicells raised, large, perforate; aperture collared, with large median tooth flanked by smaller hinge teeth; 2–4 distal spines on young; zooecia small, areolar pores large (fig. 76) .... *Smittoidea prolifica*

– Avicularia not median and suboral 19

19. Frontal with hills (colli), raised knobs, or tubercles; aperture round with truncate median tooth and hinge teeth proximally; collar low, with 2 distal spines in young; shapes of avicularia include: (1) small, raised triangles originating proximolaterally to aperture, directed distally, (2) small ovals scattered on frontal, (3) long spatulates on frontal (rare); ovicells with 6–8 variable pores (fig. 77) .... *Parasmittina collifera*

– Frontal granular, without knobs; aperture round with median tooth and hinge teeth proximally; collar high, hiding teeth, notched proximally, low distally with 2–4 spines in young; a very large, triangular avicularium originating on proximal frontal, directed distally, with raised pointed tip; often other various triangular and oval avicularia on frontal, on collar, and on margins; ovicell raised, with large pores (fig. 78) .... *Parasmittina trispinosa*

20. Ovicell imperforate, a shallow hood without frontal filled in; aperture with median tooth, collar, and raised sides; sometimes 2–3 distal spines; a tall umbo carrying a vertical avicularium just proximal to aperture, forming a notch with collar; giant interzooecial avicularium with spade-shaped mandible; colony in form of bristly, grayish brown nodules (fig. 79) .... *Celleporaria brunnea*

– Ovicell complete and mostly imperforate; aperture without median tooth 21

21. Ovicell imperforate, globular, smooth; aperture semicircular with a high collar; no median tooth or hinge teeth; 2–4 distal spines on young; avicularium median, oval, tiny, on proximal

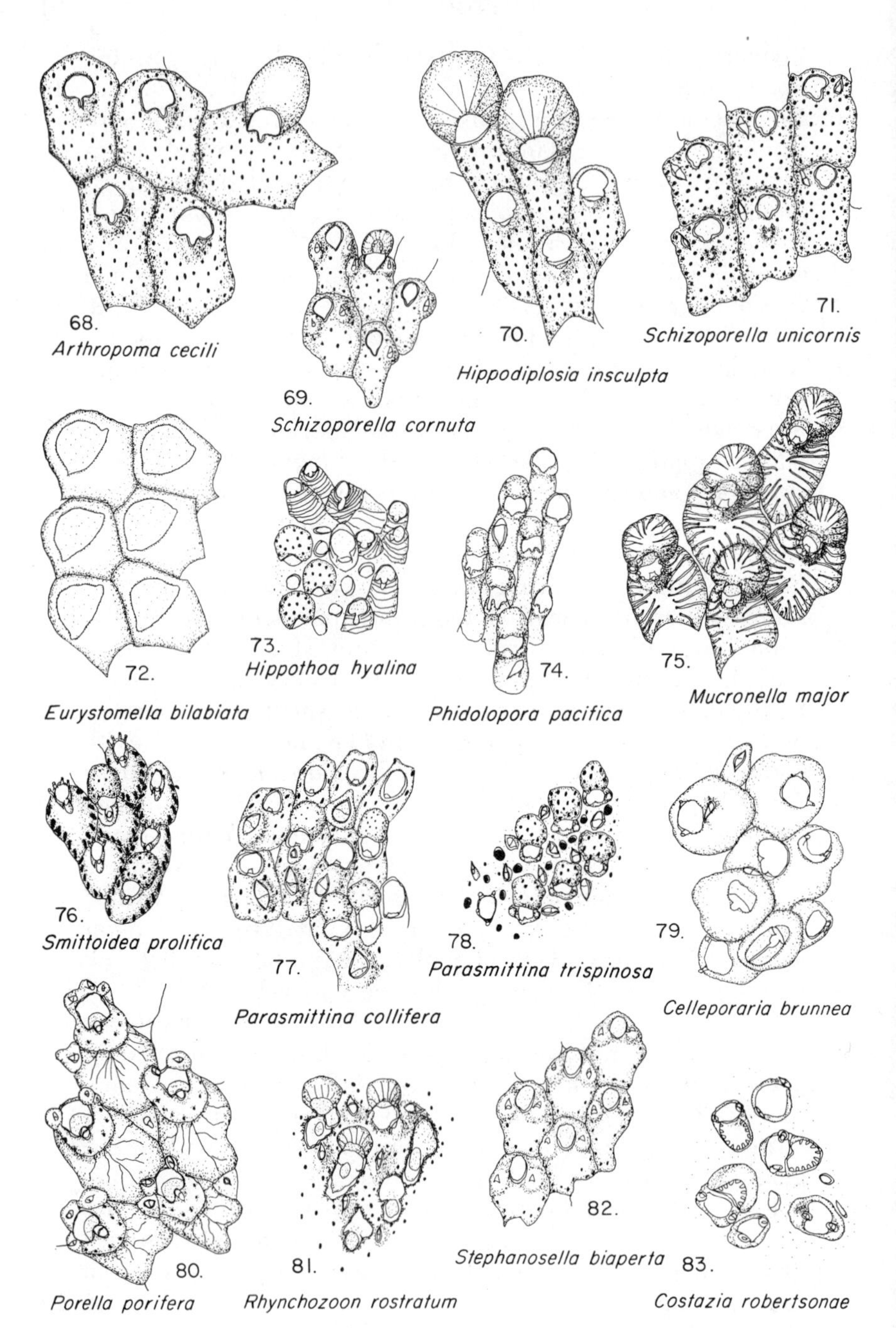

PLATE 142. **Ectoprocta** (10). Cheilostomata, Ascophora: 68–83.

lip, with raised chamber around proximal collar; zooecia small; areolae small, few; usually with other small, oval, marginal avicularia (fig. 80) ....*Porella porifera*

– Ovicells mostly imperforate but with frontal area of ovicell striated 22

22. Avicularia paired, flanking aperture 23

– Avicularia unpaired, with large bulbous base just to one side proximal to the aperture, with tip of base and mandible hooked; aperture in young rounded with sinuate proximal border; collar becomes raised in points, forming a median proximal sinus (spiramen) between merged points; ovicells with thick distal hood, front part striated, very white, descending deep into collar; pointed avicularia scattered on frontal; colony incrusting, heavily calcified, spiny; zooecia irregular, submerged (fig. 81) ......*Rhynchozoon rostratum*

23. Paired oval avicularia on mounds flanking aperture but separated from low aperture collar; a large, triangular avicularium on the lateral frontal; aperture with a rounded, proximal sinus; ovicell rounded; frontal with radiate grooves ending in distal pits between ribs; zooecial frontal imperforate, with peripheral areolar pores (fig. 82) ....*Stephanosella biaperta*

– Paired oval avicularia mounted on raised apertural collar, a similar, oval, median, distal avicularium also present; aperture with a V-shaped proximal sinus; ovicell rounded; frontal with radiating triangular pores or grooves; sometimes an oval interzooecial avicularium raised on a pivot; incrusting with erect branches or nodules; similar to *Schizoporella cornuta* and *Stephanosella biaperta* (with imperforate frontals) and *Lagenipora* spp. (with perforate frontals) (fig. 83) .... *Costazia robertsonae*

LIST OF ECTOPROCTA

Order **Ctenostomata**

Alcyonidium parasiticum (Fleming, 1828).

Alcyonidium polyoum (Hassall, 1841). On old, heavily incrusted kelp crabs, on stones or algae.

Bowerbankia gracilis O'Donoghue, 1926. Extremely abundant as a brown furry coating on rocks and pilings.

Farrella elongata (von Beneden, 1845).

Flustrellidra corniculata (Smith, 1871) (=*Flustrella corniculata*). On algal stipes, especially at bases of *Laminaria sinclairii*, in large tan colonies.

Triticella elongata (Osburn, 1912). On legs, gills, and bodies of various crabs; because of transparency and insensitivity to handling, this is excellent class material.

Victorella pavida Kent, 1870. In brackish water of bays and estuaries; introduced.

Order **Cyclostomata**

Bicrisia edwardsiana (d'Orbigny, 1839).

Crisia maxima Robertson, 1910. The Crisias occur in the low intertidal, among sponges and hydroids.

Crisia occidentalis Trask, 1857.

Crisia serrulata (Gabb and Horn, 1862) (=*C. serrata*, preoccupied).

Crisulipora occidentalis Robertson, 1910.

Diaperoecia californica (d'Orbigny, 1852) (=*Idmonea californica*).

Disporella spp.

Filicrisia franciscana (Robertson, 1910). Sparse, delicate growths on mid- and low-intertidal rocks.

Filicrisia geniculata (Milne-Edwards, 1838). The Filicrisias are delicate, whitish, and appear black-speckled; low intertidal; under rocks.

Lichenopora spp.

Stomatopora granulata (Milne-Edwards, 1836).

Tubulipora pacifica Robertson, 1910. Often on algae.

Tubulipora tuba (Gabb and Horn, 1862). Generally on shells in low intertidal.

Tubulipora tuba var. *fasciculifera* (Hincks, 1884).

Order **Cheilostomata**

Suborder **Anasca**

* *Aetea anguina* (Linnaeus, 1758). This tiny, creeping species (fig. 23) is practically cosmopolitan; at least one other species may be present in our area (Osburn, 1950).

* Not in key.

Bugula californica Robertson, 1905. Distinctive spiral form; low intertidal; excellent for demonstration of avicularia.

Bugula mollis Harmer, 1926. Probably introduced.

Bugula longirostrata Robertson, 1905.

Bugula neritina Linnaeus, 1758. Characteristic of harbors and pilings.

Bugula pacifica Robertson, 1905.

Bugula pugeti Robertson, 1905.

Callopora circumclathrata (Hincks, 1881).

Callopora horrida (Hincks, 1880). Incrusting shells, sponges, and algae.

Caulibugula ciliata (Robertson, 1905). Tiny, delicate, elm-tree like form; look for it in pools in lower midtidal, often on red sponges.

Caulibugula occidentalis (Robertson, 1905).

Cauloramphus spiniferum (Johnston, 1832). On algae.

Cellaria diffusa Robertson, 1905.

Cellaria mandibulata Hincks, 1882.

Chapperia patula (Hincks, 1881).

Colletosia radiata (Moll, 1803).

* *Conopeum commensale* Kirkpatrick and Metzelaar, 1922. In bays.

Conopeum reticulum (Linnaeus, 1767). A common bay (brackish water) form; frequently associated with *Victorella pavida* and the entoproct *Barentsia;* incrusting on rocks and shells.

Copidozoum tenuirostre (Hincks, 1880). On rocks.

Dendrobeania laxa (Robertson, 1905).

Dendrobeania lichenoides (Robertson, 1900). With the green colonial ascidian *Perophora,* common in low, shaded intertidal.

Electra crustulenta (Pallas, 1766). On rocks and shells.

Electra crustulenta arctica Borg, 1931. On rocks, shells, and occasionally on algae.

Figularia hilli Osburn, 1950. Incrusting rocks and shells.

Hincksina velata (Hincks, 1881). On shells.

Lyrula hippocrepis (Hincks, 1882). Incrusting rocks and shells.

Membranipora fusca Osburn, 1950. Incrusting rocks and shells.

Membranipora membranacea (Linnaeus, 1767). Commonly incrusting the floating kelp *Macrocystis*.

Membranipora perfragilis (MacGillivray, 1881). Usually in harbors.

Membranipora serrilamella Osburn, 1950. On *Macrocystis* in late summer.

Membranipora tuberculata (Bosc, 1802). Widely distributed on California coast; large, coalescing colonies common on algae.

Membranipora villosa Hincks, 1880. On *Macrocystis* in spring and early summer.

Reginella nitida Osburn, 1950. Incrusting rocks and shells.

Scrupocellaria californica Trask, 1857. Scrupocellarias are brownish, greenish, or white bushy growths, very common under rocks and floats.

Scrupocellaria diegensis Robertson, 1905. See Osburn, 1950, pp. 131–132.

Scrupocellaria varians Hincks, 1882.

Tegella robertsonae O'Donoghue, 1926. Excellent for large avicularia; incrusting on shells, sponges, and algae.

Thalamoporella californica (Levinsen, 1909).

Tricellaria occidentalis (Trask, 1857).

Tricellaria ternata (Solander, 1786).

Suborder **Ascophora**

Arthropoma cecili (Audouin, 1826).

Celleporaria brunnea (Hincks, 1884) (=*Holoporella brunnea* in Osburn, 1952). Thick, brown or gray brown, multilaminar colonies.

Coleopora gigantea (Canu and Bassler, 1923).

Costazia robertsonae Canu and Bassler, 1923.

Cryptosula pallasiana (Moll, 1803). Common, incrusting any available surface.

Eurystomella bilabiata (Hincks, 1884). Common; the rose color is distinctive.

Fenestrulina malusii (Audouin, 1826). Widespread in shallow water.

Fenestrulina malusii var. *umbonata* O'Donoghue, 1926.

Hippodiplosia insculpta (Hincks, 1882). Common on stipes of red algae and under rocks.

Hippothoa spp. Colonies disc-shaped, commonly forming delicate, silvery patches on red-algal fronds; see Pinter, 1973.

Lagenipora spp. Several species; see Osburn, 1952, for key.

Microporella californica (Busk, 1856).

Microporella ciliata (Pallas, 1766).

Microporella cribrosa (Osburn, 1952).

Mucronella major (Hincks, 1884).

Parasmittina collifera (Robertson, 1908).

Parasmittina trispinosa (Johnston, 1838).

Phidolopora pacifica (Robertson, 1908). Subtidal or in deep, sheltered pools; the lacy form and orange pink color are distinctive.

Porella porifera (Hincks, 1884). Incrusting.

Rhynchozoon rostratum (Busk, 1856).

Schizoporella cornuta (Gabb and Horn, 1862).

Schizoporella unicornis (Johnston, 1847). In bays and harbors; see Powell, 1970, J. Fish. Res. Bd. Canada 27: 1847–1853.

Smittoidea prolifica Osburn, 1952. Incrusting.

Stephanosella biaperta (Michelin, 1845). Incrusting.

REFERENCES ON ECTOPROCTA

(For references before 1950, see Osburn's monographs.)

Banta, W. 1966. Systematic list of Bryozoa identified from Farnsworth Bank, Santa Catalina Island, in samples of Velero IV, Stations numbered 3594, 3595 and 10334. Allan Hancock Pac. Exped. 19: 419–422.

Bassler, R. S. 1953. Bryozoa. Part G. *In* R. C. Moore, ed., *Treatise on Invertebrate Paleontology,* Geol. Soc. Amer. and Univ. Kansas Press.

Humason, G. 1972. *Animal Tissue Techniques.* 3rd ed., xiv, 641 pp. San Francisco, Freeman.

Hyman, L. H. 1959. *The Invertebrates: Smaller Coelomate Groups.* Vol. 5. Phylum Ectoprocta, pp. 275–515. McGraw-Hill.

Nielsen, C. 1971. Entoproct life-cycles and the entoproct/ectproct relationship. Ophelia 9: 209–341.

Osburn, R. C. 1950. Bryozoa of the Pacific Coast of America. Part 1, Cheilostomata—Anasca. Allan Hancock Pac. Exped. 14 (1): 1–269.

Osburn, R. C. 1952. Bryozoa of the Pacific Coast of America. Part 2, Cheilostomata—Ascophora. Allan Hancock Pac. Exped. 14 (2): 271–611.

Osburn, R. C. 1953. Bryozoa of the Pacific Coast of America. Part 3, Cyclostomata, Ctenostomata, Entoprocta and Addenda. Allan Hancock Pac. Exped. 14 (3): 613–841.

Osburn, R. C. and J. D. Soule 1953. Suborder Ctenostomata. *In* Bryozoa of the Pacific Coast of America. Part 3, Cyclostomata, Ctenostomata, Entoprocta and Addenda. Allan Hancock Pac. Exped. 14 (3): 726–758.

Pinter, P. A. 1969. Bryozoan-algal associations in southern California waters. Bull. So. Calif. Acad. Sci. 68: 199–218.

Pinter, P. A. 1973. The *Hippothoa hyalina* (L.) complex with a new species from the Pacific coast of California. In *Living and Fossil Bryozoa*, G. P. Larwood, ed., pp. 437–446. Academic Press.

Ross, J. P. R. 1970. Keys to the Recent Cyclostomata Ectoprocta of marine waters of northwest Washington State. Northwest Sci. 44: 154–169.

Ryland, J. S. 1970. *Bryozoans*. London: Hutchinson Univ. Library, 175 pp.

Soule, D. F. and J. D. Soule 1964. Clarification of the family Thalamoporellidae (Ectoprocta). Bull. So. Calif. Acad. Sci. 63: 193–200.

Soule, D. F. and J. D. Soule 1964. The Ectoprocta (Bryozoa) of Scammon's Lagoon, Baja California. Amer. Mus. Nat. Hist. Nov. 2199: 1–56.

Soule, D. F. and J. D. Soule 1973. Morphology and speciation of Hawaiian and eastern Pacific Smittinidae (Bryozoa, Ectoprocta). Amer. Mus. Bull. 152: 365–440.

Soule, J. D. 1959. Results of the Puritan-American Museum of Natural History Expedition to western Mexico. 6. Anascan Cheilostomata (Bryozoa) of the Gulf of California. Amer. Mus. Nat. Hist. Nov. 1969: 1–54.

Soule, J. D. 1961. Results . . . 13. Ascophoran Cheilostomata . . . *Ibid.* 2053: 1–66.

Soule, J. D. 1963. Results . . . 16. Cyclostomata (Ectoprocta) and Entoprocta . . . *Ibid.* 2144: 1–34. (The above group of papers correct or update certain information in Osburn's volumes.)

Soule, J. D. and D. F. Soule 1969. Systematics and biogeography of burrowing bryozoans. Amer. Zool. 9: 791–802.

Woollacott, R. M. and W. J. North 1971. Bryozoans of California and northern Mexico kelp beds. *In* The biology of giant kelp beds (*Macrocystis*) in California, pp. 455–479. Nova Hedwigia 32: 1–600.

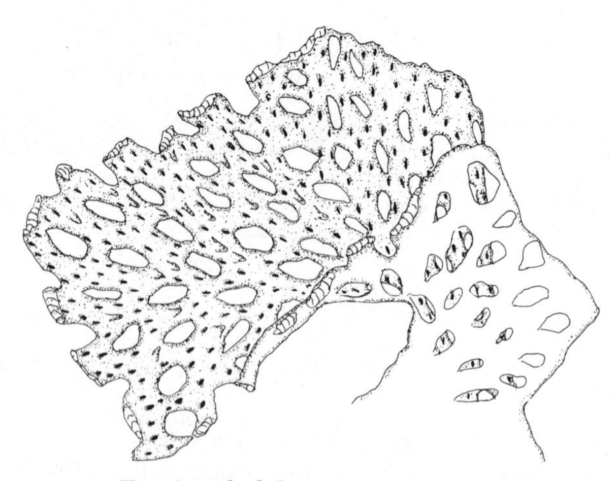

Fig. 84. *Phidolopora pacifica:* (x 6)

LESSER LOPHOPHORATES AND ENTOPROCTA

The lophophorate phyla—**Brachiopoda, Ectoprocta,** and **Phoronida**—are characterized by a circle, crescent, or double spiral of straight, ciliated tentacles containing extensions of the coelom. The Ectoprocta have been treated in the previous chapter; the remaining two groups are covered below with the **Entoprocta,** a small pseudocoelomate phylum formerly lumped with the ectoprocts, but actually not closely related in the opinion of most zoologists (but see Nielsen, 1971).

PHYLUM ENTOPROCTA

Richard N. Mariscal
Florida State University, Tallahassee

(PLATE 143)

Members of the phylum **Entoprocta (Kamptozoa, Calyssozoa)** may be either solitary or colonial and have a cuplike head or **calyx,** bearing a circlet of tentacles, perched at the end of a stalk of variable length. In the past, entoprocts have generally been united with ectoprocts in the phylum Bryozoa (Polyzoa). However, Entoprocta differ fundamentally from Ectoprocta in lacking a true coelom, in possessing protonephridia with flame cells, and especially in having the anus within the circlet of tentacles which also surrounds the mouth, hence "entoproct." In contrast, ectoprocts are coelomate, lack excretory organs, and have the anus outside the circlet of tentacles ("ecto-proct").

Entoprocts, because of their small size and relative obscurity, are frequently mistaken for certain of the stalked protozoans, hydroids, or ectoprocts. Colonial entoprocts can generally be recognized by the characteristic jerky bending movements of the stalk and calyx, which has led to their common name of "nodding heads." Stimulation of the tentacles of an entoproct causes them to be curled toward the center of the calyx. An ectoproct colony, on the other hand, is not capable of bending, and the tentacles are withdrawn by being pulled base first back into the basal portion without curling inwards.

Free-living colonial entoprocts are generally found at very low tide levels in sheltered pools or on the underside of rocks, often attached to algae or hydroids. However, in bays and other sheltered areas such as the Palo Alto Yacht Harbor, entoprocts may comprise a major part of the fauna at certain times of the year, forming extensive mats on any available hard substrate.

KEY TO ENTOPROCTA

1. Colonial forms with individuals arising at regular intervals from a ramifying stolon .... PEDICELLINIDAE 3
– Noncolonial, living singly attached to, or in close association with, tube-dwelling polychaetes, sponges, ectoprocts, sipunculans, etc. .... LOXOSOMATIDAE 2
2. Basal portion of the stalk consists only of a muscular sucking disc, lacking differentiation into a pedal gland, pedal groove, etc.; adults not permanently attached, capable of movement on host (fig. 1) ....*Loxosoma* spp.
– Basal portion of stalk of the buds possesses a well-differentiated pedal gland, generally reduced or lost in adults, which may become permanently attached to host (fig. 2) .... *Loxosomella* spp.
3. Thick, muscular stalk arising directly from the stolon, without a distinct enlargement at the base 4
– Thin, delicate stalk with a well-defined, muscular enlargement at the base where it arises from the stolon .... *Barentsia* 5
4. Calyx perched at an oblique angle to the thick, muscular stalk; both stalk and calyx possess a well-developed musculature; stalk also with a well-developed, diagonal musculature (fig. 3) ....*Myosoma spinosa*
– Calyx terminal on stalk; calyx without a well-developed musculature, and stalk without a strongly developed, diagonal musculature (fig. 4) ....*Pedicellina cernua*
5. Stalk with muscular enlargements where bending occurs ... 6
– Stalk without muscular enlargements; walls of stalk chitinized and inflexible 7
6. Stalk commonly branched, with one or more new individuals often arising from the region of muscular enlargements; stalk with pores throughout length (fig. 5) ....*Barentsia ramosa*
– Stalk almost always unbranched; rarely a new stolon or stalk may arise from a muscular enlargement; stalk may have many

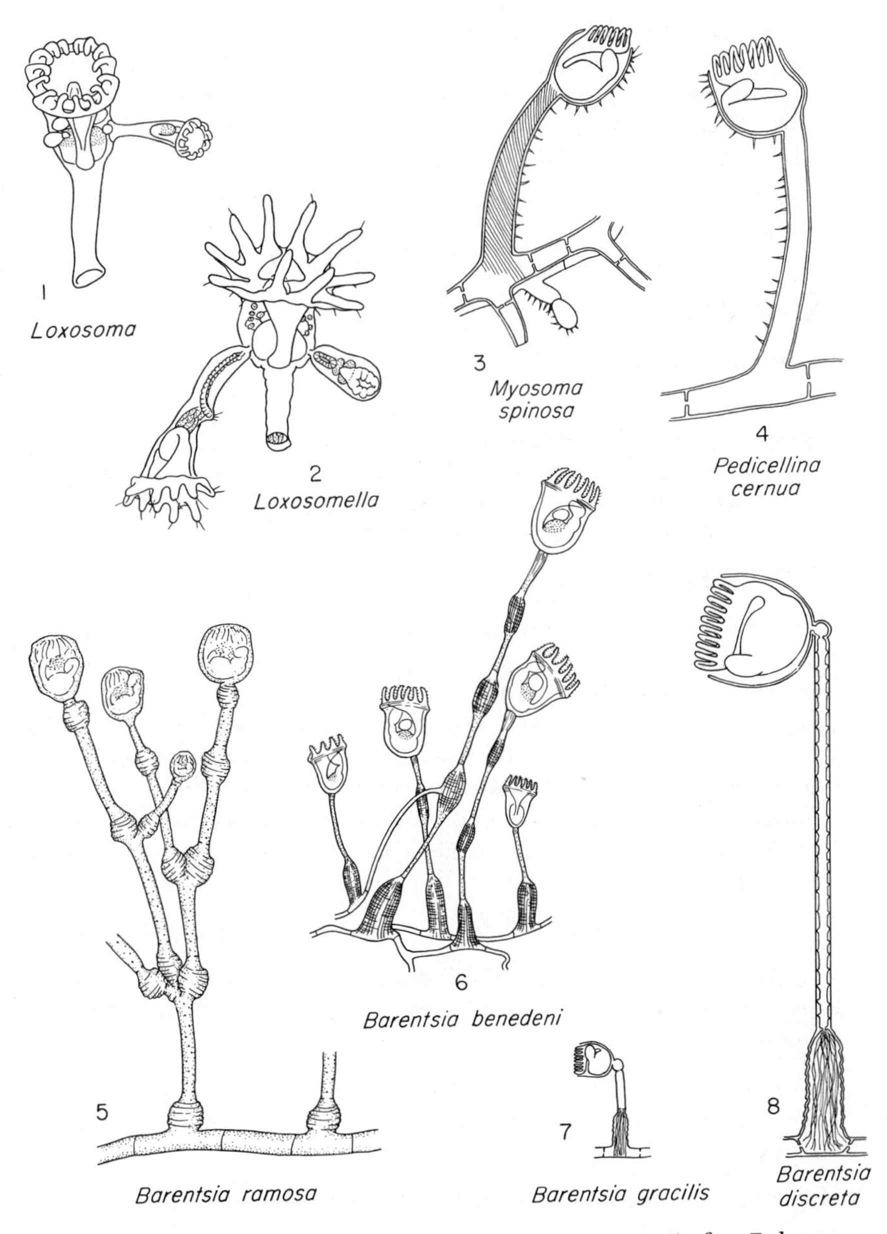

PLATE 143. **Entoprocta.** 1, 2, and 6 after Nielsen, 1964; 5 after Robertson, 1900; 3, 4, 7, and 8 to same scale after Osburn, 1953; rest to various scales.

(up to 12) muscular enlargements; pores absent (fig. 6) .... *Barentsia benedeni*

7. Usually less than 1 mm in total height; stalk with few to no pores (fig. 7) ....*Barentsia gracilis*
– About 2–4 mm in total height; stalk with many pores (fig. 8) .... *Barentsia discreta*

LIST OF ENTOPROCTA

LOXOSOMATIDAE

Loxosoma spp. Commensals in worm tubes and with other invertebrates.

Loxosomella spp. An undescribed species is found on the branchiae of the onuphid polychaete *Diopatra ornata.* Another undescribed loxosomatid has been found embedded in an encrustation of sand grains at the base of the introvert of the sipunculan *Phascolosoma agassizii.* Probably many loxosomatids associated with invertebrates on the Pacific coast have yet to be discovered and described.

PEDICELLINIDAE

Barentsia benedeni (Foettinger, 1887). A European estuarine species preferring silty harbors with much suspended detritus. This species may be very common in certain parts of San Francisco Bay, such as the Palo Alto Yacht Harbor, on pilings, floats, and barnacles, shells of the mussels *Ischadium, Mytilus,* etc. This is the *B. gracilis* of Mariscal, 1965. Because the young of this species are nearly identical to *B. gracilis* in size and in lacking muscular enlargements on the stalk, the two species have been frequently confused. However, it is probable that any estuarine *Barentsia* with muscular swellings on the stalks of at least some members of the colony is *B. benedeni.*

Barentsia discreta (Busk, 1886). See Rogick, 1956, pp. 224–227, pl. 1–2.

Barentsia gracilis (M. Sars, 1835). Prefers fairly clean water; although sometimes described as having muscular enlargements on stalk (var. *nodosa*), such are probably referrable to *B. benedeni.*

Barentsia ramosa (Robertson, 1900). Barentsias occur intertidally and to moderate depths on algae, hydroids, ectoprocts, chitons, and on rocks and other hard substrates.

Myosoma spinosa (Robertson, 1900). Intertidal on hard substrates.

Pedicellina cernua (Pallas, 1771). Under rocks, on algae, hydroids, ectoprocts, etc., in the mid to low intertidal, also on pilings.

REFERENCES ON ENTOPROCTA

Brien, P. 1959. Classe des Endoproctes ou Kamptozoaires. *In* P-P. Grassé, ed., *Traité de Zoologie,* Vol. 5, pp. 927–1007. Paris: Masson et Cie.

Hyman, L. H. 1951. *The Invertebrates, Acanthocephala, Aschelminthes, and Entoprocta,* Vol III. McGraw-Hill, pp. 521–554.
Mariscal, R. N. 1965. The adult and larval morphology and life history of the entoproct *Barentsia gracilis* (M. Sars, 1835). J. Morph. 116: 311–338.
Nielsen, C. 1964. Studies on Danish Entoprocta. Ophelia 1: 1–76.
Nielsen, C. 1971. Entoproct life-cycles and the entoproct/ectoproct relationship. Ophelia 9: 209–341.
Osburn, R. D. 1953. Phylum Entoprocta. *In* Bryozoa of the Pacific Coast of America, Allan Hancock Pac. Exped. 14 (3): 759–773, pl. 82. Los Angeles: University of Southern California Press.
Prenant, M. and Bobin, G. 1956. *Bryozoaires. Première Partie. Entoproctes, Phylactolèmes, Cténostomes.* Faune de France 60: 1–398.
Robertson, A. 1900. Studies in Pacific coast Entoprocta. Proc. Calif. Acad. Sci. (3) 2: 323–348.
Rogick, M. D. 1956. Bryozoa of the U.S. Navy 1947–1948 Antarctic Expedition, I–IV. Proc. U.S. Nat. Mus. 105: 221–317.
Soule, D. F. and J. D. Soule 1965. Two new species of *Loxosomella,* Entoprocta, epizoic on crustacea. Allan Hancock Found. Occas. Pap. 29: 1–19.

PHYLUM PHORONIDA

Russel L. Zimmer
University of Southern California, Santa Catalina Marine Biological Laboratory

(PLATE 144)

Phoronids are a small phylum with only two genera, both represented intertidally on the central California coast. Of eight species known from the west coast, only three are locally abundant in the intertidal, but the actinotroch larvae (fig. 10) of others may be taken in the plankton and adults of other species may be collected inter- or subtidally.

The adult has a wormlike body terminating in the crescentic or doubly coiled lophophore (fig. 9). The body is unsegmented, smooth and soft, and secretes a chitinous tube to which sand grains, shell fragments, or detritus may adhere, except in those species burrowing in shell or limestone. The digestive tract is U-shaped with the mouth and anus close together at the base of the lophophore.

KEY TO COMMON PHORONIDA

1. With delicate epidermal collar overlapping base of lophophore; tentacles pale green; separate, straight, sandy tubes, vertical in muddy sand flats, commonly in vast numbers .... *Phoronopsis viridis*

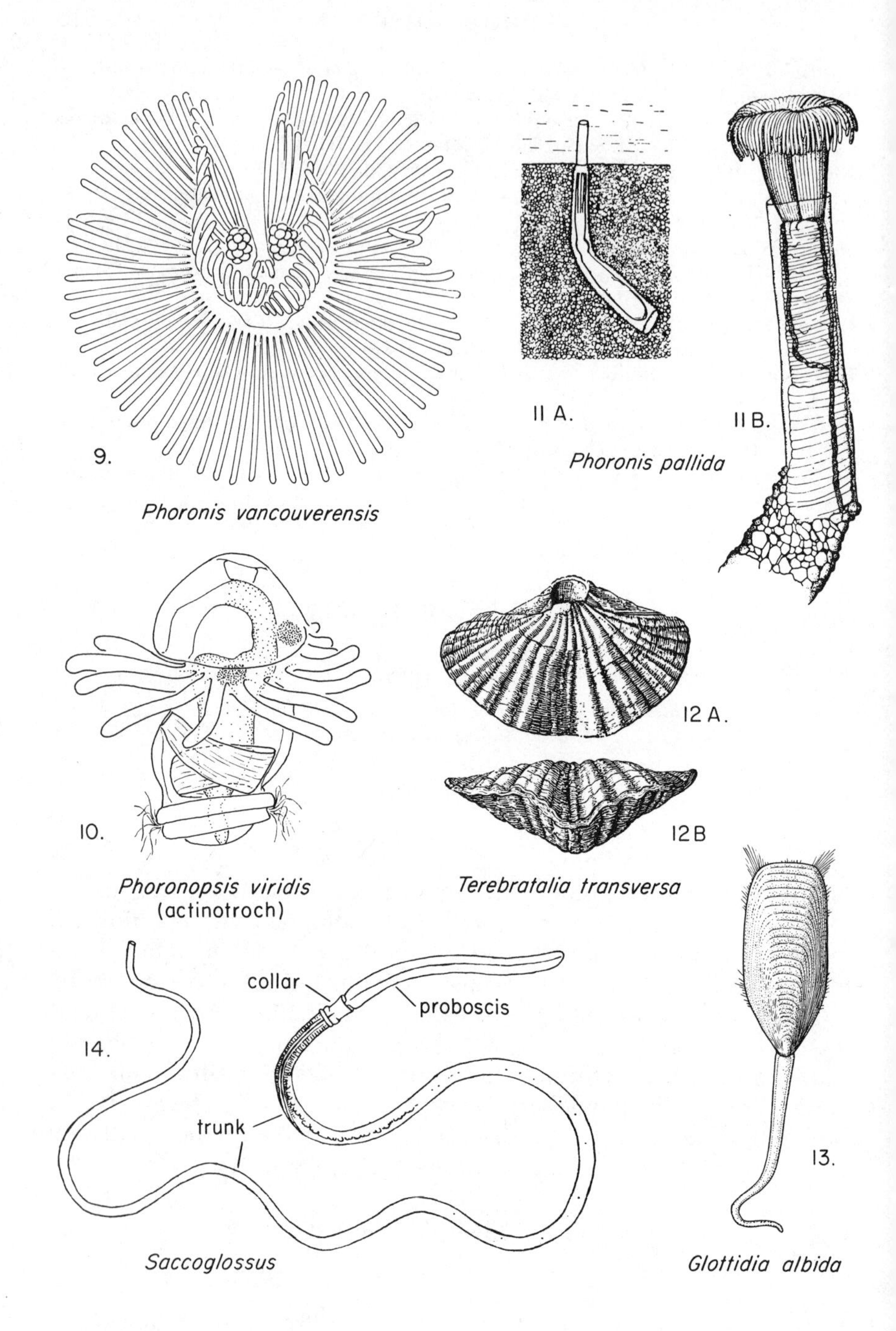
9.
Phoronis vancouverensis
11 A.
11 B.
Phoronis pallida
10.
Phoronopsis viridis
(actinotroch)
12 A.
12 B
Terebratalia transversa
collar
proboscis
14.
trunk
Saccoglossus
13.
Glottidia albida

- Lacking collar at base of lophophore; tentacles white; tubes and habitat otherwise .... *Phoronis* 2
2. Tubes without sand grains, but often with adherent diatoms and detritus; commonly in tangled masses on rocks, floats, and pilings, but may be more or less separate, burrowing into shells or rocks; during spring and summer, masses of embryos may be held among tentacles of adult (fig. 9) .... *Phoronis vancouverensis*
- Very tiny; tubes usually bent in obtuse angle (fig. 11), with distal part membranous, proximal sand-encrusted; may be commensal in walls of *Upogebia* burrows .... *Phoronis pallida*

LIST OF PHORONIDA

* *Phoronis architecta* Andrews, 1890. Locally abundant on Atlantic coast and subtidally in southern California; probably ranges length of west coast; occasionally intertidal, with sand-encrusted tubes, straight and vertical in sand, or meandering and adherent to rocks and shells.

* *Phoronis ovalis* Wright, 1856. Boring in shell and limestone; known from Washington on west coast, but possibly cosmopolitan.

Phoronis pallida (Schneider, 1862). This tiny form was found in the firm mucus-cemented walls of burrows of *Upogebia* by Rogene K. Thompson (Ph.D. thesis, University of California, Berkeley, 1972). It is known from Europe and probably occurs along our entire coast in association with burrows of thallasinid crustaceans, and perhaps of other animals forming comparable firm-walled burrows. Hermaphroditic; breeds summer and fall, possibly year round; has large spermatophoral organ; sheds eggs directly into sea.

* *Phoronis psammophila* Cori, 1889. Rare in central and southern California; also known from Europe.

* Not in key.

PLATE 144. **Phoronida, Brachiopoda, Hemichordata.** 9, *Phoronis vancouverensis*, adult: lophophore showing paired masses of a few embryos each, adjacent to paired nephridial ridges and median anus, mouth covered by a flap (epistome) within indented ring of tentacles; 10, *Phoronopsis viridis*, actinotroch larva: often has same green cast as adult, digestive tract sparsely stippled, metasomal sac looped around gut, two densely stippled areas represent blood corpuscle masses which are incorporated into circulatory system at metamorphosis; 11A, *Phoronis pallida* in its tube; 11B, *P. pallida*, partly emerged from tube; 12A, brachiopod *Terebratalia transversa*, dorsal view; 12B, *T. transversa*, apical view; 13, brachiopod *Glottidia albida;* 14, hemichordate, *Saccoglossus* sp. 9, 10, Zimmer, original; 11A,B, Silén, 1952; 12, Keep, 1904; 13, original, Emily Reid; 14, Assheton in van der Horst, 1934–39.

Phoronis vancouverensis Pixell, 1912. Known from length of west coast; sometimes referred to *P. hippocrepia* Wright, a European species (Marsden, 1959) or to *P. ijimai* Oka, a Japanese form (Emig, 1971).

* *Phoronopsis californica* Hilton, 1930. With large, bright orange, elaborately spiralled lophophore; originally described from intertidal, but now known only subtidally from Point Conception south.

* *Phoronopsis harmeri* Pixell, 1912. The common *Phoronopsis* of Oregon and Washington; united with *P. viridis* by Marsden (1959) and Emig (1971) but probably a distinct species on the basis of spermatophore and larval differences; lacks green pigmentation of *P. viridis*.

Phoronopsis viridis Hilton, 1930. Forms massive beds on sand- and mudflats of central California; dioecious; males with large spermatophoral organs during spring and early summer; females shed eggs directly into sea.

REFERENCES ON PHORONIDA

Emig, C.-C. 1971. Taxonomie et systématique des Phoronidiens. Bull. Mus. National Hist. Nat., Zool. 8: 473–568.

Marsden, J. C. R. 1959. Phoronidea from the Pacific coast of North America. Canad. J. Zool. 37: 87–111.

Rattenbury, J. C. 1953. Reproduction in *Phoronopsis viridis*. . . . Biol. Bull. 104: 182–196.

Rattenbury, J. C. 1954. The embryology of *Phoronopsis viridis*. J. Morph. 95: 289–349.

Silén, L. 1952. Research on Phoronidea of the Gullmar Fiord area (West coast of Sweden). Ark. Zool. 4: 95–140.

Zimmer, R. L. 1967. The morphology and function of accessory reproductive glands in the lophophores of *Phoronis vancouverensis* and *Phoronopsis harmeri*. J. Morph. 121: 159–178.

PHYLUM BRACHIOPODA

(PLATE 144)

Brachiopods (lamp shells) are entirely marine and sessile, with a bivalved shell generally attached to the substrate by a stalk (pedicel or peduncle). Some, at first encounter, may be mistaken for bivalved molluscs. There are two classes: **Articulata,** whose valves are hinged by interlocking teeth and sockets, and whose pedicel emerges through a hole in the ventral valve; and **Inarticulata,** whose valves are not hinged, but are held together by muscles. The pedicel in inarticulates may be long and used to retract the shell into a burrow in soft substrate, or may be reduced or absent.

The Inarticulata are represented locally by *Glottidia albida* (Hinds, 1844) (fig. 13), which has been taken by shallow dredging in Tomales, San Francisco, and Monterey Bays. *Glottidia* has a rather fragile, whitish to brownish, smooth, elongate shell, borne on a long, muscular pedicel, which it uses to burrow into fine, sandy or muddy substrates. The Articulata are represented by several species, all generally in deeper waters, but *Terebratalia transversa* (Sowerby, 1846) (fig. 12) has been taken at very low tides under rocks; it has a reddish, subquadrate shell, and heavily ribbed forms have been given the subspecific name *caurina* (Gould, 1850).

REFERENCES ON BRACHIOPODA

Hertlein, L. G. and U. S. Grant IV 1944. The cenozoic Brachiopoda of western North America. Publ. Univ. Calif. Los Angeles, Math. Phys. Sci. 3: 236 pp.

Jones, G. F. and J. L. Barnard 1963. The distribution and abundance of the inarticulate brachiopod *Glottidia albida* (Hinds) on the mainland shelf of southern California. Pac. Nat. 4: 27–52.

Rudwick, M. J. S. 1970. *Living and Fossil Brachiopods.* London: Hutchinson Univ. Library, 199 pp.

Smith, A. G. and M. Gordon, Jr. 1948. The marine mollusks and brachiopods of Monterey Bay, California, and vicinity. Proc. Calif. Acad. Sci. (4) 36: 147–245 (see pp. 209–210).

Williams, A. and A. J. Rowell 1965. Brachiopoda, Part H. *In* R. C. Moore, ed. *Treatise on Invertebrate Paleontology,* 927 pp., Geol. Soc. Amer. and Univ. Kansas Press.

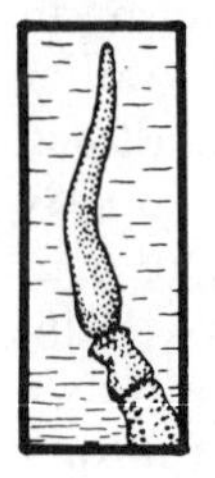

PHYLUM HEMICHORDATA

Theodore H. Bullock
University of California, San Diego

(Plate 144)

Hemichordates are a small group, somewhat apart from, but showing affinities with both echinoderms and chordates. The free-swimming larva of some hemichordates, the "tornaria," is strongly reminiscent of an echinoderm larva. On the other hand, hemichordates possess gill slits comparable to those of amphioxus. Other alleged similarities are questionable.

The phylum is divided into two classes: **Pterobranchia** are minute, deep-sea, sessile organisms; **Enteropneusta** are wormlike forms that are only occasionally a conspicuous part of the fauna. Many well-worked localities have never yielded an enteropneust; nevertheless, the group is worldwide, and about two dozen species are known from the western North American intertidal zone and adjacent waters.

Enteropneusts inhabit sand or mud, occasionally living under rocks or among seaweed holdfasts. They may be identified in the field by the presence of an anterior, bulbous to elongated, nonretractile proboscis, protruding from a distinct collar; the rest of the body is soft and wormlike, but unsegmented (fig. 14). Some burrowing species may be located by a coiled casting of sand thrown up in a cone at one end of the burrow, but most are encountered by chance digging. In some forms the small, pale or orange proboscis may be exposed and conspicuous on substrate exposed at low tide. Being soft and fragile, enteropneusts are usually collected in fragments, and should be taken back to the laboratory in separate containers.

Good preservation soon after collection is essential, as all serious taxonomic study depends upon serial sectioning and histologic staining. It is usually not necessary to narcotize these animals, but it is desirable to leave them in clean sea water for a few hours to allow evacuation of sand from the gut; frequent removal of the mucous sheath with its sand grains prevents reingestion. Fixation without coiling is aided if each worm is lifted into the air by a match stick under the middle of the specimen and killed by dropping fixative solution over it for a few minutes. It may then be lowered into the fixative. A general histologic fixative such as Bouin, Heidenhain's Susa, formol-acetic-alcohol or 10% formalin is satisfactory. Changing the fluid twice in the first 24 hours and again after several days is important.

One locally common species in our area is possibly *Mesoglossus* sp., in shelly shale gravel in protected tidal channels, as at Duxbury Reef, where it is easily located by its small piles of threadlike castings and its orange proboscis, but collected with difficulty because of its extreme delicacy. Several species of *Saccoglossus* (fig. 1), recognized by their long proboscides, more than 6 times longer than wide, are known from the west American coast, but none is common except *S. pusillus* (Ritter, 1902) in southern California, and *Saccoglossus* sp. on the coast of Oregon and Washington. At least one species each of *Balanoglossus, Glossobalanus, Schizocardium, Stereobalanus,* and two still-unpublished genera, besides others further north and south, occur in scattered localities but are rare. Specimens found may be of great scientific interest and should be carefully preserved. The West Coast fauna is unusually rich in variety, with representatives of half the known genera and all three of the known families of enteropneusts. The most recent, but still unpublished, treatment of local forms is that of Rao, Ritter, and Bullock and the class has been monographed by van der Horst (1934–1939) and Dawydoff (1948).

REFERENCES ON HEMICHORDATA

Barrington, E. J. W. 1965. *The Biology of Hemichordata and Protochordata.* Freeman, San Francisco, 176 pp.

Davis, B. M. 1908. The early life-history of *Dolichoglossus pusillus* Ritter. Univ. Calif. Publ. Zool. 4: 187–226.

Dawydoff, C. 1948. Stomocordes. In P. Grassé, ed., *Traité de Zoologie.* Masson, Paris, pp. 367–532.

Horst, C. J. van der 1934–39. Hemichordata. *In* Bronn, *Klassen und Ordnungen des Tierreichs* 4: 4: 2: 2: pp. 1–737.

Rao, K. P., W. E. Ritter, and T. H. Bullock. The Enteropneusta of the eastern Pacific. (in manuscript)

Ritter, W. E. and B. M. Davis 1904. Studies on the ecology, morphology, and speciology of the young of some Enteropneusta of western North America. Univ. Calif. Publ. Zool. 1: 171–210.

PHYLUM ECHINODERMATA

Echinoderms are diversified animals having, in the adult, an imperfect radial symmetry and a skeleton of calcareous plates, sometimes reduced to small, concealed ossicles. A unique feature of this phylum is the water-vascular system, which consists of a ring canal, radial canals, tube feet, and certain derivatives from the ring canal, such as Tiedemann's bodies and the polian vesicles. The tube feet occur either with or without terminal suckers and are used for locomotion, feeding, and respiration.

Of the several living classes, the **Crinoidea** are not represented in the local intertidal zone. This group is characterized by the upward direction of the mouth, which is located in a disc bearing a number of branching rays or arms. The rays bear, on their oral surface, ciliated ambulacral grooves leading to the mouth. The anus is usually on the oral surface of the disc. Of more than 600 living species of crinoids, some 80 are stalked, sessile forms, found only in depths below 50 fathoms; the remainder are vagile as adults and cling to rocks and seaweeds or swim about, often in shallow water.

Of the remaining classes, two have arms or rays: the **Asteroidea** or sea stars and the **Ophiuroidea** or brittle stars. (These groups, together with the **Somasteroidea,** are treated as subclasses of the class **Stelleroidea** by most modern echinoderm systematists.) The **Echinoidea,** or sea urchins and sand dollars, lack arms and have a rigid test with movable spines. The **Holothuroidea,** or sea cucumbers, lack spines and are generally soft-bodied; the skeletal plates are usually reduced to tiny ossicles in the body wall.

GENERAL REFERENCES ON ECHINODERMATA

Boolootian, R. A., ed. 1966. *Physiology of Echinodermata.* Wiley Interscience, 822 pp.

Holland, N. D. and L. Z. Holland 1969. A bibliography of echinoderm biology, continuing Hyman's 1955 bibliography through 1965. Publ. Staz. Zool. Napoli 37: 441–453.

Hyman, L. H. 1955. *The Invertebrates:* Vol. IV. *Echinodermata.* McGraw-Hill, 763 pp.

Johnson, M. W. and L. T. Johnson 1950. Early life history and larval development of some Puget Sound echinoderms, with special reference to *Cucumaria* spp. and

Dendraster excentricus. In M. H. Hatch, ed.; *Studies Honoring Trevor Kincaid*, pp. 74–84. University of Washington Press.
Mortensen, T. 1921. Studies of the development and larval forms of echinoderms. Copenhagen: G. E. C. Gad, 261 pp., 33 pl.
Nichols, D. 1966. *Echinoderms.* London: Hutchinson University Library, 220 pp.

CLASS ECHINOIDEA

John S. Pearse
University of California, Santa Cruz

Echinoids are globose to flattened echinoderms covered with spines. The tube feet are arranged in five paired ambulacral rows extending from the mouth on the oral surface to the opposite, aboral pole. The ambulacral rows alternate with five interambulacral areas. Calcareous plates in the ambulacral and interambulacral areas occur in orderly patterns, and are usually bound into a rigid test over which the movable spines and pedicellariae are arranged. Echinoids are traditionally divided into the **Regularia** (sea urchins) and the **Irregularia** (sand dollars, cake urchins, and heart urchins), but can alternatively be divided into two subclasses: the **Perischoechinoidea** (an ancient stock represented today by the Order Cidaroidea, occurring mainly in the tropics and deep sea) and the **Euechinoidea,** including all other living echinoids.

Sea urchins are globular, with the anus, madreporite, and five gonopores at the aboral pole, and are usually covered with long spines. They possess an intricate jaw apparatus, with five teeth, called Aristotle's lantern. Most are omnivorous grazers and scavengers that scrape algae and encrusting animals from rocks. Many, including our west-coast species, also eat drift algae caught with their extended tube feet.

Only two sea urchins, the purple urchin *Strongylocentrotus purpuratus* (Stimpson, 1857) and the red urchin *S. franciscanus* (Agassiz, 1863) are found intertidally in central California. Both range from Alaska to central Baja California, and *S. franciscanus* is also found in Asia as far south as northern Japan. *S. purpuratus* can occur in tremendous numbers on surf-swept reefs at low tide. Often the animals are in small depressions and crevices, especially on soft shales and sandstones. Most individuals are 3 to 8 cm in diameter, but they may reach nearly 10 cm. The spines are usually a uniform rich purple, while the body is dark reddish purple. Juveniles (less than about 1 cm in diameter) tend to be pale greenish, and even large adults which have remained under rocks can retain this pale green color. *S. francis-*

canus is mainly subtidal, especially in kelp beds, but does occur intertidally in low tidepools and tidal channels. Adults are relatively large, usually between 5 and 10 cm in diameter. The spines are long and range from pale, nearly white, to a dark, nearly black, reddish or maroon. Even juveniles, down to a few millimeters in diameter, have orange or reddish spines.

Four other sea urchins may be found intertidally between Canada and Baja California. These include the primarily subtidal *Lytechinus pictus* and *L. anamesus* (which may be synonymous); the black, long-spined *Centrostephanus coronatus*, occasionally intertidal in southern California; and, on Washington shores and northward, the circumpolar green sea urchin *Strongylocentrotus droebachiensis*. The most numerous sea urchin off the coast of California, *Allocentrotus fragilis*, is not intertidal; it often occurs in enormous numbers at depths of 50 to 1200 meters.

Several irregular echinoids occur intertidally along the Pacific coast; the heart urchin *Lovenia cordiformis* is occasionally seen at low tide on southern California beaches. Only the purplish gray sand dollar *Dendraster excentricus* (Eschscholtz, 1829) occurs on the central California coast. It may be found on flat, sandy bottoms from low-tide mark to 100 meters, often in densely packed beds with each animal nearly vertical in the sand, perpendicular to the current. In life *Dendraster* is covered with a bristly plush of short spines, and its bleached tests are common objects on exposed sandy beaches.

REFERENCES ON ECHINOIDEA

Birkeland, C. and F. -S. Chia 1971. Recruitment risk, growth, age and predation in two populations of sand dollars, *Dendraster excentricus* (Eschscholtz). J. Exp. Mar. Biol. Ecol. 6: 265–278.

Durham, J. W. and R. V. Melville 1957. A classification of echinoids. J. Paleontol. 31: 242–272.

Ebert, T. A. 1968. Growth rates of the sea urchin *Strongylocentrotus purpuratus* related to food availability and spine abrasion. Ecology 49: 1075–1091.

Grant, U. S. IV and L. G. Hertlein 1938. The west American Cenozoic Echinoidea. Publ. U.C.L.A. Math. Phys. Sci 2: 1–225.

McCauley, J. E. and A. G. Carey, Jr. 1967. Echinoidea of Oregon. J. Fish Res. Bd. Canada 24: 1385–1401.

Mortensen, T. 1928–1951. *A Monograph of the Echinoidea*. Copenhagen: C. A. Reitzel, 5 volumes.

Philip, G. M. 1965. Classification of echinoids. Paleontol. 39: 45–62.

Swan, E. F. 1953. The Strongylocentrotidae (Echinoidea) of the northeast Pacific. Evolution 7: 269–273.

CLASS ASTEROIDEA

James E. Sutton
California Academy of Sciences, San Francisco

Starfishes or sea stars are familiar animals of the intertidal zone. Most are predators although some, like *Patiria*, are also herbivorous. They possess strong arms with well-developed ambulacral grooves in which are numerous tube feet, bearing suckers in all local forms. Mouth and ambulacral grooves are on the ventral or oral surface; anus and madreporite are on the dorsal or aboral surface. Among characters important in classification are the spines, the pincerlike pedicellariae (absent in some, such as *Henricia*), the radius (the distance from the center of the disc to the tip of an arm), and the shape and arrangement of skeletal plates, usually only observable after removal of the surface tissues.

Monographic treatments include those of Verrill (1914) and Fisher (1911–1930). Hopkins and Crozier (1966) cover the area south of the present treatment. For recent higher classifications, see Spencer and Wright (1966). The following key is based upon that by Frances Weesner Lechleitner in the second edition of this manual.

KEY TO ASTEROIDEA

1. With 8–20 or more arms 2
– Normally with 5 or 6 arms (as a result of injury and regeneration, individuals with other than the normal number of arms are common) 3
2. With 8–13, typically 11 or 12, firm and rigid arms, without fleshy covering; color in life reddish brown *Solaster dawsoni*
– With 15–24 or more limp and flexible arms, with soft fleshy covering; color in life pink to purple .... *Pycnopodia helianthoides*
3. Lateral margins of disc and arms with row of large, conspicuous, granulated plates; dorsal surface covered by rounded granulated plates; disc large; arms tapering; disc and arms distinctly flattened; color in life bright orange .... *Mediaster aequalis*
– Lateral margins of disc without large, rectangular, granulated plates 4
4. Arms short, thick, broadly attached to disc, producing webbed appearance 5

– Arms long, not broadly attached to disc, not producing webbed appearance 6

5. Dorsal surface covered with smooth, slippery, leathery skin, with embedded, flat, round lime deposits; color in life gray and orange mottling; has a garlic or sulfurous odor .... *Dermasterias imbricata*

– Dorsal surface not covered by leathery skin; flat skeletal plates arranged in crescent-shaped, concentric clusters mixed with tiny granules, giving scaled appearance; color in life highly variable, solid or mottled red, orange, brown, yellow, and green; number of arms typically 5 but ranges from 4–9 .... *Patiria miniata*

6. With 6 arms; radius in adults rarely larger than 4 cm; broods eggs under disc .... *Leptasterias* 7

– With 5 arms; radius in adults much larger than 4 cm; does not brood eggs 8

7. Radius less than 2.5 cm; arms slender, not usually swollen at base; dorsal skeletal plates not in distinct radial rows, middorsal row not conspicuous; color in life green or dark red mottled with light yellow .... *Leptasterias pusilla*

– Radius up to 4 cm; arms often somewhat swollen at base; dorsal skeletal plates typically in distinct radial rows, middorsal row conspicuous; color olive green or red, solid or mottled (a highly variable species; small specimens easily confused with *L. pusilla*) .... *Leptasterias hexactis*

8. Arms long, thin; pedicellariae absent; disc small; dorsal surface superficially smooth without distinct spines; skeleton forming conspicuous, flat, reticulated meshwork; color in life variable, but generally red, orange, or yellow .... *Henricia leviuscula*

– Arms thick, with distinct spines; pedicellariae present; disc generally large 9

9. Arm spines generally small, low, rounded, beadlike; patchy in distribution 10

– Arm spines large, distinct, clubshaped or sharp; uniform in distribution 12

10. Adambulacral spines (long thin spines on edge of ambulacral grooves, immediately adjacent to tube feet) with clusters of pedicellariae; color extremely variable *Evasterias troschelii*

– Adambulacral spines without clusters of pedicellariae, although some pedicellariae may be visible around bases of spines, attached to skeletal plates 11

11. Spines on lateral and dorsal surfaces of arms forming exten-

sive irregular, reticulated pattern, or separate convex, curved groups at arm tips; distinct middorsal radial row of spines not typically present, or at least not straight; color purple, red, brown, or ochre yellow .... *Pisaster ochraceus*

– Spines on lateral and dorsal surfaces of arms not forming reticulated pattern (except rarely); spines usually single or in groups of 2 or 3, separated by areas of soft tissue; middorsal radial row of spines typically distinct, straight; color in life consistently pink, sometimes mottled with gray-green or maroon-purple .... *Pisaster brevispinus*

12. Dorsal spines typically blunt, club-shaped, surrounded at base by ring of flesh (blue in life), with ring of pedicellariae outside fleshy ring; tiny pedicellariae thickly scattered over surface between spines; spines dense, but typically not arranged in distinct radial or concentric rows; color in life bluish gray .... *Pisaster giganteus*

– Dorsal spines typically sharp, surrounded by distinct ring of large pedicellariae; fleshy ring absent; few large pedicellariae on surface between spines; spines arranged in rather distinct radial rows; color in life bright to brick red, often with yellow banding or mottling on arms ....*Orthasterias koehleri*

LIST OF ASTEROIDEA

Order **Valvatida**

Mediaster aequalis Stimpson, 1857. Common subtidally, occasional in low intertidal zone, Monterey area (formerly in old order Phanerozonia).

Order **Spinulosida**

Dermasterias imbricata (Grube, 1857). Leather star; preys on sea urchins, starfish, sponges, sea anemones, bryozoans, etc.; characteristic of seaweedy boulder flats; formerly in old order Phanerozonia.

Henricia leviuscula (Stimpson, 1857). A ciliary small-particle feeder; may also feed on sponges and bryozoans (see Mauzey *et al.*, 1968); characteristic of boulder areas on exposed coast.

Patiria miniata (Brandt, 1835). Bat star; a particulate feeder, also eats starfish, ascidians, and algae; characteristic of protected rocky areas; the polychaete *Ophiodromus pugettensis* is common in the ambulacral grooves (see page 183).

Solaster dawsoni Verrill, 1880. Largely subtidal, rarely very low intertidal zone; primarily northern, uncommon in central California.

Order **Forcipulatida**

Evasterias troschelii (Stimpson, 1862). Very low intertidal, generally scarce; superficially similar to *Pisaster ochraceus* but with more slender arms and orange-red coloration, although color variable; see Christiansen 1957, Limnol. Ocean. 2: 180–197 (feeding).

Leptasterias hexactis (Stimpson, 1862). A highly variable species, including what was formerly known as *L. aequalis* on the California coast; mainly low intertidal zone; broods mass of large yellow eggs beneath disc; feeds on barnacles, sea cucumbers, chitons, snails, etc. (see Chia, 1966a, b).

Leptasterias pusilla (Fisher, 1930). A delicate form of the mid-intertidal zone; like *L. hexactis,* it broods its eggs.

Orthasterias koehleri (de Loriol, 1897). Very low intertidal; common in subtidal zone.

Pisaster brevispinus (Stimpson, 1857). Feeds on clams on sand bottoms (Smith 1961, Behaviour 18: 148–153), sand dollars, etc.; and on barnacles on rocks and pilings in bays; unlike other local *Pisaster,* occurs on soft bottoms.

Pisaster giganteus (Stimpson, 1857). Low intertidal zone; mainly southern; see Landenberger 1968, Ecology 49: 1062–1075 (feeding); 1969, Physiol. Zool. 42: 220–230 (distributional ecology).

Pisaster ochraceus (Brandt, 1835). Ochre star; the common predator of the lower *Mytilus californianus* beds; see Feder 1959, Ecology 40: 721–724 (food); 1970, Ophelia 8: 161–186 (growth, predation); Mauzey 1966, Biol. Bull. 131: 127–144 (feeding, reproductive cycles).

Pycnopodia helianthoides (Brandt, 1835). Sunflower star; a large and very active opportunistic predator; common in low intertidal and subtidal zones.

REFERENCES ON ASTEROIDEA

Chia, F. -S. 1966a. Systematics of the six-rayed sea star *Leptasterias* in the vicinity of San Juan Island, Washington. Syst. Zool. 15: 300–306.

Chia, F. -S. 1966b. Brooding behavior of a six-rayed starfish, *Leptasterias hexactis.* Biol. Bull. 130: 304–315.

Fisher, W. K. Asteroidea of the North Pacific and adjacent waters. Bull. 76, U.S. Nat. Mus. Part 1, Phanerozonia and Spinulosa, 419 pp., 122 pls., 1911. Part 2, Forcipulata (part), 245 pp., 81 pls., 1928. Part 3, Forcipulata (concluded), 356 pp., 93 pls., 1930.

Hopkins, T. S. and G. F. Crozier 1966. Observations on the asteroid echinoderm fauna occurring in the shallow water of southern California. Bull. So. Calif. Acad. Sci. 65: 129–145. Covers area south of Point Conception.

Mauzey, K. P., C. Birkeland, and P. K. Dayton 1968. Feeding behavior of asteroids and escape responses of their prey in the Puget Sound region. Ecology 49: 603–619.

Spencer, W. K. and C. W. Wright 1966. Asterozoans. Part U, Echinodermata. *In* R. C.

Moore, ed., *Treatise on Invertebrate Paleontology,* Univ. Kansas Press and Geol. Soc. America 3 (1): 4–107.
Verrill, A. E. 1914. Monograph of the shallow-water starfishes of the North Pacific coast from the Arctic Ocean to California. Smithsonian Inst., Harriman Alaska Series, 14: Part 1, text, 408 pp.; Part 2, plates 1–110. Much of nomenclature is obsolete.

CLASS OPHIUROIDEA

James E. Sutton
California Academy of Sciences, San Francisco

(PLATES 145, 146)

Ophiuroids, brittle or serpent stars, are inconspicuous but numerous in the intertidal zone. They are relatively small, occurring in more or less concealed habitats, buried in sand, under rocks, or in the fronds and holdfasts of algae. Brittle stars have a central disc and 5 long, narrow, segmented arms (rays), of great flexibility in the horizontal and, to a lesser extent, in the vertical plane. Ophiuroids differ from asteroids in several respects: they lack ambulacral grooves, the gonads are limited to the central disc, and the tube feet lack suckers.

Certain structures are common to nearly all ophiuroids. Each arm segment has a large upper (aboral, dorsal) and a lower (oral, ventral) arm plate (shield) (figs. 1, 3). The upper arm plate may consist of many small plates (fig. 6), or may take the form of a large central plate with small supplementary pieces about its margin (figs. 7, 8), or may be a single entire piece (fig. 3). Between the upper and lower arm plates, on either side of the arm segment, there is a lateral (side) arm plate which bears a vertical row of arm spines. These spines differ in size, structure, and function according to species, and also according to the size of the animal, the distance of the segment from the base of the arm, and the extent of regeneration after autotomy. At the margin of the lower arm plate, a pair of podia (tentacles, tube feet) project through openings in the plate called podial pores (tentacle pores). The contractile podia may be simple or feathery, or may possess glands. Alongside the podial pores may be one or more small, flat tentacle scales; in some species these scales may close over the pores when the podia are retracted.

The mouth is directed toward the substrate, but is not in direct contact with it. Rather, it is located at the upper end of a pre-oral cavity, which contains a variety of structures (fig. 2). There are usually 5 triangular structures, the jaws, each composed of 2 plates, the half-jaws (oral plates, jaw plates). The outer edge of each jaw may have protrusions called oral papillae (mouth papillae): those lying along the sides of the jaw are lateral oral papillae; those at the pointed tip of the jaw are the terminal (apical, infradental) oral papillae. Further into the pre-oral cavity, there may be small rounded structures called tooth (dental) papillae. Still further in are larger blocklike structures, the teeth. At the upper (dorsal) end of the cavity lies the mouth, a muscular sphincter, which leads into the gut, a simple sac without anus or diverticula. On either side of each jaw there are one or two buccal podia (oral tentacles) and, on some species (although not on any species in this key), a modified oral tentacle scale, which is often quite similar in shape and position to a lateral oral papilla.

The base of the jaw is composed of 3 plates, a large central oral shield (buccal plate) and a smaller adoral shield on either side. These are located in the interradius, the area of the disc between any two adjacent arms. Also found in the oral surface of the interradius are the genital slits, opening into the genital bursae where eggs and/or sperm are produced and where, in viviparous species, the young are brooded.

The aboral or upper surface of the disc may be smooth, or may be covered with granules, spines, papillae (small rounded nodules), or scales. Most species possess radial shields, 5 pairs of large plates lying on the outer margin of the disc at the base of each arm (fig. 3). In some species the radial shields may be partially or totally covered by overlying structures. Virtually all structures show some variability. The greatest variability is seen in: length of the arms, number and shape of arm spines and of oral papillae, types and extent of disc covering. Color and pattern, although characteristic in some species, cannot be used as key characters in others, notably in the genera *Ophiothrix* and *Ophiopholis*. Enough characters are provided in the key that, should one or more differ from the description, the remaining characters should suffice to permit identification.

One of the characters used in the key is the arm:disc ratio, the ratio of the length of the arm to the diameter of the disc. For example, a specimen with arms 75 mm long and a disc 15 mm in diameter has an arm:disc ratio of 5:1. Because many species readily autotomize, or break off portions of arms when disturbed, and then regenerate slowly from the point of the break, the arm:disc ratio as a distinguishing character must be used with caution.

Ophiuroids have a variety of feeding habits. Some are detritus feeders, consuming small bits of organic matter in the substrate. Others are suspension feeders, clinging to algal holdfasts or other objects and removing particles and small animals from the water by means of their arms, spines, and podia. Still others are ciliary feeders, producing a current through burrows they form around themselves in

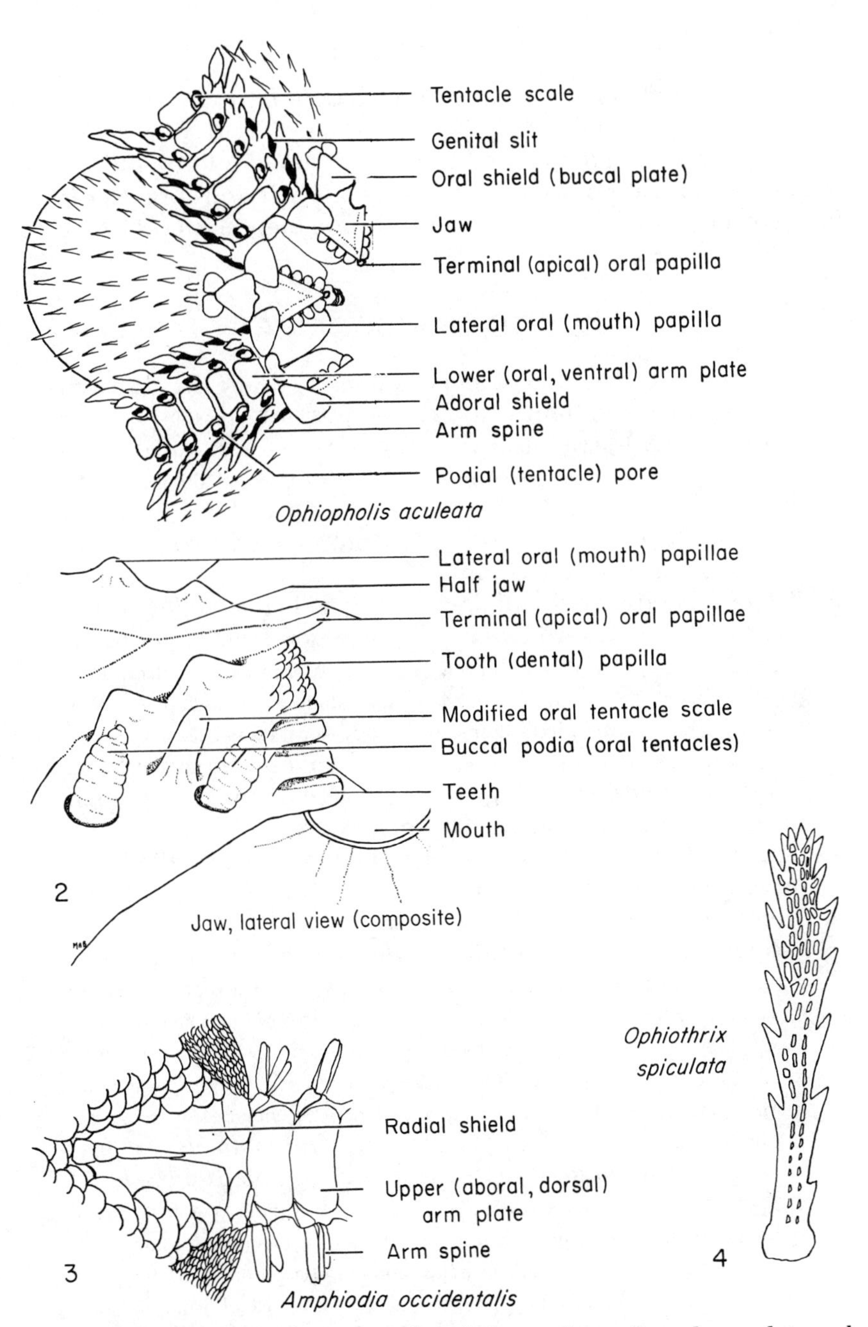

PLATE 145. **Ophiuroidea** (1). 1, *Ophiopholis aculeata* form *kennerlyi,* oral view of part of central disc; 2, composite diagram of lateral view of an ophiuroid jaw; 3, *Amphiodia occidentalis,* aboral view of part of disc, showing radial shields and two entire upper arm plates; 4, *Ophiothrix spiculata,* arm spine.

the substrate. Others move water through their burrows by a pumping motion of their central discs.

"Basket stars," a group of ophiuroids not found intertidally, are occasionally brought up from deeper waters. Their arms branch dichotomously and may be recurved toward the mouth, forming a tangled, basketlike shape.

KEY TO OPHIUROIDEA

1. Oral papillae absent; base of jaw pierced by hole (fig. 10); arm spines glassy and thorny (fig. 4); 6–9, typically 7, arm spines per vertical row; disc and radial shields covered by similar spines; 20–40 tooth papillae per jaw; 4–6 teeth per jaw; 1 tiny tentacle scale per podial pore .... *Ophiothrix spiculata*
– Oral papillae present; base of jaw not pierced by hole; arm spines not glassy and thorny 2
2. 2–6 oral papillae per half-jaw; 2 genital slits per interradius 3
– 8–9 oral papillae per half-jaw (fig. 5); 4 genital slits per interradius; oral papillae small, blunt, adjacent papillae in close contact; tooth papillae absent; 2–3 teeth per jaw; 4–5 short, stout arm spines per vertical row; disc granulated; radial shields small, oval, separated; 2 tentacle scales per podial pore .... *Ophioncus granulosus*
3. Upper arm plate entire (fig. 3), not broken up into many small plates, not with supplementary pieces 7
– Upper arm plate not entire, broken up into many small plates, or with supplementary pieces, or both 4
4. 1 tentacle scale per podial pore 5
– 2–5 tentacle scales per podial pore; typically 2 large plus 1 or 2 (rarely 3) small tentacle scales per podial pore; upper arm plate broken up into many small plates (fig. 6); 4–6 lateral oral papillae per half-jaw; terminal papilla may be single or paired; tooth papillae absent; 5–6 teeth per jaw; 2 (rarely 3) very short, tapered, blunt arm spines per vertical row; disc smooth, leathery; radial shields small, oval, widely separated, may be covered; viviparous .... *Ophioplocus esmarki*
5. Tentacle scale large, round, capable of completely covering the podial pore; a small, triangular supplementary piece on either side of upper arm plate; 4–5 irregular, blunt, lateral oral papillae per half-jaw; 2 terminal oral papillae per jaw; may be a few tooth papillae per jaw; 4–5 teeth per jaw; 3 long, tapered arm spines per vertical row (4 at base of arm); disc covered by tiny scales, larger near radial shields; radial shields small, oval, widely separated .... *Ophionereis eurybrachyplax*
– Tentacle scale small, variable in shape, not capable of completely covering the podial pore; upper arm central plate surrounded by 6–20 supplementary pieces, of variable shape; upper arm central plate

typically entire, or occasionally broken up into several plates; 2–5, typically 3, lateral oral papillae per half-jaw; 1 small terminal oral papilla per jaw (rarely absent); tooth papillae absent; 8–10 teeth per jaw .... *Ophiopholis* 6

6. Upper arm central plate surrounded by 6–12 large angular supplementary pieces in close contact (fig. 7); 4–6, typically 5, short, blunt, slightly flattened arm spines per vertical row; disc and radial shields covered by short flattened papillae, grading to longer spinelike papillae on oral surface; this species may intergrade with following .... *Ophiopholis aculeata* form *kennerlyi*
– Upper arm central plate surrounded by 14–20 small, rounded supplementary pieces not in close contact (fig. 8), and may be covered by tiny spinelets; 5–7, typically 5, long tapered arm spines per vertical row; disc and radial shields covered by short sharp spines ending in several points .... *Ophiopholis bakeri*

7. 5–7, typically 6, arm spines per vertical row; arm spines large, much flattened, blunt; 3–5, typically 4, oral papillae per half-jaw; 15–20 tooth papillae per jaw; 2–3 teeth per jaw; disc and radial shields covered by knobs and papillae; 1 large flat tentacle scale per podial pore .... *Ophiopteris papillosa*
– 3–4 arm spines per vertical row; arm spines variously shaped; 3 (rarely 4) oral papillae per half-jaw; tooth papillae absent; 3–5 teeth per jaw; disc covered by flat scales; radial shields not covered, closely appressed or separated by single row of scales; 0–2 tentacles scales per podial pore ... 8

8. Oral papillae subequal in size; disc diameter up to 10 mm; 3 arm spines per vertical row, blunt and flattened; 3 (rarely 4) oral papillae per half-jaw; 4–5 teeth per jaw; radial shields typically separated for much or all of length by single row of scales (fig. 3); 0–2 tentacle scales per podial pore; arms characteristically long and coiling .... *Amphiodia* 9
– Oral papillae unequal, those closest to base of jaw much longer radially than others (fig. 9); disc diameter up to 5 mm; 3 arm spines per vertical row (often 4 at base of arm), shape variable; 3 (rarely 4) oral papillae per half-jaw; 3–4 teeth per jaw; radial shields small, typically closely appressed for entire length except for proximal tips; 2 tiny tentacle scales per podial pore; arms characteristically short and not coiling .... *Amphipholis* 10

9. Arm:disc ratio greater than 10:1; typically 2 tentacle scales per podial pore; oviparous ....*Amphiodia occidentalis*
– Arm:disc ratio 7 or 8:1; typically 1 (occasionally 0 or 2) tentacle scale per podial pore; may be viviparous .... *Amphiodia* sp.

10. Arm:disc ratio 3 or 4:1; arm spines equal, tapered, blunt; viviparous; very small .... *Amphipholis squamata*
– Arm:disc ratio 7 or 8:1; arm spines unequal, middle spine in middle

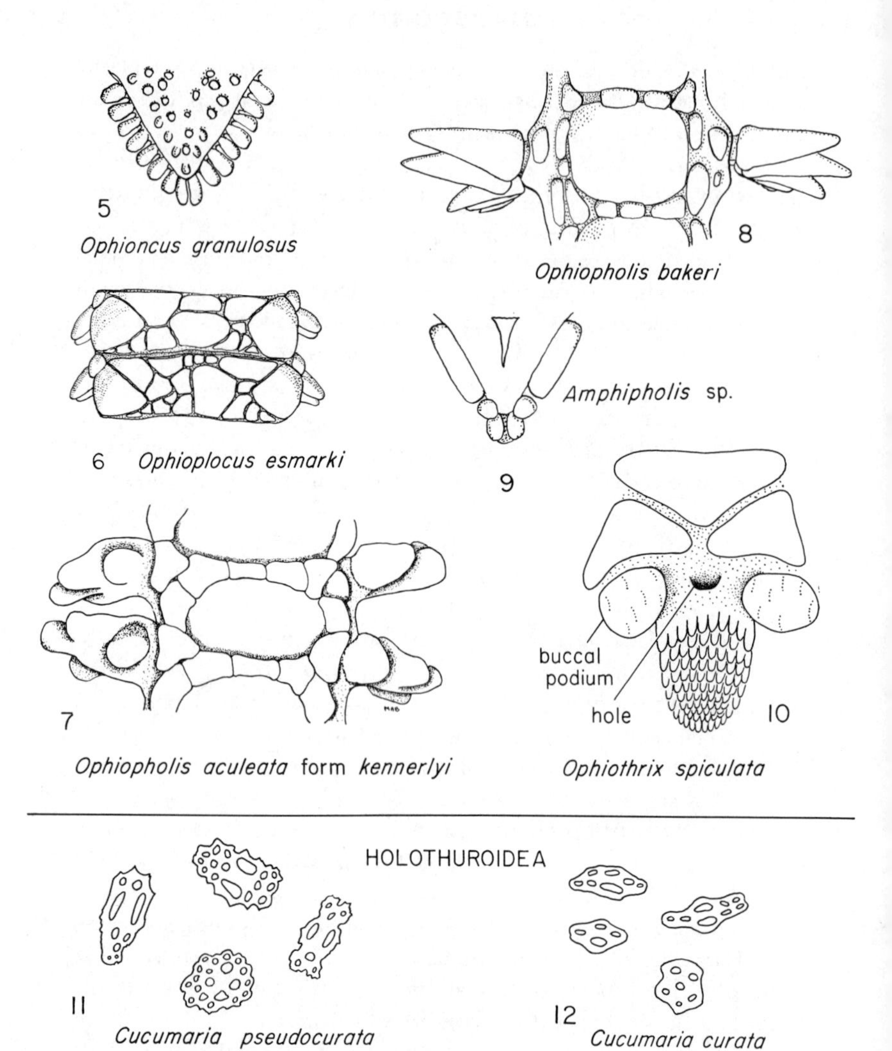

PLATE 146. **Ophiuroidea** (2), **Holothuroidea.** 5, *Ophioncus granulosus,* oral view of jaw; 6, *Ophioplocus esmarki,* two upper arm plates; 7, *Ophiopholis aculeata* form *kennerlyi,* upper arm plate, showing central plate surrounded by 12 large angular supplementary pieces; 8, *Ophiopholis bakeri,* upper arm plate, showing central plate surrounded by 16 small rounded supplementary pieces; 9, *Amphipholis,* oral view of jaw, showing elongate oral papillae at base of jaw; 10, *Ophiothrix spiculata,* oral view of jaw, showing absence of oral papillae, hole in base of jaw, and large oral tentacles; 11–12, **Holothuroidea,** body-wall ossicles: 11, *Cucumaria pseudocurata;* 12, *C. curata.* 2, 7, 10, Sutton, original, drawn by Melissa A. Barbour; 11, 12, Rutherford, original; rest, Weesner, in Light *et al.,* 1954.

part of arm longer than others, less tapered, more clublike; oviparous; small specimens easily confused with *A. squamata* *Amphipholis pugetana*

LIST OF OPHIUROIDEA

Amphiodia occidentalis (Lyman, 1860) (=*Diamphiodia occidentalis*). Abundant in mid-intertidal pools, under rocks or buried 1–20 cm in sand; also common in muddy sand among *Zostera* roots, where its coiling arms resemble tangled worms at collection; aggregates readily; autotomizes easily.

Amphiodia sp. Same habitat as above species in Monterey Bay; uncommon. May be a variety of *A. occidentalis.*

Amphipholis pugetana (Lyman, 1860) (=*Axiognathus pugetanus*). Similar habitat to following species; may be somewhat larger; rare in Monterey area.

Amphipholis squamata (Delle Chiaje, 1829) (=*Axiognathus squamatus*). Very common in coarse sand and gravel of upper middle and midtidal pools; smallest ophiuroid on Pacific coast; viviparous.

Ophioncus granulosus Ives, 1889. Very low intertidal zone, on lower side of rocks; extremely rare.

Ophionereis eurybrachyplax Clark, 1911. Generally subtidal, rarely in low intertidal zone.

Ophiopholis aculeata (Linnaeus, 1767) form *kennerlyi* (Lyman, 1860). Middle and low tidal pools, in areas of strong tidal currents or wave action; often attached to red algae such as *Gigartina;* does not burrow.

* *Ophiopholis aculeata* (Linnaeus, 1767) form *typica* Clark, 1911. The North Pacific form, rare in California; differs from *kennerlyi* in having smaller and more numerous supplementary pieces, somewhat longer arm spines, and the disc may be covered with spines.

Ophiopholis bakeri McClendon, 1909. More southern in distribution than *O. aculeata;* generally subtidal.

Ophioplocus esmarki Lyman, 1874. Middle and lower intertidal zone, under rocks; generally conspicuous, slow-moving; rarely autotomizes; common in Monterey area.

Ophiopteris papillosa (Lyman, 1875). Lowest tidepools, under rocks and in kelp holdfasts; uncommon intertidally.

Ophiothrix spiculata Le Conte, 1851. Middle and low intertidal pools, in

* Not in key.

algal fronds and holdfasts, under rocks on hard substrates; does not aggregate; autotomizes easily; very agile; brilliant, highly variable color patterns; very common.

REFERENCES ON OPHIUROIDEA

Austin, W. C. and M. P. Haylock 1973. British Columbia Marine Faunistic Survey Report: Ophiuroids from the northeast Pacific. Fish. Res. Bd. Canada, Tech. Rept. no. 426: 36 pp.

Boolootian, R. A. and D. Leighton 1966. A key to the species of Ophiuroidea (brittle stars) of the Santa Monica Bay and adjacent areas. Los Angeles County Museum Contr. in Science no. 93: 1–20. Covers area south of that in Light's manual, but many species found in both areas; contains errors, use with caution.

Clark, H. L. 1911. North Pacific ophiurans in the collection of the U.S. National Museum. Bull. 75, U.S. Nat. Mus., xvi + 302 pp.

Kyte, M. A. 1969. A synopsis and key to the recent Ophiuroidea of Washington State and southern British Columbia. J. Fish. Res. Bd. Canada 26: 1727–1741.

May, R. M. 1924. The ophiurans of Monterey Bay. Proc. Calif. Acad. Sci. ser. 4, 13: 261–303.

Nielsen, E. 1932. Papers from Dr. Th. Mortensen's Pacific Expedition, 1914–16. LIX. Ophiurans from the Gulf of Panama, California, and the Strait of Georgia. Vidensk. Medd. Dansk Naturh. Foren. Bd. 91: 241–346. Excellent review of chronology of discovery of new species, references for each species, synonyms, descriptions, and figures of most species.

CLASS HOLOTHUROIDEA

James C. Rutherford

University of California, Berkeley, and Bodega Marine Laboratory

(Plate 146)

Holothurians, or sea cucumbers, are soft-bodied and elongated in the mouth-anus axis. The skeleton is reduced to minute ossicles embedded in the body wall, which are often of distinctive form (figs. 11, 12). Ossicle preparations may be made by dissolving a piece of body wall in household bleach on a depression slide. The ossicles can be observed under a microscope after the bleach is flushed away with water. Some species show a clear, pentamerous arrangement of five longitudinal bands of tube feet; others have a more marked bilateral symmetry and a distinct "sole" bordered by tube feet. In still others, the tube feet are scattered over the body surface and, in some, tube feet are lacking. The mouth is surrounded by branching, contractile tentacles, which represent much modified tube feet. The madreporite is internal, opening into the coelom. The following key is based upon that by Frank P. Filice and James C. Cannan in the second edition of this manual.

KEY TO HOLOTHUROIDEA

1. Tube feet absent 2
– Tube feet present 3
2. Wormlike, delicate, white or semitransparent .... *Leptosynapta albicans*
– Not wormlike; stout body tapering to a "tail"; color in life purplish to dark brown .... *Caudina chilensis*
3. True tube feet with suckers found only on "ventral" surface; venter easily distinguished from dorsum by form and color 4
– Tube feet not confined to venter; venter not sharply different from rest of surface 6
4. Body cylindrical; large (up to 50 cm), with prominent papillae on dorsal surface; color in life mottled reddish brown .... *Stichopus californicus*
– Venter a distinct, flattened "sole"; smaller (less than 12 cm) 5
5. Dorsal surface with firm, granulate, easily visible plates; color orange; up to 12 cm long ....*Psolus chitonoides*
– Dorsum without visible plates; color usually bright red; rarely exceeds 2 cm in length ....*Lissothuria nutriens*
6. Color in life white, without spots; tube feet rigid and nonretractile, giving animal a bristly appearance; tube feet with spicules, and restricted to ambulacra; long, branched yellow tentacles ....*Eupentacta quinquesemita*
– Body color variable (white, black, orange, etc.); tube feet retractile 7
7. (*Note 3 choices*) Basic body color white or cream with small brown spots, which may be widespread or restricted to tentacular crown ....*Cucumaria piperata*
– Body color white to light gray to black 8
– Body color other (dark brown, pink, yellow, orange, etc) ... 10
8. Body color white to black, with ventral surface straw colored; tube feet scattered in interambulacral areas especially in ventral ones .... *Cucumaria lubrica*
– Body color light gray to black; tube feet absent in interambulacral areas; 10 equal-sized tentacles (except that some populations in Monterey area have 8 large and 2 small); often in considerable aggregations 9
9. Body-wall ossicles with jagged edges (fig. 11) .... *Cucumaria pseudocurata*
– Body-wall ossicles similar to above but with smooth edges (fig. 12) ....*Cucumaria curata*
10. Anterior and posterior tips reddish; usually white ventrally with dorsal, red or orange red stripe ....*Pachythyone rubra*

– Body color orange, pale yellow, pink, reddish to dark brown, and variously spotted; coloration uniform over dorsal and ventral surfaces 11

11. Body reddish to dark brown or mottled; large (up to 25 cm); tentacles orange red to purple ....*Cucumaria miniata*

– Body yellow with brown spots and tube feet absent in interambulacral areas, or body orange yellow without spots and with tube feet in interambulacral areas; 5 small anal papillae; up to about 3 cm long ....*Cucumaria fisheri*

LIST OF HOLOTHUROIDEA

Order **Aspidochirotida**

Stichopus californicus (Stimpson, 1857). Usually subtidal; contains interesting ciliates in respiratory trees; see Stevens, 1901, Proc. Calif. Acad. Sci. (3) 3: 1–42.

Order **Molpadida**

Caudina chilensis (J. Müller, 1850). A sand dweller; rare in this area.

* *Caudina* sp. See Clark, 1936.

Order **Apodida**

Leptosynapta albicans (Selenka, 1867). See Glynn, 1965.

* *Leptosynapta* spp. See Heding, 1928.

Order **Dendrochirotida.** See Pawson and Fell, 1965.

Cucumaria curata Cowles, 1907. Occurs in great numbers among corallines and mussels on exposed ledges; broods eggs under body. See Filice, 1950; Smith, 1962.

Cucumaria fisheri Wells, 1924. Low intertidal and subtidal; characteristically in kelp holdfasts.

Cucumaria lubrica Clark, 1901. Low intertidal and subtidal; broods its eggs and may form large subtidal aggregations; rare in this area, more common in Friday Harbor area.

Cucumaria miniata Brandt, 1835. Subtidal, in algal holdfasts on Monterey Peninsula; low rocky intertidal north of San Francisco.

Cucumaria piperata (Stimpson, 1864).

Cucumaria pseudocurata Deichmann, 1938. Commonly aggregated, in

* Not in key.

habitats similar to those of *C. curata* but range extends further north; *C. curata* and *C. pseudocurata* are probably conspecific. See Rutherford, 1973.

Eupentacta quinquesemita (Selenka, 1867) (=*Cucumaria chronhjelmi* Theel, 1886).

Lissothuria nutriens (Clark, 1901) (formerly *Thyonepsolus;* see Pawson, 1967). Broods eggs and young upon back; intertidal among rocks and in kelp holdfasts; common at Monterey, rare at Bodega.

Pachythyone rubra (Clark, 1901) (formerly *Thyone*). In kelp holdfasts.

* *Pentamera* sp. From tide pools at Monterey; similar to *Eupentacta;* see Deichmann, 1938.

Psolus chitonoides Clark, 1902. A northern species; has been taken subtidally at Bodega Head.

REFERENCES ON HOLOTHUROIDEA

Clark, H. L. 1901. The holothurians of the Pacific coast of North America. Zool. Anzeiger 24: 162–171.

Clark, H. L. 1924. Some holothurians from British Columbia. Can. Field Nat. 38: 54–57.

Clark, H. L. 1936. The holothurian genus *Caudina*. Ann. Mag. Nat. Hist. (ser. 10) 15: 267–284.

Deichmann, E. 1938. New holothurians from the western coast of North America, and some remarks on the genus *Caudina*. Proc. New England Zool. Club 16: 103–115.

Deichmann, E. 1941. The Holothuroidea collected by the *Velero* III during the years 1932 to 1938. Part I. Dendrochirota. Allan Hancock Pac. Exped. 8 (3): 61–195.

Filice, F. P. 1950. A study of some variations in *Cucumaria curata* (Holothuroidea). Wasmann J. Biol. 8: 39–48.

Glynn, P. W. 1965. Active movements and other aspects of the biology of *Astichopus* and *Leptosynapta* (Holothuroidea). Biol. Bull. 129: 106–127.

Heding, S. G. 1928. Synaptidae. Papers from Dr. Th. Mortensen's Pacific Exped. 1914–16, no. 46. Vidensk. Medd. Dansk Naturhist. Foren. 85: 105–323.

Pawson, D. L. 1967. The psolid holothurian genus *Lissothuria*. Proc. U.S. Nat. Mus. 122: 1–17.

Pawson, D. L. and H. B. Fell 1965. A revised classification of the dendrochirote holothurians. Breviora 214: 1–17.

Rutherford, J. C. 1973. Reproduction, growth and mortality of the holothurian *Cucumaria pseudocurata*. Marine Biol. 22: 167–176.

Smith, E. H. 1962. Studies of *Cumaria curata* Cowles, 1907. Pac. Nat. 3: 233–246.

Wells, H. W. 1924. New species of *Cucumaria* from Monterey Bay, California. Ann. Mag. Nat. Hist. (ser. 9) 14: 113–121.

PHYLUM CHORDATA: INTRODUCTION AND UROCHORDATA

Donald P. Abbott
Hopkins Marine Station of Stanford University, Pacific Grove

Chordates are characterized by possessing, at least in embryonic stages, three distinctive features: (1) a notochord, (2) pharyngeal clefts or pouches ("gill slits"), and (3) a tubular dorsal nerve cord. In addition, all chordates possess either an endostyle or its evolutionary derivative, a thyroid gland, and all manufacture the hormone thyroxin.

The best known subphylum of chordates, the **Vertebrata,** is properly beyond the scope of this manual, but intertidal pools so frequently contain fishes that a key to the commoner species is included (p. 656). The student of invertebrates is also likely to be rewarded by the sight of such marine mammals as sea lions, seals, whales, porpoises, and sea otters, as well as a variety of shore birds. The otters and many species of birds feed in good part on shore invertebrates, and some knowledge of such marine vertebrates is highly desirable for any seashore biologist.

The remaining chordates (subphyla **Urochordata** and **Cephalochordata**) are invertebrates. They were once lumped with the **Hemichordata** and spoken of as "protochordates," but it is now generally held that the hemichordates are sufficiently distinct to be recognized as a separate phylum. Cephalochordates, typified by the well-known "amphioxus," are represented on the central California coast by *Branchiostoma californiense* Andrews, 1893, which is taken subtidally by trawl or grab sampler in Monterey and Tomales bays, and has occasionally been encountered in sand in the very low intertidal (Elkhorn Slough). Urochordates (tunicates) are abundant in our intertidal zone, and receive detailed treatment below.

SUBPHYLUM UROCHORDATA (=TUNICATA)

(PLATES 147–149)

Urochordates are divided into the classes **Larvacea, Thaliacea,** and **Ascidiacea.** The first two groups are pelagic and are not encountered intertidally unless washed ashore. The Ascidiacea are sessile animals, especially abundant on the sides and undersurfaces of rocks low in the intertidal zone where they occur with hydroids, sponges, and bryozoans.

Simple (or "solitary") ascidians are ovoid, elongate, or somewhat irregular in shape. Most species are attached directly to the substratum by one side or by the base, but some are borne on a conspicuous stalk. The body lies encased in a protective outer tunic, or test, provided with a pair of apertures which are often borne on tubular extensions called siphons. When the animal is undisturbed, a current of water enters through the oral aperture (mouth) and leaves by the atrial aperture. This current brings food to the ascidian and carries off waste products. Simple ascidians reproduce only by sexual means, each egg developing into a swimming tadpole larva which settles, metamorphoses, and develops into a single adult. Development is typically pelagic, but in some species the eggs are retained and tadpoles brooded in the atrial cavity surrounding the pharynx (e.g., *Styela truncata*). Adults are usually solitary, but may occur gregariously in clusters.

Colonial ascidians, although basically similar to simple ascidians in structure and sexual reproduction, also produce offspring by budding. As a result, several to many individuals are usually found connected together in a colony or clone. The form of the colony varies greatly. In "social" ascidians, the individuals, while joined at the base, are largely distinct from one another. In "compound" ascidians, the small individuals or zooids are embedded in a continuous mass of common tunic which frequently has a characteristic growth form of its own. The separation of ascidians into simple, social, and compound types is convenient, but has no taxonomic significance. Some colonial species form colonies which are intermediate between the social and the compound types, and some simple ascidians are closely related to colonial species.

In some compound ascidians, the zooids are scattered more or less continuously throughout the common test, and both apertures of each zooid open independently to the outside. In others the zooids are arranged in recognizable clusters or "systems" within the colony. Where this occurs, the atrial apertures of all zooids in a system empty into a common cloacal cavity or pit in the test, which in turn opens to the outside through a common cloacal aperture.

The key uses external features as far as is possible, but ascidians are so variable in size, shape, and color that it is often necessary to refer to the internal anatomy to confirm identifications. Internal features are best seen in preserved individuals. Ascidians taken in the field should be allowed to relax in a bowl of sea water. When apertures are expanded, the sea water may be replaced by isotonic $MgCl_2$, or Epsom

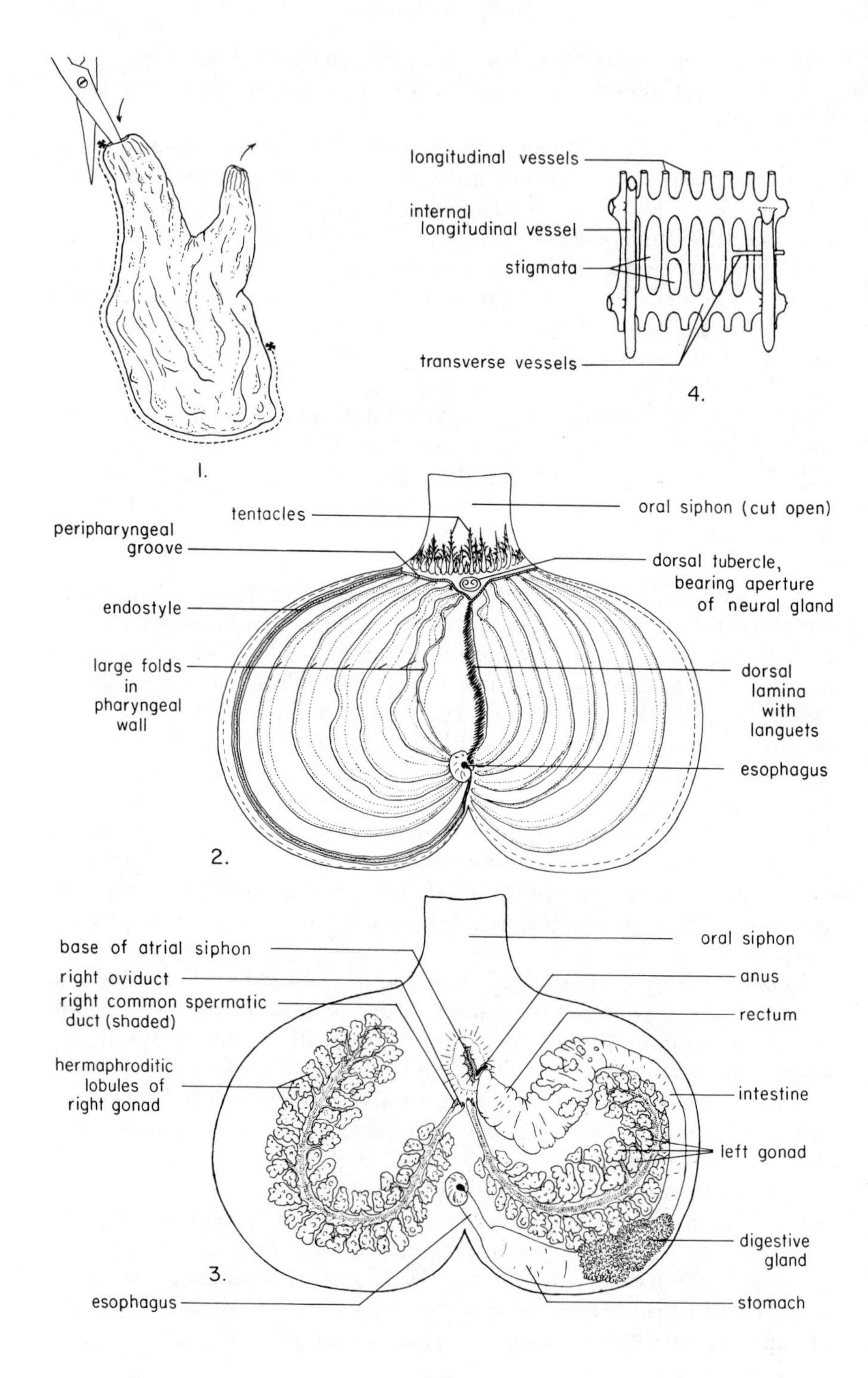

1.
longitudinal vessels
internal longitudinal vessel
stigmata
transverse vessels
4.
tentacles
oral siphon (cut open)
peripharyngeal groove
dorsal tubercle, bearing aperture of neural gland
endostyle
large folds in pharyngeal wall
dorsal lamina with languets
esophagus
2.
oral siphon
base of atrial siphon
anus
right oviduct
rectum
right common spermatic duct (shaded)
hermaphroditic lobules of right gonad
intestine
left gonad
digestive gland
3.
esophagus
stomach

salts may be added to the sea water (1–2 heaping teaspoons per liter). A few crystals of menthol added to the surface of the water may enhance narcosis. Complete narcotization takes 4–24 hours. After the animals no longer close their apertures when probed, they should be fixed for an hour or more in 10% formalin. Such treatment permits easy observation of features that are difficult or impossible to see in contracted, living or preserved specimens.

For internal anatomy, simple ascidians are conveniently dissected in the following manner (fig. 1). Insert one point of a pair of scissors into the oral siphon and cut downward along, or very slightly to the right of, the median line (for most simple ascidians the median sagittal plane is defined roughly by the positions of the apertures: a plane passing downward through the centers of the two apertures will divide the body bilaterally). Continue the cut around the base of the body, cutting through the tissues of the tunic, body wall (mantle), and pharynx, until the two attached halves can be spread apart like an opened book. Remove the body from the tunic, pin the opened animal down in a wax-bottomed pan, and cover it with water. The inner surface of the pharynx thus exposed shows many features of taxonomic importance (figs. 2, 4). In most simple ascidians, the stomach, intestinal loop, and gonads lie lateral to the pharynx (fig. 3). To expose them, cut the numerous fine tissue strands which attach the pharynx to the mantle and other organs on each side, and fold back the pharynx. In *Ciona* the gut loop and gonads lie posterior to the pharynx and can be seen without further dissection.

Colonial ascidians with large zooids may be dissected like simple ascidians. For those with small zooids completely buried in a massive common test, it is usually only necessary to remove the zooids from the tunic. Slice the colony parallel to the long axes of the zooids, observe how the zooids lie in the tunic, then remove several for study. Select well-expanded individuals in which the rows of pharyngeal stigmata are clearly visible. In most cases, zooids can be removed intact by grasping them with fine forceps by their anterior ends (oral siphons) and *gently* pulling them out of the tunic. Place them in a small dish of sea water and observe under a dissecting microscope in a good light. Body organization and anatomy are shown in figures 5–8. Colonies containing sexually mature individuals are much more eas-

---

PLATE 147. **Ascidiacea** (1). Method of dissection and anatomy of taxonomic importance in a representative simple ascidian (*Pyura haustor*): 1, method of dissection: cut along dotted line; 2, specimen cut along midventral line and spread open to show the inner surfaces of oral siphon and pharynx; 3, same view as in 2, but with pharynx removed, tentacles and dorsal tubercle not shown; 4, a small area (one "mesh") of a pharyngeal wall, much enlarged (diagrammatic).

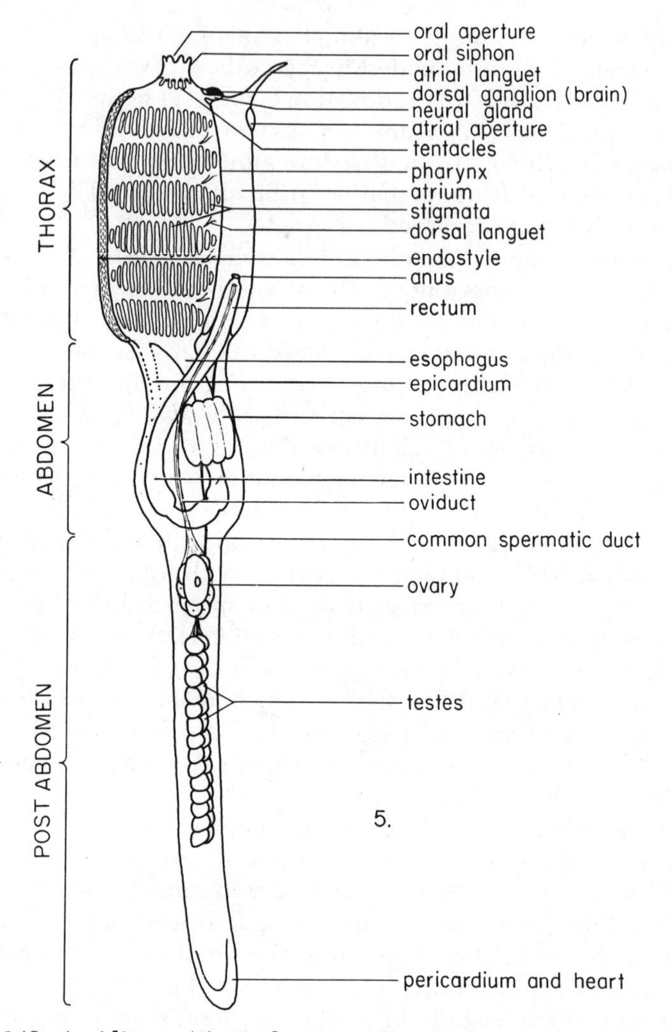

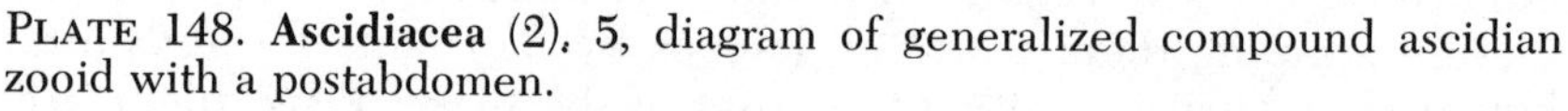

PLATE 148. **Ascidiacea** (2). 5, diagram of generalized compound ascidian zooid with a postabdomen.

ily identified than juvenile colonies or old, degenerating colonies. Sexually ripe zooids possess well-developed gonads. Most colonial ascidians retain the eggs after fertilization; developing tadpole larvae are most often brooded in the atrial cavity, and are released only when ready to swim.

KEY TO ASCIDIACEA

1. Individuals usually occur singly, unattached to others of their own kind; where occasionally found settled on top of one

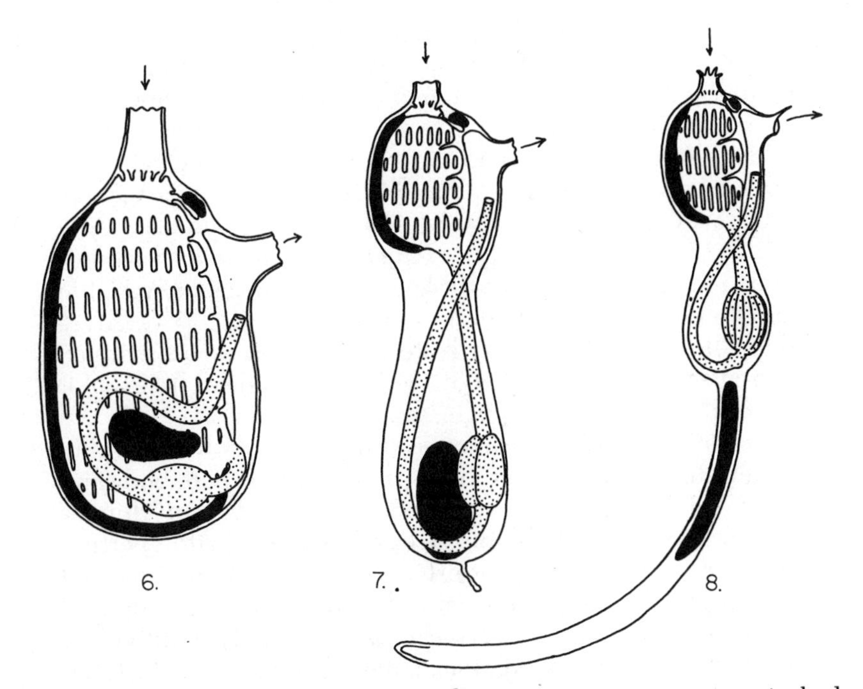

PLATE 149. **Ascidiacea** (3). Diagrams showing common variations in body form and arrangement of gut and gonads: gut stippled; brain, endostyle, and gonads solid black. 6, intestinal loop lying beside and at left of the pharynx, body not divided into regions (holosomatous), gonads on one or both sides, attached to gut or to mantle. This condition is common in simple ascidians and is found in numerous colonial species. 7, 8, zooids of colonial ascidians showing the subdivision into more or less distinct regions (merosomatous): 7, a generalized polycitorid zooid (e.g., *Clavelina, Archidistoma*) showing division of body into two regions, thorax and abdomen; 8, a generalized synoicid zooid (e.g., *Aplidium, Synoicum*) showing division of body into three regions, thorax, abdomen, and postabdomen.

another in groups (e.g., *Ascidia ceratodes* on floats or pilings), adjacent individuals are not organically joined .... simple or solitary ascidians 2

– Individuals more or less separate but always occurring in clusters or groups, their bodies interconnected basally by stolons or a sheet of tunic .... "social" colonial ascidians 18

– Individuals small, numerous, and completely embedded in a common test, which may vary in shape from a thin, flat sheet to a tall, stalked lobe ...... "compound" colonial ascidians 22

2. Test surface provided with numerous projecting spines or stiff bristles 3

– Test surface variable; smooth, wrinkled, lumpy, papillate, or with fine, soft tendrils, but without distinct spines or stiff bristles 4

3. Body with a distinct stalk; shafts of test spines either unbranched or with numerous, tiny, secondary spinelets irregularly distributed; test orange or orange brown, apertures in life reddish .... *Boltenia villosa*
– Body unstalked; shaft and tip of each spine typically encircled by regular rings of recurved, thornlike secondary spinelets; test dark brown, apertures reddish or orange .... *Halocynthia hilgendorfi igaboja*
4. Body elongate, borne on a distinct, narrowed stalk of variable length; color pale tan to dark brown 5
– Body elongate, globular, or flattened, not borne on a distinct, narrowed stalk but attached directly by the side or base; color various 6
5. Test usually lacking conspicuous tubercles, but with regular, longitudinal ridges and grooves (running along both body and stalk) which represent, respectively, thicker and thinner regions of test; oral siphon distinctly curved, with aperture directed ventrally or posteriorly; esophageal aperture located near posterior end of elongate pharynx; ovaries tubular, attached to body wall, 1–3, typically 2 on each side; individual testes separate, arranged along sides of ovaries, not clustered into compact lobes; common, outer coast and open bays .... *Styela montereyensis*
– Test usually with distinct tubercles, at least anteriorly near siphons; longitudinal folds in tunic, if present, consisting of wrinkles rather than structural features; both siphons straight or nearly so, directed anteriorly; esophageal aperture located halfway down length of elongate pharynx; ovaries tubular, attached to body wall, typically 4–9 on right side and 2–5 on the left; individual testes clustered into numerous compact lobes arranged along sides of ovaries; in protected bays *Styela clava*
6. Free (unattached) surface of body flattened into a conspicuous disc bearing the 2 apertures and a geometrical arrangement of plates on the test; apertures each surrounded by a ring of 6 triangular plates ....*Chelyosoma productum*
– Free surface of body without such a disc or plates on test 7
7. Test clear, glasslike, transparent, generally colorless but sometimes with very small, scattered, discrete flecks of pigment 8
– Test partially transparent to translucent, never glasslike, and always tinted with gray-brown, olive, or green, occasionally orange 10
– Test opaque; color red, yellow, tan, brown, or white 12
8. Test bearing large, conspicuous, well-spaced, blunt, conical,

transparent papillae over body and siphons; nearly always subtidal in California .... *Ascidia paratropa*

– Test not so, smooth or with minute papillae or processes .. 9

9. Intestine curving upward (anteriorly) after leaving stomach, and forming an S-shaped loop; pharyngeal stigmata straight, never arranged in spirals; common intertidally on pilings and on protected floats .... juvenile *Ascidia ceratodes*

– Intestine curving downward (posteriorly) after leaving stomach, and forming a wide, C-shaped loop; pharyngeal stigmata curved, arranged in spirals; nearly always subtidal in California .... *Corella willmeriana*

10. Body elongate, cylindrical, often 10–15 cm long, attached by one end, siphons close together projecting from free end; body soft, somewhat gelatinous to touch; viscera often visible through tunic; common in bays on floats and pilings .... *Ciona intestinalis*

– Body not as above; shape globular or oval and somewhat flattened; usually less than 7 cm in greatest dimension 11

11. Body globular; surface of test with numerous soft projecting tendrils (often somewhat encrusted with debris); tentacles inside base of oral siphon branched; pharyngeal stigmata curved and arranged in spirals; gonads present on both sides of body; introduced, in protected bays *Molgula manhattensis*

– Body oval, laterally compressed, attached by one side; test without projecting tendrils; tentacles inside base of oral siphon unbranched; pharyngeal stigmata straight, never in spirals; gonads confined to left side of body, associated with intestinal loop .... *Ascidia ceratodes*

12. Tunic usually bright red or rose, occasionally white, smooth and pearly in appearance; body hemispherical; at least 5 tubular gonads attached to body wall on left side (side bearing intestinal loop); uncommon intertidally in California .... *Cnemidocarpa finmarkiensis*

– Test whitish, tan, or brown; red color, if present, confined to apertures of siphons; 1–3 gonads on left side of body 13

13. Body globular, generally less than 1.5 cm in greatest diameter; test surface bearing fine, soft tendrils; entire tunic completely and evenly coated with sand, rendering it opaque; some pharyngeal stigmata distinctly curved and in some areas arranged in spirals; prominent renal organ (a clear oval sac) present on body wall on right side, just below gonad .... *Molgula verrucifera* and possibly another species of *Molgula*

– Body hemispherical to elongate; test without tendrils and not evenly encrusted with sand; pharyngeal stigmata straight, never arranged in spirals; renal organ absent 14

14. Body elongated, tubular, attached by one side; apertures borne on siphons arising from opposite ends of the body; test whitish to pale brown; rare in California ...*Pyura mirabilis*

– Not as above 15

15. Tentacles at base of oral siphon branched; pharynx with 6 or more folds on each side; free margin of dorsal lamina toothed or cleft to form small tentacles (languets); test very tough and hard, yellowish to dark brown, often partly encrusted with debris; in expanded specimens both apertures are bright red and borne on prominent siphons (fig. 1) .... *Pyura haustor*

– Tentacles at base of oral siphon unbranched; pharynx with only 4 folds on each side; free margin of dorsal lamina smooth 16

16. Test whitish to tan and bearing, at least in some areas, numerous large, rounded elevations, separated by deep grooves, which give test a "cobblestone" appearance; usually 3–9 tubular ovaries attached to body wall on right side (side opposite intestinal loop) ....*Styela plicata*

– Test usually brown, smooth or wrinkled, but lacking a "cobblestone" appearance; only 2 tubular ovaries attached to body wall on right side 17

17. Body elongate and cylindrical, up to about 60 mm long and usually 2–4 times as long as broad; tunic tough, leathery, usually conspicuously wrinkled; typically 4–6 internal longitudinal vessels on areas of pharynx between adjacent pharyngeal folds; oviparous, tadpole larvae not brooded in atrial cavity; rare intertidally, commoner subtidally in California ....*Styela gibbsi*

– Body hemispherical, globular, or elongate-oval, usually less than 30 mm long and less than twice as long as broad; tunic tough but thin, not conspicuously wrinkled; typically only 1–2 internal longitudinal vessels on areas of pharynx between adjacent pharyngeal folds; ovoviviparous, tadpole larvae brooded in atrial cavity; common intertidally in California .... *Styela truncata*

18. (*Note 3 choices*) Test bright red or orange, opaque; individual zooids rounded or ovoid, up to about 6 mm in greatest diameter, clearly demarcated from one another, or more or less laterally fused 19

– Test pale green, transparent; individual zooids rounded, 2–3 mm in diameter, resembling tiny green grapes growing in loose sheets or compact clusters; common on rocks and algae ....*Perophora annectens*

– Test colorless or tinged with olive or brown, usually transparent at least at distal end (through which colored parts of internal body may be visible), basally sometimes encrusted with sand; individual zooids elongate, usually 2.5–18 times as long as broad 20

19. Adjacent individuals distinct from one another, joined only at their bases by fine stolons or a thin sheet of test; moderately common on sides and undersurfaces of low-intertidal rocks .... *Metandrocarpa taylori*

– Adjacent individuals laterally fused, giving colony the appearance of a compound ascidian; usually on algae, generally subtidal but occasionally washed up on beaches .... *Metandrocarpa dura*

20. Test not sand-encrusted or only lightly so; 2 bright pink or orange pink bands on thoracic region, clearly visible through transparent tunic; individual zooids 2.5–5.5 times as long as broad; 16–20 rows of stigmata .... *Clavelina huntsmani*

– Test generally heavily encrusted with sand except near distal tip, occasionally free of sand entirely; thoracic regions of zooids usually visible through test, orange or gray; individual zooids tubular, 7–18 times as long as broad; 7–13 rows of stigmata 21

21. Pharynx bright orange in life; tests of individual zooids are sandy tubes up to 20 mm long and 1.5 mm in diameter; in life the colony appears as a series of well-spaced orange specks on a sandy background; 7 rows of stigmata; stomach smooth-walled; common on sandy rocks and among roots of *Phyllospadix* .... *Pycnoclavella stanleyi*

– Pharynx pale gray or colorless in life; tests of individual zooids are tubular, usually sand-encrusted, up to 50 mm long and 2–4 mm in diameter; usually 12–13 rows of stigmata; stomach wall with distinct longitudinal ridges .... *Euherdmania claviformis*

22. Colonies in the form of continuous, thick or thin, encrusting sheets, or one or more rounded mounds broadly attached to the substratum; free surface of colony smooth or irregular, sometimes with projecting bumps or ridges 23

– Colonies in the form of one to many upright or pendant lobes borne on narrowed or tapering peduncles (stalks) arising from a basal sheet or network of stolons 36

23. Zooids clearly arranged in clusters ("systems") associated with common cloacal apertures in test; systems irregularly oval, star-shaped, or consisting of elongate double rows of zooids 24

– Zooids not clearly arranged in clusters in test; systems either

absent or complex and obscure 29
24. Zoods (removed from test) with 3 body regions (fig. 8) 25
– Zooids with 2 body regions (fig. 7) 26
– Zooids with only one body region (fig. 6) 27
25. Zooids with 8–12 rows of stigmata; colonies generally pale yellowish or tan, occasionally orange brown, growing in encrusting sheets with smooth to irregular surfaces; very common .... *Aplidium californicum*
– Zooids with 13–15 rows of stigmata; colonies generally red, orange, or pink, growing in massive, thick, flat-topped slabs or cakes; fairly common .... *Aplidium solidum*
26. Atrial siphon elongate, tubular, terminating in a circular aperture whose margin is cleft into about 6 small, equal teeth; no atrial languet; pharynx with 3 rows of stigmata; test very tough and leathery, often impregnated with sand grains; free surface of colony uniformly colored; usually bearing numerous small, shallow-rimmed, craterlike depressions (common cloacal cavities) .... *Archidistoma psammion*
– Atrial siphon and aperture not so; atrial languet prominent; pharynx with 4 rows of stigmata; test soft, not impregnated with sand; free surface of colony often exhibiting 2 or more colors and lacking craterlike pits ..... *Distaplia occidentalis*
27. Systems mainly rounded, oval, or star-shaped 28
– Systems mainly elongate, sometimes branched or somewhat irregular; most zooids appear to be arranged in straight or curved, parallel rows; color highly variable .... *Botrylloides* spp.
28. Test dark purple to blackish, with anterior tips of zooids golden in life; pharynx with 4 rows of stigmata .... *Botryllus tuberatus*
– Test variously colored, usually yellow, orange, reddish, or bluish; pharynx with at least 8 rows of stigmata *Botryllus* spp.
29. Test containing numerous calcareous spicules (examine slice or torn fragment of colony with hand lens or dissecting microscope) 30
– Test lacking spicules 33
30. Spicules shaped like minute, spiny globes, very abundant in both superficial and deeper regions of test; colony opaque, whitish or gray, sometimes tinged with pink or lavender ... 31
– Spicules mainly disc-shaped, present almost entirely in deeper regions of test, especially clustered around abdominal portions of zooids; superficial regions of test free of spicules but containing many transparent bladderlike cells 32
31. Colony grayish or white, often tinged with lavender in life;

atrial siphon usually with margin flared out like a funnel, and projecting from base of thorax; 3 rows of stigmata; common .... *Trididemnum opacum*

– Colony dense white, sometimes tinged with pink in life; atrial siphon small, without flaring margin, sometimes reduced to a simple perforation in the mantle; 4 rows of stigmata; abundant .... *Didemnum carnulentum*

32. Colony flat, encrusting, up to 10 mm thick; tunic translucent and pinkish to gray-white; common .... *Cystodytes* sp.

– Colony encrusting but massive, up to 30 mm thick, with free surface of test often sculptured into low, smooth, irregularly arranged ridges or convolutions; test translucent, pinkish to gray in life .... *Cystodytes lobatus*

33. Colony delicate, thin (1–3 mm thick), gelatinous, encrusting, often overgrowing sponges and other ascidians; test transparent, colorless to olive green, speckled with small zoooids 0.5–2mm long .... *Diplosoma macdonaldi*

– Colony and test not so 34

34. Test dense opaque white, containing zooids with bright red anterior ends and apertures; living colony appears pink speckled with bright red spots; colony consisting of one or more rounded, smooth-surfaced mounds or slabs up to 25 mm thick and rather soft to touch .... *Archidistoma molle*

– Test transparent to translucent, at least in superficial layers; colony yellowish or gray, zooids yellowish to colorless; colony form variable 35

35. Test and zooids colorless; colony encrusting, forming a flat sheet or one or more low rounded mounds; zooids averaging about 3 mm long .... *Archidistoma diaphanes*

– Test colorless to pale yellow, but yellow zooids give color to the whole colony; larger colonies forming great encrusting sheets, produced at intervals into projecting lobes of variable size and shape; small colonies in midtide zone consisting of groups of small, clear knobs borne on stubby, sandy stalks .... *Archidistoma ritteri*

36. Zooids in systems (more or less regular rows or clusters, associated with common cloacal apertures), which may be clearly visible or partly obscured by encrusting sand; larger zooids always with a distinct atrial languet on one margin of the atrial aperture 37

– Zooids never in systems; both siphons of each zooid open independently at surface of colony; atrial siphon much like oral siphon, with a lobulated or toothed margin but never with an atrial languet 42

37. Colonies form clusters of shovel- or paddle-shaped lobes, each lobe consisting of a somewhat flattened head or blade attached by one edge to a stalk; functional zooids present only in the blades, and here arranged in parallel rows visible only on one flat surface of the blade; colony pale cream to light orange brown, never encrusted with sand *Distaplia smithi*
– Colonies and systems not so 38
38. Zooids lacking a postabdomen (see slice through colony cut parallel to long axes of zooids); gonad when present lying within and beside intestinal loop, and extending only slightly below bottom of loop when fully developed; 4 rows of stigmata; colony never sand-encrusted; smallest colonies club-shaped and pedunculate, larger colonies mushroom-shaped or forming low mounds; largest colonies prostrate and encrusting; free surface of colony variously colored, often showing 2 contrasting hues .... *Distaplia occidentalis*
– Zooids with a conspicuous postabdomen, usually at least as long as abdomen, containing the heart and, during the breeding season, all the gonads; 5–20 or more rows of stigmata; colony form, sand encrustation, and color variable 39
39. Colony orange, tan, or brown, with or without encrusting sand; stomach wall of zooid with a single longitudinal groove down one side, but otherwise smooth and lacking obvious longitudinal ridges and grooves 40
– Colony colorless or tinged with red or pink, always heavily encrusted with sand; stomach wall of zooid with several large, obvious longitudinal ridges separated by grooves 41
40. Typical larger colony consisting of a single, thick, flattened lobe, up to several inches across, attached at one margin by a short, stout stalk; smaller colonies forming globular, pedunculate heads; tunic not sand-encrusted except in smallest colonies; color brown to orange; zooids seldom more than 5 mm long, with their long axes always perpendicular to the surface of the colony regardless of their position (see slice through colony); common .... *Polyclinum planum*
– Typical colony consisting of a cluster of orange, club-shaped lobes up to 60 mm long, tapering toward the attached bases, and frequently somewhat sand-encrusted; zooids 10–40 mm long, with their long axes always parallel to the long axis of the colony lobe and perpendicular to the surface only at the expanded tip of the club .... *Synoicum parfustis*
41. Club-shaped lobes of colony bulbous, usually with flattened and corrugated free ends, and containing many zooids; sand grains encrusting colony surface but not usually deeply em-

bedded in test matrix; 17–21 rows of stigmata; common .... *Aplidium propinquum*

– Club-shaped lobes of colony narrow, often rounded distally, and containing only a few zooids; test heavily encrusted and impregnated with sand; 5 rows of stigmata .... *Aplidium arenatum*

42. Colony a cluster of bright scarlet or crimson lobes, borne on short peduncles; stomach wall with a single longitudinal groove down one side, otherwise covered with small tubercles or occasionally almost smooth; esophagus curves and enters stomach from one side ....*Ritterella rubra*

– Colony a cluster of lobes, either bright orange and at least partly free of sand, or heavily sand-encrusted; stomach wall with several conspicuous longitudinal ridges separated by grooves, never tuberculate; esophagus enters anterior end of stomach 43

43. Larger lobes of colony up to 15 mm wide across the top and reaching 30 mm in height; upper surfaces of lobes (and sometimes whole lobes) relatively free of sand; bright orange zooids clearly visible in transparent tunic *Ritterella pulchra*

– Larger lobes seldom more than 5–7 mm wide across the top, often less; lobes slender, tapering toward the base, usually occurring in close-packed clusters that form low mounds on rocks; lobes completely encrusted with sand, obscuring both tunic and zooids ....*Ritterella aequalisiphonis*

LIST OF ASCIDIANS

Order **Enterogona**

Suborder **Aplousobranchia**

Aplidium arenatum (Van Name, 1945) (=*Amaroucium arenatum*).

Aplidium californicum (Ritter and Forsyth, 1917) (=*Amaroucium californicum*). Very common.

Aplidium propinquum (Van Name, 1945) (=*Amaroucium propinquum*). Common.

Aplidium solidum (Ritter and Forsyth, 1917) (=*Amaroucium solidum*). Fairly common.

Archidistoma diaphanes (Ritter and Forsyth, 1917) (=*Eudistoma diaphanes*).

Archidistoma molle (Ritter, 1900) (=*Eudistoma, Distoma molle*).

Archidistoma psammion (Ritter and Forsyth, 1917) (=*Eudistoma psammion*). Common.

Archidistoma ritteri (Van Name, 1945) (=*Eudistoma ritteri*). Common; see Levine, 1962.

Clavelina huntsmani Van Name, 1931. Common; very favorable for class study.

Cystodytes lobatus (Ritter, 1900) (=*Distoma lobatus*).

Cystodytes sp.

Didemnum carnulentum Ritter and Forsyth, 1917. Abundant.

Diplosoma macdonaldi Herdman, 1886 (=*D. pizoni* Ritter and Forsyth, 1917).

Distaplia occidentalis Bancroft, 1899.

Distaplia smithi Abbott and Trason, 1968.

Euherdmania claviformis (Ritter, 1903). See Trason, 1957.

Polyclinum planum (Ritter and Forsyth, 1917) (=*Glossophorum planum*). Common.

Pycnoclavella stanleyi Berrill and Abbott, 1949. Common on sandy rocks and among roots of *Phyllospadix;* see Trason, 1963.

Ritterella aequalisiphonis (Ritter and Forsyth, 1917) (=*Amaroucium, Sigillinaria aequalisiphonis*).

Ritterella pulchra (Ritter, 1901) (=*Distoma, Sigillinaria pulchra*).

Ritterella rubra Abbott and Trason, 1968.

Synoicum parfustis (Ritter and Forsyth, 1917).

Trididemnum opacum (Ritter, 1907) (=*Didemnum opacum, T. dellavallei* Ritter and Forsyth, 1917). Common.

Suborder **Phlebobranchia**

Ascidia ceratodes (Huntsman, 1912) (=*A. californica* Ritter and Forsyth, 1917; *Phallusia ceratodes*). Common intertidally on pilings and on protected floats.

Ascidia paratropa (Huntsman, 1912) (=*Ascidiopsis paratropa*). Nearly always subtidal in California.

Chelyosoma productum Stimpson, 1864.

Ciona intestinalis (Linnaeus, 1767). Common in harbors on floats and pilings; in marine and brackish water.

Corella willmeriana Herdman, 1898. Nearly always subtidal in California; see Lambert, 1968.

Perophora annectens Ritter, 1893. Common on rocks and algae, forming pale green sheets in lower midtidal zone.

Order **Pleurogona**

Suborder **Stolidobranchia**

Boltenia villosa (Stimpson, 1864).

Botrylloides spp. May be conspicuous on harbor floats.

Botryllus tuberatus Ritter and Forsyth, 1917.

Botryllus spp. Sometimes conspicuous on harbor floats.

Cnemidocarpa finmarkiensis (Kiaer, 1893). Uncommon intertidally in California.

Halocynthia hilgendorfi igaboja Oka, 1906. Rare intertidally in California.

Metandrocarpa dura (Ritter, 1896) (=*Goodsiria dura; M. dermatina* Huntsman, 1912). Usually on algae; generally subtidal but occasionally washed up on beaches.

Metandrocarpa taylori Huntsman, 1912 (=*M. michaelseni* Ritter and Forsyth, 1917). Moderately common on low intertidal rocks. See Abbott, 1953, 1955; Haven, 1971; Newberry, 1965.

Molgula manhattensis (DeKay, 1843). Introduced from Atlantic; in brackish waters and protected bays.

Molgula verrucifera Ritter and Forsyth, 1917.

Molgula sp.

Pyura haustor (Stimpson, 1864) (=*Halocynthia johnsoni* Ritter, 1909).

Pyura mirabilis (von Drasche, 1884) (=*Herdmania, Halocynthia mirabilis*). Rare in California.

Styela clava Herdman, 1881 (=*S. barnharti* Van Name, 1945). In protected bays; see Abbott and Johnson, 1972, for review of *Styela*.

Styela gibbsi (Stimpson, 1864). Rare intertidally; commoner subtidally in California.

Styela montereyensis (Dall, 1872). Common; outer coast and open bays.

Styela plicata (Lesueur, 1823) (=*S. barnharti* Ritter and Forsyth, 1917). Introduced.

Styela truncata Ritter, 1901. Common intertidally in California.

REFERENCES ON UROCHORDATA

Abbott, D. P. 1953. Asexual reproduction in the colonial ascidian *Metandrocarpa taylori* Huntsman. Univ. Calif. Publ. Zool. 61: 1–78.

Abbott, D. P. 1955. Larval structure and activity in the ascidian *Metandrocarpa taylori*. J. Morph. 97: 569–594.

Abbott, D. P., and W. B. Trason 1968. Two new colonial ascidians from the west coast of North America. Bull. So. Calif. Acad. Sci. 67: 143–154.

Abbott, D. P., J. V. Johnson 1972. The ascidians *Styela barnharti, S. plicata, S. clava,* and *S. montereyensis* in Californian waters. Bull. So. Calif. Acad. Sci. 71: 95–105.

Berrill, N. J. 1950. *The Tunicata. With an Account of the British Species.* London: Ray Society, 354 pp.

Brien, P. 1948. Embranchement des tuniciers. Morphologie et reproduction. *In* P. Grassé, ed., *Traité de Zoologie,* Tome 11: *Echinodermés, Stomocordés, Procordés.* Paris: Masson, 1077 pp.

Eakin, R. M., and A. Kuda 1971. Ultrastructure of sensory receptors in ascidian tadpoles. Z. Zellforsch. 112: 287–312.

Fay, R. C., and J. V. Johnson 1971. Observations on the distribution and ecology of the littoral ascidians of the mainland coast of southern California. Bull. So. Calif. Acad. Sci. 70: 114–124.

Haven, N. D. 1971. Temporal patterns of sexual and asexual reproduction in the colonial ascidian *Metandrocarpa taylori* Huntsman. Biol. Bull. 140: 400–415.

Hirai, E. 1968. Tunicata. *In* M. Kumé and K. Dan, eds., *Invertebrate Embryology* (English edition), pp. 538–577. Publ. for Nat'l. Library of Medicine . . . NOLIT, Belgrade, Yugoslavia. 605 pp.

Huntsman, A. G. 1912. Holosomatous ascidians from the coast of western Canada. Contrib. Canad. Biol. 1906–1910; 103–185.

Kott, P. 1969. Antarctic Ascidiacea. Antarc. Res. Ser. 13: 239 pp. (Contains an outline of modern ascidian classification.)

Lambert, G. 1968. The general ecology and growth of a solitary ascidian, *Corella willmeriana.* Biol. Bull. 135: 296–307.

Levine, E. P. 1962. Studies on the structure, reproduction, development, and accumulation of metals in the colonial ascidian *Eudistoma ritteri* Van Name, 1945. J. Morph. 111: 105–138.

MacGinitie, G. E. 1939. The method of feeding in tunicates. Biol. Bull. 77: 443–447.

Millar, R. H. 1966. Evolution in ascidians. *In* H. Barnes, ed., *Some Contemporary Studies in Marine Science,* pp. 519–534. London: Allen and Unwin.

Millar, R. H. 1971. The biology of ascidians. Adv. Mar. Biol. 9: 1–100.

Newberry, A. T. 1965. The structure of the circulatory apparatus of the test and its role in budding in the polystyelid ascidian *Metandrocarpa taylori* Huntsman. Mem. Acad. Roy. Belg., Cl. Sci. (2) 16: 1–57.

Reverberi, G. 1971. Ascidians. *In* G. Reverberi, ed., *Experimental Embryology of Marine and Fresh-Water Invertebrates,* p. 507–550. New York: American Elsevier, 587 pp.

Ritter, W. E. 1913. The simple ascidians from the northeastern Pacific in the collection of the United States National Museum. Proc. U. S. Nat. Mus. 45: 427–505.

Ritter, W. E., and R. A. Forsyth 1917. Ascidians from the littoral zone of southern California. Univ. Calif. Publ. Zool. 16: 439–512.

Tokioka, T. 1963. The outline of Japanese ascidian fauna as compared with that of the Pacific coasts of North America. Publ. Seto Mar. Biol. Lab. 11: 131–156.

Trason, W. B. 1957. Larval structure and development of the oozooid in the ascidian *Euherdmania claviformis.* J. Morph. 100: 509–545.

Trason, W. B. 1963. The life cycle and affinities of the colonial ascidian *Pycnoclavella stanleyi.* Univ. Calif. Publ. Zool. 65: 283–326.

Van Name, W. G. 1945. The North and South American ascidians. Bull. Amer. Mus. Nat. Hist. 84: 1–476.

PHYLUM CHORDATA: INTERTIDAL FISHES

The late Rolf L. Bolin
Hopkins Marine Station of Stanford University, Pacific Grove

Revised by
Margaret G. Bradbury
California State University, San Francisco
and
Lillian J. Dempster
California Academy of Sciences, San Francisco

(PLATES 150–151)

Certain fishes are characteristic of the intertidal and of tidepools and are seldom found elsewhere, hence a key to them is useful even in a manual devoted primarily to invertebrates. Some larger fishes, even if never present at low water, may enter the intertidal zone at high tide. Still others, characteristic of offshore waters (or even of the open sea, like the giant sunfish *Mola mola*) are seen from time to time intertidally or in harbors. The following key covers fishes which may be taken along the shores with dip nets or by guddling, and includes a few rare forms from tidepools, but not the common fishes that are caught only by angling or with seines.

Terms requiring explanation are illustrated in figure 1 or explained in the following glossary. Most definitions are from Walford (1931).

GLOSSARY

adipose fin: a soft, fleshy, usually small, second dorsal fin that has no supporting rays running through it.
anal fin: the unpaired fin on the midline of the underside of the body, just behind the vent.
anal origin: the point where the anterior margin of the anal fin arises from the ventral surface of the body.
anal spine: the spine or spines at the front of the anal fin.
caudal fin: the tail fin.

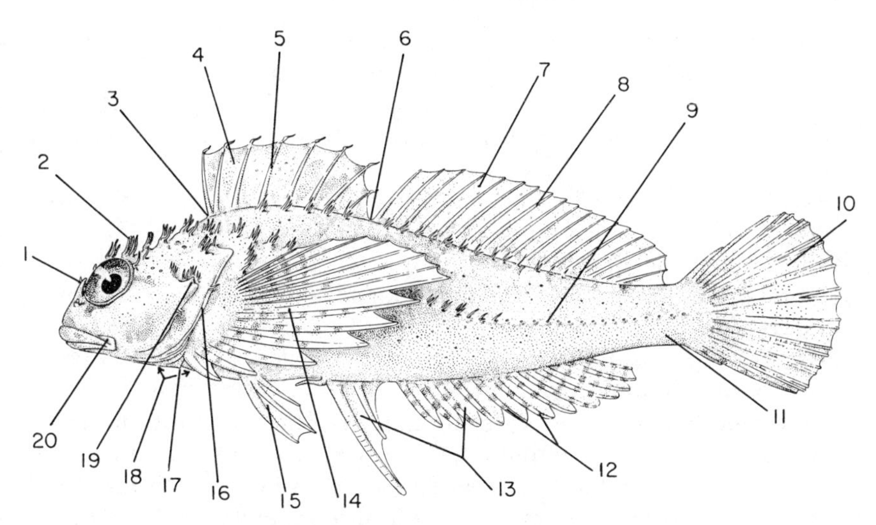

Figure 1. *Oligocottus snyderi*, after Bolin (1944)

PLATE 150. 1, *Oligocottus snyderi,* a common tidepool fish (from Bolin, 1944), labeled to show structures of systematic importance: 1, nasal spine; 2, cirri; 3, origin of first dorsal fin (dorsal origin); 4, first dorsal, or spinous dorsal fin; 5, spine; 6, junction of first and second dorsal fins; 7, second dorsal, or soft dorsal fin; 8, fin ray; 9, lateral line; 10, caudal fin; 11, caudal peduncle; 12, rays of anal fin; 13, first and second fins, often contiguous to form a single anal fin; 14, pectoral fin; 15, pelvic or ventral fin; 16, operculum or gill cover; 17, gill membrane; 18, isthmus—the region, not a specific structure; 19, preopercular spine; 20, maxillary.

caudal peduncle: that part of the body, behind the anal fin, which holds the tail fin.

ctenoid scale: a bony fish scale with the posterior margin and/or surface bearing teeth or spines.

depth: the great vertical diameter of a fish.

dorsal fin or "**dorsal:**" the unpaired fin or fins on the midline of the back.

dorsal scale band: in some fishes (e.g., some cottids), a band of scale rows extending from the nape backward along the body on either side of the dorsal fins.

first dorsal fin or "**first dorsal:**" the most anterior of the dorsal fins, where more than one is present.

gill membrane; the fleshy extension of the lower part of the gill covers or opercula. These membranes are supported by hard parts, and may be connected to each other across the isthmus or be free from it.

isthmus: the region between the gill openings ventrally on the chest and throat.

lateral line: a longitudinal line of modified scales, bearing pores, running horizontally along the side of a fish, and overlying a sensory canal; not to be confused with colored stripes.

maxillary: one of a pair of bones which form the upper part of the upper jaw; the rear ends move downward when the mouth is opened.

opercula: the somewhat movable structures composed of several bones, which cover the gill openings on either side.

orbit: the eye socket.

pectoral fins or "**pectorals:**" the more dorsal, and also usually the more anterior, of the paired fins.

pelvic fins or **ventral fins:** paired fins placed below and usually posterior to the pectoral fins.

preopercular margin: the posterior margin of the most anterior of the bones forming the operculum. The margin is often visible externally as a ridge on the opercular surface, running more or less parallel to the opercular margin but anterior to it, and often bearing one or more spines.

preopercular spine: a spine on the preopercular margin.

rays: see **soft rays.**

second dorsal fin or "**second dorsal:**" the more posterior of the dorsal fins, when two are present.

soft dorsal fin or "**soft dorsal:**" a dorsal fin supported by soft rays; usually a second dorsal fin.

soft rays or **rays:** the biramous supporting rods of a fin, which are usually branched distally and are composed of many small parts placed end to end. (Compare with **spines.**)

snout: that part of the head in front of the eyes.

spines: the stiff, usually sharply pointed rods, not composed of separate parts placed end to end, which support part or all of some fins. Also, any sharp, projecting point.

standard length: the length of the body, measured from the most anterior part of the head to the tip of the hypural. "The latter point . . . may be determined with fair accuracy in larger individuals by bending the caudal fin from side to side and noting the position of the abrupt wrinkle formed at the base of the rays." (Bolin, 1944)

suborbital stay: a projection from one of the bones forming the lower margin of the orbit. (See key, no. 15)

ventral fins or "**ventrals:**" see **pelvic fins.**

Counts of the spines and rays supporting particular fins are often given in abbreviated notations; examples are explained below. "Dorsals VIII (VII–IX)—19 (17–20)" means: there are two dorsal fins which are completely separated from one another. The heavy dash, indicating complete separation of the two fins separates the counts for the first and second dorsals (if the dorsals are connected by a membrane, the dash is replaced by a comma). The first dorsal is supported entirely by spines (in roman numerals), the number of spines averaging eight (VIII), but showing a variation of from seven (VII) to nine (IX). The second dorsal is supported entirely by soft rays (always indicated in arabic numerals), the number present averaging nineteen, but varying from seventeen to twenty in the series of specimens examined. "Anal III,6" means: the anal fin is supported anteriorly by a series of three spines, followed by a row of six soft rays.

KEY TO INTERTIDAL FISHES

1. Eyes unsymmetrical, both on same side of head 2
– Eyes symmetrical, one on each side of head 4
2. Dorsal and anal fins marked by conspicuous blackish bars; dorsal with fewer than 63 rays, anal with fewer than 47 .... *Platichthys stellatus*
– No conspicuous blackish cross bars on fins; dorsal with more than 73 rays, anal with more than 53 3
3. Eyes on right side of head (except in rare reversed individuals); anterior dorsal rays elongated and connected by membrane only at their bases .... *Psettichthys melanostictus*
– Eyes on left side of head (except in rare reversed individuals); anterior dorsal rays not elongated, connected by membrane to their tips .... *Citharichthys stigmaeus*
4. Adipose fin present behind first rayed dorsal 5
– No adipose fin 6
5. Mouth large, maxillary reaching vertical of hind margin of orbit .... *Spirinchus starksi*
– Mouth small, maxillary scarcely reaching vertical of anterior margin of pupil .... *Hypomesus pretiosus*
6. A sucking disc on belly 7
– No sucking disc on belly 12
7. Pelvic fins alone forming sucking disc 8
– Basal parts of pectoral fins together with pelvics forming the sucking disc 11
8. Sucking disc closely applied to belly, the pelvic rays appearing as short, heavy, quadrangular pads .... *Liparis florae*
– Sucking disc in form of a conical cup attached to belly only at its base; pelvic rays normal 9
9. Eyes obsolescent, under skin .... *Lethops connectens*
– Eyes well developed, fully functional 10
10. Scales rather large, about 25 along lateral line .... *Coryphopterus nicholsi*

– Scales too small to count .... *Clevelandia ios*

11. Body broad, tadpole shaped; dorsal rays about 13 .... *Gobiesox maeandricus*

– Body slender, elongate; dorsal rays 4–6 *Rimicola eigenmanni*

12. Body encased in a series of bony rings; mouth at end of long, tubular snout .... *Syngnathus californiensis*

– Body not encased in bony rings; mouth not as above 13

13. Pelvic fins present 14

– Pelvic fins absent 50

14. First dorsal composed of very short spines entirely unconnected by membrane .... *Aulorhynchus flavidus*

– No separate first dorsal composed of isolated spines 15

15. Suborbital stay present (this bony bridge may be felt by running a needle down across cheek between eye and preopercular margin) 16

– No suborbital stay developed 42

16. Sides of body clearly marked by several longitudinal lateral lines 17

– None or only a single longitudinal lateral line (transverse branches present in one species) 18

17. Scales covering suborbital stay; 2 pairs of cirri on head, the 2nd pair on nape far behind the conspicuous supraorbital flaps, minute and difficult to see .... *Hexagrammos decagrammus*

– Area over suborbital stay naked; only supraorbital pair of cirri present .... *Hexagrammos superciliosus*

18. Pelvics with 4 or 5 soft rays 19

– Pelvics with 3 soft rays 28

19. Anal with 3 spines at its anterior end 20

– Anal composed entirely of soft rays 24

20. Dorsal XVI,15; anal III,13; mouth small, maxillary reaching about to vertical from anterior edge of orbit *Oxylebius pictus*

– Dorsal XIII,13–16; anal III,6–9; mouth large, maxillary reaching to or beyond vertical from hind margin of pupil 21

21. Anal III,6; color black and yellow, tending toward vertical bars .... *Sebastes chrysomelas*

– Anal III,8–9; color uniform, no bars developed 22

22. Lower jaw much projecting, its tip continuing dorsal profile of head when mouth is closed; maxillary extending beyond posterior margin of orbit .... *Sebastes paucispinus*

– Lower jaw not markedly projecting; maxillary not reaching beyond vertical of posterior margin of orbit 23

23. Peritoneum white .... *Sebastes melanops*

– Peritoneum black .... *Sebastes mystinus*

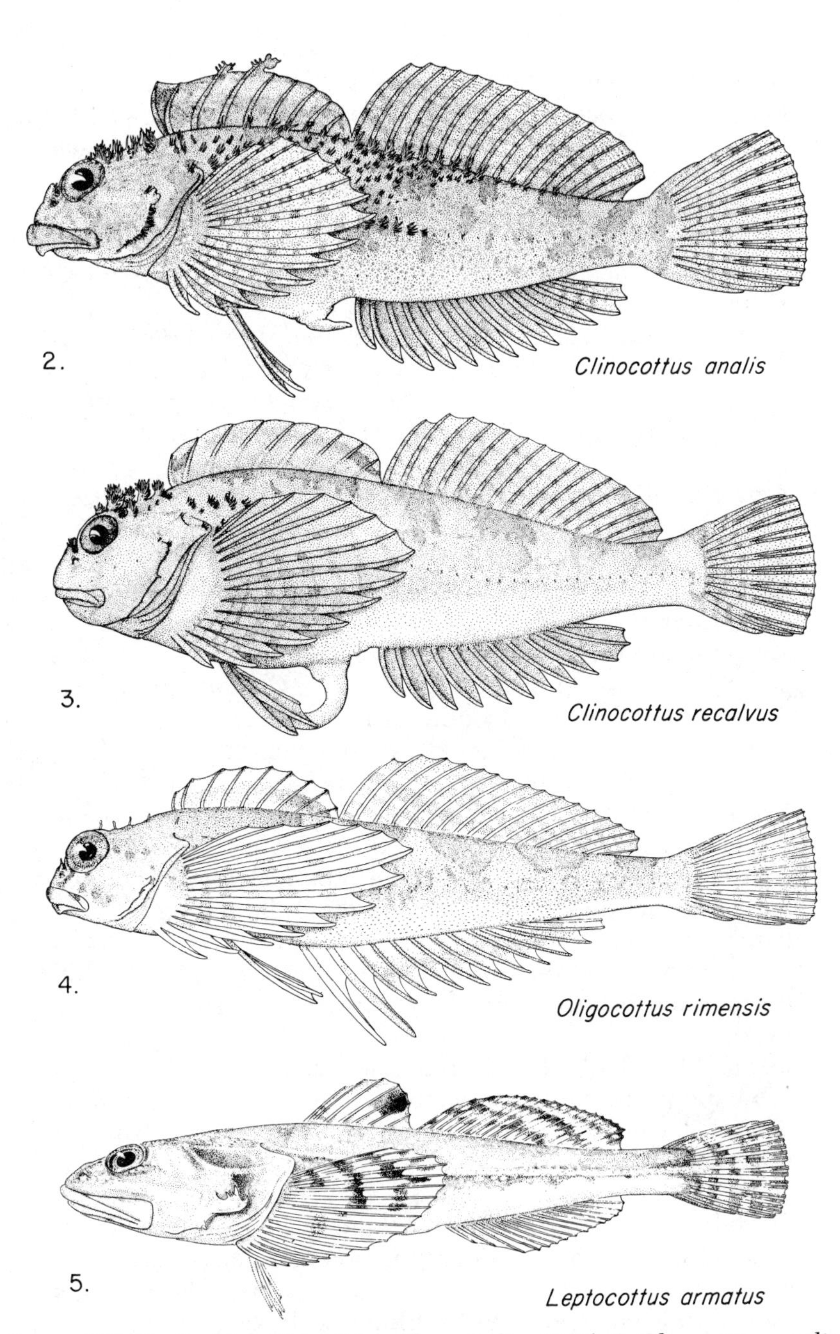

PLATE 151. **Typical intertidal fishes** (from Bolin, 1944): 2, *Clinocottus analis*, one of the commonest tidepool species, occurring even in high pools; 3, *C. recalvus*, another common tidepool species; 4, *Oligocottus rimensis*, less common in tidepools; 5, *Leptocottus armatus*, common in bays and brackish waters.

24. Body with strongly ctenoid scales 25
– Body entirely naked 27
25. First dorsal XVII–XVIII, separate from second dorsal; anal 22–24 .... *Jordania zonope*
– First dorsal X–XI, strongly joined to second dorsal; anal 18–20 26
26. Dorsal scale band of 7–8 rows ..... *Hemilepidotus spinosus*
– Dorsal scale band of 4–5 rows .... *Hemilepidotus hemilepidotus*
27. Gill membranes free from isthmus; pelvics I,5 .... *Scorpaenichthys marmoratus*
– Gill membranes joined to isthmus; pelvics I,4 (fig. 5) .... *Leptocottus armatus*
28. Area between dorsal fins and lateral line with well-developed scales in oblique or longitudinal bands or covering entire area 29
– Area between dorsal fins and lateral line naked or with minute, prickly scales not arranged in definite bands.......... 34
29. Anus much nearer pelvic base than anal origin .... *Orthonopias triacis*
– Anus not notably advanced in position 30
30. Conspicuous ctenoid scales on top of head 31
– No scales on top of head 33
31. No cirrus on upper anterior margin of orbit; dorsal scale band originating about under base of 3rd dorsal spine, separated from scales of head by a naked area or by scales so minute and scattered that they do not obscure the definite origin of the band .... *Artedius fenestralis*
– A well-developed cirrus on upper anterior margin of orbit; dorsal scale band more or less merging with squamation of head 32
32. Second dorsal 12–14; anal 10; scales extending under entire orbit and present even on snout .... *Artedius creaseri*
– Second dorsal 16–18; anal 12–14; scales extending only under posterior part of orbit, if at all; no scales on snout .... *Artedius harringtoni*
33. Dorsal scale band with 24–29 oblique scale rows, 6–11 scales in longest row; no scales behind opercular flap .... *Artedius lateralis*
– Dorsal scale band with 39–49 oblique scale rows, 10–18 scales in longest row; a few small scales just behind opercular flap between pectoral base and lateral line *Artedius corallinus*
34. Pectoral fins united ventrally, of 21–24 rays *Synchirus gilli*
– Pectoral fins entirely separate, of 12–17 rays 35

35. Anus immediately in advance of anal origin 36
- Anus in middle 1/3 of the distance between pelvic base and anal origin 39
36. Body covered with minute, prickly scales; preopercular spine simple (fig. 4) .... *Oligocottus rimensis*
- Body without visible scales; preopercular spine bifid to quadrifid except in very young 37
37. No cirri on nasal spines and none on body above lateral line .... *Oligocottus maculosus*
- A well-developed cirrus on nasal spine and tufts of cirri along base of dorsal fin 38
38. No cirri on maxillary and none on suborbital stay; preopercular spine usually bifid in adults (fig. 1) .... *Oligocottus snyderi*
- One to 4 cirri on end of maxillary and a small tuft of cirri on suborbital stay; preopercular spine usually trifid in adults .... *Oligocottus rubellio*
39. Cirri and minute, prickly scales present between dorsal fins and lateral line; preopercular spines bifid or trifid (fig. 2) .... *Clinocottus analis*
- Neither cirri nor scales between dorsal fins and lateral line; preopercular spines simple 40
40. Head moderately pointed and angular, not hemispherical; upper lip strictly terminal; a small, fleshy tubercle in median line of groove which limits upper lip dorsally; no cirri behind opercular flap between pectoral base and lateral line .... *Clinocottus embryum*
- Head very bluntly rounded, hemispherical; upper lip inferior except in juveniles; no fleshy tubercle in groove bordering upper lip; a patch of cirri behind opercular flap between pectoral base and lateral line 41
41. No cirri in anterior 1/2 of interorbital space (fig. 3) .... *Clinocottus recalvus*
- Cirri in anterior 1/2 of interorbital space in specimens more than 35 mm. in standard length (juveniles impossible to differentiate from *C. recalvus* with key) *Clinocottus globiceps*
42. Pelvics with 5 soft rays 43
- Pelvics with fewer than 5 soft rays 45
43. Dorsal and anal fins evenly rounded, without any stiff and clearly differentiated spines anteriorly *Icichthys lockingtoni*
- Dorsal clearly differentiated into 2 parts; both dorsal and anal with stiff spines anteriorly 44
44. Dorsal VIII–IX, 12–16; distance from dorsal origin to pelvic base 42–50% of standard length; distance from upper end of

pectoral base to dorsal origin 7.2–9.1% of length of dorsal base .... *Micrometrus minimus*
- Dorsal VII–IX, 16–19; distance from dorsal origin to pelvic base 36–41% of standard length; distance from upper end of pectoral base to dorsal origin 5.5–7.1% of length of dorsal base .... *Micrometrus aurora*

45. Head and body with several rows of conspicuous photophores .... *Porichthys notatus*
- No photophores developed 46

46. Pectoral girdle with a small, upturned hook on anterior margin (readily seen by lifting operculum) 47
- No upturned hook on pectoral girdle 49

47. Caudal fin forked; anal II, 31–35 .. *Heterostichus rostratus*
- Caudal truncate or rounded; anal II, 24–28 48

48. Soft dorsal with rounded profile, with 7–10 rays all evenly spaced .... *Gibbonsia metzi*
- Soft dorsal with angular profile, with 5–8 rays of which 2 or 3 in posterior of the fin are separated by markedly enlarged interspaces .... *Gibbonsia montereyensis*

49. Well-defined transverse rows of pores at right angles to main longitudinal lateral line; no dense, hairlike growth of cirri on head .... *Plagiogrammus hopkinsi*
- No transverse lines of pores as above; a dense, hairlike growth of cirri covering entire top of head .... *Chirolophis nugator*

50. Gill membranes free, gill openings extending forward ventrally; a prominent longitudinal fold of skin extending just below pectoral base and slightly above base of anal .... *Ammodytes hexapterus*
- Gill membranes joined to each other or to isthmus; no longitudinal fold of skin as above 51

51. Gill membranes attached to isthmus, gill openings restricted to sides of head; body in front of anal origin naked, behind anal origin scaled .... *Anoplarchus purpurescens*
- Gill membranes connected to each other, forming a fold across isthmus; body either completely naked or scaled 52

52. A well-developed naked spine at anterior end of anal fin (this spine, although fitting into a sheath formed in anterior end of fin, is clearly visible without dissection by forcing the membrane back with a needle and pulling spine forward) 53
- No naked spine at anterior end of anal, all skeletal elements completely covered by membrane 54

53. Anterior surface of anal spine channeled .... *Apodichthys flavidus*

– Anterior surface of anal spine convex .... *Xererpes fucorum*

54. Posterior part of dorsal fin composed of needlelike spines which may be detected by drawing a finger forward across their tips (fairly hard pressure may be required in some cases) 55

– Posterior part of dorsal composed of soft rays 57

55. Dorsal origin over pectoral fin; pectoral fin slightly larger than eye .... *Phytichthys chirus*

– Dorsal origin behind pectoral fin; pectoral fin slightly smaller than eye 56

56. Two or 3 color bands with light centers and dark margins radiating from the eye onto the head and cheek; distance from snout to dorsal origin relatively short, about 5.1 to 6.6 times standard length .... *Xiphister mucosus*

– Three dark color bands, each bordered by pale lines, radiating from the eye onto the head and cheek; distance from snout to dorsal origin relatively long, about 4.1 to 5.1 times standard length .... *Xiphister atropurpureus*

57. Body covered by readily visible scales .... *Cebidichthys violaceus*

– Body naked .... *Scytalina cerdale*

LIST OF INTERTIDAL FISHES

OSMERIDAE **Smelts**

Hypomesus pretiosus (Girard). Surf smelt.

Spirinchus starksi (Fisk). Night smelt.

BOTHIDAE **Lefteye flounders**

Citharichthys stigmaeus Jordan and Gilbert. Speckled sand-dab.

PLEURONECTIDAE **Righteye flounders**

Platichthys stellatus (Pallas). Starry flounder.

Psettichthys melanostictus Girard. Fringe sole.

STROMATEIDAE **Butterfishes**

Icichthys lockingtoni Jordan and Gilbert. Medusafish.

EMBIOTOCIDAE **Surfperches**

Micrometrus aurora (Jordon and Gilbert). Reef perch.

Micrometrus minimus (Gibbons). Dwarf perch.

SCORPAENIDAE **Scorpion fishes**

Sebastes chrysomelas (Jordan and Gilbert). Black and yellow rockfish.

Sebastes melanops Girard. Black rockfish.

Sebastes mystinus Jordan and Gilbert. Blue rockfish.

Sebastes paucispinis Ayres. Bocaccio.

HEXAGRAMMIDAE **Greenlings**

Hexagrammos decagrammus (Pallas). Kelp greenling.

Hexagrammos superciliosus (Pallas). Rock greenling.

Oxylebius pictus Gill. Painted greenling.

COTTIDAE **Sculpins**

Artedius corallinus (Hubbs). Coralline sculpin.

Artedius creaseri (Hubbs). Roughcheek sculpin.

Artedius fenestralis (Jordan and Gilbert). Padded sculpin.

Artedius harringtoni (Starks). Scalyhead sculpin.

Artedius lateralis (Girard). Smoothhead sculpin.

Clinocottus analis (Girard). Wooly sculpin.

Clinocottus embryum (Jordon and Starks). Calico sculpin.

Clinocottus globiceps (Girard). Mosshead sculpin.

Clinocottus recalvus (Greeley). Bald sculpin.

Hemilepidotus hemilepidotus (Tilesius). Red Irish lord.

Hemilepidotus spinosus Ayres. Brown Irish lord.

Jordania zonope Starks. Longfin sculpin.

Leptocottus armatus Girard. Staghorn sculpin.

Oligocottus maculosus Girard. Tidepool sculpin.

Oligocottus rimensis (Greeley). Saddleback sculpin.

Oligocottus rubellio (Greeley). Rosy sculpin.

Oligocottus snyderi Jordan and Evermann. Fluffy sculpin.

Orthonopias triacis Starks and Mann. Snubnose sculpin.

Scorpaenichthys marmoratus (Ayres). Cabezon.

Synchirus gilli Bean. Manacled sculpin.

LIPARIDAE **Snailfishes**

Liparis florae (Jordon and Starks). Tidepool snailfish.

GASTEROSTEIDAE **Sticklebacks**

Aulorhynchus flavidus Gill. Tubesnout.

SYNGNATHIDAE **Pipefishes** and **Seahorses**

Syngnathus californiensis Storer. Kelp pipefish.

GOBIIDAE **Gobies**

Clevelandia ios (Jordan and Gilbert). Arrow goby.

Coryphopterus nicholsi (Bean). Blackeye goby.

Lethops connectens Hubbs. Halfblind goby.

BATRACHOIDIDAE **Toadfishes**

Porichthys notatus Girard. Plainfin midshipman.

GOBIESOCIDAE **Clingfishes**

Gobiesox maeandricus (Girard). Northern clingfish.

Rimicola eigenmanni (Gilbert). Slender clingfish.

CLINIDAE **Clinids**

Gibbonsia metzi Hubbs. Striped kelpfish.

Gibbonsia montereyensis Hubbs. Crevice kelpfish.

Heterostichus rostratus Girard. Giant kelpfish.

STICHAEIDAE **Pricklebacks**

Anoplarchus purpurescens Gill. High cockscomb.

Cebidichthys violaceus (Girard). Monkeyface prickleback.

Chirolophis nugator (Jordan and Williams). Mosshead warbonnet.

Plagiogrammus hopkinsi Bean. Crisscross prickleback.

Phytichthys chirus (Jordan and Gilbert). Ribbon prickleback.

Xiphister atropurpureus (Kittlitz). Black prickleback.

Xiphister mucosus (Girard). Rock prickleback.

PHOLIDIDAE **Gunnels**

Apodichthys flavidus Girard. Penpoint gunnel.

Xererpes fucorum (Jordan and Gilbert). Rockweed gunnel.

SCYTALINIDAE **Graveldivers**

Scytalina cerdale Jordan and Gilbert. Graveldiver.

AMMODYTIDAE **Sandlances**

Ammodytes hexapterus Pallas. Pacific sandlance.

REFERENCES ON INTERTIDAL FISHES

Baxter, J. L. 1966. *Inshore Fishes of California.* Sacramento: Calif. Dept. Fish and Game, 80 pp.

Bolin, R. L. 1944. A Review of the Marine Cottid Fishes of California. Stanford Ichthyol. Bull. 3: 1–135.

Clemens, W. A. and G. V. Wilby 1961. Fishes of the Pacific Coast of Canada. 2nd ed. Bull. Fish. Res. Bd. Canada 68: 443 pp.

Fitch, J. E., and R. J. Lavenberg 1971. Marine Food and Game Fishes of California. California Natural History Guides no. 28. University of California Press, 179 pp.

Herald, E. S. 1972. *Fishes of North America.* Doubleday and Co., 255 pp.

Hubbs, C. 1952. A Contribution to the Classification of the Blennioid Fishes of the Family Clinidae with a Partial Revision of the Eastern Pacific Forms. Stanford Ichthyol. Bull. 4: 41–165, illus.

Macdonald, C. K. 1972. A key to the fishes of the family Gobiidae (Teleostomi) of California. Bull. So. Calif. Acad. Sci. 71: 108–112.

Miller, D. J., D. Gotshall, and R. Nitsos 1965. *A Field Guide to Some Common Ocean Sport Fishes of California.* 2nd rev. Sacramento: Calif. Dept. Fish and Game, 87 pp.

Miller, D. J. and R. N. Lea 1972. Guide to the coastal marine fishes of California. Calif. Dept. Fish and Game, Fish Bull. 157: 235 pp.

Phillips, J. B. 1957. A review of the rockfishes of California. (Family Scorpaenidae). Calif. Dept. Fish and Game, Fish Bull. 104: 158 pp.

Roedel, P. M. 1953. Common Ocean Fishes of the California Coast. Calif. Dept. Fish and Game, Fish Bull. 91: 184 pp.

Walford, L. A. 1931. Handbook of Common Commercial and Game Fishes of California. Calif. Dept. of Fish and Game, Fish Bull. 28: 181 pp.

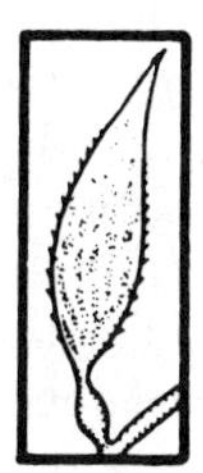

INTERTIDAL PLANTS

Isabella A. Abbott
Hopkins Marine Station of Stanford University, Pacific Grove

No zoologist can study the intertidal zone profitably without an acquaintance with the common and conspicuous plants that so plainly mark the zones of the shore. Most of these are algae—red, green, or brown—but a few flowering plants also occur in the marine habitat. Chief among these are the seagrasses (Zosteraceae). The broad-leaved eelgrass *Zostera marina* grows rooted in muddy bottoms of bays and estuaries. Two species of surfgrass, *Phyllospadix*, occur in the rocky intertidal zone of the outer coast. *P. scouleri* on rocks and ledges is a useful indicator of the "zero" tidal level.

KEY TO THE SEAGRASSES (ZOSTERACEAE)

1. Rooted in soft bottom in protected bays; leaves 6–12 mm broad ....*Zostera marina* Linnaeus
– On exposed rocks of low intertidal, or in tidepools and channels of rocky shores; leaves relatively narrow (1–4 mm broad) .... *Phyllospadix* 2
2. Leaves narrow (1–2 mm), firm, relatively thick, eliptical in cross section; blades long (up to 3 m); inflorescence multiple on elongate peduncle; rhizomes commonly buried in sand ....*Pyllospadix torreyi* Watson
– Leaves broader (2–4 mm), thinner and flatter than above; blades shorter (up to 1 m); inflorescence solitary, nearly basal; rhizomes commonly exposed ....*Phyllospadix scouleri* Hooker

Intertidal algae are extremely useful indicators of intertidal zonation, and a knowledge of the commonest larger forms is indispensable in ecological studies. An excellent discussion of intertidal zonation by J. W. Hedgpeth is found in chapter 14 in the fourth (1968) edition of *Between Pacific Tides*.

The algal flora of the Pacific Coast of North America is one of the richest in the world, rivaled only by those of the coasts of Japan, Australia, New Zealand, and South Africa. For the Monterey Peninsula alone, Smith (1969) lists 440 species, representing an estimated 80 percent of the species known for the coast. Of this bewildering variety, only those forms most abundant, commonly noticed, strikingly different, or ecologically important have been selected for the following key, which is based on external morphological characters and attempts to use only obvious features and to avoid specialized terms. This leaves much to be desired for critical determinations (the names of certain genera likely to be confused with those in the key are occasionally included in parentheses). As a check on identifications, reference may be made to the excellent figures and keys in G. M. Smith, *Marine Algae of the Monterey Peninsula, California* or Abbott and Hollenberg (1974). Our figures are reproduced from Dr. Smith's monograph, and do not duplicate those reprinted in *Between Pacific Tides*. Positive identification of many red algae requires detailed microscopic examination of sections by a specialist.

KEY TO GENERA OF CONSPICUOUS INTERTIDAL ALGAE

(PLATES 152–156)

1. Plants bright grass green or dark spinach green 2
– Plants olive-green, tan, brown, red, purple, or pink 6
2. Forming clumps of free, smooth, expanded sheets of varying width and height, with ruffled margins (figs. 1, 2); occurring in upper midtidal zone or in bays; "sea lettuce" .... *Ulva*
– Form not as above 3
3. Plants forming groups of convoluted or straight, soft tubes (figs. 3, 4); especially common in estuarine and brackish-water habitats .... *Enteromorpha*
– Plants growing in tufts or appressed to rock surface; not characteristic of brackish water 4
4. In low, compact, grass green, pincushionlike tufts, occurring well up in intertidal zone .... *Cladophora*
– Not as above; surface of thallus feltlike; very dark green .... *Codium* 5
5. Thallus prostrate, adhering closely to rock surfaces (fig. 6); frequently so dark green as to appear almost black. .... *Codium setchellii*
– Growing in erect tufts; branches slender, cylindrical, and repeatedly branched (fig. 5) .... *Codium fragile*
6. Thallus prostrate, encrusting the substrate 7
– Not prostrate or encrusting 10

7. Plants conspicuously and heavily calcified to form a brittle crust; usually in pastel shades of pink or purplish (encrusting coralline algae) .... *Lithothamnium*
   (see also *Melobesia*, *Fosliella*, and *Lithophyllum*)
– Plants not so calcified; color darker 8
8. Exposed surface bearing concentric rings or ridges; yellow brown to brownish black .... *Ralfsia*
– Exposed surface without concentric rings or ridges; surface smooth or convoluted 9
9. On rock and on shells of turban snails (*Tegula*); dark purplish red to black .... *Peyssonellia*
– Always on rocks; very dark red to reddish black; surface leathery to waxy; never convoluted unless on a convoluted substrate .... *Petrocelis* or *Hildenbrandia*
– Always on rocks; light to dark brown, never red; surface smooth; convoluted in irregular ridges and folds (fig. 20) .... *Cylindrocarpus*
10. Plants conspicuously calcified, branched and stiffly jointed; usually reddish pink, pink, lavender, or chalky white; plants generally less than 30 cm (non-encrusting coralline algae) 11
– Plants not calcified; color and size variable 13
11. Plants slender, stiff, with segments of most smaller branches more or less cylindrical (fig. 23); conceptacles (appearing as small, hemispherical bumps) only terminal to segment of the branchlets in fertile plants, and only 1 conceptacle to a segment .... *Corallina*
– Plants thicker, heavier, more robust, with segments of most smaller branches not cylindrical but conspicuously flattened and flared laterally into small "wings;" conceptacles not confined to tips of branchlets, and frequently more than 1 to a segment (figs. 21, 22, 24) 12
12. Conceptacles occurring only on the flattened faces of segments (figs. 21, 22) .... *Bossiella*
– Conceptacles occurring both on the margins of lateral wings and on the flattened faces of segments (fig. 24) .... *Calliarthron*
13. Plants tan, yellow-brown, or brownish, sometimes slightly tinged with green (as in *Fucus*) but predominant color always brown or tan; size variable, up to about 40 m long (large kelps and other brown algae) 14
– Plants olive green (with purplish bases), or red, pinkish, or purplish (including very dark purple); plants always less than 1 m long 26
14. Plant with an unbranched, whiplike stipe up to about 40 m long, terminating in hollow, bulbous float bearing elongate

PLATE 152. **Green and Brown Algae.** 1, *Ulva taeniata* (x 1/3); 2, *U. lobata* (x 1/3); 3, *Enteromorpha intestinalis* (x 1/3); 4, *E. compressa* (x 2/3); 5, *Codium fragile* (x 1/3); 6, *C. setchellii* (x 1/3); 7–9, *Macrocystis integrifolia;* 7, holdfast (x 1/3); 8, entire plant (x 1/12); 9, upper part of a branch (x 1/3) (from G. M. Smith, *Marine Algae of the Monterey Peninsula*, 1969; used with permission of Stanford University Press).

blades distally (figs. 10, 11); commonly in offshore kelp beds or washed ashore; occasional intertidal individuals are much smaller .... *Nereocystis*
– Plant otherwise; habitat variable 15
15. Plant resembling small palm tree; with erect, trunklike stipe bearing terminal cluster of blades; on exposed, surf-beaten rocks .... *Postelsia*
– Plant not as above; habitat variable 16
16. Plant consisting of numerous stout straps up to several meters long anchored by common holdfast; hundreds of small blades borne along sides of each strap, some blades modified as small floats; very common in rocky intertidal ...... *Egregia*
– Plant otherwise 17
17. Whole plant regularly dichotomously branched 18
– Whole plant not dichotomously branched; size variable, but up to 25 m or more in length; restricted to lower intertidal (+1 ft and below) and kelp beds 21
18. Stipe massive and spreading, repeatedly branched; blades narrowly linear, 6–12 mm broad, completely smooth, 100–500 per plant; plants exposed to full force of surf (fig. 14) ....*Lessoniopsis*
– Stipe conical and discrete; blades 1–2 cm broad, frequently with air bladders in upper portions; on rocks 2–6 ft above zero tide level, protected from full surf 19
19. Branches bladelike, conspicuously thin and flattened, and with distinct midrib .... *Fucus*
– Branches cylindrical or somewhat flattened but fleshy and thick; no midrib 20
20. Plants usually 4–8 cm tall (occasionally up to 15 cm); lower branches distinctly flattened but rather thick and fleshy ... . *Pelvetiopsis*
– Plants 15–40 cm tall; lower branches slender and cylindrical or slightly flattened .... *Pelvetia*
21. With a single, distinct, elongate, swollen float at base of each blade; blades large, corrugated, and with toothed margins (figs. 7–9); plants found in low intertidal and kelp beds; up to 20 m long .... *Macrocystis*
– Floats otherwise or absent; blades lacking regularly toothed margins; habitat variable 22
22. Terminal branches of thallus slender, many bearing linear series of rounded, close-set floats (each series resembling row of peas in pod; lost in winter); plants up to 8 m long; in low intertidal and kelp beds .... *Cystoseira*
– No floats of any sort; plants not in kelp beds 23

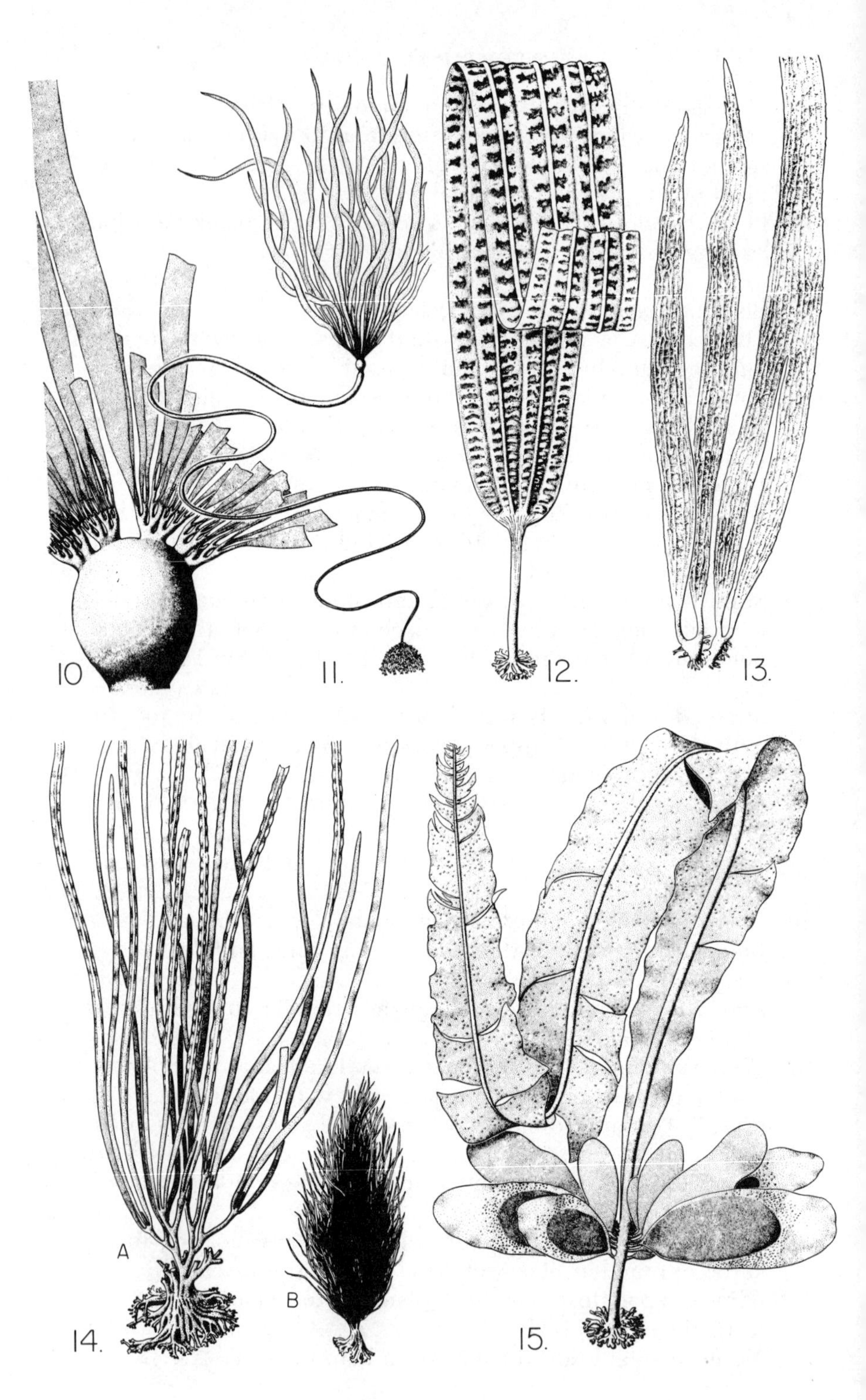

10
11.
12.
13.
A
B
14.
15.

23. Axis (arising from a holdfast) bearing lateral blades or branches 24
– Axis or axes arising from a holdfast or prostrate rhizome, never bearing lateral branches or blades .... *Laminaria* (see also *Costaria* and *Dictyoneurum*, figs. 12, 13)
24. Axis terminating in, or bearing distally, a single very prominent blade, larger than any of lateral blades 25
– Axis not terminating in single blade conspicuously larger than lateral blades (figs. 16–18); with sharp, distinctive odor (liberates strong acid when detached) .... *Desmarestia*
25. Terminal blade with midrib (fig. 15) .... *Alaria*
– Terminal blade lacking midrib (fig. 19) ...... *Pterygophora*
26. Plants consisting of elongate, unbranched, hollow sacs or bladders, up to 25 cm long; olivaceous to purplish in color; in midtide zone .... *Halosaccion*
– Plants not so formed 27
27. Plants with a short, stout, erect stipe centrally perforating one or more disc-shaped or circular, concave blades, the whole resembling a red mushroom or shallow red goblet (fig. 25); low intertidal .... *Constantinea*
– Plants not so shaped, generally bladelike or bushy 28
28. Plant form variable, but blades (if any) or branches of the thallus always bearing many elongate papillae or conical, pointed spines 29
– Plant form variable, but never bearing either elongate papillae or conical, pointed spines; fertile female plants in some may form small, low, less-than-hemispherical bumps on blades or branches 30
29. Plants 4–6 cm high; erect, profusely branched; all branches cylindrical and bearing conical, pointed spines (fig. 26); dark red to nearly blackish; occurring on rocks 3–6 ft above zero tide level .... *Endocladia*
– Plants larger, up to 1 m long; thallus of broad blades or narrow, compressed branches, bearing elongate papillae—in some species so thickly grouped as to suggest a turkish towel (fig. 29–33); usually found below the + 3.5 ft tidal level; color variable, usually some shade of red or purple tinged with green or brown; several species .... *Gigartina*

PLATE 153. **Brown Algae.** 10, 11, *Nereocystis luetkeana:* 10, float and bases of blades (x 1/40); 11, entire plant (x 1/60); 12, *Costaria turneri* (x 1/6); 13, *Dictyoneurum californicum* (x 1/6); 14, *Lessoniopsis littoralis:* A, plant with most of blades removed (x 1/8); B, entire plant (x 1/24); 15, *Alaria marginata* (x 1/4) (from G. M. Smith, *Marine Algae of the Monterey Peninsula*, 1969; used with permission of Stanford University Press).

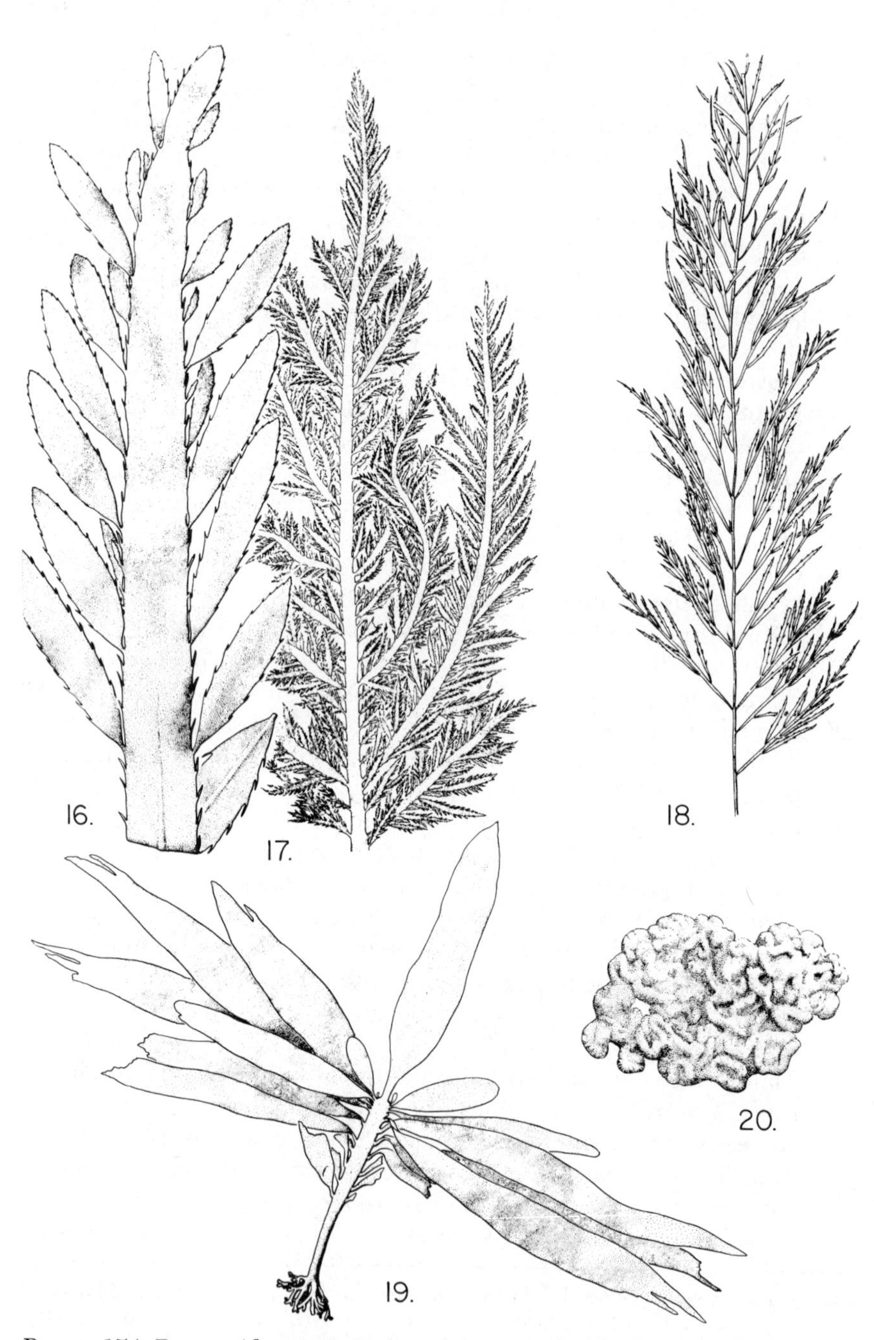

PLATE 154. **Brown Algae.** 16, *Desmarestia munda* (x 2/9); 17, *D. herbacea* (x 1/3); 18, *D. latifrons* (x 2/9); 19, *Pterygophora californica* (x 1/2); 20, *Cylindrocarpus rugosus* (x 1/9) (from G. M. Smith, *Marine Algae of the Monterey Peninsula*, 1969; used with permission of Stanford University Press).

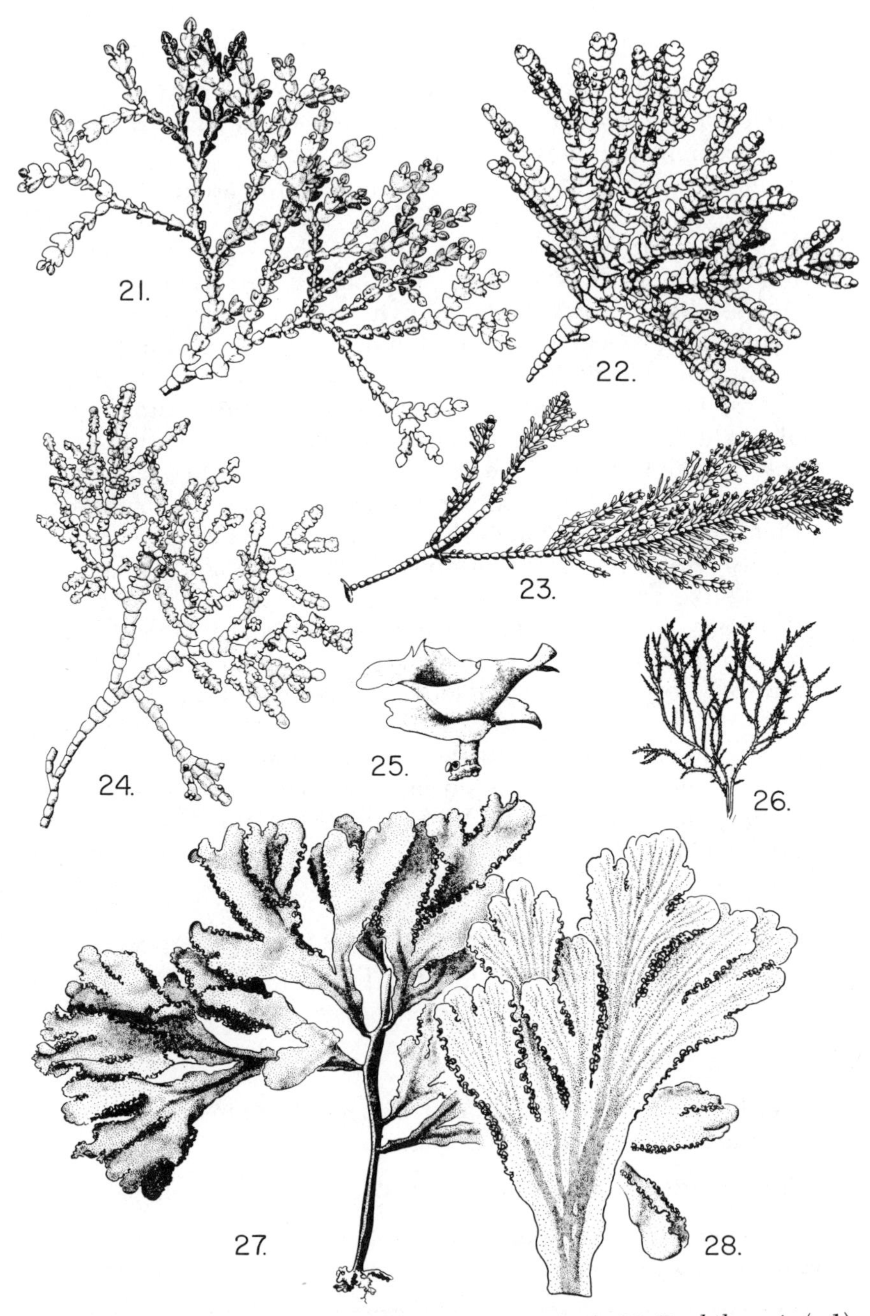

PLATE 155. **Red Algae.** 21, *Bossiella orbigniana* (x 1); 22, *B. chiloensis* (x 1); 23, *Corallina officinalis* var. *chilensis* (x 2/3); 24, *Calliarthron tuberculosum* (x 2/3); 25, *Constantinea simplex* (x 1/2); 26, *Endocladia muricata* (x 1 1/3); 27, 28, *Botryoglossum farlowianum:* 27, entire plant (x 1/3); 28, segment of blade (x 2/3) (from G. M. Smith, *Marine Algae of the Monterey Peninsula*, 1969; used with permission of Stanford University Press).

30. Thallus "leafy," consisting of broad, simple or partly dissected or subdivided blades; blades always bearing patterns of conspicuous ridges or "veins" 31
- Thallus variable; blades or branches never bearing ridges or veins 32

31. Blades simple, without frilled margins, but with anastomosing network of veins or ridges (fig. 35) .... *Polyneura*
- Blades distally dissected and with margins narrowly and tightly frilled; veins not anastomosing, but branching and fanning out distally (figs. 27, 28) .... *Botryoglossum*

32. Plants forming clumps of paper-thin, smooth, broad or elongate sheets with ruffled margins (fig. 34); in form (but not color) much like *Ulva;* species occur on rocks up to 4 ft above zero-tide level, or as epiphytes on larger plants ... *Porphyra*
- Plants otherwise, not like *Ulva* 33

33. Plants rarely branched, forming clumps of tough, leathery, paddlelike or lanceolate blades, each arising from a stubby stipe (fig. 36); olive green to purplish and often iridescent; blades of fertile plants bearing numerous small bumps (cystocarps); very common in middle and low intertidal zones .. .. *Iridaea*
- Plants much branched; blades either absent or narrow and lanceolate, borne laterally on the main axis or on branches; margins of branches or blades frequently bearing tiny bladelets; thallus surface smooth .... *Prionitis*

LIST OF COMMONER CENTRAL CALIFORNIA INTERTIDAL ALGAE

(Not all taken to species in the key)

Chlorophyta (Green algae)

Cladophora columbiana Collins (=*C. trichotoma* in Smith). Common in high intertidal as small, cushionlike tufts.

Codium fragile (Suringar) Hariot. Dark green, branching tufts (fig. 5).

Codium setchellii Gardner. Dark emerald green, velvetlike patches, closely adherent to rock (fig. 6).

PLATE 156. **Red Algae.** 29, *Gigartina californica* (x 1/3); 30, *G. agardhii* (x 1/3); 31, *G. papillata* (x 1/3); 32, *G. canaliculata* (x 1/2); 33, *G. harveyana* (x 1/4); 34, *Porphyra perforata* (x 1/3); 35, *Polyneura latissima* (x 1/2); 36, *Iridaea flaccida* (x 1/4) (from G. M. Smith, *Marine Algae of the Monterey Peninsula*, 1969; used with permission of Stanford University Press).

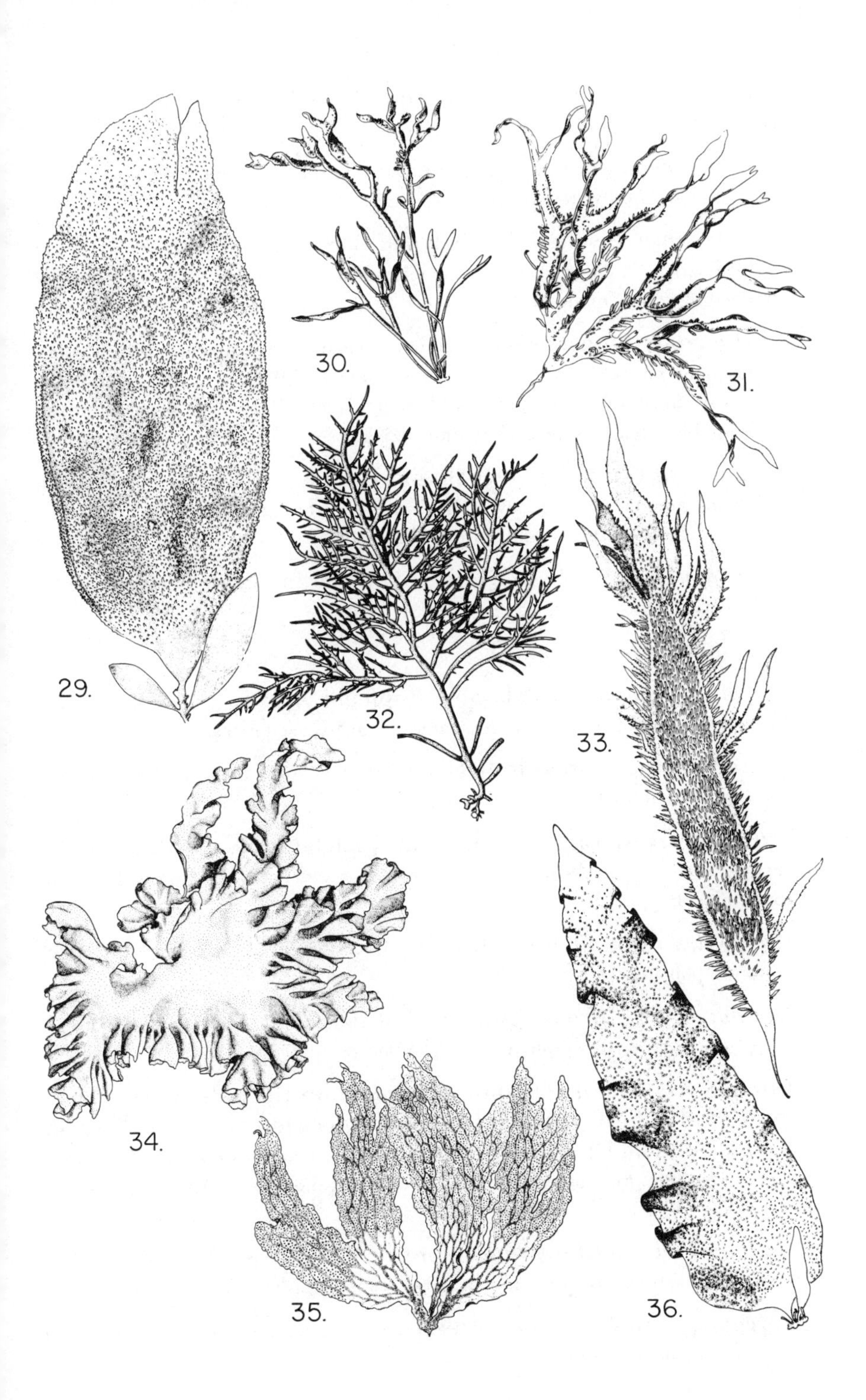
29.
30.
31.
32.
33.
34.
35.
36.

Enteromorpha compressa (Linnaeus) Greville. Often floating in brackish bays, or stranded on flats (fig. 4).

Enteromorpha intestinalis (Linnaeus) Link. Frequently in brackish habitats and below freshwater seeps (fig. 3).

Ulva expansa (Setchell) Setchell and Gardner. Often very large (up to 1 m in diameter); pale green; common in sheltered bays and harbors, where it may be stranded in masses on mudflats.

Ulva lactuca Linnaeus. Although often stated to be common, is actually rare; blades more transparent than *U. expansa* or *U. lobata;* "When . . . spread out in the hand one's fingers are plainly visible through blades of *U. lactuca* but not through blades of the other two species" (Smith, 1969).

Ulva lobata (Kützing) Setchell and Gardner. Dark green, lobed (fig. 2); frequently produces reproductive cells during lowest tidal period of month. Up to a foot long; characteristic of pools in rocky intertidal.

Ulva taeniata (Setchell) Setchell and Gardner. Usually elongate, with curly margins (fig. 1).

Phaeophyta (Brown algae)

Alaria marginata Postels and Ruprecht. Very long, smooth, straplike blades with central rib (fig. 15); low intertidal on exposed rocks.

Costaria turneri Greville (=*C. costata*). With 5 conspicuous longitudinal ribs (fig. 12); subtidal.

Cylindrocarpus rugosus (Okamura) Setchell and Gardner (=*Petrospongium rugosum*). Small, discrete, elevated patches on higher intertidal rocks (fig. 20).

Cystoseira osmundacea (Menzies) C. A. Agardh. A favorite habitat of spirorbid polychaetes.

Desmarestia latifrons (Ruprecht) Kützing. The narrow, shiny fronds frequently carry small, epiphytic red algae (fig. 18).

Desmarestia ligulata (Lightfoot) Lamouroux (includes the common low intertidal somewhat feathery *D. herbacea* and the broader bladed subtidal *D. munda*). Sulfuric acid is produced by decaying *Desmarestia* species; keep from other algae, and avoid using them for packing animal specimens (figs. 16, 17).

Dictyoneurum californicum Ruprecht. With a creeping basal portion; characteristic of surfy areas; bases harbor rich assortment of invertebrates (fig. 13).

Egregia menziesii (Turner) Areschoug. The long, flat stipes are the typical home of *Notoacmea insessa.*

Fucus distichus Linnaeus. "Rockweed" or "Bladder Wrack;" in mid-intertidal beds usually below *Pelvetia.*

Laminaria dentigera Kjellman (formerly widely known as *L. andersonii;* also as *L. setchellii*). Very common; a marker (indicator) of −1.0 ft tide level. Stipes are the typical habitat of *Collisella instabilis.*

Laminaria farlowii Setchell. Very rough, large blades.

Laminaria sinclairii (Harvey) Farlow. With creeping rhizomes and slender, very mucilaginous blades and stipe.

Lessoniopsis littoralis (Farlow and Setchell) Reinke. A perennial with tough, woody stipe and hundreds of blades; characteristic of very exposed, surf-beaten low intertidal (fig. 14).

Macrocystis integrifolia Bory. Has a creeping holdfast; characteristic of shallow water and tidal channels, 0.5 to –0.5 ft level; blades and floats often bear *Membranipora* and *Obelia* colonies (figs. 7–9).

Macrocystis pyrifera (Linnaeus) C. A. Agardh. Has a conical holdfast; characteristically attached in 6–80 m of water, forming extensive floating beds; supports a rich invertebrate epifauna.

Nereocystis luetkeana (Mertens) Postels and Ruprecht. This giant "Bull Kelp" forms extensive floating beds in 6–14 m of water; small individuals occur in intertidal pools (figs. 10, 11).

Pelvetia fastigiata (J. G. Agardh) De Toni. Forms conspicuous mid-intertidal beds.

Pelvetiopsis limitata (Setchell) Gardner. Upper midtidal, on exposed vertical surfaces.

Postelsia palmaeformis Ruprecht. The "Sea Palm;" characteristic of midtidal rocks exposed to direct surf.

Pterygophora californica Ruprecht. Perennial in subtidal beds, with extremely woody stipe showing growth rings; drift specimens are frequently riddled by boring isopods (fig. 19).

Ralfsia pacifica Hollenberg. The commonest olive brown crust; with discrete margins; at all tidal levels.

Rhodophyta (Red algae)

Bossiella chiloensis (Decaisne) Silva (=*Bossea corymbifera*). Fig. 22.

Bossiella orbigniana (Decaisne) Silva (=*Bossea orbigniana*). Fig. 21.

Bossiella plumosa (Manza) Silva.

Botryoglossum farlowianum (J. G. Agardh) De Toni. Forms large patches at low intertidal levels (figs. 27, 28).

Calliarthron cheilosporioides Manza.

Calliarthron tuberculosum (Postels and Ruprecht) Dawson (=*C. setchelliae*). The commonest erect coralline of low intertidal in central California (fig. 24).

* *Callophyllis violacea* J. G. Agardh.

Constantinea simplex Setchell. Common north of Dillon Beach (fig. 25).

Corallina officinalis var. *chilensis* (Harvey) Kützing (=*C. chilensis*). The commonest erect coralline in tidepools at every level (fig. 23).

Corallina vancouveriensis Yendo. Common on high rocks, exposed to surf.

Endocladia muricata (Postels and Ruprecht) J. G. Agardh. Frequently the highest macroscopic alga found; very resistant to drying (fig. 26). For an account of the diverse animal community of the *Endocladia* zone, see Glynn, 1965, Beaufortia 12: 1–198.

Gigartina agardhii Setchell and Gardner. Usually below *Endocladia* zone (fig. 30).

Gigartina californica J. G. Agardh. Large simple blades with close-set papillae (fig. 29).

Gigartina canaliculata Harvey. Mostly smooth branches growing in grass-like clumps (fig. 32).

Gigartina corymbifera (Kützing) J. G. Agardh. Common at −1.0 ft level and subtidally; blades broader at apex than near base.

Gigartina harveyana (Kützing) Setchell and Gardner. More irregularly dissected and with softer spines than *G. californica* (fig. 33).

Gigartina papillata (C. A. Agardh) J. G. Agardh. The most variable of *Gigartina* species (fig. 31); high to mid-intertidal, below *Endocladia*.

* *Gracilaria verrucosa* (Hudson) Papenfuss. Common, half-buried in coarse sand.

Halosaccion glandiforme (Gmelin) Ruprecht. Groups of hollow, finger-like sacs, containing fluid, in which may be found the copepod *Diarthrodes cystoecus*.

Hildenbrandia occidentalis Setchell. Forms dark red to dark brown, thin, continuous crusts on rocks; common everywhere.

* Not in key.

Iridaea cordata var. *cordata* (Turner) Bory (=*Iridophycus cordata*). Dark violet to purple, low intertidal to subtidal, 15 ft. of water.

Iridaea cordata var. *splendens* (Setchell and Gardner) Abbott (=*Iridophycus splendens*). Dark purple, clustered, with stipes if on exposed rocks; without stipes and larger blades if lining channels.

Iridaea flaccida (Setchell and Gardner) Silva (=*Iridophycus flaccida*). Forming conspicuous light green band between + 1.5 and + 3.0 ft tide level; plants usually purplish and iridescent near bases, but yellowish green throughout most of length (fig. 36).

Iridaea heterocarpa Postels and Ruprecht (=*Iridophycus heterocarpa*).

Lithothamnium californicum Foslie. An encrusting, dark violet, stony alga in low intertidal, especially on vertical rock surfaces.

Lithothamnium pacificum Foslie. An encrusting, rose to violet, stony alga especially on low rocks.

Petrocelis franciscana Setchell and Gardner. Reddish brown crust with waxy surface; spore-bearing phase of *Gigartina papillata* and/or *G. agardhii*, see West, 1972, Brit. Phycol. J. 7: 299–308.

Peyssonellia meridionalis Hollenberg and Abbott. Common on rocks and on shells of *Tegula;* reddish purple; low intertidal.

Polyneura latissima (Harvey) Kylin. Distinctive for its seersuckerlike surface, and found in the low intertidal, this species is one of the commonest subtidal red algae from Washington to Baja California (fig. 35).

* *Polysiphonia pacifica* Hollenberg. Common in midtidal; frequently woolly and intertangled when dry.

* *Polysiphonia paniculata* Montagne. More common in exposed places than *P. pacifica.*

Porphyra lanceolata (Setchell and Hus) G. M. Smith. Sexes usually on separate plants forming creamy margins ("males") or red margins ("females").

Porphyra occidentalis Setchell and Hus. A deep-water species, sometimes cast ashore in large numbers.

Porphyra perforata J. G. Agardh. Commonest California *Porphyra* (fig. 34); upper midtidal.

Prionitis andersonii Eaton.

Prionitis lanceolata Harvey. Very common about low tide level and in pools.

* *Schizymenia pacifica* (Kylin) Kylin.

* *Smithora naiadum* (C. L. Anderson) Hollenberg. Common on leaves of surfgrass *Phyllospadix*.

REFERENCES ON ALGAE

Abbott, I. A. and G. J. Hollenberg 1975. *Marine Algae of California.* Stanford University Press. 832 pp.

Dawson, E. Y. 1956. *How to Know the Seaweeds.* Pictured Key Nature Series. Dubuque: W. C. Brown, 197 pp.

Dawson, E. Y. 1966. *Seashore Plants of Northern California.* California Natural History Guides no. 20. University of California Press, 103 pp. Dawson's books are useful beginners' guides.

Smith, G. M. 1969. *Marine Algae of the Monterey Peninsula, California.* Revised ed., including Supplement of Hollenberg and Abbott, Stanford University Press, 752 pp.

INDEX

Boldface numbers indicate illustrations.

Abalones. *See Haliotis*
Abarenicola claparedii oceanica, 226
pacifica, 226, **227**
vagabunda oceanica, 226
Abbott, Donald P., 638
Abbott, Isabella A., 669
Abietinaria spp., 77, 83
Acanthina, 10, 11
punctulata, 495, 509
spirata, 495, **496**, 509
Acanthocephala, 125, 281
Acanthochitonidae, 464
Acanthodoris brunnea, 529, 537
hudsoni, 529, 537
lutea, 529, 537
nanaimoensis, 529, 537
pilosa, 529, 537
rhodoceras, 529, 537
Acanthomysis sculpta, 273
spp., 273
Acanthonotozomatidae, 324, 345
Acanthoptilum gracile, 91, 93
Acari, 425
Acaridiae, 425
Acarnus erithacus, 38, 42, 54, 61
Acartia tonsa, 251, **252**, 255
Accedomoera, 319
vagor, 345, 358
Achelia chelata, 418, **419**, 423
echinata, **419**, 421, 423
gracilipes, **419**, 421, 423
nudiuscula, **419**, 421, 423
simplissima, **419**, 421, 423
spinoseta, **419**, 421, 423
Acila, 543
Acmaea mitra, **476**, 480, 500
paradigitalis, 501
rosacea, **477**, 480, 500
testudinalis scutum, 502
Acmaea. See also Collisella, Notoacmea
Acmaeidae, 500–502
Acoela, 100, 107
Acotylea, 104, 108
Acrothoracica, 259, 267
Actacarus, 425, 428
Acteocina, 516
culcitella, **484**, 485, 512, 519, 521
harpa, 482, **484**, 512, 519, 521
inculta, 472, 512, 519, 521
Acteocinidae, 512
Acteon punctocaelatus, 512, 521
Acteonidae, 469, 512
Actiniaria, 85, 87, 91
Actiniidae, 91
Actinotroch, 613, 614
Adephaga, 447
Adocia gellindra, 60
Adociidae, 60
Adula californiensis, **553**, 554
diegensis, 554
falcata, **553**, 554
Aedicira, 207
Aega lecontii, 297, 306
microphthalma, 297, 306
symmetrica, 297, 307
tenuipes, 297, 307
Aegialites, 447
californicus, 447, **448**
fuchsii, 450, 451
subopacus, 450, 451
Aegidae, 293, 297, 306
Aeginellidae, 367, 368, 372
Aegires albopunctatus, 524, 537
Aeolidia papillosa, 536, 537
Aeolidiella oliviae, 541
takanosimensis, 535, 538
Aepophilus bonairei, 432
Aequorea spp., **75**, 77, 82
Aequoreidae, 82
Aetea, 581
anguina, **590**, 604
Agaue, 430
longiseta, **429**
Agauopsis, 427, **429**, 430
productus, **429**
Aglaja diomedea, 519, **520**, 521
nana, 521
ocelligera, 519, **520**, 521

Aglaophenia, 26, 306, 424
inconspicua, 83
spp., 76, **78**, 83
struthionides, 83
Aglauropsis aeora, 77, 83
Alaria, 675
marginata, **674**, 680
Albuneidae, 410
Alcyonaria, 86, 93
Alcyonidium, 43, 581
parasiticum, 584, **585**, 603
polyoum, 584, **585**, 603
Alderia modesta, 523, 537
Aldisia sanguinea, 528, 538
Alepas pacifica, 262, 267
Aletes squamigerus, 505
Alexia setifer, 22
Algae, 669–684
Allocentrotus fragilis, 622
Alloeocoela, 100, 108
Allogaussia, **327**, **336**, **341**
recondita, 345, 358
Alloioplana californica, 105, 108
Alloniscus perconvexus, 303, 304, 310
Allopora, 66
californica, 214, 268
porphyra, 84
venusta, 214
Allorchestes, **319**, 323, **337**, **341**
angusta, 343, 358
Alpheidae, 403–404
Alpheus, 384, 404
bellimanus, 403
dentipes, **392**, 403
spp., 388, 403
Alvania, 504
Alvinia spp., **489**, 490, 504
Amaeana occidentalis, 232, **233**, 234
Amage, **165**
Amaroucium aequalisiphonis, 652
arenatum, 651
californicum, 651
propinquum, 651
solidum, 651
Amblopusa, 447
borealis, 449, 450
Amblyosyllis sp., 187
Ammodytes hexapterus, 664, 668
Ammodytidae, 668
Ammothella menziesi, 418, **419**, 423
setosa, 418, **419**, 423
tuberculata, 417, **419**, 423
Ammotrypane aulogaster, 222, 224
Ampelisca, **329**, **337**, 340
agassizi, **319**, **327**, 341, 358
compressa, 358
cristata, **337**, 340, 358
lobata, **319**, 342, **350**, 358
macrocephala, **319**, 342, 358
milleri, 340, **350**, 358
pugetica, **319**, 342, **350**, 358
Ampeliscidae, 326, 340, 342
Ampharete labrops, 231, **233**
Ampharetidae, **165**, 168, 171, 231–232, **233**
Amphicteis spp., 231
Amphilochidae, 324, 326, 344
Amphilochus, 323, **327**, **336**, **350**
litoralis, 344, **347**, 358
neapolitanus, 344, **347**, 358
Amphinomidae, 162, **173**, 177–178
Amphiodia occidentalis, **629**, 631, 633
sp., 631, 633
Amphioxus. *See* Cephalochordata
Amphipholis, 122, **632**
pugetana, 633
squamata, 631, 633
Amphipoda, 272, 313–376
Amphiporus bimaculatus, 115, 117, **119**
formidabilis, 114, 117, **119**
imparispinosus, 114, 117
Amphissa columbiana, 493, **494**, 497, 510
versicolor, **494**, 497, 510
Amphithalamus tenuis, **484**, 490, 504
Amphithoe. *See Ampithoe*
Amphitrite, **157**, **167**
Amphoriscidae, 64
Ampithoe, 323, 331, **336**, **337**
aptos, 335, 337, **341**, 358
dalli, 358
humeralis, 334, **341**, 358
lacertosa, **319**, 334, **347**, 358
lindbergi, 334, 358
mea, 334, 358
plumulosa, 335, **350**, 358
pollex, 335, **347**, 358
simulans, 334, 335, **346**, 358
sp., 358
valida, **319**, 334, 335, 358
Ampithoidae, 322, 324, 330, 331, 334–335
Anaata brepha, 61
spongigartina, 40, 42, 54, 61
Anaitides groenlandica, **179**, 180, 182
medipapillata, 180, 182
mucosa, **179**, 180, 182
williamsi, **179**, 180, 182
Anamixidae, 324, 342
Anamixis, **347**
linsleyi, 342, 358
Anasca, 582, 583–584, 589–596, 604–606
Anaspidea, 515, 517, 518, 521–522
Anatanais normani, 279, 280
Anchicolurus occidentalis, 275
Ancinus, 307
daltonae, 307
Ancula lentiginosa, 524, 538
pacifica, 524, 538

Anisodoris nobilis, 530, 538
Anisogammarus, **323**
  confervicolus, **319, 336,** 351, 358
  pugettensis, **319, 336,** 351, 358
  ramellus, **319,** 351, 359
Annelida, 135–243
Anomalodesmata, 546
Anomalohalacarus, 425, 428
Anomia peruviana, 549
Anomiidae, 545, 549
Anomura, 378, 379, 386, 399–402, 408–411
Anopla, 113, 116
Anoplarchus purpurescens, 664, 667
Anoplodactylus erectus, **420,** 421, 423
  oculospinus, **420,** 421, 423
Anorostoma grande, 445
  maculatum, 445
  wilcoxi, 445
Anostraca, 245, 246, **249**
Anotomastus gordiodes, 225
Antho lithophoenix, 42, 55, 60
Anthomedusae, 80
Anthomyiidae, 437, 445
Anthopleura, 424
  artemisia, 90, 91
  elegantissima, 11, 86, **88,** 90, 91, 358, 506
  xanthogrammica, **88,** 90, 91, 506
Anthozoa, 85–94
Anthuridae, 306
Anthuridea, 283, 291, **292,** 306
Antias hirsutus, 298, 309
Antiasidae, 298, 309
Antinoe sp., 175
Antinoella sp., 175
  aureocincta, 538
Antiopella barbarensis, 533, 538
Anurida maritima, 434, **435**
Anuropidae, 291
Aoridae, 322, 326
Aoroides, 323, **336, 341**
  californica, 359
  columbiae, 333, 359
Aphrodita, **161**
  armifera, 174
  castanea, 174
  refulgida, 174
Aphroditidae, 160, **161,** 162, 174
Aphrosylus direptor, 442, 443
  grassator, 442, 443
  praedator, **438,** 440, 442, 443
  wirthi, 442, 443
Aplidium, 255, 363, 643
  arenatum, 651
  californicum, 648, 651
  propinquum, 651
  solidum, 648, 651
Aplousobranchia, 651
Aplysia californica, 521
  vaccaria, 521
Aplysilla glacialis, 42, 44, 59
  polyraphis, 42, 44, 59
Aplysillidae, 59
Aplysiopsis oliviae, 523, 537
  smithi, 523, 537
Apodichthys flavidus, 664, 667
Apodida, 636
Apomatus, 156, **157**
Aporobopyrus, 410
  muguensis, **285,** 286, 305
Arabella iricolor, **154, 161,** 203, 204, 205
  semimaculata, 203, **204,** 205
Arabellidae, 152, **161,** 162, 166, 203–205
Archaeogastropoda, 453, 468, 499–503
Archaeomysis grebnitzkii, 273
  maculata, 273
  spp., 273
Archiannelida, 140, 147–150
Archidistoma, 255, 643
  diaphanes, 649, 651
  molle, 649, 651
  psammion, 648, 652
  ritteri, 649, 652
Archidoris montereyensis, 33, 530, 538
  odhneri, **520,** 531, 538
Arcoida, 545
Arctonoe fragilis, **168,** 174, 175
  pulchra, **168,** 174, 175
  vittata, **168,** 174, 175, 500
Arcturidae, 287, 305
Arcuatula demissa, 554, 555
Arenicola, **157, 167, 227**
  brasiliensis, 226
  cristata, 226
Arenicolidae, 160, **167,** 169, 226, **227**
Argeia, 404
  pauperata, 287, 305
  pugettensis, **285,** 287, 305
Arguloidea, 253
Argulus japonicus, 253
  pugettensis, 253, **256**
Aricidea, 207
  suecica, 207
Armadillidiidae, 303, 310
Armadillidium vulgare, 303, 310
Armadilloniscus coronacapitalis, **301,** 304, 311
  holmesi, **301,** 304, 311
  lindhali, **301,** 303, 304, 311
  tuberculatus, 311
Armandia bioculata, 224
  brevis, 222, 224
Armina californica, 536, 538
Arnold, Zach M., 26
Artedius corallinus, 662, 666
  creaseri, 662, 666
  fenestralis, 662, 666

Artedius (*continued*)
harringtoni, 662, 666
lateralis, 662, 666
Artemia salina, 245, 247, **249**, 434
Arthropoda, 244–452
Arthropoma cecili, 599, **602**, 606
Articulata, 616, 617
Aruga spp., 361
Aschelminthes, 122–127
Ascidia californica, 652
ceratodes, 643, 644, 652
paratropa, 644, 652
Ascidiacea, 638–655
Ascidians, 638–655
Ascidiopsis paratropa, 652
Ascophora, 582, 584, **597**, 598–603, 606–607
Ascophyllum, 20
Asellidae, **282**, 298, 308
Aselloidea, 298, 308
Asellota, 283, 298–302, 308–310
Asellus tomalensis, 298, 308
Asilidae, 437, 444
Aspidobranchia, 468
Aspidochirotida, 636
Assiminea californica, **484**, 490, 504
translucens, 504
Assimineidae, 504
Astacillidae, 305
Astacura, 378, 386
Astacus, 386
Asteroidea, 620, 623–627
Astraea, 513
gibberosa, 485, 489, 503
inaequalis, 503
Astylinifer arndti, 42, 48, 61
Asychis amphiglypta, 228
elongata, 19, 228
Atagema quadrimaculata, 528, 538
Athecata, 80
Atyidae (Crustacea), 402
Atyidae (Gastropoda), 512
Atylus, **329**, 340
levidensus, **337**, 340, 359
tridens, **319**, 340, **350**, 359
Aulorhynchus flavidus, 660, 667
Aurelia, **88**
aurita, 95
labiata, 95
Aurellia, 95
Austrobilharzia variglandis, 19
Austrodoris odhneri, 538
Autolytinae, 187
Autolytus spp., 184, **186**, 187
Axiognathus pugetanus, 633
squamatus, 633
Axiothella, 407
rubrocincta, **227**, 228
Axocielita hartmani, 60
originalis, 34, 42, 51, **57**, 60

Babaina festiva, 538
Babakina festiva, 533, 538
Balanidae, 264
Balanoglossus, 619
Balanomorpha, 262, 268–269
Balanophyllia elegans, 86, 93
Balanus, 214, 260, **261**, 286, 504, 540
Balanus (*Armatobalanus*) *nefrens*, 264, **265**, 268
Balanus (*Balanus*) *amphitrite amphitrite*, 18, 263, 265, **266**, 268
aquila, **265**, 267, 268
balanoides, 305
crenatus, **266**, 267, 268
flos, 268
glandula, 264, **266**, 268, 269, 442
improvisus, 18, 23, 261, **266**, 267, 268
nubilus, **263**, **265**, 267, 268
pacificus, 264, **265**, 268
trigonus, 264, **266**, 268
Balanus (*Conopea*) *galeatus*, 264, **265**, 268
Balanus (*Megabalanus*) *tintinnabulum californicus*, 264, **266**, 268
Balanus (*Semibalanus*) *cariosus*, 264, **265**, 268
Balanus (*Solidobalanus*) *hesperius laevidomus*, 264, **266**, 269
Balcis rutila, 506
spp., 488, **489**, 506
thersites, 506
Bamboo worms. *See* Maldanidae
Bankia setacea, 575, **576**
Barentsia, 605, 610
benedeni, 18, 23, **611**, 612
discreta, **611**, 612
gracilis, **611**, 612
gracilis nodosa, 612
ramosa, 610, **611**, 612
Barleeia, **489**
acuta, 490, 504
dalli, 504
haliotiphila, 490, 504
marmorea, 504
oldroydi, 504
sanjuanensis, 504
subtenuis, 504
Barnacles. *See* Cirripedia
Barnard, J. Laurens, 314
Barnea pacifica, 573
subtruncata, 572, 573, **574**
Baseodiscus punnetti, 115, 117
Basiliochiton heathii, 461, 464
Basket stars. *See* Ophiuroidea
Basommatophora, 469, 513–514
Bathycopea daltonae, 294, 307

Batillaria attramentaria, 19, **489,** 492, 505
cumingi, 505
multiformis, 505
zonalis, 505
Batrachoididae, 667
Bdellonemertea, 113, 118
Beach hoppers. *See* Talitridae
Beetles. *See* Coleopotera
Beroe cucumis, 98, 99
forskali, **97,** 98, 99
Beroida, 98
Berthella californica, 522
Betaeus, 382, 384, 385
ensenadensis, 403
gracilis, **389,** 390, 403
harfordi, **389,** 390, 403
harrimani, 388, **389,** 403
longidactylus, 388, **389,** 404
macginitieae, 388, **389,** 404
setosus, **389,** 390, 404
Bicrisia edwardsiana, 586, **588,** 604
Bimeria franciscana, 80
Bittium attenuatum, 492, 505
eschrichtii, 492, **494,** 505
eschrichtii montereyense, 505
interfossa, 492, 506
purpureum, 492, 506
Bivalvia, 543–578
classification, 545–546
terminology, 547, **548**
Blake, James A., 151
Blepharipoda occidentalis, 399, 410, 546, 559
Boccardia berkeleyorum, 208, **209,** 213
columbiana, 208, **209,** 214
hamata, 208, **209,** 214
proboscidea, 208, 214
tricuspa, 208, **209,** 214
truncata, 208, 214
uncata, 214
Bolin, Rolf, 98, 656
Bolinopsis microptera, 98, 99
Boltenia villosa, 644, 653
Bopyridae, 284, 286, 305, 382
Bossea corymbifera, 681
orbigniana, 681
Bossiella, 671
chiloensis, **677,** 681
orbigniana, **677,** 681
plumosa, 681
Bothidae, 665
Botrylloides spp., 648, 653
Botryllus spp., 648, 653
tuberatus, 648, 653
Botryoglossum, 678
farlowianum, **677,** 682
Botula californiensis, 554
diegensis, 554
falcata, 554
Bougainvillia mertensi, 80
ramosa, 80
spp., 74, 79, 80
Bougainvilliidae, 80
Bousfield, E. L., 363
Bowerbankia, 581
gracilis, 540, 584, **585,** 603
Bowers, Darl E., 355
Brachiopoda, 609, 616–617
Brachycera, 436
Brachyura, 378, **380,** 386, **387,** 393–399, 405–408
Brada sp., 221
villosa, 222
Bradbury, Margaret G., 656
Bradyagaue, 428
bradypus, **427**
Branchellion, 146
Branchiobdellids, 386
Branchiomaldane vincentii, 226
Branchiopoda, 245, 246–247
Branchiostoma californiense, 638
Branchiura, 245, 253, **256**
Brania sp., 185, 187, **188**
Brine flies. *See Ephydra*
Brine shrimp. *See Artemia*
Brinkhurst, Ralph O., 136
Brittle-stars. *See* Ophiuroidea
Bryobiota, 446
bicolor, 449, 450
Bryopsis spp., 537
Bryozoa. *See* Ectoprocta, Entoprocta
Buccinidae, 510
Bugula, 580, 582
californica, 596, **597,** 605
longirostrata, 596, **597,** 605
mollis, 596, **597,** 605
neritina, 596, **597,** 605
pacifica, 540, 596, **597,** 605
pugeti, 596, **597,** 605
Bulla, 516
gouldiana, 482, **483,** 512, 519, 521
Bull kelp. *See Nereocystis*
Bullidae, 512
Bullock, Theodore H., 618
Bunodactis elegantissima, 91
Busycon canaliculatum, 509
Busycotypus canaliculatus, 19, **496,** 497, 509–510

Cactosoma arenaria, 87, 92
Cadlina, 540
flavomaculata, 530, 538
limbaughi, 530, 538
luteomarginata, 530, 538
modesta, 531, 538
sparsa, 529, 538
Cadulus fusiformis, 455

Caecianiropsis psammophila, 297, 302, 309
Caecidae, 505
Caecum californicum, 474, **479,** 505
Cafius, 446
canescens, 450
seminitens, 450
spp., 450
Calanoida, 251, **252,** 255
Calanus finmarchicus, 251
spp., **252,** 255
Calappidae, 405
Calcarea, 34, 63
Caligoida, 251, 253, **256,** 257
Caligus clemensi, **256,** 257
Callianassa, 10, 175, 176, 251, 255, 286, 379, 384, 403, 407
californiensis, 305, **383, 400,** 401, 408
gigas, **400,** 401, 408
longimana, 408
Callianassidae, 408–409
Calliarthron, 671
cheilosporioides, 682
setchelliae, 682
tuberculosum, **677,** 682
Callophyllis violacea, 682
Calliopiella, **317, 319,** 324, **337, 346**
pratti, 348, 359
Calliopiidae, 322, 323, 326
Calliopius, **337, 350**
laeviusculus, 359
sp., 348, 359
Calliostoma annulatum, 487, 502
canaliculatum, **486,** 487, 502
costatum, 502
ligatum, **486,** 487, 502
splendens, 502
supragranosum, **486,** 487, 502
Callistochiton crassicostatus, 463, 464
palmulatus mirabilis, 463, 464
palmulatus palmulatus, 463, 464
Callistoplacidae, 464
Callopora, 580
circumclathrata, **592,** 593, 605
horrida, **592,** 593, 605
Calycella syringa, 82
Calyptoblastea, 82
Calyptraeidae, 507
Calyssozoa, 609
Campanulinidae, 76, 82
Campanularia compressa, 82
spp., 76, 82
Campanulariidae, 82
Canace aldrichi, 442, 443
Canaceidae, 436, 439, 442, 443
Canaceoides nudatus, 442, 443
Cancer antennarius, **383,** 396, **397,** 406
anthonyi, 396, 406
branneri, 406
gibbosulus, 406
gracilis, 396, 406
jordani, 396, 406
magister, 117, 380, 396, 406
oregonensis, 406
productus, 396, **397,** 406
Cancridae, 406
Candelabridae, 80
Candelabrum sp., 74, 80
Capellinia rustya, 540
Capitella capitata, **223,** 225
Capitellidae, 140, 169, **223,** 225–226
Caprella angusta, 376
brevirostris, 370, 372
californica, 370, 372, **375**
equilibra, 23, 370, 372, **374**
ferrea, 372, **374**
gracilior, 370, 372, **375**
greenleyi, 367, 370, 372, **375**
incisa, 372, **374**
laeviuscula, 370, 372, **374**
mendax, 370, 372, **373**
natalensis, 370, 372, **373**
penantis, 23, 370, 372
pilipalma, 372
uniforma, 370, 376
verrucosa, 370, **373,** 376
Caprellidae, 367, 368, 370, 372
Caprellidea, 367–376
Caprogammarus, 367
Carabidae, 447, 450
Carcinonemertes, 384
epialti, 112, 113, 117
Cardiidae, 546, 551
Carditidae, 546, 551, 563
Cardium corbis, 551
Caridea, 378, **380,** 385, 386–392, 402–404
Carinoma mutabilis, 114, 116
Carleton, James T., xvii, 17, 385, 453, 467
Carmia macginitiei, 60
Catriona alpha, 535, 538
Caudina chilensis, 635, 636
sp., 636
Caulibugula ciliata, 596, **597,** 605
occidentalis, 596, 605
Caulleriella spp., 219, **220**
Cauloramphus spiniferum, 591, **592,** 605
Cebidichthys violaceus, 665, 667
Cellaria diffusa, 594, **595,** 605
mandibulata, 593, **595,** 605
Celleporaria brunnea, 108, 601, **602,** 606
Centrostephanus coronatus, 622
Cephalaspidea, 469, 512, 515, 516, 517, 518–519, 521
Cephalocarida, 245, 246
Cephalochordata, 638
Cephalopoda, 455–457
Ceradocus spinicaudus, 359
Cerapus, **332, 336, 337**

abditus, 359
tubularis, 333, 359
Ceratonereis tunicatae, 190, **191**, 193
Ceratostoma, 10
foliatum, 497, 508
inornatum, 19, 497, 508
Cercyon, 446
fimbriatus, 447, 450
luniger, 447, 450
sp., **448**
Cerebratulus californiensis, 113, 115, 117, **119**
Ceriantharia, 86, 87, 93
Cerithidea californica, 488, **489**, 505
Cerithiidae, 505–506
Cerithiopsidae, 469, 506
Cerithiopsis montereyensis, **494**, 506
spp., 493, 506
Cestoda, 100
Chaceia ovoidea, 572, **574**
Chaetogaster diaphanus, 141, 143, 145
limnaei, **139**, 141, 142, 145
Chaetomorpha, 537
Chaetopleura gemma, 462, 464
Chaetopleuridae, 464
Chaetopteridae, **157**, 169, 216–218
Chaetopterus, 9, **157**
variopedatus, **217**, 218
Chaetozone setosa, 219, **220**
Chalinula, 59
Chama, **544**
pellucida, 556, 558
Chamidae, 546, 549, 556, 558
Chapperia patula, **592**, 593, 605
Cheilonereis cyclurus, 190, **191**, 193
Cheilostomata, 581, 582, 583, 589–603, 604–607
Chelicerata, 245, 413
Chelidonura inermis, 518, **520**, 521
Chelifera, 277–280
Chelonethida, 431, **435**
Chelura, 323, **332**, **336**
terebrans, 331, 359
Cheluridae, 322, 326, 330
Chelyosoma productum, 644, 652
Chersodromia cana, 444
inchoata, 444
insignita, 444
magacetes, 444
Childia groenlandica, 19, 100, 107
Chioraera leonina, 540
Chirolophis nugator, 664, 667
Chironomidae, 436, 437, 439–440, 443
Chironomus oceanicus, 432
Chitons. *See* Polyplacophora
Chlorophyta, 678, 680
Chondrophora, 66, 84
Chone ecaudata, 215, 236, 238, **240**
gracilis, 238
minuta, 238
mollis, 238, **240**
Chordata, 638–668
Choristida, 63
Chromodoris californiensis, 540
macfarlandi, **520**, 528, 538
porterae, 529, 538
Chromolepida bella, 444
Chrysaora melanaster, 95
Chrysopetalidae, 162, **173**, 177
Chrysopetalum occidentale, **173**, 177
Chthamalidae, 264
Chthamalus, 260, **261**, 286, 442
dalli, 264, **265**, 269
fissus, 264, **265**, 269
microtretus, 269
Cidaroidea, 621
Ciliocincta sabellariae, **119**, 121
Ciona, 641
intestinalis, 23, 645, 653
Cirolana harfordi, 269, 307
Cirolanidae, **282**, 293, 296, 307
Cirratulidae, 169, 218–219, **220**
Cirratulus cirratus cingulatus, 219
cirratus cirratus, 219, **220**
cirratus spectabilis, 219
Cirriformia luxuriosa, 219, **220**
spirabrancha, 219, **220**
Cirripedia, 259–269
Cirrophorus, 207
Cistenides brevicoma, 230, **231**
Citharichthys stigmaeus, 659, 665
Cladocera, 245, 247
Cladonema californica, 69, **73**, 74, 79, 80
Cladonemidae, 80
Cladophora, 280, 670
columbiana, 678
trichotoma, 678
Clam shrimps. *See* Conchostraca
Clams. *See* Bivalvia
Clathria sp., 42, 53, 60
Clathriidae, 60
Clathrina sp., 42, **57**, 58, 63
Clathrinida, 63
Clathrinidae, 63
Clathriopsamma pseudonapya, 42, 53, 60
Clathromangelia interfossa, 491, 511
Clathurella canfieldi, 491, **494**, 511
Clausidium, 408
vancouverense, 251, **254**, 255
Clava leptostyla, 18, 76, 80
Clavelina, 643
huntsmani, 647, 652
Clavodoce splendida, 180, **181**, 182
Clavularia, 11
sp., 91, 93, **97**
Claw shrimps. *See* Conchostraca

Cleantis heathi, 305
Clevelandia, **131**
ios, 660, 667
Clinidae, 667
Clinocardium nuttallii, 551, **557**
Clinocottus analis, **661,** 663, 666
embryum, 663, 666
globiceps, 663, 666
recalvus, **661,** 663, 666
Cliona celata var. *californiana,* 38, 42, 47, 62
spp., 19, 33, 41, 42, 48, 62
Clionidae, 33, 62
Clunioninae, 443
Clymenella, **167**
californica, 228
complanata, 228
sp., **227**
Clytia, 82
Cnemidocarpa, 255
finmarkiensis, 645, 653
Cnidaria, 65–97
Cnidopus ritteri, 90, 91
Coan, Eugene V., 543
Cockles. *See* Cardiidae, Veneridae
Codium fragile, 537, 670, **672,** 678
setchellii, 670, **672,** 678
Coelenterata. *See* Cnidaria
Coelopa (Neocoelopa) vanduzeei, 445
Coelopidae, 437, 445
Colanthura squamosissima, 291, **292,** 306
Coleoidea, 456
Coleopora gigantea, 599, **600,** 606
Coleoptera, 434, **435,** 446–452
Collastoma pacifica, 100, 107, 131
Collembola, 433, 434, **435**
Colletosia radiata, 594, **595,** 605
Collisella, **483**
asmi, **476,** 480, 501
digitalis, **477,** 478, 501
instabilis, 9, **477,** 478, 501, 681
limatula, **477,** 480, 501
limatula moerchii, 501
ochracea, **476,** 481, 501
pelta, **477,** 478, 480, 481, 501
scabra, **476,** 478, 501
strigatella, **477,** 480, 481, 501
triangularis, **477,** 478, 501
Coloboneura, 444
Colobranchiata, 293, 294
Columbellidae, 510
Colurostylis occidentalis, 275
Comb jellies. *See* Ctenophora
Conchostraca, 245, 247
Conidae, 511
Conopeum commensale, 605
reticulum, 23, **590,** 591, 605
Constantinea, 675
simplex, **677,** 682
Conualevia alba, 524, 538
Conus californicus, 491, 511
Cook, David G., 136
Coon-stripe shrimp. *See Pandalus*
Copepoda, 245, 250–258
Copidognathus, 428
curtus, **427, 429**
pseudosetosus, 428, **429**
Copidozoum tenuirostre, **592,** 593, 605
Corallimorpharia, 85, 87, 93
Corallina, 671
chilensis, 682
officinalis var. *chilensis,* 677, 682
vancouveriensis, 682
Corambella bolini, 539
Corambe pacifica, 516, 537, 538
Corbicula fluminea, 551–552
manilensis, 20, 551–552
Corbiculidae, 546, 551–552
Cordylophora lacustris, 19, **73,** 76, 80
Corella willmeriana, 645, 653
Corixidae, 433
Cornuspira lajollaensis, **27,** 28
Corophiidae, 322, 326, 330, 338–340
Corophioidea, 322, 324
Corophium, 323, 324, **332,** 333, **336, 337,** 338, **339, 347**
acherusicum, 338, **339,** 340, 359
baconi, 338, **339,** 359
brevis, **339,** 340, 359
californianum, 338, 359
insidiosum, 338, **339,** 340, 359
oaklandense, 338, **339,** 359
spinicorne, **339,** 340, 359
stimpsoni, 340, 359
uenoi, 19, 338, **339,** 340, 359
Corynactis californica, 86, 87, 93
Coryne brachiata, 81
spp., 74, 81
Corynidae, 81
Coryphella cooperi, 534, 538
fisheri, 539
pricei, 534, 539
trilineata, 533, 539
Coryphopterus nicholsi, 659, 667
Cossura pygodactylata, **220,** 221
Cossuridae, 169, **220,** 221
Costaria, 675
costata, 680
turneri, **674,** 680
Costazia robertsonae, 599, **602,** 603, 606
Cottidae, 657, 666
Cotylea, 104, 109
Crabs. *See* Brachyura
Crago. See Crangon
Crangon, 381, 384
alaskensis elongata, 388, **394,** 404

franciscorum, 305, **383**, 386, **387**, **394**, 404
munitella, 404
nigricauda, 388, **394**, 404
nigromaculata, 388, **394**, 404
sp., 388, 404
stylirostris, 386, **394**, 404
See also Alpheus
Crangonidae, 404
Craniella arb, 63
Craspedacusta sowerbii, 77, 83
Crassostrea gigas, 19, 215, 547, 549, 556
virginica, 19, 556, **557**
Cratena, 541
rutila, 541
spadix, 538
Crayfish. *See* Astacura
Crepidula, 471
adunca, 475, **484**, 507
convexa, 19, 22, 475, **484**, 507
fornicata, 19, 475, **484**, 507
glauca, 507
nivea, 507
nummaria, 475, 507
perforans, 475, 507
plana, 19, 475, **484**, 507
Crepipatella lingulata, 474, **479**, 507
Cribrina elegantissima, 91
Crimora coneja, 526, 539
Crinoidea, 620
Crisia maxima, 589, 604
occidentalis, 589, **590**, 604
serrata, 604
serrulata, 589, **590**, 604
Crisulipora occidentalis, 588, **588**, 604
Crucigera zygophora, 239, **240**, 241
Crustacea, 244–246
Cryptochiton, 407
stelleri, 458, 460, 464
Cryptodonta, 545
Cryptolithodes sitchensis, 399, 409
Cryptomya californica, 546, **548**, 570
Cryptoniscidae, 284, 286, 305
Cryptosula pallasiana, 598, **600**, 606
Cryptothir balani, 286, 305
Cryptothiria balani, 305
Ctenobranchia, 468
Ctenodrilidae, 169, **220**, 221
Ctenodrilus serratus, **220**, 221
Ctenophora, 98–99
Ctenostomata, 581, 583, 584, **585**, 603–604
Cubaridae, 303, 310
Cubaris affinis, 310
californica, 310
microphthalma, 303, 310
spp., 303
Cucumaria chronhjelmi, 637
curata, 9, 11, **632**, 635, 636, 637
fisheri, 636
lubrica, 635, 636
miniata, 636
piperata, 635, 636
pseudocurata, **632**, 635, 636–637
Cumacea, **249**, 272, 273–276
Cumanotus beaumonti, 536, 539
Cumella vulgaris, **249**, 275
Cumingia californica, **565**, 568
Curculionidae, 449, 451
Cuthona rosea, 541
Cuttle-fish. *See* Sepioidea
Cyamidae, 367–368
Cyamus gracilis, **375**
scammoni, **375**
Cyanea capillata, 95
Cyanoplax dentiens, **459**, 462, 464
fackenthallae, 464
hartwegii, 462, 465
raymondi, 464
Cyathura munda, 291, **292**, 306
Cyclaspis sp., 275
Cyclopoida, 251–252, **254**, 255
Cyclostomata, 582, 583, 584–589, 604
Cycloxanthops novemdentatus, 398, 406
Cydippida, 98
Cylindroberis sp., **249**
Cylindrocarpus, 671
rugosus, **676**, 680
Cymadusa, **319**, **325**
uncinata, 359
Cymakra aspera, 493, **494**, 511
gracilior, 495, 511
Cymbulidae, 516
Cymothoidae, 293, 296, 307
Cypraeolina pyriformis, 511
Cyprideis, 248
Cystiscus jewettii, 488, 511
Cystodytes lobatus, 649, 652
sp., 649, 652
Cystoseira, 673
osmundacea, 680
Cythereis aurita, **249**

Dajidae, 284
Daly, Howell V., 432
Daphnia, 247
Dasybranchus glabrus, 225
lumbricoides, 225
Decachela discata, 421, 422, 423
Decapoda, 377–412
biology, 378
Decorator crabs. *See* Majidae
Demonax media, 239
Demospongiae, 32, 34, 59
Dempster, Lillian J., 656
Dendraster, 107
excentricus, 622

Dendrobeania laxa, 594, **595**, 605
lichenoides, 594, **595**, 605
Dendroceratida, 59
Dendrochirotida, 636
Dendrochiton thamnoporus, 465
Dendrodoris albopunctata, 539
fulva, 539
nigromaculata, 539
sp., 528, 539
Dendronotus albus, 532, 539
frondosus, **527**, 532, 539
iris, 532, 539
subramosus, 532, 539
venustus, 539
Dendropoma lituella, 473, **483**, 505
rastrum, 474, 505
Dendrostomum. *See Themiste*
petraeum, 131
Dentalium hexagonum, 455
pretiosum, 455, **498**
pretiosum berryi, 455
Dermasterias imbricata, 624, 625
Desmarestia, 675, 680
herbacea, **676**, 680
latifrons, **676**, 680
ligulata, 680
munda, **676**, 680
Deutella californica, 368, 372, **373**
Dexaminidae, 326, 340
Diadumene franciscana, 22, 89, 92
leucolena, 19, 89, 92
lighti, 89, 92
Diadumenidae, 92
Diala acuta, 504
Dialineura melanophleba, 444
Diamphiodia occidentalis, 633
Diaperoecia, 582
californica, 586, **587**, 604
Diaphana, 516
californica, 482, 512, 516, 519, 521
Diaphanidae, 512
Diaptomus spp., 251
Diarthrodes cystoecus, 253, **254**, 255, 682
Diastylopsis dawsoni, 275
Diaulota, 447
densissima, **448**, 449, 450
fulviventris, 449, 450
vandykei, 449, 450
Diaulula sandiegensis, 530, 539
Dicranomyia signipennis, 443
Dictyoceratida, 36, 59
Dictyoneurum, 675
californicum, **674**, 680
Dicyema apollyoni, **119**
Dicyemida, **119**, 121
Didemnum carnulentum, 649, 652
opacum,652
Dikonophora, 278, 280
Dimecoenia, 444
Dinophilidae, 149
Dinophilus gyrociliatus, **148**, 149
Diodora arnoldi, 500
aspera, 474, **479**, 500
Diogenidae, 409
Diopatra ornata, **198**, 199, 612
Diotocardia, 468
Diplodonta orbella, 552, **561**
Diplodontidae, 552
Diplosoma macdonaldi, 649, 652
pizoni, 652
Diplostraca, 247
Diptera, 433, 436–446
Dipurena sp., 79, 81
Dirona albolineata, 533, 539
picta, 533, 539
Discodoris heathi, 530, 539
Disoma franciscanum, 216, **217**
Disomidae, 216, **217**
Disporella spp., 586, **587**, 604
Distaplia, 577
occidentalis, 648, 650, 652
smithi, 650, 652
Distoma lobatus, 652
molle, 651
pulchra, 652
Diurodrilus, **148**, 149
Dodecaceria concharum, 219
fewkesi, 218, 219, **220**
Dogielinotidae, 326, 343
Dogielinotus, **317**, **329**, **336**, **337**
loquax, 343, 359
Dolichopodidae, 436, 437, 440, 442, 443
Doridella steinbergae, 516, 537, 539
Doriopsilla albopunctata, **528**, 539
Doris sp., 539
tanya, 539
Dorvillea annulata, 202
articulata, 202
moniloceras, 202
rudolphi, **200**, 202
Dorvilleidae, 152, 166, **200**, 202
Doto amyra, 534, 539
columbiana, 534, 539
kya, 534, 539
sp., **525**
varians, 539
Doyen, John T., 446
Drilonereis falcata, 203, **204**, 205
nuda, 203, **204**, 205
Dryomyzidae, 436, 437, 442–443, 444
Dryopidae, 446
Dungeness crab. *See Cancer magister*
Duvaucelia, 541
gilberti, 541
Dynamenella dilatata, 295, 307
"*Dynamenella*" *benedicti*, 295, 308

glabra, 294, 308
sheareri, 295, 308
Dytiscidae, 446

Echiniscoides sigismundi, 432, **435**
Echinodera, 124
Echinodermata, 17, 620–637
Echinoidea, 620, 621–622
Echiura, 132–133
Ectoprocta, 579–608, 609
anatomy, 579–581
methods of study, 582–583
Edotea sublittoralis, 287, 305
Edwardsia sp., 87, 92
Edwardsiidae, 92
Eel-grass. *See Zostera*
Egregia, 410, 478, 501, 673
menziesii, 502, 680
Elasmopus, 323, **325**, **336**, 351
antennatus, 351, 359
mutatus, **337**, **346**, 351, 359
rapax, **337**, 359, 360
serricatus, **346**, 351, 360
Electra crustulenta, 591, **592**, 605
crustulenta arctica, 591, **592**, 605
Ellobiidae, 513
Elmidae, 446
Elphidium crispum, **27**, 28
Elysia bedeckta, 537
hedgpethi, **520**, 522, 537
Emarcusia morroensis, 535, 539
Embiotocidae, 665
Emerita, 9, 379, 381, 384
analoga, 399, **400**, 410
Emphyastes, 446
fucicola, 449, 451
Empididae, 437, 444
Emplectonema gracile, 115, 118
Emplenota, 446
arenaria, **448**, 449, 450
Enchytraeidae, 136, 137, 138, 140, 141, 142, 144
Enchytraeus, **139**
albidus, 141, 144
cryptosetosus, 144
multiannulatus, 144
pugetensis, 144
Endeodes, 446
collaris, **448**, 449, 451
rugiceps, 449, 451
Endocladia, 118, 271, 442, 501, 504, 521, 675
muricata, 555, **677**, 682
Enopla, 113, 117
Enterogona, 651
Enteromorpha, 439, 442, 502, 512, 537, 670
compressa, **672**, 680
intestinalis, **672**, 680
Enteropneusta, 618
Entodesma inflatum, 577
saxicola, **571**, 577
Entoniscidae, 284, 286, 305
Entoprocta, 582, 609–613
Eohaustorius, 323, **325**, 330, **336**, **337**
spp., 348, 360
Eophliantidae, 324, 330
Ephelota gemmipara, 28, **29**
Ephydra, 444
gracilis, 434
Ephydridae, 439, 444
Epiactis prolifera, 86, 90, 92
Epialtus productus, 405
Epicaridea, 283, 284–287, 305
Epilabidocera amphitrites, 251, 255
longipedata, 251, **252**, 255
Epilucina californica, 551, **571**
Epinebalia pugettensis. *See Nebalia pugettensis*
Epitoniidae, 468, 506
Epitonium cooperi, 506
tinctum, 488, **489**, 506
Erato, 469
columbella, 492, 508
vitellina, 492, 508
Eretmoptera browni, 439, 443
Ericthonius, 323, 324, **332**, 333, **336**, **337**
brasiliensis, 334, **341**, 360
hunteri, 334, **341**, 360
Errinopora pourtalesia, 268
Erycinidae, 546, 550, 558–559
Esperiopsis originalis, 60
Eteone californica, 180, 182
dilatae, 178, **179**, 182
lighti, 178, **179**, 182
pacifica, 178, **179**, 182
Eubranchiata, 293, 295
Eubranchus misakiensis, 19, 535, 539
occidentalis, 540
olivaceus, 540
rustyus, 535, 540
Eucarida, 271, 377–412
Eucopella, 82
Eudendriidae, 81
Eudendrium californicum, 74, 81
spp., 81
Eudistoma diaphanes, 651
molle, 651
psammion, 652
ritteri, 652
Eudistylia polymorpha, 236, **237**, 238
vancouveri, 236, **237**, 238
Eudorella pacifica, 275
Euechinoidea, 621
Euherdmania claviformis, 647, 652
Eulalia aviculiseta, **181**, 182

Eulalia (*continued*)
bilineata, **181**, 182
quadrioculata, 182
viridis, 182
Eulamellibranchia, 545
Eulimidae, 468, 506
Eumida bifoliata, 180, **181**, 182
sanguinea, 180, 183
Eunereis longipes, 190, **191**, 193
Eunice antennata, **200**, 201
kobiensis, 201
valens, **200**, 201
Eunicea, **154**
Eunicidae, 152, **157**, **163**, 166, 201
Eunoe senta, **170**, 175
Eupentacta quinquesemita, 635, 637
Euphausia, 377
pacifica, **249**
Euphausiacea, **249**, 377
Euphrosine aurantiaca, 178
sp., **173**, 178
Euphrosinidae, 162, 178
Eupolymnia, 407
crescentis, 234
Eupomatus gracilis, 239, **240**, 241
Euplokamis californiensis, 99
Eurylepta aurantiaca, 107, 109
californica, 107, 109, **110**
Eurypon asodes, 60
Eurystomella bilabiata, 599, **602**, 606
Eurytemora hirundoides, 251, **252**, 255
Eurythoe complanata, **173**, 177, 178
Eusiridae, 322, 326, 328, 345
Euthyneura, 468, 469
Eutimidae, 83
Eutonina indicans, **75**, 77, 83
Euzonus dillonensis, **223**, 224
mucronata, **223**, 224
williamsi, **223**, 224
Evactis artemisia, 91
Evadne, 247
Evasterias troschelii, 624, 626
Excirolana chiltoni, 296, 307
kincaidi, 296, 307
linguifrons, 296, 307
Excorallanidae, 291
Exogone, 162
gemmifera, 187, **188**
lourei, 187, **188**, 189
uniformis, 187, 189
verugera, 187, **188**, 189
Exogonella sp., 164, 166, 185, 189
Exosphaeroma amplicauda, 295, 308
inornata, 295, 308
octoncum, 295, 308
rhomburum, 295, 308

Fabia, 379, 381
subquadrata, **394**, 398, 407
Fabricia berkeleyi, 236, 238, **240**
brunnea, 236, 238, **240**
sabella, 236, 238, **240**
Fabricinae, 235
Facelina stearnsi, 534, 540
Farrella elongata, 584, **585**, 603
Fasciolariidae, 511
Figularia hilli, 594, **595**, 605
Fenestrulina malusii, 598, **600**, 606
malusii var. *umbonata*, 598, 606
Filibranchia, 545
Filicrisia franciscana, 589, **590**, 604
geniculata, **588**, 589, 604
Fiona pinnata, 516, **525**, 534, 540
Fish, 656–668
Fissurella volcano, 474, **479**, 500
Fissurellidae, 500
Flabelliderma commensalis, 221, 222
Flabellifera, 283, 291–297, 306–308
Flabelligera infundibularis, 221, 222
Flabelligeridae, **157**, **165**, 171, 221–222, **223**
Flabellina iodinea, 540
Flabellinopsis iodinea, 533, 540
Flatworms. *See* Platyhelminthes.
Flies. *See* Diptera
Flosmaris grandis, 87, **88**, 93
Flukes. *See* Trematoda
Flustrella corniculata, 604
Flustrellidra, 581
corniculata, 584, **585**, 604
Folliculina, 28, **29**, 123
Foraminifera, 26–29, 274, 455
Fosliella, 671
Fouling, 18
Freemania litoricola, 106, 108
Fresnillo, **327**, **337**
fimbriatus, 344, 360
Fucellia antennata, 445
assimilis, 445
costalis, 445
fucorum, 445
pacifica, 445
rufitibia, 445
separata, 445
thinobia, 445
Fucus, 20, 439, 673
distichus, 681
Fusinus luteopictus, **498**, 499, 511

Gadinia reticulata, 513
Gadinidae, 513
Galatheidae, 410
Galatheidea, 378, 410
Galeommatacea, 550, 558–559
Gammaridacarus brevisternalis, 425
Gammaridae, 322, 328, 349, 351–352
Gammaridea, 313–366
Basic Key, 328, 330

Dissection, 320–321
Identification, 314–320, 321–324
Key to Families, 324, 326, 328
Morphology, 314–320
Gammaropsis, 323
mamolus, 333, **350**, 360
thompsoni, **319**, 333, 360
tenuicornis, 360
Gammarus, 323, **336**
Gaper clam. *See Tresus*
Gari californica, 566, 568, **576**
Garidae, 568
Garveia annulata, 74, 80
franciscana, 18, 80
Garypus californicus, 431, **435**
Gasterosteidae, 667
Gastrodelphys dalesi, 238
Gastropoda, 467–542
classification, 468–469
terminology, 470–471
Gastrotricha, **119**, 123–124
Gemma gemma, 19, 22, 547, 559, **561**, 562, 563
Genetyllis castanea, **179**, 180, 183
Geodia mesotriaena, 41
Geonemertes sp., 112, 118
Gephyrea, 128
Gerridae, 433
Geukensia demissa, 554
Ghost shrimp. *See Callianassa*
Gibbonsia metzi, 664, 667
montereyensis, 664, 667
Gigartina, 633, 675
agardhii, **679**, 682, 683
californica, **679**, 682
canaliculata, **679**, 682
corymbifera, 682
harveyana, **679**, 682
papillata, 682, 683
Gitanopsis vilordes, 344, **347**, 360
Gladfelter, William B., 273
Glans carpenteri, 563
subquadrata, **557**, 563
Glossobalanus, 619
Glossodoris macfarlandi, 538
Glossophorum planum, 652
Glottidia albida, **614**, 617
Glycera americana, 196, **196**, 197
capitata, 194, **195**, **196**, 197
convoluta, 196, **196**, 197
dibranchiata, 196, **196**, 197
robusta, 195, **196**, 197
tenuis, 195, **196**, 197
Glyceridae, 20, 164, 194–197
Glycinde armigera, 194, **195**
polygnatha, 194, **195**
Gnathia crenulatifrons, 284, 305
Gnathiidae, 284, 305
Gnathiidea, 283, 284, 304–305
Gnathostomula, **119**, 122
Gnathostomulida, **119**, 122
Gnorimosphaeroma lutea, **292**, 294, 308
noblei, 294, 308
oregonense, **292**, 294, 308
rayi, 294, 308
Gobiesocidae, 667
Gobiesox maeandricus, 660, 667
Gobiidae, 667
Golfingia hespera, 131
margaritacea californiensis, 131
Golfingiidae, 131
Goniada brunnea, 194, **195**
Goniadidae, 164, 194, **195**
Gonionemus vertens, 77, 83
Gonothyraea clarki, 82
spp., 76, 82
Gonyaulax polyedra, 26, **27**
Goodsiria dura, 653
Gooeyduck. See *Panopea*.
Gordiacean worms. *See* Nematomorpha
Gosliner, Terrence M., 517
Gracilaria verrucosa, 682
Grandidierella japonica, 19, 24, 333, **356**, 360
Grantia sp., 41, 42, 58, 63
Grantiidae, 63
Granula subtrigona, 492, 511
Granulina margaritula, 488, **489**, 511
Grapsidae, 408
Grass shrimp. *See* Hippolytidae
Gribble. *See Limnoria*
Gromia oviformis, 26, **27**
Gymnoblastea, 80
Gymnophila, 469, 515
Gyptis arenicola glabra, 183
brevipalpa, 183
brunnea, 183

Haderlie, Eugene C., 100, 112
Hadromerida, 62
Hadrotes, 446
crassus, 450
Halacaridae, 425–431
Halacarus, 428
frontiporus, **427**, **429**
Halcampa crypta, 87, 92
decemtentaculata, 87, 92
Halcampidae, 92
Haleciidae, 83
Halecium spp., 76, **78**, 83
Halichondria bowerbanki, 42, 44, **49**, 62
panicea, 19, 33, 42, 44, **49**, 62, 538
Halichondrida, 62
Halichondriidae, 62
Haliclona ecbasis, 59
lunisimilis, 59
permollis, 59
spp., 33, 35, 38, 42, 45, **49**, 59

Haliclonidae, 59
Haliclystus auricula, **88**, 96
salpinx, 96
sanjuanensis, 96
stejnegeri, 96
Haliotidae, 499–500
Haliotis, 403, 573
cracherodii, 481, 499
kamtschatkana, 482, 499–500
kamtschatkana assimilis, 499–500
rufescens, 482, 500
walallensis, 481, 500
Haliplanella luciae, 23, **88**, 89, 92, 541
Haliplanellidae, 92
Halisarca sp., 33, 42, 43, 59
Halisarcidae, 59
Hallaxa sp., 529, 540
Halobates sericeus, 433
spp., 433
Halobisium occidentale, 431, **435**
Halosaccion, 253, 255, 675
glandiforme, 682
Halosoma compactum, **422**, 423
viridintestinale, **422**, 423
Halosydna brevisetosa, 175
johnsoni, 175
Halocynthia hilgendorfi igaboja, 644, 653
johnsoni, 653
mirabilis, 653
Haminoea, 516
vesicula, 482, 512, 519, 521
virescens, 482, **483**, 512, 519, 521
Hancockia californica, **525**, 531, 540
Hand, Cadet, 65, 85
Hapalogaster, 379
cavicauda, 399, 409
Haplophragmoides columbiensis var. *evolutum,* **27**, 28
Haplosclerida, 59
Haploscoloplos elongatus, 205, **206**, 207
Haplosyllis spongicola, 185, 189
Harenactis attenuata, 92
Harmothoe hirsuta, 175, 176
imbricata, **161**, **170**, 175, 176
lunulata, **170**, 175, 176
Harpacticoida, 251, **254**, 255, 257
Hartman, Willard D., 32
Haustoriidae, 322, 328, 348
Heart-urchins. *See Lovenia*
Hedgpeth, Joel W., 413
Helcomyzidae, 437, 445
Heleomyzidae, 439, 445
Helix, 468
Hemibranchiata, 293, 294
Hemichordata, 618–619, 638
Hemicyclops, 408
thysanotus, 255, **256**
Hemigrapsus, 286, 305
nudus, 2, 393, **397**, 408
oregonensis, 112, 117, **383**, 393, 408
Hemilepidotus hemilepidotus, 662, 666
spinosus, 662, 666
Hemioniscus balani, **286**, 305
Hemipodus borealis, 196, **196**, 197
californiensis, 196, **196**, 197
Hemiptera, 433, 434, **435**
Henricia leviuscula, 372, 623, 624, 625
Heptacarpus brevirostris, 391, **392**, 402
carinatus, **387**, 391, 402
cristatus, 391, **392**, 403
franciscanus, 403
gracilis, 391, **392**, 403
palpator, 391, **392**, 403
paludicola, 390, 391, **392**, 403
pictus, 391, **392**, 403
taylori, 391, **392**, 403
Herdmania mirabilis, 653
Hermaea ornata, 537
vancouverensis, 523, 537
Hermaeina oliviae, 537
smithi, 537
Hermans, Colin O., 147
Hermissenda, 255
crassicornis, 517, 533, 540
Hermit crabs. *See* Anomura, Paguridae
Hesionidae, 166, **173**, 183–184
Hesionura coineaui difficilis, 183
sp., 178, 183
Hesperonoe, **131**, 408, 409
adventor, **170**, 175, 176
complanata, **170**, 175, 176
Heterocrypta occidentalis, 393, **394**, 405
Heterodonta, 545
Heteromastus filobranchus, 225, 226
filiformis, 225
Heteronemertea, 113, 117
Heterophlias, **317**, **327**, **332**, **336**
seclusus, 330, 360
Heterosaccus californicus, 260, 267
Heterostichus rostratus, 664, 667
Hexacorallia, 85, 86, 91
Hexactinellida, 34
Hexagrammidae, 666
Hexagrammos decagrammus, 660, 666
superciliosus, 660, 666
Hiatella arctica, **571**, 572
pholadis, 572
Hiatellidae, 546, 550, 552, 570, 572
Hildenbrandia, 671
occidentalis, 682
Hincksina velata, 591, **592**, 605
Hinnites, 544
giganteus, 549, **557**
multirugosus, 549
Hippidae, 411

Hippidea, 378, 410–411
Hippodiplosia insculpta, 599, **602**, 607
Hippolyte, 385
californiensis, 390, **392**, 403
Hippolytidae, 402
Hippomedon, **317, 336**
denticulatus, 345, 360
Hipponicidae, 507
antiquatus, 507
cranioides, 475, 507
tumens, 475, **483**, 507
Hippothoa hyalina, **602**
spp., 599, 607
Hirudinea, 146–147
Holopagurus pilosus, 409
Holoporella brunnea, 606
Holothuroidea, 620, 634–637
Homalopoma baculum, 485, **489**, 503
carpenteri, 503
fenestratum, 503
luridum, 485, **489**, 503
paucicostatum, 485, 503
radiatum, 485, 503
Homarus americanus, 20
Hopkinsia rosacea, 526, 540
Hoplocarida, 246, 271
Hoplonemertea, 113, 117
Hoploplana californica, 104, 108, **110**
Hormiphora sp., 99
Horsehair snakes. *See* Nematomorpha
Horseneck clam. *See Tresus*
Horseshoe crab. *See Limulus*
Hyadesia, 425
Hyale, **317, 319**, 323, **327**, 331, **336, 337**, 343
anceps, 344, **350**, 360
frequens, 344, **350**, 360
grandicornis californica, 343, 360
nigra, 360
plumulosa, 344, 360
rubra, 360
Hyalidae, 318, 320n., 323, 326, 343
Hyalinoecia sp., 199
Hydractinia spp., 71, **73**, 74, 80, 541
Hydraenidae, 447, 450
Hydrobiidae, 504
Hydrocoral, 66
Hydrocoryne sp., **73**, 74, 79, 81
Hydrocorynidae, 81
Hydroida, 65–85
Hydroides norvegicus, 241
pacificus, 239, **240**, 241
Hydrophilidae, 446, 447, 450
Hydrozoa, 65–85
glossary, 71–72
key, 72–79
terminology, 66–69
Hymedesanisochela rayae, 42, 52, 61
Hymedesmia brepha, 42, 53, 61
spp., 42, 51, 52, 61
Hymenamphiastra cyanocrypta, 38, 41, 42, 51, 61
Hymeniacidonidae, 62
Hymendectyon lyoni, 42, 52, 61
Hymeniacidon sp., 42, 46, 62
ungodon, 38, 42, 46, 62
Hyperiidea, 313
Hypomesus pretiosus, 659, 665
Hypselodoris californiensis, 528, 540
porterae, 538

Iais californica, 18, 21, 23, **301**, 302, 309
Ianiropsis, 300
analoga, 302, 309
epilittoralis, 302, 309
kincaidi derjugini, 302, 309
kincaidi kincaidi, 302, 309
magnocula, 302, 309
minuta, 302, 309
montereyensis, 302, 309
tridens, 302, 309
Icichthys lockingtoni, 663, 665
Idanthyrsus ornamentatus, **229**, 230
Idarcturus hedgpethi, 287, 305
Idmonea californica, 604
Idotea, 287
(*Idotea*) *fewkesi*, 289, **290**, 305
(*Idotea*) *rufescens*, 289, 305
(*Idotea*) *urotoma*, 289, **290**, 305
(*Pentidotea*) *aculeata*, **290**, 291, 306
(*Pentidotea*) *gracillima*, 306
(*Pentidotea*) *kirchanskii*, **290**, 291, 306
(*Pentidotea*) *montereyensis*, **290**, 291, 306
(*Pentidotea*) *resecata*, 290, **290**, 306
(*Pentidotea*) *schmitti*, 289, 306
(*Pentidotea*) *stenops*, **290**, 291, 306
(*Pentidotea*) *whitei*, 306
(*Pentidotea*) *wosnesenskii*, 289, **290**, 306
Idoteidae, 282, 287, 305
Idothea. *See Idotea*
Illg, Paul L., 250
Illyodrilus frantzi capillatus, 143, 145
Ilyanassa obsoleta, 510
Immergentia, 581
Inachidae, 405
Inachoides tuberculatus, 405
Inarticulata, 616, 617
Insecta, 432–452
Introduced species, 17–25
Ione, 408
brevicauda, 305
cornuta, **285**, 286, 305
Iridaea, 678

cordata var. *cordata*, 683
Iridaea (*continued*)
cordata var. *splendens*, 683
flaccida, **679**, 683
heterocarpa, 683
Iridia serialis, **27**, 28
Iridophycus cordata, 683
flaccida, 683
heterocarpa, 683
splendens, 683
Irregularia, 621
Irus lamellifer, 560, **561**, 562
Isaeidae, 322, 326
Isanthidae, 92
Ischadium, 612
demissum, 19, 24, 547, **553**, 554–555
Ischnochiton fallax, 465
heathiana, 465
interstinctus, 465
radians, 463, 465
regularis, 463, 465
Ischnochitonidae, 464
Ischyroceridae, 322, 323, 326, 330, 331, 335
Ischyrocerus, **336**
anguipes, 335, **336**, 360
minutus, 360
spp., 335, **336**, 360
Iselica fenestrata, 512
ovoidea, **483**, 488, 512
Isobactrus, 425, 426, **429**, 430
Isocheles pilosus, 409
Isociona lithophoenix, 60
Isodictya quatsinoensis, 35, 42, 50, 60
Isodictyidae, 60
Isophelliidae, 93
Isopoda, 272, 281–312

Jaeropsidae, 298, 300, 309
Jaeropsis dubia dubia, 300, **301**, 309
dubia paucispinis, 300, 309
davisi, **301**, 302, 309
lobata, 300, 309
occidentalis, **301**, 302, 309
rajata, 302, 309
solasteri, 300, 309
triangulata, 300, 309
Janiridae, 298, 300, 309
Janthina spp., 506
Janthinidae, 506
Jassa, **332, 336**
falcata, 335, 360
Jellyfish. *See* Scyphozoa
Jordania zonope, 662, 666

Kaburakia excelsa, 105, 108
Kamptozoa, 609
Katharina tunicata, 458, 461, 465
Kellia, 546
laperousii, **557**, 558, 559
suborbicularis, 559
Kelliidae, 546, 550, 558, 559
Kelp. *See* Phaeophyta
Kelp crab. *See Pugettia*
Kelp flies. *See* Coelopidae, Anthomyiidae
Keyhole limpet. *See* Fissurellidae
Kinorhyncha, **119**, 124
Krill. *See* Euphausiacea
Kuris, Armand M., 378, 385

Labiata, 245
Lacrymaria olor var. *marina*, 28, **29**
Lacuna carinata, 503
marmorata, **489**, 491, 503
porrecta, 491, 503
solidula, 503
unifasciata, **489**, 491, 503
vincta, 503
Lacunidae, 503
Lagenipora punctulata, **600**
spinulosa, **600**
spp., 599, 603, 607
Laila cockerelli, 526, 540
Lamellaria, 469, 471, **479**
spp., 473, 487, 508, 515–516
stearnsi, 508
Lamellariidae, 508
Lamellibranchia. *See* Bivalvia
Laminaria, 404, 410, 456, 478, 501, 675
andersonii, 681
dentigera, 501, 681
farlowii, 681
setchellii, 681
sinclairii, 604, 681
Lamp shells. *See* Brachiopoda
Lamprops quadriplicata, 275, 276
sp., 275, 276
Laodicea spp., 77, 83
Laodiceidae, 83
Laonice cirrata, **211**, 213, 214
Lar, 69
Larvacea, 638
Lasaea, 546
cistula, 558–559
rubra, 558, 559
sp., 19, 547, 558, 559
subviridis, **557**, 558, 559
Lasaeidae, 558
Lasiopogon actius, 444
Lebbeus lagunae, 388, 403
Lecythorhynchus hilgendorfi, 418, **419**, 423–424
marginatus, 423
Leeches. *See* Hirudinea
Leioptilus spp., 93
Leiosella idia, 59

Lembos, **341**
  spp., 333, 360
Lepadomorpha, 262, 267–268
Lepas, 260, **261**, 408, 540
  (Dosima) fascicularis, 262, 267
  (Lepas) anatifera, **261**, 263, 267
  (Lepas) pacifica, **261**, 262, 267
Lepidasthenia interrupta, **168**, 174, 176
Lepidonotus squamatus, 174, 176
Lepidopleuridae, 465
Lepidozona cooperi, **459**, 463, 465
  mertensii, **459**, 463, 465
  sinudentata, 463, 465
Leptasterias, 11
  aequalis, 626
  hexactis, 624, 626
  pusilla, 624, 626
Leptochelia dubia, **279**, 279, 280
Leptochiton rugatus, 462, 465
Leptoclathria asodes, 42, 53, 60
Leptocottus armatus, **661**, 662, 666
Leptomedusae, 71, 82
Leptoplana chloranota, **103**, 108
Leptostraca, **249**, 270, 271
Leptosynapta albicans, 635, 636
  spp., 636
Lernaeodiscus, 410
Lessoniopsis, 673
  littoralis, **674**, 681
Lethops connectens, 659, 667
Leucandra heathi, 32, 42, 58, 63
Leucilla nuttingi, 33, 41, 42, 58, 64
Leuckartiara octona, **73**, 74, 79, 81
Leuconia heathi, 63
Leucosiidae, 405
Leucosolenia, 40
  eleanor, 32, 42, **57**, 58, 63
  sp., 43, **57**, 58, 63
Leucosoleniida, 63
Leucosoleniidae, 63
Leucothea sp., 99
Leucothoe, **327**, **337**, **346**
  alata, **341**, 342, 360
  spinicarpa, **341**, 342, 360
Leucothoidae, 326, 342
Leucothoides, **346**, **350**
  pacifica, 342, 361
Lichenopora, 582
  spp., 586, **587**, 604
Light, S. F., ix–x
Lightiella serendipita, 246
Ligia (Ligia) pallasii, **301**, 303, 310
  (Megaligia) occidentalis, **301**, 303, 310
Ligidium gracilis, 303, 310
  latum, 303, 310
  sp., **301**
Ligiidae, 303, 310
Lignophliantis, **329**, **337**
  pyrifera, 331, 361
Liljeborgia, **325**, **346**
  spp., 349, 361
Liljeborgiidae, 326, 349
Limnodrilus hoffmeisteri, **139**, 141, 143, 145
Limnomedusae, 83
Limnoria, 281, 295
  (Limnoria) quadripunctata, 18, **292**, 296, 307
  (Limnoria) lignorum, 23, 296, 307
  (Limnoria) tripunctata, 18, 296, 307
  (Phycolimnoria) algarum, **292**, 295, 307
Limnoriidae, 293, 295–296, 307
Limonia (Idioglochina) marmorata, **437**, 439, 443
Limpets. *See* Acmaeidae
Limulus, 413
Lineus pictifrons, 116, 117
  ruber, 114, 117
  vegetus, 115, 116, 117
Liotia fenestrata, **483**, 487, 503
Liotiidae, 503
Liparidae, 667
Liparis florae, 659, 667
Liparocephalus, 447
  cordicollis, **448**, 449, 450
Lipochaeta, 444
Liriope balani, 305
Lironeca californica, 297, 307
  vulgaris, 296, 307
Lirularia funiculata, 487, 502
  parcipicta, 502
  succincta, 487, **494**, 502
Lissodendoryx firma, 40, 43, **49**, 52, 61
  noxiosa, 61
  topsenti, 43, 47, **49**, 61
Lissothuria nutriens, 11, 635, 637
Listriella spp., 349, 361
Listriolobus pelodes, **130**, 133
Lithodidae, 409
Lithophaga plumula kelseyi, 553, **553**, 555
  subula, 555
Lithophyllum, 500, 671
Lithothamnium, 500, 671
  californicum, 683
  pacificum, 683
Littleneck clam. *See Protothaca, Tapes*
Littorina, 10
  littorea, 20, **489**, 503
  planaxis, **489**, 491, 504
  scutulata, **489**, 491, 504
Littorinidae, 503–504
Littorophiloscia richardsonae, **301**, 304, 310
Livoneca. *See Lironeca*
Lobata, 98
Lobsters. *See* Palinura, *Homarus*

Lohmannella, 430
falcata, **427**
Loimia medusa, **233**, 234
Loligo, 121
opalescens, 456
Lophopanopeus bellus, **397**, 399, 406
leucomanus heathii, **397**, 399, 406
Lophophorates, 609, 613–617
Lottia, 500
gigantea, 478, **483**, 501
Lovenia cordiformis, 662
Loxorhynchus crispatus, 33, 395, **397**, 405
Loxosoma spp., 610, **611**, 612
Loxosomatidae, 610, 612
Loxosomella spp., 610, **611**, 612
Lucinidae, 546
Lumbricillus belli, 144
georgiensis, **139**, 144
mirabilis, 144
qualicumensis, 144, 145
santaeclarae, 145
vancouverensis, **139**, 144, 145
Lumbrineridae, 152, 162, 164, 202–203
Lumbrineris erecta, 203, **204**
japonica, 202, 203
latreilli, **154**, 202, 203, **204**
tetraura, 203, **204**
zonata, 203, **204**
Lycastopsis pontica neapolitana, 193
Lyonsia californica, **571**, 577
Lyonsiidae, 546, 550, 575, 577
Lyrodus pedicellatus, 18, 22, 547, 575, **576**
Lyrula hippocrepis, 594, **595**, 605
Lysianassa, **317**, **336**, **337**, **341**
spp., 344, 361
Lysianassidae, 323, 324, 326, 344–345
Lysidice sp., **200**, 201
Lytechinus anamesus, 622
pictus, 622

Macoma, 9, 507, **548**
acolasta, **567**, 568
balthica, 23, **567**, 568, 569
expansa, **567**, 569
inconspicua, 569
indentata, 566, **567**, 569
inquinata, **567**, 568, 569
irus, 569
nasuta, **565**, **567**, 568, 569
secta, 118, **565**, 566, **567**, 569
yoldiformis, 566, **567**, 569
Macrocystis, 83, 96, 108, 109, 290, 306, 358, 359, 502, 538, 539, 540, 606, 673
integrifolia, **672**, 681
pyrifera, 681
Mactra californica, 564
dolabriformis, 564
Mactridae, 546, 551, 563–564
Madreporaria, 86, 93
Maera, **317**, **336**
inaequipes, 352, 361
simile, 351, 361
sp., **346**, 352, 361
vigota, **350**, 351, 361
Magelona pitelkai, 216, **217**
sacculata, 216, **217**
Magelonidae, 169, 216, **217**
Majidae, 385, 405
Malacobdella grossa, 112, 113, 118, **119**
Malacostraca, 270–412
Maldanidae, **157**, **167**, 169, **227**
Manania sp., 96
Manayunkia speciosa, 236, 238, **240**
Mandibulata, 244
Mandibulophoxus, **325**
gilesi, 348, 361
Mangelia interlirata, 511
Mantis shrimps. *See* Hoplocarida
Margarites parcipictus, 502
pupillus, 502
salmoneus, 487, **494**, 502
succinctus, 502
Marginellidae, 511
Marionina subterranea, 143, 145
Mariscal, Richard N., 609
Market crab. *See Cancer*
Marphysa, **163**
stylobranchiata, **200**, 201
Marsenina, 508
Masking crabs. *See* Majidae
Mayerella banksia, 368, 372, **373**
McCain, John C., 367
McDonald, Gary R., 522
Mediaster aequalis, 623, 625
Mediomastus californiensis, **223**, 225, 226
Megalomma splendida, 235, **237**, 238
Megaluropus, **325**, **336**, **337**, **346**
spp., 351, 361
Megamphopus effrenus, 333, **347**, 361
martesia, 333, 361
Megatebennus bimaculatus, 474, **479**, 500
Megathura, 407
crenulata, 474, **479**, 500
Melampidae, 513
Melanderia (Melanderia) crepuscula, 440, 442, 443
mandibulata, **438**, 440, 442, 443
Melibe leonina, **520**, 532, 540
Melita, 323, **325**, **336**, 352
appendiculata, **347**, 352, 361
californica, 352, 361
dentata, 352, 361
fresneli, 361
sp., 361
sulca, **341**, 352, 361
Melobesia, 671

Melongenidae, 509–510
Melyridae, 446, 449, 451
Membranipora, 538, 539, 581, 681
fusca, 591, **592,** 606
membranacea, 589, **590,** 606
perfragilis, 589, **590,** 606
serrilamella, 591, **592,** 606
tuberculata, **590,** 591, 606
villosa, 591, **592,** 606
Mercenaria mercenaria, 20, 547, 560, **561,** 562
Mercierella enigmatica, 18, 21, 239, **240,** 241
Mermaid, 1
Mertensia ovum, 99
Mesochaetopterus, 131
taylori, **217,** 218
Mesocrangon, 404
Mesodinium pulex, 28, **29**
Mesogastropoda, 468–469, 503–508
Mesoglossus sp., 619
Mesometopa esmarki, 343, 361
sinuata, 343, **350,** 361
Mesonerilla, **148,** 149
intermedia, **148**
"Mesozoa," **119,** 121–122, 456
Metacaprella anomala, 370, **371,** 376
kennerlyi, 370, **371,** 376
Metandrocarpa dermatina, 653
dura, 647, 653
michaelseni, 653
taylori, 647, 653
Metaphoxus frequens, **341,** 349, 361
fultoni, **341,** 349, 361
Metapseudidae, 279
Metaxia convexa, 492, 506
diadema, 506
Metopa cistella, **317,** 343, 361
Metoponorthus pruinosus, 311
Metridiidae, 93
Metridium, 10, 424
exilis, 89, 93
senile, 22, 86, 89, 93
senile fimbriatum, 93
Metriochemini, 443
Microcerberidae, 308
Microcerberidea, 283, **292,** 297, 308
Microcerberus abbottii, **292,** 297, 308, 309
Microciona, 60, 61
microjoanna, 38, 43, 55, **57,** 61
parthena, 43, 56, **57,** 61
prolifera, 19, 33, 40, 43, 55, **57,** 61
Microdeutopus, **341**
schmitti, 333, 362
Microdriles, 136
Microhydra, 83
Microjassa litotes, 335, 362
Micrometrus aurora, 664, 665
minimus, 664, 665
Micronereis sp., 178
Microphthalmus sczelkowii, 183
Microporella californica, 598, **600,** 607
ciliata, 598, **600,** 607
cribrosa, 598, **600,** 607
Micrura, 113
alaskensis, 115, 117
pardalis, 116, 117
verrilli, 116, 117, **119**
Midges. *See* Diptera
Miller, Milton A., 277
Milneria minima, 563
Mimulus foliatus, **394,** 395, 405
Mitella polymerus, 268
Mites. *See* Halacaridae
Mitrella aurantiaca, 493, 510
carinata, 493, **494,** 510
gausapata, 510
tuberosa, 493, **494,** 510
Mitromorpha aspera, 511
carpenteri, **494,** 495, 511
filosa, 511
Modiolus capax, **553,** 554, 555
carpenteri, 554, 555
demissus, 554
flabellatus, 555
fornicatus, 555
rectus, 554, 555
sacculifer, 555
senhousia, 555
Mola mola, 656
Molgula, 255
manhattensis, 19, 23, 645, 653
sp., 645, 653
verrucifera, 645, 653
Mollusca, 453–578
Molpadida, 636
Monocelis cincta, **102,** 108
Monokonophora, 278, 279
Monopylephorus irroratus, **139,** 143, 145
Monotocardia, 468
Montacutidae, 546, 550, 558, 559
Moon snail. *See Polinices*
Mopalia ciliata, 461, 465
hindsii, 101, 108, 462, 465
hindsii recurvans, 465
imporcata, 465
lignosa, 462, 465
lowei, 461, 462, 466
muscosa, **459,** 461, 465, 466
porifera, 461, 466
Mopaliidae, 465
Mucronella major, 601, **602,** 607
Mud shrimp. *See Upogebia*
Munna (Munna) halei, **299,** 300, 310
(Neomunna) chromatocephala, **299,** 300, 310
(Neomunna) stephenseni, **299,** 300, 310
(Uromunna) ubiquita, 298, **299,** 310

Munnidae, 298–300, 310
Muricidae, 470, 508–509
Mursia gaudichaudii, 393, **394**, 405
Musculus pygmaeus, 552, **553**, 555
senhousia, 19, 23, 547, 554, 555
Mussels. *See* Mytilidae
Mya arenaria, 19, 22, 24, 547, **548**, 570, 571
hemphillii, 22
Mycale macginitiei, 40, 43, 50, 60
richardsoni, 43, 50, 60
Mycalidae, 60
Myidae, 546, 550, 570
Myoida, 546
Myosoma spinosa, 610, 611, 612
Myriapoda, 245
Mysella, 546
golischi, 559
pedroana, 559
spp., 558, 559
tumida, 559
Mysidacea, **249**, 272–273
Mysis sp., **249**
Mytilicola orientalis, 19, 22, 251, 255, **256**
ostrea, 22
Mytilidae, 545, 550, 552–555
Mytilimeria nuttalii, 546, **571**, 575, 577
Mytiloida, 545
Mytilus, 131, 512, 554, 573, 612
californianus, 33, 268, 407, 423, 552, **553**, 555, 626
edulis, 20, 22, 251, 255, 552, **553**, 555
Myxicola infundibulum, 235, **237**, 238
Myxicolinae, 235
Myxilla agennes, 51, 61
Myxillidae, 61

Naididae, 136, 138, 140, 141, 142, 145
Naineris dendritica, 205, **206**, 207
Nais communis, 143, 145
Najna, **317**
consiliorum, 362
sp., 362, 433
Najnidae, 324, 343
Namanereinae, 193
Nassa, 510
Nassariidae, 510
Nassarius, 469
cooperi, 510
fossatus, 493, **496**, 510
mendicus, 81, 493, **496**, 510
obsoletus, 19, 23, 495, **496**, 507, 510
perpinguis, 510
Natantia, 378, 402–404
Naticidae, 507–508
Nautilus, 468
Navanax inermis, 521
Neanthes brandti, 192, **192**, 193
caudata, 18
diversicolor, 193
lighti, 193
limnicola, 11, **192**, 193
succinea, 19, 192, **192**, 193
virens, 192, 193
Nebalia, 123
bipes, **249**, 271
pugettensis, **249**, 271
Nebaliacea, **249**, 271
Nectonema, 122, 125
Nectonemertes, 112
Nematocera, 436
Nematoda, **119**, 124–125
Nematomorpha, 125
Nematostella vectensis, 87, **88**, 92
Nemertea, 112–120
Nemertopsis gracilis, 116, 118
Neoamphitrite robusta, 234
Neocyamus physeteris, **375**
Neogastropoda, 469, 508–511
Neomachilis halophila, 434
Neomysis awatschensis, 273
mercedis, 273
Neoturris spp., 79, 81
Nephtyidae, **163**, 164, 197–199
Nephtys caeca, 163, 197
caecoides, 197, **198**
californiensis, 197, **198**
cornuta, **198**
cornuta franciscana, 197, **198**, 199
parva, 197, 199
Nereidae, 20, **157**, 159, 166, 178, 190–194
Nereis, 151, 152, **153**, 155
eakini, 190, **191**, 193
grubei, 190, **192**, 193
latescens, 191, **192**, 193
mediator, 193
natans, 194
pelagica neonigripes, 190, **192**, 194
procera, 190, **191**, 194
vexillosa, 190, **191**, 194
Nereocystis, 673
luetkeana, **674**, 681
Nerilla antennata, 149
Nerillidae, 149
Nerillidium, **148**, 149
simplex, **148**
Nerinides acuta, **211**, 213, 214
tridentata, 213, 214
Nerocila californica, 296
Netastoma rostrata, **549**, 572, 573, **576**
Nettastomella rostrata, 573
Newell, Irwin M., 425
Newman, William A., 259
Nexilis epichitonius, 100–101, **102**, 108, 465
Nocticanace arnaudi, 442, 444
Noctiluca scintillans, 26, **27**

Notaspidea, **498,** 512, 516, 517, 518, 522
Nothria elegans, **198,** 199, 201
iridescens, **198,** 199, 201
stigmatis, 199, 201
Notoacmea fenestrata, **476,** 480, 501
fenestrata cribraria, 501
insessa, 9, **476,** 480, 501, 502, 680
paleacea, **476,** 480, 502
persona, **477,** 481, 502
scutum, **476,** 481, 502
Notomastus magnus, 225, 226
tenuis, 225, 226
Notophycidae, 162, **173,** 178
Notoplana acticola, 108, **110**
inquieta, 105, 108
rupicola, 106, 108
saxicola, 106, 109
Notostraca, 245, 247
Nototropis tridens, 359
Nucella, 10, 11
canaliculata, **496,** 497, 509
canaliculata compressa, 509
emarginata, **496,** 497, 509
lamellosa, **496,** 497, 509
Nucula, 543
Nuculoida, 543, 545
Nuda, 98, 99
Nudibranchia, 515, 518, 523–542
Nuttallia nuttallii, 566, 568
Nuttallina californica, **459,** 462, 464, 465
thomasi, 464
Nybakken, James W., 515
Nymphon, 414, **415**
Nymphonidae, 414
Nymphopsis spinosissima, 417, **419,** 424

Obelia, 26, **67,** 69, 681
geniculata, 82
longissima, 82
spp., 18, 23, 76, 77, 82
Ocenebra atropurpurea, 499, 508–509
circumtexta, **498,** 499, 509
clathrata, 508–509
gracillima, 499, 509
interfossa, **498,** 499, 509
japonica, 508
lurida, **498,** 499, 509
munda, 509
sclera, 509
subangulata, 509
Ochthebius, 447
vandykei, 447, 450
Ocosingo, **327, 336**
borlus, 344, 362
Octocorallia, 86, 93
Octolasmis californiana, 262, 267
Octopoda, 456
Octopus apollyon, 456
bimaculatus, 456
bimaculoides, 456
dofleini martini, 456
rubescens, 456
Odontosyllis, **161**
parva, 187, 189
phosphorea, 187, **188,** 189
Odostomia bisuturalis, 513
columbiana, 513
spp., 488, 513
(Evalea) phanea, **484,** 513
(Evalea) tenuisculpta, 513
(Menestho) fetella, **484,** 513
Oedicerotidae, 326, 345
Oedignathus, 379
inermis, 399, 410
Oedoparena glauca, **438,** 442, 444
Okenia angelensis, 527, 540
plana, 19, 526, 540
Oligochaeta, 136–146
Oligochinus, 323, **337**
lighti, 345, 362
Oligocottus maculosus, 663, 666
rimensis, **661,** 663, 666
rubellio, 663, 666
snyderi, **657,** 663, 666
Olindiadidae, 83
Olivella baetica, 495, **496,** 511
biplicata, 81, 495, **496,** 511
pycna, 495, **496,** 511
Olividae, 511
Omalogyra atomus, 504
sp., 482, 504
Omalogyridae, 504
Oncoscolex pacificus, 222, **223**
Onchidella, 469
borealis, 515, **520**
Onchidiacea, 515
Onchidoris aspera, 540
bilamellata, 526, 540
hystricina, 530, 540
muricata, 531, 540
Oniscidae, 303, 310
Oniscoidea, 281, 283, 302–304, 310–311
Onuphidae, 152, 156, **198,** 199, 201
Onuphis spp., 201
Opalia chacei, **489,** 490, 506
montereyensis, **489,** 490, 506
wroblewskyi, 506
Ophelia assimilis, 222, 224
limacina, **223,** 224
pulchella, 224
Opheliidae, 159, 170, 222–224
Ophiodromus pugettensis, **173,** 183–184, 625
Ophioncus granulosus, 630, **632,** 633
Ophionereis eurybrachyplax, 630, 633

Ophiopholis, 628
  aculeata, **629**
  aculeata form *kennerlyi*, 631, **632**, 633
  aculeata form *typica*, 633
  bakeri, 631, **632**, 633
Ophioplocus esmarki, 630, **632**, 633
Ophiopteris papillosa, 631, 633
Ophiothrix, 628
  spiculata, **629**, 630, **632**, 633–634
Ophiuroidea, 620, 627–634
  anatomy, 627–628, **629**
Ophlitaspongia pennata, 33, 34, 38, 40, 43, **49**, 50, 61
Ophryotrocha puerilis, 166, **200**, 202
Opisthobranchia, 467, 469, 512, 515–542
Opisthopus transversus, 398, 407, 500
Opossum shrimps. *See* Mysidacea
Orbinia johnsoni, 205, **206**, 207
Orbiniidae, **165**, 171, 205–207
Orchestia, 9, 364
  chiliensis, 19, 22, 353, **356**, 363
  enigmatica, 22, 363
  georgiana, 353, 363
  traskiana, 355, **356**, 363
Orchestoidea, 9, 314, 352
  benedicti, **354**, 355, 357, 363
  californiana, **354**, 355, **356**, 357, 363
  columbiana, **354**, 355, **356**, 357, 364
  corniculata, **354**, 355, **356**, 357, 364
  pugettensis, **354**, 355, **356**, 357, 364
  spp., 425
Orchomene, **317**, **336**
  pacifica, 345, 362
Oregonia gracilis, 395, 405
Oregoniplana opisthopora, 101, 108
Oriopsis gracilis, 236, 238
Orobitella rugifera, 382, 409, 546, 558, 559, 576
Orthasterias koehleri, 625, 626
Orthocladiinae, 443
Orthonectida, **119**, 121
Orthonopias triacis, 662, 666
Osmeridae, 665
Ostracoda, 245, 247–248, **249**
Ostrea, 544
  edulis, 556
  lurida, 20, 556
Ostreidae, 545, 550, 556
Ovatella, 28
  myosotis, 18, 22, 24, 469, **484**, 488, 513
Owenia, **165**
  collaris, 228
  fusiformis, 228
Oweniidae, **165**, 171, 228
Oxylebius pictus, 660, 666
Oxyrrhis marina, 26, **27**
Oysters. *See* Ostreidae
Pachycerianthus fimbriatus, 87, 93, 539
  plicatulus, 93
  torreyi, 93
Pachychalina lunisimilis, 43, 45, **49**, 59
Pachycheles, 384
  pubescens, 399, **400**, 410
  rudis, 286, 305, 399, **400**, 410
Pachygrapsus, 28, 381, 384
  crassipes, 2, 10, **383**, 393, 408
Pachythyone rubra, 635, 637
Pacifastacus leniusculus, 386
  nigrescens, 386
  trowbridgii, 386
Pacificides psammophilus, 101, **102**, 108
Pagurapseudes sp., 277, 278, 280
Pagurapseudidae, 280
Paguridae, 409
Paguridea, 378, 409–410
Pagurus, 379
  beringanus, 402, 409
  granosimanus, **400**, 402, 409
  hemphilli, 401, 409
  hirsutiusculus, 305, **383**, 402, 409
  quaylei, 409
  samuelis, **400**, 401, 409
Palaemon macrodactylus, 18, 24, 385, **387**, 390, 402
Palaemonidae, 402
Palaeotaxodonta, 545
Paleanotus bellis, 177
Paleonemertea, 113, 116
Palinura, 378
Palola paloloides, 201
Pancolus californiensis, 279, 280
Pandalidae, 402
Pandalus danae, 390, 402
Pandeidae, 81
Panope. *See Panopea*
Panopea generosa, **561**, 572
Panoploea, **329**, **341**
  spp., 345, 362
Paphia, 562
  bifurcata, 22
  staminea, 562
Papyrula saccharis, 56, 63
Paracerceis cordata, 295, 308
Paraclunio alaskensis, **438**, 439, 440, **441**, 443
  trilobatus, 440, 443
Parajassa, **337**
  angularis, 335, 362
Parallorchestes, 323, **336**
  ochotensis, 343, 362
Paramoera, **325**
  mohri, 348, 362
Paranais frici, 143, 145
  litoralis, 141, 143, 145

Paranaitis, **163**
Paranemertes peregrina, 115, 118
Paranerilla, 149
limicola, **148**
Paranthessius columbiae, **256**
spp., 251, 255
Paranthura elegans, 291, **292**, 306
Paraonidae, 171, 207, **217**
Paraonides, 207
Paraonis, 207
gracilis, 207, **217**
Parapholas californica, 573, **574**
Paraphoxus, **325, 336, 350**
cognatus, **319**, 349, 362
epistomus, **319, 341**, 349, 362
milleri, 349, 362
obtusidens, **319**, 349, 362
spinosus, 349, 362
tridentatus, 349, 362
Parapleustes, **317**
nautilus, **346**, 348, 362
pugettensis, **346**, 348, 362
subglaber, 362
Paraprionospio pinnata, **211**, 213, 214
Paraselloidea, 298, 309
Parasitiformes, 425
Parasmittina collifera, 601, **602**
trispinosa, 601, **602**, 607
Paratanaidae, 280
Parathalassius aldrichi, 444
melanderi, 444
Paraxanthias taylori, **397**, 398, 406
Pareurythoe californica, 177, 178
Parthenopidae, 405
Parvatrema borealis, 19
Parydra, 444
Patellacea, 516
Patiria, 10
miniata, 184, 623, 624, 625
Pea crabs. *See* Pinnotheridae
Peanut worms. *See* Sipuncula
Pearse, John S., 621
Pebble crabs. *See* Xanthidae
Pectinaria, 407
californiensis, 230–**231**
Pectinariidae, 171, 230–231
Pectinibranchia, 468, 469
Pectinidae, 545, 549
Pedicellina cernua, 610, **611**, 612
Pedicellinidae, 610, 612
Peisidice aspera, 176
Peisidicidae, 162, 176
***Pelagia colorata*, 95**
***noctiluca, panopyra*, 95**
Pelecypoda. *See* Bivalvia
Peloscolex, **139**, 141
apectinatus, 141, 143, 145
gabriellae, 141, 143, 145
nerthoides, 141, 143, 145
Peltogasterella gracilis, 260, 267
Pelvetia, 673, 681
fastigiata, 681
Pelvetiopsis, 673
limitata, 681
Penaeidea, 378
Penares cortius, 34, 43, 56, 63
saccharis, 34, 43, 56, **57**, 63
Penetrantia, 581
Penitella conradi, 573, **574**
gabbii, 573
penita, 573, **574**, 575
turnerae, 575
Pennatulacea, 86, 91, 93
Pentadibranchiata, 293, 295
Pentamera sp., 637
Pentidotea. See Idotea
Peracarida, 271, 272
Perigonimus repens, 81
Perinereis monterea, 191, 194
Perischoechinoidea, 621
Periwinkles. *See Littorina*
Perophora, 605
annectens, 646, 653
Perotripus brevis, 368, **371**, 372
Petaloconchus, 9
montereyensis, 473, **483**, 505, 513
Petelodoris spongicola, 538
Petricola, 28
carditoides, **561**, 563
pholadiformis, 19, 547, **561**, 563
Petricolidae, 546, 551, 552, 563
Petrocelis, 671
franciscana, 683
Petrolisthes, 9, 379, 381, 385
cabrilloa, 410
cinctipes, **383, 400**, 401, 410
eriomerus, 401, 410
gracilis, 410
manimaculis, 401, 410
rathbunae, 399, 410
Petrospongium rugosum, 680
Peyssonellia, 671
meridionalis, 683
Phacellophora, 267
camtschatica, 95
Phaeophyta, 680–681
Phallusia ceratodes, 652
Phanerozonia, 625
Phascolosoma agassizii, 129, **130**, 131, 612
Phascolosomatidae, 131
Phasianellidae, 503
Pherocera sp., 444
Pherusa, **165**

Pherusa (*continued*)
inflata, 221, 222, **223**
papillata, 221, 222, **223**
Phialella sp., **75**, 79, 83
Phialellidae, 76, 83
Phialidium gregarium, 82
spp., **75**, 76, 79, 82
Phidiana nigra, 540
pugnax, 533, 540
Phidolopora, 582
pacifica, 599, **602**, 607
Philobrya setosa, 545, 550, **576**
Philobryidae, 545, 550
Philodina, **119**, 123
Phlebobranchia, 652
Phliantidae, 324, 330
Pholadidae, 546, 549, 572–575
Pholadidea ovoidea, 573
penita, 575
Pholadomyoida, 546
Pholeterides furtiva, 255, **256**
Pholididae, 667
Pholoe glabra, **172**, 176, 177
minuta, 177
tuberculata, **172**, 176, 177
Phoronida, 609, 613–616
Phoronis, 615
architecta, 615
hippocrepia, 616
ijimai, 616
ovalis, 615
pallida, **614**, 615
psammophila, 615
vancouverensis, **614**, 615, 616
Phoronopsis, 92
californica, 616
harmeri, 616
viridis, 613, **614**, 616
Photidae, 322, 326
Photis, 323, **332**, 333, **336**
bifurcata, 333, **346**, 362
brevipes, 333, **347**, 362
californica, 333, **347**, 362
conchicola, 333, **347**, 362
Phoxichilidium femoratum, 422, 423, 424
quadridentatum, 422, 424
Phoxocephalidae, 316, 322, 328, 348–349
Phragmatopoma, 521
californica, **229**, 230
Phreatoicidea, 281
Phtisicidae, 367, 368, 372
Phyllaplysia taylori, 521, 522
zostericola, 522
Phyllobranchopsis enteromorphae, 537
Phyllochaetopterus prolifica, 216, **217**, 218
Phyllodoce, **153**
sp., 180, 183
Phyllodocella bodegae, **173**, 178
Phyllodocidae, **157**, 159, 162, **163**, 178–183
Phyllodurus abdominalis, **285**, 286, 305, 409
Phyllospadix, 9, 92, 117, 177, 226, 255, 291, 306, 404, 423, 480, 503, 647, 652, 684
scouleri, 669
torreyi, 502, 669
Physcosoma agassizii, 131
Phytia setifer, 513
Phytichthys chirus, 665, 667
Piddocks. *See* Pholadidae
Pilargiidae, 166, 184, 186
***Pilargis berkeleyae*, 184, 186**
***maculata*, 184, 186**
Pill bugs. *See* Oniscoidea
Pinnixa, 379
barnharti, 407
faba, **394**, 396, 407
franciscana, 398, 407
littoralis, **394**, 396, 407
longipes, **394**, 398, 407
occidentalis, 407
tubicola, 398, 407
schmitti, 407
spp., 398
weymouthi, 407
Pinnotheres concharum, 407
Pinnotheridae, 385, 407
Pinter, Penny A., 579
Pionosyllis gigantea, 187, 189
Pisaster brevispinus, 625, 626
giganteus, 625, 626
ochraceus, 625, 626
Pisidiidae, 546, 551
Pisidium casertanum, 551
Pismo clam. *See Tivela*
Pista, 407
brevibranchiata, 234
elongata, **233**, 234
pacifica, **233**, 234
Pistol shrimp. *See Alpheus*
Placida dendritica, 523, 537
Placiphorella velata, **459**, 461, 466
Plagiogrammus hopkinsi, 664, 667
Planes cyaneus, 408
marinus, 408
minutus, 408
Platichthys stellatus, 659, 665
Platybranchiata, 293, 294
Platydoris macfarlandi, 528, 540
Platyhelminthes, 100–111
Glossary, 101, 104
Key, 104–107
Platynereis bicanaliculata, 190, 191, 194
Platyodon cancellatus, 570, **571**
Pleonexes aptos, 358
Pleurobrachia bachei, **97**, 98, 99

Pleurobranchus californicus, 552
strongi, 522
Pleurogona, 653
Pleuroncodes planipes, 410
Pleuronectidae, 665
Pleusirus, **341, 347**
secorrus, 348, 362
Pleustes, **329**
depressa, 348, 362
Pleustidae, 322, 323–324, 326, 345
Pleusymptes subglaber, **346**, 348, 362
Plocamia karykina, 38, 43, 54, 62
Plocamiidae, 62
Plocamissa igzo, 40, 43, **49**, 54, 62
Plumularia plumularoides, 83
setacea, 83
spp., 76, **78**, 83
Plumulariidae, 71, 83
Podarke pugettensis, 183
Podoceridae, 322, 324, 330
Podocerus, **319**, 324, 331, **332, 337**
brasiliensis, 331, **346**, 363
cristatus, 331, 363
falcatus, 360
spongicolus, 331, **346**, 363
Pododesmus cepio, 213, 214, 549
macrochisma, 549
Podon, 247
Poduridae, 433
Poecillastra tenuilaminaris, 38
Poecilosclerida, 60
Polinices, 409, 469
lewisii, **484**, 490, 507–508
Pollicipes, 501
polymerus, 260, **261**, 262, 268
Polycera atra, 524, 540
hedgpethi, 524, 540
sp., **527**
tricolor, 524, 540
zosterae, 524, 540
Polychaeta, 151–243
Collection and Preservation, 157–158
Dissection, 158
Errantia, 151
Glossary, 159–160
Index to Families, 160
Key to Families, 162
Morphology, 151–157
Sedentaria, 151, 155
Polycheria, **329, 350**
antarctica, 363
osborni, 340, 363
Polychoerus carmelensis, 100, **102**, 107
Polycirrus spp., 232, 234–235
Polycladida, 101, 108–109
Polyclinum planum, 650, 652
Polydora, **157**, 213, 409
alloporis, 214
brachycephala, **209**, 212, 214
commensalis, **209**, 210, **211**, 214
convexa, 212, 214
elegantissima, **209**, 210, 214
giardi, 212, 214
ligni, 19, **209**, 210, 214
limicola, **211**, 212, 215
nuchalis, **209**, 210, 215
pygidialis, **209**, 212, 215
socialis, **211**, 212, 215
spongicola, **211**, 212, 215
tridenticulata, 214
websteri, 212, 215
Polygordiidae, 147
Polygordius, 147, **148**, 149
Polymastia pachymastia, 33, 43, 46, 62
Polymastiidae, 62
Polyneura, 678
latissima, **679**, 683
Polynoidae, 159, **161**, 162, **168, 170**, 174–176
Polyophthalmus pictus, 222, 224
Polyorchidae, 81
Polyorchis haplus, 81
montereyensis, 81
pacifica, 82
penicillatus, 81, **97**
spp., 77, 81
Polyphaga, 447
Polyplacophora, 457–466
Polysiphonia, 512, 537
pacifica, 683
paniculata, 683
Polyzoa. *See* Ectoprocta
Pontobdella, 146
Pontogeneia, **319, 337**
inermis, 345, 363
intermedia, **332**, 345, 363
rostrata, **332**, 345, 363
Pontogeneiidae, 322, 323, 326
Pontomalota, 446
californica, 449, 450
luctuosa, **448**
nigriceps, 449, 451
Pontomyia natans, 432
Porcelain crabs. *See* Porcellanidae
Porcellanidae, 385, 410
Porcellidium sarsi, **254**
spp., 253, 255
"Porcellio" littorina, 304, 310
Porcellio (*Mesoporcellio*) *laevis*, 301, 304, 310
(*Porcellio*) *dilatatus*, **301**, 304, 310
(*Porcellio*) *scaber americanus*, 304, 310
(*Porcellio*) *scaber scaber*, **301**, 304, 311
(*Porcellio*) *spinicornis occidentalis*, 310
Porcellionides pruinosus, 304, 311
Porella porifera, **602**, 603, 607

Porichthys notatus, 664, 667
Porifera, 32–64
Glossary, 41
Key, 43–59
Spicules, 37–41, **49, 57**
Techniques, 35
Porphyra, 678
lanceolata, 683
occidentalis, 683
perforata, **679,** 683
Portunion conformis, **285,** 286, 305
Postelsia, 464, 501, 673
palmaeformis, 681
Potamididae, 505
Prawns. *See* Natantia
Praxillella affinis pacifica, 228
Precuthona divae, 536, 541
Priapula, 133–134
Priapulus caudatus, **130,** 133
Prionitis, 678
andersonii, 683
lanceolata, 683
Prionospio cirrifera, **211,** 213, 215
pinnata, 214
pygmaeus, **211,** 213, 215
Pristes oblongus, 559
Proales, 123
Proboscidactyla, 239
circumsabella, **75,** 84
flavicirrata, **75,** 84
spp., 72, 79, 84
Proboscidactylidae, 84
Procambarus clarkii, 386
Procerodes pacifica, 108
Prosobranchia, 467, 468–469, 499–511, 515
Prostheceraeus bellostriatus, 107, 109
Prostoma rubrum, 112, 118
Prosuberites sp., 19, 43, 47, 62
Protoaricia sp., 205, 207
Protobranchia, 545
Protochordates, 638
Protodorvillea gracilis, **200,** 202
Protodrilidae, 149
Protodriloides, **148,** 149
Protodrilus, **148,** 149
Protonymphon, **422**
Protothaca, 255, 407, 562
laciniata, 562
restorationensis, 562
staminea, 560, 562
staminea orbella, 562
tenerrima, 560, 562
Protozoa, 26–31
Protrichoniscus heroldi, 304, 311
Psammobiidae, 546, 550, 552, 564, 566, 568
Psettichthys melanostictus, 659, 665
Pseudione sp., 287, 305
Pseudocella triaulolaimus, **119**
Pseudoceros canadensis, 106, 109, **110**
luteus, 106, 109
montereyensis, 106, 109
Pseudochama exogyra, 556, **557,** 558
Pseudodiaptomus euryhalinus, 251, **252,** 255
Pseudomelatoma torosa, 491, 511
Pseudopallene sp., 421, 424
Pseudopolydora kempi, 19, **209,** 210, 215
paucibranchiata, 19, **209,** 210, 215
Pseudopotamilla, 513
intermedia, 236, 239
occelata, 84, 236, **237,** 239
socialis, 236, **237**
Pseudopythina rugifera, 559
Pseudoscorpionida, 431, **435**
Pseudostylochus burchami, 105, 109
ostreophagus, 21
Psolus chitonoides, 635, 637
Pterioida, 545
Pteriomorpha, 545
Pterobranchia, 618
Pteropods, 516
Pterygophora, 675
californica, **676,** 681
Ptilosarcus gurneyi, 91, 93
Pugettia, 381, 382
gracilis, 395, 405
producta, 9, 117, 260, 395, **397,** 405
richii, 395, 405
Pulmonata, 467, 469, 513–514, 515
Purpura foliata, 508
Pycnoclavella stanleyi, 647, 652
Pycnogonida, 413–424
Pycnogonum, 414
rickettsi, **415,** 417, 424
stearnsi, **415,** 417, 424
Pycnopodia helianthoides, 623, 626
Pygodelphys aquilonaris, 255, **256**
Pygospio californica, **211,** 213, 215
elegans, **211,** 212, 215, 238
Pyramidellida, 469, 471, 512, 516, 518
Pyramidellidae, 512–513
Pyromaia tuberculata, 393, 405
Pythodelphys acruris, 255, **256**
Pyura, 255
haustor, **640,** 646, 653
mirabilis, 646, 653

Quinqueloculina angulostriata, **27,** 28
Quohog. *See Mercenaria*

Ralfsia, 671
pacifica, 681–684
Ramex californiensis, 232, **233,** 235
Randallia ornata, 393, **394,** 405
Razor clams. *See* Solenidae

Red crab. *See Pleuroncodes*
Red tides, 26
Rees, John T., 65
Reginella nitida, 594, **595,** 606
Regularia, 621
Reniera spp., 43, 45, **49,** 59
Renieridae, 59, 60
Renilla, 538
Reptantia, 378, 405–411
Retusa harpa, 512, 521
Rhabdocoela, 100, 107–108
Rhabdodermella nuttingi, 64
Rhithropanopeus harrisii, 19, 385, 398, 407
Rhizocephala, 259, 267, 382, 405
Rhizoclonium, 537
Rhodophyta, 681–684
Rhombognathus, 425, **429,** 430
Rhopalura, 122
Rhyacodrilinae, 140
Rhynchocoela, 112–120
***Rhynchospio arenincola,* 211,** 213, 215
Rhynchothorax philopsammum, 418, **420,** 421, 424
Rhynchozoon rostratum, **602,** 603, 607
Ribbon worms. *See* Nemertea
Rice, Mary E., 128
Rictaxis punctocaelatus, **483,** 488, 512, 516, 519, 521
Rimicola eigenmanni, 660, 667
Rissoidae, 504
Rissoina spp., 504
Rissoinidae, 504
Ritterella aequalisiphonis, 651, 652
pulchra, 651, 652
rubra, 651, 652
Rochefortia spp., 559
Rocinela angustata, 297, 307
belliceps, 297, 307
Rock crabs. *See Cancer*
Rosalina columbiensis, 28, **29**
Rostanga pulchra, 33, **527,** 528, 541
Rotatoria, 123
Roth, Barry, 467
Rotifera, **119,** 123
Roundworms. *See* Nematoda
Ruditapes, 562
philippinarum, 562
Runcina sp., 469, 518, 521
Rutherford, James C., 634

Sabella crassicornis, 236, **237,** 239
media, 236, **237,** 239
Sabellaria cementarium, 122, **229,** 230
gracilis, **229,** 230
nanella, **229,** 230
spinulosa, **229,** 230
Sabellariidae, **157,** 171, 228–230
Sabellidae, 155, **157,** 159, 160, **167,** 171, 235–239, **240**
Sabellinae, 235
Saccocirridae, 149
Saccocirrus, **148,** 149
Sacoglossa, 515, 518, 522–523, 537
Saccoglossus, **614,** 619
pusillus, 619
Sagartia leucolena, 92
luciae, 92
Salicornia, 92, 408, 431, 490, 504, 513, 537
Salmacina tribranchiata, 239, 241
Salpingidae, 447, 449, 451
Sand crabs. *See* Albuneidae, Hippidae
Sand dollars. *See* Echinoidea
Sanguinolaria nuttallii, 568
Sarcophagidae, 437, 445
Sarcoptiformes, 425
Sarsia spp., **73,** 79, 81
Sarsiella tricostata, 22
zostericola, 22, 248
Saxicava arctica, 572
Saxidomus, 251, 255
giganteus, 560
nuttalli, 560, **561,** 562
Scale worms. *See* Polynoidae
Scalibregmidae, 171, 222, **223**
Scallops. *See* Pectinidae
Scaphopoda, 455
Scaptognathus, 425, **427,** 430
Scatella, 444
Scatophila, 444
Schistocomus hiltoni, 231, 232
sp., **233**
Schizobranchia insignis, 235, **237,** 239
Schizocardium, 619
Schizoporella cornuta, 599, **602,** 603, 607
unicornis, 599, **602,** 607
Schizothaerus capax, 564
nuttallii, 564
Schizymenia pacifica, 683
Schlinger, Evert I., 436
Scissurellidae, 500
Scleractinia, 93
Scleroplax granulata, **131,** 396, 407
Sclerospongiae, 34
Scolelepis squamatus, 213, 215
Scoloplos acmeceps, 205, **206,** 207
armiger, **165,** 205, **206,** 207
Scorpaenichthys marmoratus, 662, 666
Scorpaenidae, 666
Scrippsia pacifica, 77, 82
Scrupocellaria, 582
californica, 594, 606
diegensis, 594, **597,** 606
varians, 596, **597,** 606
Scypha spp., 41, 43, 58, 63
Scyphacidae, 304, 311

Scyphistoma, **88,** 94
Scyphozoa, 94–97, 516
Scyra acutifrons, 395, 405
Scytalina cerdale, 665, 668
Scytalinidae, 668
Sea anemones. *See* Actiniaria
Sea cucumbers. *See* Holothuroidea
Sea lettuce. *See Ulva*
Sea mice. *See* Aphroditidae
Sea palm. *See Postelsia*
Sea pens. *See* Pennatulacea
Searlesia dira, **494,** 497, 510
Sea slugs. *See* Opisthobranchia
Sea spiders. *See* Pycnogonida
Sea squirts. *See* Urochordata
Sea urchins. *See* Echinoidea
Seagrasses. *See Zostera*
Sebastes chrysomelas, 660, 666
melanops, 660, 666
mystinus, 660, 666
paucispinus, 660, 666
Seila montereyensis, 492, 505
Seison, **119,** 123, 271
Semaeostomeae, 95
Semele rubropicta, 568
rupicola, **565,** 568
Semelidae, 546, 550, 564, 568
Sepioidea, 456
Septibranchia, 546
Septifer bifurcatus, 552, **553,** 555
Serolidae, 291
Serpula, **157**
vermicularis, 239, **240,** 241
Serpulidae, 155, **157,** 159, 160, 171, 239–242, 505
Serpulorbis squamigerus, 473, 474, **483,** 505
Sertularia, 26, *furcata,* 83
spp., 76, **78,** 83
Sertulariidae, 83
Sertularella spp., 77, **78,** 83
turgida, 83
Shipworms. *See* Teredinidae
Shore crabs. *See* Grapsidae
Shrimps. *See* Caridea
Sigalionidae, **157,** 162, **172,** 176–177
Sigambra bassi, 184, **186**
Sige montereyensis, 180, 183
Sigillinaria aequalisiphonis, 652
pulchra, 652
Sigmadocia edaphus, 43, 48, **57,** 60
sp., 48, **57,** 60
Siliqua lucida, 569, 570
patula, 118, 569, 570, **571**
Sinezona rimuloides, 482, 500
Simognathus, **427,** 430
Siphonariidae, 513
Siphonodentalium quadrifissatum, 455
Siphonophora, 66
Siphonosoma ingens, **130,** 131
Sipuncula, 128–132
Sipunculus nudus, 129, 131
Sipunculidae, 131
Skeleton shrimps. *See* Caprellidea
Slipper lobsters. *See* Palinura
Slipper shells. *See* Calyptraeidae
Smith, Allyn G., 455
Smithora naiadum, 684
Smittia, 440
clavicornis, 440, 443
marina, 440, 443
pacifica, **438,** 440, 443
Smittoidea prolifica, 601, **602,** 607
Solaster dawsoni, 623, 625
Solemya, 545
Solemyoida, 545
Solenidae, 546, **548,** 550, 569–570
Solen rosaceus, 570
sicarius, 569
Soleolifera, 469
Somasteroidea, 620
Soule, Dorothy F., 579
Soule, John D., 579
Sow bugs. *See* Oniscoidea
Sphaeriidae, 551
Sphaerium patella, 551
Sphaeroma pentodon, 22, 295, 302, 308, 309
quoyana, 18, 22, 23, 308
Sphaeromatidae, **282,** 293–295, 307
Sphaerosyllis californiensis, 185, **188,** 189
hystrix, 185, **188,** 189
pirifera, 189
Spheciospongia confoederata, 43, 46, 62
Spider crabs. *See* Majidae
Spinosphaera oculata, 232, **233,** 235
Spio filicornis, **211,** 213, 215
Spinulosida, 625
Spiny-headed worms. *See* Acanthocephala
Spiny lobsters. *See* Palinura
Spiochaetopterus costarum, 218
costarum pottsi, 218
Spionidae, 158, **165,** 169
Spiophanes bombyx, **211,** 212, 215
fimbriata, **211,** 212, 215
missionensis, **211,** 212, 215
Spirastrellidae, 62
Spirinchus starksi, 665
Spiroglyphus lituellus, 505
Spirontocaris, 385, 402
prionota, 390, 403
See also Heptacarpus
Spirophorida, 40, 63
Spirorbis borealis, **240,** 241
eximius, 241

moerchi, 241
spirillum, 239, **240**, 241
spp., 19, 242
Spisula, 544
catilliformis, 564
spp., 564
Sponges. *See* Porifera
Spongia idia, 43, 44, 59
Spongiidae, 59
Spoon worms. *See* Echiura
Springtail. *See* Collembola
Spurilla chromosoma, 534, 541
oliviae, 534, 541
Squid. *See Loligo*
Staphylinidae, 446, 447, 449, 450–451
Starfishes. *See* Asteroidea
Stauridiosarsia japonica, **73**, 81
spp., 74, 81
Stauromedusae, 95, 96
Steggoa californiensis, 180, **181**, 183
Stelleroidea, 620
Stelletta clarella, 38, 40, 43, 48, 58, 63
Stellettidae, 63
Stenoglossa, 469
Stenoplax fallax, 461, 464, 465
heathiana, 464, 465, 487, 504, 559
Stenothoe, **329, 336**
valida, 343, 363
Stenothoidae, 323, 326, 342–343
Stenothoides burbanki, 342, 363
Stenula incola, 343, **350**, 363
Stephanosella biaperta, 599, **602**, 603, 607
Stereobalanus, 619
Sternaspidae, 166, **167**, 224
Sternaspis, **167**
fossor, 224
Sthenelais berkeleyi, **172**, 176, 177
fusca, **172**, 177
Stichaeidae, 667
Stichopogon coquilletti, 444
Stichopus californicus, 635, 636
Stiliger, 518
fuscovittatus, 523, 537
Stolidobranchia, 653
Stolonifera, 86, 91, 93
Stomotoca spp., 79, 81
Stomatopoda, 246
Stomatopora granulata, 586, **587**, 604
Stone crabs. *See* Lithodidae
Stony corals. *See* Madreporaria
Streblospio benedicti, 19, 22, **211**, 212, 215–216
lutinicola, 22
Streblosoma crassibranchia, 232, 235
Streptoneura, 468
Stromateidae, 665
Strongylocentrotus, 108, 404
droebachiensis, 622
franciscanus, 9, 621–622
purpuratus, 9, 221, 222, 621
Styela barnharti, 22, 653, 654
clava, 644, 653
gibbsi, 646, 653
montereyensis, 644, 654
plicata, 22, 646, 654
truncata, 639, 646, 654
Stylantheca, 11
porphyra, 66, 84
Stylasterina, 66, 84
Stylatula elongata, 91, 93
Stylochoplana agilis, **103**
gracilis, 105, 109, **110**
Stylochus atentaculatus, 105, 109
californicus, 105, 109
exiguus, 105, 109
franciscanus, 104, 109
tripartitus, 104, 109, **110**
Stylommatophora, 469
Stylostomum lentum, 107, 109
Suberites sp., 43, 47, 62
Suberitidae, 62
Surf clams. *See* Mactridae
Surf-grass. *See Phyllospadix*
Sutton, James E., 623, 627
Sycettida, 63
Sycettidae, 63
Syllidae, **157**, 159, **161**, 162, 166, 184–189
Syllis elongata, 185, **186**, 189
gracilis, 185, **186**, 189
spenceri, 189
spp., 189
Synalpheus lockingtoni, 404
Synapseudes intumescens, 278, 279, **279**
Syncaris pacifica, **387**, 390, **392**, 402
Synchelidium, 324, **329, 337, 350**
rectipalmum, **337**, 345, 363
shoemakeri, **337, 341**, 345, 363
Synchirus gilli, 662, 666
Syncoryne eximia, 74, 81
mirabilis, 18, 23, 74, 81
Syndesmis dendrastrorum, 100, 107
franciscana, 107–108
Syndisyrinx franciscanus, 100, **102**, 107–108
Syngnathidae, 667
Syngnathus californiensis, 660, 667
Synidotea berolzheimeri, **288**, 289, 306
bicuspida, **288**, 289, 306
consolidata, 306
harfordi, **288**, 289, 306
laticauda, 22, 23, 287, **288**, 306
pettiboneae, **288**, 289, 306
ritteri, **288**, 289, 306
Synoicum, 643
parfustis, 650, 652
Synopiidae, 328, 349

Tadpole shrimps. *See* Notostraca
Taenioglossa, 468
Tagelus californianus, **565**, 566, 568
Talitridae, 314, 318, 326, 328, 343, 352–357, 363–364
Talitroidea, 326
Talitroides alluaudi, 353, 364
  topitotum, 353, **356**, 364
"Talitrus" sylvaticus, 353, 364
Tanaidacea, 272, 277–280
Tanaidae, 280
Tanais sp., 279, 280
  vanis, 280
Tanystylum californicum, 418, **420**, 424
  duospinum, 418, **420**, 424
  intermedium, 418, 424
  occidentalis, 418, 424
Tapes japonica, 19, 22, 24, 547, 560, 562
  philippinarum, 562
  semidecussata, 562
Tapeworms. *See* Cestoda
Tardigrada, 413, 431–432, **435**
Tealia coriacea, 90, 92
  crassicornis, 90, 91, 92
  lofotensis, 90, 92
Tectibranchs, 515
Tecticeps convexus, 295, 308
Tedania topsenti, 61
  toxicalis, 43, 52, 61
Tedanione obscurata, 43, 48, 61
Tegella robertsonae, **592**, 593, 606
Tegula, 259, 410, 513, 671, 683
  brunnea, 485, **486**, 501, 502
  funebralis, 480, 485, **486**, 501, 502, 503
  montereyi, 485, **486**, 502
  pulligo, 485, **486**, 503
Telepsavus costarum, 218
Tellina bodegensis, **565**, 566, 569
  buttoni, 569
  carpenteri, **565**, 569
  modesta, **565**, 566, 569
  nuculoides, **565**, 566, 569
  salmonea, 569
Tellinacea, 552, 564–569
Tellinidae, 546, 552, 564, 566, 568–569
Telmatogeton macswaini, **438**, 440, 443
Telmatogetonini, 443
Tenellia adspersa, 23, 534, 541
  pallida, 541
Tentaculata, 98, 99
Terebella californica, **233**, 234, 235
Terebellidae, 159, **167**, 171, 232–235
Terebratalia transversa, **614**, 617
  transversa caurina, 617
Terebripora, 581
Teredinidae, 546, 549, 575
Teredo, 28
  beachi, 22
  diegensis, 22, 575
  navalis, 18, 22, 547, 575, **576**
  townsendi, 22
Tethya aurantia var. *californiana*, 32, 33, 38, 40, 41, 43, 46, 62
Tethyidae, 62
Tethymyia aptena, 440, **441**, 443
Tetilla arb, 38, 40, 43, 56, 63
  spp., 40, 43, 56, 63
Tetillidae, 63
Tetraclita, 260
  (*Tetraclita*) *squamosa rubescens*, 264, **266**, 269
  (*Tetraclita*) *squamosa rubescens* var. *elegans*, 269
Tetrastemma nigrifrons, 115
Teuthoidea, 456
Thaididae, 509
Thais. See Nucella
Thalamoporella, 582
  californica, 593, **595**, 606
Thalassacarus, 425, 430
  commatops, **429**
Thalassarachna, 430
  capuzinus, **429**
Thalassinidea, 378, 408–409
Thaliacea, 638
Thallasotrechus, 447
  barbarae, **435**
  nigripennis, 447, 450
Tharyx multifilis, 219, **220**
  parvus, 219
  spp., 218
Thecata, 82
Thelepus crispus, 232, **233**, 235
Themiste dyscritum, 129, 131
  perimeces, 129, 131
  pyroides, 100, 107, 129, **130**, 131
  zostericola, 129, 131
Thereva hirticeps, 444
  pacifica, 444
Therevidae, 437, 444
Thinodromia, 444
Thinopinus, 446
  pictus, 451
Thinusa, 446
  maritima, 449, 451
Thoracica, 259, 267–269
Thoracophelia mucronata, 224
Thordisa bimaculata, 530, 541
Thyone rubra, 637
Thyonepsolus nutriens, 637
Thysanoessa, 377
Tigriopus californicus, 253, **254**, 257
  triangulus, 257
Tintinnopsis nucula, 28, **29**
Tipulidae, 436, 439, 443
Tiron, **327**, 330, **332, 341**

biocellata, 349, 363
Tisbe furcata, 253, **254**, 257
Tivela, 251, 544
stultorum, 214, 560, **561**, 562
Tochuina tetraquetra, 531, 541
Tonicella lineata, 461, 465
Tooth shells. *See* Scaphopoda
Top shells. *See* Trochidae
Tornaria, 618
Toxadocia sp., 43, 48, 60
Trachylina, 66
Transennella sp., 560, 563
tantilla, 560, **561**, 562, 563
Trapania velox, 523, **525**, 541
Travisia gigas, 222, 224
Trematoda, 100
Tresus, 251, 255, 407, 544
capax, 564
nuttallii, 564, **576**
Tricellaria occidentalis, 596, **597**, 606
ternata, 596, **597**, 606
Trichobranchus, **157**
Trichocorixa, 432, 434
reticulata, 434, **435**
verticalis, 434
Trichoniscidae, 304, 311
Trichotropis, 513
Tricladida, 100–101, 108
Tricolia pulloides, 490, 503
Trididemnum dellavallei, 652
opacum, 508, 649, 652
Trilobodrilus, **148**, 149
Trimusculidae, 513–514
Trimusculus reticulatus, 469, 475, **484**, 513–514
Trinchesia abronia, 535, 541
albocrusta, 536, 541
flavovulta, 535, 541
fulgens, 536, 541
lagunae, 535, 541
sp., 19, 536, 541
virens, 536, 541
Triopha aurantiaca, 541
carpenteri, 526, **527**, 541
grandis, 526, 541
maculata, 516, 526, 541
Triphora, 469, **494**
montereyensis, 506
pedroana, 506
spp., 492, 506
Triphoridae, 506
Tritella laevis, 368, **371**, 372
pilimana, 368, **371**, 372
tenuissima, 368, 372
Triticella elongata, 584, **585**, 604
Tritonia diomedea, 531, 541
exsulans, 532, 541
festiva, **525**, 531, 541
Trivia, 469
californiana, 491, 508
Triviidae, 508
Trochidae, 502–503
Trochochaeta franciscanum, 216, **217**
Trochochaetidae, 169, 216, **217**
Trombidiformes, 425
True crabs. *See* Brachyura
Tryonia imitator, 504
Trypanosyllis gemmipara, 185, **186**, 189
ingens, 185, **186**, 189
Trypetesa, **263**
lateralis, 259, 267
Tubificidae, 136, 137, 138, 140, 141, 142, 145
Tubulanus pellucidus, 114, 116
polymorphus, 115, 116
sexlineatus, 116, 117
Tubularia, 66
crocea, 18, 23, 72, 82, 538, 541
marina, 72, 82
Tubulariidae, 82
Tubularioidea, 84
Tubulipora pacifica, 586, **588**, 604
tuba, 586, **588**, 604
tuba var. *fasciculifera*, 586, 604
Tunicata. *See* Urochordata
Turban shells. *See* Trochidae
Turbanella, **119**
Turbellaria, 100–111
Turbinidae, 503
Turbonilla, **489**
franciscana, 513
spp., 488, 513
Turridae, 511
Turritopsis nutricula, 76, 80
Tusk shells. *See* Scaphopoda
Tylidae, 283
Tylodina fungina, 472, **498**, 512, 516, 522
Typosyllis aciculata, 185, **186**, 189
adamanteus, 185, 189
armillaris, 185, **186**, 189
hyalina, 185, **186**, 189
pulchra, 185, 189

Ulva, 100, 107, 182, 271, 439, 442, 502, 670, 678
expansa, 680
lactuca, 680
lobata, **672**, 680
taeniata, **672**, 680
Umbraculidae, 512
Umbrella crab. *See Cryptolithodes*
Ungulinidae, 546, 552
Upogebia, 251, 255, 379, 382, 384, 403, 404, 407, 546, 558, 559, 570, **576**, 615
pugettensis, 286, 305, 383, 409

Urechis, 9, 10, 28, 175, 176, 404, 407, 546, 570
caupo, **130**, 132, 133
Urochordata, 638–655
Urosalpinx cinerea, 19, **498**, 499, 509
"Urosalpinx" subangulata, **498**, 499, 509

Vaucheria, 123, 537
Velella, 262
lata, 84
velella, 66, 84, **97**
Vallentinia adherens, 77, 83
Valvatida, 625
Valvifera, 283, 287–291, 305–306
Velutina, 469
laevigata, 508
prolongata, 488, 508
velutina, 487, 508
Velutinidae, 508
Veneridae, 546, 551, 552, 559–563
Veneroida, 546
Venerupis, 562
philippinarum, 562
staminea, 562
Venus mercenaria, 562
Vermetidae, 468, 505
Vermiliopsis multiannulata, 239, **240**, 242
Vertebrata, 638
Victorella pavida, 19, 584, **585**, 604, 605
Vitrinella oldroydi, 465, **484**, 487, 504
Vitrinellidae, 504
Volsella capax, 555
demissa, 554
recta, 555
senhousia, 555

Washington clam. See *Saxidomus*
Water bears. *See* Tardigrada
Water boatmen. *See* Corixidae
Water fleas. *See* Cladocera
Water strider. *See Halobates*
Whale lice. *See* Cyamidae
Wheel animalcules. *See* Rotifera
Williamia peltoides, 469, 475, 513
vernalis, 513
Williams, Gary C., 517

Xanthidae, 406–407
Xererpes fucorum, 665, 667
Xestospongia vanilla, 43, 45, **49**, 60
Xiphister atropurpureus, 665, 667
mucosus, 665, 667
Xiphosura, 245, 413

Zanclea costata, 82
Zancleidae, 82
Zaolutus actius, 91, 92
Zimmer, Russel L., 613
Zirfaea pilsbryi, 572, **574**, 575
Zoantharia, 85, 86, 91
Zoothamnium, 28, **29**
Zostera, 66, 83, 96, 108, 131, 133, 182, 189, 207, 218, 219, 255, 271, 290, 306, 402, 403, 405, 423, 502, 507, 522, 633
marina, 669
Zosteracea, 669
Zygeupolia rubens, 113, 114, 117
Zygherpe hyaloderma, 43, **49**, 50, 60

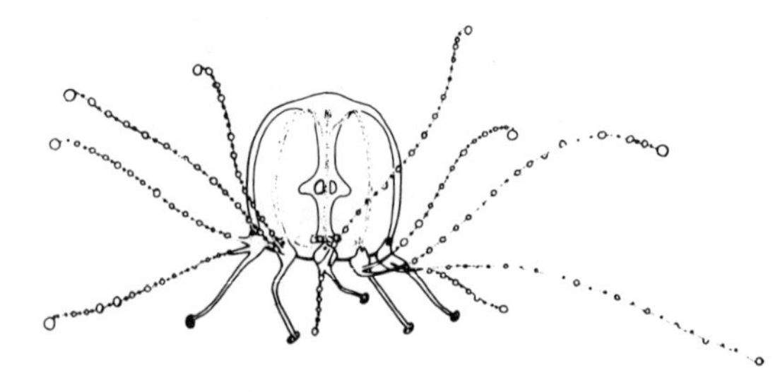

ERRATA AND ADDENDA

The following changes should be noted, in addition to the corrections made in the text for the third printing, 1980.

pages 18, 22, 23, 308, 312, 712: *quoyana* should read *quoyanum*

pages 22, 295, 302, 308, 309, 712: *Sphaeroma pentodon* Richardson, 1904 appears to be a synonym of *S. quoyanum* H. Milne Edwards, 1840; see Rotramel, 1972.

pages 292, 294, 308, 697: *Gnorimosphaeroma lutea* Menzies, 1954 appears to be a synonym of *G. insulare* (Van Name, 1940); see Hoestlandt, 1977, Crustaceana, 32: 45–54.

pages 69, 72, 73, 74, 79, 80, 691: *Cladonema californica* should read *C. californicum.*

pages 72, 73, 74, 79, 81, 699: *Hydrocoryne* sp. should read *H. bodegensis* Rees, Hand, and Mills, 1976; see Wasmann J. Biol. 34: 108–118.

pages 77, 83, 686: For *Aglauropsis* sp. read: *Aglauropsis aeora* Mills, Rees, and Hand, 1976. P, M Medusa an open water form. See Wasmann J. Biol. 34: 23–42.

pages 251, 252, 255, 696: *Eurytemora hirundoides* (Nordquist, 1888) appears to be a synonym of *E. affinis* (Poppe, 1880).

pages 269, 688: The paper by Newman and Ross, 1976, makes several changes of names used in our key by raising subgenera to generic rank:
Balanus (Armatobalanus) nefrens becomes *Armatobalanus (Armatobalanus) nefrens* (Zullo, 1963)
Balanus (Conopea) galeatus becomes *Conopea galeata* (Linnaeus, 1771)
Balanus (Megabalanus) tintinnabulum californicus becomes *Megabalanus californicus* (Pilsbry, 1916)
Balanus (Semibalanus) cariosus becomes *Semibalanus cariosus* (Pallas, 1788)
Balanus (Solidobalanus) hesperius laevidomus becomes *Solidobalanus (Hesperibalanus) hesperius laevidomus* (Pilsbry, 1916)